Electron Microscopy and Analysis, 1981

Electron Microscopy and Analysis, 1981

Proceedings of the Institute of Physics Electron Microscopy and Analysis Group conference held at the University of Cambridge, 7–10 September 1981 (EMAG 81)

Edited by M J Goringe

Conference Series Number 61

The Institute of Physics
Bristol and London

CODEN IPHSAC 61 1—563

British Library Cataloguing in Publication Data

Electron microscopy and analysis, 1981. —
    (Conference series/Institute of Physics,
    ISSN 0305-2346; no. 61)
    1. Electron microscopy — Congresses
    I. Goringe, M. J.
    502'.8   QH212.E4

    ISBN 0-85498-152-7
    ISSN 0305-2346

The conference was organised by the Electron Microscopy and Analysis Group of the Institute of Physics in association with the Royal Microscopical Society.

Programme Committee
    J N Chapman (*Chairman*), A C Brown, L M Brown, J R A Cleaver, A G Cullis, P Doig, J A Eades, P J Goodhew, D J Smith, G J Tatlock and D Vesely

Exhibition and Local Arrangements Committee
    L M Brown and J R A Cleaver (*Co-Chairmen*), A W Agar, A C Brown, W R Clarke, A Everid, D A Jefferson, R A Jones, L G P Jones, W C Nixon, W O Saxton, W M Stobbs and M Waterworth

Electron Microscopy and Analysis Group Committee 1980—81
    A G Cullis (*Chairman*), J N Chapman (*Secretary*), A C Brown, L M Brown, J R A Cleaver, P Doig, J A Eades, P J Goodhew, G J Tatlock and D Vesely

Honorary Editor
    M J Goringe

Published by The Institute of Physics, Techno House, Redcliffe Way, Bristol BS1 6NX and 47 Belgrave Square, London SW1X 8QX.

Printed in Great Britain by J W Arrowsmith Ltd, Bristol.

Preface

Ten years ago the Electron Microscopy and Analysis Group held its twenty-fifth anniversary meeting in Cambridge. To mark that occasion the proceedings of the meeting were published and such was the popularity of this venture that all subsequent meetings in the biennial EMAG series have been accompanied by substantial proceedings. This volume bears evidence of the continuation of this tradition and it is fitting that after an absence of ten years the 1981 meeting should return to Cambridge.

Since the last meeting in Cambridge, the Cavendish Laboratory has moved from its original site in the centre of the city to a new site in West Cambridge. The new buildings provided not only excellent lecture theatres, but also a splendid exhibition site for the 33 firms whose products were displayed. The meeting itself followed a largely familiar format with sessions on instrumentation, electron scattering and diffraction, microanalysis, surface studies, image formation and analysis, high-resolution electron microscopy, beam-sensitive materials and studies of defects. However, in addition to these topics there were sessions on subjects not dealt with extensively in previous EMAG meetings. Foremost among these were data acquisition and processing, multi-signal analytical electron microscopy and semiconductor materials. Another venture introduced into the 1981 meeting was the incorporation of a half-day workshop on 'STEM — the current and future state of the art' in place of one of the parallel sessions. No formal presentations were given, but a summary of the discussions which took place is included at the end of the volume.

Topics in the individual sessions were introduced by 14 invited speakers describing work carried out in Australia, USA, Canada, Germany and France, as well as in the United Kingdom. To open the conference Professor E Zeitler tackled one of the most difficult problems in electron microscopy, that of acquiring a quantitative description of specimens which damage under the electron beam. Despite the constraints this imposes on the experimentalist, the emphasis here and in later papers in the chapter is not on how little can be done but rather, with ingenuity and care, just how much information can be derived! Turning to specimens which are more stable under the electron beam, Dr J C H Spence and Dr P Goodman both show how detailed crystallographic information can be obtained using techniques far removed from the conventional imaging and selected area diffraction modes. These include convergent beam electron diffraction, beam rocking modes and the use of inelastically scattered electrons. Further papers describe theoretical beam—specimen interactions and the exploitation of carefully selected portions of the scattered radiation field to yield crystallographic data.

The instrumentation contributions fall into a number of categories. Dr K C A Smith describes how recent advances in computer technology, together with a decrease in cost of semiconductor memory, have led to a revolution in the collection and analysis of data in microscopy and related analytical techniques. Rather closer to the specimen itself are the many advances in the design of everything from individual microscope components to complete instruments. The state of the art here, together with a glimpse of what the future might hold, is provided by Dr E D Boyes.

Instrumental developments also play a large part in the contributions in the chapter on microanalysis where, despite competition from techniques such as cathodoluminescence, x-ray microanalysis and electron energy loss spectroscopy remain the most popular techniques. Dr G W Lorimer and Dr C Colliex provide excellent reviews of recent progress in the two techniques and many of the themes introduced here are developed in more detail later in the chapter. In many instances it is desirable to utilise both techniques simultaneously, along with the cathodoluminescent signal and different parts of the elastically scattered electron distribution, for complete characterisation of samples. This approach is introduced by Dr H Dexpert, who describes the need for it in a number of examples from solid state chemistry.

The advantages in using carefully selected portions of the scattered electron intensity distribution are described in the chapter on image formation and analysis. Two very different examples of what can be achieved are presented by Professor R E Burge and Dr R Sinclair, who show how innovative imaging methods can lead to information inaccessible using normal techniques. The large number of high-resolution investigations now under way highlighted in the paper by Dr D J Smith, and his subsequent papers in the chapter show the enormous advances that have been made in recent years in this field.

Finally, there are chapters on arguably the most important aspect of electron microscopy, namely its application to solving particular problems of widely varying natures. Dr A Howie introduces a chapter on surface studies in the electron microscope where he shows how information can be obtained which is not generally available using the conventional techniques of the surface physicist. The study of defects continues to be of great interest to electron microscopists and recent developments here are contained in the chapter headed by Professor G C Weatherly's paper on interfaces. A growth area within electron microscopy is the study of semiconductors, and Dr P M Petroff shows how a wide range of signals may be used to study these materials and in particular to provide information on electrically active defects.

In all, the topics contained in this volume cover a wide range of subjects of interest primarily, but not exclusively, to the physical scientist who wants to understand his specimen on a microscopic scale. It is hoped that those reading it will find it not only informative, but also a stimulus to the development and application of further advances in electron microscopy and analysis.

J N Chapman
Chairman
Programme Committee

Contents

Chapter 3: Data acquisition and processing

Chapter 4: Microanalysis

Chapter 5: Multi-signal analytical electron microscopy

Chapter 6: Electron scattering and diffraction

Chapter 7: Image formation and analysis

Chapter 8: High resolution

Chapter 11: Semiconductors

Plenary Lecture:
Radiation damage in beam-sensitive material

E Zeitler

Fritz-Haber-Institut der Max-Planck-Gesellschaft
Faradayweg 4-6, D-1000 Berlin 33

1. Introduction

Williams and Fisher (1970) have clearly demonstrated that the production
of a micrograph with high energy electrons degrades the structure of the
(biologic) object to be recorded. As a remedy they propose applying the
minimum dose required by the photographic recording medium. As a logical
consequence, there exist two possibilities for improving this given
situation--by (a) finding ways to make the specimen *less* sensitive to
the interacting radiation and (b) making the photographic emulsion *more*
sensitive to the interacting radiation. This dilemma is enhanced by the
fact that the primary interaction of the radiation with matter--specimen,
emulsion or any recording medium--is the same. The dilemma is also
expressed in the nomenclature; i.e., the microscopist whose structures
disappear speaks pessimistically of "radiation damage", while the chemist
speaks optimistically of the "speed" of his emulsion (Mees, 1954).

In this context, we should not forget the many radiation-matter
interactions which are considered beneficial and therefore investigated
with vigor. For example, the shortage of energy has awakened new
interest in photosynthesis with the aim of improving the conversion of
light to energy which takes place in plants by means of man-made systems.
The radiation-induced degradation of polymers (Tsuji, 1973) has become
an inherent part of "lithographic" processes in the fabrication of micro-
circuits. Here "radiation damage" is partly responsible for the advent
of the microprocessor age. And there is radiation therapy. Of course
we cannot overlook the other radiation damage problem more formidable
than that of electron microscopy--namely that of mankind, where the
awesome aspects are genetic and carcinogenic effects.

Radiation chemistry and radiobiology (Huttermann, 1978; Dertinger, 1969)
have developed into vast and fast-moving fields with very serious problems to solve. The intention of this contribution is to vitalize the
radiation damage studies in electron microscopy. (At present they are
too descriptive; they are phenomenological and, especially, too isolated
from the related above mentioned scientific endeavors.) Therefore I
mention these active fields, which are sponsored by rich industries or
responsible public health agencies, because there is a wealth of pertinent
knowledge to be tapped and applied to our problems in electron microscopy.

In turn, our findings will become knowledge in radiation chemistry. After
all, the final purpose of *all* these studies must be to find a way to
deduce the unadulterated structure of the microscopic specimen.

2.   State of the art

In electron microscopy the term "radiation damage" has two meanings.
For most of the users, it means destruction of the structure which the
electron microscopist sets out to determine. To the materials scientist,
it often means the displacement of atoms by electrons, simulating
neutrons and other particles. Electrons offer the possibility for
direct observation of the structure damage caused. We deal only with
the first aspect, that of structure loss--which, except for newer low
temperature efforts, is still covered adequately by Isaacson's review
of four years ago! (Isaacson, 1977).

The loss of structure has not been quantified, although this is possible
with modern image processing (Linfoot, 1956; Hawkes, 1980). Instead,
standard procedures have been developed which record, as "analog signals",
as it were, the appearance or disappearance of certain phenomena during
the interaction of the electron beam with the specimen. The charge
density, mostly measured in number of electrons per square Angstrom,
is considered the cause of the fading signal and is therefore plotted on
the abcissa. Over a wide range the law of reciprocity holds--that is,
the effects depend only on the product of the current density multiplied
by the exposure time. The customary phenomena are mass loss, fading of
diffraction peaks, fading of energy loss peaks, and fading of optical
absorption peaks. And, of course, a loss of visible structure. Most of
the decay follows an exponential (single hit) law, so that the resistivity to radiation can be characterized by the charge density, which
reduces the phenomenon under investigation to 37 percent (1/e) of its
virgin value.

Probably for the sake of convenience, the fading of diffraction patterns
is most often applied to the studies. It is found that the spots of
higher-order reflexes fade more quickly than those of lower order. It
is also found that at 4.2 K the diffraction spots of L-valine resist
a charge density fifteen times larger than the critical charge density
at room temperature (Müller, 1981). Although no relation between the
various phenomena and the loss of structure information has been made,
it is of advantage to have hard numbers signifying one aspect of
radiation damage. For example, the success of attempts to reduce
radiation damage can thus also be quantified. Many such studies can be
found in the literature in which the effect of special specimen treatments
--embedding and staining, freezing and cooling--lead to moderate success.
As long as we do not know the chemical events which are triggered by the
primary assault, however, all attempts to influence the radiation sensitivity will fall into the category of trial and error. In other words,
enough measurements have been made within the microscope. We must now
turn to radiation chemistry.

3.   What goes on in the specimen?

In his classical contribution to radiation chemistry, Platzman (1967)
distinguishes three phases in the response of molecules to irradiation.

The distinction is based on the response times involved. Determined by the velocity v of the projectile and its distance b (impact parameter) from the target molecule, the time b/v of the initial energy transfer is 10^{-13} to 10^{-16} seconds. In the ensuing second phase of 10^{-10} seconds, the dislodging of electrons from their regular orbits may render free radicals--that is, molecules with unpaired electrons. In the third phase, the truly chemical phase, with reaction times of 10^{-6} seconds, these radicals will react with their surroundings. This scheme shows that radiation damage can only be controlled by controlling the reactions of the radicals. The radiation damage is, first of all, the chemical response of the target.

4.   What is known about radicals?

Whereas regular molecules are diamagnetic, free radicals are paramagnetic. This fact does not suffice to distinguish molecules from radicals in the electron microscope. However, concentrations of radicals can be determined very readily in well established electron spin resonance (ESR) spectrometers with great sensitivity (Box, 1977). It is interesting that a plot of radical concentration versus dose follows the same single hit law as the optical density curve for photographic material or the damage curve of diffraction spots, namely $R = R_\infty(1 - \exp[1-D/D_o])$ (Wyards and Cook, 1969).

In chemical parlance, the removal of one electron is called "oxidation"; the addition of one electron, "reduction". Both processes are induced by radiation. The electrons needed for reduction might stem from traps or from oxidized molecules. Most of these products are unstable at room temperature, so that for their determination the ESR measurements and the irradiation must be performed at liquid nitrogen or lower temperatures. As an example, the ionization process of water forms two charged radicals, H_2O^+ and e^-. If the electron can escape recombination and react with an additional water molecule, three highly reactive species derive--a hydroxyl radical, a hydrogen atom, and a solvated electron.

In solid materials of molecules MH similar processes occur, including the homolytic fission of a highly excited molecule into two radicals. Hydrogen atoms may diffuse to other molecules and form addition radicals at unsaturated bonds or abstract hydrogen from saturated sites:

$$MH + H^\cdot \rightarrow MH_2^\cdot \qquad \text{or} \qquad MH + H^\cdot \rightarrow M^\cdot + H_2$$

The above variety of reactions shows that a prediction of the actual pathway is possible only if the nature of the specimen is known and taken into account.

5.   Direct and indirect effects

In electron microscopy the specimen is either embedded into a matrix or it consists of particles or microcrystals supported by a thin film. In radiation chemistry the experiments are performed both in the dry state and in aqueous solutions--perhaps solidified at low temperature. In all the cases distinction must be made as to whether the energy is

transferred directly from the irradiating beam to the investigated
structure or indirectly from an excited molecule of the "solvent" which
acts as an intermediate energy depository. In condensed matter the
dissociated molecular fragments tend to recombine because they are
encaged at their place of birth. This so-called cage effect (Franck and
Rabinowitsch, 1934) may be reduced by the particular geometry of the
electron microscope specimen which may allow the escape of electrons and
other reaction products. These facts must be observed when experimental
results are to be compared.

6. G-values

The Bethe formula describes the loss which a charged particle experiences
while going through matter. The material to which this energy is transferred enters the formula only in the guise of two averaged numbers,
namely, the effective number of electrons per volume and the mean
ionization potential. With our interest in the target response, more
specific and detailed information about the specimen must be applied.
As Platzman (1962) has shown, the proper function would be the so-called
oscillator strength, which is related to the dielectric function.

In order to quantize experimental yields and compare them with results
of radiation chemistry, the appropriate G values are recommended. The
G value is the number of observed events which are produced by a deposit
energy of 100 eV. Typical G values are of the order of unity--or,
one unit corresponding to 1.0364×10^{-7} mol per joule. These values
can be readily related to the internationally standardized dose unit,
that is, 1 rad = 6.24×10^{13} eV per gram. A proper dose should be
defined as an amount per mass or per volume (two tablets per adult,
one per child). In electron microscopy it is customary to call the
charge density the "dose". This seems to be permitted because the
samples are so thin that the change in energy of the passing electron
due to the energy loss is negligible. Furthermore, the primary energies
are all of the same magnitude, say, 100 kV, and deviations can be
corrected (β^2 law).

7. Perspective

The phenomenological cataloging of radiation damage will not lead to an
understanding, let alone a reduction, of radiation damage in electron
microscopy. Radiation chemistry has to be performed, or at least con-
sulted, in order to come up with even educated guesses. Embedding a
sample into hydrogen-rich material such as sugars does not seem wise
on account of the indirect yet high reactivity of atomic hydrogen
(Tobias et al., 1960). Certain types of scavenger molecules react with
the prime radiation products and thus check the danger of radicals.
Best known are the sulfhydryl compounds cysteine and cysteamine. Also
hydrogen atoms might be "donated" from harmless protector molecules to
damaged molecules. This scheme of protection has not been tried in
electron microscopy. It is obvious that the efficacy of low temperature
schemes will depend on the ability to freeze the reactive components
into place. The disagreements between the various authors concerning
their experimentally found protection factors at low temperatures might
be due to the fact that the diffusibility of the reactants is critically

dependent on the temperature and on the composition of the matrix. The
success of creating an effective cage--and that is the idea of cryo-
protection--is determined by diffusion-limited and competing processes.
In the end it all becomes radiation chemistry. But in electron
microscopy, reaction kinetics and the rate constants for radical decay
have been studied only in very few instances (Fryer, 1979). Arrhenius
plots and thus a knowledge of activation energies are not available.

8. <u>Conclusion</u>

As an appropriate ending for this contribution, mention should be made
of those studies on radiation damage in which the authors apply chemical
reasoning to explain their findings.

As preparation for radiation chemistry, Egerton (1980) monitors with his
energy loss spectrometer the escape of the individual elements carbon,
nitrogen and oxygen from amorphous polymer films. (Hydrogen, unfortunately, escapes this method.) He was able to show that a change of the
specimen temperature from 300 K to 77 K alters the escape rates for
nitrogen and oxygen from cellulose nitrate by a factor of 100. Irradiating at low temperature, and measuring the content at room temperature,
leads to a loss of the elements which depends on the time spent for the
warm-up. These experiments prove again that the damage must be
envisaged as a two-stage process whereby the primary step leads to break-
age of chemical bonds and inter- and intramolecular disorder, whereas the
secondary step involves diffusion and migration of chemical species.
Therefore, low temperature will be helpful, since the diffusion coefficients, for example, are typically a factor of 10^{20} lower at 77 K than
at room temperature.

The change of the diffraction pattern of di-para-anthracene under the
influence of the electron beam was explained by Jones (1976) as a beam-
induced monomerization, i.e., as an in situ chemical reaction.

Chemical arguments are also used by Clark et al. (1980) to explain the
difference in radiation sensitivity of copper phthalocyanine and its
chlorinated derivatives. The authors refine the diffraction method in
following the decay of specific spots (hkl) in both materials. Although
the chlorine atoms are liberated by the radiation, their diffusion is
restricted on account of their size, which is not true in the case of
freed hydrogen. Thus it is the cage effect which makes the halogen
compound more resistant.

Another fine example for the interplay of structure, chemical bonding
and diffusibility is provided by the low temperature study of para-
Terphenyl by Parkinson et al. (1980). They show that hydrogen bridges
can be formed between the molecules at temperatures around 15 K. These
bonds lock the molecules in place to form a superlattice--which, however,
can again be destroyed by radiation to transform into the lattice
observed at 300 K.

Finally, the theorists also engage in electron radiation chemistry.
Schnabl (1980) performed quantum chemical calculations on the radicals
of DNA bases to compare them with known orbital results of the intact
molecules. He can demonstrate that the removal of hydrogen leads, in

most cases, to the loss of one π-electron. In his view this effect is responsible for the change in the energy loss spectra of the bases as observed by Isaacson et al. (1973).

This list of papers is small, but they represent an important step towards the understanding of radiation damage. I conclude with the hope that this article may contribute to the encouragement of future work in the direction of radiation chemistry--but always with the goal of reducing the damage in electron microscopy.

References

Box H C 1977 Radiation effects: ESR and ENDOR analysis (New York: Academic Press)

Clark W R K et al 1980 Radiation damage mechanisms in copper phthalo-cyanine and its chlorinated derivatives. Ultramicroscopy **5** 195

Dertinger H and Jung H 1969 Molekulare Strahlenbiologie (Berlin: Springer)

Egerton R F 1980 Chemical measurements of radiation damage in organic samples at and below room temperature. Ultramicroscopy **5** 521

Frank J and Rabinowitsch E 1934 Some remarks about free radicals and the photochemistry of solutions. Trans. Faraday Soc. **30** 120

Fryer J R 1979 The chemical applications of transmission electron microscopy (New York: Academic press)

Hawkes P W ed. 1980 Computer processing of electron microscope images (Berlin: Springer)

Huttermann J et al ed 1978 Effects of ionizing radiation on DNA (Berlin: Springer)

Isaacson M 1977 Specimen damage in the electron microscope. In Principles and techniques of electron microscopy ed. M A Hayat (New York: van Nostrand-Reinhold) p 1

Isaacson M, Johnson D and Crewe A V 1973 Electron beam excitation and damage of biological molecules: its implication for specimen damage in electron microscopy. Rad. Res. **55** 205

Jones W 1976 The electron-beam-induced monomerization of di-p-anthracene. J. Microscopy **107** 151

Linfoot E M 1956 Information theory and optical images. JOSA **46** 740

Mees C E K 1954 The theory of the photographic process (New York: Macmillan)

Müller K-H et al 1981 Cryoprotection of electron-irradiated organic crystals. Proc. Electron Microscopy Society of America p 26

Parkinson G M et al 1980 Electron microscopy at liquid helium tempera-tures. In Electron microscopy at molecular dimensions ed Baumeister and Vogel (Berlin: Springer) p 208

Platzman R L 1962 Superexcited states of molecules. Rad. Res. **17** 419

Platzman R L 1967 Energy spectrum of primary activations in the action of ionizing radiation. In Radiation research ed G Silini (Amsterdam: North-Holland) p 20

Schnabl H 1980 Does removal of hydrogen change the electron energy loss spectra of DNA bases? Ultramicroscopy **5** 147

Tobias C A et al 1960 in The initial effects of ionizing radiation on cells ed J. C. Harris (New York: Academic Press) p 257

Tsuji K 1973 ESR study of photodegradation of polymers. Adv. Pol. Sci. **12** 131

Williams R C and Fisher H W 1970 J. Mol. Biol. **52** 121

Wyards S J and Cook J B 1969 Solid State Biophysics (New York: McGraw-Hill)

Electron microscopy of polymer blends

D Vesely and H Lindberg

Department of Non-Metallic Materials, Brunel University, Uxbridge Middx.
and Department of Materials, University of Luleå, Sweden.

1. Introduction

It is possible to modify the properties of polymers by blending one or
more polymer, as with metals. It is necessary to control the size and
distribution of the phases, ideally by thermal treatment, or more
realistically by the control of processing conditions. To study the
phases, which are usually smaller than 1µm, it is necessary to use
electron microscopy. Most commercial polymers have similar electron
scattering properties and the contrast between the phases is very weak.
To identify the phases and to distinguish them from numerous artefacts,
a range of analytical techniques and their careful application is
required. The importance of analysis is shown in Fig 1. where the
structure of POM/SAN is compared with three different artefacts:
contamination, thickness contrast and voids.

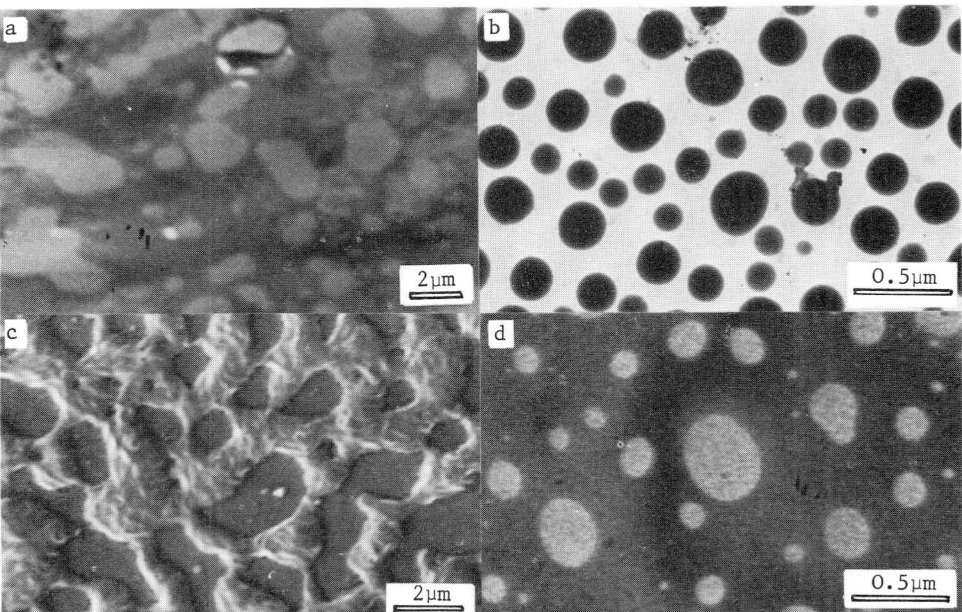

Fig. 1 a – an ultramicrotome section of POM with 20% SAN b – carbon film
with contamination spots, c – thickness contrast on PE/PDMS solvent cast
film – only PE is present d – voids formed during solvent casting of
PS/PDMS system. Only PS is present.

0305-2346/82/0061-0007$01.50   © 1982 The Institute of Physics

The purpose of this paper is to compare the available techniques, namely, electron energy loss analysis and energy dispersive x-ray analysis, and to discuss the difficulties associated with their application to polymers.

2. Electron beam damage

The major limiting factor in utilisation of analytical techniques in the electron microscope is beam damage. The highly ionizing electron beam will cause rapid changes in synthetic polymers, which are not only structural as in the case of loss of crystallinity, but also compositional as in the case of chain scissions, crosslinking and the formation of free radicals. Most important however from the point of view of analysis are the decomposition of molecules, the formation of gases and the loss of mass. These effects are illustrated in Figure 2 and Figure 3.

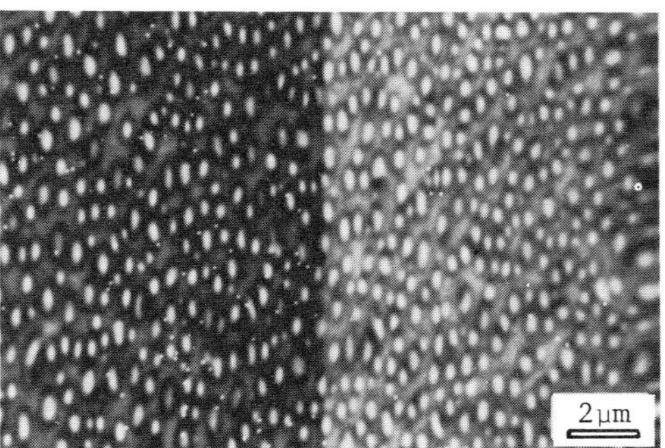

Figure 2 – STEM image of PVC: the right hand portion of the specimen was exposed to the beam for much longer periods. The reduction in specimen thickness due to loss of mass is apparent. The superimposed Cl X-ray signal (small white dots) shows the change in composition.

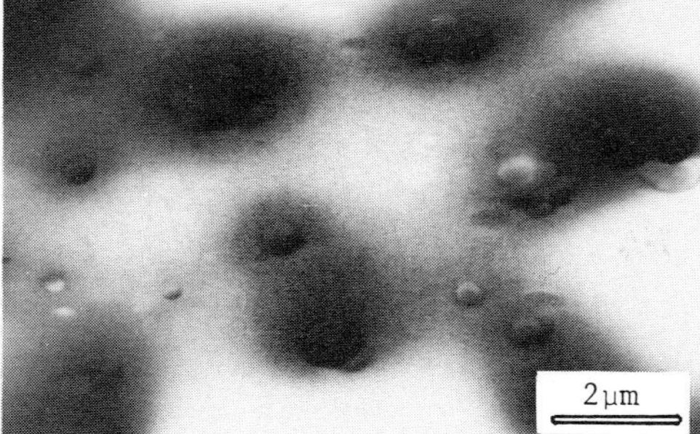

Figure 3 – CTEM image of solvent cast film of PE/PDMS. The formation of voids due to gas evolution is apparent in the darker regions.

The ionizing energy for most organic molecules is very low (10–15 eV) and the energy transferred from an incident electron can propagate along the molecular chain. The probability of the outer shell electrons excitation is greater than the inner shell excitation and therefore the degradation is a more rapid process than the collection of the analytical information. The analysis should thus be considered as a dynamic process with the specimen continuously changing its thickness and chemical composition. The recording time and the dose rate are therefore the most important parameters. Knowledge of the history of specimen area is also essential. The electron beam damage properties vary from polymer to polymer and it is important to evaluate the individual components separately, as well as the polymer blend system. The differences between the polymers are shown in the table below, where the number of gas molecules or cross links formed by absorbing an energy of 100 eV are shown (Chapiro 1962).

Polymer	PS	PB	PVC	PAN	PCL	PIB	PMMA	PE	PDMS	PM(Ph)S
gas mol/100 eV	0.03	0.2	0.3	0.4	0.84	0.81	1.5	2.1	3.1	0.8
cross links/ 100 eV	0.04	–	0.3	–	0.35	1	1.1	2	3.1	0.8

The above table can explain the localised formation of voids in PS/
PDMS systems in the regions containing PDMS. Another example is shown
in Fig. 4 where the phase of SAN in POM matrix changes contrast. This
is most probably due to loss of the matrix with little change in the
SAN.

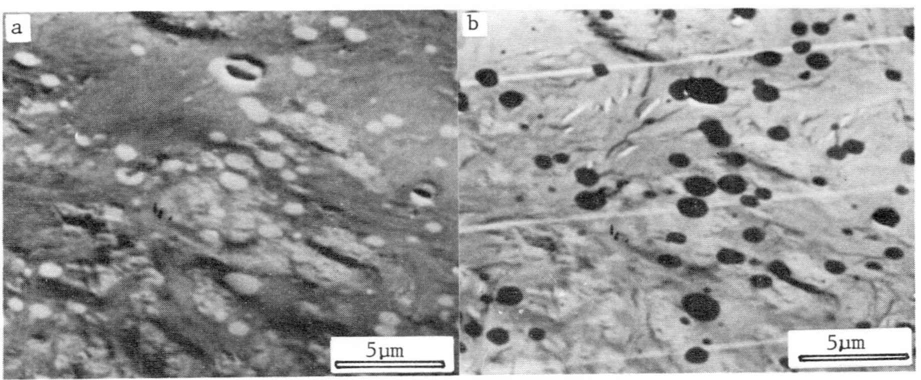

Fig. 4 – STEM micrographs of a POM/SAN system a – first image, b – second
image after 3 min irradiation. The contrast of the SAN phase is
reversed.

3. Analytical techniques

Few polymers contain heavier elements than C, namely N, O, Cl or Si.
The application of EDAX is therefore limited to special cases only.
As the efficiency of this analytical technique is rapidly decreasing
with atomic number, then high electron beam currents are needed to
gain useful results. The spatial resolution might be limited by the
loss of mass due to the beam damage and an attempt was made to prove
this. Several spectra were recorded after a lapse time and it was
found that in the case of PVC the Cl peak decreases rapidly. However
in the case of PDMS, the Si peak increased. This is illustrated
in Fig. 5.

The explanation of this effect could be that the loss of mass is
mainly due to the weakly bonded side groups as in the case of Cl in
PVC, leaving the stronger molecular chain backbone intact. This would
not be true for elevated temperatures, e.g. when high current
densities are used, imposing thus another limitation on a high
spatial resolution.

In an attempt to gain more information on the decay of the Cl peak,
the change in the count rate was measured. It was found that the
count rate does not change much initially, but then falls rapidly and
levels off at a lower value. This behaviour is very similar to the
loss of crystallinity and suggests that the local disorder and the
chemical changes are closely related. The loss of mass could be to

some extent prevented, e.g. by sealing the specimen surfaces with a
carbon layer (Vesely et al. 1975), but it is an open question as to
how the spatial resolution will be affected. It is obvious that a
spectrum from an undamaged specimen could be obtained only for very
small irradiation doses and in this case an x-ray map is much more
valuable for the study of polymer blends

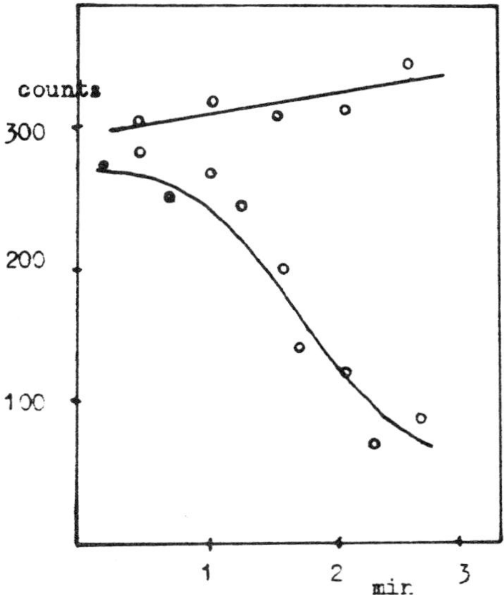

Fig. 5. The changes in Cl and Si
peaks counts with time are
shown.

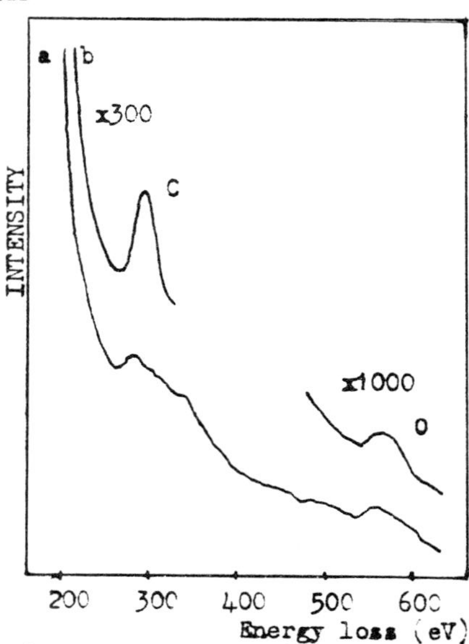

Fig. 6 – EELA spectra for PDMS
a, recorded on the film, 5 sec.
exposure time.
b, sequential recording, 30 sec.
scan. Vertical scale is
expanded.

The main advantage of energy loss analysis is its ability to detect
light elements. This makes this technique very valuable for polymer
studies, especially when only O or N are present besides C and H. The
high detection efficiency is unfortunately outweighed by poor signal
to noise ratio. The recording of the spectrum is usually done by
scanning the electron beam across the exit slit. This method has the
advantage that the signal can be processed electronically and therefore
the large change in the background intensity is not a matter of concern
as in the case of film recording. On the other hand the spectrum can
be recorded on a film with fewer electrons (10x). This is shown
in Fig. 6.

The multi channel detector can combine the advantages of both previous
recording techniques, i.e. the short exposure time and the electronic
signal processing. Moreover it should be possible with this detector
to subtract the background from the selected peak and use it for
elemental mapping. With such a technique it should then be possible
to identify any polymer and perhaps even to study the structural
differences on a finer scale. However the spatial resolution, which
is limited by the beam sensitivity of polymers will remain rather poor.
Chapiro A, 1962, Rad.Chem. of Poly. Systems (Interscience,Wiley)pp385–539
Vesely D. Low A. and Bevis M 1975, Proc. EMAG 75 (Acad. Press)pp333–5.

Inst. Phys. Conf. Ser. No. 61: Chapter 1
Paper presented at EMAG, Cambridge, 7–10 September 1981

Radiation damage imposed limits to the quantitative characterisation of an organic molecular crystal

M J Drummond, J N Chapman and W A P Nicholson

Department of Natural Philosophy, University of Glasgow, Glasgow G12 8QQ

1. Introduction

In electron microscopy a wide variety of signals are available to aid the experimenter in his attempt to fully describe a specimen. Thus,for example, the position and intensity of spots in the diffraction pattern relate to the structure of the material whilst the presence and strength of lines in an x-ray spectrum enable information on elemental composition to be obtained. When dealing with radiation sensitive specimens, such as organic molecular crystals, a problem arises because the characteristic signal and the usually uninformative background signal both vary as electron irradiation proceeds. This may arise due to disruption of the molecules within the crystal structure, loss of a particular element from the sample or a number of other causes. A necessary consequence of this is that if reliable information on undamaged material is to be obtained the dose incident on unit area of the specimen must be constrained to be less than a certain value, which differs not only from one material to another but may differ for different characteristic signals from the same specimen.

To obtain the required information about the specimen, the statistical accuracy of the characteristic signal (detected signal minus background signal) must surpass a prescribed value. This in turn imposes limits on the minimum electron dose to which the specimen must be subjected. Thus the requirements of a sufficiently large dose for statistical significance in the characteristic signal, together with a limitation on the dose/unit area before significant specimen damage occurs, define a minimum specimen area which is required for a particular investigation. It follows that meaningful experiments cannot be carried out if the individual regions of interest within samples are smaller than this critical size.

A specimen of scientific interest and commercial importance which serves to illustrate the way an investigation is limited by the radiation sensitivity of the material is the pigmentary form of chlorinated copper phthalocyanine (Cl-CPC). The individual particles which result from the pigmentation of a particular industrially prepared Cl-CPC are typically 50nm in size and are known to contain an average 14.5 Cl atoms/molecule (figure 1a). Examination of the material in the CTEM reveals that (a) one dimensional lattice fringes of separation $\sim$1.3nm are present in many of the particles,(b) spot and streak selected area diffraction patterns which are difficult to interpret are generally observed and (c) the structure in the diffraction pattern and image fringes essentially fade to zero after a dose/unit area of typically 25 C.cm^{-2}. These observations suggest that not only is the material more susceptible to radiation damage than epitaxially prepared specimens of fully chlorinated CPC (16 Cl atoms/molecule)(Clark et al 1980) but that the molecular ordering is significantly lower. To proceed further with a structural investigation and to assess the variation of composition between particles it is necessary to record diffraction patterns and x-ray spectra from individual particles. The remainder of this paper concerns a theoretical and experimental assessment of the possibility of achieving this successfully.

2. Diffraction Patterns

As a guide to determining the minimum particle size required for a diffraction pattern with information extending to spacings $\sim$0.1nm to be recorded using a dose < 10 C.cm^{-2}, we assume initially that the pigments have the same structure as epitaxially deposited Cl-CPC (Uyeda et al, 1972). Knowledge of the atomic scattering factors for the constituent atoms within the molecule then enables ready computation of the general form of the diffraction pattern together with an estimate of the size of particles required to yield, say 10^4 electrons, in a weak diffraction spot. The number 10^4 is selected as being the number required to easily detect a spot using Kodak 4407 electron emulsion normally processed, or to detect a spot using the detector system employed in this work and described later in this section. Use of nuclear or x-ray emulsions with the electrons incident directly on them and with the optimum camera length reduces this number by $\sim$10^3. The calculated minimum particle size is $\sim$10nm suggesting that even though the structure in the pigments may be less ordered than in epitaxial films, diffraction patterns from individual particles should be readily obtained using restricted doses.

To confirm that this is true we have recorded diffraction patterns from individual particles using an extended VG Microscope HB5. The area from which the pattern was recorded was defined by the size of the incident probe which was varied over the range 15-50nm. A probe convergence angle of 0.5 mradian was used and the current density adjusted so that spots remained visible for a period $\sim$10s. Two of the three post specimen lenses with which the microscope is equipped were used to project the diffraction pattern onto a 20mm diameter viewing screen in the detector chamber of the instrument. As the detector and specimen chambers are directly connected and at a pressure < 5x10^{-9} mbar it is impossible to use film inside the vacuum system and a prism was used to transfer the photon image through a window in the detector chamber. The window size limits the photon collection angle so that an F2.8, 100mm focal length macro lens is adequate for viewing and recording purposes. To compensate as much as possible for the poor photon collection efficiency ($\sim$ 9x10^{-2}%) it is important to use as fast a film as possible. Ilford XPI chromogenic film push processed to beyond 1600 ASA yields good results. With this system diffraction spots containing > 10^4 electrons over an area $\sim$1.3x10^{5}μm^2 on the phosphor screen were detected successfully.

Figure 2a shows an image, recorded in the HB5 with a small probe, of a series of individual pigmentary particles of Cl-CPC in which one-dimensional lattice fringes are clearly visible. The range of diffraction patterns obtained from particles similar to these is shown in figure 2b, c and d. Exposure times of 1-4s were used and in all cases the pattern was still clearly visible after the end of the recording. The pattern in figure 2b was observed only on a few occasions whilst those in the other two figures were commonly seen. They may be regarded as arising from increasingly faulted versions of the structure which yielded the pattern in figure 2b.

The spot pattern in figure 2b arises from a structure with spacings of 1.3nm and 0.35nm. These agree well with the (020) and (001) spacings

present in epitaxially prepared Cl-CPC although the angle between the major
axes is not as expected. The (OO1) spacing corresponds to the separation
of molecules when stacked on top of one another as shown in figure 1b. That
this spacing should be present in pigmentary material is not surprising for
there is a natural tendency for the planar molecules to stack together with
their delocalised ring structures overlapping as shown in figure 1c. Thus,
in the absence of a polar substrate to encourage ordered growth from a well
defined base plane there is a tendency for long oblique columns of molecules
to grow. Eventually the columns will be bonded together by weak residual
forces and the streaking commonly observed in the diffraction patterns
suggests that the binding is insufficiently strong to cause the molecules
in individual columns to align to form well defined sheets parallel to
their own plane. This is represented schematically in figure 1d. Thus using
diffraction patterns from individual particles we have deduced a structure
for non-epitaxially grown Cl-CPC which differs from the perfect crystal
structure in a manner consistent with the different growth mechanism.

3. <u>X-Ray Microanalysis</u>

In the same way it was possible to calculate a theoretical diffraction
pattern from an individual Cl-CPC particle it is possible to calculate the
x-ray spectrum from it. For the latter purpose a knowledge of cross-
sections for characteristic and bremsstrahlung x-rays is required (Gray et
al, 1980) and these together with the approximate specimen composition and
the efficiency and resolution of the detector system enable the spectral
shape to be calculated. Once this has been done it is a simple matter to
calculate the number of photons in any characteristic peak arising when a
particle of a specific size is irradiated with a specified dose/unit area.
It is also important to know the error in this quantity particularly if a
peak of interest is on a high background.

Assuming a detector solid angle appropriate to the HB5 of O.04sr, a
particle size of 5Onm and a dose/unit area of 1O $C.cm^{-2}$ the number of Cu
and Cl Kα photons detected would be $\sim$1 and $\sim$15 respectively. From these
figures we can see that x-ray microanalysis of individual Cl-CPC particles
is impractical. We should note, however, that when recording x-ray spectra
there is no reason to restrict the dose/unit area to that appropriate to
the occurance of significant structural disruption. Nonetheless, experiments involving the recording of spectra from a large area of pigments show
that there is a significant loss of chlorine for a dose/unit area of
2O $C.cm^{-2}$ (Clark et al, 1980). Thus the value used above could not be
significantly exceeded with safety.

We conclude, therefore, that whilst structural information may readily be
obtained from individual Cl-CPC particles, compositional information is not
so easy to derive. To determine the Cu:Cl ratio at the 10% level > 1OO
particles would be required whilst variations of 3% would necessitate
recording spectra from > 1OOO particles. It is unclear whether such
analyses would provide useful information about the pigments themselves,
but the ability to perform analyses on platelets of area $\sim$1μm^2 and thickness $\sim$5Onm is of distinct value when examining the larger blocks of
Cl-CPC which exist prior to pigmentation.

One of us (MJD) would like to acknowledge a CASE studentship with ICI.

Clark W R K, Chapman J N, MacLeod A M and Ferrier R P 1980 Ultramicroscopy
  <u>5</u> 195
Gray C C, Chapman J N, Nicholson W A P, Ferrier, R P 1980 Proc EMAG 79 425
Uyeda N, Kobayashi T, Suito E, Harada Y, Watanabe M 1972 J Appl Phys <u>43</u> 12

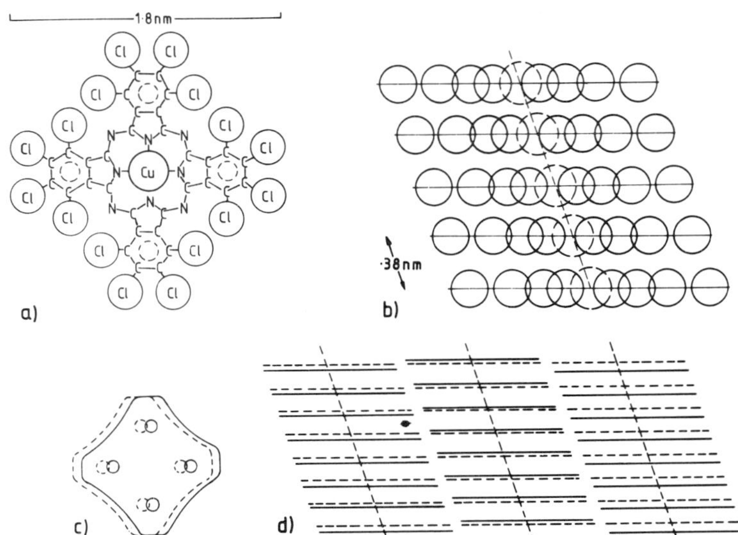

Figure 1: a) Atomic disposition in Cl-CPC molecule. In the material used here on average 1.5 of the Cl sites are occupied by H atoms;
b) Stacking of planar molecules in a column with Cl atoms represented by solid circles and Cu atoms by dashed circles; c) Projection of 2 molecules indicating π-orbital overlap. The π-orbitals are represented by circles; d) Schematic diagram illustrating displacement of columns of molecules from perfect crystal positions.

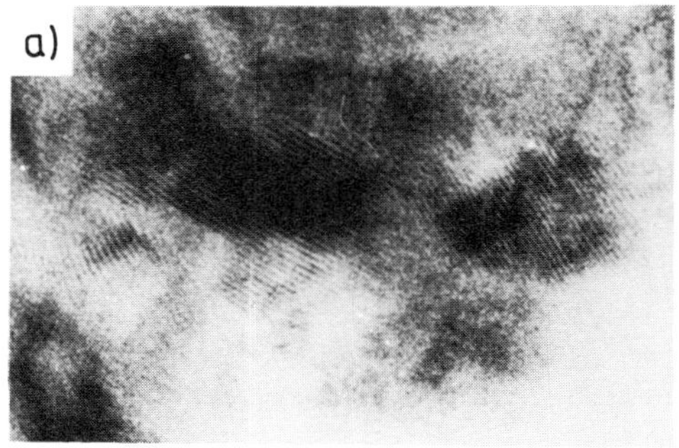

Figure 2: a. Bright field image from pigmentary Cl-CPC showing 1.5 nm lattice fringes.
b,c,d. Diffraction patterns recorded from ≃ 20 nm diameter areas within individual Cl-CPC pigments. c and d show increasing disorder.

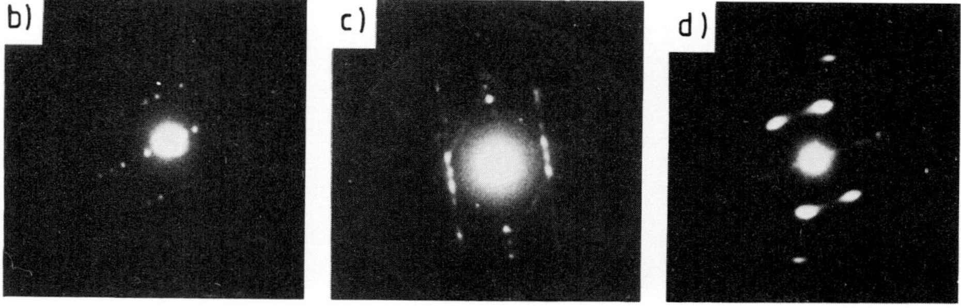

Lattice imaging of radiation-sensitive materials

J R Fryer,

Chemistry Department, University of Glasgow, Glasgow G12 8QQ.

1. Introduction

Molecular crystals are not only radiation sensitive but also do not adopt high symmetry space groups. In consequence, the natural orientation of the crystal on a major zone axis is unusual, and to obtain a molecular projection extremely difficult. Furthermore, the radiation sensitivity limits the opportunities to tilt the specimen to the desired orientation.

Lattice imaging provides less information than molecular imaging about non-periodic regions of crystals but is still of far higher resolution than electron diffraction. For molecular crystals the lattice spacings are normally large so that contrast reversals are rarely a problem, and the lower signal to noise ratio necessary for resolution of a lattice increases the effective lifetime of the specimen in the electron beam by a factor of 3 or 4.

An example of the quantitative application of lattice imaging is provided by study of the $\beta \rightarrow \alpha$ phase change in copper phthalocyanine. This phase change – against the thermodynamic inclination of these phases – is brought about by crushing and grinding in a ball mill during the formation of pigmentary size particles, and valid examination can only take place on a dispersion of the fine particles in a random orientation. Therefore accurate lattice imaging is the only means of differentiating between the phases of the particles and this entails distinguishing between the 1.28 and 1.21nm of the α phase and the 1.25nm of the β phase.

A more typical problem of molecular crystals is in the aromatic hydrocarbon series coronene, ovalene and hexabenzocoronene   . A projection of a typical unit cell is shown in Fig.1 but the correct orientation for molecular imaging is rarely achieved. Therefore crystal behaviour and non-periodic structure interpretation relies on lattice imaging.

2. Experimental

The copperphthalocyanine samples were retrieved from an industrial milling process at specific intervals, dispersed in isopropylalcohol and sprayed onto a carbon coated grid. Isopropylalcohol has been shown not to cause any structural changes during brief immersion at room temperature.

The coronene, ovalene and hexabenzocoronene were purified by vacuum sublimation and this material evaporated onto cleaved KCl maintained at 250°C.

The film was backed with carbon, floated off and collected on a clean grid.

A similar preparation on a heated carbon substrate was also carried out.

All specimens were examined in a JEOL 100C using minimal exposure techniques and Kodak Industrex C X-ray film.

3. Results

3.1 Copper phthalocyanine β→α phase change.

Accurate calibration of each micrograph was essential and this was done by measuring either the 0.98nm planes of the β phase or 0.57nm of the α phase. These spacings were sufficiently different from all other spacings not to be confused and normally some crystals were in the correct orientation for these periodicities to be observed. With accurate calibration the α and β phases could be identified by the images of lattice planes of the phases that occurred on most crystals.

Initially the β phase consisted of very large crystals but grinding for 3 hours produced some break-up of the particles via the formation of small α phase crystals on the outside of the particle whilst a core of β phase remained inside - Fig.2. This micrograph shows the random nature of the process with crystal at A being β phase, at B being α phase showing the molecular columns and at all other places being α phase in random orientation. Subsequent grinding extended the α phase throughout the particle with consequent dispersion of the small crystals as in Fig.3. In some cases small remaining crystals of β phase are also dispersed as the particle disintegrates.

3.2 Aromatic hydrocarbons

The situation with the aromatic hydrocarbons is quite different in that all the crystals are prepared in the same orientation but it is normally an incorrect one for molecular imaging. Coronene and ovalene grow normal to the molecular plane on KCl so that the molecular axis is at 45° to the electron beam and grows with the molecular axis parallel to the substrate on carbon which is equally inconvenient as shown in Fig.4. Fine lattice structure is given that provides information about boundaries of these crystals as shown in Fig.5. Hexabenzocoronene follows its smaller analogues in many ways but shows an unusual tendency to spiral growth as shown in Fig.6. There is molecular imaging in the core of the screw dislocation but the molecular projection axes are inclined for the surrounding molecules suggesting a lattice tilt in the strain field of this dislocation.

Lattice images have been obtained from pyrene, perylene, anthanthrene, naphthacene, and quaterrylene showing that the techniques are generally applicable.

4. Discussion

Lattice imaging has two main advantages over molecular or structure imaging for radiation sensitive materials. Firstly, the improved resolution with respect to radiation damage, secondly, the specimen does not require specific orientation. The copper phthalocyanine showed the

application of phase identification from which it was possible to determine the nature and mechanism of the phase transformation - Fryer et al.
1981. At discontinuities, interpretation of molecular crystals is aided
by the large periodicities involved that avoid aberration complications
from the microscope - Fryer 1980.

The aromatic hydrocarbons are much more radiation sensitive than phthalo-
cyanines but the lattice images show that molecular rows either bend at
junctions to minimise strain energy or have a rigid angle which comprises
a regular stacking sequence again avoiding strain energy from mismatch as
shown for coronene.

References.

Fryer J R, McKay R B, Mather R R, Sing K S W 1981 J. Chem. Tech. Bio.tech.
   31 371
Fryer J R 1980 J. Micros. 120 1

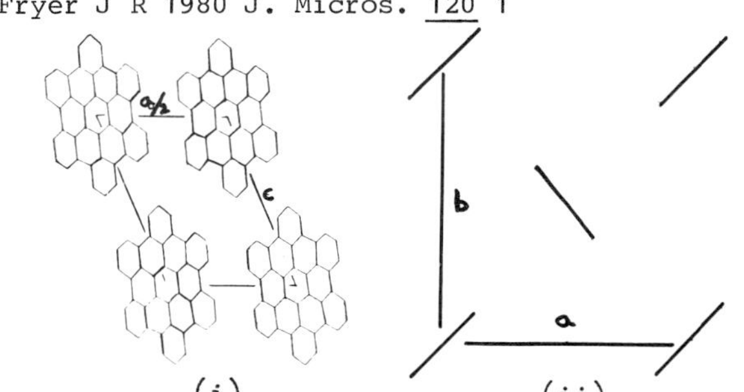

Fig 1 (i) Projection of one half of a
   hexabenzocoronene unit cell along b axis
   (ii) Projection along c axis.

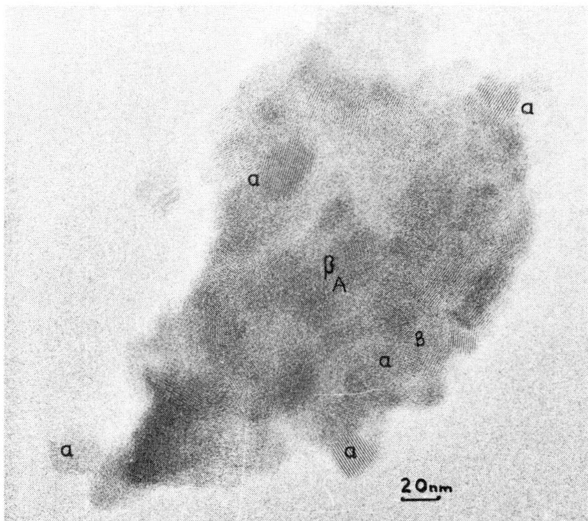

Fig.2 Copper phthalocyanine after grinding
   for 3 hours.

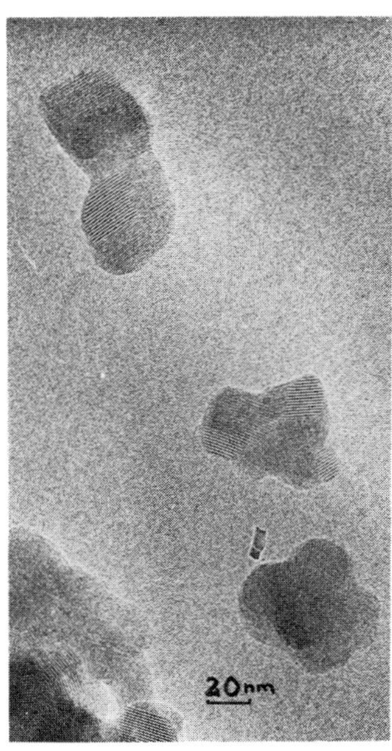

Fig.3 Dispersion of small
   crystals of α phase.

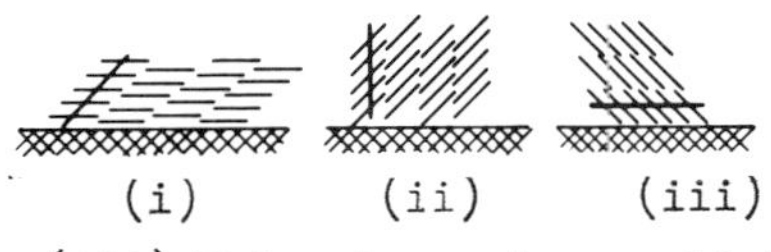

(iii) Molecular axis parallel to substrate.

Fig.4 Growth of planar aromatic molecules on a substrate.

(i) Molecular plane parallel to substrate.

(ii) Molecular axis normal to substrate. Suitable for molecular imaging.

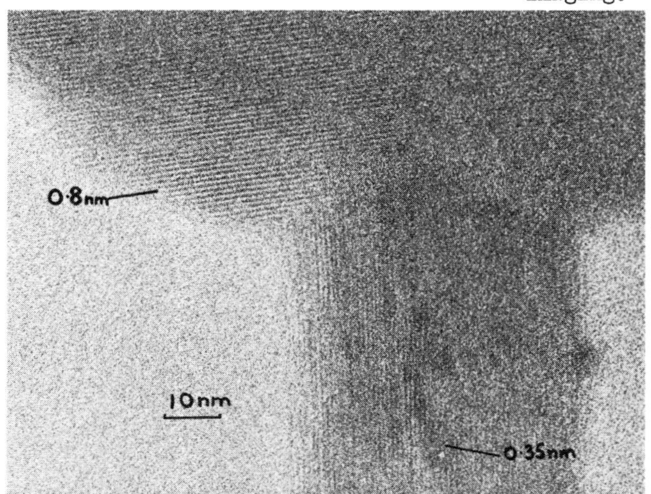

Fig.5 Junction between two coronene crystals.

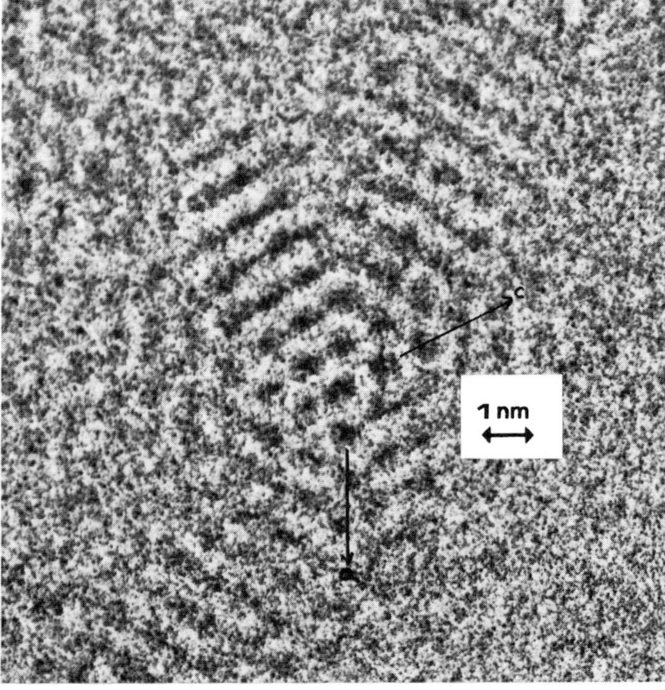

Fig.6 Molecules in spiral growth structure in hexabenzocoronene.

Crystallisation and lattice resolution of a straight chain paraffin $nC_{36}H_{74}$ and its adducts

J R Fryer

Chemistry Department, University of Glasgow, Glasgow, G12 8QQ.

1. Introduction

The two major difficulties in the study of organic and biological crystals lie in their radiation sensitivity and often unfavourable orientation for the acquisition of information. The use of minimal dose techniques and high speed photographic film has overcome the radiation limitation for some of the more stable aromatic systems such as poly-nuclear aromatic hydrocarbons - Fryer 1978, Smith and Fryer 1981 - but these specimens are 10-100 times more stable than aliphatic hydrocarbons. However, the majority of systems of biological interest are composed of aliphatic bonds so that it is desirable that some degree of high resolution imaging is achieved to permit the study of boundaries and other non-periodic regions of crystalline arrays. This work describes the extent to which low dose techniques have been successful in imaging structure in an aliphatic hydrocarbon, the specimen preparation techniques necessary to maximise information and some possible adducts of the system with an aromatic hydrocarbon.

1.1 Radiation sensitivity

According to the statistical assessment of resolution - Rose 1949, Glaeser 1975 - the point resolution that could be achieved with a paraffin ($>C_{26}$) should be approximately 10nm. The molecule of $C_{36}H_{74}$ is 4.75nm in length and approximately 0.25nm wide so that molecular resolution is unlikely but the much lower value of signal to noise ratio that can be achieved with lattice resolution should permit lattice resolution of the 4.75nm periodicity. This resolution could be achieved by large (>150nm) defocus values that would enhance large periodicities at the expense of fine structure which is lost in radiation damage.

1.2 Paraffin structure and orientation

The zig-zag chain of n-paraffins stacks - for even numbers of carbon atoms - in an orthorhombic unit cell structure - Muller 1932. Various polymorphic forms have been reported but the normal structure is orthorhombic and has been extensively studied by X-ray and electron diffraction - Dorset (1978). The n-paraffin $nC_{36}H_{74}$ is soluble in most hydrocarbon solvents and Dawson and Vand (1951) showed that it normally crystallised with the long c axis vertical to the substrate and exhibited spiral growth. This orientation precluded lattice resolution of the 4.75nm periodicity.

A similar problem had been encountered in work on polyesters and Willems (1955) had shown epitaxial growth of polymers on aromatic hydrocarbons whilst Wittmann and St. John Manley (1978) had found that the paraffin $C_{32}H_{66}$ crystallised with its c axis parallel to the substrate on naphthalene and other aromatic hydrocarbons.

2. Experimental

2.1 Electron microscopy

The JEOL 100C microscope was operated at a magnification of 10-20K to achieve resolution of the 4.75nm lattice without undue radiation damage. Minimal exposure techniques were employed - using the dark-field switching mode to remove the beam from the specimen and control exposure - and Kodak Industrex C X-ray film.

2.2 Specimen preparation

A dilute (0.1%) solution of $nC_{36}H_{74}$ was made up in molten naphthalene, and a drop of this solution put onto a carbon film on a grid or the grid dipped into the melt. Vacuum sublimation at room temperature removed the naphthalene leaving the paraffin mostly in the (001) orientation parallel to the carbon film.

When coronene was added for co-crystallisation the coronene was added to the melt in the same proportion as the paraffin and specimens prepared in a similar manner.

3. Results

Fig.1 shows a low resolution micrograph of the paraffin crystallised from naphthalene with inset diffraction pattern showing (00ℓ) reflections. The lattice in Fig.2 corresponds to the (001) of 4.75nm although the optical diffraction (inset) extends to ~1.2nm suggesting a resolution of 2.37nm in the image although the signal to noise ratio prevents its observation.

When coronene is present the paraffin is much less ordered, except when the paraffin co-crystallises with the coronene as shown by the diffraction pattern in Fig.3 with the paraffin c axis normal to the coronene surface. Other regions gave diffraction patterns similar to double diffraction Fig.4 and lattice images such as Fig.5 of 12nm periodicity. However, in all cases this same periodicity pertained so that it is suggestive of some order between the coronene and the paraffin.

4. Discussion

The crystallisation technique to change orientation could have much wider application and has recently been successfully applied to many phospho-lipid systems - Dorset 1980. It appears that staggered aliphatic chain moieties of a molecule will align on the (110) plane of an aromatic hydrocarbon such as naphthalene, giving acicular crystals in this unusual orientation.

The resolution of the crystal lattice and the crystal shape show even growth with the (001) being the slowest growth direction and it is this plane that comprises the sides of the needles. The growth is clearly controlled by the initial orientation of the crystal nucleus and in the case of the aromatic additives this orientation is affected. The coronene and paraffin have at least an epitaxial relationship but the much greater radiation stability of the coronene does not appear to confer extra stability on the paraffin as judged by the rates of fading of their respective diffraction patterns. This would suggest that there is little or no chemical interaction between the two molecular species.

References

Dawson I M and Vand V 1951 Proc. Roy. Soc. 206A 555.
Dorset D L 1978 Z. Naturforsch 33a 964
Dorset D L 1980 Private Comm.
Fryer J R 1978 Acta Cryst. A34 603
Glaeser R M 1975 In "Physical Aspects of Electron Microscopy and Micro-
  beam Analysis" ed. B M Siegel and D R Beaman (New York, Wiley) 205-230
Muller A 1932 Proc. Roy. Soc. 138A 514
Rose A 1948 Advances in Electronics 1 131
Smith D J and Fryer J R 1981 Nature 291 481
Willems J 1955 Naturwiss. 42 176
Wittmann J C and St. John Manley R 1978 J. Polym. Sci. 16 1891

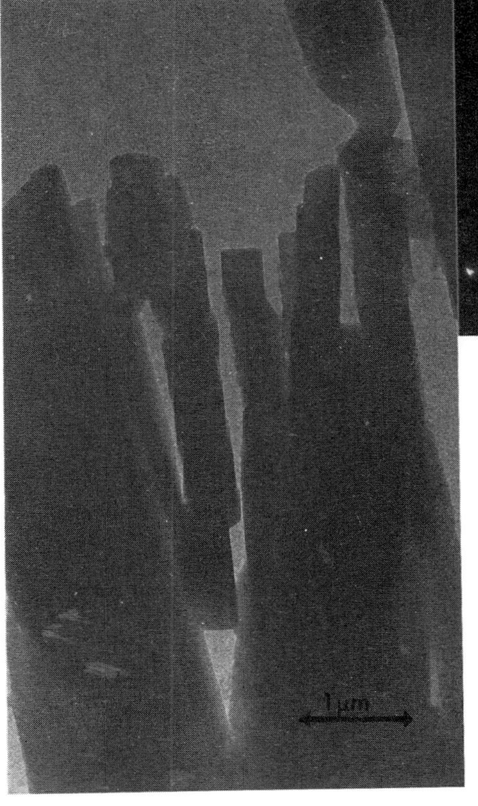

Fig.1 Paraffin crystals and their electron diffraction pattern, in (001) orientation.

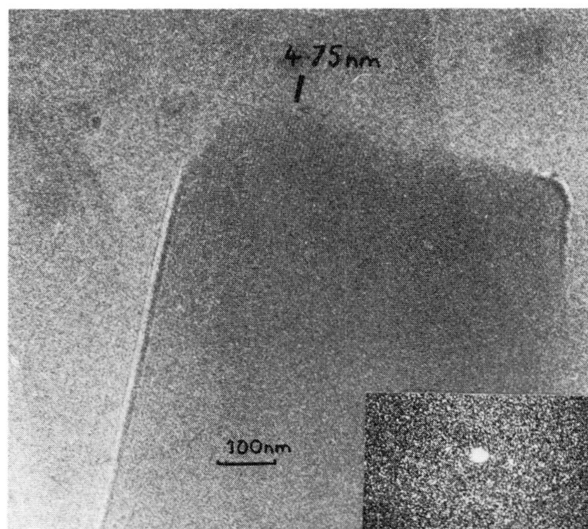

Fig.2 4·75nm lattice of
    paraffin in (601)
    orientation and the
    optical diffraction.

Fig.3 Coronene diffraction pattern
together with (hk0) paraffin
reflections.

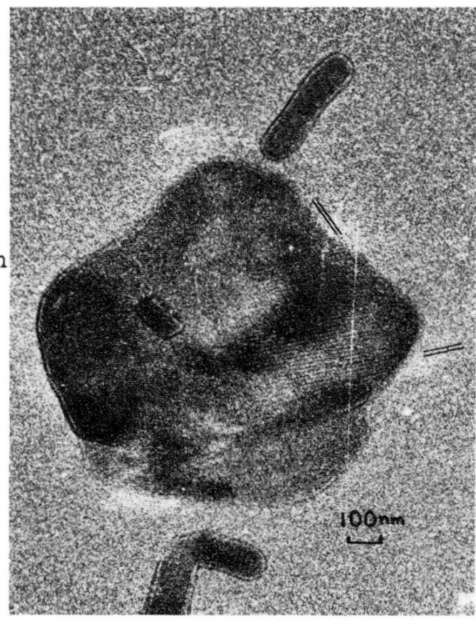

Fig.5 Typical coronene/paraffin
    crystal exhibiting 12nm
    lattices.

Fig.4 (hk0) paraffin multiple
    diffraction.

Electron irradiation damage mechanisms in sodium β''-alumina

R Hull, D Cherns, C J Humphreys and J L Hutchison

University of Oxford, Department of Metallurgy and Science of Materials,
Parks Road, Oxford OX1 3 PH, England.

Introduction

Sodium β''-alumina ($Na_2O.5Al_2O_3$) is a superionic conductor, used as the
solid-state electrolyte in the sodium-sulphur cell. The structure, shown
schematically in Fig.1, consists of the alternate stacking of spinel-like
blocks and 'conduction planes' of composition Na_2O. The spinel-like blocks
are separated by Al-O-Al bridges (the O lying in the conduction planes) and
the loosely bound Na^+ ions are able to move in the conduction planes. A
single crystal is therefore a two-dimensional ionic conductor. The space
group is R3m and the conventional hexagonal unit cell, with a = 5.614 and
c = 33.85Å. contains 3 spinel-like blocks separated by conduction planes
(normal to $\underline{c}$ and 11.3Å apart).

The material damages rapidly in the electron microscope. The effect of the
damage is that conducting planes 'disappear' so that adjacent spinel-like
blocks join up (see Fig.2) (De Jonghe 1977; Matsui and Horiuchi 1977,1981;
Bovin 1979; Humphreys et al. 1980). Similar missing conduction planes
have been observed in battery material both before and, at greater density,
after use, suggesting that the electron irradiation damage process may be
a vastly speeded-up version of the degradation of the material which occurs
during cycling in batteries. An understanding of the electron irradiation
damage mechanisms may therefore be of importance in improving the lifetime
of the electrolyte in the Na/S cell.

The primary damage mechanism

Since electron irradiation causes the conduction planes to 'disappear' the
damage mechanism results in the displacement of the oxygen atoms in the
Al-O-Al bridges separating the spinel blocks. Experiments we have
performed over a range of incident electron energies (100keV to 1MeV) show
that the damage is reduced as the incident electron energy increases. The
primary damage mechanism therefore cannot be displacement damage of the
oxygen atom, but is likely to be ionisation damage (the cross-sections for
ionisation and displacement damage decrease and increase, respectively,
with increasing electron energy). The ionisation damage leads to
displacement of the conduction plane oxygen atoms by a mechanism which is
at present unclear.

The production and growth of crystallographic shear planes

When a conduction plane is eliminated the spinel-like blocks on either
side of it collapse and shear. This has been studied by De Jonghe (1977),

Bovin (1979) and, in particular, by Matsui and Horiuchi (1981). However it
has not previously been recognised that the defect formed is a crystallographic shear (CS) plane and no mechanism for the growth of the defect has
been proposed. In the perfect crystal a conduction plane oxygen atom is at
the corner-sharing site of two AlO_4 tetrahedra. When this O atom is
removed the structure collapses and shears by the two tetrahedra sliding
towards each other along an edge and becoming edge sharing. The fault
vector thus corresponds to an edge direction of one of these tetrahedra,
i.e. $R = 1/30$ [$55\bar{0}2$] or one of the two other crystallographically equivalent $\bar{d}$irections. This fault vector is consistent with fig.2 and other
micrographs. Some subsequent rearrangement of atoms may occur.

Electron microscopy reveals that the mechanism for the growth of the CS
plane is similar to that proposed by Anderson and Hyde (1967) for binary or
complex oxides. Some oxygen vacancies, produced by the primary damage
mechanism, segregate into a disc. The central part of the disc collapses
and shears, and this disc is bounded by a dislocation loop of Burgers
vector $\underline{b} = 1/30$ [$5\bar{5}0\bar{2}$]. (This dislocation is partial and of mixed edge and
screw character.) In our case, the CS plane always appears to nucleate at
either the top or bottom surface of the crystal. We therefore observe not
a complete dislocation loop, but a crescent in the plane of the conduction
plane. An expanding crescent is shown in Fig.3 obtained by tilting the
crystal. The crescent expands because the elastic strain at the dislocation
makes it a sink for absorbing further oxygen vacancies. Thus the conduction plane is eliminated by the growth of the dislocation crescent. In
fig.2 the bounding dislocation segment is being viewed end-on. Fig.4 shows
a tilted crystal demonstrating that a dislocation crescent is associated
with the end of each disappearing conduction plane.

The nucleation of CS planes at the crystal surfaces may be simply understood. Unlike the case of metals, the vacancies cannot be absorbed at the
crystal surfaces except by structural collapse. In this case, the lower
line energy of a half loop at the surface compared with a complete loop in
the bulk will favour surface nucleation. In battery material we would
expect the CS planes to nucleate at the electrolyte surfaces and possibly
at grain boundaries.

Conclusion

After the primary damage event, secondary damage proceeds by the production
of CS planes. These grow by a dislocation mechanism and the nucleation is
at crystal surfaces. If a similar damage mechanism occurs in batteries
(the primary ionisation event being caused by the moving sodium ions interacting with the oxygen atoms) the lifetime of the electrolyte might be
improved by adding impurities to be located in the conduction plane. A
suitable impurity should pin dislocations without adversely affecting the
conductivity.

It may be relevant that the material used in this work was lithia doped,
which is believed to perform better than magnesia doped material. Neutron
diffraction indicates that whereas all the magnesium is located within the
spinel blocks, some of the lithium may be located within the conduction
planes. If this is the case, one way in which the lithium may improve the
material is by pinning the movement of dislocations along the conduction
planes and thus minimising the loss of these planes. If a sufficient
number of conduction planes are eliminated the resultant strain may lead to
cracking and fracture.

The authors are grateful to Dr. W. Bugden and Dr.J.H.Duncan of British Railways Board for their encouragement and for supplying the specimens.

Anderson J S and Hyde B G 1967 J.Phys.Chem.Solids 28 1393
Bovin J O 1979 Acta Cryst. A35 572
De Jonghe L C 1977 Mater.Res.Bull. 12 667
Humphreys C J, Hutchison J L, Wickens V J, Bugden W G and Duncan J H
    Proc.Eur.Cong. on Electron Microsc., The Hague 1980 1 382
Matsui Y and Horiuchi S 1977 Proc. 5th Int.Conf. on HVEM Kyoto p.321,
    1981 Acta Cryst. A37 51.

The authors thank Professor Sir Peter Hirsch for very useful discussions, and the British Railways Board (R.H.), the Central Electricity Generating Board (D.C.) and the Science and Engineering Research Council (J.L.H., R.H.) for financial support.

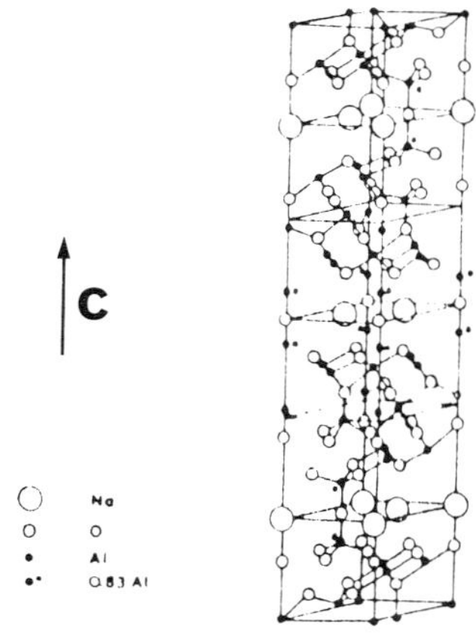

FIG.1   (a) Diagram of one unit cell of the sodium β''-alumina structure.
(b) HREM lattice image, electron beam along [11$\bar{2}$0], rows of bright white spots correspond to conduction planes 11.3Å apart, the dark space between each bright spot corresponds to a bridging oxygen atom (c) HREM lattice image, beam along [21$\bar{3}$0]. Both micrographs taken at Scherzer defocus on the JEOL 200CX.

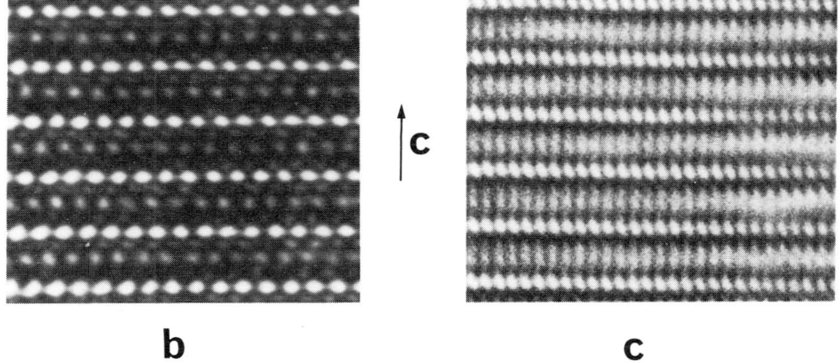

b   c

FIG.2 Sequence of lattice images from the same area taken at approximately 15sec intervals showing a disappearing conduction plane (arrowed). Scherzer defocus, [11$\bar{2}$0].

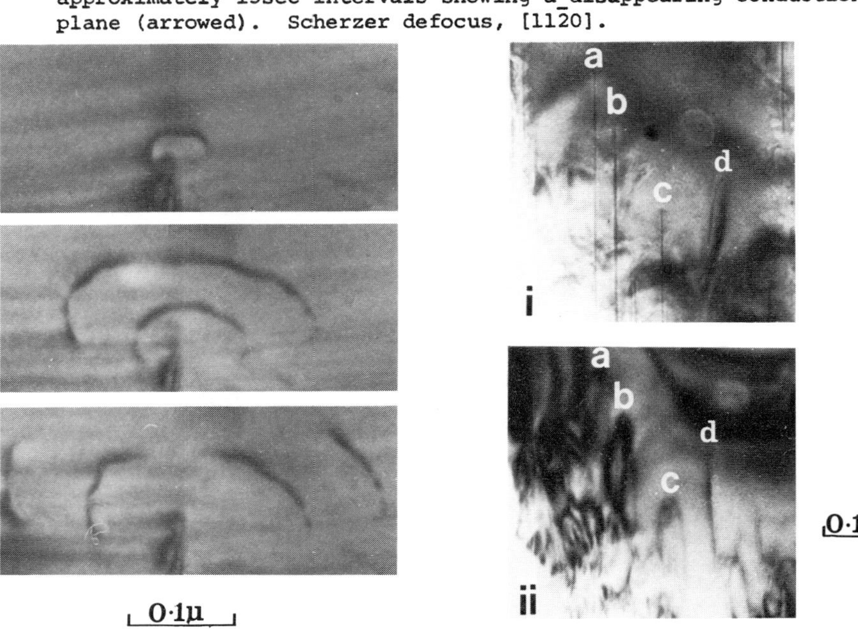

FIG.3 Bright field micrographs of the same area of crystal with conduction planes obliquely inclined to the beam. The sequence shows dislocation half-loops expanding on different conduction planes.

FIG.4 Bright field micrographs (i) end-on view of partially collapsed conduction planes (dark lines) terminating at (a), (b) and (c). (ii) the same conduction planes obliquely inclined, the dislocation half-loops associated with the edge of the collapsed planes are seen. (d) is a dislocation not lying in a conduction plane.

Instrumentation today and tomorrow

E D Boyes

Department of Metallurgy & Science of Materials, University of Oxford.

1. Introduction   Instrumentation comes between the specimen and the micrograph and separates the concept from the paper. It makes a contribution to many results which is often imperfectly understood and the perception of it significantly influences the design of experiments and hence the range and success of applications of electron-optical techniques. Advances can be strategic, in expanding the range of the possible; or tactical, which involves practical implementation. The properties of individual components are a necessary condition for performance but to be useful they must be integrated into a complete instrument, however simple.

2. Objective Lenses   The most critical component in any microscope is usually the objective lens. Earlier designs were restricted both by electronics of limited output and stability, which diminished the value of proposed improved lens geometries, and by a desire to avoid the non-linearities associated with saturation and which were difficult to calculate. Originally low excitations of $NI/V_R^{\frac{1}{2}} \sim$ 10-15 were used with the specimen placed almost out of the objective lens field (an advantage with magnetic specimens). Crewe-type developments of the symmetrical condenser-objective require an excitation nearer to 25 for $C_S \sim$ 0.5mm, or about 8000 Amp turns at 100kV (12000 at 200kV). Second-zone or higher order operation requires more power, and wider gap designs less, but even now few microscopes provide the 10,000 At at 100kV or up to 15,000 at 200kV necessary for full flexibility.

At very high voltages lens saturation may become a genuine problem and restrict the minimum C_S value attainable with normal pole-piece materials and coils. In this regime or others where greater excitation is required, the additional capabilities and complexity of superconducting systems can become useful, but costs may remain prohibitive except where a very cold specimen is also required and especially if this is confirmed to reduce radiation damage. Nevertheless C_S values of $\sim$ 1mm seem destined to become common and could be obtained at up to 600kV in a symmetrical condenser-objective lens with a 5-6mm gap, which is sufficient for a medium tilting goniometer and a low take-off angle X-ray detector. Without appreciable tilt and at lower voltages, $C_S \leq$ 0.5mm is quite realistic as was pointed out many years ago and is the basis of the single atom imaging STEMs. No more than an 8-9mm gap or equivalent dimension (C-O $C_S \sim$ 1.6mm) would be required for a $\pm$ 60° high tilt goniometer.

A series of papers has considered the high excitation objective lens and it can be concluded that the symmetrical type has fundamental advantages. However in a complete microscope system the integration within the whole machine is also important and here the symmetrical lens has some problems. Firstly, the relationship between image focus and illumination conditions through the pre-field is only satisfactory with a suitable condenser system and if the specimen is at all times exactly at the lens centre. Secondly, most existing high resolution goniometer stages are for top entry configurations which are not suitable for the relatively narrow bores of a classic C-O lens configuration; and finally alignment appears to be particularly critical. The illumination can be modified with

a coupled condenser lens or a 'twin-lens' which has essentially the same
function for CTEM. Alignment and other aspects of flexible operation are
in the process of undergoing a revolution as they become computer controlled, which can also be used to cycle lenses and minimise the effects of
hysteresis. Although this will allow the conventional condenser lens to be
focussed correctly on the front focal plane of the C-O lens, there are other
reasons for desiring the flexibility provided by an additional "condenser"
lens. This or some other method of obtaining diffraction patterns from
small (< 50nm) diameter areas is going to be important in HREM.

3. Specimen Stages   The mechanical stability of the specimen stage is
probably the least satisfactory aspect of many modern microscopes. High
resolution non-tilting stages and goniometer versions of them with very
restricted tilt have drift rates of $\sim 5\text{Å}$ per minute, with considerable ir-
reproducibility and vast variations between individual microscopes of the
same type. It could be argued that the positioning accuracy of $\leq$ 10nm
possible with a good stage is itself rather improbable, given normal
instrument engineering tolerances, but this indicates that stability has to
be built into the design and cannot be added by workmanship (but can be
reduced by it). For example the substitution of bellows vacuum seals for
"O"-rings on the linear motions improves both the vacuum and the mechanical
properties. The HR stages of both types and top entry goniometers are
relatively compact and have symmetrical mechanical arrangements. The
specimen holder is detached from the airlock etc. as it is loaded and
located in the stage which then surrounds it, so that its influence on the
stability of the basic stage, but not necessarily the specimen, should be
minimal.

Side-entry goniometers on the other hand are large, asymmetrical and pro-
trude through the side wall of the microscope, leading to a considerable
sensitivity to vibration and variable drift problems. The goniometer
holders, in particular, are a fair approximation to an audio resonator, with
a large mass of motors, etc, at the extreme outer end. They were designed
for a previous generation of medium to low resolution TEMs and have been
impressively developed for the current TEM/STEMs. The single axis eucen-
tricity is useful but necessitates several O -rings even in the 'clean'
versions. Both the mechanical stability and vacuum could be improved by
going over to a detachable tip system, similar to the HR side-entry stages.
Full bellows sealing would require independant tilt and shift compensation
with computer controlled external motor drives. The requirements are
mechanically no better than the best existing stages with which external
control inputs are converted into internal motions with good integrity,
minimum backlash and high reproducibility.

Several existing stages meet these criteria and it is very unfortunate that
the only attempt so far to achieve this highly desirable goal was fundamentally flawed in concept by the specification of a tilt-rotate geometry
which is very much more difficult to engineer but which nevertheless very
nearly worked satisfactorily. A double tilt, doubly eucentric version of
the same thing would have worked well and facilitated the use of more
advanced lens types, by maintaining a constant plane of focus. Long-rod
specimen holders are suitable for the many applications in which lower
stability is acceptable and for the features of heating, cooling, EBIC etc.
connections.

4. Electron Sources and Illumination Systems   Some improvements to standard thermionic guns continue to be made, it would appear almost always as
a consequence of practical experience. Pointed filaments have largely been
replaced by plug-in LaB_6 units which are brighter, more stable and have a

much longer life of 1000+ hours. They do require, however, a better vacuum of < 1×10^{-6} torr. Systems are not optimised to fully exploit the smaller source sizes and lower work function of LaB_6 tips that have been demonstrated with both single and polycrystals on other, generally lower voltage, test instruments. Practical brightness seems to be up 3-5X over tungsten, as measured by TEM phototimes and makes practical microscopy at and above 1,000,000X with direct phosphor image viewing and standard photography. With image intensifiers shot noise is reduced.

Much bigger gains e.g. $\beta\times1000$, should be available by using a field-emission gun, but there are problems of stability, total current and coupling to an HVEM accelerator stack. For high resolution probe systems at medium and low voltages, the FEG has obvious advantages and since some demagnification of 5-50X is almost always involved, the engineering is in the realm of the possible. For CTEM however the advantages are much less obvious and the coupling problems more acute since magnification and higher total current may be required from a simple adaptation of an existing CTEM illumination system much of which is then redundant.

Cold field emission in either <310> or <111> orientation has a good stability short-term but an extended emission decay over multiple minutes to hours and is only satisfactory if a genuinely clean UHV vacuum is used. TF emission at $\sim$ 1800K is less attractive than "warm" emission at $\sim$ 1200K in which the temperature is held just above that required to avoid long-term adsorption of contaminants. As a consequence long term stability (DC) can be good but there is a ripple of ± 1-15%, which must be removed by video compensation via an emission sampling aperture the reference from which is used to divide the imaging signals. Longer term data collection, such as X-ray microanalysis is improved and a refinement involves build-up treatments to localise the emission about the tip axis which increases both the available beam current and the brightness, with potential advantages in reduced outgassing from the surrounding electrodes. Coupling of FEGs to columns is presently somewhat rudimentary. Satisfactory purely electro-static systems have yet to be demonstrated above about 50kV. Electrode geometry is less critical than previously supposed, but voltage ratio restrictions remain. The simple triode geometry has been used up to 120kV and the focussing properties are insensitive but not completely independant of the voltages used. Modifications with electrostatic einzel lenses mounted on the EHT insulator immediately in front of the tip are advantageous but the biggest improvements can be expected from incorporating low aberration magnetic pre-condenser lenses either on the EHT insulator or more practically immediately behind a reshaped anode with field leakage to cover the tip. Reducing the distance between the tip and a conventional condenser lens to about 75mm is possible and can increase usefully the current density in the intermediate cross-over by improving the aberrations of the illumination system and hence its apparent brightness.

5. <u>Contamination Avoidance</u> Construction is now with liner tubes and alignment is electronically controlled but anyway more accurate by design and construction, with bellows sealed linear drives on stage, apertures etc. Outer magnetic and thermal shields contribute to the substantial mechanical stiffness of columns but a tight site specification is still necessary for the best results.

Ion pumps for the specimen chambers of TEM/STEMs are coming in but semi-UHV oil diffusion pumps are almost as clean if UK sourced. Specimen contamination is particularly serious with microdiffraction or CL and is introduced primarily by the specimen which may also have an oxide polishing residue. Pretreatment must be under high vacuum, usually in the airlock with access to the specimen for heating, ion-sputtering etc. Heating to $\sim$ 50°C at 10^{-9} torr is used to eliminate contamination on the dedicated STEMs used to image single atoms but high vacuum airlock prepump systems are not yet common.

6. <u>High Resolution Electron Microscopy (TEM and STEM)</u> Some microscopes now achieve a resolution which up to at least 200kV is similar to the theoretical predictions, indicating adequate control of the engineering and environmental factors. The important parameters are then the electron energy or wavelength; the spherical aberration of the imaging system, primarily the objective lens, and the characteristics ($\Delta\theta$, ΔE) of the illumination system which are related to source brightness. High resolution performance is most usefully described by the contrast transfer function (CTF) two aspects of which are of particular importance. The first zero (d_1) of the CTF at optimum defocus (D_{opt}) slightly beyond the Scherzer value, defines the range of spatial frequencies for a structure image which can in principle be reliably interpreted from a single micrograph of a suitably thin specimen. Beyond this, oscillatory contrast containing phase inversions and zeroes in the spatial frequency spectrum may be preserved down to a resolution (d_{min}) limited by the illumination and stability or envelope terms, but no single micrograph can contain a complete range of fourier components and a through-focus series is mandatory for any serious interpretation. Down to d_1 insights into completely unknown structures can be obtained for interactive interpretation of the images, but below d_1 it is only possible to confirm, or otherwise, matching to reasonably well defined structural models. The d_1 resolution is given by $AC_s^{1/4}\lambda^{3/4}$ where 0.6 < A < 0.7 and values for a series of real ($*$), proposed (ϕ) and hypothetical specifications are given (for A = 0.64) in the table. From this it will be seen that similar results can be obtained with various combinations of wavelength, kV and C_s value. In general it is better to optimise C_s, unless there are overriding requirements of access to the specimen for particular modes of operation. Adding new pole-pieces and if necessary excitation no longer involves critical technology, so that optimum performance can reasonably be expected. This is not the case with improving λ through increasing the kV where considerable problems still remain. For example the combined AC and DC drift per minute stability of EHT systems in the 0 to 200kV range is now of the order of one ppm and this is maintained through the cables and accelerator systems to be transferred to the beam incident on the specimen. At higher voltages no better than 5 ppm is achieved and transmission etc. problems can become acute. The disproportionate size of higher voltage instruments also requires special and costly building provisions and exposes them to a greater sensitivity to environmental factors. Worldwide there have been rather few HVEMs, perhaps 60, over which to spread development. As a consequence none of the HVEMs is believed to produce a full theoretical performance, but this may not be strictly necessary since the CTF first zero (d_1) may be more useful and attainable. In fact a case can be made for diffraction-limiting images to this level to remove confusing and probably uninterpretable fourier components which have been shown in some cases to lead to image distortions.

The immediate future of high resolution electron microscopy probably lies with medium high voltage instruments incorporating low aberration lenses ($C_s \lesssim$ 1mm), having great stability and working routinely at magnifications of 1,000,000X or more, but as far as possible based on existing technology. The engineering of the STEM and particularly of the essential high voltage FEG to achieve the calculated CTF performance may be more difficult but the fundamental optical limits will be very similar for CTEMs and STEMs.

For crystal structure determinations under coherent or partially coherent illumination conditions CTEM imaging will be preferred but a number of STEM functions are compatible with the condenser-objective lens and would extend usefully the information available from HREM images as well as providing assistance in the interpretation of them, for example by providing a CBDP to determine independantly the thickness and exact orientation of selected small regions of the specimen.

STEM imaging systems have possible advantages for incoherent imaging; for the use of a multiplicity of categories of electrons including those scattered inelastically; for complex dark-field techniques based on energy filtering or annular and other shaped geometries of detectors producing parallel signals, especially from amorphous or SRO materials and because of the continuous availability of the complementary CBDP. The limitations are likely to be ones of S/N as the available signal is further subdivided, and even at very high magnifications the brightness of the FEG is still not sufficient to provide as good a basic BF S/N as the conventional or LaB_6 gun of a CTEM used with a static beam. However the STEM provides a much greater flexibility in handling images or other data which are always collected serially, at a freely selectable rate and in principle at any magnification appropriate to a particular scientific problem. The use of more appropriate signals e.g. at defects, may require only modest resolution to be effective. In either type of instrument an interface for on-line analysis and processing of images would be available.

Individual surface atoms, holes in surfaces and single atom steps on them have been imaged with both techniques and either type of instrument could be developed to image under the rigorous UHV conditions appropriate to a definitive surface physics experiment, but the STEM can be more readily controlled to restrict radiation damage.

The most pressing needs in HREM are now in a better understanding of the basic physical process, and the use of carefully controlled experimental procedures to assist image interpretation, particularly where strong phase objects are also present. It is therefore timely to consider whether the value of the enhanced information available in a STEM system is more important than any probably quite small differences in BF image quality which may result as a consequence of the reduced S/N.

7. <u>Microanalysis</u> Microanalysis of thin specimens is most conveniently performed using EDS X-ray methods for which neatly packaged accessories are readily available. However the spatial resolution of X-ray methods is substantially inferior to that which can be obtained with microdiffraction or electron spectroscopy (EELS). It can be improved by using very thin specimens, which is not always possible if real problems e.g. grain boundary precipitates are to be contained, and at higher voltages. Major problems of extraneous electrons and hard X-rays from the illumination system producing primary excitations outside the main probe can be overcome in both TEM/STEMs and dedicated STEMs by careful attention to opaque aperturing close to the gun and well before the specimen. This "hole-count" problem is most serious since thick regions of the outer rim of a disk specimen and the grid or specimen holder may be more efficiently excited than the low X-ray yield from the thin region of the specimen area which is under the probe. Low Z specimen-holders of carbon or beryllium are also useful in reducing secondary fluorescence, the interaction of which with other regions of the specimen and with its environment would seem to be a rather fundamental limitation of the X-ray microanalysis of thin specimens.

In modern EPMA instruments normal incidence and a take-off angle (γ) of

30-50° is used to reduce corrections. However other considerations are also important with thin specimens: bremsstrahlung is increasingly forward peaked at higher voltage   and the peak to background in X-ray spectra is improved at higher γ; the projected specimen thickness, important for beam spreading at high resolution is minimised and quantitation generally is improved at or near zero tilt; insensitive geometrical conditions for X-ray collection allow the tilt to be chosen for the narrowest projected width of planar features e.g. grain boundaries and plate-like precipitates, improving both the accuracy and sensitivity of the analysis, or to set appropriate diffracting conditions for image contrast.

Integration into high performance, C_S < 10mm, objective lenses can be between the pole-pieces (γ = 0-40°) with a close approach detector or above the upper pole-piece (γ > 60°). Thin window/windowless detectors improve the detection of low Z materials but the resolution (ΔE > 100eV) is in-adequate to resolve overlaps of L and M lines from heavier elements. In addition the very low energy X-rays can be heavily absorbed and unlike EELS this is worse with the thicker specimens which are practical at higher kV. The major advantage is access to the comprehensive and convenient EDS data-handling routines and the detectors seem to be surprisingly robust, but still for rather specialised applications.

The least probe current (pAs) is required for microdiffraction because the elastic scattering cross-sections are much higher than for the single-electron excitations. Recent improvements have included parallel detection using low light level TV cameras equipped with an external high gain micro-channel-plate image intensifier so that the limiting detection efficiency is that of the primary scintillator at about 20%. With an appropriate short persistance phosphor such a system has the possibility of masked or otherwise selected imaging conditions. Comparable resolution Grigson scanning systems have a detection $\sim$ 0.001% but allow quantitation over a wide dynamic range. Continuous VTR tape recording can produce a diffraction pattern from one or more frames each of .033 or .04 seconds using a frame-store and utilising more electrons per frame than a pattern obtained by scanning for multiple seconds. This is particularly useful if the specimen is unstable due to charging, beam heating, radiation damage or contamina-tion as are many catalyst species. A resolution of better than 5nm is obtained with an FEG system. It is inferior to the image resolution since the phase lock of the scanning system converts probe position diffusion due to AC fields into image distortion for this type of instability.

Analytical information may be contained over a wide angular range and it is necessary for this to be transmitted and displayed. However many TEM/STEM pole-piece types seriously constrict the maximum angle. A symmetrical C-O lens or an inverted asymmetrical design is suitable. An angle which is variable up to at least ± 10° at 100kV is desirable but optimum results, especially in the use of higher order layer lines may require a liquid helium cooled specimen holder which is also likely to become desirable to stabilise easily damaged specimens.

The design of electron spectrometers has advanced rapidly to combine high resolution of $\leq$ 1eV with useful transmission. Nevertheless at higher losses S/N may be more limiting than the instrument and the most urgent require-ment is now for a system of parallel signal detection to replace the serial scanning systems over at least a limited range e.g. to look at edge detail. Post-spectrometer optics may be required to magnify the dispersion of 1-5eV/μm to match the dimensions of available detector arrays (diode and CCD). Multiple scattering complicates the interpretation of the spectra and reduces sensitivity: either exceptionally thin specimens or higher kV

are required. Potentially the extended fine structure (EXELFS) may usefully complement EXAFS for thin specimens and light elements.

The maximum acceptance angle of the spectrometer is small at $<10^{-2}$ radians.
A wider angle of scattering from the specimen can be collected by inter-
posing lenses which need to magnify the size of the source. The "TEM" part
of a TEM/STEM lens already performs this function but a considerable re-
organisation of a dedicated STEM, possibly including an objective lens with
an appreciable post field would be required.

8. <u>Electronics for Scan Conversion, Distribution Maps etc.</u>   Little short
of a revolution is underway in the organisation of microscope electronics.
Control functions are increasingly digital with considerable interaction
possibilities as they become software based; making major modifications
simple to implement with a high probability of success. New ways of operating columns can be explored easily and will be particularly applicable to
probe scanning modes with selection of resolution, current and convergence
angle combinations appropriate to specific problems.

Analogue conversion of the primary slow scan of $\sim$10 seconds per frame to TV
standard are an important feature of the STEM instruments used for imaging
single atoms. More flexible digital versions incorporate programmeable
signal processing including spatially resolved time averaging of images etc.
and remove the need for direct TV rate scanning which is not originally
provided on many TEM/STEMs.

Distribution images obtained by mapping the X-ray intensities for a single
element can be very unrepresentative of the actual composition and in many
cases are of limited use for thin foils unless some method of normalisation, e.g. a ratio method, is employed. Firstly there is the major problem
of obtaining adequate statistics with the limited through-put capability of
current EDS detector systems ($\sim$5000 cps) over a reasonable number of resolv-
able pixels in an acceptable period of time; and secondly, variations in
specimen thickness are combined with changes in the relative composition.
In some important applications which relate to stress corrosion susceptibility the specimen preparation process may actually couple these two
effects and the problem of differential polishing or ion thinning is common
in many types of multi-phase materials.

A digital scanning system has been developed specifically for profiling
variations in chemical composition related to mechanical, embrittlement and
corrosion properties. This approach has the advantage of producing at each
of a restricted number of selected points or small areas, a quantitative
spectrum which to first order is independent of thickness or of any beam
current fluctuations. For a selected analysis time the relative importance
of spatial resolution (number of points) and precision in determining composition (dwell time) can be adjusted appropriately for each problem. It
is important to be able to organise the analysis arrays in relation to the
microstructures through high quality electron images. These require a
faster digital scan clock rate than is available from a directly computer
generated scan, implying a dedicated scanning system which can be computer
controlled, for example a 204.8kHz PLL clock for a 512x512 visual image in
1.28 seconds. A digital scan converter/framestore with interactive micro-
computer based graphics which in the standard TV format can be developed by
the microscopist, would provide even greater flexibility. However scan
rotation is more efficiently organised as an analogue function.

As computer hardware costs are reduced there is little advantage in maximum
system integration, particularly where this has to be done by the microscopist. Separate or modular mini/microcomputers for different functions and

with appropriate communications interfaces, are generally to be preferred.

9. <u>Conclusions</u>   Combined TEM/STEM instruments will continue to be the main commercial product, with genuinely clean vacuum and completely revised electronics but some continuity of column engineering. The main type of objective lens will have higher excitation and operate in strong condenser-objective mode with low aberrations, primarily to improve the current into small probes for microanalysis applications but also a high magnification in TEM. Higher stability stages will contribute to a usefully enhanced high resolution imaging capability in this type of instrument.

Higher voltages are in general advantageous with thin specimens, but 1MeV HVEM microanalysis is only just beginning: 400-600kV may be more practical. Very powerful tools for characterising heterogeneous catalysts will be provided by combining analysis, including multi-beam microdiffraction, with a controlled atmosphere environmental cell for in situ studies of reactions.

The future of the FEG is assured in low voltage SEMs etc. and for the highest resolution STEM images, but it may face considerable competition for microanalysis from higher voltage TEM/STEMs equipped with improved LaB_6 guns. Higher voltages should improve the brightness and reduce the beam spreading in the specimen, which may be the determining factor on both the usefulness and accuracy of analysis in many practical situations. Conversely, in the SEM of bulk specimens the range of interaction is reduced at the lower voltages towards which development will accordingly be directed; to surpass TEM/STEMs for imaging surfaces.

For HREM, improvements to extend the CTF first zero (C_s, λ and stability) are to be preferred to more highly coherent illumination systems incorporating FEGs which, if there are no other limiting conditions, modify only the envelope terms. On-line image analysis with diffraction patterns, including CBDPs, and chemical analysis from selected small areas 10-50nm in diameter, will advance interpretation but require a small probe capability in future instruments.

The potential role of transmission techniques in surface science is considerable and the heterogeneity of many surface processes will require improved SEM-style resolution from bulk samples. It remains to be seen whether a new type of combined surface physics/microscopy instrument will be required or whether separate experiments can be related satisfactorily. The superconducting lens PSEM is an outstanding example of a new approach.

Overall electron-optical instrumentation will become substantially more productive and capable of tackling increasingly complex problems, the needs of which will largely determine the direction of future developments. They must be based on a better understanding of the basic physics of electron/specimen interactions and will continue to require a challenging combination of innovation and careful engineering.

The SERC is thanked for continuing support of several projects.

Structure Image Resolutions ($d_1$nm) for Combinations of kV and/(C_s)mm	
kV = 40/(0.5), d_1 = 0.37*	kV = 500/(1.1), d_1 = 0.15*
100/(0.7) = 0.28*	600/(3.3) = 0.18*
200/(4.5) = 0.33φ	1000/(8) = 0.17*
200/(2.0) = 0.27*	1000/(2.8) = 0.13φ
200/(1.2) = 0.24*	1000/(1.5) = 0.11

High temperature *in situ* experimentation in HVEM: instrumentation and application to materials science

R Valle, B Genty, A Marraud and J Cadoz*

Laboratoire de Microscopie THT CNRS-ONERA, 92320 Châtillon (France)

* Laboratoire de Physique des Matériaux, CNRS Bellevue, 92190 Meudon (France)

1. Introduction

High temperature in situ investigation in HVEM necessitates a very specific equipment such as a high vacuum specimen chamber, a low light level high resolution visualization system, a eucentric goniometer stage as well as a double tilting heating holder (1500 K) and a high temperature double tilting straining holder. In each case, the physical phenomena which may limit the capabilities of those systems were evidenced through a physical approach of the problems. The optimization of these physical parameters has then guided the design and construction of these high performance instruments.

Most of the high temperature double tilting or straining holders described in the litterature were designed to perform in situ experiments at a temperature of 1000 K. However, the ever increasing importance of the in situ experimentation on superalloys and ceramics necessitates the design and construction of very high temperature specimen holders (1500 K). The present paper essentially deals with the design and construction of the high temperature specimen holders and the corresponding heat transfer investigation.

2. The double tilting heating specimen holder

This new specimen holder is based on a commercially available furnace (Philips) which may attain a temperature of 1300 K at an input power of 5 W. The platinum furnace is held into a pivoting bushing which not only ensures the second tilt movement but also acts as a thermal shield. With such an optimized construction, a maximum temperature of 1500 K is attained at an input power of 7.5 W (Genty and Gervais 1980).

3. The high temperature double tilting straining holder

In situ deformation tests on ceramics and superalloys not only necessitate very high temperatures (1300 K - 1500 K) but also a complex and strong stretching device to deform these high strength materials. The straining holder designed in Grenoble (Pélissier, Lopez and Debrenne 1980) may be operated up to 1500 K ; however, the maximum load to which the specimen is subjected is only 0.6 N. In order to apply a load of several daN on the specimen, a strong mechanical device is required. Such a strong device should be similar in size to the one previously designed at ONERA which comprizes tungsten jaws with linear ball-bearings directly machined in the jaws (Valle and Martin 1974). Unfortunately, a system of this size would require an enormous power (150 W) to reach a temperature of 1500 K. The radiant energy would contribute elevating the temperature of the surrounding elements such as the pole pieces, the aperture holder, the cryogenic pump and the tip of the specimen holder. Due to the length of the rod of this side-entry specimen holder, this would result in an important thermal drift. The present investigation is therefore aimed at limiting to the lowest possible value the input power required for a given temperature eg 1500 K. This result may be attained by optimizing the heating system and reducing the loss of energy due to thermal radiation and thermal conduction.

The loss of energy by thermal radiation is proportional to the fourth power of absolute temperature (Stefan-Boltzmann law) whereas it is proportional to the temperature gradient in the case of heat conduction (Fourier conduction law). At high temperature levels the loss of energy is therefore essentially due to thermal radiation. Consequently, the high temperature straining holder must be designed in order to reduce the loss of radiant energy.

3.1 Composition and surface finish

The chemical composition and the surface roughness have a great effect on the radiative properties of materials ie the reflectivity and the emissivity.

The total hemispherical radiant energy of a non-black body of emissivity ϵ is given by the equation :

$$E = \epsilon \sigma \, T^4 \quad (\text{Unit} : W\, m^{-2}) \tag{1}$$

where σ is the Stefan-Boltzmann constant ($\sigma = 5.7 \times 10^{-8}\ W\, m^{-2}\, K^{-4}$). (The term total denotes that radiation from all wavelengths is included).

With regards to the chemical composition of the high temperature jaws, tungsten seems to be the most appropriate material since it presents superior mechanical properties at elevated temperatures. Consequently, optically smooth tungsten jaws are used. Experiments have demonstrated that thermal radiation is thus reduced by a factor 1.5.

3.2 Exchange of radiant energy between the jaws and the tip of the specimen holder

If the jaws were radiating inside a perfectly black enclosure, then the radiant energy would be given by equation 1. This energy may be reduced, provided that the emissivity of the enclosure would be lower than unity. In this case, radiant energy would be reflected by the surface of the enclosure ie the tip of the specimen holder. Consequently, the material of the tip must be chosen in order to optimize this reflection phenomenon.

The reflectivity of a surface depends on the chemical composition, the surface finish and the wavelength of the incident radiation (fig. 1). Reflectivity rises as wavelength is increased.

The spectral distribution of hemispherical spectral emissive power E_λ is described by Planck's radiation law :

$$E_\lambda\, (\lambda) = \frac{2\,\pi\, C_1}{\lambda^5\, (e^{\, C_2 /\lambda T} - 1)} \quad (\text{Unit} : W\, m^{-3}) . \tag{2}$$

The wavelength λ_{max}, at which the emissive power E_λ is a maximum for a given temperature, shifts towards shorter wavelengths as the temperature is increased, and is given by the Wien's displacement law :

$$\lambda_{max}\ T = 2.9 \times 10^3\ \mu m\ K. \tag{3}$$

At the temperature of the jaws (1500 K), the emissive power is a maximum for a wavelength of approximately 2 μm. Consequently, the most appropriate metal is gold (Fig. 1). In fact, molybdenum was chosen, since good mechanical properties are required. However, this molybdenum tip has been electropolished and gold coated.

Due to the exchange of energy between the jaws (T_1 = 1500 K) and the tip of the specimen holder, the tip will attain an equilibrium temperature T_2. If a concentric cylindrical configuration is assumed, then the net radiant energy is given by the relation :

$$Q_{12} = \sigma\, S_1\, \epsilon_e\ (T_1^4 - T_2^4) \quad (\text{Unit} : W) \tag{4}$$

where the effective emissivity ϵ_e is given by the relation :

$$\epsilon_e = \frac{\epsilon_1\, \epsilon_2}{1 - (1 - \epsilon_2)\,(1 - (S_1/S_2)\,\epsilon_1)} \simeq \frac{\epsilon_1\, \epsilon_2}{\epsilon_1 + \epsilon_2 - \epsilon_1\, \epsilon_2} \tag{5}$$

where ϵ_1, S_1 and ϵ_2, S_2 are the emissivities and the surfaces of the jaws and of the tip. For example, if $\epsilon_1 = \epsilon_2 = 0.1$, then $\epsilon_e = 0.05$. The radiant energy is therefore reduced by a factor 2. However, the exchange of radiant energy between the jaws and the tip of the specimen holder is limited to only two sides of the jaws. Nevertheless, thermal tests on experimental jaws have demonstrated that the use of optically smooth tungsten jaws, held in an optically smooth golden molybdenum tip, already reduces the thermal radiation by a factor 2.

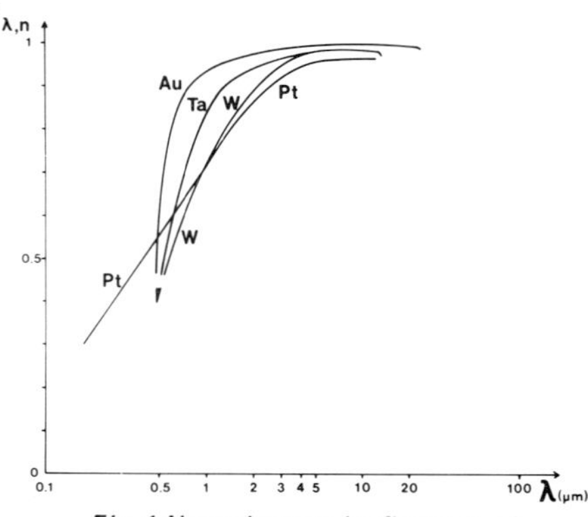

Fig. 1 Normal spectral reflectance of polished metals.

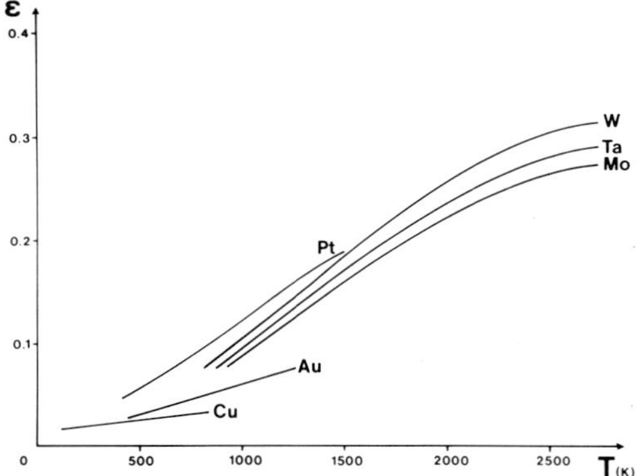

Fig. 2 Effect of temperature on the hemispherical total emissivity of polished metals.

3.3 The radiation shield

When shields are placed between mutually radiating surfaces, the effective emissivity of the system is substantially reduced. If two gray surfaces such as the jaws (ϵ_1, T_1) and the tip of the holder (ϵ_2, T_2) are separated by one radiation shield having emissivity ϵ_s on both sides, then the radiation shield will reach an equilibrium temperature T_s so that the net radiant energy Q_1 from the jaws towards the shield would be equal to the radiant energy Q_2 from the shield towards the tip of the holder. If a concentric cylindrical configuration is assumed, then the quantities Q_1 and Q_2 are given by the equations :

$$Q_1 = \sigma S_1 \frac{\epsilon_1 \epsilon_s}{1 - (1 - \epsilon_s)(1 - (S_1/S_s)\epsilon_1)} (T_1^4 - T_s^4) = \sigma S_1 \epsilon_{1s} (T_1^4 - T_s^4) \tag{6}$$

$$Q_2 = \sigma S_s \frac{\epsilon_s \epsilon_2}{1 - (1 - \epsilon_2)(1 - (S_s/S_2)\epsilon_s)} (T_s^4 - T_2^4) = \sigma S_s \epsilon_{s2} (T_s^4 - T_2^4) \tag{7}$$

where ϵ_1, ϵ_s, ϵ_2, S_1, S_s, S_2, T_1, T_s, T_2 are the emissivities, the surfaces and the temperatures of the jaws, the thermal shield and the tip of the specimen holder. The temperature of the thermal shield may be determined from equations (6) and (7) :

$$T_s^4 = \frac{\epsilon_{1s}(S_1/S_s) T_1^4 + \epsilon_{s2} T_2^4}{\epsilon_{s2} + \epsilon_{1s}(S_1/S_s)} \tag{8}$$

For example, if $\epsilon_1 = 0.3$, $\epsilon_s = \epsilon_2 = 0.1$, $T_1 = 1500$ K and $T_2 = 300$ K, then $T_s = 1200$ K. The reduction of radiant energy is given by the ratio Q_3/Q_1, where Q_3 is the radiant energy in the absence of a radiation shield.

$$Q_3 = \sigma S_1 \epsilon_1 T_1^4 \tag{9}$$

In this case, the radiation shield would reduce the energy loss due to thermal radiation by a factor 5.

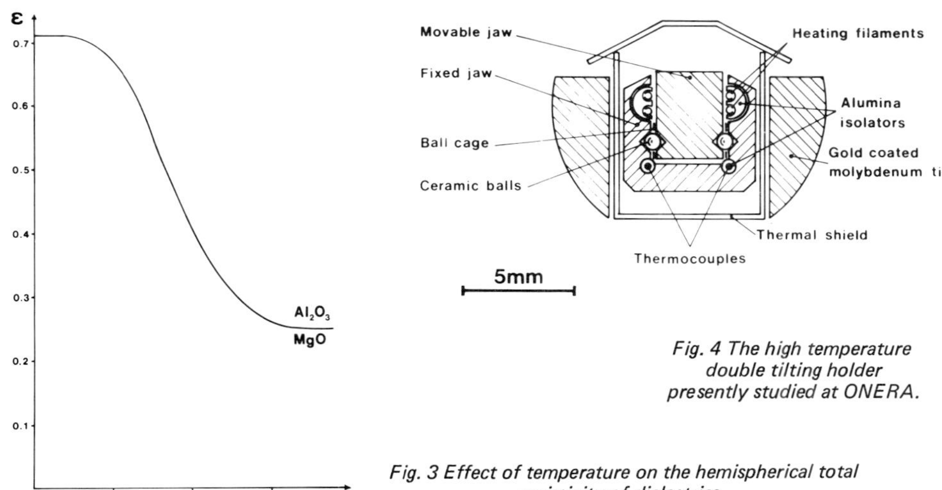

Fig. 4 The high temperature double tilting holder presently studied at ONERA.

Fig. 3 Effect of temperature on the hemispherical total emissivity of dielectrics.

The thermal shield must offer good radiating properties (low emissivity) as well as good mechanical properties in order to remain flat at high temperature. As regards the emissivity (Fig. 2), gold is the most appropriate metal. However, since pure gold does not offer sufficiently good mechanical properties at high temperature, a gold alloy foil must be used. Unfortunately, the emissivity of the alloy may be much higher than the emissivity of gold. For instance, gold-palladium alloys may have an emissivity of 0.5. In order to combine good mechanical properties at high temperature and low emissivity, gold coated tantalum or molybdenum shields seem to be more appropriate. However, due to the high diffusion rate of gold in tantalum or molybdenum at high temperature (1200 K), the use of a gold coated Ta or Mo shield is impossible. As regards metal shields, the only possible solution is thus a tantalum shield ($\epsilon_s = 0.2$ at 1200 K). In fact, the most appropriate solution seems to be a golden silica shield (Fig. 4) which offers superior mechanical properties and would remain flat at high temperature. This solution is in fact similar to those of infrared optics where gold coated front mirrors are used.

The various thermal shields were tested on an experimental set-up. For a given temperature, the reduction of the input power is barely a factor 1.3 in the case of a tantalum shield, whereas it is higher than a factor 2 in the case of pure gold. Consequently, a temperature of 1500 K may be attained with an input power lower than 30 W.

3.4 The heating system

The heating system must be designed in order to minimize the thermal gradient between the jaws and the heating element, thereby limiting the temperature of the tungsten heating filament. This is both aimed at preventing the embrittlement of the tungsten filament and also at minimizing the loss of energy by conduction through connecting wires of a large cross-section. The heat transfer between the tungsten filament and the jaws may be achieved either by heat conduction or by thermal radiation. Thermal conduction through an isolating cement seems to be the most efficient but it presents a major drawback since changing the heating elements necessitates breaking the cement. Moreover, the quality of the thermal contact between the heating elements and the jaws is of outmost importance. In fact, the interface is never perfect and in the case of non-ideal contact the phenomenon is interface controlled. Due to interface resistance there will be an important temperature discontinuity at the interface. If there is no contact, the exchange of energy is confined to radiation. Unfortunately, the emissivity of ceramics decreases at the temperature is increased (Fig. 3); the isolating cement will thus act as a thermal shield. Consequently, as concerns the heating system, radiation heat transfer seems to be the most appropriate solution. The system consists of tungsten heating coils radiating in long holes directly machined in the jaws (Fig. 4). As regards the heat transfer process, the system may be represented by the equation :

$$Q_{12} = \sigma\, S_1\, \frac{\epsilon_1\, \epsilon_2}{1 - (1 - \epsilon_2)\,(1 - (S_1/S_2)\,\epsilon_1)}\quad (T_1^4 - T_2^4) \qquad (\text{Unit}:\text{W}) \tag{10}$$

where ϵ_1, S_1, T_1 and ϵ_2, S_2, T_2 are the emissivities, the surfaces and the temperatures of the filaments and of the holes. Therefore, the system will be optimized by maximizing the exchange surface and increasing the emissivities. At high temperature, the emissivity of the as-extruded tungsten wire is nearly that of a blackbody. Since perfectly black surfaces do not exist in nature, the fixed jaw has been hollowed out to form two long cylindrical cavities having a rough surface and small openings thereby providing a very close approximation to a blackbody. The tungsgen filaments are held in alumina isolators which are machined in such a way that the tungsten filaments directly radiate towards the fixed jaw and the movable jaw (Fig. 4). Due to the large exchange surface between these two nearly-black bodies, the thermal gradient between the wires and the jaws is quite small eg T_1 = 1600 K, T_2 = 1300 K. In order to limit the loss of energy by conduction, the connections are made on isolating alumina pins directly fixed on the jaws. The temperature of the connections, which directly determine the loss of energy by conduction is thus the temperature of the jaws (1500 K) instead of that of the filaments (2000 K). This loss of energy is thus reduced by a factor 1.5. This system works perfectly and, for a given temperature, the input power may be reduced by a factor 2 as compared to a system consisting of filaments embedded in alumina.

4. Applications to materials science. Conclusion

The high temperature (1500 K) double tilting specimen holder is already operating in a satisfactory manner for the last two years and high temperature in situ recrystallization, irradiation and thermal cycling tests on superalloys and ceramics have already demonstrated the efficiency and the reliability of this new specimen holder.

As concerns the high temperature double tilting straining holder, this basic study on heat transfer has open the way to the possibility of designing a straining holder which permit to attain both a very high operating temperature (1500 K) and a heavy load on the specimen. Moreover, the polished tungsten jaws, the golden molybdenum tip and the thermal shield have all been satisfactorily tested on an experimental set-up.

In situ experiments in HVEM on NiO single crystals have been performed and published recently by Cadoz, Hokim, Pélissier and Valle (1981). This investigation has shown that it is possible to deform NiO single crystals at room temperature since dislocation movements have been observed. Moreover, the evolution of a dislocation dipole into dislocation loops was observed and recorded during high temperature in situ experiments (1400 K), thereby confirming the already published models explaining the formation of dislocation loops in ceramics. These recent experiments already demonstrate the possibilities of high temperature in situ experimentation in HVEM.

5. References

Cadoz J, Hokim D, Pélissier J and Valle R Journal Mat. Sc. to be published.
Genty B and Gervais H 1980 Proc. 7th Eur. Cong. Elect. Micr. 528.
Pélissier J, Lopez JJ and Debrenne P 1980 Proc. 6th Int. Conf. HVEM 30.
Valle R and Martin JL 1974 Proc. 8th Int. Cong. Elect. Micr. 180.

Inst. Phys. Conf. Ser. No. 61: Chapter 2
Paper presented at EMAG, Cambridge, 7–10 September 1981

Wehnelt modulation beam blanking in the scanning electron microscope

S M Davidson

The General Electric Co Ltd   Hirst Research Centre   Wembley   England

Beam blanking is essential for much SEM assessment of semiconductors. Typical applications are listed in Table 1, together with the necessary modulation waveforms. Our requirements were for beam pulsing at up to 20 MHz at 2 kV for voltage contrast operation, and blanking at up to ∿5 MHz at 20 kV for EBIC and CL measurements. The aim of this paper is to demonstrate that this is possible on the JSM–35 (and other SEMs) using Wehnelt modulation.

The gun geometry must be optimised so that small changes in gun bias cause large changes in emission current. The most important gun parameter is the Wehnelt to filament distance, sometimes called the filament height. Also of significance are the Wehnelt aperture diameter and the Wehnelt to anode spacing. Figure 1 shows how the emission current varies with gun bias (the gun characteristics) for different filament heights at 20 kV. Ideally the characteristic must be as steep as possible. The gun characteristics as a function of kV are illustrated in Figure 2, for a filament height of 1 mm. These are roughly linear, with a pseudo–exponential tail close to cut–off. Assuming an operating emission current of 200 μA, the blanking voltage is typically 30% of the steady state bias at 2 kV, reducing to ∿20% at 20 kV. If we wish to blank with 50V at 20 kV, and 10V at 2 kV,the steady state bias must therefore be roughly 250V at 20 kV, and 30V at 2 kV. Wehnelt beam blanking should be possible provided that the bias resistor values lie in the range 0.2   to 2 MΩ. On the JSM–35, the standard bias resistor values lie well above this range, and they must be altered. Table 2 shows the standard and modified bias resistor values as a function of bias setting. The optimum gun geometry and operating conditions are summarised in Table 3.

Capacitative coupling was used to feed the modulation signal to the Wehnelt, the coupling components being mounted inside a new lower anode chamber (Figure 3). Rather than breaking the connection from HT cable to Wehnelt, a new Wehnelt was constructed, with an isolated lower section. A spring strip assembly connects the inner end of the coupling capacitor (500 pF/ 30 kV epoxy encapsulated ceramic) to the lower section of the Wehnelt, and the upper section to the lower via a 10 M resistor. In this configuration the capacitance of the gun driven part of the Wehnelt is 15 pF, low enough not to cause loading of the drive circuitry. Isolation is adequate to 25 kV. For stroboscopic operation at up to 5 kV, a 100 MΩ bleed resistor connected from cathode to earth biasses the gun to cut–off with a 2 MΩ cathode bias resistor.

Figure 4 shows the input/output characteristics of the system at 2 kV and 20 kV. At 2 kV, the blanking is driven directly from the 10V output of a

pulse generator; at 20kV a single transistor (PN2369) boosts the amplitude
up to a maximum of 50V. The upper trace is the beam current (taken from
the head amplifier); the lower trace is the modulation signal. The modulation is non-linear; at 2 kV a 2V change in bias can cause a 50% change in
beam current, but 10V is needed for complete cut off (c.f. Fig 2). At
20 kV the behaviour is similar, and in practice the non-linearity is of
little consequence. Figure 5 shows EBIC decays recorded at 20 kV from two
regions of a BFY51 transistor. The minimum decay time is ∿10 nS, limited
by the pulse driver transistor. The longer decay is characteristic of the
device.

In summary, electron beam modulation by Wehnelt blanking is a practical
alternative to the more usual deflection blanking. Advantages over the
latter include the lack of spot movement during switch-off (although a
small amount of defocussing does occur), and the lack of contamination/
charging effects associated with closely spaced blanking plates. The
blanking sensitivity does not vary with lens settings and hence probe
current; no modifications to the column are necessary, and any modulation
waveform is possible, although some distortion may occur. With suitable
gun geometry and bias conditions, electron beam modulation can be implemented using standard pulse generators and/or simple drivers.

Table 1

Application	Mode	kV	Modulation	Waveform
Lock-in amplification	EB,EBIC,CL	5–30	1 kHz– 1 MHz	
Transient analysis	EBIC,CL	10–30	1 kHz– 5 MHz	
Stroboscopy	VC	1–3	50 kHz–50 MHz	

Table 2

Bias setting	0	1	2	3	4	5
Standard (M)	5	6	8.5	11	13	16
Modified (M)	0.5	1	2	3	5	7

Table 3

	2 kV		20 kV
Wehnelt aperture diameter		3 mm	
Wehnelt – filament spacing		1 mm	
Wehnelt – anode spacing		20 mm	
Bias resistor	0.5 M		1 M
Emission current	60 µA		250 µA
Bias voltage	30V		250V
Blanking voltage	10V		45V

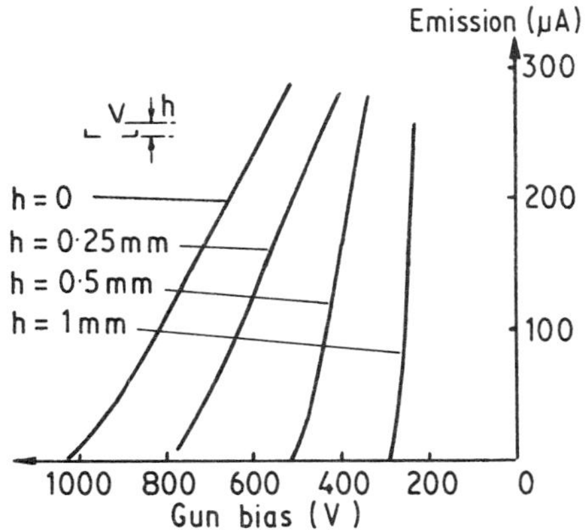

Figure 1 : Emission current v gun
bias for **various** Wehnelt
to filament distances
(h), 20 kV

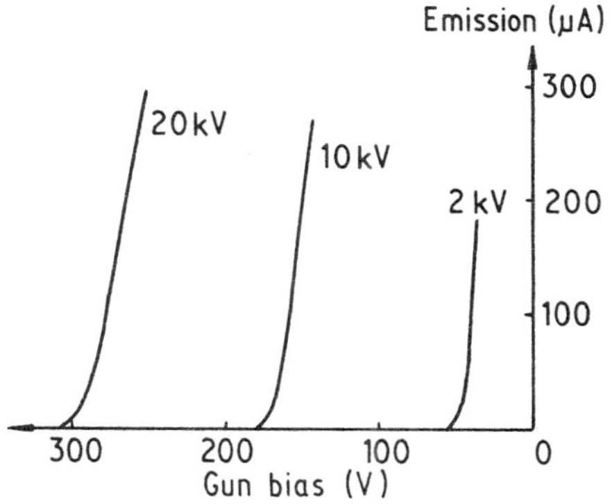

Figure 2 : Emission current v gun
bias for 2, 10, 20 kV
h = 1 mm

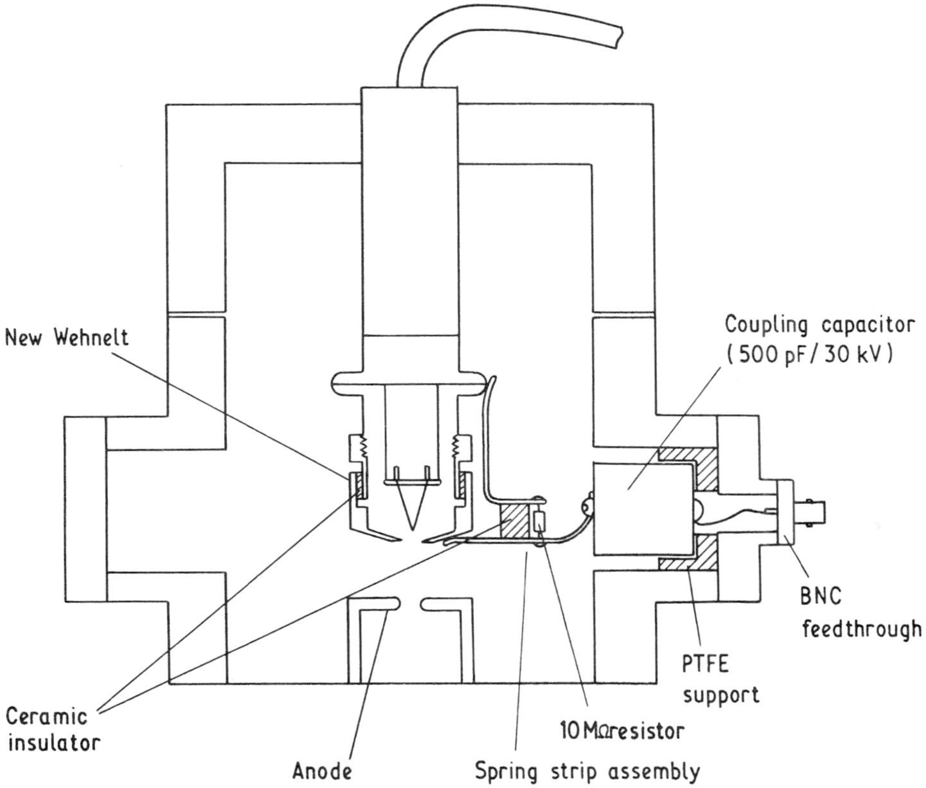

Figure 3 : Modified Wehnelt and gun housing

↑ 2V/DIV

(a) 2 kV

↑ 5V/DIV

↑ 2V/DIV

(b) 20 kV

↑ 20V/DIV

2 μs/DIV

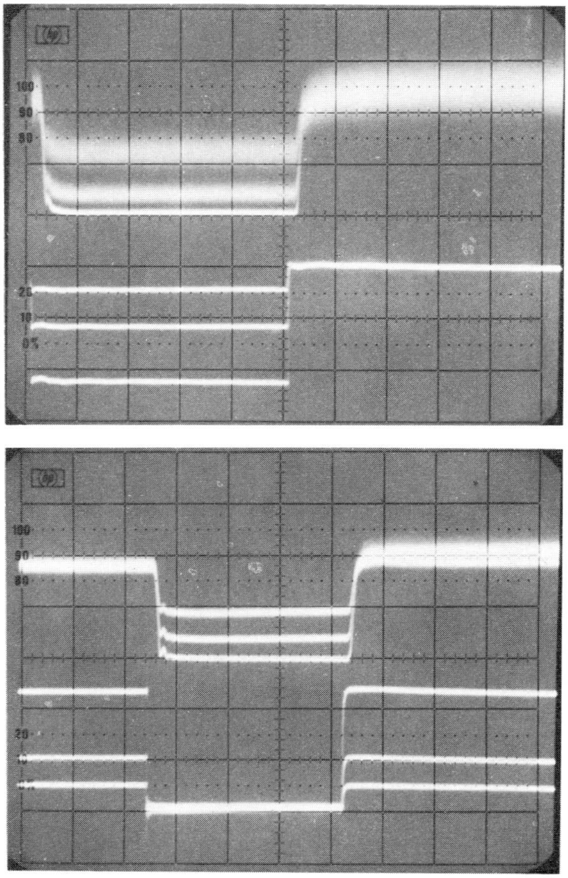

Figure 4 Modulation characteristics of electron gun. Upper trace-signal; lower trace-modulation.

20 ns/DIV

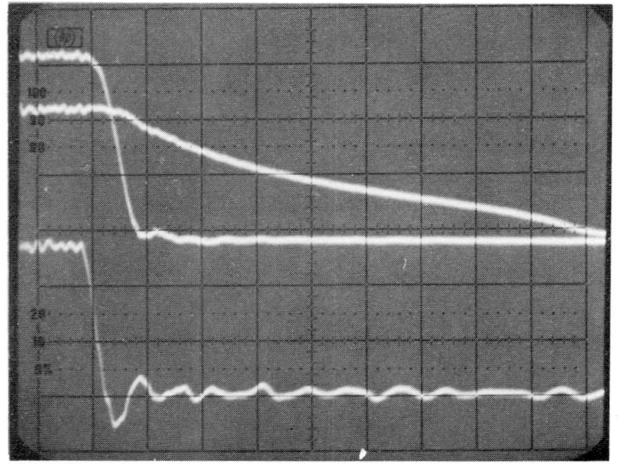

Figure 5 EBIC decays recorded from BFY51.

Uses of contamination in STEM: projection electron lithography

M.T. Browne, P. Charalambous and R.E. Burge
Physics Department, Queen Elizabeth College, University of London,
Campden Hill Road, LONDON W8 7AH.

Contamination in STEM is generally a nuisance and considerable effort is
made to avoid it.   However contamination can be used for beam writing
(Muller, 1971) and can also be used to determine the spherical aberration
coefficient (C_s) for the objective lens of a VG HB5 STEM.

A source of contamination is hydrocarbons which are adsorbed on the specimen surface.   If hydrocarbons are present in the vacuum, new material is
constantly being supplied.   In a CTEM a fairly uniform contamination film
builds up on the specimen in the irradiated area.   In STEM contamination
occurs where the electron probe impinges on the specimen.   Using a
stationary probe, contamination spots a few tens of Å in diameter can be
made.   Contamination rates in this situation can be very high.   The beam
current can be sufficient to cross-link all material which moves across the
surface through the probe.   Muller (1971) explains this movement in terms
of diffusion but Fourie (1979) considers that drift due to locally generated electric fields is also necessary to explain this effect.   Fourie
found that contamination rates in STEM could be reduced by using a low
energy electron beam to irradiate the surrounding area of specimen.   He
assumed that the reduction of contamination was due to the discharging
effect the low energy electrons would have.   We have found that contamination rates in STEM can be greatly reduced by irradiating the surrounding
area by high energy electrons. This fixes the contamination on the surrounding surface and thus reduces the diffusion of material to the area of
the probe.

In the VG HB5 STEM a simple method of irradiating the surrounding area is
to remove the objective aperture while maintaining normal focus.   Then,
due to spherical aberration, a probe with a circle of confusion of approximately 30μm diameter results.   The probe has a well focussed high intensity centre.   Thus contamination free high resolution bright field imaging
but with reduced contrast is made possible by removing the objective
aperture.

If a series of objective apertures are used at normal focus the radii (r)
of the circles of confusion Fig.(1) can be measured.   The angle (α)
that each aperture subtends at the specimen can be determined by calibration against a known diffraction pattern.   From a plot of $(r)^{1/3}$ v α
Fig.(2) the value of $C_s = 3.4 \pm 0.1$ mm for the objective lens was determined
Thick contamination layers can be produced in STEM, and Fig.(3) shows a
contamination 'spot' formed on a 250Å thick carbon film.   After formation
the 'spot' has been rotated through a large angle by allowing the film on
which it is formed to curl.   The 'spot' is seen to be cone-shaped with
height-to-width aspect ratio of approximately 5 to 1.   The overall height
is 2000Å.   If the probe is moved across the film lines can be drawn.

Fig.(4) shows a circle generated using analogue signals to deflect the spot. The circle of diameter 0.5μm was tilted to show the large aspect ratio. Contamination patterns can be used as masks for electron lithographic projection.   A CTEM photograph of a mask generated in STEM by an analogue method is shown in Fig.(5).   This shows the high contrast obtained in bright field projection.

The generation of contamination patterns in a more controlled way has been obtained by using a microprocessor with two 8 bit D/A converters to generate the X and Y scans.   One problem is specimen-stage drift of about $3\mathring{A}s^{-1}$. To overcome this a registration cross was drawn on the specimen.   A registration procedure which used two more D/A converters enabled the drift to be checked and corrected to a best accuracy of $\pm40\mathring{A}$.   An example of this form of pattern generation is shown in Fig.(6).   This is part of a 256x256 dot matrix circular pattern of outside dia. 5μm which took about 25 mins.to generate.

Using more accurate D/A converters and using correcting factors to allow for lens and scan distortion, computer generated patterns of greater accuracy could be obtained.

Fig.1

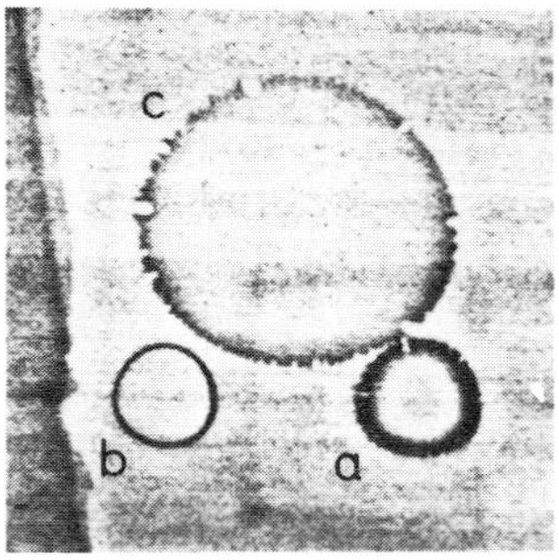

Confusion circles:a at focus, b under focus, c over focus

Fig.2

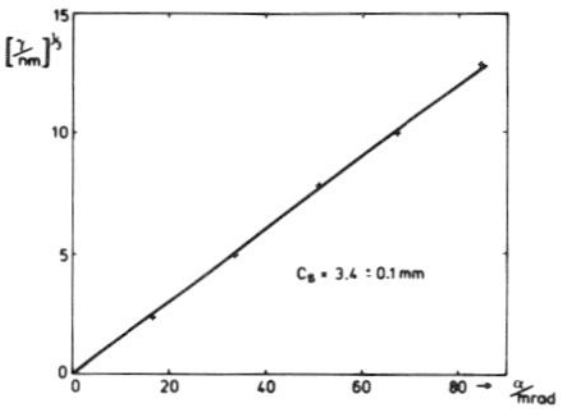

Determination of Cs

Fig.3

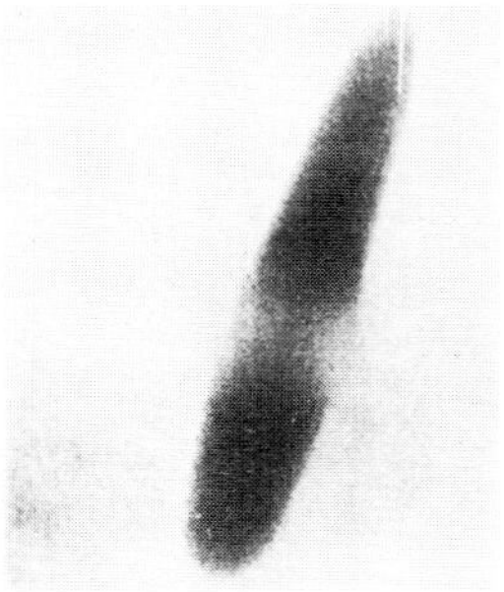

Tilted 'spot of contamination on $250\mathring{A}$ thick carbon film

Fig.4

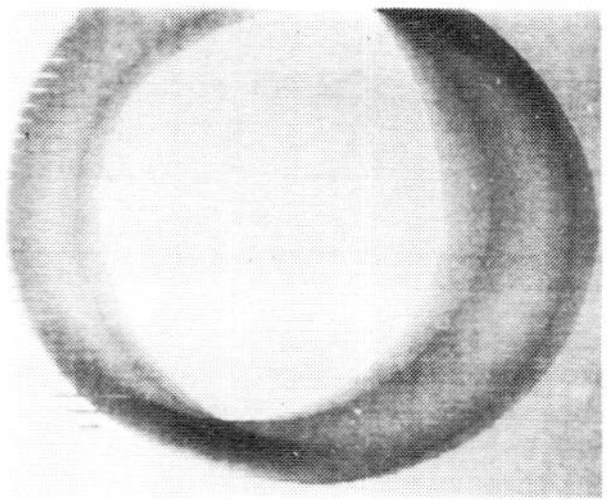

Tilted circle of contamination

Fig.5

CTEM projection of STEM generated mask: line width about $500\mathring{A}$

Fig.6

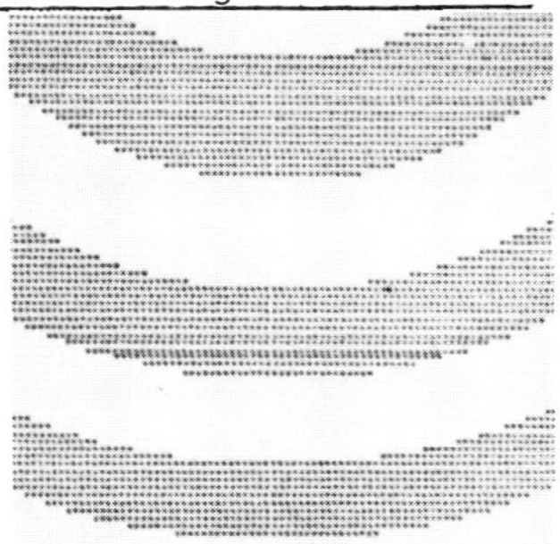

CTEM projection of part of STEM generated mask: spot spacing $200\mathring{A}$

References

Fourie ,J.T. 1979 Scanning Electron Microscopy, Vol.II, SEM.Inc., AMF O'Hare pp.87-102.

Muller, K.H. 1971 Optik **33**, 296-311; pp.331-343.

Practical aspects of the operation of a field emission probe

M M El Gomati, R Browning and M Prutton

Department of Physics, University of York, Heslington, York YO1 5DD, UK

Over the last decade, field electron emitters have been used increasingly as high brightness sources in electron probes. As a result, different electron lenses to those used in conventional probes have had to be adopted. One of the simplest lens types used is the two element lens introduced by Butler (1966). In addition to its simplicity in construction and operation, this lens has the advantage of a low energy focus (usually a few hundred volts). This gives the operator a larger field of view, beneficial in locating an area of interest, and extra possibilities of applications such as electron loss spectroscopy. We have designed an all electrostatic high resolution probe with a simplified two element lens. The electron optics of the lens was computed with programs provided by Munro (1975). Figure 1 is a schematic of the column, and figure 2 shows the variation of beam diameter as a function of convergence angle for a focussing ratio of 7:1.

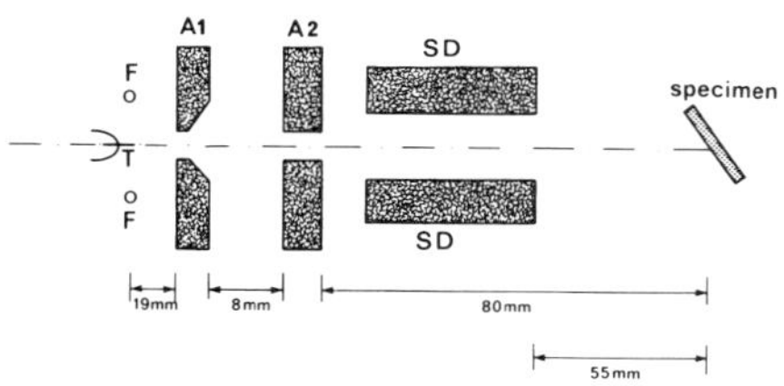

Fig 1: A schematic of the electron column. T tip, F degassing filament, A1 and A2 anodes, SD stigmator/deflector

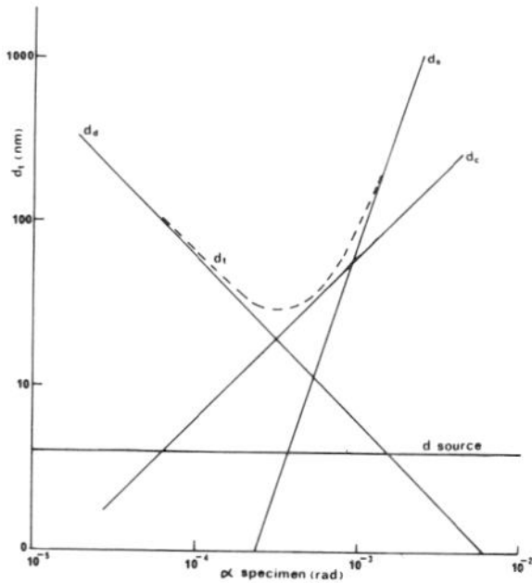

Fig 2: The variation of beam diameter as a function of convergence angle at a focussing ratio of 7:1

The field emitter is mounted on bellows which allows it to be scanned in a plane perpendicular to the first anode as well as adjusting its separation from this anode. This allows a change in the focussing ratio over the range 6–12. An eight pole electrostatic stigmator/deflector is used, Rang (1949). The angle stop is mounted on the first anode to prevent secondaries generated at the second anode from reaching the insulators and building up a charge. Thorough outgassing of the first anode is essential to reduce electron stimulated desorption. The metal–insulator junction was designed carefully so that the insulator at the negative electrode was in a low field to reduce field emission effects. Two different emitters have been

0305-2346/82/0061-0045$01.50   © 1982 The Institute of Physics

used, namely W(310) and Zr/W(100). Table 1 compares the performance of these emitters and that of a third type, built up W(100)4. Table 1 shows the advantages of the built up W(100) over W(310) and to a lesser extent the Zr/W(100). In particular, the built up W(100) has better stability than W(310) and yet a narrower energy spread than the equally stable Zr/W(100). In addition, it has the highest brightness. The built up W(100) has not been tested in this column due to power supply limitations. However it has been used by one of us (MP) in another column, Veneklasen et al (1978).

	W(310)	Built up W(100) Veneklasen (1972)	Zr/W(100) Swanson (1975)
emission angle	0.8 - 1 rad	.12 - .2	.12 - .2
$\frac{dI}{d\Omega}$ (I=10 µA)	10 µA/Sr	130 µA/Sr	100 µA/Sr
Δv	.2 eV	.4 - .5 eV	.7 - .8 eV
operating temperature	300 K	1200 K	1200 - 1500 K
emission noise at 10^{-9} torr	5%	10%	10%
useful operating time	$\sim$ 10 mins	> 10 hours	> 10 hours
emitter radius	< 100 nm	> 500 nm	> 500 nm
wf	4.3 eV	4.5 eV	2.6 eV

Table 1: A comparison between the characteristics of three different emitters (assuming 10 µA total emission)

References

Butler J W 1966 6th Int Cong Elect Microscopy, Rome
Munro E 1975 Cambridge University report no CUED/B-Elect. TR45
Rang O 1949 Optik 5 518
Veneklasen L and Siegel B 1972 J App Phys 43 1600
Swanson L 1975 J Vac Sci Tech 1228
Veneklasen G L, Todd G and Poppa H 1978 **Proc** 9th Int Cong Elect Microscopy, Toronto

Simultaneous detection techniques and computer-controlled data acquisition for spectrally and time resolved cathodoluminescence in the SEM

M. Hastenrath, K. Löhnert, E. Kubalek

Universität - Gesamthochschule - Duisburg, Fachbereich Elektrotechnik, "Werkstoffe der Elektrotechnik", Kommandantenstr. 60, 4100 Duisburg 1, West-Germany

1. Introduction

In a lot of disciplines like solid state electronics, mineralogy, biology and medicine cathodoluminescence (CL) measurements in the SEM enable the micron-scaled assessment of a variety of local material properties. In optoelectronic materials   important properties like radiative transitions or minority carrier lifetimes can be determined by recording of the spectral distribution or the decay behaviour of the CL-signal. These properties are usually evaluated by the time-consuming sequential recording of the CL-spectrum or the CL-decay for each analyzed point on the specimen. The increase of the number of measuring points for getting additional information about local variations of these quantities therefore normally excludes the sequential recording techniques and requires a faster detection method with subsequent data processing. In the following a computer-controlled CL-measurement system is presented, which utilizes the optical multichannel analysis (OMA) technique both for spectrally and, in conjunction with a preceding streak camera, for time resolved CL-measurements. The measuring efficiency of this system will be demonstrated by some experimental results in optoelectronic materials.

2. Instrumentation

The following description of the measurement system is confined only to its essential features, whereas details are given elsewhere (Hastenrath et al. 1979, 1980, Löhnert et al. 1978, 1979). The basic CL-measurement system consists of:
- an electron optical column with imaging electronics of a commercial SEM (Cambridge Stereoscan S4-10)
- a beam blanking system with electron pulse decay times of less than 50 ps (Menzel and Kubalek 1978, 1979)
- a clean, oil-vapour free vacuum system based on turbomolecular vacuum pumps
- a specially modified liquid helium flow cryostat (Thor cryogenics)
- an ellipsoidal mirror for CL-collection with about 30% efficiency (Balk and Kubalek 1973) and
- a vacuum grating monochromator (McPherson 218) for spectrally resolved CL-measurements.

In conjunction with a suitable photomultiplier this system can be used for
- recording of integral or spectrally resolved CL-micrographs or linescans
- sequential recording of CL-spectra
- sequential recording of CL-decays by use of a boxcar integrator.

In the last two cases the CL-signal is recorded either by a chart recorder or transmitted to a desktop computer (Hewlett
Packard 9845 S) for data storage and evaluation. Significant improvements in measurement speed are achieved in this system
by use of the OMA technique, which enables
the simultaneous recording of a complete
CL-spectrum or, in conjunction with a
preceding streak camera as time resolving
photodetector, of a complete CL-decay in a
few seconds for a single measuring point.
This CL-measurement system together with
the interfaced computer control system is
shown in Fig. 1.
In spectral analysis the silicon inten-
sified target (SIT) vidicon camera of the
OMA system is directly adapted to the
vacuum monochromator (dashed connection
instead of adapted streak camera in
Fig. 1). The exit slit of the mono-

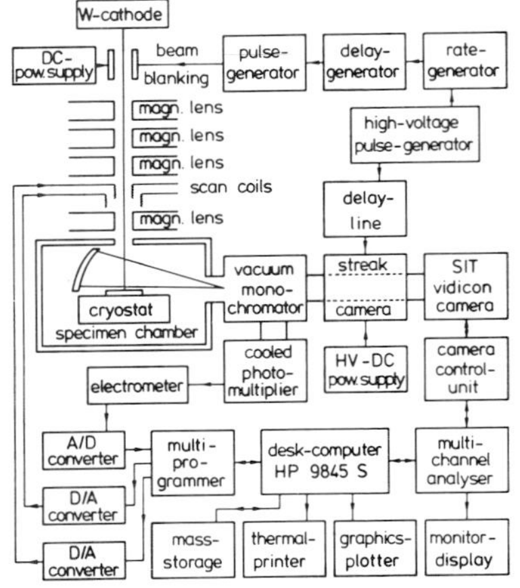

Fig. 1 Computer controlled CL
measurement system

chromator must be removed in this mode. The different wavelengths in the
spatially dispersed CL-spectrum correspond then to a definite line on the
vidicon multidiode target. After integration of the complete video signal
of each line in the subsequent multichannel analyser (MCA) the CL-spectrum
is displayed on a TV monitor together with the direct video image. Gating
circuits in the MCA enable to integrate only over the really illuminated
parts of the vidicon target. The dark current of the target is automatically subtracted from each spectrum. By use of a 300 lines/mm grating a
spectral resolution of about 1 nm is typically achieved within a simultaneously detected wavelength range of about 125 nm (Löhnert et al. 1978,
1979).
In time resolved analysis a streak camera as time resolving photodetector
is inserted between the monochromator and the SIT vidicon camera. The
streak camera (Hamamatsu C 979) transforms the temporal intensity distribution of a CL-pulse, arriving at the streak camera photocathode, into a
lateral intensity distribution at the camera output phosphor screen. This
lateral intensity distribution is then read out and displayed by the same
OMA system as used for the simultaneous registration of CL-spectra. Time
ranges between 10 ns and 300 ns with a time resolution of about 1/100 of
the selected time range are available in this operation mode (Hastenrath
et al. 1979, 1980, Hastenrath 1981).
Computer controlled data acquisition is obtained by interfacing the desktop computer to the OMA system and to the scanning circuits of the SEM,
so that all measurement steps, which are necessary for the automatic recording of CL-spectra or CL-decays, can be performed completely under computer control. The electron beam on the specimen is positioned by external
driving of the SEM scanning coils with two DAC-cards, incorporated in a
multiprogrammer (Hewlett Packard 6940 B), the operation of the MCA is controlled by use of a relay-card and the MCA-data are transferred to the
computer via a serial RS 232 C interface. The recorded CL-spectra or CL-
decays are either directly plotted on a graphics plotter or a thermal
printer or can be stored on a floppy disc or tape cartridge.

3. Operation

Before starting an automatic measurement only the positioning of the spec-

imen and the optimizing of the SEM-performance and CL-
detection parameters must be manually established. Subsequently the CL-spectra or CL-decays of a preselected
number of specimen points are automatically recorded
and stored together with all measurement parameters in
a data file on the mass storage device. This automized
operation of the whole system is realized by extensive
software routines for system controlling and data processing, which are all included in a programm packet
written in the HP extended BASIC language. In automatic data acquisition mode measurements are performed
by running the programm sequences illustrated in Fig.
2. The main measuring subroutines of this program
are:
- Positioning of the electron beam in the preselected
  pattern
- Controlling the operation parameters of the MCA
- Entering the digitized CL-spectra or decay curves
  from the MCA
- Storing the data on floppy disc or tape cartridge.

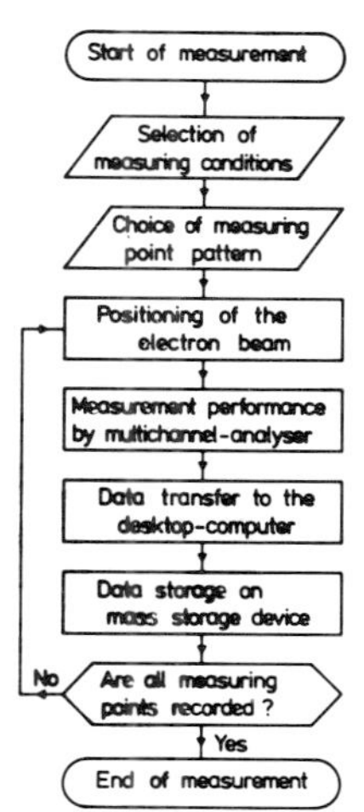

Fig. 2 Measurement
flow chart

The minimum time needed for running one such cycle
is less than 15 seconds. Evaluation and presentation of recorded CL-data is comprised in the following subroutines:
- Calibration of the wavelength axis utilizing the known line spectrum of
  a Hg/Cd spectral lamp
- Correction of the recorded spectrum with respect to the quantum efficiency characteristics of the SIT vidicon photocathode
- Plotting of the normalized spectrum (linear wavelength or energy axis)
  or decay (linear time axis) with either linear or logarithmic intensity
  scale
- Plotting of up to 11 spectra or decays in 3-dimensional presentation
- Mapping of evaluated parameters (peak position and FWHM in spectrally
  or 1/e-decay time in time resolved measurements) in the preselected
  pattern.

4. Applications

The capabilities of the system are demonstrated by the following measurements at optoelectronic materials:
Fig. 3 shows a CL-pulse recorded with
the streak camera from a $1.6 \cdot 10^{18}$ cm^{-3}
Se-doped GaAs specimen at 80 K. A mi-
nority carrier lifetime of $\tau = 2.1$ ns
is directly evaluated from the exponential part of the decay. This measurement was completed in about 30 s,
including 200 integrations for S/N
improvement. Fig. 4 shows a set of
11 CL-spectra taken at the cleavage
plane of a GaP LED, isoelectron-
ically doped with $1 \cdot 10^{18} cm^{-3}$N, at
80 K in distances of 1.6 /um from the
top $2 \cdot 10^{18} cm^{-3}$ Zn-doped p-side to the
$1 \cdot 10^{17} cm^{-3}$ S-doped n-side. A change
in recombination behaviour from domina-
ting NN$_1$ pair-band emission at the p-side to prevailing emission from

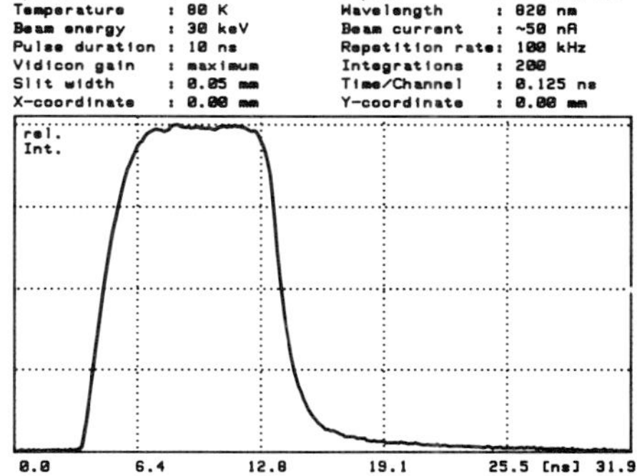

Fig. 3 Recorded CL-pulse from a
$1.6 \cdot 10^{18} cm^{-3}$ Se-doped GaAs specimen

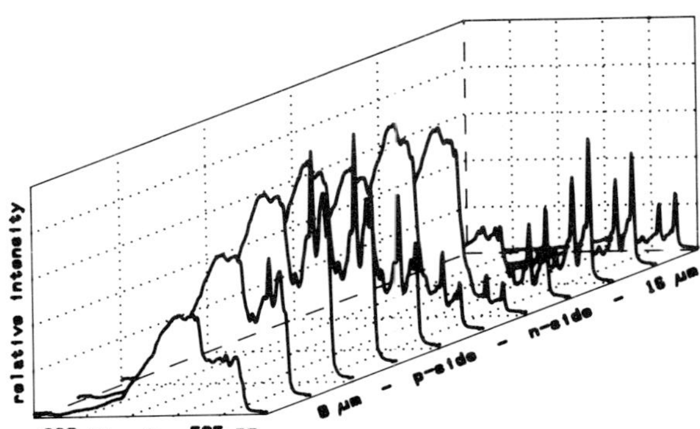

Fig. 4 Spatially resolved CL-spectra
on GaP-LED

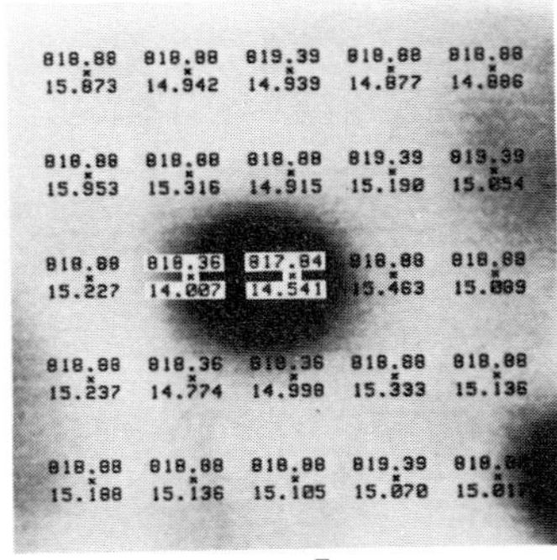

Fig. 5 Mapping of peak position
and FWHM (in nm) at dislocation
in GaAs

isolated N-centers at the n-side is clearly disclosed in this measurement
series, which was obtained in less than 3 minutes.
The last example in Fig. 5 shows a dislocation in a 3.7-$8.0 \cdot 10^{17}$cm^{-3} Si-
doped GaAs specimen. At this dislocation 25 spectra were recorded in a 5x5
point pattern with a 5 μm point distance in about 6 minutes. The directly
determined values of the peak wavelength (upper value) and FWHM (lower
value) both show a slight decrease at the dislocation core, indicating a
local change in the occuring radiative transitions.

5. Acknowledgements

The financial supports of the "Ministerium für Wissenschaft und Forschung
des Landes Nordrhein-Westfalen" and of the "Deutsche Forschungsgemeinschaft" are gratefully acknowledged.

6. References

Balk L J and Kubalek E 1973 BEDO 6 pp 559-569
Hastenrath M, Löhnert K and Kubalek E 1979 BEDO 12/1 pp 163-175
Hastenrath M, Balk L J, Löhnert K and Kubalek E 1980 J. Microscopy 118,
  Pt3, pp 303-308
Hastenrath M 1981 PhD Thesis, Duisburg University
Löhnert K, Hastenrath M, Balk L J and Kubalek E 1978 BEDO 11 pp 95-104
Löhnert K, Hastenrath M and Kubalek E 1979 SEM/I pp 229-236
Menzel E and Kubalek E 1978 BEDO 11 pp 47-66
Menzel E and Kubalek E 1979 SEM/I pp 305-318

BEDO   : Beitr. Elektronenmikroskop. Direktabbildung und Analyse von
         Oberflächen, Ed. Pfefferkorn G, Münster, W.-Germany
SEM   : SEM Inc., AMF O'Hare, IL 60666, U.S.A.

Cathodoluminescence spectrometer design for the TEM

S H Roberts

H H Wills Physics Laboratory, University of Bristol, Royal Fort,
Tyndall Avenue, Bristol BS8 1TL

1. Introduction

Cathodoluminescence (CL) is finding increasing application in the scanning
electron microscope (SEM) and the past decade has seen development of a
wide range of spectrometers (CLS). Highly efficient CLSs have been developed which can form monochromatic images and measure decay lifetimes. To
correlate CL with the crystallography of a thin foil, it is most convenient to have the CLS on a transmission electron microscope (TEM). Furthermore high spatial resolution requires thin specimens and high beam voltages to reduce beam spreading as it passes through the foil.

Adaptation of SEM designs to the TEM is not a straight-forward matter.
The CL signal from thin foils is very weak and can be swamped by CL from
thicker adjacent material that is excited by scattered electrons and X-
rays. Also the space available for an efficient collector is severely
limited.

The problem of collecting light emitted over a wide angular range and
directing it into the entrance slit of a monochromator for spectral
analysis has been overcome in all high efficiency collectors by use of a
concave mirror shaped as an ellipse of revolution about its major axis,
one form of an ellipsoid. This particular shape has the property that all
rays emitted at one focus are reflected to pass through the other focus.
When applied to the collection of CL from a microscope specimen with the
area of interest at one focus, angular compression can also be achieved
depending on the eccentricity of the ellipsoid and configuration. In the
TEM, the lower pole piece of the objective lens precludes CLS optics after
the specimen as in for example Davidson et al (1980). The design in
fig. 1 after Petroff et al (1978) has been successfully used for their
JEOL 200B. Two lead-backed planar mirrors in a periscope arrangement are
used to shield the quartz light guide on the right from X-rays that would
otherwise excite luminescence in the quartz.

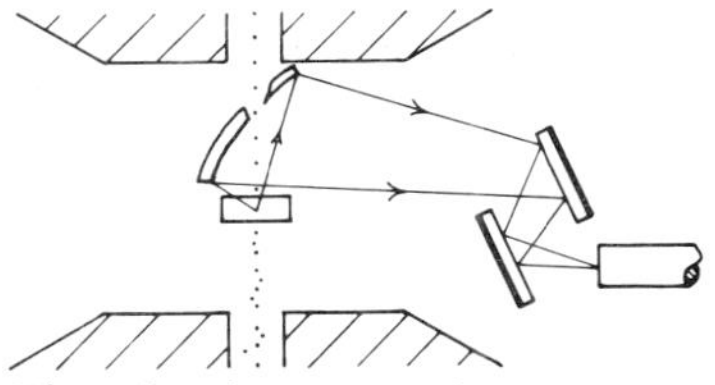

2. Evaluation of possible ellipsoidal collectors for the Philips EM400

In designing a collector for our Philips
EM400 TEM, various mirror optics based on
the arrangement employed by Petroff et al
(1978) were studied in detail. For optimising ellipsoid parameters and comparing

Fig. 1 Mirror optics of a CLS
used in a JEOL 200B TEM.

different designs, the solid angle subtended by the mirrors at the specimen was maximised. A numerical approach was taken for evaluating the
solid angle by dividing the 2π steradian emission above the specimen into
a large number of small elements of equal solid angle. Each element was
examined to see whether mirror was present in that direction. The interaction of spatial limitations in reducing the efficiency of the mirror can
be appreciated by a projection in which solid angle is proportional to
area. Fig. 4 shows examples of the projection in which characters represent the elements, although only half of the 1020 elements used are plotted because the characters are rectangular. The lozenge shape is arbitrary and results from the shearing involved in making a hemisphere flat
without stretching or tearing. The centre corresponds to the incoming
beam and the direction of the access port is on the right.

Proposed designs for a TEM are shown in fig. 2. The design in fig. 2(a)
has the image focus outside the microscope column for a high degree of
collimation. The ellipsoids in fig. 2(b) and (c) have short separations
of foci (2c) with little demagnification of aperture angle. The angle
can be further reduced by a reflecting objective, as drawn, which has no
chromatic aberration and can be lead-backed as in the periscope arrangement of fig. 1. Construction is simplified by making the small convex
mirror spherical and a semi-aplanat system is possible up to as large an
aperture as 40° semi-angle using an ellipsoidal concave mirror (Burch
1947). There is sufficient space for the mirror to be extended below the
specimen. However this was not considered because the increase in collection efficiency would be outweighed by increased back scattering of electrons from the extra mirror close to the transmitted beam (Petroff 1979).

In evaluating various designs for the Philips EM400, the following spatial
limitations were considered. 7mm spacing between the upper pole piece and
the specimen with access through a port of 15mm diameter. Full translation and tilting of the specimen holder requires clearance within a
cylinder of radius 4.6mm which is perpendicular to the beam and 45° to
the access port. This is indicated by the dashed ellipses in fig. 2.
Finally the specimen holder obstructs emission of light at low angles from
the specimen surface; below 16° for a standard holder and below 38° for a
He-cooled holder designed by Dr J A Eades. Both holders have integral
drives for specimen rotation or tilting.

Results of the numerical analysis which allow for full tilting of the
specimen holders in the arrangements shown in fig. 2(a) and (b) are shown
in fig. 3. For the first design, 2c is large and the solid angle can be
maximised with respect to the focal length (F) of the ellipsoid giving $17\frac{1}{2}$%
efficiency at F=3.5mm. Both ellipsoid parameters need to be varied for the

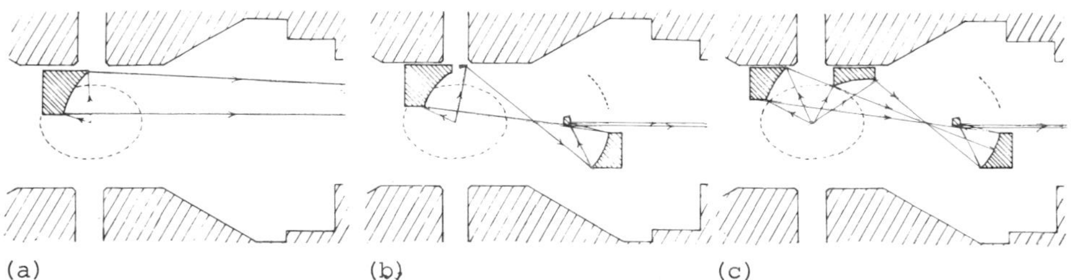

(a)   (b)   (c)

Fig. 2 Proposed collector designs. (a) image focus outside microscope
column. (b) and (c) use a reflecting objective for collimation.

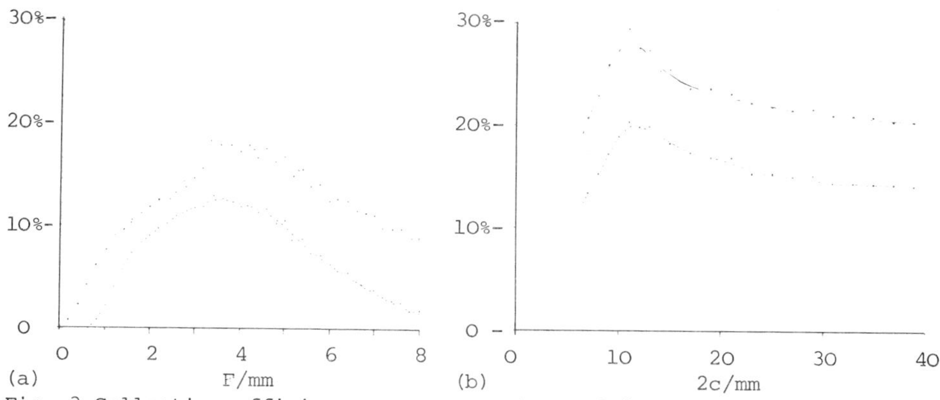

(a)   F/mm   (b)   2c/mm

Fig. 3 Collection efficiency as a percentage of 2π sr emission for the
designs in fig. 2(a) and (b), respectively, and for a standard holder
(——) and a He-cooled holder (- -).   (a) 2c is large.   (b) F=4+$\frac{1}{2}$mm for
the maximum efficiency at each value of 2c.

second design although F was found to be invariant for maximum solid angle
given any 2c, so the results are presented as a function of 2c. Maximum
efficiency is 28% at 2c=11mm.

Despite the optimisation of ellipsoid parameters, these efficiency figures
are rather low. The substantial part of the mirror that is cut away to
clear the holder is denoted by h's in the equal solid angle projections.
A novel design was considered which would fit around the holder clearance
envelope using two ellipsoids that share foci, see fig. 2(c). This introduces two extra parameters of another F and the cut-off angle between the
two mirrors. A brief analysis showed that this was not as beneficial as
hoped. Nevertheless it may be advantageous to extend this idea to many
ellipsoids in concentric bands, analogous to the thin Fresnel lens.

The analysis was repeated, this time for no tilting of the holder.
Maximum efficiency is increased to 52% and 60% respectively showing that
this is the most efficient approach so long as the mirror can be retracted
without breaking the vacuum for tilting of the holder.

```
              l l l l l l p p p p                          l l l l l l l l l v
           l l l l l l l l l p p p p p p p p            l l l l l l l l l l l l l v v v v
        l l l l l [m m m m m m]p p p p p p p p p      l l l l l [m m m m m m m]h v v v v v v
       l l l h h [m m m m m m]p p p p p p p p p p    l l l [m m m m m m m m m m]v v v v v v v
      l l l h h h h [m m m m m m]p p p p p p p p p p  l l l [m m m m m m m m m m m]v v v v v v v v
     l l l h h h h h [m m m m m m]p p p p p p p p p p p  l l l h [m m m m m m m m m m m]v v v v v v v v v
    l l l h h h h h h h [m m m m m m]p p p p p p p p p p p  l l l h h h [m m m m m m m m m m m]v v v v v v v v v v
   l l l h h h h h h h h [m m m m m]p p p p p p p p p p p p  l l l h h h h [m m m m m m m m m m]v v v v v v v v v v v
  l l h h h h h h h h h [m m m m m]t p p p p p p p p p p p p  l l h h h h h h [m m m m m m m m m]t t v v v v v v v v v v v v
 l l l l h h h h h h h h h [m m m m]t p p p p p p p p p p p p v  l l l l h h h h h h h h [m m m m m]b b m t t v v v v v v v v v v v v v
 l l l h h h h h h h h h [m m m m]t p p p p p p p p p p p p p p  l l l l h h h h h h h h h [m m m m m]t t v v v v v v v v v v v v v
 l l l h h h h h h h h h h [m m m m]t p p p p p p p p p p p p p p  l l l h h h h h h h h h [m m m m m m m]t v v v v v v v v v v v v v
  l l h h h h h h h h h [m m m m]p p p p p p p p p p p p p p  l l h h h h h h h h h [m m m m m m m]v v v v v v v v v v v
  l l l h h h h h h h h [m m m m]p p p p p p p p p p p p  l l l h h h h h h h h [m m m m m m]v v v v v v v v v v
   l l l h h h h h h h [m m m]p p p p p p p p p p p  l l l h h h h h h h [m m m m m]v v v v v v v v
    l l l h h h h h h h [m m m]p p p p p p p p p  l l l h h h h h h h [m m m m m]v v v v v v v
     l l l h h h h h h [m m m]p p p p p p p p  l l l h h h h h h [m m m m]v v v v v v
      l l l l l l h h [m m]p p p p p p p p  l l l l l l l h h h [m m m]l v v v v v
         l l l l l l l p p p p p           l l l l l l l l l l l v v
            l l l l p p                       l l l l l l
```

(a) 2c large, F=3.5mm   (b) 2c=11mm, F=4.2mm

Fig 4 Proportionate solid angle projections for the maxima in fig. 3.
Characters: v acceptance angle at image focus, p access port, t pole
piece, b beam hole, l holder obstruction, h holder clearance and m mirror.

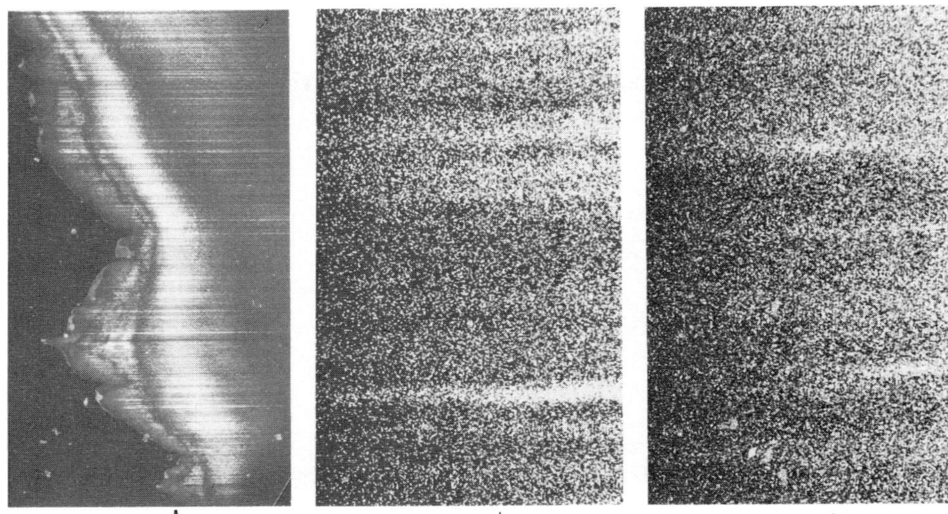

(a)   ↑ ——————1µm   (b) 329.0 $^{\pm}$2 nm   (c) 334.5 $^{\pm}$2 nm

Fig. 5 (a) TEM dark field micrograph of ZnS using 0001 reflection. (b) and (c) monochromatic CL topographs from the same area as (a), cooled to 40K nominally. The left edges in (b) and (c) coincide with the arrow.

3. Results from a CLS built for the EM400

Collection mirror optics based on fig 2(a) with restricted tilting of the holder were used in a CLS built for the EM400. The calculated collection efficiency for the mirror is 36%. This is less than the maximum possible value because of a small cut-away for the mounting bracket and generous clearance from the pole piece and holder. Details of the complete optics and preliminary results from striated ZnS are given in Roberts (1981).

The performance of the CLS in both spatial and spectral resolution is demonstrated by monochromatic topographs using a 120kV probe on ZnS, fig. 5. The thinned crystal was cooled to about 40K in order to increase preference for photon emission of peaks in the UV that show detail on a fine scale. The dark field micrograph, using the 0001 reflection present through double diffraction in hexagonal ZnS, indicates the severe stacking disorder in this specimen and the edge of the thin region.

Acknowledgements

Many people have contributed to the success of this work but particular thanks go to S Mardix for the ZnS crystal, J A Eades for use of the He-cooled holder and J W Steeds for his help and encouragement. Financial support from the SRC and IBM (UK Labs Ltd) is gratefully acknowledged.

References

Burch C R 1947 Proc. Phys. Soc. 59 41 and 47
Davidson S M, Clarke W R L, Cumberbatch T J, Huang E and Myhajlenko S 1980
   Inst. Phys. Conf. Ser. No. 52 351
Petroff P M, Lang D V, Strudel J L and Logan R A 1978 IITRI/SEM 1 325
Petroff P M 1979 private communication
Roberts S H 1981 Inst. Phys. Conf. Ser. No. 60 377

A simple cathodoluminescence detector for the STEM

S J Pennycook

Cavendish Laboratory, Madingley Road, Cambridge CB3 OHE

Design considerations

An efficient CL detector must collect light from as large a solid angle as possible. Normally, a parabolic or ellipsoidal mirror is focused on the specimen (Petroff et al. 1978, Roberts 1981) which gives high collection efficiency from the area at the focus of the mirror and low efficiency for light generated elsewhere. Such mirrors need to be accurately made and aligned and a large gap in the polepiece is needed to accommodate them. A tapered, internally reflecting silver tube provides a simple alternative, having a comparable collection efficiency, while being straightforward to manufacture and align, and it can be used with polepieces having only a 2 – 3 mm gap. Light travelling towards the larger tube diameter is bent by twice the taper angle towards the tube axis at each reflection with the walls, giving compression of the angular spread of the light. Efficient collection is only achieved from areas close to the input end of the tube.

Figure 1 shows the detector assembly. The tapered tube is mounted on the end of a straight quartz light guide which passes out of the vacuum system through a viton 'O'-ring seal, giving efficient coupling to a photomultiplier or spectrometer and avoiding losses due to reflections or divergence at a window. Alignment is achieved by three screws which tilt the whole assembly by several degrees in any direction. The quartz rod is rigidly held in the final flange, which slides on a second bellows, allowing the detector to be withdrawn by 15 mm. When retracted, no spurious X-rays can be detected from the tapered tube.

To minimise the flux of scattered electrons hitting the tapered tube, the detector is placed on the electron entrance side of the specimen (fig.2). It is placed in a port at 90° to that carrying the X-ray detector and simultaneous collection of CL and X-rays is achieved. At present the specimen is held in a special tilting cartridge to give a clear view of the specimen from both CL and X-ray detectors, but the normal tilting X-ray cartridge should also be suitable.

Alignment of the detector is simply achieved by viewing the specimen at low magnification when the tapered tube is also imaged. The tube is pointed at the region of interest, and then the tube raised towards the specimen, while monitoring the CL output. If necessary, its height can be measured by focusing onto the tube. The alignment may be preset and the detector slid in or out as required.

For spectral analysis the quartz light guide is used as an input slit to an f/4 spectrometer giving a spectral resolution of 20 nm. Higher resolutions could be achieved using a spot to slit converter, although further light losses would be incurred.

Choice of tapered tube

Silver was chosen for the tube material, since it has high reflectivity throughout the visible range. A sheet of silver was mechanically polished to 95% reflectivity (measured with a He-Ne laser). It was then rolled around a mandrel into the required taper. With a tube of constant length (30 mm in this case) the degree of compression increases as the angle of taper is increased, but the diameter of the exit end must be kept small, since it defines the diameter of the quartz light guide and hence the spectral resolution obtainable.

For the detection of total emitted CL, good results were obtained with a silver tube of input diameter as large as 2 mm. The results of ray tracing calculations for various incident angles are compared in the table to the measured performance (tube A).

TABLE

Tube A	Calculation			Experiment	
2-4 mm diameter 30 mm length	Incident angle(°)	Exit angle(°)	% transmission	Range of Exit angle (°)	% transmission
	10	6	95	0 – 10	85
	20	9	86	3 – 20	78
	30	15	81	5 – 26	71
	40	17	74	6 – 32	62
	50	23	70	7 – 35	52
	60	22	60	8 – 37	45
Tube B	10	3	95	0 – 4	
	20	0	86	0 – 5	
0.5-4 mm	30	4	81	0 – 9	
diameter	40	7	77	0 – 9	
30 mm	50	4	70	0 – 9	
length	60	7	66	0 – 9	

The actual transmission is slightly lower than expected, probably due to losses from the seam of the tube, but integrating up to an angle of 50° indicates that 13% of the CL emission into 4π sr is collected by the tube, and compressed sufficiently to fall on a 2" photocathode. The tube focuses light into an annular cone with a range of exit angles as shown in the table. Greater compression is needed for feeding the light into the spectrometer, and can be achieved by reducing the input diameter to 0.5 mm as shown in the table (tube B). It was not possible to measure the actual transmission of this tube with the simple apparatus available, since the diameter of the light-collecting end was so small, but the range of angles emerging showed that compression was almost as good as expected from the calculation. Integrating the calculated transmission over angle gives a collection efficiency of 15% (of 4πsr) while allowing for the measured angular distribution indicates that the collection efficiency into the spectrometer is 10%. This compares well to the collection efficiency of an ellipsoidal mirror (for example Steyn et al.(1976) quote an efficiency of 40% 2π sr). The small number of optical components required to remove the light from the microscope may help to make up this factor.

Results

The intensity of CL emitted from a thin specimen per incident electron can be written as $I_{CL} = n_g \eta_{CL}$ where $n_g = t/\Lambda_g$ is the number of electron hole pairs generated by the fast electron, t is the specimen thickness and Λ_g represents the mean free path for electron hole pair production by all possible routes. Contrast has been observed due to variations in Λ_g with incident beam diffraction conditions (Pennycook and Howie 1980) although great care has to be taken to avoid luminescence due to variations in the scattered electron distribution (for example by using a very small volume of specimen). Such effects are small, and are smeared by using a convergent probe, so that roughly $n_g \propto t$.

When the specimen thickness is less than the bulk diffusion length D the effect of non-radiative recombination at the surfaces can reduce η_{CL} to a value of approximately $\eta_{CL} D_{eff}/D$, where $D_{eff} = t/\pi$ (Shockley 1950), when $I_{CL} \propto t^2 \eta_{CL}$. Thus it is necessary to know the local sample thickness before interpreting the CL contrast as variations in η_{CL}. Thickness may be estimated using the annular dark field intensity, X-ray emission rate, or if thin enough by EELS.

The smaller effective diffusion length means that the CL image resolution improves linearly as thickness is reduced (Pennycook 1981). Single dislocations can be imaged in thin samples of diamond (Pennycook et al. 1980) with resolutions better than achieved on bulk diamond (Hanley et al.1977) and figure 3 shows an example.

The technique seems particularly suitable for the study of commercial phosphors (Pennycook et al. 1981). An example is shown in fig.4 of the analysis of two regions of a grain of Y_2O_2S:Tb. In each region the X-ray, EELS, and CL spectra were taken simultaneously using a Link Systems 860 analyser. The EELS spectra allow the thickness to be measured, even in the thicker region from the position of the maximum of the multiple plasmon losses. The X-ray spectra show spurious Cu peaks from the grid, Au from the cartridge and Ag from the tapered tube, as well as the specimen peaks. The CL spectra were obtained in single scans of 3 mins, but multiscanning would improve their quality. The spectra were taken using an S20 photomultiplier and require correction for spectral response before quantitative work can be undertaken.

References

Hanley P L, Kiflawi I and Lang A R   1977   Phil.Trans.Roy.Soc. A284 329
Pennycook S J   1981   Ultramicroscopy 7 No.1 in press
Pennycook S J and Howie A   1980   Phil. Mag. 41 809
Pennycook S J, Brown L M and Craven A J   1980   Phil. Mag. 41 589
Pennycook S J, Dexpert H and Charrière Y   1981   J. Micr. et Spect. Elec.
  in press
Petroff P M, Lang D V, Strudel J L and Logan R A   1978   Scanning Electron
  Microscopy 1 325 (SEM Inc, A M F O'Hare, Illinois)
Roberts S H, these proceedings
Shockley W   1950   Electrons and Holes in Semiconductors (Van Nostrand,
  Princeton) p 323
Steyn J B, Giles P and Holt D B   1976   J. Microscopy 107 107

Fig.1

STEM-CL
detector

Fig.2   Specimen geometry

Fig.3   Single dislocations in IIb diamond

Fig.4   Rare earth phosphor showing simultaneous acquisition of X-ray, EELS and CL spectra.

A   B

Acknowledgments

Thanks are due to Prof. T Evans for the diamond specimen, to Drs. H. Dexpert and Y Charrière for the rare earth materials and to the SRC for a research assistantship.

Some applications of quantitative backscattered electron imaging using a Robinson-type detector

M.D. Ball, M.P. Amor and H.J. Lamb

Alcan International Limited, Banbury, Oxford, OX16, 7SP

Quantitative measurements of the backscattered signal in an SEM can be used to determine atomic numbers of bulk specimens, or the thicknesses of thin ($\leqslant$ 2–3 μm) specimens.

The atomic number dependence of electon backscattering is well documented (e.g. Bishop 1966). By standardising the important operating conditions, and calibrating the backscattered signal variation with atomic number, a quantitative measure of the apparent atomic number of any region of a specimen can be obtained (Ball and McCartney, 1981). The speed and accuracy with which these measurements can be obtained will depend on a number of factors, including specimen quality, incident beam current and the performance of the detector.

In these investigations an I.S.I. DS130 SEM equipped with a Robinson-type backscattered electron detector (Robinson 1974) was used. This detector is a high-efficiency scintillator photomultiplier with a large collection angle. The measurements of the output signal were facilitated by the use of a "Cambridge Technology IMAS" signal processor. This device is designed to enable regions of a desired signal level and bandwidth to be selectively displayed at normal imaging rates. Thus when calibrated and used in conjunction with a backscattered detector, "iso-atomic number" images are generated.

This combination of the Robinson backscattered detector and IMAS quantitative signal processor offers a number of significant advantages over standard solid-state detector systems.

1. The high signal to noise ratio of the detector allows atomic number measurements to be made on ideally prepared specimens to better than ± 0.1, although on a typical specimen a more realistic accuracy would be ± 0.25 (limited by specimen preparation).

2. The rapid response of the detector and signal processing system allows backscattered imaging and iso-Z mapping even at T.V. scanning rates. This not only makes operation more convenient but it also allows the signal level for a phase of a particular atomic number to be selected more precisely.

3. The Robinson detector performs satisfactorily for low incident beam energies ($\leqslant$5 keV). By operating under these conditions, the volume from which backscattered electrons originate can be substantially reduced. This improvement in spatial and depth resolution is important where small particles of low atomic number are under investigation.

Applications

1. Identification and distribution mapping of phases in complex alloy systems

Second-phase particles which have a well-defined composition will necessarily have a characteristic electron backscattering coefficient. Figure 1a is a backscattered image of an area of a multiphase sample. In Figure 1 b, c and d each of the three phases present has been imaged in turn by using the IMAS to select the appropriate backscattered signal levels. Although the mean atomic number (or backscattered electron coefficient) does not uniquely define a phase, in any given alloy system the possible phases which could be present will almost invariably have distinct mean atomic numbers, so that the backscattered signal measurements can normally be used to "fingerprint" each phase unambiguously. (Where some ambiguity does exist, a qualitative x-ray microanalysis survey will usually provide the necessary extra information.

The speed and spatial resolution which is offered by this approach enables large numbers of particles with a size range down to about 0.25 μm to be identified quickly and easily.

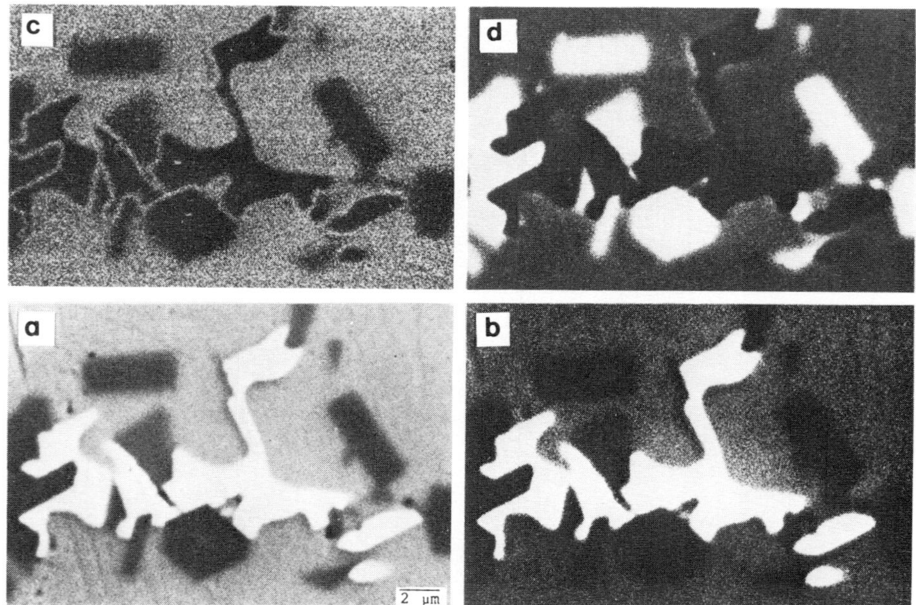

Figure 1. Backscattered electron images of an aluminium sample containing aluminium carbide particles and α-Al$_{12}$Fe$_3$Si intermetallics
(a) Normal backscattered electron image and Iso-Z maps (using the IMAS signal processor) for:
(b) Z = 17 ± 0.5, showing the α-Al$_{12}$Fe$_3$Si particles,
(c) Z = 13 ± 0.5, showing the aluminium matrix,
(d) Z = 11 ± 0.5, showing the aluminium carbide particles.

2. Measurement of the Composition of Binary Phases (or Solid Solutions)

For binary or pseudobinary systems, where the individual atomic numbers of
the two components are known and are significantly different, measurements
of the mean atomic number (or backscattered coefficient) can be used to
derive the ratio of the two components.

A number of relationships between the backscattered coefficient of a
sample and its composition have been proposed (Poole and Thomas 1961).
However, since backscattered coefficients cannot be measured directly
using backscattered detectors, the relationship proposed by Danguy and
Quivy (1956), which expresses the composition in terms of atomic numbers,
has been used.

For a phase which can be represented by the formula $A_m B_n$:

$$\frac{m}{n} = \frac{Z_B(Z_B - \bar{Z})}{Z_A(\bar{Z} - Z_A)} \quad \ldots\ldots\ldots \text{(Danguy \& Quivy 1956)}$$

where $\bar{Z}$ is the mean atomic number
   Z_A and Z_B are the atomic numbers of elements A and B respectively.

Using this expression, the concentrations of elements such as Ag, Zn or Cu
in solid solution in aluminium can be estimated to within 1 atomic %.

3. Identification of Low Atomic Number Phases

Second phase particles containing low atomic number elements ($Z < 11$) can
be difficult to identify by x-ray techniques. However, it is relatively
straightforward to measure atomic numbers down to $Z = 4$ (or lower in
principle) using the quantitative backscattered electron imaging
technique. Thus in aluminium alloys, carbides, borides, oxides and
nitrides have been distinguished. Figure 1d shows the distribution of
aluminium carbide particles ($Z \approx 11$) in an aluminium sample.

4. Thickness Measurement

For thin specimens (< 2-3 μm) the backscattered signal varies with
thickness (for a homogeneous sample). This has been used as a method of
measuring the thickness of (a) self-supporting thin films and (b) thin
films deposited on substrates of a substantially different atomic number
(see Niedrig 1978).

By using the IMAS to select a particular signal level, equal thickness
contour images can be generated. Figure 2 shows an image of a TEM
specimen of aluminium with a number of equal thickness contours superimposed. Operating at 20 keV, the thickness of an aluminium specimen can
be determined to within 50 nm (for the thickness range 0-2 μm). At lower
kV a higher accuracy can be achieved over a reduced range and 10-20 nm
thickness resolution should be attainable at about 5 keV. The speed and
convenience of this method make it attractive for the measurement of TEM
foil thicknesses, especially where x-ray microanalysis corrections are
required. The method should also facilitate the measurement of thin
(<2-3 μm) deposited coatings.

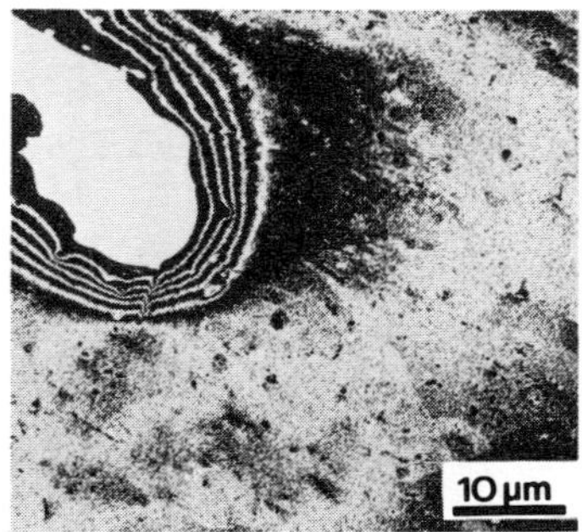

Figure 2.   An image of a TEM specimen of aluminium with a number of equal
thickness contours (by selecting several signal levels using
the IMAS and superimposing the images obtained). The
thicknesses have not yet been accurately calibrated but the
thickness increments between fringes are approximatley 300 nm
and the fringe widths are about 50 nm.

REFERENCES

Ball, M.D. and McCartney, D.G. 1981 J.Microscopy 124 (In Press)

Bishop, H.E. 1966, In Castaing et Al. "Optiques des Rayons X et
Microanalyse" (Paris. Hermann) p.153

Danguy, L. and Quivy, R. 1956 J. Phys. Radium, 16 , 320

Niedrig, H. 1978, In "Scanning Electron Microscopy Vol.1"
(Ed. by O. Johari) p.841 SEM Inc.Chicago.

Poole, D.M. and Thomas, P.M (1961) J.I.M. 90, 228

Robinson, V.N.E. 1974 J. Phys E. Scientific Instruments 7, 650

Line profiling of Kossel X-ray diffraction lines in a SEM using an energy-dispersive X-ray detector

D.J. Dingley, C. Harper and S. Long

H.H. Wills Physics Laboratory, University of Bristol, Bristol BS8 1TL.

I. Introduction

Crystal structure determination requires diffracted intensities to be
determined as well as the reciprocal lattice and accurate lattice spacings. Whereas this is relatively straightforward in X-ray diffraction
experiments on bulk crystals, it still proves very difficult when using
micro diffraction techniques whether electron or X-ray. In our recent
studies using micro Kossel X-ray diffraction to investigate micro crystals in the SEM, very precise values of lattice parameter and crystal
orientation have been obtained from a variety of materials including
metals, semi-conductors and mineralogical specimens. Dingley and
Razavidadeh (1981). It seemed appropriate therefore that efforts should
be made to determine the X-ray intensities at the same time so as to
permit structure determination. This paper reports the results of the
first experiments in this endeavour.

2. Experimental

The principle was to use an energy dispersive X-ray detector to record
the X-ray intensity coming from the specimen as the specimen was rocked
so as to cause the Kossel line to sweep across the detector window. A
Kevex detector was used with a Link Systems microprocessor. The
experiments were carried out in an S4 Cambridge Instruments SEM using a
modified series 100 specimen stage. The specimen examined was a copper
single crystal.

The detector window was suitably collimated to reduce the acceptance
angle of the incident radiation. Several designs were tried, including narrow bore hypodermic needles, but the most successful device was a
2.5 mm thick lead disc with a central hole 0.35 mm in diameter. The disc
was mounted in a plastic cap which fitted over the standard collimator
supplied with the detector. Later, a brass disc with a 0.25 mm bore was
mounted behind the lead to further increase the collimation. With a
specimen to collimator distance of 50 mm the acceptance angle was thus
reduced to 5 x 10^{-3} radians. This is comparable with the width of the
higher order Kossel lines produced from well annealed metal specimens.

The specimen was mounted on a specially constructed tilting stage which
consisted of two Osbourne worm and wheal gear boxes mounted in series.
The overall gear reduction was 625:1, so that one complete rotation of
the external control knob rotated the specimen by 0.057°. A stepper
motor with 7.5° per step was attached to this control providing an
incremental change in angle of the specimen of 0.2 x 10^{-3} radians.

The main experimental difficulty was to align the electron beam at the same level as the collimator on the detector. This was achieved by changing the specimen height and moving the electron beam laterally until a maximum count rate was recorded by the detector.

The specimen chosen was a copper single crystal. Although this metal gives excellent Kossel patterns in transmission with well defined fine structure in the diffraction lines, Fig 1, it was decided to work in back reflection, even though this gives poorer patterns, Fig 2, because the back reflection condition is of more practical application.

The specimen current in all experiments was set at 16nA providing a background count rate of 500 cps. The total number of counts arriving over the Cu kα peak for a period of 3 minutes was recorded at each setting of the crystal, i.e. at each 0.057$^\text{O}$ increment. To reduce the overall angular range sampled in this fashion the specimen was first oriented so that the required Kossel line was close to the collimator. This was achieved by recording the Kossel pattern first photographically with the film positioned centrally and in front of the collimator.

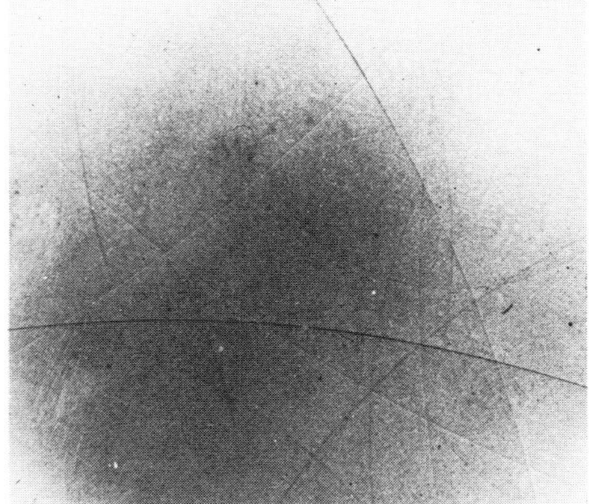

<table>
<tr><td align="center">Fig 1.</td><td align="center">Fig 2.</td></tr>
<tr><td>Transmission KDP from Cu crystal.</td><td>Back reflection KDP from Cu single crystal with Aℓ in solid solution.</td></tr>
</table>

3. <u>Results</u>

Line profiles were obtained for {020} excess lines and an {002} deficiency line. They were recorded with both positive and negative directions of rotation. The results are shown in Figs 3, 4 and 5. Further experiments were carried out with the finer collimator and an example is shown in Fig 6. The improved sharpness of the peak is evident. The solid lines in Figs 3, 4 and 5 are a 9th degree polynomial fit.

The percentage increase in intensity above background has been calculated for each line and is tabulated in Table 1 together with the half height widths. The errors quoted are a combination of standard errors.

4. <u>Conclusions</u>

The experiments to measure Kossel X-ray diffraction line intensities using an energy dispersive X-ray detector have been successful and

indicate that it would be feasible to obtain sufficently precise data
from single crystals to permit structure determination.

5. Acknowledgements

The authors would like to thank M. Welland for useful discussion and
assistance and to Professor J. Ziman for permitting this work to be
undertaken as an undergraduate research project.

6. References

Dingley, D.J. and Razavizadeh, N., 1981 Scanning Electron Microscopy.
(To be published.)

Table 1

Percentage increase in intensity of Kossel lines and half height widths.

Line	% increase	half height width
020 (positive rotation)	9.0 ± 0.7	.004 rads ($\pm$ 5 x 10^{-4} rads)
020 (negative rotation)	9.3 ± 0.7	.005 rads ($\pm$ 7 x 10^{-4} rads)
002 (deficiency)	-4.4 ± 0.8	.007 rads ($\pm$ 12 x 10^{-4} rads)
020 (finer collimator)	13.7 ± 1.8	.002 rads ($\pm$ 5 x 10^{-4} rads)

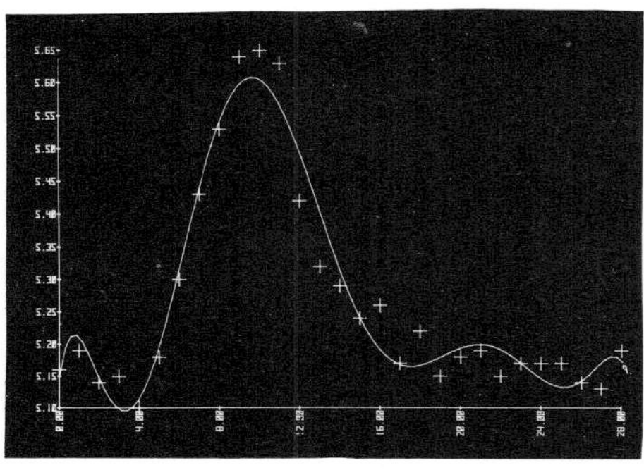
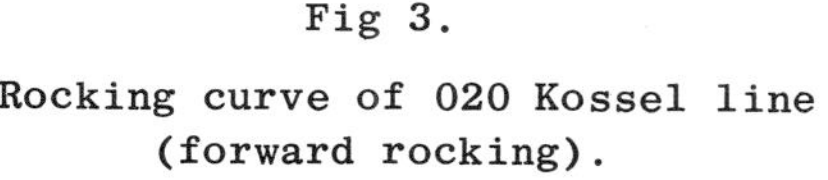

Fig 3.

Rocking curve of 020 Kossel line
(forward rocking).

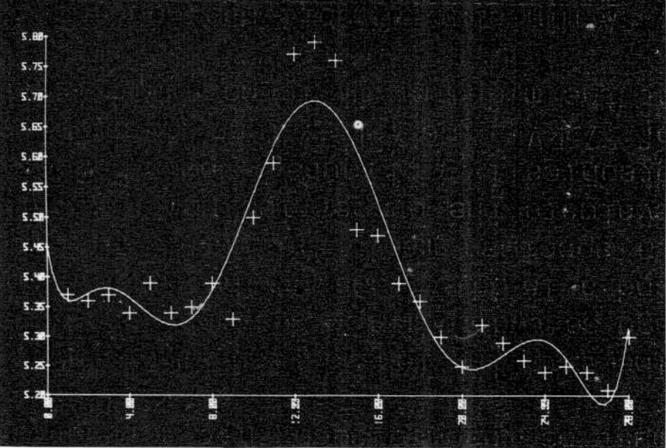

Fig 4.

Rocking curve of 020 Kossel line
(reverse rocking).

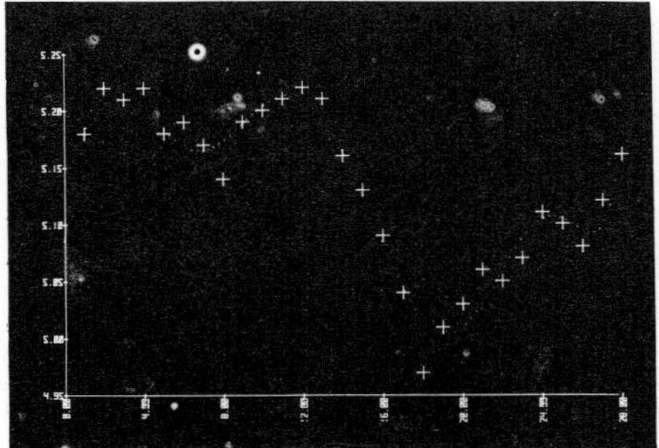

Fig 5.

Rocking curve of 002 deficiency
Kossel line.

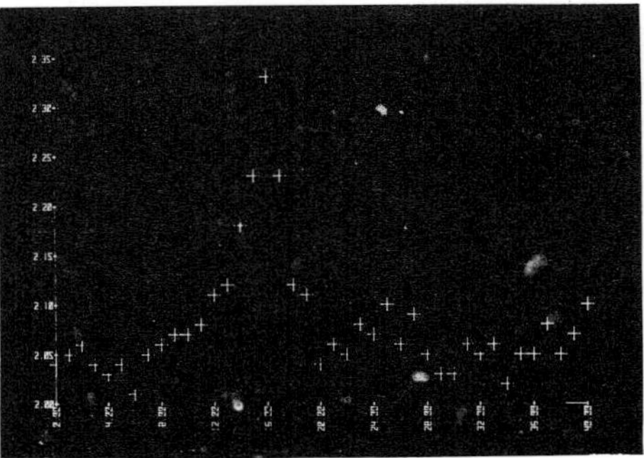

Fig 6.

Rocking curve of 020 Kossel line
0.25 mm aperture.

Microdiffraction of industrial supported catalysts

J P Lynch, E Lesage, H Dexpert and E Freund

Institut Français du Pétrole, BP 311, 92506 Rueil Malmaison, France

1. Instrumentation

Parallel recording of microdiffraction patterns in the STEM has been shown
(Cowley 1979) to be more efficient than the conventional method, using
post-specimen scan coils. Such a system has been developed   for the VG HB5
and is shown schematically in Figure 1. The diffraction pattern is formed
on a phosphor-coated glass plate positioned beyond the specimen. A mirror
angled at 45° allows the pattern to be viewed through a port. A commercial
low light level TV camera (RCA 1030H) was found to be sufficiently sensitive to record the image at TV rates. In order to allow the operation of the
normal bright field imaging system (via the electron spectrometer) the
screen and mirror are pierced by central holes, larger than the collector
apertures normally used to define the spectrometer acceptance angle. The
video signal may be viewed directly or recorded for later study. The screen
and mirror are retractable to allow the annular dark field detector (which
integrates the diffracted intensity at each image point) to be used when
required. Although less flexible than the system described by Cowley and
Spence (1979), the present arrangement was chosen so as to interfere as
little as possible with the other detection systems.

The angular size of the diffraction discs in microdiffraction patterns is
defined by the objective aperture size and specimen position and determines
the maximum lattice spacing which may be usefully examined. In the conventional specimen position (a few mm from the objective aperture) the probe
is highly convergent and rather large diffraction discs are seen. The
specimen may also be placed beyond the objective lens (in the cartridge
designed for secondary electron imaging) so that the aperture-specimen
distance is several cm. Low convergence angles and correspondingly smaller
diffraction discs are then obtained, at the expense of spatial resolution
in the image. In this work the angles of convergence and image resolutions
were, in the conventional position, 3.0 mrad and 1 nm, and in the "long
focal length" position, 0.5 mrad and 10 nm, using the same physical aperture size. Since the specimen is scanned in forming a bright field image,
the spatial resolution of the diffraction pattern is set by the area
visible on the viewing screen, or in spot mode by the image resolution
element.

2. Results

The catalysts chosen for study consist of metal particles (Pt and Pd) from
2 to 10 nm in diameter supported on highly divided alumina. The support
was in the form of platelets, typically 100 nm by 100 nm and a few nm
thick. It was prepared by heating aluminium hydroxide gel and appeared

from X-ray diffraction to be γAl_2O_3.

The post lens specimen position was found most useful in distinguishing the different types of oxide and platelet orientation present since the lattice spacings involved are large. In all cases for which zone axis patterns were obtained, the [111] direction was found to lie in the plane of the platelet, the most frequent orientations normal to the plane being [31$\bar{2}$] and [1$\bar{1}$0]. Figure 2 shows two [1$\bar{1}$0] patterns typical of (a) γAl_2O_3 and (b) δAl_2O_3 (Lippens and de Boer 1964), the latter being found in approximately 30% of cases.

For the study of relative orientations of particles and support the conventional specimen position was used to obtain the required spatial resolution. The angular definition was sufficient to identify zone axes from both components and, since the image is available simultaneously, tilting of the specimen to obtain indexable patterns is easier than when using the conventional recording system.

Local variations of support structure were again seen. Figure 3(a) shows a diffraction pattern taken whilst scanning a single platelet. When the probe is stopped, variations of intensity, presumably due to changes in defect concentration, are seen from points separated by a few nm (Fig. 3(b)-(d)). Support-particle orientations characteristic of epitaxy were found for only a small percentage of particles in this system. Figure 4 shows an example of two particles supported on the same [1$\bar{1}$0] oriented platelet (the image has been reproduced from a video recording). Although one particle is clearly also in [1$\bar{1}$0] orientation, the other does not show a well defined zone axis pattern.

3. Discussion

The above patterns reveal some of the drawbacks of the present system. The screen has suffered some damage, visible as a dark area near the centre hole, and since the screen is inside the microscope vacuum repairs are very time consuming. In addition saturation, giving rise to increased spot size, is evident in the more intense diffraction spots. Although the system is sufficiently sensitive to record the weak diffraction from particles down to 3 nm in diameter, the dynamic range (nominally 10^8) is not large enough to accomodate strong diffraction simultaneously.

References

Cowley JM 1979 Ultramicroscopy $\underline{4}$ 413
Cowley JM and Spence J C H 1979 Ultramicroscopy $\underline{3}$ 433
Lippens B C and de Boer J H 1964 Acta. Cryst. $\underline{17}$ 1312

Acknowledgements

The authors would like to thank Mssrs R Goudemand and M Bisiaux for technical support. One of us (J P L) would like to thank the Royal Society for an Exchange Fellowship.

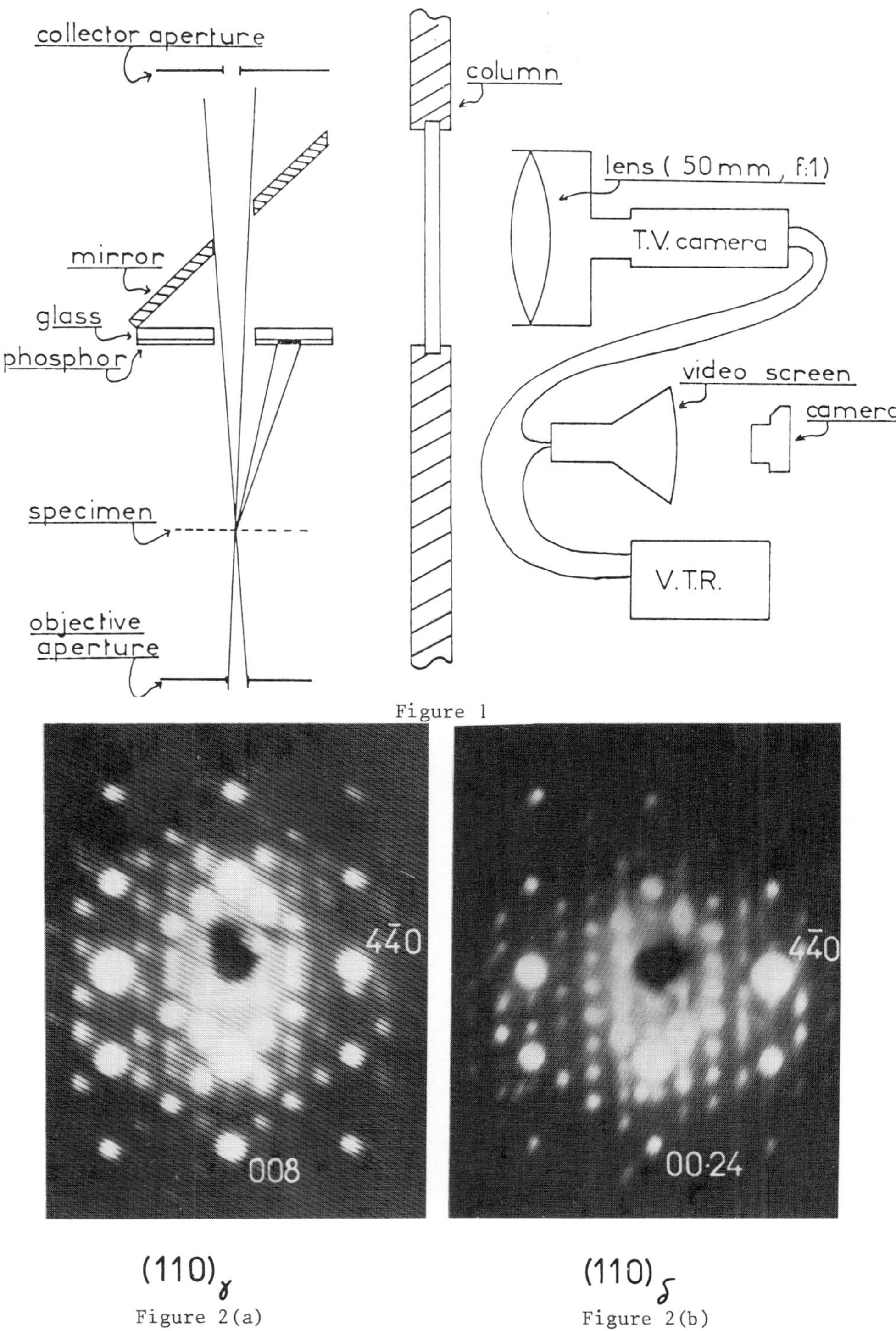

Figure 1

$(110)_\gamma$

Figure 2(a)

$(110)_\delta$

Figure 2(b)

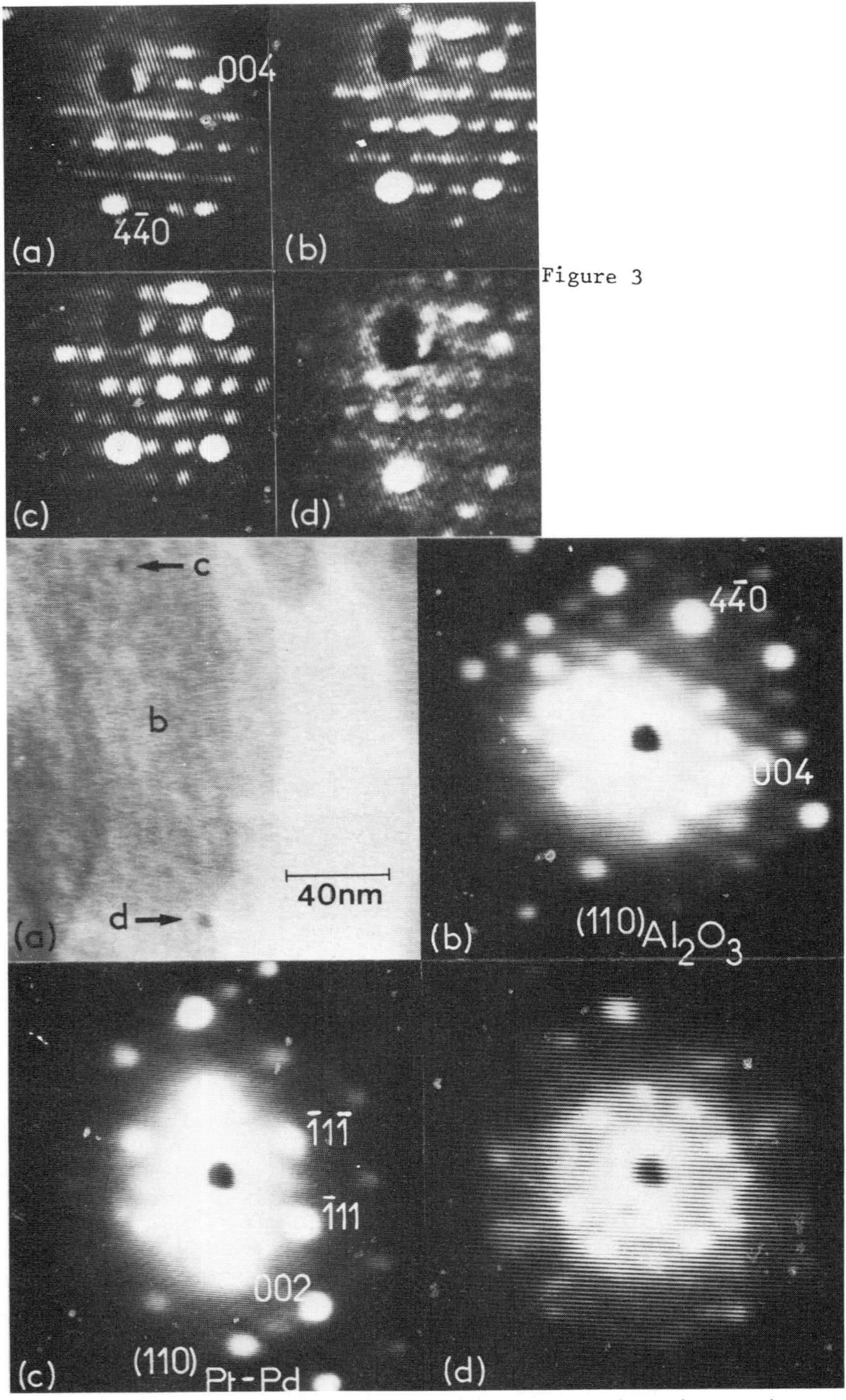

Figure 4 : (a) bright field image, (b) support microdiffraction , (c) and (d) particle microdiffractions .

Applications of the Cambridge interactive electron lens analysis system: CIELAS

R Hill and K C A Smith

University Engineering Department,Trumpington Street, Cambridge CB2 1PZ, UK

1. Introduction

An interactive electron lens analysis system (CIELAS) has been developed based on a minicomputer (Hill and Smith, 1980). It utilizes the basic finite element method of Munro (1971, 1975) to compute the fields around the lens structure, from which it is possible to further compute the electron optical properties. Various types of lens have been analyzed using this system, and the results of some of this work are presented in this paper.

2. The Pinhole Lens

A study of the parameters affecting the performance of the pinhole lens was undertaken, and the results compared directly with those published by Liebmann (1955). The main features of a pinhole lens are its back bore radius R_0, gap width S, and front bore radius R_1. Liebmann analyzed the problem by considering the lens to be split into two parts: half a symmetric lens of bore radius R_0 and gap width 2S, followed by a thin pinhole lens. He produced a set of curves, all normalized by the factor $(S+R_0+R_1)$, for the principle lens properties. The focal length f and the distance l from the inside face of the pinhole to the image plane are represented by single curves, since Liebmann concluded that f and l are little affected by the thin pinhole lens. (The curves correspond to the f and l of the half-symmetric lens.) The spherical aberration C_s is increased by the presence of the thin pinhole lens, but it is not clear how the C_s of the pinhole lens combines with that of the half-symmetric lens. Note, the chromatic aberration C_c behaves in the same way as f.

To test these conclusions the properties of a series of pinhole lenses were calculated, and the results presented in the same format as that used by Liebmann, as shown in Fig. 1. For each of the properties a lower and upper curve was found, which enclosed all the curves for the other lenses. The single curves predicted by Liebmann for f and l correspond to the lower of the respective curves.It can be concluded that the work of Liebmann on the pinhole lens is correct only to a first approximation. The parameters f and l cannot be accurately represented by a single normalized curve (note that the factor $(S+R_0+R_1)$ is the same for both the limit cases). All of the optical properties vary between two bounding curves, in a manner which is not related in any simple way to the bore and gap sizes.

3. Investigations of a High Resolution Electron Microscope Objective Lens

The properties of a range of lenses for the Cambridge University High Resolution Electron Microscope (HREM) objective lens have been investigated. The present lens was analyzed by Cleaver (1976, 1977) using the finite element method. A simplified lens model had to be used due to limits imposed by the maximum finite element mesh size then available. Using CIELAS a larger mesh size is available (7200 mesh nodes compared with the original 1250), and so a full lens model can be used in the analysis. This approach ensures that there are no unforeseen problems with stray fields, etc. The existing lens is a symmetric design, having top and bottom bore diameters of 7mm and a gap size of 14mm (referred to as a 7-14-7 lens). The specimen is placed at the midpoint of the gap, with the electrons emerging via the lower bore. Under these conditions the lens has C_S=3.33mm and C_C=3.27mm.

Initially lenses were analyzed assuming no saturation of the magnetic circuit, to determine the basic lens parameters that affect their performance. (An unsaturated condition was used to speed up the analysis, since it takes considerably longer to calculate the fields for a saturated lens.) By varying each parameter of the lens the following behaviour was found, for an object placed at the midpoint of the gap:
(i) the aberrations are little affected by changes in the top bore diameter;
(ii) smaller gap sizes produce smaller aberrations;
(iii) the aberrations vary considerably with lower bore diameter.
This last point is illustrated graphically in Figs. 2(a) and 2(b). These plots show the variation of C_S and C_C with object plane for three lenses with different lower bore diameters. For an object at the midpoint of -7.5mm (the origin is taken to be at the face of the lower pole-piece) then a large bore diameter is preferable for low C_S but gives larger C_C. Obviously a compromise must be reached such that neither aberration is particularly dominant.

The existing lens design allows only the top bore and gap to be easily modified and so two modified designs were selected for further consideration: a 7-11-7 and a 16-11-7 lens. (The larger top bore diameter offers certain advantages in specimen manipulation.) For both of these lenses it was necessary to evaluate the properties under saturated conditions. The results are shown graphically in Figure 3 for a range of excitations and assuming a beam voltage of 575kV. The 7-11-7 lens gives superior performance, and for an object placed at the midpoint it is possible to achieve C_S=2.4mm and C_C=2.9mm.

4. A Miniature Double-gap Condenser Lens

As mentioned earlier, it is possible under CIELAS to use a large number of nodes in the finite element mesh, which allows the user to analyze very complex lens geometries. To test this facility, a miniature double-gap condenser lens was selected, as shown in Figure 4, where both gaps are excited by the same coil. To achieve an acceptable accuracy in the analysis of such a geometry it is necessary to have a high node density in and around both gaps; this is possible with CIELAS, and such a finite element mesh is shown in Figure 5. Using this data it was possible to compute the axial flux density with an error of only -1.5% (see Hill and Smith, 1979 for a description of how this error is derived). From this field the properties of the lens can be calculated. The analysis of any such complex geometry can now be undertaken with confidence, using the large mesh size available with CIELAS.

5. Conclusions and Acknowledgements

In assessing the effectiveness of a design tool such as CIELAS the time taken to analyze a lens is an important consideration. The work on the pinhole lens was completed over a period of two days, whilst the bulk of the work on the HREM objective lens took about six days. The analysis of the double-gap condenser lens took two hours, but no optimization of the geometry was done.

The applications presented in this paper illustrate the power, speed, and
flexibility of CIELAS. The ability to be able to present the results of any
analysis in a convenient format, e.g. as Liebmann, makes the assimilation of
these results easier for the user, and the availability of large mesh sizes
facilitates the accurate analysis of very complex lens geometries.

We wish to thank Dr J R A Cleaver, Dr W C Nixon, Dr D J Smith and Prof T Mulvey
for many useful discussions.

6. References

Cleaver J R A 1976 Optik <u>48</u> 95
Cleaver J R A 1977 Optik <u>49</u> 413
Hill R and Smith K C A 1979 Electron Microscopy and Analysis 1979
   ed T Mulvey (Inst. of Phys. Conf. Series number 52) pp 49-52
Hill R and Smith K C A 1980 Proc. 7th European Cong. on Elec. Microscopy
   ed P Brederoo and G Boom <u>1</u> pp 60-61
Liebmann G 1955 Proc. Phys. Soc. <u>B</u> 682
Munro E 1971 Ph D Thesis University of Cambridge
Munro E 1975 Engineering Dept Technical Report CUED/B-Elect TR45
   University of Cambridge

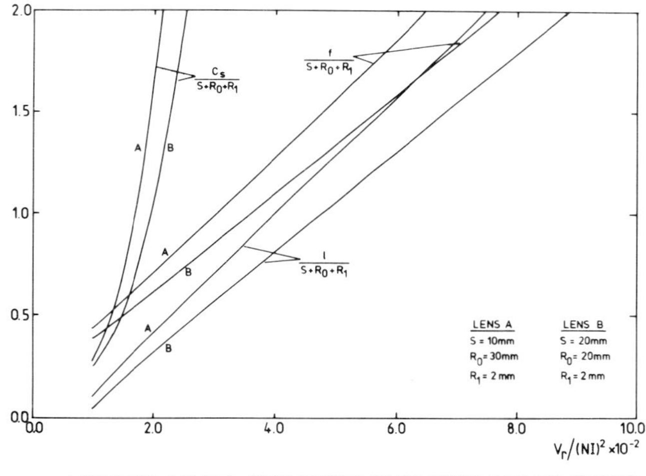

Fig. 1 The variation of
the optical properties
of a pinhole lens

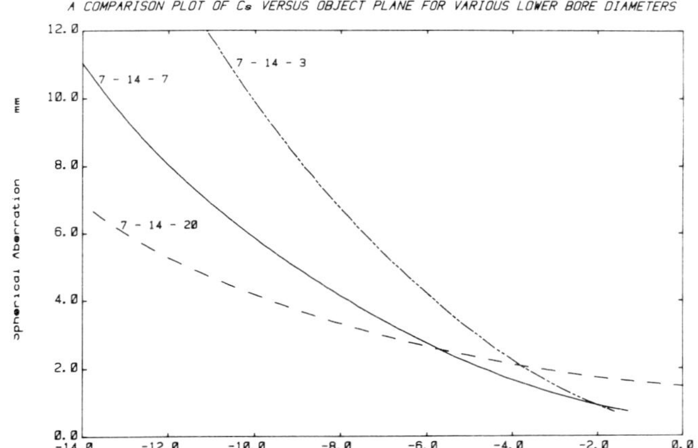

Fig. 2(a)

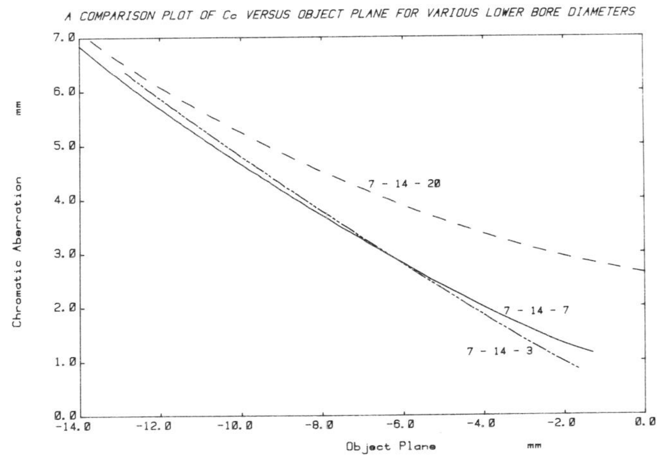

Fig. 2(b)

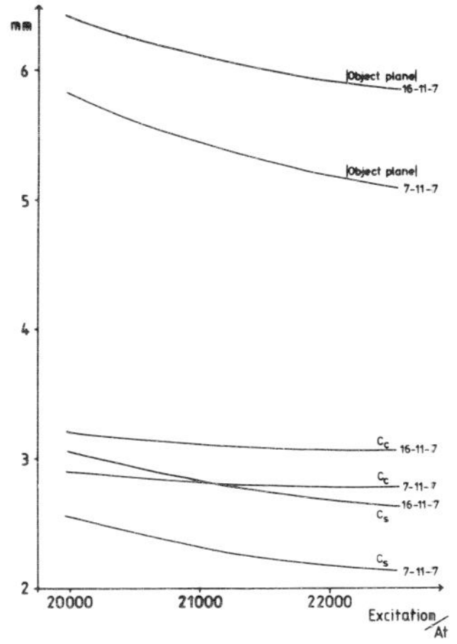

Fig. 3 The saturated
behaviour of the 7-11-7
and 16-11-7 lenses

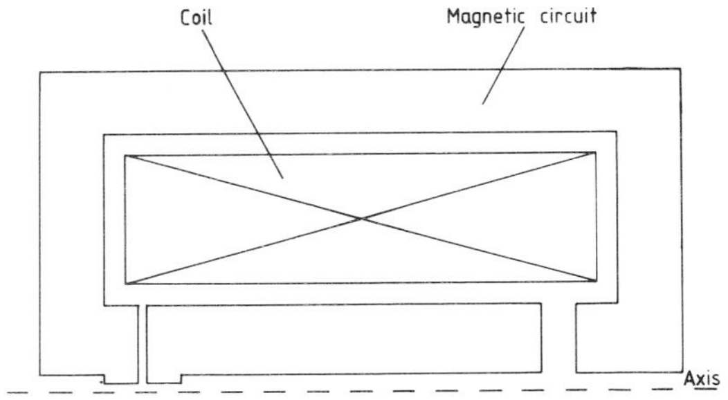

Fig. 4 Cross section through the
double gap condenser lens

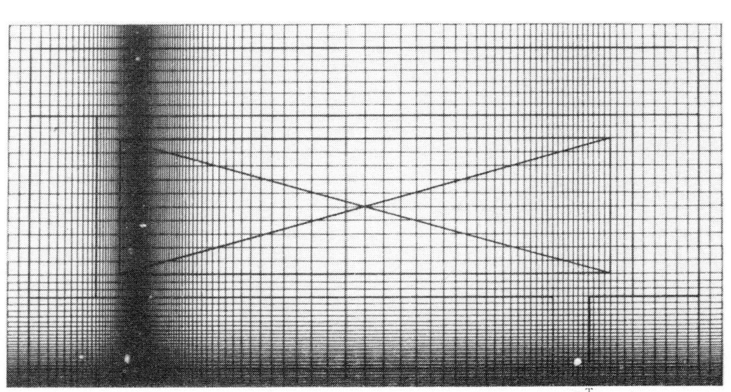

Fig. 5 The complete
finite element mesh
for the double gap
condenser lens

Improved programs for calculating optical properties of electron lenses

H Nasr, W Chen[*] and T Mulvey

 Department of Physics, The University of Aston in Birmingham, B4 7ET
*On leave from the Department of Physics, Beijing University, Chinese
Peoples' Republic

1. Introduction

The electron-optical system of a modern electron microscope has now become
too complex to design without the aid of a computer. Frequently the
optimisation of a system can be carried out efficiently only by the use of
interactive computation (Hill and Smith 1980). In this case it is desirable
to employ an immediately accessible mini-computer. In the past this has not
always been feasible because of the large computer store needed in the
calculation of the lens fields by the convenient finite element method.
The two forms of this method at present available, namely the differential
method (Munro 1971) and the integral method (Newman et al 1972), were both
designed primarily for large computers where the size of the computer store
is not a problem.

Recent developments in low-store field calculations (Mulvey and Nasr 1981,
Nasr 1981) have led to a substantial reduction in core store without
loss of accuracy. In order to take full advantage of these programs
however, additional ray tracing and aberration finding programs are also
needed that are suited to the requirements of interactive computing. This
includes the calculation of the general ray equation, which is especially
useful in aberration-correcting systems (Al-Hilly and Mulvey 1981). All
the programs described in this paper can be applied to a wide range of
mini-computers, or even micro-computers such as the Commodore PET. To
illustrate this point Figure 1 shows the axial magnetic field distribution
of a single-polepiece lens carried out on a PET using a mesh size of 29
axial x 19 radial using the selected area method developed by the authors
(Mulvey and Nasr 1981). It should be mentioned that this is quite a
demanding calculation on a main frame computer using the standard (Munro
1975) program. Figure 2 shows some diagnostic tests that can be carried
out on the accuracy and consistency of this calculation by means of the
newly-developed differential-integral method (Nasr 1981). A central
feature of this diagnostic test is that the magnetic field on the axis
due to the coil is calculated independently, as shown in Figure 2. This
is then subtracted from the total field, giving the axial field due to the
iron. The area under this field distribution must be algebraically zero
since the iron does not contribute to the lens excitation. In general the
net area under the curve will not be zero; this allows one to make an
estimate of the accuracy of the computation. Finally the field due to the
iron (smoothed if necessary) can be added to that due to the coil calculated

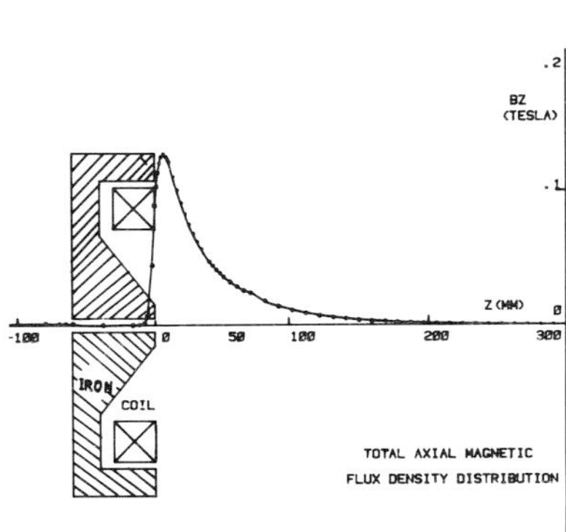

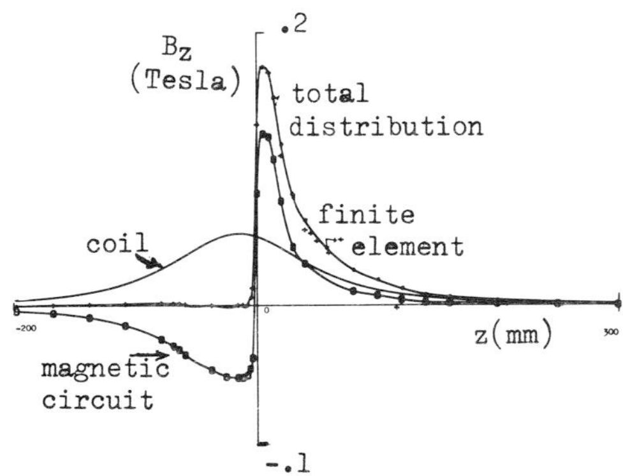

Fig. 1 Calculation of the axial
field distribution of a single-
polepiece lens on a Commodore
PET micro-computer by   the
'selected area' method.Mesh size
19x29.(Programs VPLIN, VPSAT)

Fig. 2 Diagnostic check and improvement of the axial field distribution
shown in Fig. 1 by means of the
differential-integral method.
(Program DIFINT)

by the Biot-Savart law producing a total field of improved accuracy.

2. Optimisation of saturated lenses

If the polepieces of an objective lens are allowed to saturate in a
controlled manner it is often possible to improve the spherical and
chromatic aberrations. However these aberrations do not reach their
minimum values simultaneously and so some compromise may be necessary in
the design. In order to carry out the necessary calculation of the electron
trajectories on a mini-computer, it is essential to reduce the   storage
required by the program without appreciable loss of accuracy.   One way of
achieving this is to vary the step length of the calculated   electron
trajectory along the axis of the magnetic field.   Ideally, this step length
should vary according to the distribution of the axial magnetic   flux
density in such a way that in the important steep part of the field, the
step length should be short while for other parts it can be longer. This
requirement is in fact the same as that of the division of the field into
finite elements in the initial calculation of the field. In our trajectory
and aberration program DOCOL (Design Of Condenser Objective Lenses)   the
axial region between the finite element mesh points is sub-divided   into
four equal lengths to provide extra field values by interpolation.   This
program is particularly useful in investigating the   electron-optical
properties of saturated polepieces in a mini-computer.   An alternative
program REALAB (REAL ABberration) has also been developed giving a further
reduction in core store and computing time.

To investigate the correlation between the aberration coefficients C_s
and C_c and the saturation conditions in the lens polepiece, it is useful
to display the incremental values ΔC_s and ΔC_c of the spherical   and
chromatic aberration coefficients along the axis.   This provides useful
diagnostic information in the early design stages. As an example Figure 3
shows the polepiece structure and local relative axial field distribution
B_z/B_0 in the lower polepiece of a hypothetical 600 kV high resolution
microscope objective lens (cf. Cleaver 1978).

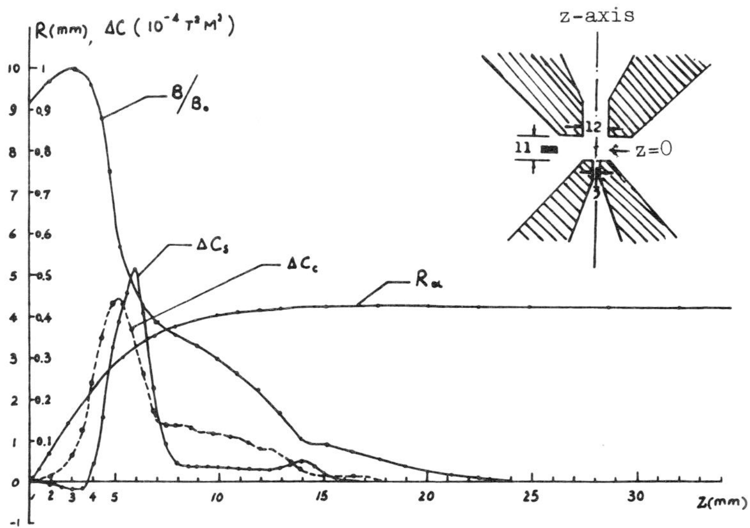

Fig. 3 Computer-aided optimisation of the chromatic and spherical aberrations of a high resolution TEM objective with saturated polepieces. For V=600 kV, best C_S=1.9 mm (C_C=2.8 mm) and best C_C=2.4 mm (C_S=3.6 mm). Program DOCOL or REALAB.

The respective areas enclosed between the ΔC_S and ΔC_C curves and the axis corresponds to the magnitude of the field aberration coefficient C_S or C_C. By suitably adjusting the polepiece structure and excitation in order to minimise C_S and C_C it is possible to reach a near-optimum lens   design entirely by the interactive use of a mini-computer. The above investigation showed, for example, that a number of compromise solutions were available between the extreme cases of the minimum C_S value of 1.9 mm (but with C_C= 2.8 mm) and the minimum C_C value of 2.4 mm (but with C_S=3.6 mm).

3. Optimisation of projector lens systems

It is frequently necessary to consider the interaction between   two lenses, for example, the intermediate and final projector lenses of a TEM especially in wide-angle projector systems (Al-Hilly and Mulvey 1981). The program   SYSLIN (SYStem under LINear conditions) or SYSSAT   (SYStem under SATuration conditions) allows one to calculate gaussian trajectories and total aberration coefficients F of a twin-lens field distribution as a function of the lens excitation parameter as shown in Figure 4. The image distortion (spiral and radial)$\Delta\rho/\rho$ is given by

$$\Delta\rho/\rho = F \tan^2\alpha_p$$

where α_p is the semi-projection angle of the final projector lens. Figure 4 shows that simultaneous correction of spiral and radial distortion is possible in this lens system.

It should perhaps be mentioned that third order aberration theory cannot be validly used however at the high gaussian semi-projection angles of 28° envisaged. For such large angles, general ray calculations are needed   and suitable programs have already been developed that make use of the improved

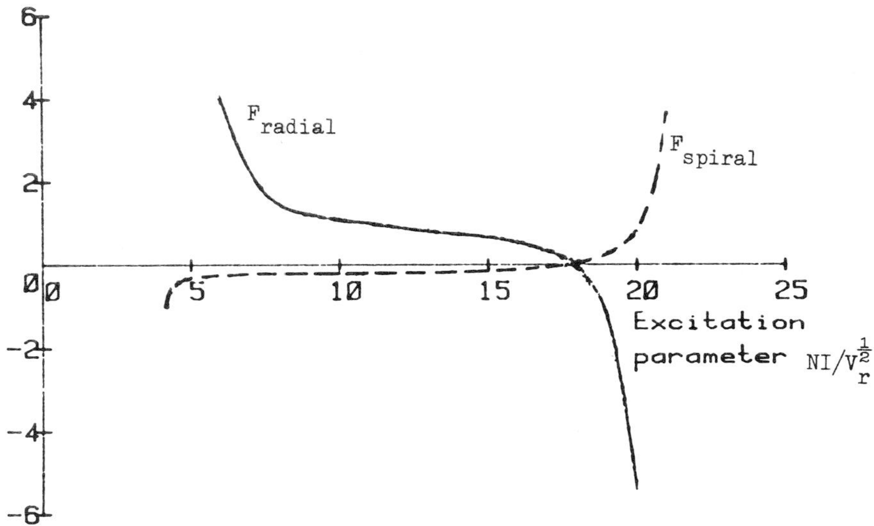

Fig. 4 Calculation of the condition for the simultaneous correction of spiral and radial distortion in a double projector lens system. (Programs SYSLIN and SYSSAT).

finite element method (Nasr 1981, Al-Hilly 1982).

The programs described above have already proved useful in the solution of a wide range of electron-optical problems, especially where   it was desirable to use a mini-computer. It is hoped to make these programs generally available in the near future.

4. References

Al-Hilly S M 1982 PhD thesis Aston University
Al-Hilly S M and Mulvey T (These Proceedings)
Cleaver J R A 1978 Optik <u>49</u> 413-431
Hill R and Smith K C A 1980 Electron Microscopy 1980 (Leiden) eds. P
    Brederoo and G Boom Vol.1 PP60-61
Mulvey T and Nasr H 1980 Nuclear Instruments and Methods (In press)
Mulvey T and Nasr H 1980 Electron Microscopy 1980 (Leiden) eds. P Brederoo
    and G Boom Vol.1 PP64-65
Munro E 1971 PhD dissertation Cambridge University
Munro E 1975 A set of computer programs for calculating the properties of
    electron lenses Department of Engineering Cambridge University
Nasr H 1981 PhD thesis Aston University
Newman M J, Trowbridge C W and Turner L R 1972 GFUN: An interactive program
    as an aid to magnet design Proc. 4th Int. Conf. on Magnet Technology
    Brookhaven pp617-626

Analysis of space charge limited electron beams

A T Stokes and K C A Smith

University Engineering Department, Trumpington Street, Cambridge CB2 1PZ,UK

1. Introduction

A computer program has been developed for the analysis of electron beams
in space charge limited conditions.   Because of the discrete nature of
digital computer programs it cannot fully emulate the continuous changes
that occur in any real problem.   The standard manner of solving electron
beam problems with space charge is to allow current emitted from cathodes
to flow through the electric field.   A charge density distribution is
determined from this current flow and a new electric field found.   This
cycle is repeated until a self-consistent solution evolves.   The effect of
charge density is to depress the potential and to reduce the initial
acceleration rate of emitted charges at the emitter surface.   This in turn
lowers the electron velocities throughout the field area, increasing the
charge density even further.   Changes in charge and potential, however,
decrease with each cycle until final values are asymptotically approached.
An example of this is given in Fig. 1 where the final potential is ϕ_{FINAL}.

For a sufficiently high current however the charge density is so great
that the potential goes negative in places before ϕ_{FINAL} is obtained,
producing a negative gradient normal to the emitter face.   This is shown
in Fig. 1 as ϕ_{OFF}, and its effect is to retard the current emanating from
the emitter.   In a real problem the potential goes negative slowly, and
as it does so the current is reduced until a final potential is established,
ϕ_{CRIT}.   However, a computer works in discrete steps and when the potential goes negative it does so by several volts.   Even the hottest thermionic emitter cannot produce electrons with initial velocities that can
surmount a negative potential barrier of more than a fraction of a volt.
The effect of a potential like ϕ_{OFF} is to cut off all emission.   Since
no current flows charge density becomes zero and the potential reverts to
the charge-free value of ϕ.   Successive cycles again build up the charge
density, lowering the potential until it again saturates and emission
stops.   This on-off situation cannot produce meaningful results.   The
difficulty is overcome by reducing the emission currents.   Whereas
previous methods of estimating the current reductions as given by Radley
(1975) and Birtles (1975) involve complex calculations, the new method
outlined below does not.

To obtain solutions by computer methods it is necessary to reduce the
emitted current artificially such that no total cut-off state can occur.
The barrier can only be a fraction of a volt high, since this is responsible for retarding part of the current.   In practice this barrier is

ignored and considered as zero, resulting in ϕ_{CRIT} on Fig. 1.   In conditions of space charge limitation therefore the current density from those parts of the cathode which are saturated, that is having negative electric field gradients normal to their surface, must be reduced to make their gradient zero.   Furthermore, different parts of the emitter may be under differing amounts of saturation and some regions not saturated at all.

2. <u>Method of Solution</u>

Space charge limited problems are solved by the implementation of the iterative sequence of events of figure 2.   It augments the standard manner of solution by terms introduced to control the current emission from source.   Electric field solutions are obtained by finite element methods, see Munro (1975).   Trajectory and charge density determination utilises the Lorentz equation and charge sharing as described by Stokes (1978).   The check at each cycle, except the first one, is a comparison of the charge dependent potential ϕ_ρ, and the current emission between successive cycles.

Potentials modified by charge density ρ, such as ϕ_{FINAL} in Fig. 1, consist of two components: a charge free component ϕ_L due to electrodes alone, and a charge dependent component ϕ_ρ.   The existence of the superposition of the two components to give the total potential of ϕ_T is used as a basis of the current reduction.   At all places of emission G_L, the gradient of ϕ_L normal to the emitting surface, must be positive; whereas G_ρ, the gradient of ϕ_ρ will always be negative and the total gradient G_T is the sum of these.

For simplicity a problem having only a single source of emission will be considered first.   Then ρ, ϕ_ρ and, more importantly, G_ρ will be directly dependent upon its emitted current.   The sequence of Fig. 2 is run and charge is allowed to cyclically build up, with no reduction in current, until saturation occurs and the potential becomes ϕ_{OFF}. In this state $|G_\rho| > |G_L|$ and, therefore, G_T is negative.   The criterion is to reduce the current emission until G_T is zero by reducing the emitted current by the ratio $|G_L/G_\rho|$.   This is readily achieved due to the dependency of ρ on the current Subsequent cycles, as in the standard technique, produce further saturation due to the additional increases in ρ, necessitating further reductions in current emission. The changes in ρ and current emission however grow less as the cycles proceed until solutions for both are asymptotically approached.

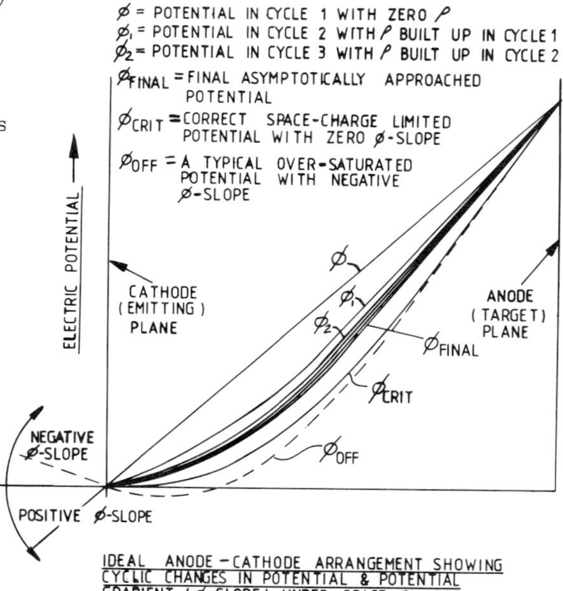

Figure 1.

Accuracy of the solution will increase with the number of emission points, so that in practice the cathode will have many points of emission. Then ρ and G_ρ will no longer be dependent upon the current emitted from any one point. Alteration of the current at any point will not alter its G_ρ by the same proportion and will affect the G_ρ values of every emission point differently. However, if the current from all the emission points are adjusted by an identical fraction, then at every point ρ and G_ρ will be altered by that same fraction.

As before, the solution proceeds with full emission current from all points. Charge density is cyclically built up until saturation conditions occur at one or more emission points. All points under saturation have their ratio $|G_L/G\rho|$ calculated (these will always be less than 1.0) and the smallest taken as a scaling factor SF. All emission current levels are scaled by SF, which results in the charge density ρ and hence the values of ϕ_ρ and G_ρ at all emission points being identically scaled. Points with a ratio equal to SF now have zero G_T gradients, and therefore satisfy the criterion. All other points have a positive value of G_T, and from their viewpoint the current reduction is too great. Indeed points previously not in saturation did not require their currents to be reduced at all, however, the reduction is necessary to eliminate all negative G_T gradients. Therefore, although the field ϕ_ρ is acceptable for the present cycle of the solution, the current levels at the emission points require adjusting for use in the next cycle's trajectory calculations. These adjustments are made on the assumption that the current at each emission point is directly and solely responsible for its G_ρ value. In this case all current is adjusted by the new ratio $|G_L/G_\rho|$, which will always be greater than or equal to 1.0. An upper limit is

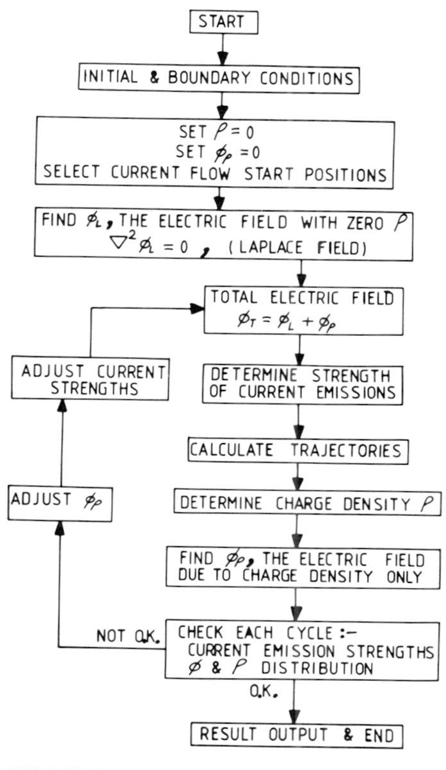

MODIFIED FLOW DIAGRAM TO ENABLE COMPUTER ANALYSIS OF PROBLEMS INVOLVING SPACE – CHARGE LIMITING OF THE EMISSION CURRENT

Figure 2.

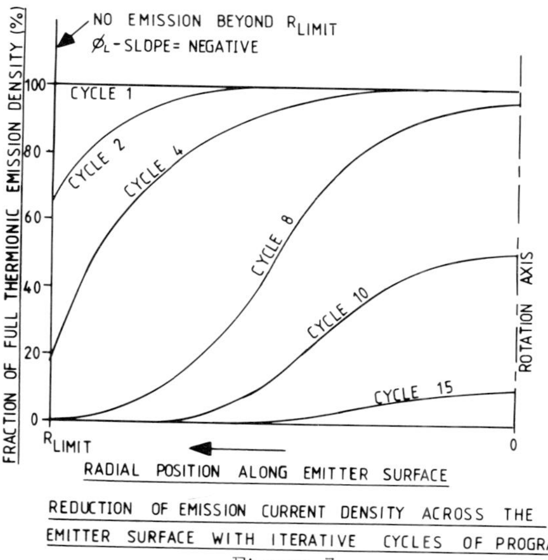

REDUCTION OF EMISSION CURRENT DENSITY ACROSS THE EMITTER SURFACE WITH ITERATIVE CYCLES OF PROGRAM

Figure 3

set since a point cannot emit more current than that predicted for purely
thermionic emission.   With this proviso three types of emission point
exist: those with zero G_T which do not have their current adjusted since
their new ratio is 1.0; those not previously saturated which regain their
maximum thermionic current; and the remainder, being previously saturated,
have their current increased to something less than maximum.   The
assumption upon which the current changes are made is not entirely true,
causing too great a current increase so that the next cycle will again
saturate, necessitating further current adjustments.   Simultaneously
charge density is still inexorably increasing cycle by cycle, the reduction in current merely reducing the amount of its increase.   However both
causes of saturation operate simultaneously and changes in current and
charge decrease cyclically until solutions for both are asymptotically
approached.

In practice ϕ_L along the emitter will vary from a positive value at the
axis to a negative at its outer limits, emission ceasing when ϕ_L crosses
zero.   The value of ϕ_L and hence the emission can be controlled by the
introduction of an extra electrode (Wehnelt) close to the emitter and
biased negatively.   Because of this differential value of ϕ_L the emission
points also saturate differentially with saturation occurring first at
areas of low ϕ_L.   In the example from which Fig. 3 is drawn saturation
occurs, in cycle 2, at the outer emitter regions and these are consequently
reduced.   As shown, subsequent cycles will not only further reduce points
already reduced once, but also reduce newly saturated points closer towards
the axis.   This proceeds until all parts are saturated with reduced self-
consistent currents.

3. Conclusions

The technique outlined here maintains the criterion of zero or positive
G_T at all times.   Ideally, when some points saturate, it would be possible
to calculate the effect of each emission point on all G_0 values and thereby deduce the exact currents needed to give zero G_T at all saturated
points.   However this would be extremely difficult for the great number
of trajectories available (up to 1000 maximum).   It would also be super-
fluous since the cyclic increases in charge density would make a new
current determination necessary.   In the present method as the value of
ρ converges to a suitable answer so do the current values.

The analytical technique outlined here has been successfully applied to
the solution of a space charge limited electron beam problem with thermionic emission and an electrostatic lens.   The predicted values are in
good agreement with those measured.

This work forms part of a general study of the analysis of electron beams
in electromagnetic fields by computer methods.   Financial support from
the Science Research Council, the National Research Development Corporation
and the Royal Aircraft Establishment in Farnborough, is gratefully acknowledged.

References

Birtles A B and Dirmikis D, 1975, Int. J Electronics Vol 38 pp 49-67
Munro E 1975 Report CUED/B-ELECT/TR45, Dept. of Engineering, Univ of
    Cambridge
Radley D E, Dirmikis D and Birtles A B, 1975, Proc IEE, Vol 122 pp 620-624
Stokes A T 1978 Thesis, Dept. of Elect. Eng., Univ. of Aston in Birmingham

Materials science applications of hybrid scanning modes

M.N. Thompson

E.O. Applications Laboratory, S&I, Philips Industries, 5600 MD Eindhoven, The Netherlands.

1. The Analytical System

A hybrid diffraction and dark field (HDD) module capable of conical scanning, descan, rocking beam and the scanning of TEM image and diffraction patterns over the STEM detector has been described by Thompson (1980). The scanning of the TEM image mode (SCIM) and the scanning in the TEM diffraction mode (SCID) provide the means of converting basic TEM information into a time sequential form suitable for input into a computer.

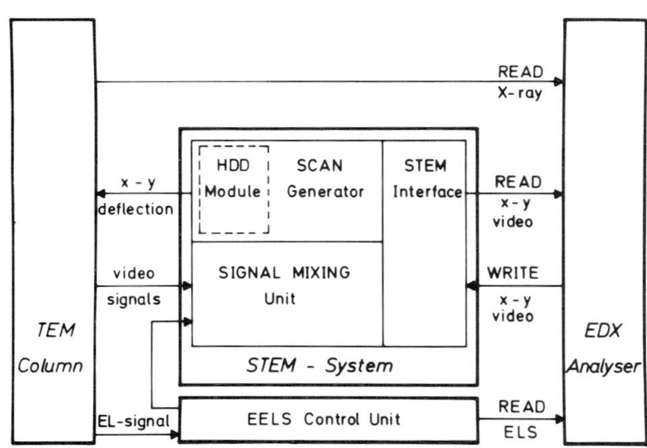

Figure 1. System comprises an EM400T and STEM unit + HDD module interfaced to an EDAX 9100/65 analyser. Electron energy analysis and energy filtering is conducted with a GATAN 607 spectrometer.

Applications of on-line quantification of TEM diffraction patterns and specimen thickness measurements have been published by Hagemann (1981).

2. Conical Scan Mode and its Applications

The conical scanning mode is an extension of the TEM mode and provides dynamic, hollow-cone illumination of the specimen. The technique has been applied to high resolution imaging by Krakow (1976,1981) and to texture studies and the differentiation of phases by van Oostrum et al (1976). The mode of operation is shown schematically in Figure 2. The beam is tilted through a semi-angle θ and subsequently conically scanned with its apex in the specimen plane. As a result all diffracted beams originally scattered through an angle θ enter the objective aperture and contribute to the image. TEM specimens which form concentric ring patterns such as extraction replicas, aerosol, pollution particles, thin films catalysts etc. can be usefully studied using the conical scan method. Despite the power of the technique to discriminate between randomly oriented different phases, and indicate their location and density using the dark field technique, little general attention has been directed to its general application.

<u>Catalyst Application</u>: In the TEM study of a catalyst sample no evidence of a random distribution of palladium particles was observed in bright field Figure 3a. The diffraction pattern Figure 3b, however indicated both the presence of the particles as a ring pattern and of the single crystal substrate. To optimise the dark field contrast of the particles a Pd diffraction ring (and objective aperture size) was selected which did <u>not</u> allow any single crystal reflection to contribute to the dark field image during conical scanning. The resultant dark field micrograph which integrates and registers <u>all</u> Pd particles diffracting into the ring is shown in Figure 3c. The selection of a lower order ring which contains both Pd diffraction spots and the single crystal substrate is not advocated as the effective contrast and recognition of the palladium particles is obstructed by the strong substrate contribution Figure 3d.

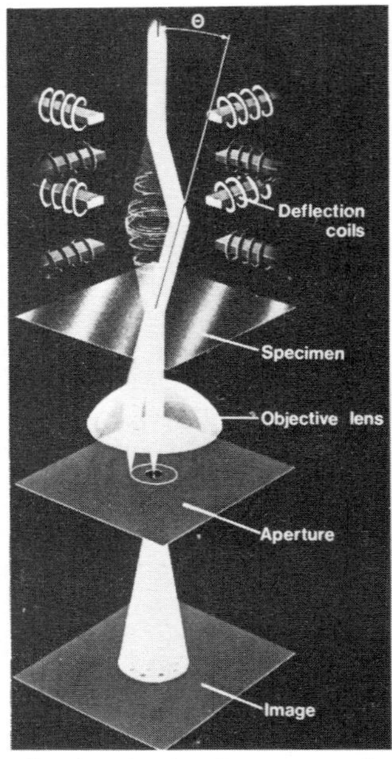

Fig.2. Conical Scan Mode

Fig.3 Catalyst specimen
a) Bright field TEM image
b) Corresponding Diffraction Pattern
c) Integrated D.F. image from Pd ring
d) Integrated D.F. image from low order Pd ring + substrate reflections

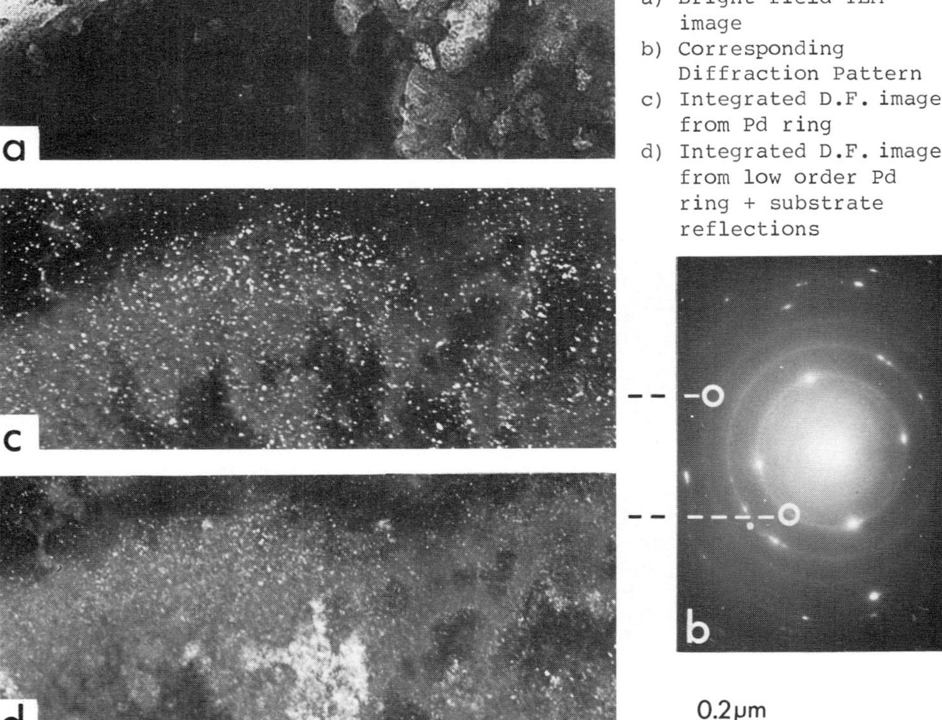

3. A new Rocking Beam Application

Applications of the rocking beam mode to microdiffraction have been
reported by van Oostrum et al (1973) and dark field analysis by Thompson
(1973). The latter methods rely on rocking semi-angles greater than 2°.
The present investigation has shown that smaller rocking angles $< 2\theta_{BRAGG}$
can be used to eradicate unwanted strong diffraction contrast arising
from bend contours and other dynamical diffraction effects in crystalline
materials.

Figure 4a shows strong diffraction contrast around a triple point in a
ceramic material. The strong contrast in the grains diverts attention
from the region of interest along the grain boundary. To improve the
bright field TEM image information content, the illumination was rocked.
As a consequence the local diffraction conditions varied causing the bend
contours to be locally displaced. Registration of the integrated images
generated by the rocking process results in an image, Figure 4b, in which
the strong diffraction contrast is smoothed out and the presence of a
grain boundary phase made evident. The diffraction pattern generated during rocking is shown in Figure 4c. The circle denotes the position of the
objective aperture and defines the angular range of the transmitted
electrons used to produce the bright field Micrograph 4b.

The above technique has a potential application in removing undesirable
contrast in micrographs of micro-crystalline materials, matrices obscured
by strain contrast, complex mineral phases, particulates etc, designated
for off-line image analysis. The benefit of the technique relative to the

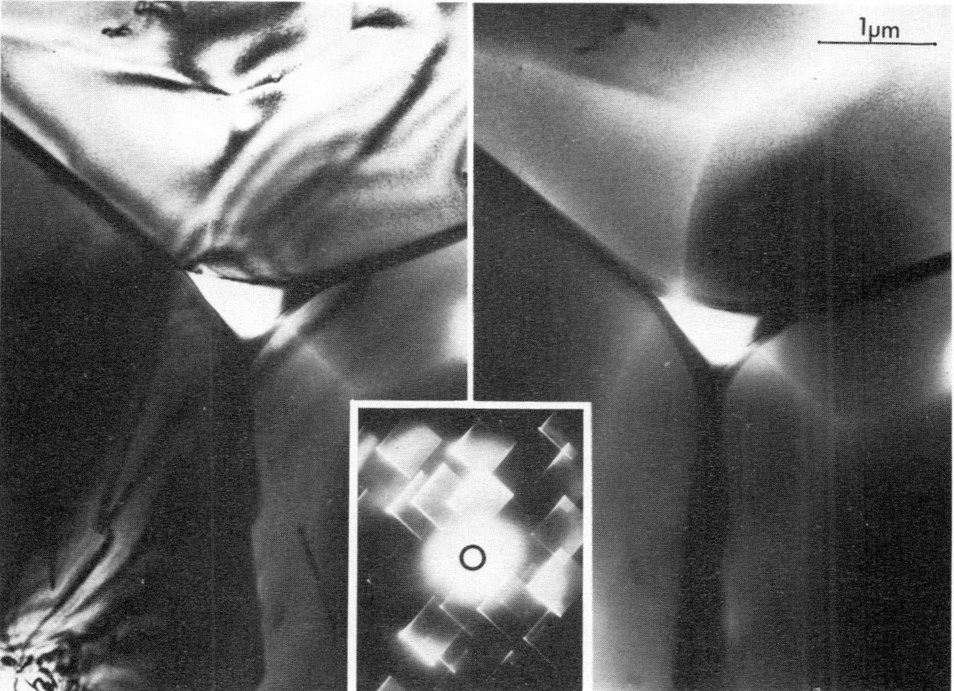

Fig 4a Conventional Bright   Fig 4c Rocking Beam   Fig 4b Rocking Beam Bright
Field image of a ceramic   D.P.   Field image. Note absence
alloy.   of bend contours.

competing STEM imaging method is in the higher image resolution and information density at low magnifications inherent in the TEM photographic process.

4. Double Rocking Beam Mode

This technique was first performed in a SEM by Möllenstedt and Meyer (1975) and later in a TEM system by Eades (1980).

The double rocking beam mode allows wide angled zone axis patterns (ZAP's) to be displayed on the STEM unit CRT. The optics require that scanning coils above the specimen rock the beam in the specimen plane while a set of scan coils beneath the specimen returns the beam to the optical axis for detection. The relative rotation and magnification differences of the rasters above and below the specimen are matched using the STEM scan rotation control and the auxiliary magnification controls of the HDD module and zoom magnification controls.

Figure 1a shows a [100] orientation in Aluminium produced by the double rocking beam method following STEM detection and Figure 5b after energy filtering to allow only no-loss electrons to contribute to the image.

Figure 5
[100] ZAP in
Aluminium

Courtesy of
P.Hagemann.

a) STEM Image   b) No-loss electron image

5. Conclusions

It is encouraging to note that some years following the introduction of the TEM/STEM system and the concept of hybrid scanning modes a variety of these modes are available for general application to materials science problems. These modes are, and will, continue to be subject to evolution in specialised areas. In the future we can anticipate further innovations involving the integration of the operational flexibility of the hybrid modes with the programming and data processing capability of the computer.

References
Eades J.W. 1980 I.of P. Conf. series No.52, Chpt 1, 9-12.
Hagemann P. 1981 Proc.39th EMSA Meeting, 250-251.
Hagemann P. 1981 Beitr.elektronenmikroskop.Direktabb.Oberfl.14, 339-346.
Krakow W. 1976 Proc.34th EMSA Meeting, 566-567.
Krakow W. 1981 Proc.39th EMSA Meeting, 224-225.
Möllenstedt G. and Meyer H.R. 1975 Optik 42, 487.
van Oostrum K.J., Leenhouts A. and Jore A. 1973 Appl.Phys.Lett 23, 283
van Oostrum K.J., Premsla H.F., Leenhouts A. and Jore A. 1976 Philips
   Bulletin EM101.
Thompson M.N. 1973 SEM systems and applications. I.of P.Conf. series
   No.18, 176-181.
Thompson M.N. 1980 Electron Microscopy 1980, Den Haag, Bd.1, 524-525.

Ion bombardment effects on field emission current stability

H Todokoro, N Saitou and S Saitou[*]

Central Research Laboratory, Hitachi Ltd., Kokubunji, Tokyo, 185, JAPAN
*Naka Works, Hitachi Ltd., Katsuta, Ibaragi, 312, JAPAN

1. Introduction

Field emission source brightness is 10^3 times as large as a thermionic emission source. Therefore, it has been used as the electron source for high resolution scanning electron microscopes (Crewe, Wall and Langmore 1970, Todokoro, Nomura and Komoda 1977) or conventional electron microscopes (Tonomura, Matsuda, Endo, Todokoro and Komoda 1979). However, stability of the field emission depends on pressure, emission current, and anode cleanliness, etc. There are two types of instability in field of emission current; one is long term instability called the drift; the other is short term instability, the fluctuation.

Fluctuation is increased by emission current increase as well as pressure rise. A large fluctuation ($\Delta I_E/I_E = 20 \sim 30\%$) ultimately results in source destruction. Thus, to apply a field emission source in electron beam analytical instruments requiring much greater probe current than scanning electron microscope, it is necessary to identify the true mechanism of the fluctuation and remove it.

Little experimental work has been done on fluctuation increase due to ion bombardment. Swann and Smith (1973) concluded that a pressure of at least 1.3×10^{-7} Pa (1×10^{-9} Torr), and a current below $10\mu A$ was necessary to ensure relative fluctuation below 5%. They also indicated that operation in a pulsed mode could reduce the fluctuation but that the gains were relatively small.

This paper clarifies the fluctuation mechanism by varying emission current and pressure over a wide range.

2. Ion Bombardment Effects

It is assumed that fluctuation increase is caused by ion bombardment of the emitting area. Relative fluctuation should be the function of N_i:

$$\Delta I_E / I_E = f (N_i) \tag{1}$$

where N_i is the number of the ions that bombard the tip per unit time. Parameter N_i is composed of ions formed by electron stimulated desorption at the anode and ions generated by electron collision with residual gas.

It is given by:

$$N_i = \frac{1}{2} n_e \beta \left(\frac{n_a r_o}{R}\right)^2 + n_e n_G \sigma_G (r_1 - r_o) \qquad 2),$$

where n_e is the number of electrons that bombard the anode surface, β is the ion ejection ratio for electron bombardment on the anode, $n_a r_o$ is apparent source radius ($n_a \sim 10$) (Saitou 1977), r_o is real source radius, R is anode distance from the tip, σ_G is the electron impact ionization cross section and r_1 (~ 1.4um) (also Saitou 1977) is the effective radius in which all ionized molecules can strike the emitting area.

By substituting pressure P and emission current I_E in equation 2), N_i is expressed by:

$$N_i = 3.1 \quad 10^{18} \beta \left(\frac{n_a r_o}{R}\right)^2 I_E$$

$$+ 1.7 \quad 10^{33} \sigma_G (r_1 - r_o) P I_E \qquad 3).$$

The first term is proportional to only emission current I_E . The second term is proportional to the product of pressure P and emission current I_E.

Ratio γ of the second term to the first term is:

$$\gamma = 5.5 \quad 10^{14} \left(\frac{R}{n_a r_o}\right)^2 \left(\frac{r_1 - r_o}{\beta}\right) P \qquad 4).$$

With R=1cm, $\beta = 10^{-5}$, and $\sigma_G = 10^{-16}$, γ is one in the above equation when the pressure is 1.8×10^{-8} Pa. This indicates that fluctuation is dominated by bombarding ions produced by electron collision with residual gas in the range larger than 1.8×10^{-8} Pa.

3. Experiment

Measurements were carried out using two types of guns in a bakable stainless steel chamber evacuated by a 160 l/s ion pump (Fig.1). Lowest pressure in the chamber was about 6×10^{-8} Pa measured with a B-A gauge. Air can be introduced by opening the slow leak valve in the chamber.

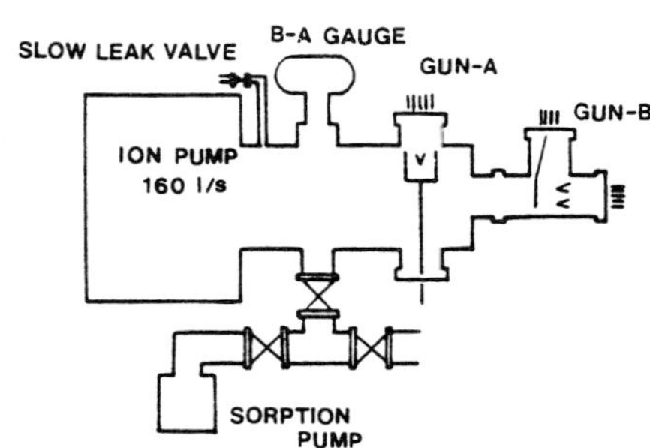

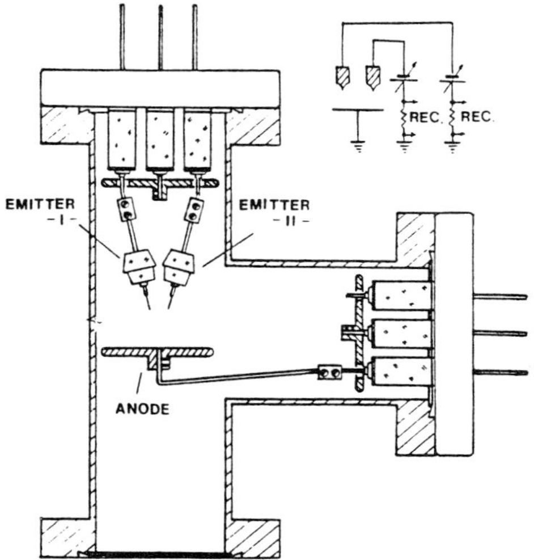

Fig.1   Schematic diagram of
        experimental apparatus

Fig.2   Experimental arrangement
        of B type gun

Fluctuation increase was measured by the A type gun. The B type gun
(Fig.2) was used to confirm the origin of the ions. In the gun, two
emitters were arranged facing the same anode at a distance of 1 cm. Each
emitter emitted different current.

4. Result and Discussion

Relative fluctuation increase as a function of P I_E is shown in Fig.3.
Pressure was controlled by air leak except at 6 x 10^{-8} Pa. Relative
fluctuation is proportional to the logarithm of the P I_E above 7 x 10^{-12}
The result shows fluctuation increase is caused by residual gas ion
bombardment of the tip. Under the above value, fluctuation is constant
and is about one percent. This small fluctuation is assumed to be caused
by absorbed gas molecule migration on the emitter surface.

The result of the two-beam experiment is shown in Fig.4. Emitter-I emitted
the current in the range 5µA to 250µA. However, emitter-II was controlled
to emit a constant 5µA current. Emitter-I fluctuation increased with
emission current. On the other hand, emitter-II fluctuation did not change.
The number of ions coming from the anode is the same for both emitters.
This result shows that bombarding ions are produced at the close region to
the emitter, not at the anode.

5. The Practical Gun

The big difficulty in getting a high emission current in a practical field
emission gun is that high emission currents result in a pressure rise due
to gas desorption from the anode due to emitted electron bombardment.
Pressure rise ΔP is related to emission current I_E by the following
equation:

$$P = 22.75 \frac{I_E}{S} \xi \qquad\qquad 5),$$

where ξ is the desorption efficiency defined by the number of gas
molecules per bombarding electron, ΔP is in Pa, I_E is in ampere, and S is
the pumping speed in l/s. Parameter ξ is about one for an anode with no
treatment. In such anodes, pressure rise by desorption is on the order of
10^{-4} Pa when I_E and S are 100µA and 10 l/s. Therefore, stable emission
is not possible in these high pressure ranges. Thus, desorption

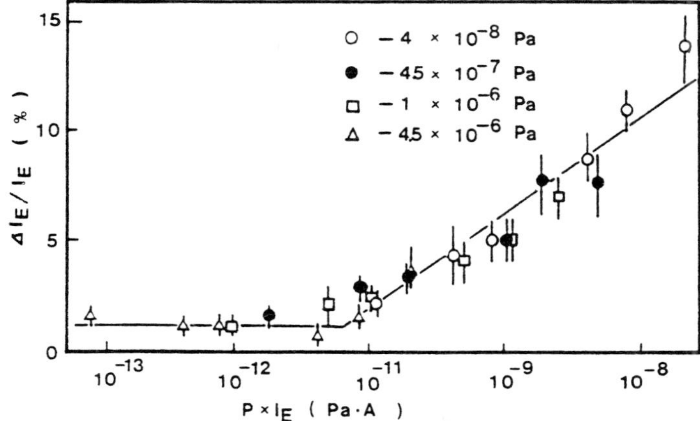

Fig.3 Relative fluctuation as a function of the product of P and I_E

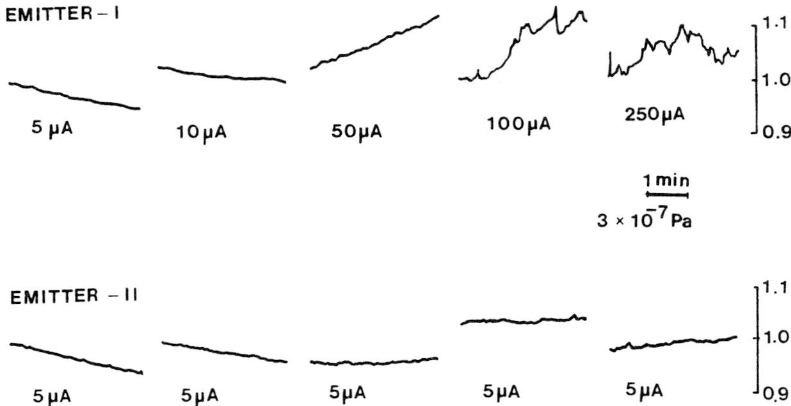

Fig.4   Result of two-beam experiment

efficiency ξ must be less than 1×10^{-4}.

To decrease ξ a heater that can heat the anode to 500°C in a vacuum
is attached. The ξ was decreased to less than 1×10^{-4} after two hours
heating. A stable 100μA emission can be obtained for more than eight
hours without an addition flashing operation. In this gun, the product
of P I_E was kept under 7×10^{-12} Pa A. Relative fluctuation was about one
percent. This type of gun has been applied to a high resolution scanning
Auger electron microscope (Todokoro, Sakitani, Fukuhara anf Okajima
1981).

6. <u>Conclusion</u>

It is found that relative fluctuation increase of total emission current
is proportional to the logarithm of the product of pressure and emission
current in the range larger than 7×10^{-12} Pa A. This fluctuation is about
one percent in the smaller region of the product. It is concluded that
fluctuation increase with product is caused by ionized gas molecule
bombardment of the emitting area of the emitter. Two-beam experiment
results show that ions are mainly produced around the emitter.

In a practical gun that has an anode baking heater to keep the product
of pressure and emission current under 7×10^{-12} Pa A, a stable 100μA
emission current has been obtained for more than eight hours.

7. <u>Refernces</u>

Crewe A V, Wall J and Langmore J 1970 Science 168 pp1338-40
Swann D J and Smith K C A 1973 proc. 6th Annual Scanning Electron
   microscope Symposium ed O Johari (Chicago: IITRI) pp42-8
Saitou N 1977 Surface Science 66 No.1 pp346-56
Todokoro H, Nomura S and Komoda K 1977 J. Electron microsc. 26 No.3
   pp213-4
Todokoro H, Sakitani Y, Fukuhara S and Okajima Y 1981 J. Electron
   Microsc. 30 No.2 pp107-13
Tonomura A, Matsuda T, Endo J, Todokoro H and Komoda T 1979 J.
   J. Electron Microsc. 28 No.1 pp1-11

Asymmetric superconducting shielding lens for TEM and STEM

G. Lefranc and K.-H. Müller

Forschungslaboratorien der Siemens AG, D-8000 München 83, Germany

1. Introduction

Two superconducting lens systems for commercial electron microscopes
(Fig. 1), each equipped with an objective and an intermediate lens
(Lefranc and Nachtrieb 1979, Lefranc et al 1981) have been tested. One of
them has been in operation in the Fritz Haber Institut Berlin for more than
one year. A resolution of 0.3 nm (Fig. 2) and a stability of the same order
as reported for the 400 kV microscope with a superconducting lens system
(Dietrich et al 1977) has been obtained. An objective lens of the asymmetric shielding type has been chosen. This lens has the advantage of easier
adjustment and room for analytical accessories.

2. Principles of the asymmetric shielding lens

In general the shielding lens (Fig. 3) consists of at least one superconducting coil for producing the required flux density. Since an axial flux
density distribution with a high peak value B_0 and steep slopes is of
advantage, hollow superconducting cylinders are aligned along the optical
axis for cutting off the flux outside the gap. The flux density distribution depends on the distance between the cylinders, on their shape, and
somewhat on the properties of the superconducting material. If a high B_0
value and a small half width are desired, the precondition is a high
shielding capability of the superconducting material. In order to avoid
the influence of external and internal stray fields a superconducting
shielding casing has to be provided. All superconducting parts should be
surrounded by liquid helium.

For this lens type it is easier to install a side entry than a top entry
specimen stage. In the case of side entry either a design is required as
described by Dietrich et al 1972 with Helmholtz coil like arrangement,
or one coil has to be located in the upper or lower part of the shielding
casing so that no interference between coil and specimen holder can occur.
The latter construction is trivial for iron core lenses, where the flux
is conducted into the gap region. In the shielding lens, however, the
flux density obtained by the coil along the optical axis is not increased
in the gap by adding the shielding cylinders.

3. Flux density distribution and electron optical data

In order to determine the achievable peak flux density for both above
mentioned arrangements, the axial flux density distributions without shielding cylinders had to be calculated and measured. We decided to operate the
objective lens with a high lens strength i.e. in the second zone mode,

where the aberration coefficients are
relatively low. In addition there is a
large gap at our disposal, which facilitates the handling of the specimen.
For beam voltages up to 250 kV a distance of more than five mm between
edge of the coil and central plane of
the gap (EC, Fig. 3) was required. The
outer diameter of the coils should not
exceed 70 mm so that a decrease of the
axial flux density due to the shielding casing can be kept in reasonable
limits (Dietrich et al 1970).

Fig. 4a and b show the axial flux density of both coil arrangements. In the
planes marked by C' in (a) and (b),
which correspond to the same distance
from the coil edge in the single and
the double coil case, there is a difference of about 25 % in the flux density if the shielding casing is applied. The insignificant influence of
the shielding casing in arrangement
(a) can be explained by the relatively
large average distance between coil
and superconducting walls.

The shielding cylinders have to be
symmetrically arranged in the double
coil design for technological reasons,
e.g. in order to locate the stigmator
coils in a suitable position. If one
wants the peak flux density B_0 in the
one coil arrangement to be equal to B_0'
(the peak value in the double coil
case) then the position of the gap

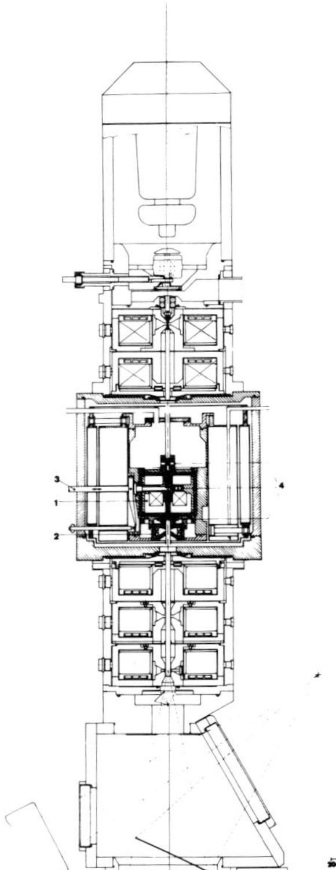

Fig. 1: Commercial microscope with
superconducting lens system:
1 objective lens, 2 intermediate lens, 3 specimen
holder, 4 correction systems.

center has to be shifted from C' to C". With this configuration, however,
the specimen transfer would be hindered by the coil. A compromise is the
construction as plotted in Fig. 3, (center of the gap at C, Fig. 4a and 3).
If the coils are excited with the maximum current applicable without causing
quenching, B_0 amounts to 3 T while B_0' would be 16 % higher. In the asymmetric case one has further to decide if the coil should be installed above
or below the specimen. For a fair evaluation the electron optical data of
the three coil arrangements have to be compared, which were calculated by
a program (Katerbau and Münchmeyer 1976) from the experimentally determined
axial flux density distribution as plotted in Fig. 5b for the asymmetric
case (Table 1). The focal lengths and aberration coefficients are essentially the same for the three designs. Due to the higher achievable peak flux
density mentioned above the symmetric arrangement could be used for beam
voltages higher than 250 kV in the second zone mode. In our opinion,
however, the advantages of easier construction and ample space for the
specimen in the asymmetric case have in general more weight than the
possibility to achieve higher beam voltages. The construction where the
coil is mounted below the specimen was mainly chosen because the magnification is increased due to the larger image distance.

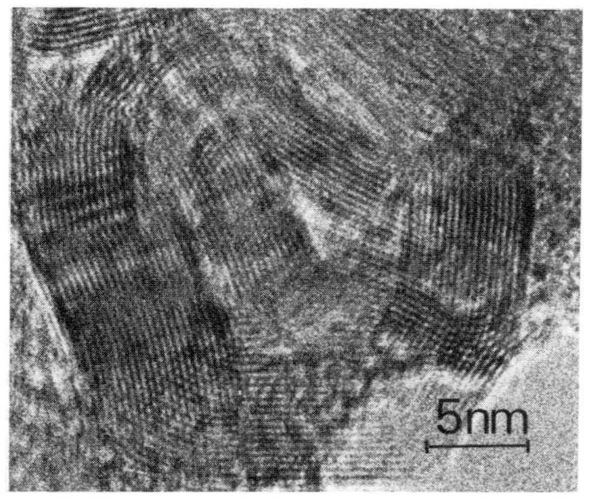

Fig. 2: Graphite fringes taken with the lens system of the MPI

Fig. 3: Principle of the asymmetric shielding lens. 1 coil, 2 shielding cylinders, 3 shielding casing. C center of the gap, E edge of field coil.

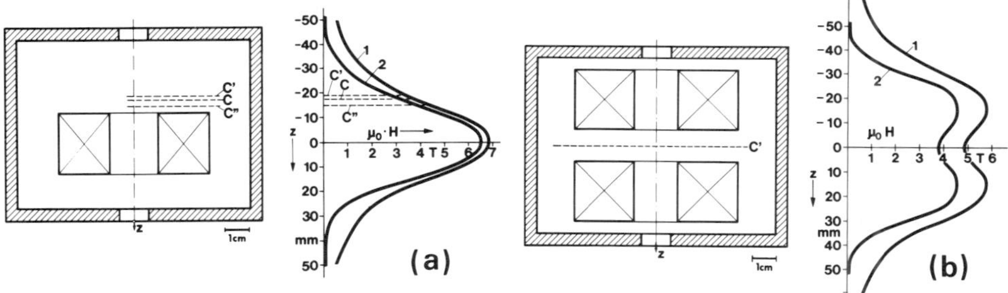

Fig. 4: Coil arrangements and corresponding axial flux density distributions without (1) and with (2) shielding casing. (a) single coil design, (b) double coil design.

	f_0	c_s	c_c	z_0
symmetrical case	1.7	1.2	1.3	1.2
coil above the gap	1.9	1.2	1.4	1
coil below the gap	1.6	1.1	1.2	1.6

Table 1: Electron optical data for the three coil arrangements. Operation in the second zone mode. Axial peak flux density 2.7 T for a beam voltage of 250 kV. f_0, c_s, c_c, z_0, focal length, spherical and chromatical aberration constants, and distance between lens center and object plane in mm.

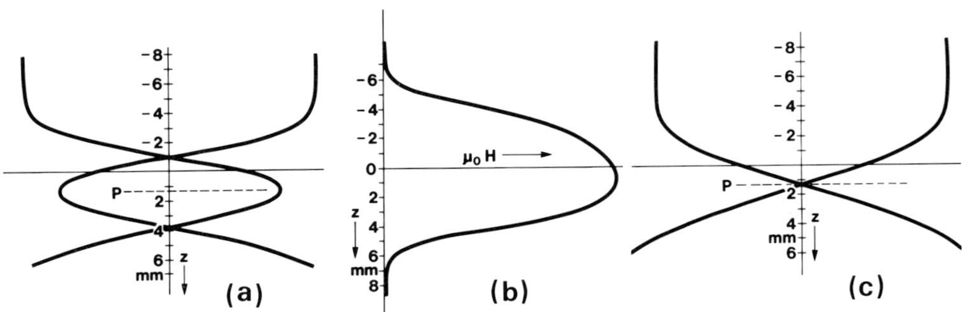

Fig. 5: Electron trajectories for the axial flux density distribution (b)
of the asymmetric shielding lens (a) in the fixed beam (TEM) and
(c) in the scanning (STEM) mode. P specimen plane.

Fig. 5a demonstrates the course of beam in the gap of the asymmetric lens
operated in the second zone mode. The objective lens has also been used in
the scanning mode without changing the position of the specimen and in this
case an approximately telecentric course of beam is adjusted (Fig. 5c). The
data for 100 kV beam voltage are given in table 2.

	B_o	f_o	c_s	c_c	z_o	I
TEM	1.6	1.7	1.1	1.3	1.4	9600
STEM	0.73	2.9	2.7	2.3	1.4	5000

Table 2: Electron-optical data for operation in TEM and STEM mode. B_o axial
peak flux density in T, $I = \int Bdz$ excitation in the gap in A;
beam voltage 100 kV.

4. Electron microscopical experiments

The main advantages of superconducting lens systems are favourable electron
optical data, mechanical and electrical stability and last not least the
strong reduction of radiation damage of organic specimens due to cryoprotection. With the superconducting lens system in Berlin, which contains
an asymmetric shielding lens besides other investigations experiments on
cryoprotection were carried out. Results as obtained with the Munich four
lens superconducting system were confirmed and in addition further cryoprotection factors determined (Müller et al, 1981).

References:

Dietrich I, Maier R G, Weyl R and Zerbst H 1970 Proc. ICEC 3

Dietrich I, Fox F, Knapek E, Lefranc G, Nachtrieb K, Weyl R and Zerbst H,
1977 Ultramicroscopy 2 241-249

Dietrich I, Koller A and Lefranc G 1972 Optik 35 468-478

Katerbau K-H and Münchmeyer W 1976 Private Communication

Lefranc G and Nachtrieb K 1979 Electron Microscopy and Analysis, ed.
T. Mulvey, Inst. of Physics 31-34

Lefranc G, Müller K-H and Dietrich I 1981 Ultramicroscopy 6 81-84

Müller K-H, Zemlin F and Zeitler E 1981 Proc. of 39th EMSA Meeting,Atlanta

Calculation of field distribution and optical properties of saturated asymmetrical objective lenses

K Tsuno and Y Harada

JEOL Ltd., Nakagami, Akishima, Tokyo 196, Japan

1. Introduction

The finite element method and associated computer programs introduced by Munro (1975) for calculating the axial magnetic field distribution are very useful for designing the objective lens of a high resolution electron microscope. The method was reportedly satisfactory for the calculation of the symmetrical condenser objective lens and probe forming lens; however, difficulties arose in the calculation of a single pole piece lens (Mulvey 1980) and the triple pole piece projector lens (Tsuno and Harada 1981). In this investigation, optical properties calculated from field distribution were compared with measured optical properties to clarify the conditions for giving accurate field distribution.

2. Lens shape and calculation method

Finite element calculation was made using a program made by Munro (1975). The objective lens used for the calculation was that of a top entry type JEM–200CX (Naruse et al. 1980). Fig. 1 shows the lens shape and mesh division. The lens yoke was made of soft steel with 0.1%carbon and the pole piece of permendur. Fig. 2 shows the magnetization curves of these materials. Optical properties are numerically computed using axial magnetic field distributions. The calculation method for optical properties is already shown in much literature (for example Kaminga 1975).

3. Effect of mesh on axial field distribution and optical properties

In order to estimate the mesh size required for obtaining the accurate result, calculation was made by changing the mesh size from "R=20, Z=40" mesh to "50, 100" mesh for the lens shown in Fig. 1. Mesh numbers are listed in Table I. Mesh was coarse in the yoke and fine near the pole face and in the gap. The computing time required for calculating axial field distribution for the lens divided by "50, 100" mesh is about 1 h for 7 steps of ampere-turns using FACOM 230-60/75. Fig. 3 shows calculated axial field distributions excited at 11 kAT for various meshes with the increase of the mesh, the maximum field strength decreases. This decrease is very significant between "20, 40" mesh and "30, 60" mesh, however, the decrease from "40, 80" mesh to "50, 100" mesh is small. Moreover, leakage field distribution inside the lower pole piece bore changes largely.

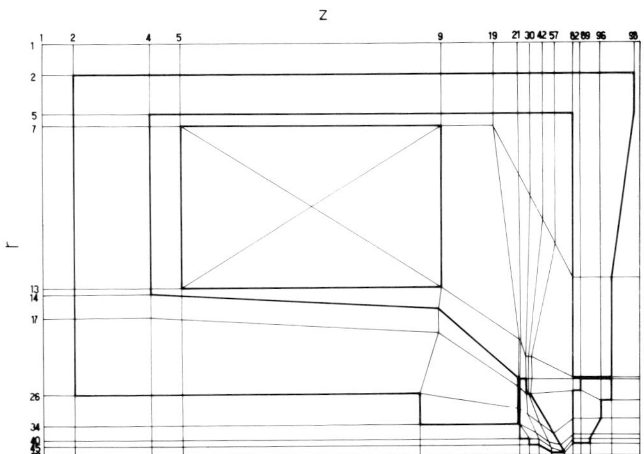

Fig. 1   Lens shape and division of mesh.

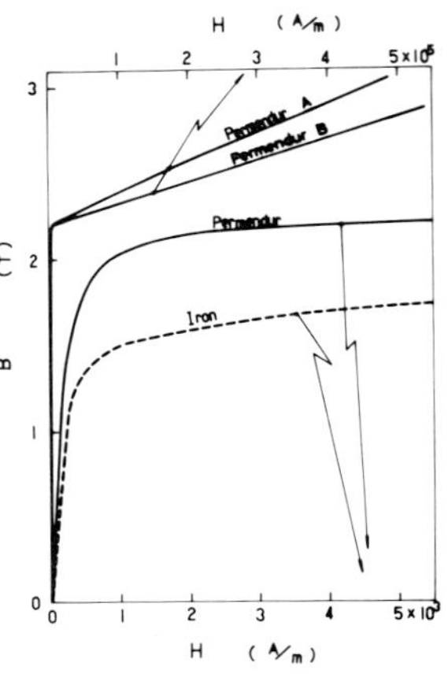

Fig. 2   Magnetization curves of iron (yoke material) and permendur (pole piece material). Two curves are shown for permendur.

Table I   Mesh numbers of R and Z directions

(50, 100)	1	2	5	7	13	14	17	26	34	40	45	50
(40, 80)	1	2	4	5	8	9	11	18	25	30	35	40
(30, 60)	1	2	4	5	7	8	9	14	19	23	26	30
(20, 40)	1	2	3	4	5	6	7	11	14	16	18	20

(50, 100)	1	2	4	5	9	19	21	30	42	57	82	89	96	98	100
(40, 80)	1	2	4	5	8	14	15	21	28	42	62	69	76	78	80
(30, 60)	1	2	4	5	7	11	12	16	22	32	47	52	57	59	60
(20, 40)	1	2	3	4	6	9	10	13	17	23	32	35	38	39	40

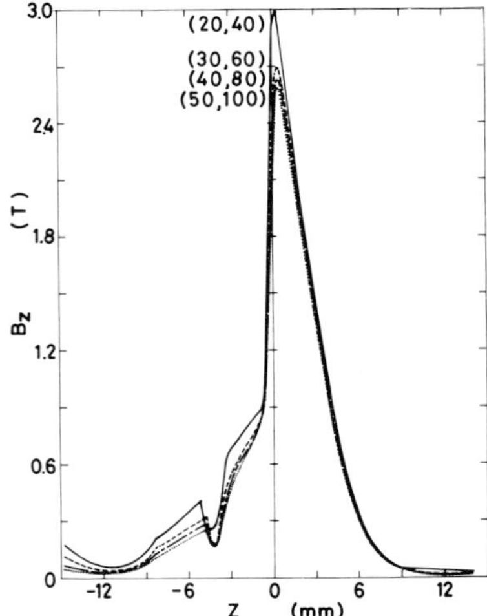

Fig. 3   Calculated axial field distributions for various meshes at 11 kAT.

Fig. 4 shows the excitation dependence of the specimen position Z_0 (distance between top face of lower pole piece and specimen) when changing the mesh. Excitation dependence of measured Z_0 is also shown. Calculated Z_0 approaches measured one with the increase of the mesh. Discrepancies of calculated Z_0 from measured Z_0 are 0.64, 0.24, 0.13 and 0.06 mm for the four meshes, respectively. The discrepancies are nearly the same as one division of mesh in the gap, except the case of "20, 40" mesh. (One division of mesh in the gap is 0.33, 0.2, 0.15 and 0.12 mm for the four meshes, respectively).

Fig. 5 shows Z_0 dependence of spherical aberration C_s for various field distributions obtained by changing the mesh. The calculated C_s also approaches the experimental value when the mesh becomes large. Thus, it is concluded that in order to obtain an accurate result, it is

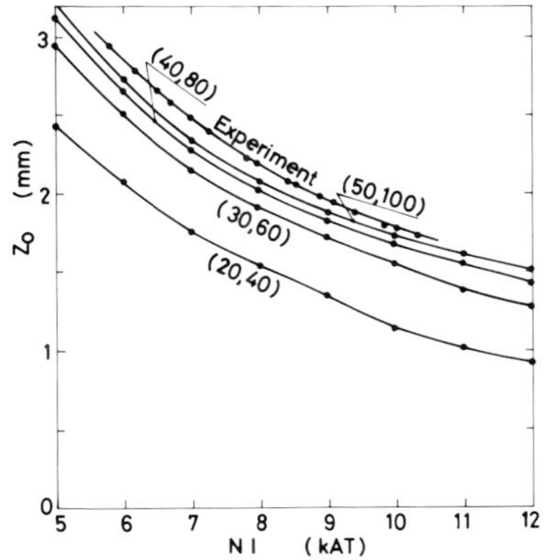

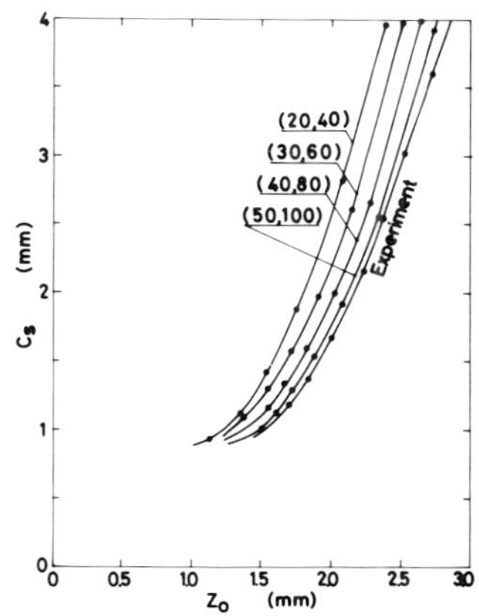

Fig. 4   Comparison of calculated and measured results of
excitation NI dependence of
specimen position Z_0.

Fig. 5   Comparison of calculated
and experimental results for the
specimen position Z_0 dependence of
spherical aberration Cs.

necessary to increase the mesh. Even in (50, 100) mesh, the discrepancy
of calculated Cs from measured one is about 0.1 mm.

4. Effect of magnetization characteristics of pole piece material

Fig. 6 shows the shape of a newly designed 30°-70° two-stage tapered pole
piece. This pole piece was proposed at EMSA '80 (Harada et al. 1980).
The calculated optical properties are shown by curves A of Fig. 7. Spherical aberration Cs of 0.8 mm and chromatic aberration Cc of 1.2 mm are
estimated at Z_0=1.3 mm and NI=11 kAT. Contrary to an expectation, the
discrepancy between calculated and measured results (curves C) is large.
Measured values of Cs and Cc at Z_0=1.3 mm are 1.06 mm and 1.55 mm, respectively. Results did not approach the experimental values, in spite of

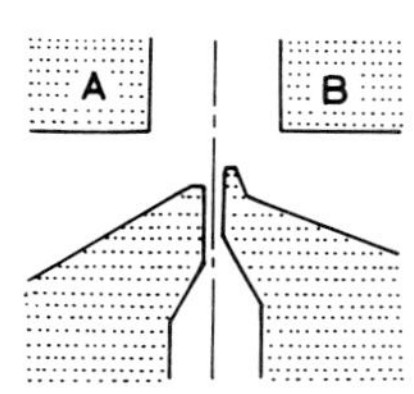

Fig. 6   Cross
section of pole
pieces.
A: Standard
shape (Fig. 1)
B: Two-stage
tapered pole
piece.

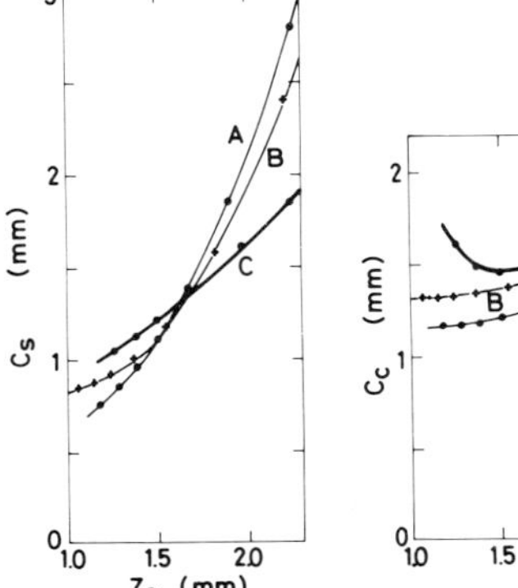

Fig. 7   Comparison of calculated
optical properties with measured
ones.
A: Calculated results using the
magnetization curve A of Fig. 2.
B: Calculated results using the
magnetization curve B of Fig. 2.
C: Measured optical properties.

several calculations made by changing the method
of dividing the mesh. Curves B of Fig. 7 show
another calculation of Cs and Cc based on the
different magnetization curve (curve B of Fig. 2)
of pole piece material. At high Z_0, Cs decreases
and Cs increases at low Z_0. In the case of Cc,
curve B is close to the experimental result. So,
the curve B gives better result than curve A.

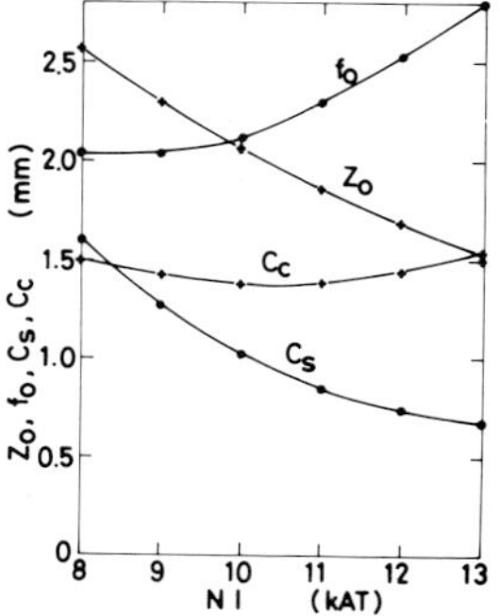

Fig. 8   Cross section
of pole piece.
A: Standard shape
(Fig. 1),
B: Large bore pole
piece.

When the taper angle of the lower pole piece is
reduced, the magnetic flux strongly concentrates
at the top face of the pole piece, and the flux
density becomes large. So, the magnetization
characteristics at the high field region greatly
influence the axial field distribution.

Thus, the origin of discrepancy between calculated and measured results is the difference of
magnetization curves of pole piece material.
Magnetization curve of permendur greatly changes
not only with the differences in the forging
method, and heat treatment conditions but also
in machining. This change of magnetization
curve by machining seems to be great near the
surface of material and small at the inner
region. So, when the thickness of the pole
piece is thin near the top face, the effect of
machining seems to be significant.

5. Design of new pole piece with improved resolution

Fig. 8 and Fig. 9 show another new shape pole
piece and calculated optical properties. Calculated values of Cs, Cc, f_0 and Z_0 are 0.9 mm,
1.4 mm, 2.2 mm and 1.9 mm at 11 kAT. The pole
piece has a large bore (2 mm) and wide upper
face (4 mm). So, it is considered to be free from the reduction of magnetic properties by machining. Due to the large bore diameter of the
lower pole piece, axial field distribution can be measured by using the
Hall probe. The measured axial field distribution is in fairly good
agreement with the measured one. So, it is expected that the resolution
at 200 kV may be improved by this pole piece.

Fig. 9   Calculated
optical properties of
new lens shown in
Fig. 8-B.

Harada Y, Tsuno K and Arai Y 1980 "38th Ann. Proc. Electron Microscopy
    Soc. Amer." San Francisco, California, G.W. Bailey (ed.) pp 266-9
Kamminga W 1975 Optik 45 39
Mulvey T and Nasr H 1980 Inst. Phys. Conf. Ser. No.52, 53
Munro E 1975 A set of computer programs for calculating the properties
    of electron lenses (CVED/B-Elect TR 45, University of Cambridge)
Naruse M, Watanabe E, Harada Y, Sakurai S and Etoh T 1980 J. Electron
    Microscopy, 29 54
Tsuno K and Harada Y 1981 J. Phys. E: Sci. Instrum. 14 (in press)

Improvement of resolution of the high voltage electron microscope by means of an asymmetrical lens

T Honda, K Tsuno and H Watanabe

JEOL Ltd., Nakagami, Akishima, Tokyo 196, Japan

1. Introduction

In the past the principal requirement for the high voltage electron microscope was for the "observation of thick specimens". Recently the applications have been extended to include "crystal structure image information". We are now in the process of constructing a high voltage (1000 KV) and high resolution (1.8Å at 400 KV) electron microscope with $C_s\lambda = 0.025$Åmm from 400 to 1000 KV (here, C_s is a spherical aberration coefficient and λ the wave length of electrons) for University of California Lawrence Berkeley Laboratory (UCLBL).

For constructing this high resolution electron microscope, we re-examined the already developed objective lenses and then produced the test lens. First, the characteristics of two objective lenses of Tohoku University (OALP and NALP lenses: the former is the original lens and the latter the improved lens of $C_s = 11$ mm at 1000 KV, Hiraga et al.1980) were examined. Second, after repeated calculations using the finite element method (Munro 1975), test objective lens (TOAR lens) was constructed. The field distribution and optical properties of this lens were measured. Third, the new lens (ARP) was designed so as to fit the resolution of 1.8Å at 400 KV.

2. Optical properties of 1000 kV electron microscope at Tohoku University

Fig. 1 shows the calculated optical properties by using the measured axial field distributions of lenses of OALP (gap S=7.5 mm, lower bore diameter b_2=3 mm, top face diameter of lower pole piece D_2=6 mm and taper angle of lower pole piece θ_2=60°) and the NALP (S=8 mm, b_2=4 mm, D_2=8 mm and θ_2= 60°). The shapes of the lens are already shown in HVEM conference held at Antwerp (Honda et al. 1980). The excitation of OALP and NALP are 18 and 25.6 kAT, respectively. As seen from the figure, the improvement of C_S from the OALP to NALP was mainly due to the increase in excitation, and it is impossible to achieve C_S of 3 mm or less with the more strong excitation when these pole pieces are used.

3. Construction of a new pole piece for the electron microscope at Tohoku University

Axial field distributions are calculated for various pole piece shapes in order to find the minimum C_S. After several calculations, the following tendency has been clear: 1) upper pole piece bore b_1 must be reduced as small as possible to insert and to move the specimen, 2) b_2 and D_2 have an optimum diameter within the limit of b_1 greater than D_2, 3) optimum angle

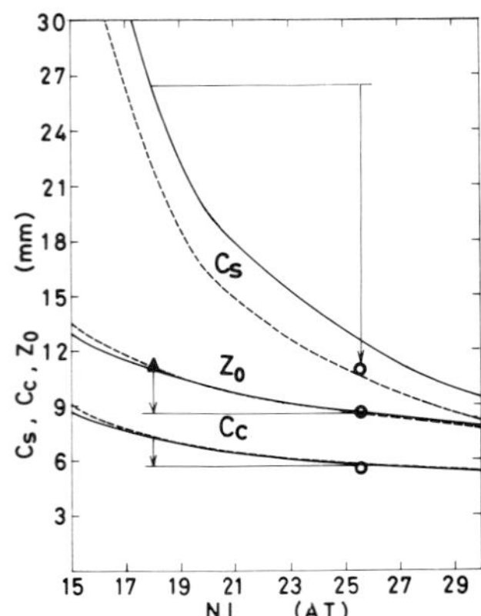

Fig. 1   Optical properties of OALP (solid lines) and NALP (broken lines) calculated from measured field distributions. A triangle and circles indicate the measured optical properties.

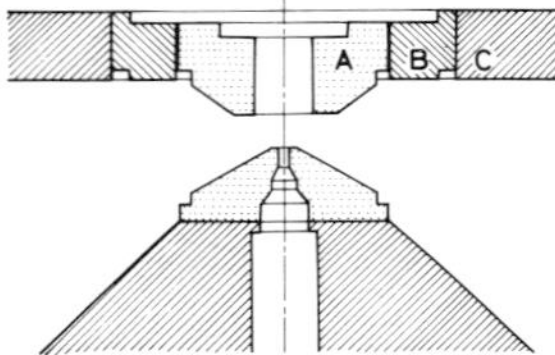

Fig. 2   Cross section of the TOAR lens. A: permendur, B: pure iron and C: soft steel.

of θ_2 is 60° and 4) S is limited by severe saturation of pole piece material. Finally, a shape of pole piece shown in Fig. 2 was obtained (TOAR lens: S =10 mm, b_2=4 mm, D_2=8 mm and θ_2=60°). After constructing the lens, axial field distributions were measured using a Hall gaussmeter (Incremental 640, Bell inc.). Calculated and measured field distributions (Fig. 3 (a) and (b)) are quite similar except the maximum field strength and the distribution in the lower pole piece bore. Solid lines of Fig. 4 show the measured and calculated maximum field strength against the total lens excitation. Broken lines show the maximum field strength against the lens excitation which is applied to the pole piece region. Therefore, difference between solid and broken lines indicates the loss excitation in the yoke. The loss excitation is large in measured distribution (about 10%), however, quite small in calculated distributions. The origin of this difference is not clear. However, several reasons are considered: 1) Small numbers of mesh in yoke (mesh for the calculation was R=47 and Z=93, here, R is the radial direction and Z the direction of optical axis. The mesh was coarse in yoke and dense in pole piece and gap). 2) Ignorance of gaps inevitably appeared on the joints of yoke and pole piece. 3) Difference of magnetic properties of yoke material.

Fig. 5 shows the comparison of measured and calculated optical properties. In this graph, horizontal axis is the excitation applied on pole piece region. The coincidence of calculated and measured results is good. As a result, good optical properties of C_s=3.7 mm, C_c=3.8 mm, and f_0=5.2 mm at Z_0=5.6 mm and NI=25.3 kAT (total NI=28.0 kAT) are obtained. Figs. 6 and 7 show

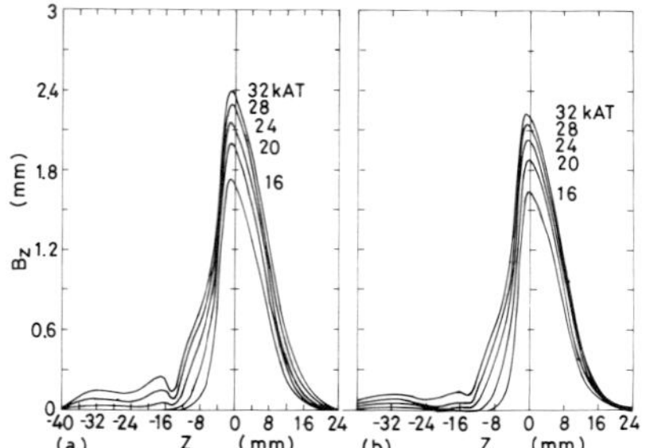

Fig. 3   Field distributions of TOAR lens for various excitations.
(a) Calculated distributions
(b) Measured distributions.

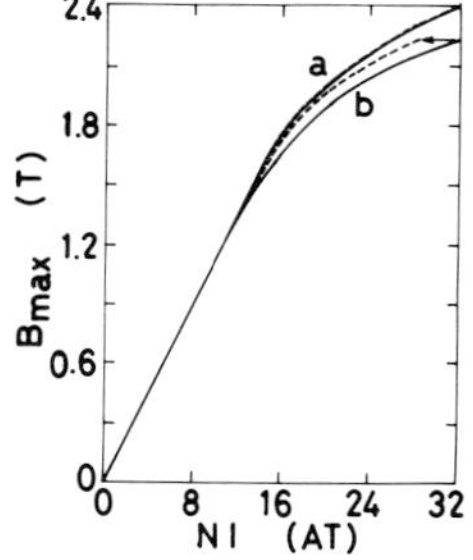

Fig. 4   Excitation dependence of the maximum field strength. a: calculation, b: measurement. Broken lines are the excitations applied to the pole piece.

the photographs of Au single crystal and β-Si_3N_4 crystal, respectively. As is proved by the photographs, good resolution is obtained. The resolution of 1.8 Å can be achieved by the C_S of 3.7 mm at 1000 kV. However, the resolution of 1.8 Å at 400 kV can not be obtained by using this lens.

4. Design of new lens for UCLBL

From the result shown in Fig. 4, TOAR lnes has a large loss excitation. In this stage, the lens yoke is severely saturated, so it is necessary to improve the lens yoke. Fig. 8 shows the shape of newly designed yoke and pole piece (ARP lens: $S=$ 10 mm, b_2=4 mm, D_2=8 mm and θ_2=60°). Calculated Z_0 dependence of C_S for accelerating voltages from 400 to 1000 kV are shown in Fig. 9. A bold line indicates the condition for $C_S\lambda$=0.025 Åmm. In order to observe the image under the condition of $C_S\lambda$=constant for various accelerating voltage, it must be moved the specimen along the bold line of Fig. 9 by means of the Z-controlled goniometer. Fig. 10 shows the phase contrast transfer function at 400 and 1000 kV. First zero of the function is located at the resolution less than

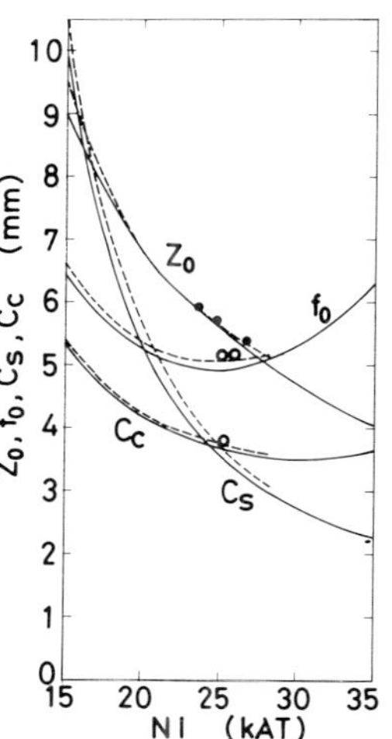

Fig. 5   Optical properties of TOAR lens. Solid lines: calculated result using the measured field distributions. Circles: measured results. The excitation *NI* is that applied to the pole piece.

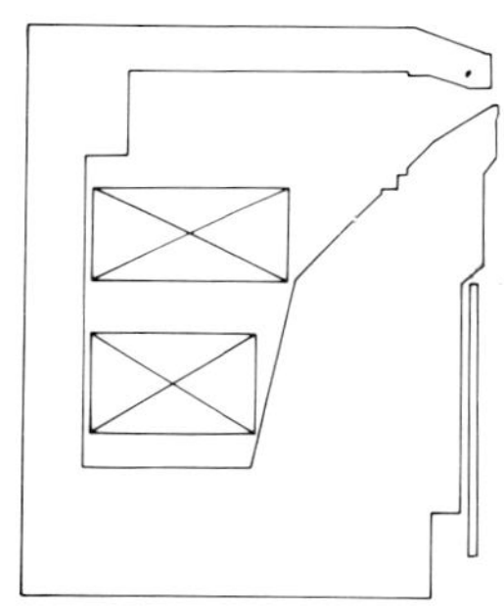

Fig. 8   Proposed shape of objective lens for UCLBL.

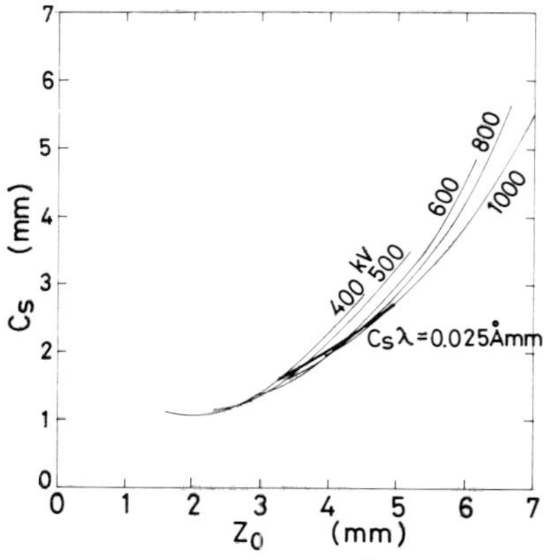

Fig. 9   Calculated C_S against Z_0 for various accelerating voltages. A bold lines shows the conditions of $C_S\lambda$=0.025 Åmm.

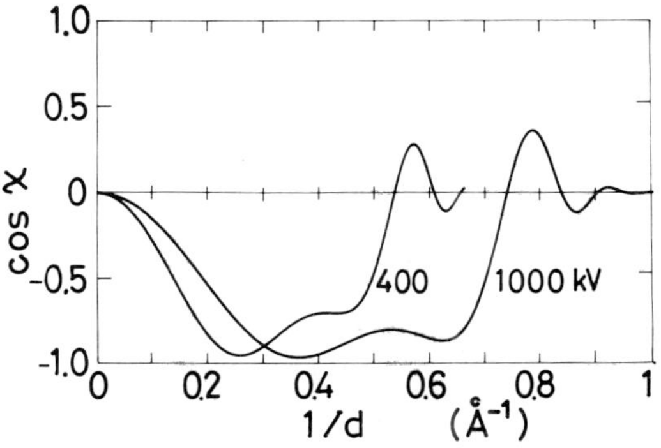

Fig. 10   Phase contrast transfer function at 400 and 1000 kV. The calculation condition is as follows.

		400 kV	1000 kV
Spherical aberr. C_S		1.5 mm	2.8 mm
Chromatic aberr. C_C		2.1 mm	3.3 mm
Defocus	Δf	580 Å	580 Å
Divergence angle	α	5×10^{-4}rad.	5×10^{-4}rad.
Energy spread	ΔE	2 eV	2 eV

1.8 Å (1/d is 0.56 Å^{-1}) for both accelerating voltages. As a result, both conditions of the theoretical resolution less than 1.8 Å at 400 kV and $C_s\lambda$=0.025 Åmm are expected to be achieved.

We acknowledge to Mr. H. Ohta and Mr. E. Aoyagi of Tohoku University for cooperating the measurement of field distributions of objective lens and photographing the images.

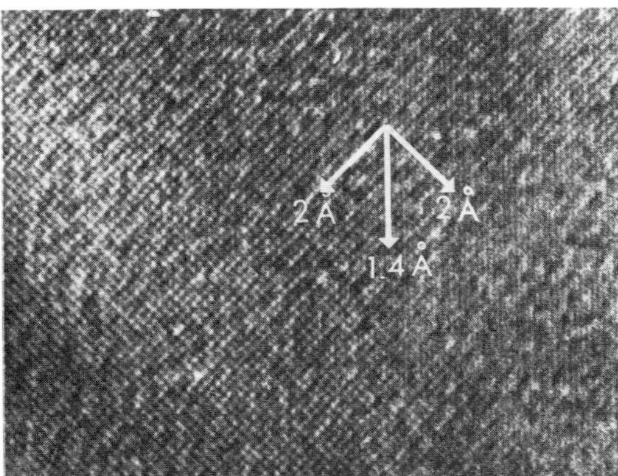

Fig. 6   Lattice fringes observed in an evaporated Au film taken at 1000 kV by axial illumination. The fringes of 2.04 and 1.44 Å corresponding to (200) and (220) planes, respectively, are clearly resolved.

Fig. 7   High resolution image of β-Si$_3$N$_4$. Large circles on model indicate the position of Si and small circles N. Si locates just on the large circle.

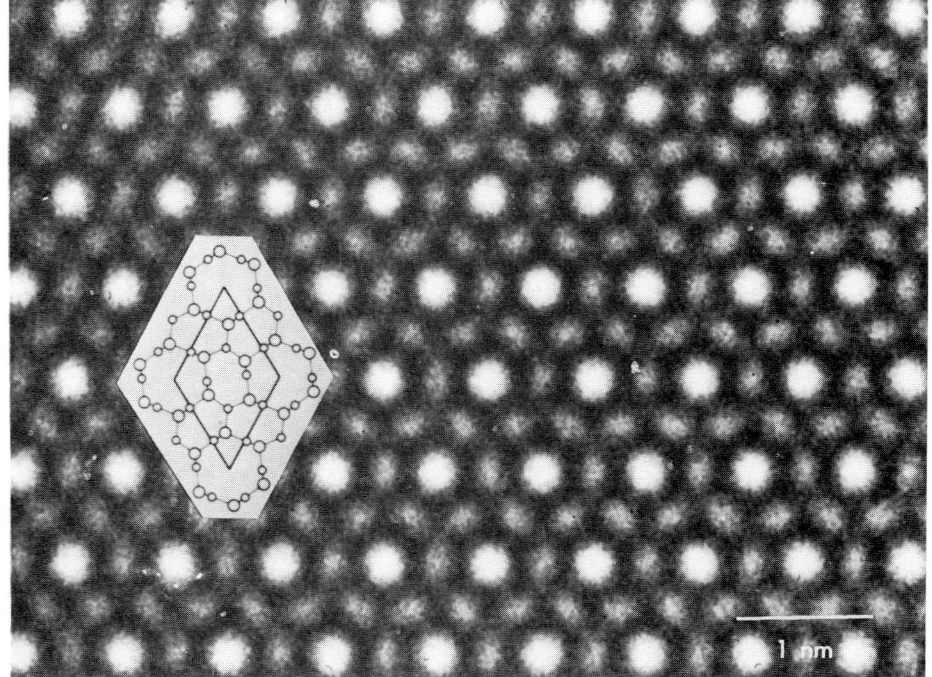

Hiraga K, Shindo D, Hirabayashi M, Ohta H and Aoyagi E 1980 HVEM Reports of Tohoku University 2 45 (in Japanese)
Honda T, Watanabe H, Etoh T and Hiraga K 1980 Electron microscopy 1980 4 18
Munro E 1975 A set of computer programs for calculating the properties of electron lenses CUED/B Elect TR, University of Cambridge

Wide-angle projector systems for the TEM

S.M.Al-Hilly and T.Mulvey The Department of Physics, The University of
Aston in Birmingham, Birmingham B4 7ET England.

In spite of the high resolving power now available in commercial electron
microscopes, the electron optical design of the projector lens and viewing
system has not changed significantly over the last forty years. This is
largely due to the difficulty of reducing the radial and spiral distortion
in the final projector to the tolerable limits of 1% radial and 2% of spiral
distortion in the micrograph. This is invariably achieved by restricting
the semi-projection angle (α_p) of the beam projected onto the fluorescent
screen to about 0.1 radian (5.7°). For a photographic image of 100mm
diameter, a minimum projection distance of 50cm is thus required between
lens and photographic plate. This leads to bulky and inconvenient viewing
chambers, especially in high voltage electron microscopes. Recently an
experimental wide-angle projection system devised by Lambrakis et al (1977)
has been modified for use in high voltage microscopes (El Kamali and Mulvey
1980) and essentially distortion-free images have been obtained at a projection semi-angle of 30°. Subsequent analysis of this system (El Kamali
1981) has shown that at such large projection angles, third order aberration
theory, although a useful preliminary guide, cannot predict with sufficient
accuracy the projected images actually observed on the microscope screen.
This, therefore, makes it difficult to optimise the electron optical system
without extensive experimental work. Recent advances in the calculation of
magnetic fields (Nasr 1981) by the finite element method has made it
convenient to use the general ray equation, and hence to determine the
form of the final image directly and without recourse to aberration theory
(Al-Hilly 1982). Since a wide-angle projector lens could find an immediate
application in a TEM with transmission fluorescent screen, it was thought
useful to employ computer-aided design in the construction of a wide-angle
projector unit that could be readily inserted into an existing JEOL "Superscope" in place of the normal internal camera ($\alpha_p=6.4^\circ$) as illustrated in
Figure 1. Because of the short projection distance (68.2mm) of the new
lens unit an image with a projection semi-angle $\alpha_p \simeq 30^\circ$ could easily be
produced on the transmission fluorescent screen and photographed externally.

Figure 2 shows the construction of the lens unit in more detail. A nearly
parallel beam of electrons from the objective lens enters the first
(corrector) lens which is an asymmetrical lens strongly excited
($NI/V_r^{\frac{1}{2}} = 17.8$) beyond its minimum focal length but still in the first
focal zone. The excitation and hence image rotation is in the opposite
sense to that of the projector. The corrector lens produces a magnification of just over three with considerable spiral distortion ($\simeq 40\%$). This
distorted image is projected onto the viewing screen by the projector lens
so that the distortion produced by the projector lens is annulled. This
arrangement is electron-optically favourable since the beam from the

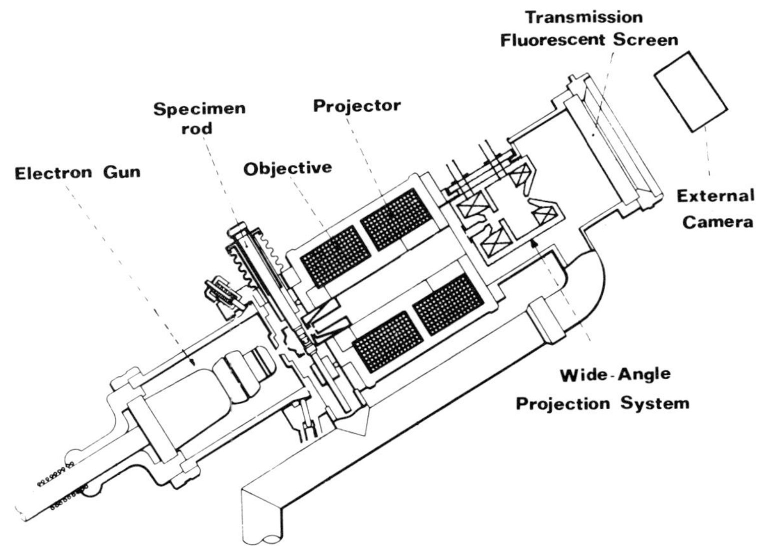

Fig.1. Cross-section of JEOL "Superscope" with transmission fluorescent screen. The internal camera (α_p = 6.4°) has been removed and a wide-angle (α_p = 28°) projector lens unit inserted in its place. Image diameter 70mm on fluorescent screen (M = 35). Photography by external camera.

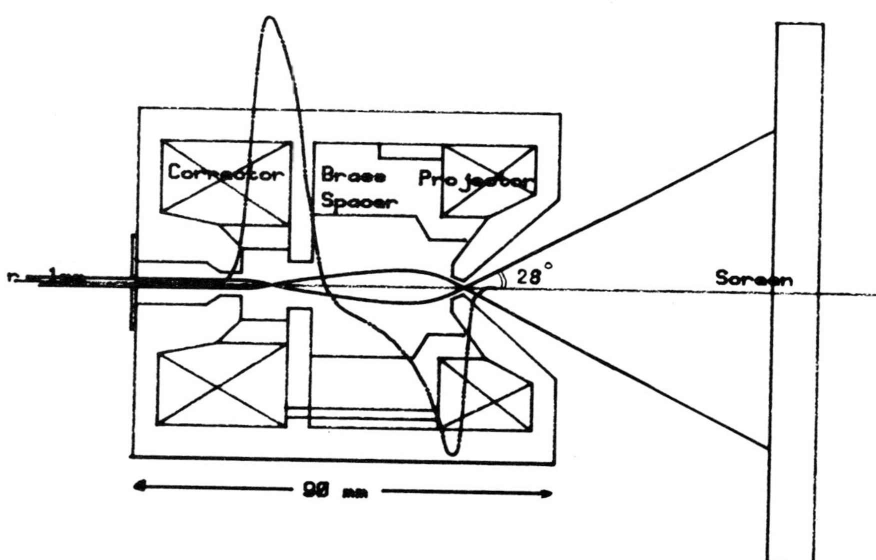

Fig.2. Cross-section of the wide-angle projector unit, showing axial field distribution and Gaussian trajectory of the extreme ray (α_p = 28°).Length of unit 90mm. Excitation of 3060A-t in the corrector and -2134 A-t in the projector at an accelerating voltage of 30KV. Average loss of ampere-turns due to field cancellation 6%. Entrance aperture 2mm dia. Effective focal length 2.2mm, projection distance 68.2mm.

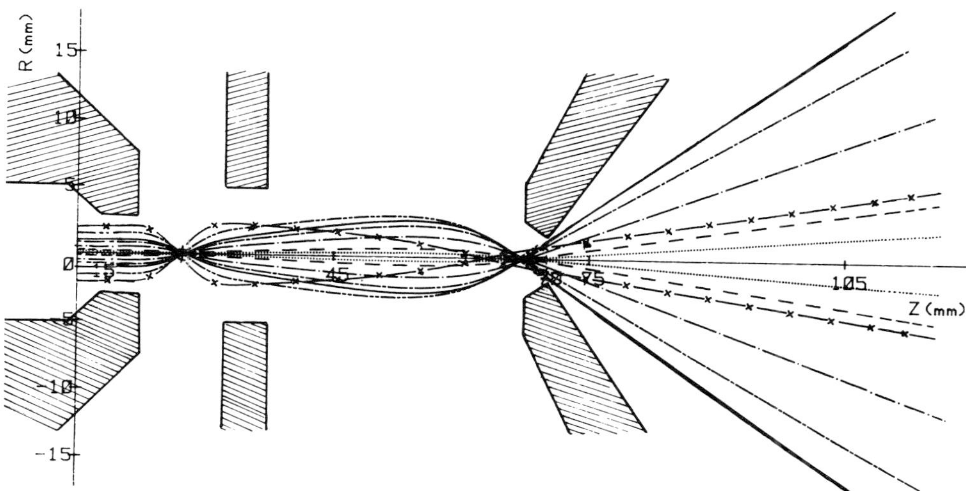

Fig.3. Computed general trajectories for the wide-angle projector unit
with excitations adjusted for negligible radial and spiral distortion
$NI/V_r^{\frac{1}{2}}$ = 17.8 (corrector) and $NI/V_r^{\frac{1}{2}}$ = 12.2 (projector).
Note that failure of correction occurs , for example, if R = 2mm.

corrector lens enters the final projector lens at nearly parallel incidence
(See Figure 2) which minimises the spiral distortion produced by this lens.
The projector lens is of the single pole-piece type which has appreciably
lower distortion than that of a conventional lens. It also has the useful
property that its spiral distortion is insensitive to excitation in the
region of minimum focal length. This is a critical feature of the present
design since the virtual image produced by the corrector produces considerable unwanted barrel distortion in addition to the wanted spiral distortion.
This means that although spiral distortion will be corrected, a considerable
amount of barrel distortion will appear in the final image. Fortunately
this can be readily compensated by reducing the excitation of the project-
or lens (Kynaston and Mulvey 1973) so as to produce an appropriate amount
of pin-cushion distortion.

Third order aberration calculations carried out with the field distribution
shown in Figure 2   (Nasr et al 1981, Al-Hilly 1982) show that the spiral
and radial distortion can indeed be corrected simultaneously. However,
such calculations give no indication of the maximum projection angle for
which the calculation remains valid, or indeed the optimum polepiece configuration. This is illustrated in Figure 3 which shows the electron
trajectories in the lens unit as calculated from the general ray equation.
For an incident beam of radius R = 1mm ($\alpha_p \simeq$ 30°) radial distortion is
virtually absent. Similarly the image rotation calculations show that the
spiral distortion is likewise absent. However, for a ray height R=2mm the
trajectory shows that the correction no longer holds. In fact, it is
necessary to provide an entrance aperture to restrict the entry of these
rays, otherwise they will appear as radial streaks in the image. It can
also be seen that the correct shaping of the projector polepiece is important and that further small improvements are possible. This can readily be

 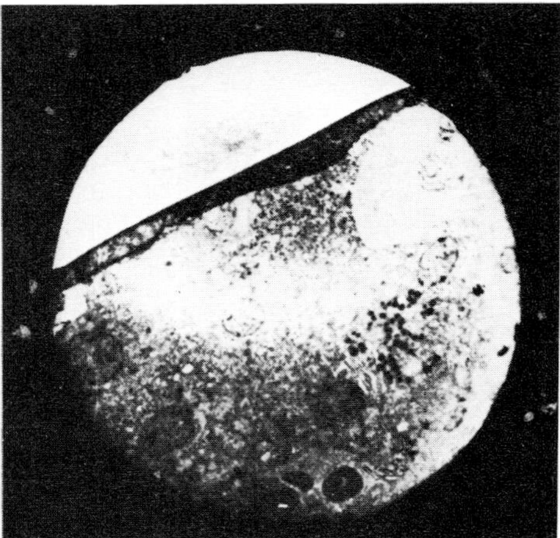

Fig.4. Wide-angle ($\alpha_p = 30^\circ$) projection images. Projection distance 68.2mm.
    (a) Specimen grid   (b) Rat pancreas

Diameter of image at standard projection distance (50cm) 58cm.

carried out by further computer-aided design calculations. Figure 4 shows
some wide-angle images obtained in the modified "Superscope". It should
perhaps be mentioned that at the standard projection distance of 50cm
these images would be 58cm in diameter.

The lens unit described here has been tested at an accelerating voltage of
30kV. However, since the axial flux density in the lens unit is low (see
Fig.2) it can be operated at accelerating voltages up to 220kV by suitably
increasing the excitation. It should also be remembered that scaling
its dimensions and excitations by a factor n leaves the distortion-free
properties unchanged but reduces the magnification by a factor n. For an
accelerating voltage of one million, for example, the appropriate factor
n is 2.7 giving a magnification of about 13.

This investigation shows that it is now feasible to construct wide-angle
projector lenses for the TEM that will allow standard size photographic
plates to be located in the optimum position for minimum exposure
techniques as well as providing a large distortion-free image for viewing.

References

Al-Hilly S M 1982 PhD thesis Aston University
El-Kamali H and Mulvey T 1980 Electron Microscopy 1980 (Leiden) eds. P
   Brederoo and G Boom Vol.1 pp74-75
El-Kamali H 1981 PhD thesis Aston University
Kynaston D and Mulvey T 1963 B. J. Appl. Phys. 14 199-206
Lambrakis E, Marai F Z and Mulvey T 1977 Inst. Phys. Conf. Ser. No. 36
   PP35-38
Nasr H 1981 PhD thesis Aston University
Nasr H, Chen W and Mulvey T (These proceedings)

Ultra-high vacuum analytical electron microscope

Y Harada, Y Ishida, Y Arai, H Hirano and N Yoshimura

JEOL Ltd., Nakagami, Akishima, Tokyo 196, Japan

Analytical electron microscopes (AEM), which provide the functions of the TEM, SEM, EDS and ELS, have been in practical use for analysis of the elements in microareas. Meanwhile, surface studies have recently been conducted with an ultrahigh vacuum (UHV) electron microscope by utilizing reflection electron microscopy and R-HEED (Yagi K 1980). In view of this trend, we have developed an ultrahigh vacuum analytical electron microscope (UHVAEM) intended for surface studies.

The present paper describes the outstanding features of the UHVAEM. Fig. 1 (b) shows a schematic diagram of its vacuum system. The specimen and gun chambers are evacuated by a sputter ion pump (SIP) after being rough pumped by a turbo-molecular pump (TMP). The viewing and camera chambers, in which much outgassing takes place from photographic films, are evacuated by the TMP. The image forming lens system is designed to provide constant projector lens excitation over the entire magnification range (Arai Y et al. 1981). Therefore, a very small aperture (orifice) can be inserted at the back focal position of the projector lens, as shown in Fig. 1(a). Consequently, gas molecules in the camera chamber are prevented from flowing into the specimen chamber. Fig. 2 shows the differential pumping characteristics between the specimen and camera chambers, which are estimated by using the resistor network simulation technique (Ohta S et al., in preparation). The pressure in the specimen chamber can be kept to within 2×10^{-7} Torr even if the pressure in the camera chamber is 1×10^{-3} Torr. In other words, the orifice makes possible a perfect differential pumping system. Moreover, the microscope column is designed to minimize outgassing sources by using stainless steel liner tubes in the column and flexible metal bellows in the variable aperture mechanisms, vacuum valves and specimen stages. The column and specimen stage can be baked out by a single touch of a pushbutton to reduce residual hydrocarbon gases and water vapor. Fig. 3 shows typical pump-down characteristics of the specimen chamber and evacuation tube. When the vacuum pressures of the specimen and gun chambers reach 1×10^{-4} Torr, the SIP is started up. The vacuum pressure reaches the 10^{-7} Torr order after its pumping for about 20 min. The pressure is further lowered when the anti-contamination device (ACD) is filled with liquid nitrogen. The ultimate vacuum pressure in the specimen chamber can thus be lowered to the 10^{-8} Torr order. Fig. 4 shows spectra of the residual gas in the specimen chamber obtained by use of the ACD, where the total pressure was 8.9×10^{-8} Torr. The partial pressure of the hydrocarbon gases with mass numbers from 41 to 43 is less than 3×10^{-10} Torr, and that with mass number of more than 50 is of a negligible order. Moreover, the partial pressure of water vapor is less than 1×10^{-8} Torr. Therfore, a clean ultrahigh vacuum is achieved in the specimen chamber.

Yagi K Proc. 38th EMSA (1980) 290
Arai Y, Ishida Y, Watabe T and Harada Y Proc. 39th EMSA (1981) 374
Ohta S and Yoshimura N in preparation

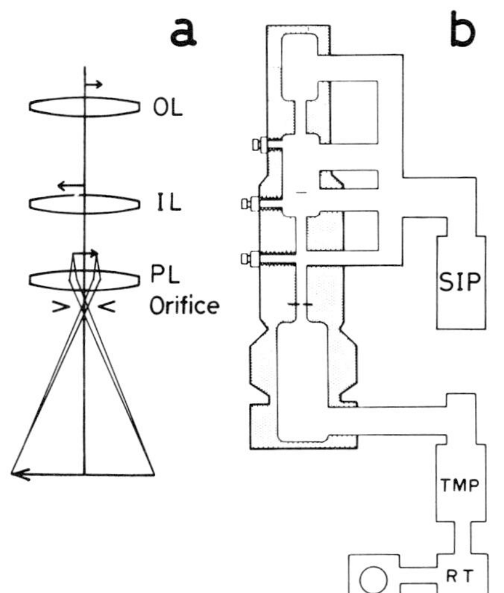

Fig. 1 Schematic diagram of an ultra-high vacuum system (b), and the basic three-stage image forming lens system showing the orifice position (a).

Fig. 2 Differential pumping characteristics between the specimen and camera chambers.

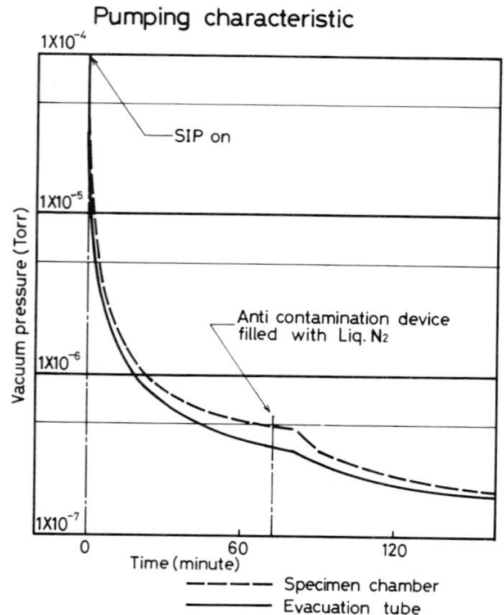

Fig. 3 Typical pump-down characteristics of the specimen chamber and evacuation tube.

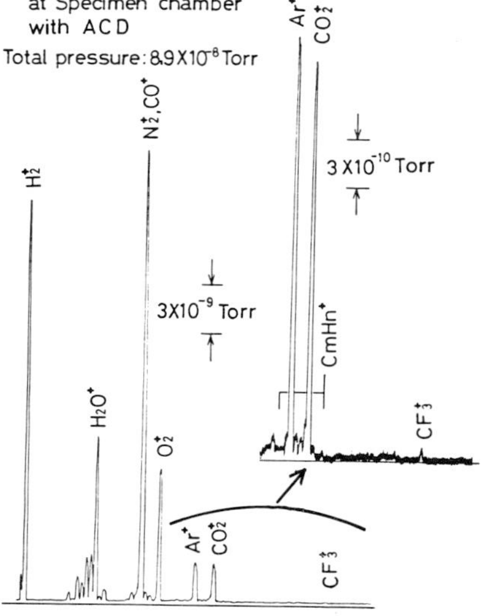

Fig. 4 Spectra of the residual gas in the specimen chamber obtained by use of the ACD.

On-line digital computer techniques in electron microscopy

K C A Smith

University Engineering Department, Trumpington Street, Cambridge CB2 1PZ

1. Introduction

In electron microscopy information concerning the nature of the specimen is
produced in the form of images and spectra. The on-line acquisition,
manipulation and storage of spectral information using standard, inexpensive minicomputer technology has become a common practice in all laboratories using electron microscopes, particularly of the SEM or STEM variety.
The on-line approach towards the collection of image information is, however, still relatively uncommon. The principal reason for this is that it
is technically more difficult, and considerably more costly, to collect
and manipulate image data. The advantages to be gained from the technique
have also yet to be fully and convincingly demonstrated. Recent advances
in computer technology have, however, significantly enhanced the feasibility of using the on-line technique in both scanning and conventional
microscopy, and its general usefulness is also becoming increasingly apparent.

The fundamental problem of manipulating image data in a small computer is
associated with the very large information content of the average image.
For example, an image consisting of 512 x 512 pixels (an image size that
fits conveniently into the standard TV picture format) with 256 grey levels, i.e. 8-bit quantization, contains 256 Kbytes of information (Note:
1 byte = 8 bits; K = 1024). For comparison a typical EDXA spectrum will
contain only 6 Kbytes of information. A standard minicomputer has a
directly addressable memory space of up to 128 Kbytes, which means that
while there is ample room for the storage and manipulation of spectra,
there is insufficient room for even a single image of normal size.

A further difference regarding the facility with which a minicomputer can
handle the two kinds of information arises in connection with the rate at
which data can be transferred. Information can be input to, or output
from, a computer at a rate of about one data word every few microseconds.
This means that complete images may be transferred at a rate of about one
per second; complete spectra can, because of their lower bit content, be
transferred between 10 and 100 times faster. The consequence of this is
that the acquisition and display of images must necessarily proceed at
'slow scan' rates, whilst spectra can be accumulated and displayed at
rates compatible with normal TV scan speeds.

Various methods have been developed for circumventing the difficulties
associated with on-line acquisition and manipulation of images. One
method suitable for SEM or STEM instruments is to transfer the image data

during acquisition to a fast-disk backing store. The data input and transfer is controlled by the computer, which determines the maximum rate of
transfer, but there is virtually no limit to the size of the image that can
be accommodated. For display purposes the image data is read off the disk
continuously and transferred via the computer to a refresh CRT display.
This display procedure is not satisfactory for normal-size images,however,
since the refresh frame rate is inordinately low.   Small images, e.g.
128x128 pixels (occupying 16 Kbytes) can be stored in computer memory and
read out to a CRT display at a sufficiently high refresh rate to provide a
flicker-free display.   This technique can be used to display selected
portions of a larger image stored on disk, but as a means of examining a
complete image it has obvious disadvantages.   Another possible technique
for displaying a normal-size image is to use a scan converter storage tube.
The image is transferred from disk in a single scan and stored in the
converter tube in analogue form.   The image is then read out continuously
at normal TV frame rate.   The scan converter tube has the disadvantage
that the image fades gradually during read out and therefore has to be
occasionally updated.   There is also often a loss of image quality.

The most satisfactory solution to the display problem is that provided by
the digital framestore; this can be used to serve the same function as
the analogue scan converter tube but suffers from none of its disadvantages.   The digital framestore is a relatively new development which
promises to remove most of the difficulties discussed above.   Furthermore, it can be used as a means for image acquisition, as well as display,
at data input rates extending up to the normal TV scan rate.   It thus
provides a way of acquiring CTEM images via a low light level TV camera.
A facility for integrating the incoming image signal can also be provided
with the digital framestore so that image noise can be reduced to any
desired level.   The framestore is linked closely to the computer so that
it acts virtually as an extension of the computer's own memory.   Thus,
any selected portion of the image, or even a single pixel, may be
addressed and modified.   A cursor system allows the operator to identify
and select image areas and pixels with high precision.   A system
incorporating such a framestore is described below.

2.   <u>On-line system hardware</u>

Fig 1 shows a comprehensive, integrated system for on-line operation,
capable of handling image and spectral signals both of which may be
displayed by means of the framestore.   TV image signals from SEM, STEM,
CTEM or video tape recorder are input directly to the framestore; slow
scan image signals from SEM or STEM are input either to the framestore or,
if the image is too large to be accommodated completely within the framestore, directly to the computer.   Analogue-to-digital converters or their
equivalent (not shown in the figure) are used in all input signal lines.
Spectra are accumulated using appropriate interfaces and input directly to
the computer.   In the case of SEM or STEM instruments, scanning is
controlled by the computer to achieve synchronisation during image
acquisition.   Preferably a separate digital scan generator is used rather
than the microscope's internal analogue scan generator (Unitt and Jones
1978).   Alternatively the computer itself may be used to generate the
scanning waveforms.   At any stage after acquisition, images or spectra
may be filed on disk or tape, the latter being more suitable for permanent
files, since retrieval is much slower than from disk.   An image may be
filed or retrieved in the order of 2-3 sec. using a fast disk.   For SEM
or STEM instruments, images may be output from the computer to the instru-

ments own recording CRT; for the CTEM, images are recorded by photographing the TV monitor screen. A printer and a plotter are essential requirements for the system, while a link to an external computer or computer
network can enhance the scope of the system and its range of applications
considerably. A system of the type shown in Fig.1 can, in addition to
image or spectral processing, serve also to control the microscope to which
it is attached; some of the possible control functions are indicated in
Fig.1.

3.   Advantages of on-line operation

The difficulties associated with the operation of a scanning or conventional electron microscope are well known: the directly viewed image is noisy
and difficult to interpret, and the operator has to labour under adverse
viewing conditions; the specimen is subject to continuous irradiation
while the image is observed; the photographic process introduces a delay
before the final results can be assessed, and the yield of satisfactory
micrographs is usually very low. All of these difficulties may be reduced
or eliminated by the application of on-line digital computer techniques.
With a system such as that described above, it is possible to record and
store an image with the minimum electron exposure necessary to achieve a
given signal-to-noise ratio. The image may then be viewed on a high-
quality TV monitor for as long as required without further irradiation of
the specimen. The quality of the directly-viewed image can be as good as
that seen in a final print using the conventional photographic recording
process. In addition, powerful image processing techniques are immediately
available to enhance contrast and image detail. If adequate file space is
available on disk or tape, photography may be dispensed with for a large
number of applications in scanning microscopy. For the CTEM, however, the
limited resolution and field of view of the normal TV camera pickup system
means that the final image must be recorded on plate or film in most instances. Another advantage of the digital on-line system is that image
manipulation can be carried out on a quantitative basis. For example,
when acquiring an image, maximum and minimum grey-level values together
with a histogram can be displayed (Fig.2). This completely eliminates the
incorrect recording of images which so often occurs when using analogue
methods and photographic film.

4.   Applications

An on-line digital system can be used to carry out all types of image processing operations, but it is particularly effective for the analysis and
extraction of quantitative information. For example, topographical analysis of surfaces examined in the SEM can be performed on-line using the
conventional method of stereometry, but working with two stored images instead of micrographs. A further example of on-line topographical analysis
is illustrated in Fig.3. This method is based on an automatic focusing
technique in which the computer is used to control the objective lens current (Holburn and Smith 1979). By means of a functional relationship stored within the computer, the working distance w can be determined for any
value of lens current. A small sub-raster is scanned over the specimen at
a point of interest (point 1 in Fig.3a) and the system is used to automatically focus the beam at that point, thus, w_1 is determined. A similar
procedure is carried out at a second point of interest to determine w_2.
The difference, w_1-w_2, then gives the height difference between the two
points. The system can be preset to carry out an analysis at any desired
number of locations; in this way line profiles may be plotted out.

Interest in applying on-line digital techniques to the STEM is growing
rapidly, and several new or proposed systems, some of them incorporating
framestores, have been reported recently. The main advantage in this case
is the facility with which the multiple signals provided by a STEM may be
recorded simultaneously. An integrated system for analytical purposes has
been discussed by Craven et al (1980), while Burge and van Toorn (1980)
have emphasised the advantages to be gained from the recording of multiple
signals from bright-field and dark-field detector arrays.

On-line digital techniques are also proving to be highly effective in the
case of the CTEM (Herrman et al 1978). Images are transferred from a low
light level TV camera directly to a framestore. A large number of frames
may be accumulated and averaged to reduce the image noise to any desired
level; the effectiveness of the technique being limited only by instrumental drift. The method can be extended to allow the detection and recording of single electrons providing the current density at the input of
the image pickup system does not exceed about $10^{-13}A/cm^2$. Framestores can
be configured to hold two or more images and it is possible, with suitable
digital circuitry to add, subtract or multiply these images in real time.
By taking the difference between a currently acquired image and an initial
reference image, changes may be detected; for example, those caused by
drift in the microscope (Pawley 1980). This method may also have important applications in the study of dynamic events occurring within specimens.
A related technique, involving image multiplication, is that of background
correction (see Erasmus and Smith; these proceedings). The TV pickup
system used to capture a CTEM image inevitably suffers from shading and
other undesirable effects which cause a non-uniform background to be superimposed on the image. This may be removed by storing a 'background image',
acquired in the absence of a specimen, which is then combined suitably with
the normal specimen image. In the CTEM the background is multiplicative,
so that the procedure involves multiplication by the reciprocal of the
background image. The procedure is illustrated in Fig.4, and an application is shown in Fig.5. An on-line system can also be used for the
rapid assessment of CTEM performance from diffractograms calculated by
means of the Fast Fourier transform. Other novel applications of the
framestore in electron microscopy have been suggested and demonstrated by
Harland and Venables (1980). In particular, these authors have shown how
the framestore can be used for the real-time dynamic analysis of diffraction patterns. The author gratefully acknowledges the assistance of
S.J. Erasmus, D.M. Holburn and W.J. Tee in preparing this paper.

5.   References

Burge R E and van Toorn P 1980 Scanning Electron Microscopy ed O Johari
    (Chicago: SEM Inc.) 1 pp81-91
Craven A J, MacLeod A M and Ferrier R P 1980 Proc. 7th Euro. Conf. on Elec.
    Mic. ed P Brederoo and V E Cosslett (Leiden: 7th Cong. Foundation 3
    pp170-171
Harland C J and Venables J A 1980 Proc. 7th Euro. Conf. on Elec. Mic.
    ed P Brederoo and G Boom (Leiden: 7th Cong. Foundation) 1 pp502-503
Herrmann K-H, Kranl D and Rust H-P 1978 Ultramicroscopy 3 227
Holburn D M and Smith K C A 1979 Scanning Electron Microscopy ed O Johari
    (Chicago: SEM Inc.) 2 pp47-52
Pawley J 1980 Proc. 38th Ann. EMSA Meeting ed G W Bailey (Baton Rouge:
    Claitor's) pp8-11
Unitt B M and Jones A V 1978 Scanning Electron Microscopy ed O Johari
    (Chicago: SEM Inc.) 1 pp27-31

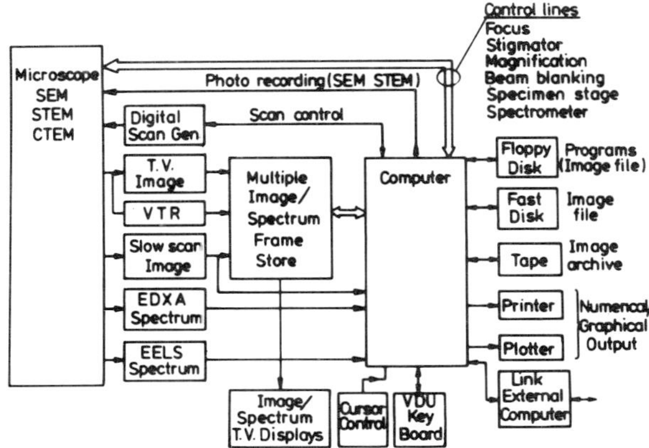

Fig.1. Schematic diagram of an on-line computer system for electron microscopy.

Fig.2. (below), Interactive display for image acquisition.

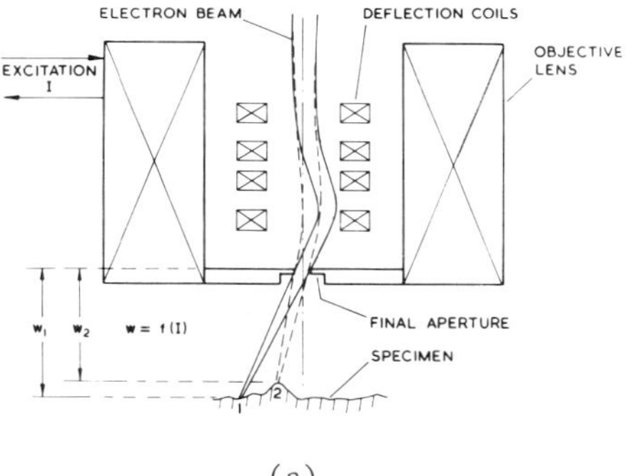

(a)

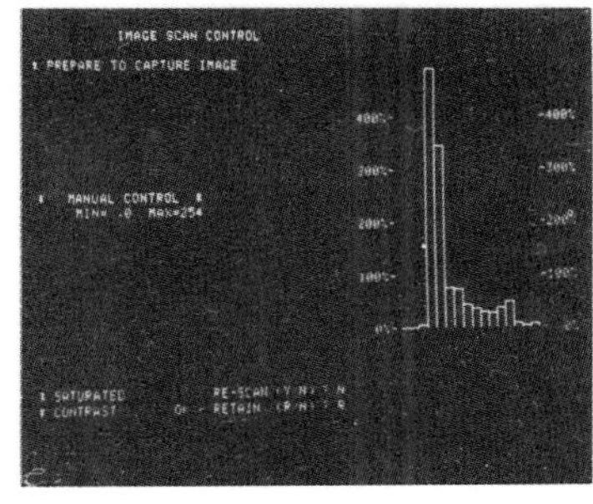

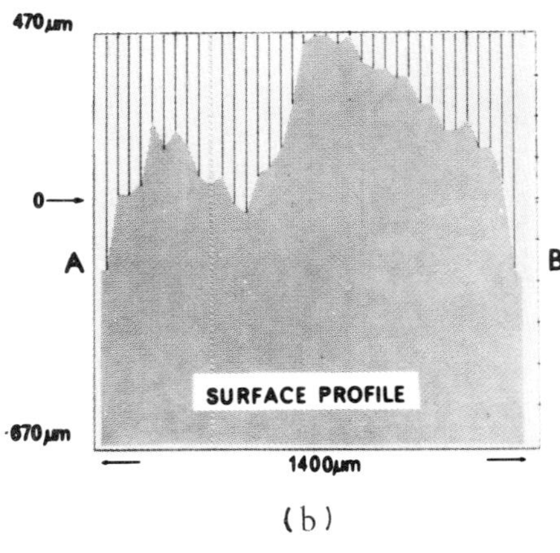

(b)

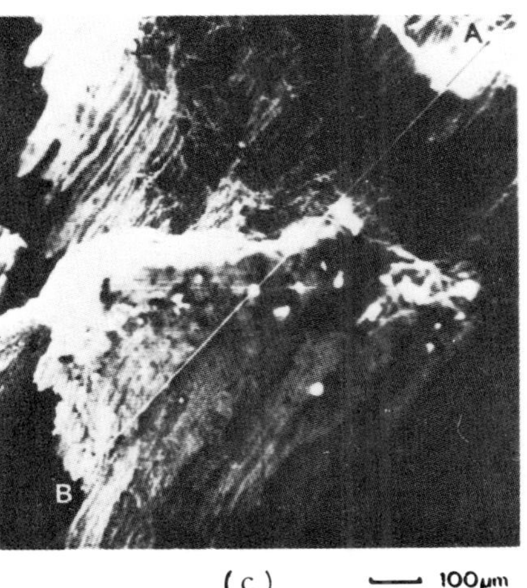

(c)

Fig.3. Determination of surface profiles in the SEM by automatic focusing.   (a) principle of method;   (b) profile on fractured steel specimen;   (c) trace of profile.

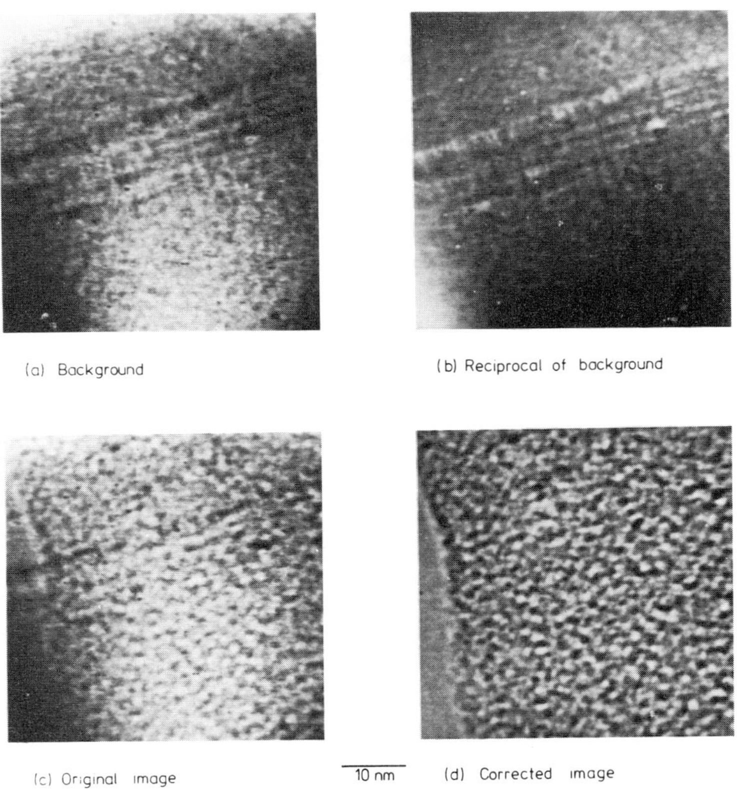

Fig.4. Illustrating background correction of am image obtained with a low-light-level image pickup system on the Cambridge HREM. Specimen: amorphous carbon film.

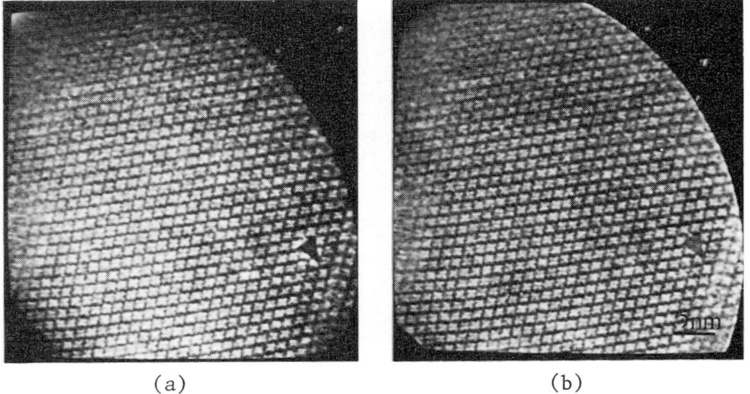

Fig.5. Images of chlorinated-Cu-phthalocyanine: (a) without background correction; (b) with background correction.

Real-time digital image processing in electron microscopy

S J Erasmus and K C A Smith

University Engineering Department, Trumpington Street, Cambridge CB2 1PZ

1. Introduction

The digital image processing system shown in Fig. 1 and known by the acronym GEMS, has been interfaced to both a SEM and a CTEM. Television rate video signals are produced directly by the SEM, while for the CTEM the video signal is produced by a phosphor screen and low light level TV camera (Catto et al 1981). The video signal is converted to digital form by an 8-bit analogue to digital converter (ADC), processed by the arithmetic unit and then stored in one of the 3 image memories, each of size 512 x 1024 points x 8 bits. The contents of each memory can be processed by the computer or by the correlator and can also be examined on a television display after background correction, which corrects for TV camera shading and phosphor screen irregularities. The correlator is capable of finding the maximum, minimum, mean, variance and cross-correlation function of a pair of images much faster than is possible by computer. The computer is also able to control the stigmator and objective lens currents so that the system can be used for automatic focusing and astigmatism correction.

2. Reduction of noise

The arithmetic unit in Fig. 1 performs the following operation at TV rate:
$$y_i = a y_{i-1} + b x_i$$

where: x_i = one frame of incoming video signal; y_{i-1} = old value of stored image; y_i = new value of stored image; a,b = constants. In order to avoid rounding errors, y is stored to 16-bit precision using two of the 8-bit image memories. By choosing appropriate values for a and b the following filters can be implemented:

(a) Averaging filter.
If we set $y_0=0$; $a=1$; $b=1/n$; then after n frames, y_n is the average of $x_1,\ldots,x_n$. The improvement in signal to noise ratio (SNR) is $\sqrt{n}$.

(b) Recursive filter.
Here we set $a=k$; $b=1-k$; $0<k<1$; then it can be shown that:
$$\frac{SNR_{out}}{SNR_{in}} = \frac{1}{\sqrt{k^{2i-2} + \frac{1-k}{1+k}(1 - k^{2i-2})}}$$

and
$$\lim_{i \to \infty}\left(\frac{SNR_{out}}{SNR_{in}}\right) = \sqrt{\frac{1+k}{1-k}}$$

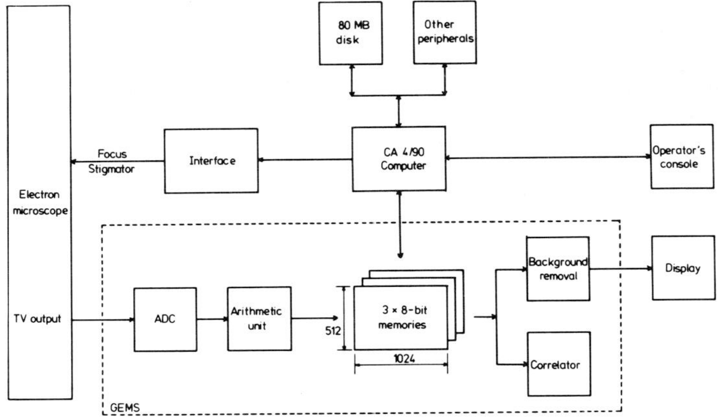

Fig. 1   Digital image processing system

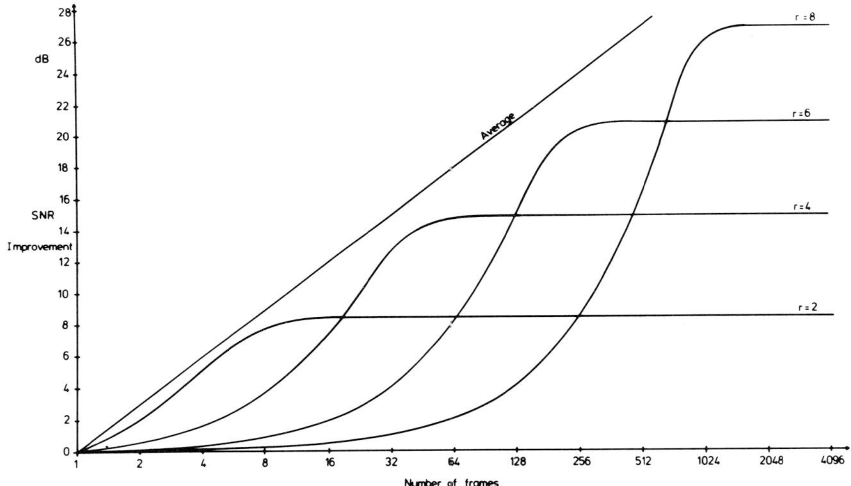

Fig. 2   Improvement ir SNR by filtering

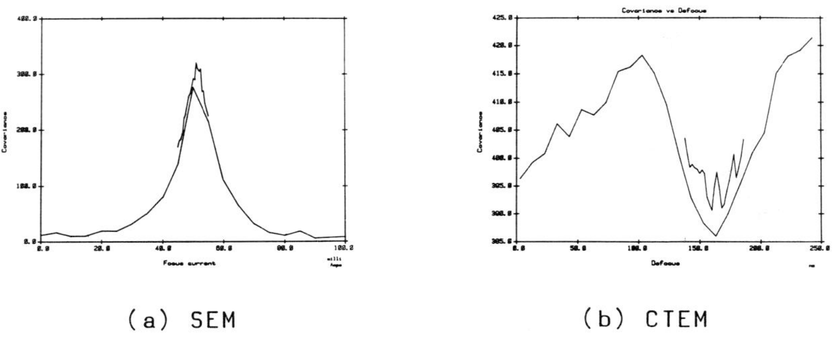

(a) SEM   (b) CTEM

Fig. 5   Variation of contrast during automatic focusing

(c) Optimal recursive filter.

The averaging filter can be shown to be optimal and Kalman filter theory produces the following recursive implementation. Here a and b are no longer constants, but are made time dependent:

$$a_i = \frac{i-1}{i}; \qquad b_i = \frac{1}{i}$$

Then y_i is the average of $x_1,\ldots,x_i$, and the improvement in SNR is $\sqrt{i}$.

These results are shown in Fig. 2. The straight line represents the ideal result for the averaging and Kalman filters, while the curves predict the results for different values of r, where $k = 1 - 2^{-r}$. Typical SEM images produced by filtering are shown in Fig. 3. Fig. 3(a) is a single frame of an image of video tape while (b), (c) and (d) are the average of 10, 100 and 1000 frames respectively. Fig. 4 shows the effect of averaging on a CTEM image of chlorinated copper phthalocyanine (specimen courtesy of Dr J Fryer).

3. Automatic focusing and correction of astigmatism

The system has been used to focus both a SEM and a CTEM automatically. The point of focus in the SEM was defined as the point of maximum contrast; in the CTEM as the point of minimum contrast. Contrast was measured by the covariance of corresponding points in a pair of images and can be calculated in 160ms by the correlator in GEMS. The point of best focus was found by sweeping the focus current under computer control initially in large steps over a wide range and then in smaller steps over a smaller range centred on the point of best focus found in the first current sweep. If necessary, a third finer current sweep can be used. Fig. 5 shows results obtained from large and small current sweeps in both the SEM and CTEM, the whole process taking about 20 seconds. In the SEM the specimen was video tape and the SNR was 0.5 (the same as in Fig. 3(a)), while in the CTEM a thin carbon film was used. In both cases, the system focused at least as well as an experienced operator could by eye. The system shown in Fig. 1 has also been used to correct the astigmatism of a SEM by analysing the dependence of the image correlation function on defocus. Automatic correction of the astigmatism of a CTEM is currently being investigated.

4. Background correction

Shading in the TV camera and irregularities in the phosphor screen used to produce a TV signal from a CTEM, superimpose an unwanted background on the image. This can be described by:

$$I(x,y) = A(x,y).E(x,y) + B(x,y)$$

where: I = image, E = incident electron current density, A = multiplicative shading, B = additive shading. In the authors' system, B was negligible and the image could be corrected by dividing by A. Since A is time invariant, its reciprocal can be calculated once by computer and then stored in one of the image memories, leaving two memories free for filtering the incoming video signal. A digital multiplier finds the product of the two images at TV rates and the result is then scaled and displayed. Typical images produced by this system are shown in Smith (1981).

References

Catto C J D, Smith K C A, Nixon W C, Erasmus S J and Smith D J 1981: these
    proceedings
Smith K C A 1981: these proceedings

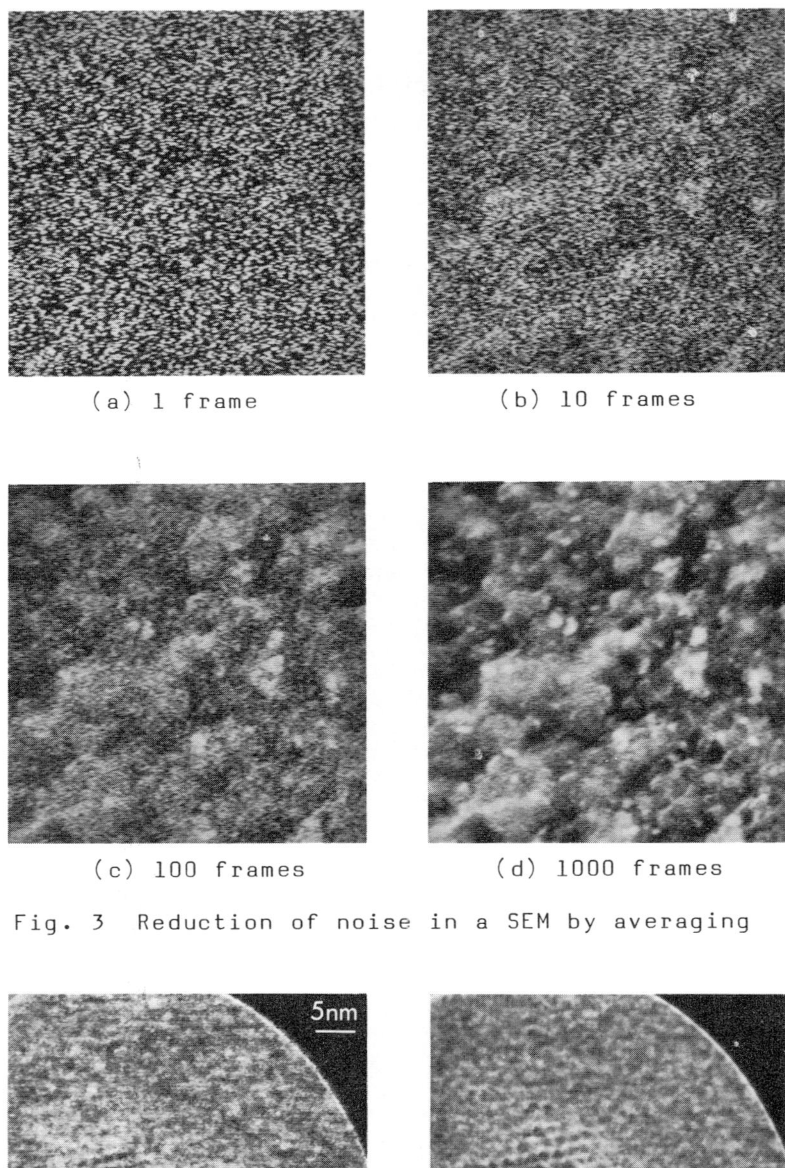

(a) 1 frame   (b) 10 frames

(c) 100 frames   (d) 1000 frames

Fig. 3   Reduction of noise in a SEM by averaging

(a) 1 frame   (b) 30 frames

Fig. 4   Reduction of noise in a CTEM by averaging

On-line image analysis and processing for very high resolution electron microscopy

E D Boyes, B J Muggridge, M J Goringe, J L Hutchison and G Catlow*

Department of Metallurgy & Science of Materials, University of Oxford
*Microconsultants Ltd, Newbury, Berks

1. Introduction   Advances in high resolution electron-microscopy (HREM) have improved the performance sufficiently to include many interatomic distances. However, even the images from very thin crystals, although superficially very attractive, are not always simple to interpret[1]. A proper understanding of the high resolution imaging process requires a knowledge of the electron amplitude and phase information on the exit surface of the specimen, and of its subsequent convolution with the contrast transfer function (CTF), which is dependent upon the properties of the microscope and on the conditions under which it is being operated.

Optimising and controlling HREM experiments to obtain reliable information about unknown structures requires a solution to the phase problem and therefore independent knowledge of the specimen thickness, microscope aberrations and the operating defocus, as well as practical assistance with correcting astigmatism. Previously this has necessitated comprehensive matching procedures between images calculated from models incorporating various combinations of parameters, and those obtained from the microscope. Experimentally, the properties of the CTF have been investigated by using a thin amorphous specimen to approximate a weak phase object with a semi-continuous range of spatial frequencies, and examining the spectrum transmitted through the lens by forming optical diffraction patterns[2] (ODM) from the micrographs on a laser optical bench. Of particular importance is the position of the first zero or phase inversion in the CTF, and the influence of defocus on it. In practice it is usually desirable to obtain images at close to optimum defocus[3] ($D_{opt} \sim 1.2 C_s^{\frac{1}{2}} \lambda^{\frac{1}{2}}$), and to have a through focal series centred on it. In the microscopy of specimens demonstrating higher symmetry the situation is more complicated since, depending on thickness, it may not be possible readily to identify the focus condition nor to correct astigmatism sufficiently precisely. Accordingly a system for on-line image acquisition and analysis is required, and this paper describes the coupling of a Microconsultants Intellect 200 image analysis system based on a double 512x512x8-bit digital video framestore, to the first ultra high resolution version of the JEOL 200CX electron microscope which has a CTF d_1 of 0.24nm and a cut-off at less than 0.17nm [4]. A related system has been used to convert slow scan SEM signals to television standard[5] and to demonstrate the feasibility of the present experiments off-line, whilst Hashimoto[6] has described a simple 64x64 point system and Herrmann et al[7] a more sophisticated one: both of which are quite slow in operation. Speed is important since to be used interactively the image acquisition and analysis together with any corrective action found to be necessary must all be accomplished within a realistic estimate of the natural stability of the microscope, which is typically of the order of one minute. Faster operation is very desirable.

2. <u>Methods and Results</u>   A transmission phosphor screen has been placed in
the base of the HREM camera chamber which was modified to provide a suitably screened viewing point. The magnification on this screen is $\sim$1.3 times
the nominal value. A low-light level television camera (SIT or micro-
channel-plate based) is focussed on the image thus provided via a pair of
large aperture lenses mounted nose-to-nose. The video output is taken to
the Intellect 200 where it is digitised to 8-bit precision at television
rate and stored in one of the digital framestores. During this process,
simple algorithms can be applied in the recursive video processor (RVP) to
improve the signal to noise ratio by averaging, or integrating the information contained in a sequence of input frames. This is spatially resolved
time averaging. Typically an eightfold improvement may be achieved after
3 seconds processing of the relatively noisy television images obtained at
high magnification. The running average mode is useful for providing a
continuous display, and the integration mode is prefered for its better
noise reduction and intensity acquisition where image stability allows acquisition without continuous display. These are comparable to the integration
also applied in the photographic exposure on very fast film where specimen
instability and stage drift also limit the useful total time.

The system is software controlled through the LSI 11/23 minicomputer which
also has random access to the framestores allowing further processing of
the stored images including their possible transfer to magnetic storage
media. Within the system, a dedicated hardware unit is available for the
production of two-dimensional fast Fourier transforms (FFTs), initially
from selected 128 by 128 pixel areas and now using 256 by 256 points.
These are an important element in many signal processing procedures and are
in many respects directly analogous to the ODM results obtained off-line,
but are available for interactive control and image selection on-line. It
is therefore possible to establish the degree of astigmatism present and
the defocus conditions from images of amorphous materials. Pseudo-micro-
diffraction is also possible but generates artefacts in addition to those
usually experienced in lattice imaging. Early results have been obtained
at 1,000,000X magnification which corresponds to 0.05nm sampling for a
$(0.1nm)^{-1}$ capability in transform space. The spatial frequency scale was
calibrated using graphite and other lattice images. The direct results are
more noisy than those obtained by ODM, which can be explained by the much
smaller area being sampled, and could be improved by the radial averaging
of properly stigmated data or a higher resolution interface combined with a
greater sample size. The latter approach is being pursued as an important
development of this essentially evaluation system, the next stage being to
determine the degree of sensitivity of the methods and the precision with
which the parameters, in particular the defocus, can be established.

Other forms of data acquisition and image analysis and processing will be
capable of implementation on this very flexible and powerful system which
represents a new approach to more systematic methods in high resolution
microscopy experiments, and should greatly simplify subsequent image
analysis.

1. Cowley J M 1979 Chemica Scripta <u>14</u> 279
2. Thon F 1966 Z Naturforsch <u>21a</u> 476
3. Scherzer O 1949 J.Appl.Physics <u>20</u> 20
4. Boyes E D et al 1980 Inst. Phys. Conf. Ser. <u>52</u> 445
5. Boyes E D et al 1980 Proc. EMSA <u>38</u> 226
6. Hashimoto et al 1980 Electron Microscopy <u>1</u> 118
7. Herrmann K-H 1978 Ultramicroscopy <u>3</u> 227

The S.E.R.C. is thanked for a grant in support of this project.

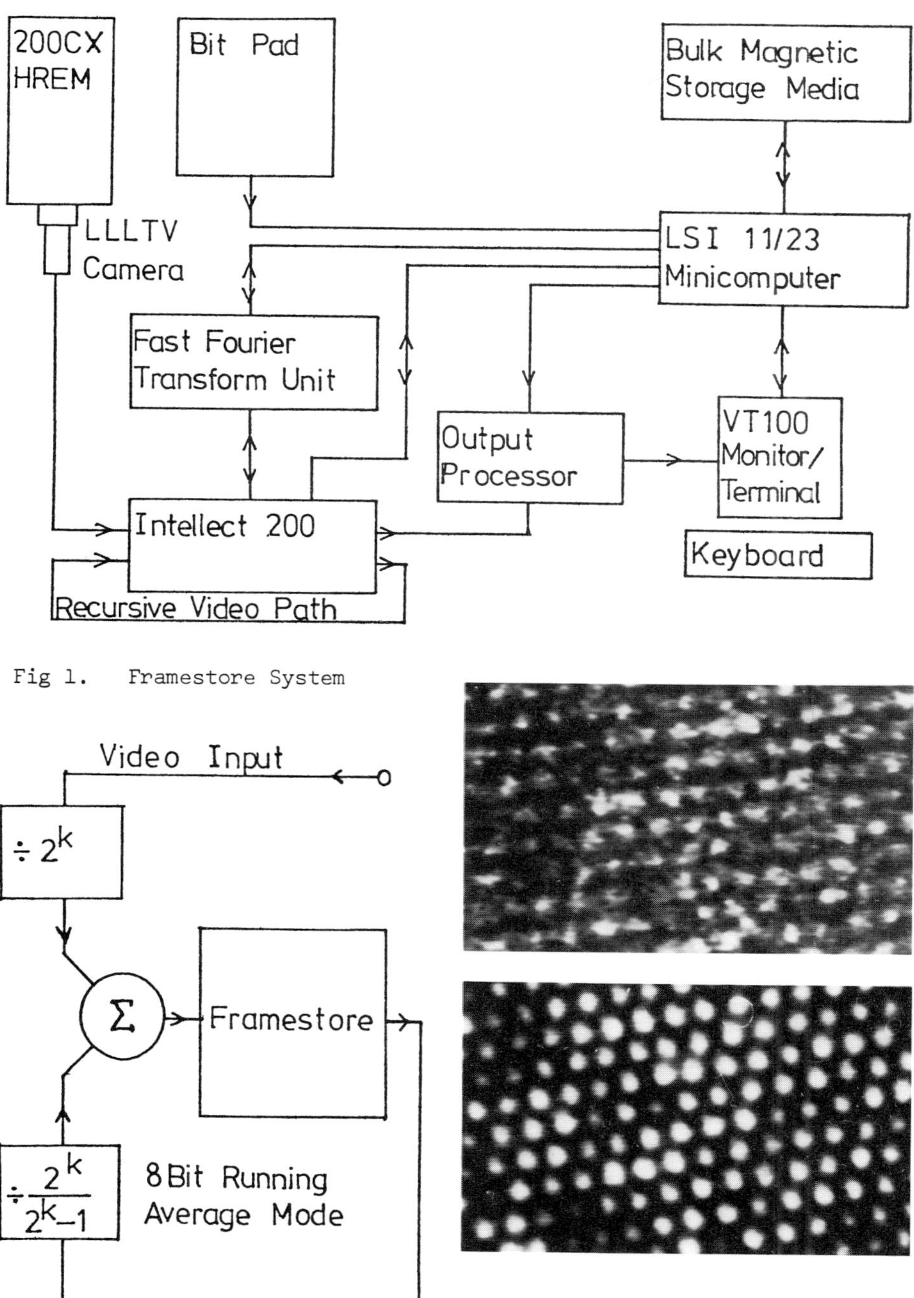

Fig 1.   Framestore System

Fig 2.   Noise reduction method applied to silicon lattice images

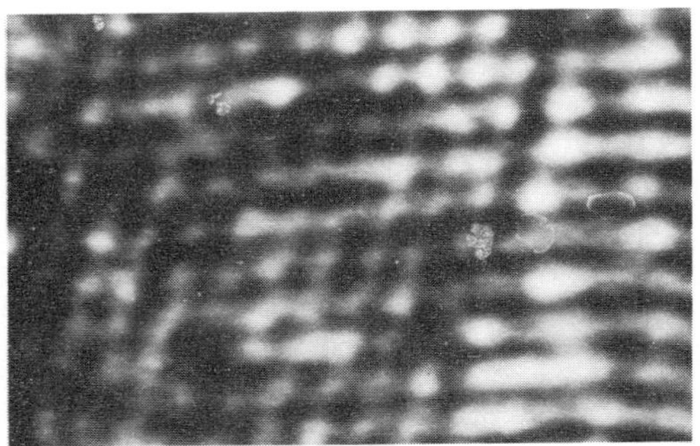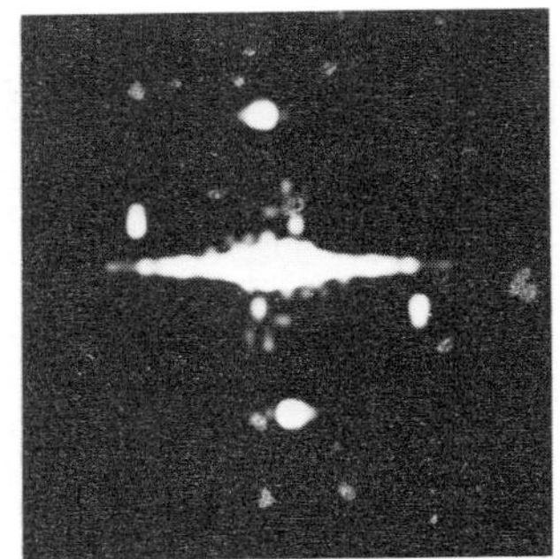

Fig 3.   Transform of selected small area of graphite lattice image

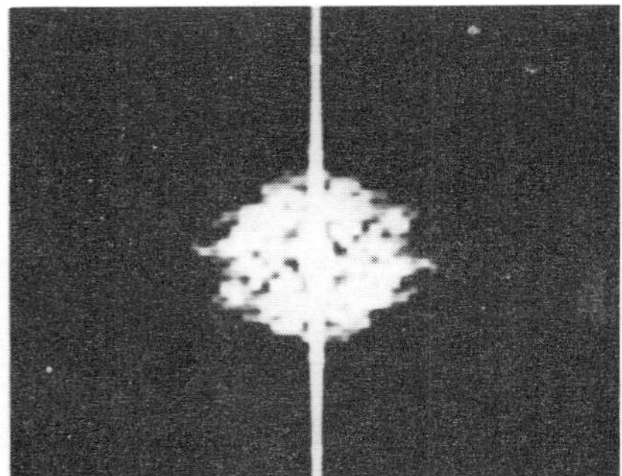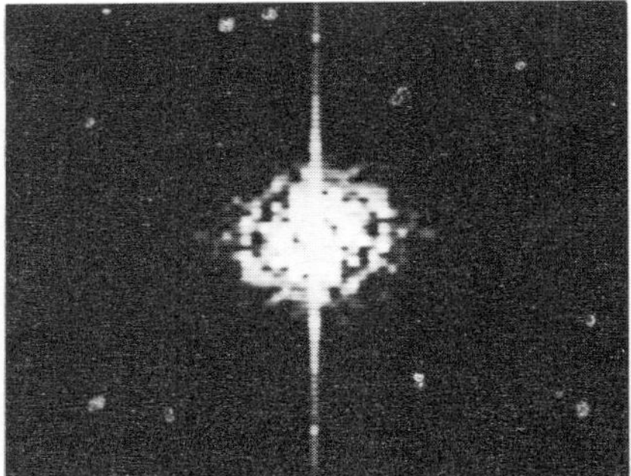

Fig 4.   Preliminary transforms of 128 x 128 images of amorphous silicon
          at different defocus settings of (a) D ∿ 1   (b) D ∿ 2, obtained
          on-line in ∿ 10 seconds.

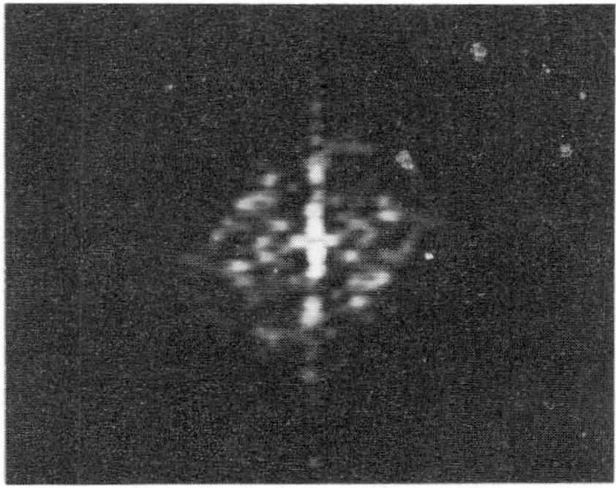

Fig 5.   Astigmatism analysis by transform

An image pickup and display system for the Cambridge University HREM

C J D Catto, K C A Smith, W C Nixon, S J Erasmus and David J Smith[*]

University Engineering Department, Trumpington Street, Cambridge CB2 1PZ.
*High Resolution Electron Microscope, Free School Lane, Cambridge CB2 3RQ.

1. Introduction

The image pickup and display system fitted to the Cambridge University
HREM was designed with the following uses in mind: (i) as an aid to the
operator when observing images at low levels of illumination, in
particular for focusing and astigmatism correction; (ii) for the observation of beam sensitive specimens; (iii) for the observation and recording of dynamic events; (iv) as a means of providing an image signal
suitable for on-line digital computer processing. To cater for this wide
range of applications the system was required to possess a high absolute
sensitivity; a wide dynamic range, including resistance to signal overload;
a high resolution in terms of lines per field; and uniformity of response
over the field of view.

To meet these requirements as far as possible a system has been constructed incorporating an Image-Isocon low light level camera tube (type P8041).
The camera is fibre-optically coupled to a fluorescent screen at unity
magnification. This type of coupling was chosen in preference to lens
coupling because of its higher transfer efficiency and freedom from
vignetting. The theoretical performance of the system has been
discussed by Catto and Smith (1975). Sufficient operating experience
has now been gained with the system to enable an assessment of its
performance to be made with respect to the design criteria outlined above.

2. Description

A schematic diagram of the system is shown in Fig. 1. The fluorescent
screen is deposited on a 40mm diameter fibre optic disk, which is sealed
to a tube at the base of the microscope by means of an O-ring. The
fibre optic faceplate of the camera tube is held in contact with the under-
side of the disk by light spring pressure. By lowering the camera on its
guide rails, access to the disk may be gained thus enabling screens to be
readily replaced. When the pickup system is not in use, the screen is
covered by a moveable shutter. A digital picoammeter is used to measure
the current falling on a known area of the normal viewing screen; from
this reading the current density in the plane of the screen may be
inferred. The image is viewed on a monitor having a long persistence
screen with an effective integration time of about 2 sec. Video signal
and black level are set with the aid of an oscilloscope. Images may be
recorded on video tape; a microphone channel is also available on the
recorder for commentary and data input. The computer interface takes the

form of a GEMS image storage and display system (see Erasmus and Smith; these proceedings).   The processed image from this system may also be displayed for comparison with the direct image.

3.   Experimental

Using the facilities described above, the sensitivity and dynamic range have been measured.   The lowest current density on the screen at which a usable image may be observed on the monitor is about 2×10^{-10} A/m² and the highest is about 10^{-8} A/m².   This latter figure corresponds to a knee in the camera transfer characteristic.   Images are obtainable at higher current densities but suffer from halation effects.   The uniformity of response was found to be to within ±5% referred to the centre of the field over the major portion of the field area.   At the extreme corners the figure was ±20%.

The modulation transfer function (MTF) of the system has been determined using a method described by Dainty and Shaw (1974): the edge response is differentiated to get the line response, which is then Fourier transformed to produce the MTF.   The edge response was found by plotting intensity contours across the edge of the image of an opaque object, in this case the diffraction stop support arm.   These measurements and calculations were conveniently performed using the on-line computer facility. Various phosphors and screen thicknesses have been tested; the best results so far being obtained with a P22 phosphor at a coating weight of 10mg/cm².   The screen is coated with approximately 20nm of aluminium to prevent ingress of stray light.   Results for this screen, obtained at different current densities, are shown in Fig. 2.   The hump in the curve for the highest current density is thought to be due to the camera being operated close to the knee of its transfer characteristic.   From the curve for the **medium** current density it can be seen that the 50% MTF point occurs at approximately 3 line pairs/mm.   Since the active area of the target is 31 x 23mm the system resolution at this value of MTF is 93 x 69 line pairs per field – equivalent to about 190 x 140 resolved image points.

4.   Operation and Applications

The system can be used to view images in subdued room lighting at screen current densities about ½% of that normally required for direct viewing. Focusing and astigmatism correction may be accomplished without the necessity for dark adaptation, although partial dark adaptation is sometimes necessary for selection of the field on the normal viewing screen. Stage drift can be checked at the very high magnifications made possible with the system, and diffractograms can be calculated, using the on-line computer facility, to check microscope performance.   The camera has proved to be extremely resistant to overload and it is possible to observe diffraction patterns, including the central spot, without damage to the tube.

The system has already established its usefulness in observations of beam-sensitive materials and dynamic events.   Figs. 3 and 4 for example, show images of chlorinated-Cu-phthalocyanine recorded under moderate dose conditions ($\sim$1.4 x 10^3 A/m² at specimen; $\sim$2 x 10^{-9} A/m² at screen) at a deliberately-chosen over-focus condition where the molecular shape is well defined. (These figures are obtained at present by photography of the monitor screen during playback of the video tape recorder – exposure typically 2 sec.)   Fig. 3 shows an area recorded soon after initial

exposure to the electron beam whilst Figs. 4(a) and (b) show a different
area recorded during a study of radiation damage.   The calculated diffraction patterns, shown inset, indicate the progressive loss of structural
integrity due to beam damage.   At very low beam current levels, very
little change in appearance was apparent even after prolonged exposure.
Fig. 5 shows stages in a progressive change of appearance, induced by
slight beam heating, in the semiconducting material cadmium telluride.

5.   Conclusions

The system has proved to be effective in meeting most of the requirements
outlined above.   Although its sensitivity cannot match that of a single-
electron counting system (Herrmann et al 1978) it is sufficient to allow
the specimen dosage during observation to be reduced by at least two
orders of magnitude.   This, coupled with the wide dynamic range and
resistance to overload, has rendered it an extremely practical system to
use.   A variety of applications, which also utilize the high-resolution
capabilities of the HREM, are envisaged.

6.   Acknowledgements

The authors are indebted to R A Camps, P Carter, L A Freeman, T L Koch
and L R Peters for their assistance with the construction and operation
of the system.   Thanks are also due to R Nixon, English Electric Valve
Co., for advice and assistance.   Specimens were kindly supplied by
Drs. J R Fryer and R Sinclair.   The system was constructed with the aid
of a grant from the Science Research Council to which the authors are
grateful for continuing support.

7.   References

Catto C J D and Smith K C A 1975 J. Microscopy <u>105</u>, 223
Dainty J C and Shaw R 1974 Image Science (London: Academic Press) p241
Herrmann K-H, Krahl D and Rust H-P 1978 Ultramicroscopy <u>3</u> 227

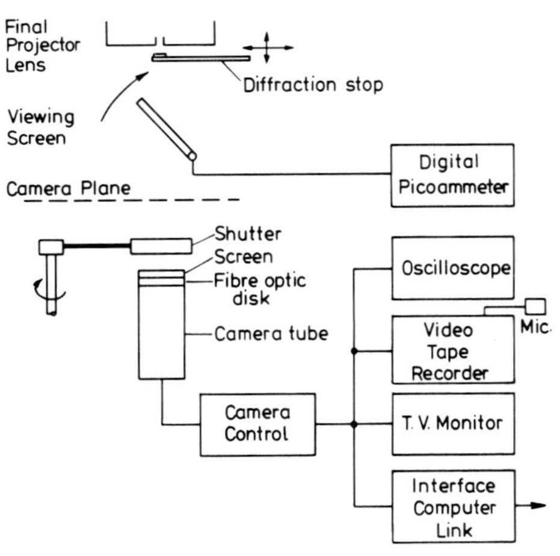

Fig. 1   Schematic diagram
of the image pickup and
display system

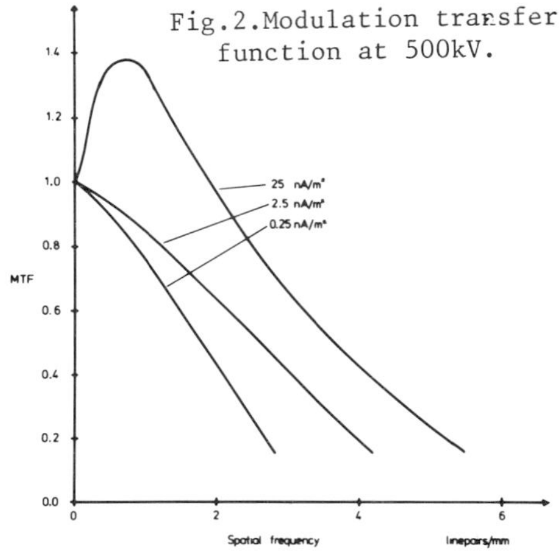

Fig.2.Modulation transfer
function at 500kV.

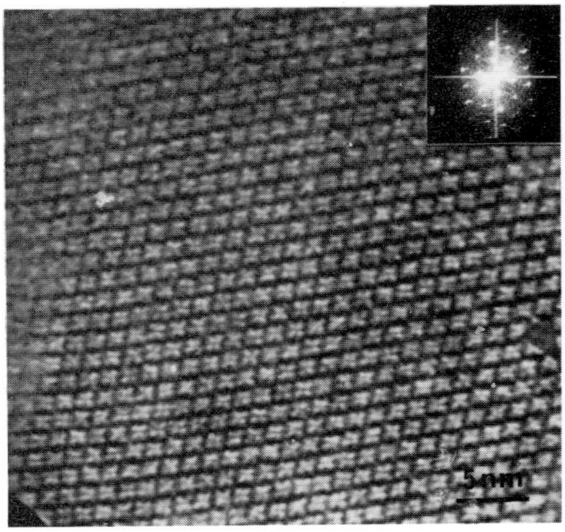

Fig.3.Molecular image of chlorinated
-Cu-phthalocyanine with calculated
diffraction pattern inset.

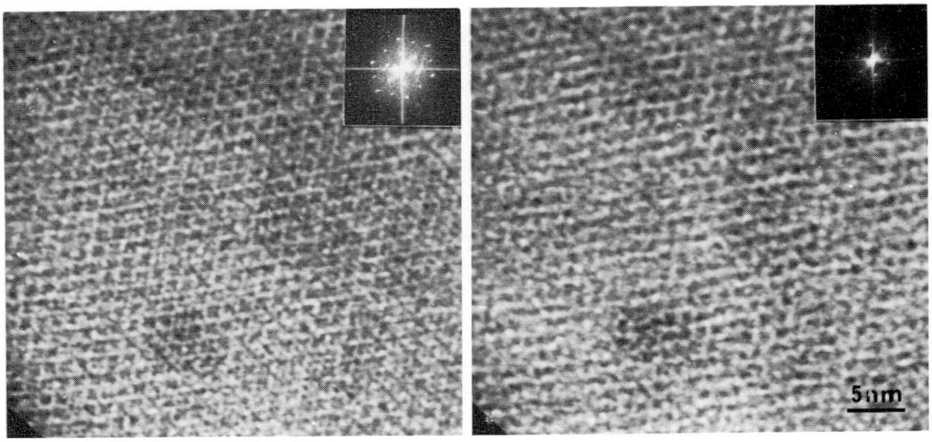

Fig.4.Images recorded during a study of radiation damage in chlorinated
-Cu phthalocyanine - time difference: 4 minutes.

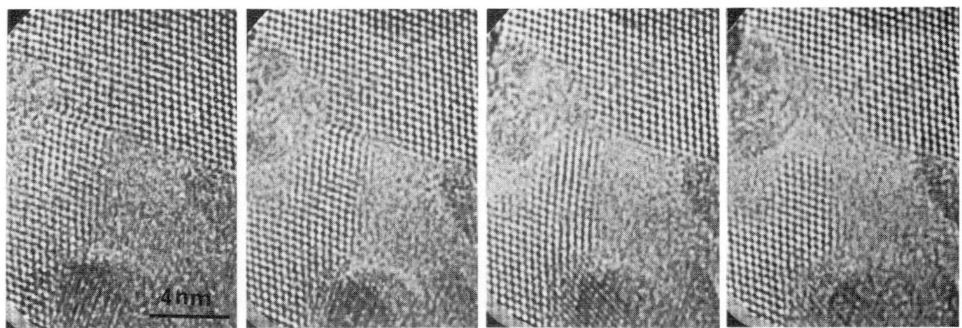

Fig.5.Cadmium telluride showing change due to slight beam heating. Images
recorded on video tape and computer processed to correct background -
total time difference: 3 secs.

An image intensifier for high-voltage electron microscopy of beam-sensitive materials

B L Jones, R C Doole and G R Booker

Department of Metallurgy & Science of Materials,
University of Oxford, Parks Road, Oxford OX1 3PH

A microchannel plate (MCP) is a compact array of channel electron multi-pliers (Fig.1) which can be used to amplify a beam of incident electrons while retaining the relative spatial intensity distribution present. The present paper describes the assessment of an MCP as an image intensifier for high-voltage electron microscopy. The MCP used was an EMI-Varian type VUW-8911ZS with the following geometry:- useful plate diamter - 25mm, plate thickness - 0.5mm, channel diameter - 12µm, channel centre-to-centre spacing - 15µm, channel bias angle - 8°. The MCP was mounted in an AEI EM7 HVEM below the standard viewing screen and photographic plate. It was uniformly irradiated with electrons of known current density, and the total output electron current was collected using a positively biassed metal plate located below the MCP. The incident electron beam was varied in the range 100 to 1000kV and 10^{-13} to 10^{-11} A (MCP irradiated area $\sim$ 3cm^2). The measurements showed that the gain increased rapidly as the MCP bias voltage was increased in the range 700 to 1100V. For a fixed bias of 1100V, the output current was closely proportional to the input current (Fig.2a) up to an input current of 10^{-11} A, and the gain decreased progressively from 7×10^4 at 100kV to 3×10^4 at 1000kV (Fig.2b). The MCP was operated at room temperature and the dark current was 0.2nA cm^{-2}. No significant degrada-tion in terms of gain or dark current occurred for sustained irradiation over the full electron beam voltage range used.

For imaging purposes, the collector plate was replaced by a 4µm thick, type P11, blue phosphor screen deposited on a pyrex glass disc. The screen was positioned 2mm below the MCP and was biassed 2.3kV positively with respect to the output side of the MCP. The screen was viewed in transmission by a low-light-level, silicon-image-tube (S.I.T.) camera and the image displayed on a TV monitor screen and simultaneously recorded using a video cassette recorder (Fig.3). The displayed image was photographed using a single frame-scan (40ms) with standard 35mm film. The recorded image could also be noise-reduced with the aid of a Quantel digital frame-store processor. With this imaging system, the optimum MCP bias voltage proved to be 800V (1100V gave too much gain), and this bias was used for all of the examples given in Figs.4 to 6. The values of electron current density, spatial resolution, etc., given below refer to the values at the MCP. A cross-grating replica (2160 lp mm^{-1}) was imaged for 1.5×10^{-14} Acm^{-2}, 100 and 1000kV, and two different magnifications, all without noise reduction, and is shown in Fig.4. A Cu-Ni thin foil specimen, onto which small Au islands up to 300Å in diameter had been deposited, was imaged for 5×10^{-14} Acm^{-2} and 1000kV, with and without noise reduction, and is shown in Fig.5. The same area recorded in the standard manner using a photographic plate, but with the incident electron current increased to 1×10^{-10} Acm^{-2}, is also shown.

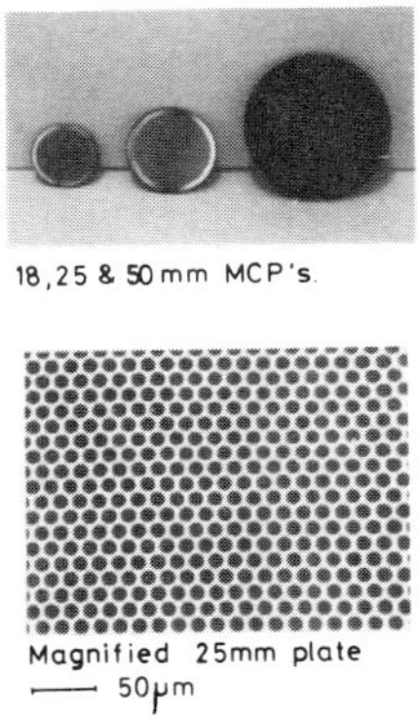

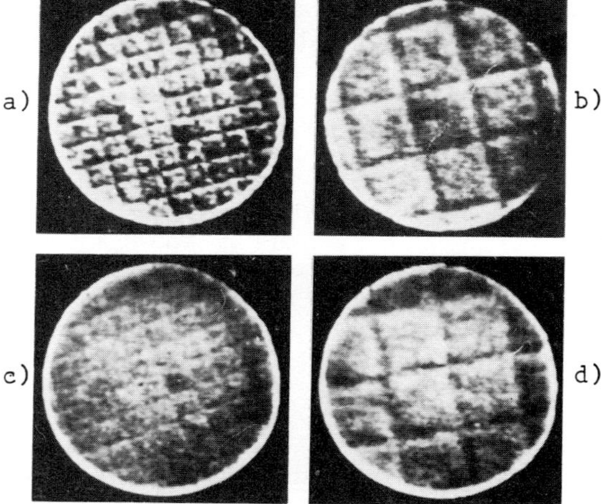

18,25 & 50 mm MCP's.

Magnified 25mm plate
—— 50μm

Fig.1.

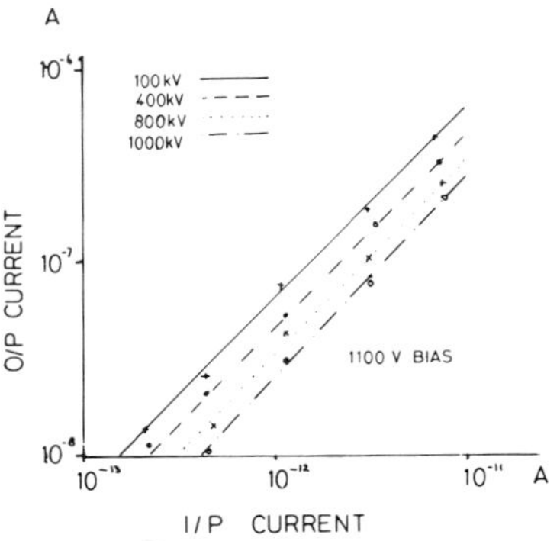

Fig.2a.

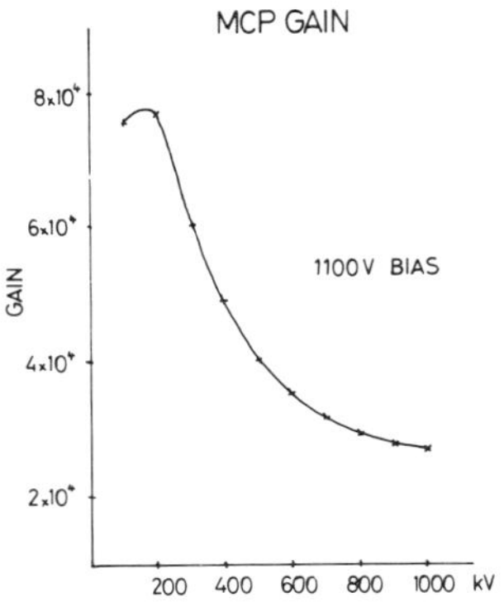

Fig.2b.

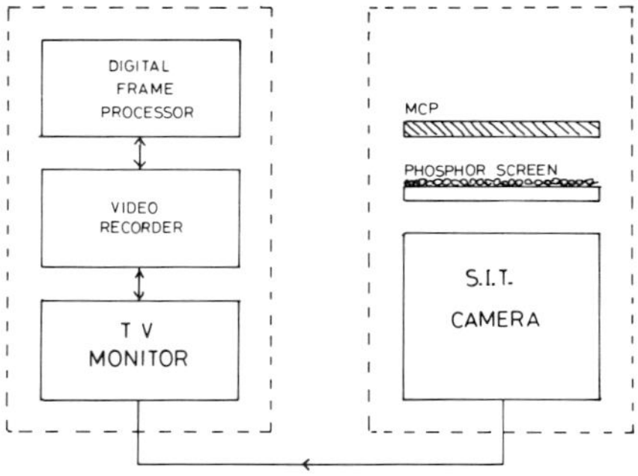

Fig.3.

Fig.4. Cross-grating replica, 2160 lp mm^{-1}, imaged using MCP, 1.5x10^{-14} Acm^{-2}. a) and b) at 100kV, c) and d) at 1000kV.

If the current had not been increased, the calculated equivalent plate
exposure time would have been ∿40min. The area imaged with the MCP shows
individual dislocations (the dark lines). The definition is better in the
noise-reduced image, where the larger of the Au islands can also be distinguished. A thin portion of a leaf specimen was imaged for 1×10^{-13} Acm^{-2} in
Fig.6. Typical areas are compared at 100 and 1000kV, without noise reduction, in Figs.6b and c. A single nucleus is compared at 100kV, with and
without noise reduction, in Figs.6d and e. A typical area recorded using a
photographic plate at1000kV (current increased to ∿1×10^{-10} Acm^{-2} or the
exposure time would have been ∿20min.) is also shown. It was not possible
to obtain an analogous photographic plate image at 100kV because the contrast was too low.

From these images obtained using the MCP it can be seen that the spatial
resolution is ∿500μm point-to-point (∿150μm on the 'edge' definition) at
both 100 and 1000kV, and that the contrast decreases on going from 100 to
1000kV. This spatial resolution is determined by (a) the primary beam
exciting additional channels as it penetrates the MCP due to the channel
bias angle of 8°, (b) the primary beam spreading as it penetrates the MCP,
(c) secondary electrons emitted from the input surface of the MCP entering
adjacent channels, and (d) spreading effects associated with back-scattered
electrons and X-rays. For a 25mm diameter MCP, this resolution corresponds
to ∿ 50 pixels, and this would be improved with a 75mm diameter MCP to ∿150
pixels. Experiments are in-hand to improve the present resolution so that
the performance is equivalent or better than a 312-line TV camera. The
contrast decreases on going from 100 to 1000kV because of (a) reduced electron scattering in the specimen, (b) primary electrons penetrating the MCP
and producing background, and (c) high-energy X-rays generated at the MCP
and phosphor screen producing background. A particular kV can often be
selected to give optimum contrast. For example, for the leaf specimen of
Fig.6, ∿300kV would probably have been optimum.

Although the quality of the images obtained at present using an MCP are
inferior to those recorded using standard photographic procedures, the
increased sensitivity enables images to be observed with incident electron
current densities ∿10^{3}x less than normally used, and this has considerable
advantages for beam-sensitive materials such as polymers and biological
specimens, e.g. the leaf specimen of Fig.6. In its present form, the MCP
system could be used with such specimens for preliminary observations,
focussing, etc., followed by conventional photographic recording. Alternatively, the MCP could be used directly in combination with a photographic
plate, e.g. an MCP in conjunction with a Kodak 4489 plate would be ∿10^{2}x
faster than Kodirex X-ray film. Later developments with the MCP system may
provide equivalent recorded image quality in addition to the much improved
sensitivity.

The authors wish to thank EMI-Varian for supplying the MCPs,
Dr. C.R.M. Grovenor for the Cu-Ni alloy specimen, and Dr. C.R. Hawes for
the leaf specimen.

A
0.1µm
Cu–Ni alloy
PHOTOGRAPHIC PLATE
B
C
1MeV
5x10⁻⁴ Acm⁻²
0.1µm
MCP
MCP + NOISE REDUCTION

Fig.5.

A
LEAF
1µm
PHOTOGRAPHIC PLATE 1MeV
B
C
1µm
MCP 1MeV
MCP 100keV
10⁻¹⁸ Acm⁻²
D
E
1µm
MCP 100keV
MCP 100keV + NOISE REDUCTION

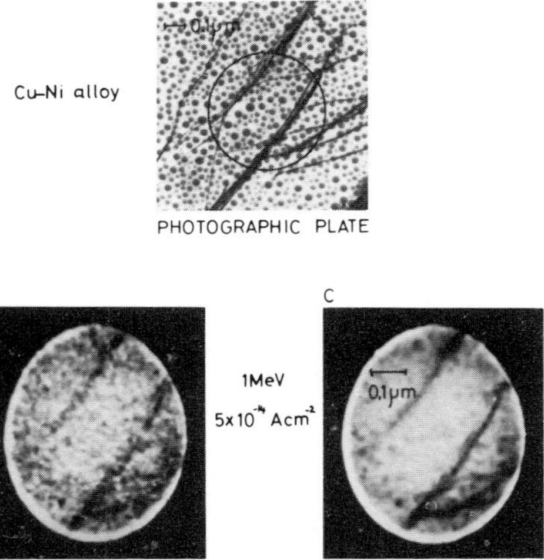

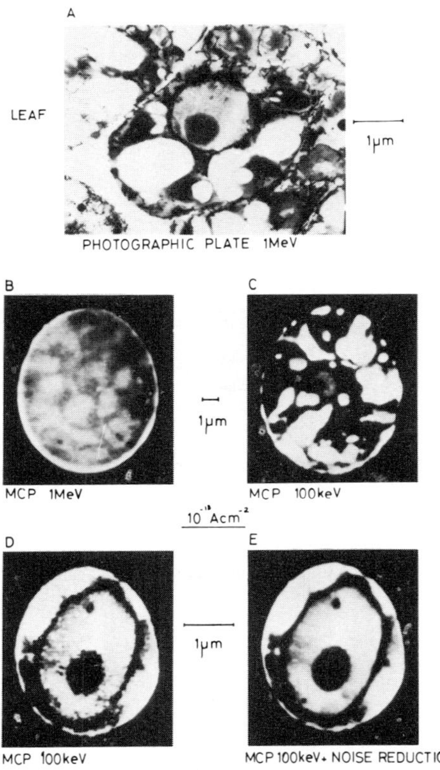

Fig.6.

Two CCD-based detector systems for use in electron microscopy and related techniques

J N Chapman, F Glas[†] and P T E Roberts

Department of Natural Philosophy, University of Glasgow, Glasgow G12 8QQ

1. Introduction

In previous papers (Chapman, Roberts, MacLeod and Ferrier, 1980a and b) we have described the use of a charge coupled device (CCD) for recording images in the transmission electron microscope. The device used was a Fairchild CCD 202 with its protective window removed and, in the experiments we have performed, the electron image was directly incident on the front face of the device. When operated in this way the CCD offers a recording system with a high sensitivity, excellent linearity up to a well defined saturation limit and direct accessibility of quantitative image intensities in the form of a series of voltage levels. It has two main disadvantages: the first is its limited dynamic range reflecting the modest storage capacity of the cells of the device and the second is that after prolonged irradiation by the electron beam the performance of the device deteriorates. The rate at which this degradation proceeds is a function of the incident electron energy and for a 100 keV beam the device is rendered effectively useless after a dose of $\sim$ 1.5 x 10^7 electrons/photocell (or $\sim$ 5 x 10^{-7} C.cm^{-2}). Before such a dose is reached it should be possible to record 10^3-10^4 electron images with reasonable statistical accuracy. However, if the device is to be used not only to record images but also to examine the specimen generally and as an aid to focussing and astigmatism correction, a device with a longer lifetime is required.

In this paper we describe briefly the way the performance of the CCD worsens with radiation dose and the nature of the damage mechanisms. It is then shown that by annealing the device an almost complete recovery of performance is possible and details of the temperatures and times required for the CCD 202 are given. Whilst the fact that the damage is reversible greatly extends the use of the CCD for image observation in the electron microscope, the problem of limited dynamic range still remains. This is particularly acute if diffraction patterns rather than electron images are to be recorded. For a device with a fixed cell capacity this problem may only be overcome by reducing the gain of the detector system as a whole. Following the method used by Jones et al (1977) we have attempted this by operating the CCD in an 'indirect' mode in which the electron signal is converted, using a transmission phosphor screen, to its photon counterpart. This is subsequently imaged onto the CCD. The device used in this aspect of the work is a one dimensional array, the Fairchild CCD 122, and details of a detector system based on this device and suitable for recording diffraction patterns are given in the final section. As the CCD is not

† Permanent address: CNET, 196 rue de Paris, 92220 Bagneux, France

exposed to ionising radiation in the 'indirect' mode there is no deterioration in performance with extensive use.

2. Radiation Damage and its Removal in the CCD 202

Two main damage effects are seen in the Fairchild CCD 202 as a result of electron irradiation in the electron microscope. The first is an increase in dark current with total electron dose whilst the second and more serious effect is a decrease in the maximum usable output voltage of the device. With an undamaged CCD 202 the dark current is negligible at room temperature for a cycle time of 40ms. By cycle time we mean the time interval between successive removals of charge from the potential wells which make up the device. However, after an electron dose $\sim$ 5 x 10^6 100 keV electrons/ photocell ($\sim$ 1.5 x 10^{-7}C.cm.$^{-2}$) the dark current generated in 40ms is sufficient to half-fill the photocells. It arises from an increased number of interface states within the energy gap of silicon. Despite the magnitude of the dark signal at room temperature, its effect is of no great consequence, for by cooling the device so that its temperature lies within the range 250-275K the dark current is again reduced to negligible proportions.

The second effect, which ultimately limits the useful lifetime of a device, arises from a decrease in transfer efficiency, the efficiency with which charge packets are removed from individual cells and transferred to the output amplifier. It manifests itself as a decrease in output signal from cells which have suffered radiation damage but is only apparent if the content of the cell exceeds a critical value. That critical value decreases with increasing electron dose and falls to one half of the initial saturation value after a dose $\sim$ 1.5 x 10^7 100 keV electrons/photocell. The cause of the increasing transfer inefficiency is related to the build-up of positive charge (mainly holes) in the oxide layer close to the silicon-silicon dioxide interface which has the effect of changing the operating conditions of the device. Application of a greater external voltage to the electrodes overcomes the adverse effect initially, but as the positive charge builds up so do electric fields within the insulator and ultimately field induced channelling causes a drop in device efficiency.

To overcome the deleterious effects described in the preceding paragraphs it is necessary to remove the interface states and the build-up of positive charge. Following reports in the literature (e.g. Zaininger and Holmes-Siedle, 1967) that the effects of radiation damage can be completely removed in MOS devices by thermal annealing, the performance of two initially completely damaged CCD 202s was studied after one hour anneals at a range of temperatures. The variation of the room temperature dark voltage and of the maximum usable output voltage, both expressed as fractions of the initial saturation voltage, are shown after each anneal in figure (1). Little change was seen until annealing temperatures $\approx$473K were used, but thereafter the dark current fell rapidly so that after annealing at 513K it was unmeasurable (< $\frac{1}{2}$% of saturation). The maximum usable voltage also improved although in this case the lightly damaged device required an annealing temperature of 573K before a complete restoration of performance was achieved and the heavily damaged area only showed a 70% recovery after annealing at 643K. Subsequent exposure of annealed devices in the electron microscope suggests that damage recurs in a similar way to that initially experienced and that further annealing for several hours at 643K again restores the performance. Such results are encouraging and indicate that only the limited dynamic range is a fundamental drawback to the use of CCDs in the direct bombardment mode in the electron microscope.

3. A CCD Based Detector System for Electron Diffraction Patterns

When the CCD 202 is irradiated with 100 keV electrons the mean output sig-
nal/incident electron is approximately 2% of the saturation voltage. This
is clearly unsatisfactory if diffraction patterns, which frequently have a
dynamic range in excess of 10^3, are to be recorded. Even if the gain of
the system were reduced, the CCD 202 would not prove ideal as its measured
dynamic range using the drive circuitry provided by the manufacturers is
limited by noise to 300:1. An intrinsically more suitable device is the
one dimensional Fairchild CCD 122 where the equivalent figure is 1500:1.
Furthermore, by use of more sophisticated drive and detector circuitry a
dynamic range in excess of 10^4 has been reported (Wen, 1975).

The simplest way of reducing the gain of the detector system is to introduce a transmission phosphor screen between the electron beam and the CCD.
As the cell size in the CCD 122 is 13μm x 13μm and the resolution of most
phosphors used in conjunction with 100 keV electrons is ≈50μm a demagnification of the optical image by a factor of 3-4x is required. This may be
achieved most readily by using a lens. The fraction of photons generated
in the phosphor which reach the CCD may be estimated to be $(M/2\sqrt{2}F(M+1))^2 t$
where F is the ratio of the focal length to diameter of the lens, M is the
magnification and t is the product of the transmission coefficients for the
lens and glass plate on which the phosphor is deposited. With an F1 lens
this figure is typically 5×10^{-3}. The phosphor used in this work is a
P22 Y_2O_2S;Eu phosphor with an estimated energy efficiency ≈15%. Combining
this information with the expected CCD responsivity of approximately
$4V/(\mu J.cm^{-2})$ suggests that the output signal from the device should be
30μV/ electron incident on the phosphor. Such a figure is rather lower than
the optimal value where the signal from a single electron would be comparable to the noise level. Its magnitude reflects primarily the rather poor
photon collection efficiency which in turn is a consequence of the demagnification required to match the phosphor resolution to the CCD cell size.
Clearly a device with a larger cell size would be desirable.

Despite these reservations a recording system with the specification described above should be of use in recording quantitative electron diffraction
intensities. Experimentally, the output from the CCD was found to vary
linearly with the current incident on the phosphor screen over a wide range
of current densities and integration times. The response/incident electron
was measured to be ≈30% of the value calculated above which suggests that
the device used has a below average responsivity, that the thickness of the
phosphor was chosen incorrectly or that losses in the optical system were
greater than expected. Integration times ranging from 14ms to 5s could be
used and provided the device was cooled to 273K the peak to peak noise level
remained < 1mV. Cooling was provided by a Peltier element.

Figures (2) and (3) show experimental results obtained with the system. Use
of a one dimensional CCD is not restrictive provided that the diffraction
patterns of interest are radially symmetric which is the case for small
grain polycrystalline and amorphous films. In the polycrystalline cobalt
film, indexing of the clearly visible rings shows that its structure is fcc
rather than hcp. The amorphous silicon specimen is of rather more interest
and illustrates the need for a recording system with a wide dynamic range.
A full structural characterisation of these films for correlation with
their preparation conditions requires both low and high angle scattering
information. Collection of this information using either photographic
emulsion or the very limited dynamic range of a CCD in the 'direct' mode

Would be both difficult and prone to error.

Chapman J N, Roberts P T E, MacLeod A M and Ferrier R P 1980a, Proc
  EMAG 79, pp77-80; 1980b, Proc 7th European Congr on EM Vol 1, pp90-91
Jones B L, Jenkins D G, Booker G R and Fry P W 1977, Proc EMAG 77, pp73-76
Wen D D 1975, Proc CCD 75 pp109-119
Zaininger K H and Holmes-Siedle A G 1967, RCA Review 28 108

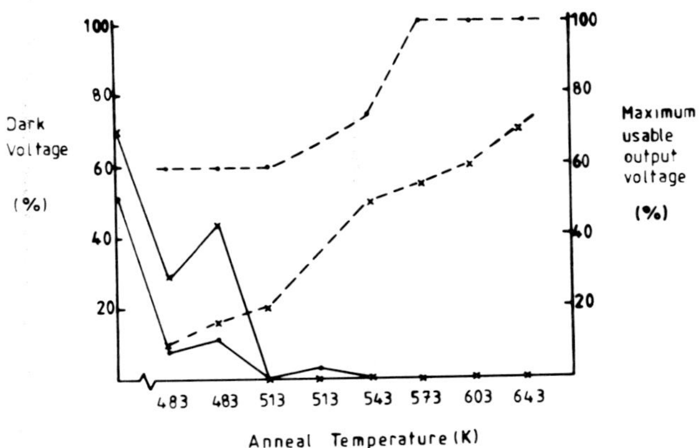

Figure (1): Variation of dark voltage (solid lines) and maximum
usable output voltage (dashed lines), expressed as percentages of
the initial saturation voltage, with anneal temperature. The dots
and crosses refer to lightly and heavily damaged devices respectively.

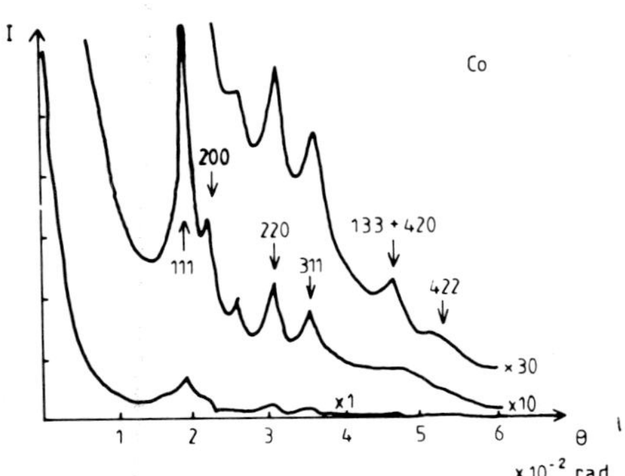

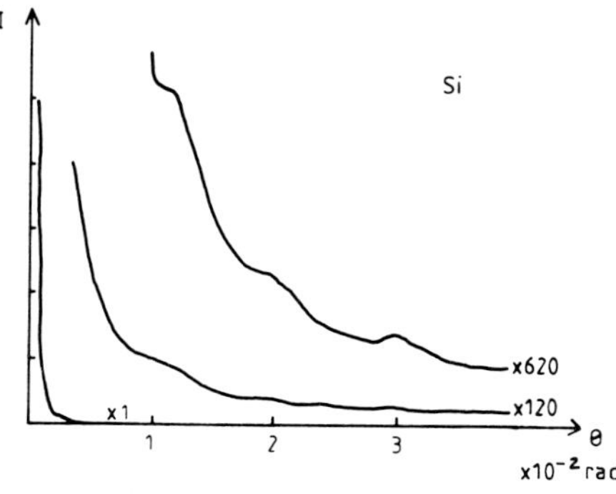

Figure (2): Variation of intensity
with scattering angle from an fcc
polycrystalline cobalt foil.
The Bragg reflections are indexed.

Figure (3): Variation of
intensity with scattering angle
from an amorphous silicon foil.
The pattern was obtained from a
foil area $\approx 1\mu m^2$ and the maximum
integration time used was 4.4 s.

Developments in the use of one- and two-dimensional self-scanned silicon photodiode arrays as imaging devices in electron microscopy

B L Jones, D M Walton and G R Booker
Department of Metallurgy & Science of Materials,
University of Oxford, Parks Road, Oxford OX1 3PH

Work on the use of 1-D and 2-D self-scanned Si photodiode arrays for
directly displaying electron microscope images (Jones et al 1977, Jenkins
et al 1978, 1980, and Walton et al 1980) has continued and the present
paper summarises some of the results. The 1-D arrays consisted of either
64, 96 or 128 diodes, and the 2-D arrays of 4096 (64x64) diodes (Table 1).
Much of the scanning circuitry was fabricated on the Si array chips. Individual arrays were positioned in the electron microscope so that on raising
the conventional viewing screen, the electrons fell directly onto the array.
The arrays were generally cooled to -40°C to reduce the diode dark current.
Each diode of the array produces an electrical output signal corresponding
to the incident local electron current density. Each diode collects charge
for a particular integration time τ, the charge for each of the diodes is
then rapidly read-out, and the process automatically repeats. The 1-D
arrays display a Y-modulation image (e.g. Fig.3), and the 2-D arrays display a brightness-modulated image (e.g. Fig.4), on an external monitor
viewing screen. The arrays were continuously operated while being electron
irradiated. For high electron current densities, τ was typically 1 to
100ms, giving continuous TV-type images. For low current densities, τ was
typically 0.1 to 10s, giving either intermittent images on a standard
oscilloscope, or continuous images on a storage oscilloscope.

Sensitivity   Values of τ required just to produce a 100% diode output
signal (saturation) for a current density of 10^{-11} Acm^{-2} are given in Fig.1a
as a function of beam voltage for both the 1-D and 2-D arrays. The corresponding values of diode collection efficiency β calculated from $\beta=3.6q/IAV\tau$,
where q is the diode saturation charge, I is the current density, A is the
diode area and V is the beam voltage, are given in Fig.1b. The decrease in
β at low voltages is due mainly to losses occurring in the surface oxide
layer, and at high voltages to the deep generation of carriers in the Si
chip, with some beam penetration of the chip occurring for V > $\sim$300kV.
Linearity   For individual 1-D and 2-D arrays, when operating with fixed τ
and V values, the output signal was closely proportional to the I value.
When operating with a fixed V value, the τ value required just to produce
saturation was closely inversely proportional to the I value over a range
of several orders of magnitude (Fig.2).
Dynamic range   For 1-D and 2-D arrays, when operating with fixed τ and V
values, the dynamic range was $\sim$100:1. By operating at different τ values,
the dynamic range could be further increased by factors of either $\sim$10^{3}X
with the array at room temperature, or $\sim$10^{5}X with the array at $\sim$-40°C.
The greater range for the cooled array occurred because the decreased diode
dark current enabled τ to be extended to larger values.
Resolution   The spatial resolution of the displayed image generally
corresponded to the centre-to-centre diode spacing of the array. The full

TABLE 1

ARRAY	TYPE	NUMBER OF DIODES	DIODE SIZE (μm)	CENTRE-TO-CENTRE SPACING (μm)	SATURATION CHARGE (pC)
IPL4064	1-D	64	100x80	100	7
IPL156	1-D	96	800x80	100	30
IPL4128	1-D	128	100x80	100	7
IPL2D1	2-D	64x64	38x30	75	2

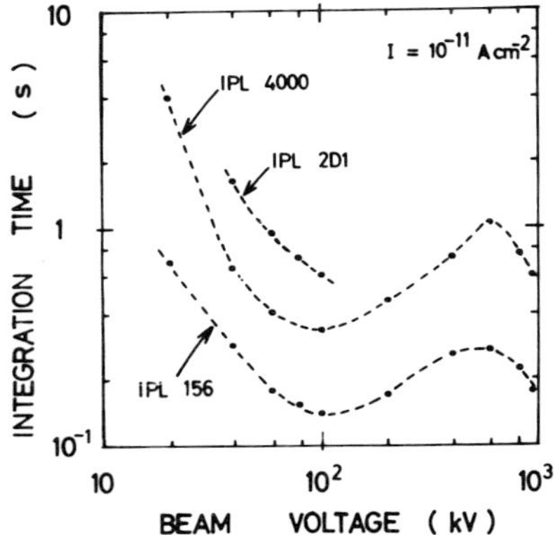

Fig 1a. τ_{sat} v V for I=10^{-11} Acm^{-2}.
Type 4064 and 4128 (same), 156 and
2D1 arrays.

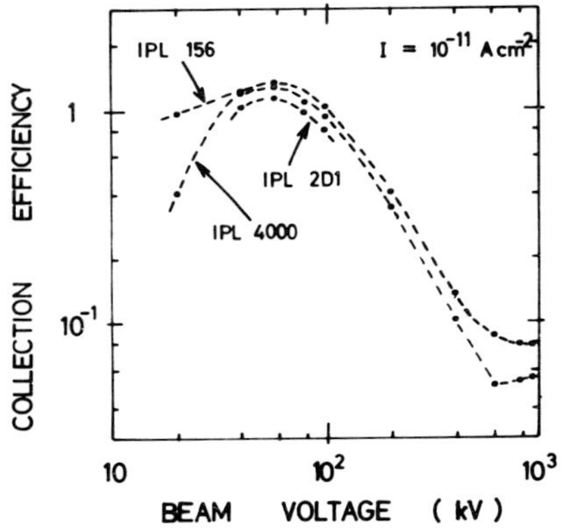

Fig 1b.   Corresponding β v V.

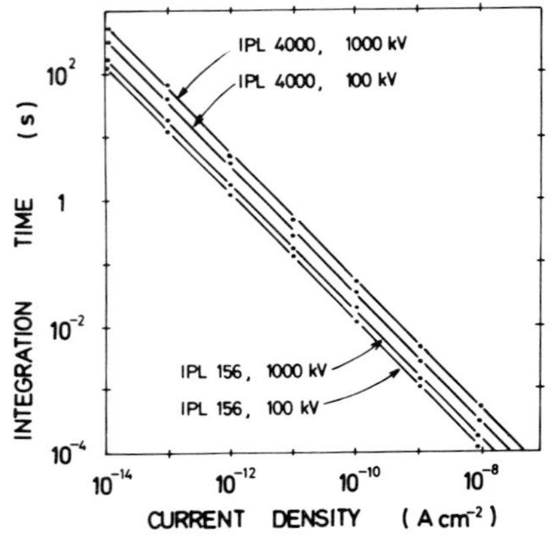

Fig 2a. τ_{sat} v I for V = 100 and
1000kV. Type 4064 and 4128
(same), and 156 arrays.

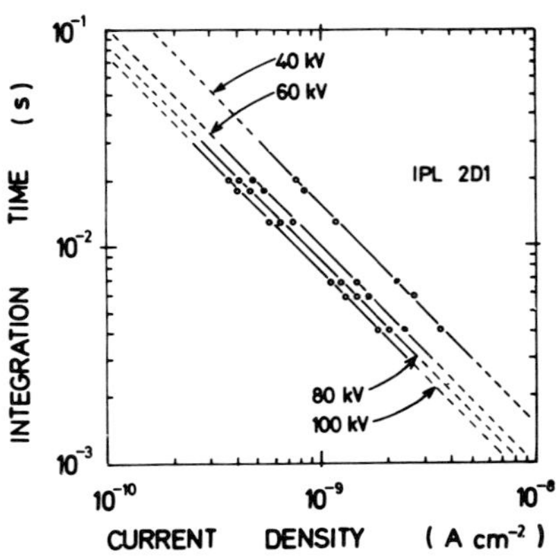

Fig 2b. τ_{sat} v I for V = 40, 60,
80 and 100kV.   Type 2D1 array.

resolution of the electron microscope image could always be revealed if required by electron optically increasing the magnification of this image at the array, although this caused a decrease in I and hence an increase in τ, and also a decrease in the field of view. For the arrays used, beam spreading effects in the Si chip caused no significant loss of spatial resolution over the beam voltage range 20 to 1000kV.

<u>Degradation</u> For the 1-D arrays, when the scanning circuitry (which included MOSTs) on the Si chip was masked with metal strips from the incident electron beam, no significant degradation occurred. When the scanning circuitry was not masked, no significant degradation occurred with 'average' daily use if $I < \sim 10^{-10}$ Acm^{-2}, and any degradation that did then arise could be subsequently annealed-out, e.g. 5 min at 300°C. For the 2-D array, the scanning circuitry at the edges of the Si chip was masked with metal strips, but the MOSTs fabricated in between the diodes (1 MOST per diode) were not masked. These arrays could be irradiated up to a dose of $\sim 5 \times 10^{11}$ electrons cm^{-2} (e.g. ~ 8000s at 10^{-11} Acm^{-2}) before any significant degradation occurred. The degradation that did then arise could be either subsequently annealed-out, e.g. 5 min at 300°C, or alternatively electron-beam 'annealed' in-situ by continuing the electron beam irradiation but with the electrical power supply disconnected from the array. Irreversible degradation occurred after only a few seconds for the 1-D arrays when $I > \sim 10^{-9}$ Acm^{-2}, and for the 2-D arrays when $I > \sim 10^{-8}$ Acm^{-2}. Degradation effects were not significantly greater at 1000 than 100kV.

<u>Applications</u> Typical τ values required for the 1-D and 2-D arrays to obtain different types of electron microscope images at 100kV are given in Table 2 and compared with corresponding typical photographic plate exposure times. A 1-D array image (1000kV, 10^{-11} Acm^{-2}, 0.4s) of bend contours in an Al-Cu alloy foil is compared with conventional photographic images of the same area in Fig.3a. 1-D array images of an electron energy loss spectrum (EELS) for a Si foil are shown in Fig.3b (80kV). The two τ values of 11ms and 2.7s corresponding to 5×10^{-10} and 2×10^{-12} Acm^{-2} respectively reveal the zero-loss and first plasmon-loss peak, and the L_{23} - shell edge, respectively. 2-D array images (100kV, 3×10^{-10} Acm^{-2}, 30ms) of a hole in a carbon film showing a through-focus series, and of a line diffraction-grating replica, are compared with conventional photographic images in Fig.4.

<u>Advantages</u> These arrays directly display electron microscope images and have high sensitivity, good linearity, large dynamic range and reasonable spatial resolution. With simple masking and sensible use, degradation effects are not a serious problem, and are insignificant for medium- to-low electron current densities. Because the output is in the form of an electrical signal, images can be stored on magnetic tape for later processing or 'on-line' processed and stored.

The authors wish to thank Integrated Photomatrix Ltd, Dorechester, Dorset, England, for supplying the arrays and for valuable discussions.

References

Jenkins D G, Rossouw C J, Jones B L, Booker G R and Fry P W 1978 Electronics Letters <u>14</u> 174
Jenkins D G, Rossouw C J, Booker G R and Fry P W 1980 Inst. Phys. Conf. Ser. No. 52, p 81
Jones B L, Jenkins D G, Booker G R and Fry P W 1977 Inst. Phys. Conf. Ser. No. 36, p 73
Walton D M, Jones B L, Booker G R and Fry P W 1980 Inst. Phys. Conf. on Data Collection in Electron Microscopy & Analysis, City University London

TABLE 2

ELECTRON CURRENT DENSITY (Acm^{-2})	TYPE OF IMAGE	ARRAY INTERGRATION TIME	PHOTOGRAPHIC PLATE EXPOSURE TIME
10^{-9}	zero and plasmon-loss spectra	3ms	0.3s (slow film)
10^{-10}	bright-field micrographs	30ms	2s
10^{-12}	weak-beam micrographs, K- and L-loss spectra	3s	30s (fast film)

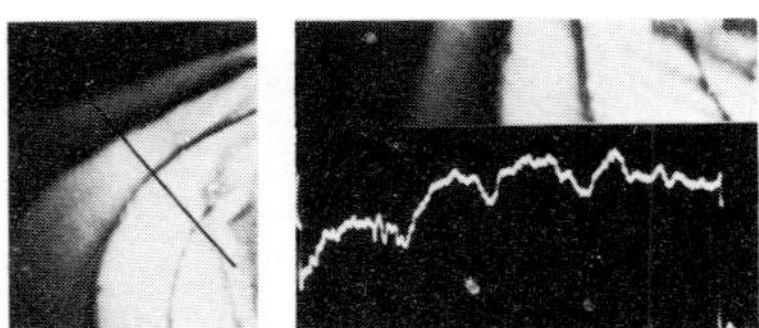

Fig.3a. Bend contours in Al-Cu foil. Type 4128 1-D array (bottom right).

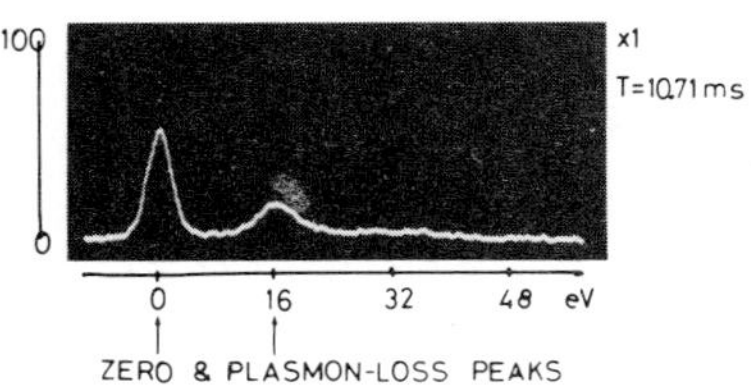

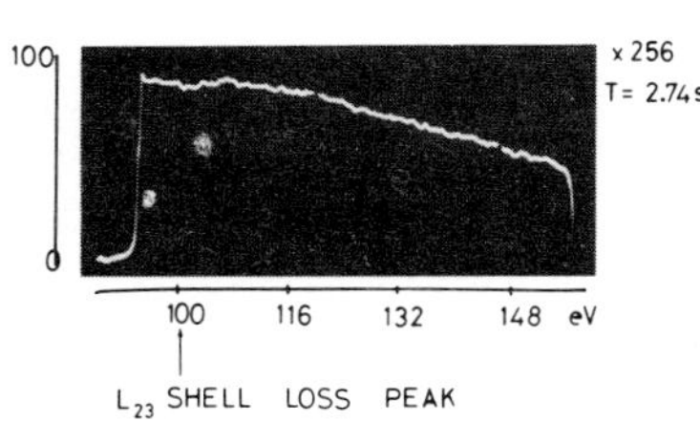

Fig.3b. EELS from Si foil. Type 4128 1-D array.

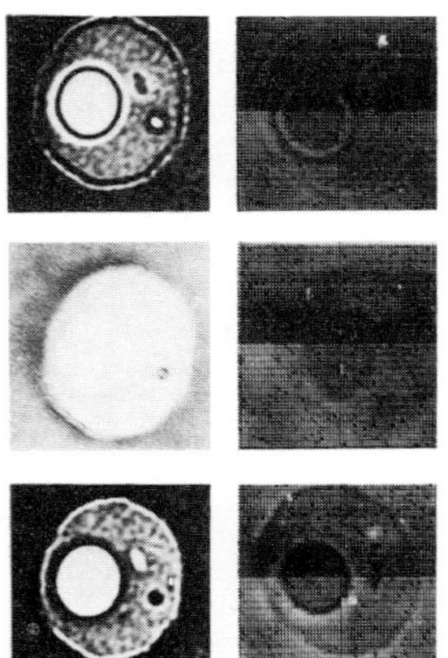

Fig.4a. Hole in carbon film. Type 2D1 2-D array (right).

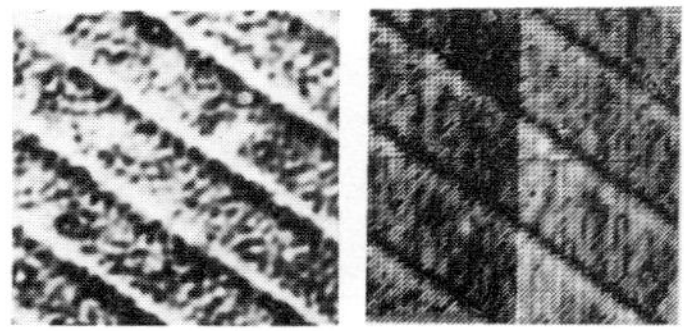

Fig.4b. Line diffraction-grating replica. Type 2D1 2-D array (right).

Computer control and instrumentation for EELS in a multi-spectrum acquisition system

P. Statham

Link Systems Ltd., Halifax Road, High Wycombe, Bucks. HP12 3SE

In an attempt to provide a flexible means of controlling and monitoring
data acquisition for a variety of signals, a multi-spectrum acquisition
system has been developed using the Link 860 analyser which has a central
processor with typical minicomputer performance and 64K byte memory
capacity. The system, described in more detail by Statham and Watson
(1981), centres on a control console which is used to establish
connections to any one of a series of data memories (fig.1). Once
connected, the acquisition details, energy calibration and vertical
scale can be altered for any one memory independently of settings for
the others. Conceptually, the central console behaves like an
oscilloscope with two channels, A and B, which can be connected to
monitor and compare the contents of memories, providing shift and gain
controls for both horizontal and vertical axes on the display as well as
logarithmic and bipolar display modes. With memories A and B defined, a
series of mathematical operations (e.g. smooth, scale, add constant,
differentiate, integrate, A+B, A-B) can be performed and background
extrapolation with SIGMAK/SIGMAL partial cross section calculations
(Egerton, 1979) are provided for core-loss quantitation. For memory A,
spectra can be stored on diskettes, transferred to other computers via a
serial link, punched on paper tape, listed to a printer or plotted in a
variety of formats. Once acquisition details have been defined, a single
key can be pressed to start analysis in all memories. Similarly, all
analyses can be immediately suspended so that, if necessary, the beam
can be repositioned to compensate for drift before an experiment is
resumed.

Signal detection
Fig. 2 shows the instrumentation involved for serial acquisition of EELS
data where a magnetic or electrostatic field is used to select the
energy of electrons which pass through a slit and strike the
photomultiplier tube (PMT) scintillator. For very low PMT currents, it is
preferable to count individual pulses derived from a threshold
discriminator and by using a high-bandwidth discriminator and ECL logic
it is possible to detect pulses separated by 10ns. However, this does
not mean that electrons can be counted at 100MHz rate because electrons
arrive randomly in time according to a Poisson distribution. For a mean
rate of n Hz and a pulse-pair separation capability of t seconds,
roughly 100nt % of pulses go undetected so if t=10ns, a count rate of
1 MHz can be recorded with only 1% losses. In practice, scintillator
rise time provides the major contribution to t (Batson, 1978) and imposes
a more severe limitation to count rate capacity. For higher currents, a

voltage-to-frequency converter (VFC) provides a pulse train with rate
proportional to current which can conveniently be counted by the same
electronic scaler. The switch over between VFC and pulse counting modes
can be controlled by the computer and arranged to occur at a specific
energy; moreover, the same control flag could be used to switch PMT
voltage to optimise gain for the different detection modes.

Ramp control
With the provision of inputs for "start","end of scan" and "channel
advance", spectra can be acquired from microscopes where image scanning
circuits are used to generate the EELS energy ramp. However, greater
flexibility is afforded if the computer can control the ramp directly.
By using a 12-bit digital-to-analogue converter (DAC), channel-to-
channel step variations for a ramp with 1024 channels can be kept to
within about 10% and the output voltage can be used to control the ramp,
provided suitable precautions are taken to minimise ground loop and r.f.
pickup. By acquiring spectra with the ramp proceeding in a direction
first of increasing, then of decreasing energy, the spectrometer can be
tested for hysteresis and distortion due to the long-term effects of the
intense zero-loss signal on scintillator and photomultiplier will be
revealed (Joy and Maher, 1980). The ramp can be fixed at any desired
energy or positioned directly on the zero-loss peak to provide a signal
for imaging. The dwell time per channel can be varied in different
regions so that more counting time can be spent counting where the
signal is weak. Alternatively, the sweep can be programmed so that the
time (in clock pulses) taken to accumulate a preset number of counts is
recorded; the relative statistical precision will then be constant from
channel to channel and the reciprocal of the recorded counts will
correspond to the input spectrum. Drift in incident beam current can be
compensated by using an external clock, with frequency proportional to
beam current, instead of the computer clock, to control dwell times.

Mains interference
Power mains interference, introduced via ground loops for example, can
appear as either a modulation of selected energy, E, or of electron
intensity, R(E). By selecting suitable dwell times and with the ability
to synchronise to the mains cycle, such interference can be revealed and
sometimes eliminated. If the dwell time per channel, d, is much less
than the mains period, T, intensity modulation is quite obvious
throughout the spectrum in a single sweep but oscillations are obscured
with multiple sweeps unless synchronisation is used. When d > T,
variations in R(E) can appear as a ragged pattern which may easily be
confused with statistical noise, especially since for N unsynchronised
sweeps, the observed magnitude of the pattern will decrease roughly as
$N^{-\frac{1}{2}}$. With d=T, or an exact multiple, intensity modulation will be
nullified. By comparison, if energy selection is affected, the result
depends on the form of the input spectrum. In a region of constant
slope, interference will be essentially indistinguishable from that
produced by intensity modulation but will have magnitude proportional to
the gradient, (dR/dE). However, for a sharp spectral feature, resolution
will be degraded for d $\gtrsim$ T whereas the position will be shifted if t<<T,
synchronisation being necessary to keep the shift constant for
sucessive sweeps to preserve resolution.

Drift correction
Energy drift can be compensated by examining the position of the zero-
loss peak after each sweep before adding the data to memory (Egerton and

Kenway, 1979). However, care is required to ensure enough counts are recorded in each sweep to accurately define a maximum; although statistical errors are easy to predict with pulse counting, with the facility to display the apparent zero position at the end of each sweep, statistical jitter can be readily observed and the experiment modified accordingly, even when using a VFC for intensity measurement.

Digital mapping

For x-ray mapping, beam position on the specimen can be digitally controlled and x-ray spectra recorded on a grid of points over the field of view, integrals from up to 18 regions of interest being stored for every grid point to give up to 18 digitised images. These images can be processed to obtain quantitative information and multiple processed images can be displayed on a colour monitor (Statham and Jones, 1980). This facility has been extended to include EELS; at each point on the specimen up to 5 regions of interest, each with appropriate dwell time, can be scanned in the EELS spectrum and the integrals for each region recorded. In order to obtain the images shown in figures 3 to 6, regions were defined to cover the zero loss, second plasmon peak and energy bands immediately before and after the Nd NIV/V edge position and the beam was scanned over a 128 x 128 grid on the specimen, covering a field of view roughly 0.5μm wide. The rare-earth oxide specimen was analysed using a Vacuum Generators HB5 microscope and spectrometer, two runs being performed, one optimised for the low loss regions taking 2 minutes and the other optimised for the edge regions taking about 4 minutes. The pseudo-grey-scale images shown were generated on a dot-matrix printer and have been subjected to several stages of copying and reduction but the salient features should still be visible. In fig. 5 the background beneath the Nd edge has been removed by subtracting a constant fraction (0.66) of the pre-edge integral for every point in the image. Although the zero-loss image (fig. 3) is markedly different from that for the raw edge integral (fig. 4), the net core-loss image (fig. 5) is rather similar, presumably because the main contrast mechanism, diffraction, is preserved by selecting mainly singly-scattered events. Although division by the zero loss image might be expected to yield a concentration map of atoms/unit area, fig. 6 shows that the possible presence of a variety of contrast mechanisms can complicate interpretation. Although this may seem discouraging, the digital format allows the relationship between images to be examined quantitatively and the danger of false interpretation of results from spot analyses can to some extent be avoided.

Acknowledgements

I am indebted to Drs. S. Pennycook and D. Macmullan for assistance in the mapping experiments and to J. Watson for designing the EELS interfaces.

References

Batson P. E. 1978 Ultramicroscopy 3 361
Egerton R. F. 1979 Ultramicroscopy 4 169
Egerton R. F. and Kenway D. 1979 Ultramicroscopy 4 221
Joy D. C. and Maher D. M. 1980 Scanning Electron Microscopy/80 1 25
Statham P. J. and Jones M. 1980 Scanning 3 168
Statham P. J. and Watson J. 1981 Proc.Quantitative microanalysis with high
    spatial resolution (to be published in Metals Soc. book 277, chap. 14)

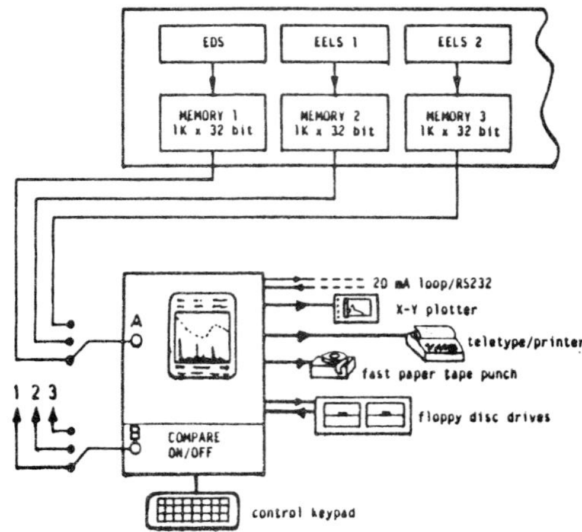

Fig 1 System layout

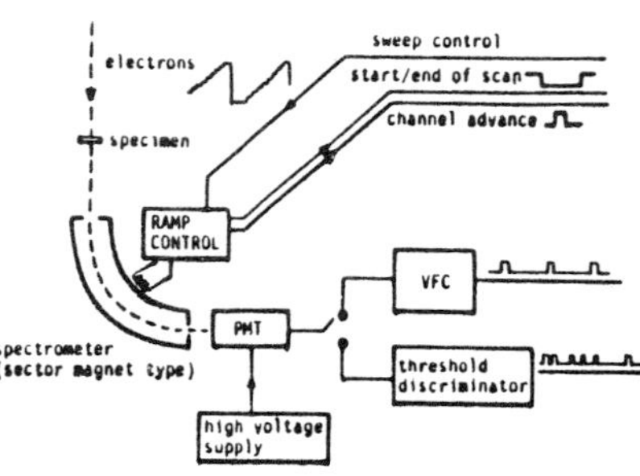

Fig 2 Instrumentation for EELS
       data acquisition

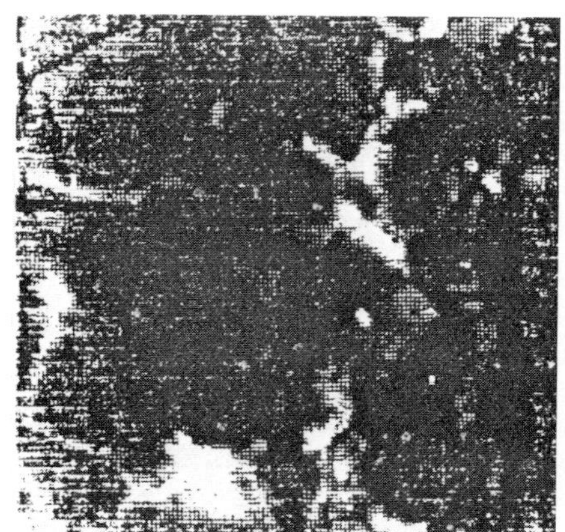

Fig 3 Zero-loss image (o.5μm field)

Fig 4 Nd N IV/V edge integral

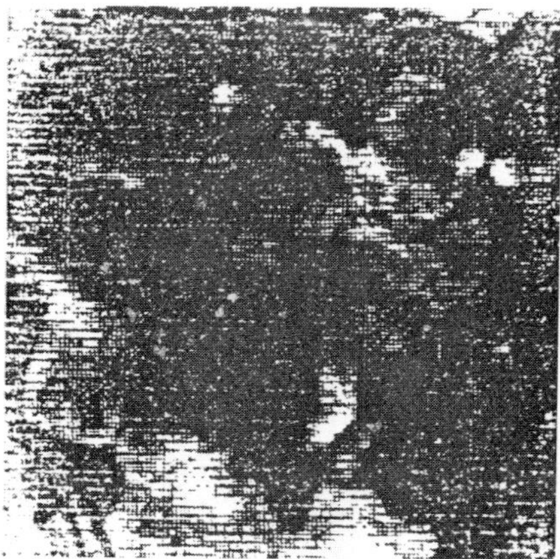

Fig 5 Nd N IV/V edge-background

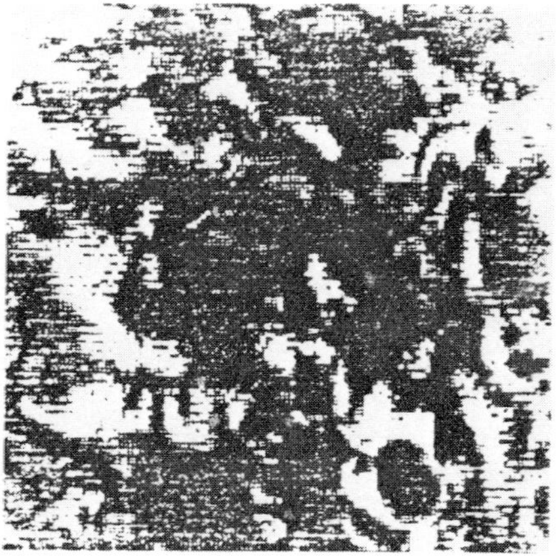

Fig 6 (Nd N IV/V edge-background)
       divided by zero-loss signal

A program suite for a computer-controlled scanning Auger electron microscope

R Browning, D C Peacock and M Prutton

Department of Physics, University of York, Heslington, York YO1 5DD, UK

1. Introduction

Because many adjustments are needed in setting up a SAM for a particular experiment it is advantageous to place such an instrument under computer control. The maximisation of the signal to noise (S/N) ratio dictates that the Auger data is collected in a digital form and it is therefore convenient to read it into a computer. For these reasons the York SAM has been interfaced to a dedicated HP9845S desktop computer and an integrated suite of programs has been written to facilitate control of the instrument as well as collection, storage and subsequent processing of spectra, line-scans and images. Although the program suite has not been designed to be easily transferred for the control of instruments other than the York SAM this paper is presented in the hope that some general principles applicable to many microscope/computer systems can be elucidated.

2. Description of the instrument

The microscope electron beam scanning and positioning and the analyser energy can be set by output from the computer which also inputs the digital electron energy analyser and photomultiplier signals (Fig 1). This is achieved with a set of transistor-transistor logic (TTL) boards designed at York and linked to the HP9845S by a 16 bit and a binary coded decimal (BCD) interface. Images can be routed to four cathode ray tube (CRT) displays, archived on a reel-to-reel tape or transferred to a multiuser computer if very time consuming processing is required. The HP9845S itself incorporates a CRT display used for communication with the user, a thermal printer for producing hard copy of results and two tape cassette drives; one used for the program suite tape; the other for data tapes.

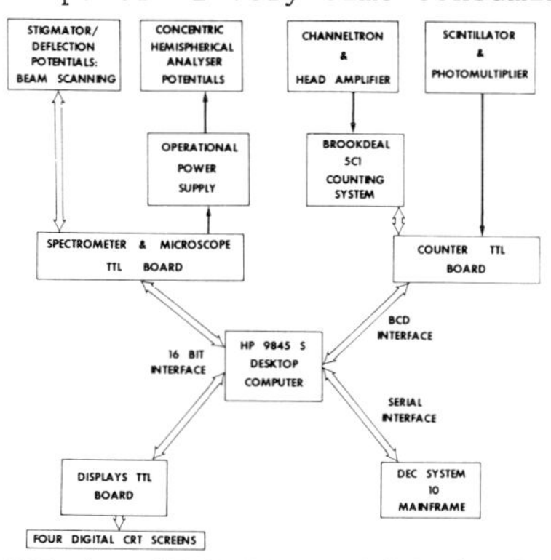

FIG 1: SIMPLIFIED SCHEMATIC DIAGRAM OF SAM CONTROL AND DATA COLLECTION ELECTRONICS

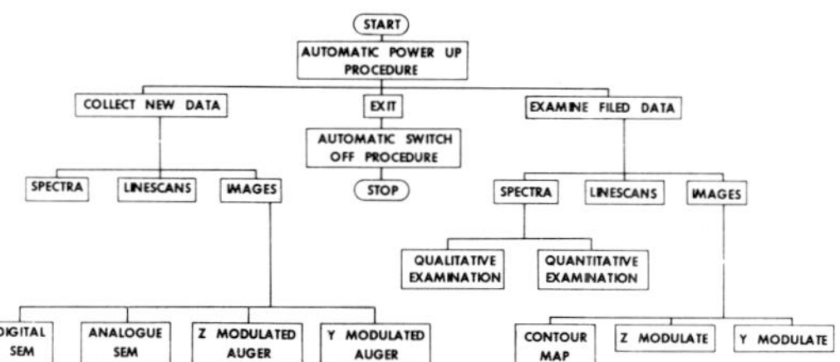

FIG 2: SIMPLIFIED PROGRAM SUITE DECISION TREE

The program suite is written in the BASIC computer language which offers powerful graphics and string handling capabilities plus easy to use input and output instructions to drive the interfaces. Relatively long compile and execution times associated with the use of BASIC, as compared with some other computer languages, are not serious compared with the dwell times necessary for acceptable S/N ratios in Auger electron images. Using the present instrument a 128 x 128 point Auger image with a 20 ms per pixel counting period can be collected in 7.5 minutes with 25% of this time being due to computation. By writing some repetitive parts of the collection routines in assembly language this percentage will be reduced in future.

3.   The structure of the program suite

At present the HP9845S has 64 k bytes of random access memory so that in order to handle image arrays of up to 256 x 256 pixels it is necessary to minimise the number of program lines in memory at a given time. This has been achieved by breaking down the program suite into logical sections which are stored as data. Each section, which constitutes an option at the end of a decision tree branch (Fig 2), is linked into memory as required while program lines that are no longer needed are written over. When the computer is first turned on a driving program, automatically loaded from tape, starts the "switch on" procedure which powers up the instrument electronics in a particular sequence to safeguard against accidental damage of any of the sensitive detection apparatus. The user is then confronted with a series of menu cards each displaying a choice of options on the HP9845S CRT. A desired option is selected by the movement of a cursor and in this way the system permits the experiment to be set up by linking into memory and executing the appropriate program lines. Advantages of this menu card technique are that the user sees all the available options laid out, is free to concentrate on physics rather than computing and can decide on an experiment without recourse to a special control language which would increase both the learning time and possibility of operator error.

Once the type of experiment, spectrum, linescan or image, has been decided the user is prompted to enter the parameters required to set up the instrument as well as any descriptive notes which are associated with the experiment and may specify the date, title, type of sample or special conditions. The software automatically interrogates the positions of the important front panel switches which select magnification and analyser mode. All these experimental conditions plus the user's notes are stored along with the data thus automating the note-taking process and ensuring that no important information necessary for later interpretation is inadvertently lost. If an error arises during data collection resulting, for example, from an exceptionally high value which would cause an integer overflow, the software automatically recognises and recovers from the error state thus avoiding the aborting of the entire experiment or the loss of the data. When one section of an experiment has been completed the user is offered another set of menu card options enabling a decision as to what to do with the result. For example, one may wish to average in another scan, repeat the experiment, move on to a completely new procedure or store the data on tape.

Spectra, linescans and images are stored on tape cartridges as integer array files each accompanied by an entry in a special directory at the start of the tape. The directory entry contains all the information needed for the interpretation of the data file including the notes made by the user, stored as a string, and the parameters defining the experiment, stored as three concatenated program lines. A typical directory entry is shown below:

(a)   IMAGE

(b) { SAMPLE #2   24hrs AFTER Ar + CLEANING   15/4/81
       FEG: 4.95, 2Ø muA, 17.6 kV, SAME AREA AS YESTERDAY
       1Ø Fℓ: READ Tp, M, N, Dt, F1, D1, Epass, Ret, Nor, Mag, Ea, Eb, Hist

(c) {2Ø DATA 3,64,64,1ØØ,11,99,5Ø,Ø,Ø,1ØØØ,5Ø,6Ø,Ø
       3Ø RETURN

where: (a) is the file type;
         (b) are the user's notes: sample, title, date and conditions;
and   (c) are three program lines stored as a string.

When a particular file is requested the corresponding directory entry is read, the notes deconcatenated and the three program lines (1Ø, 2Ø and 3Ø) stripped out. These lines are linked as a subroutine, "Fℓ", which is then called so that upon execution of lines 1Ø and 2Ø the experimental parameters are automatically assigned their correct values. As an example, for the above directory entry the variable M would have the value 64 after the RETURN statement of line 3Ø. Using some of these parameters the length of the data file is calculated so that it can then be read into an integer array redimensioned to the correct size.

Although this may seem a complex method of storing data it has been implemented because it allows for new parameters to be added to the list of those stored without the need to completely rewrite the retrieval program routines or reformat files stored in the past. This is important for an instrument where it is not possible to predict what experiments will be carried out in future. The flexible storage method also permits different types of file to be mixed up together on the same tape and enables the computer to identify the file type of the data of interest and then link in the program suite lines necessary to handle it.

Once a stored file has been retrieved it can be processed using the procedures built into the suite for this purpose. Spectra and linescans can be expanded, smoothed or printed. If required spectra can be handled using one of the quantitative processing routines such as background stripping by subtraction of an analytic form of the background (Seah 1969); unfolding of the spectrometer response function by the method of Van Cittert   (eg Carley and Joyner 1979) or integration of the area under a stripped Auger peak between energy limits specified by the user. Images retrieved from data tape can be analysed to find the frequency distribution of intensities and then rescaled (Saxton 1978), to improve the image contrast before display on one of the digitally controlled 16 grey level CRTs. Hard copy of images in a Y modulated form can be output on the HP9845S thermal printer, as shown in figure 3, which also demonstrates how the user notes and important experimental parameters are output in an easily readable form. Again codenames which may lead to confusion have been deliberately avoided.

4. Conclusion

In designing an integrated suite of programs for control of a SAM, computer memory requirements have been reduced by writing desired routines over program lines no longer needed and a flexible method of data storage

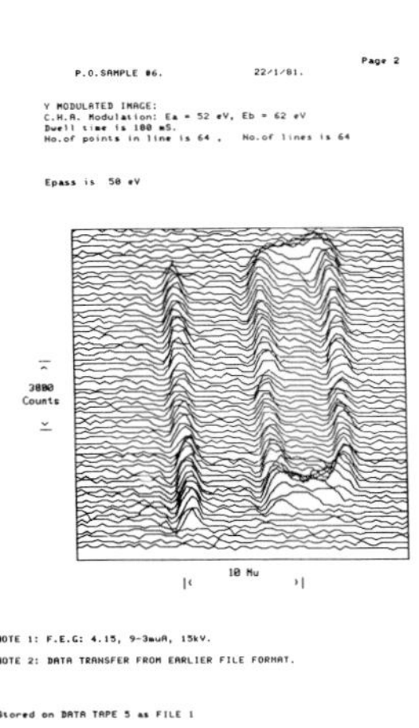

FIG 3:   EXAMPLE OF OUTPUT FROM
THE PROGRAM SUITE

has been implemented by creating a special
directory with entries which can be
executed as program subroutines. Particular
care has been taken to free the experimenter to concentrate on the experiment,
rather than operating a computer, by
making the software as user transparent as
is possible. This has been achieved by
enabling the operator to design experiments by selection of the desired course
from a "menu card" of options; automation
of routine procedures; prompting when
action or input is required; clear
presentation of results; the absence of
a special vocabulary that needs to be learnt
before using the instrument; automatic
recovery from error states and
automatic protection of the hardware
against accidental damage.

<u>References</u>

Saxton W O (1978) Computer techniques for image processing in electron
  microscopy. Academic Press (New York)
Seah M P (1969) Surf Sci <u>17</u> 132
Carley A F and Joyner R W (1979) J Elect Spect & Rel Phen <u>16</u> 1

X-ray microanalysis in the TEM

G.W. Lorimer

Joint University of Manchester/UMIST Department of Metallurgy,
Grosvenor Street, Manchester M1 7HS.

1. Introduction

In a thin specimen the spatial resolution for chemical analysis is dramatically improved, compared to the bulk; the large, bell-shaped region produced by the diffusion of the electron probe beneath the surface of the specimen, from which the majority of the X-rays are produced in a bulk sample, is absent in a thin foil, Figure 1. If the sample is sufficiently thin to carry out quantitative transmission electron-microscopy at 100kV the activated volume is approximately a cylinder equal to the beam diameter

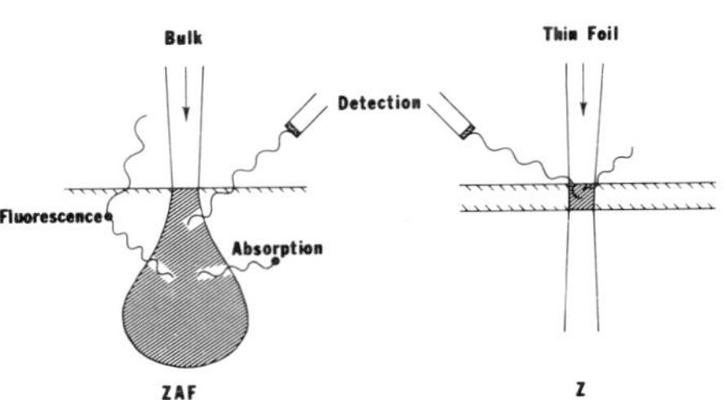

Fig.1. Schematic representation of the interaction of a high energy electron beam with a bulk and a thin specimen.

2. Thin film approximation

Characteristic X-ray fluorescence and X-ray absorption in thin specimens is less than in bulk samples, Fig.1. To a first approximation X-ray absorption and fluorescence can be neglected and the ratio of two observed X-ray intensities, I_A/I_B, can be related to the corresponding weight-fraction ratio, C_A/C_B, by the equation

$$C_A/C_B = k_{AB} \, I_A/I_B \tag{1}$$

where k_{AB} is a constant at a given accelerating voltage and is independent of specimen thickness and composition. k_{AB} values can be determined experimentally (Cliff and Lorimer (1975)) or calculated (Goldstein et al 1977)). A normalisation procedure, e.g. $\Sigma C_n = 1$, must be used to convert the ratios of the weight fractions into weight percentages. In some specimens assumptions must also be made concerning oxidation states: e.g. it is impossible to differentiate between Fe_3O_4 and Fe_2O_3 if ratios are measured, and oxygen cannot be detected. Considerable experimental effort has been spent to obtain accurate k_{AB} values, for the quality of the analysis depends directly on k_{AB}. Table 1, taken from the work of Wood et al. (1981), lists experimental and calculated k_{XSi} values at 120 kV. The calculated k_{XSi} values were obtained using the equation proposed by Goldstein et.al.(1977),

Element	Z	I	II	Element	Z	I	II
Na	11	3.57	1.83	Ti	22	1.12	1.17
Mg	12	1.49	1.31	Cr	24	1.46	1.24
Al	13	1.12	1.14	Mn	25	1.34	1.32
Si	14	1.00	1.00	Fe	26	1.30	1.36
P	15	0.99	1.04	Ni	28	1.67	1.48
S	16	1.08	1.07	Ca	29	1.59	1.65
K	19	1.12	1.05	Mo	42	4.95	4.68
Ca	20	1.15	1.04				

Table 1.   Comparison of experimental, I, and calculated, II, k_{XSi} data at 120 kV.   Calculated values assume t_{Be} = 7.5μm, t_{Au} = 0.02μm and t_{Si} = 0.1μm From Wood et al (1981).

$$k_{XSi} = A_X (Q_K \omega_K a)_{Si} e^{-\mu_{Be}^{Si} \rho t} \Big/ A_{Si} (Q_K \omega_K a)_X e^{-\mu_{Be}^{X} \rho t} \qquad (2)$$

where A_X and A_{Si} are atomic weights, Q_K the Green and Cosslett (1961) values of the K shell ionization cross sections, ω_K the fluorescence yield, a the $K_\alpha/(K_\alpha + K_\beta)$ ratio (equal to one for Z < 19) and μ_{Be}^{X} and μ_{Be}^{Si} the 'effective' mass-absorption coefficients of Si and the element X in the detector window (the Be window, Si dead layer and Au contact film) of density ρ and thickness t.   The ionization cross-section was assumed to be constant along the electron trajectory.   The agreement between the experimental and calculated k_{AB} values is satisfactory for Z > Si, and for these elements it would appear to be sufficient to calculate k_{AB} values for a given detector and accelerating voltage.   The origin of the divergence between experiment and theory at low Z may be due to a combination of inadequate theory, absorption of low energy X-rays within the specimen, contamination on the detector window and loss of light elements during irradiation.   Figure 2 is from the work of McGill and Hubbard (1981) and shows the variation in the apparent k_{NaSi} value for a thin specimen of Amelia Albite as a function or irradiation time (EM6G, 100kV, beam current

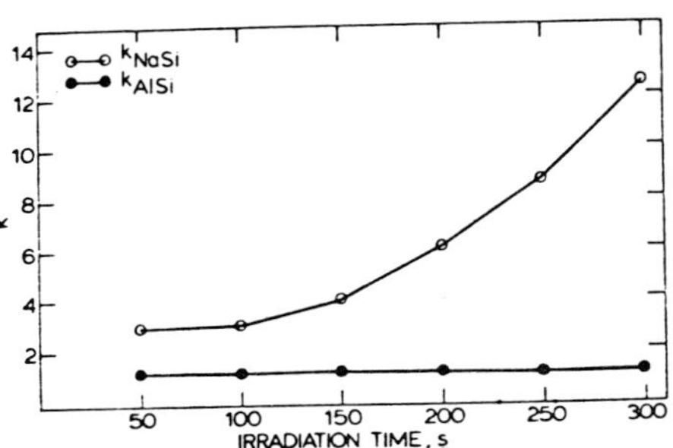

Fig.2. Variation in k_{NaSi} and k_{AlSi} with irradiation time for albite feldspar. From McGill and Hubbard (1981).

Element	Z	I	II	III
Na	11	3.2	2.85	2.02
Mg	12	1.6	1.67	1.39
Al	13	1.2	1.15	1.16
Si	14	1.0	1.00	1.00
K	19	1.0	1.10	1.08
Ca	20	1.0	1.08	1.11

Table 2   Comparison of experimental and calculated k_{XSi} data at 100 kV.

I   Cliff and Lorimer (1975)
II   McGill and Hubbard (1981)
III   Calculated values. These assume t_{Be} = 8μm, t_{Au} = 0.02μm and t_{Si} = 0.1μm.

200μA, beam diameter 3μm). Their beam current density was well below that available on modern STEM instruments and this result highlights the effects of electron beam induced elemental loss. Table 2, also from McGill and Hubbard (1981), is a list of data for Na, Mg, Al, K and Ca, as obtained from mineral specimens and, where necessary, extrapolating to zero irradiation time.

3. X-ray absorption within the specimen

X-ray absorption follows the exponential law

$$I = I_o \, e^{-\mu\rho x} \tag{3}$$

where I is the transmitted X-ray intensity, I_o the initial X-ray intensity, μ the mass-absorption coefficient of a material of density ρ and path length x. When the ratio of characteristic X-ray intensities is considered, the <u>difference</u> in X-ray absorption coefficients is the important parameter (if two characteristic X-rays have similar values of μ they will be absorbed equally and their intensity ratio will be unaffected). If t is the sample thickness and α the angle between the line of the detector and the sample surface, the X-ray intensity ratio I_A/I_B recorded by the detector will be modified by absorption in the specimen from the ratio recorded for an infinitely thin specimen, I_{oA}/I_{oB} (no absorption), as given in equation 4.

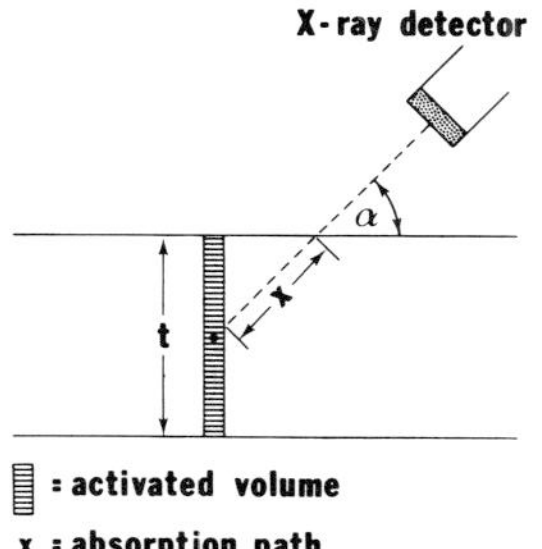

Fig.3. Geometry used to calculate absorption within the specimen x is the X-ray path length.

$$\frac{I_A}{I_B} = \frac{I_{oA}}{I_{oB}} \; \frac{\mu_B}{\mu_A} \; \frac{[1-e^{-\mu_A \rho t \csc \alpha}]}{[1-e^{-\mu_B \rho t \csc \alpha}]} \tag{4}$$

where μ_A and μ_B are the mass absorption coefficients of the characteristic radiation in the sample of density ρ. This equation has been derived with the assumption that the ionization cross section is independent of sample thickness. Recent work by Stenton et al (1981) has indicated that $\phi(\rho t)$ for thin foils of nickel is approximately unity up to a thickness of 80 nm but then increases. They report a value of $\phi(\rho t)$ of ∿ 1.3 at a thickness of 380 nm. If this result is substantiated it has important implications for making absorption corrections, and will specify a thickness limit for the validity of equation 4. In order to make an absorption correction it is necessary to know the X-ray path length within the specimen. There are various methods of determining the thickness of a sample, but it may be difficult to determine the X-ray path length with any degree of accuracy. This is particularly true for electrolytically or ion-beam-thinned samples which can be irregularly wedge-shaped. This problem has been discussed in detail by Goldstein and Williams (1981) who illustrated the large errors which could arise during the analysis of a NiAl sample if the shape of the foil and its orientation relative to the X-ray detector were not known.

Figure 4 shows the observed variation in $\ln(I_{SiK}/I_{FeK\alpha})$ as a function of specimen thickness for a 2 wt% Si eutectoid steel which had been quenched to produce a martensitic microstructure of uniform composition. The line has a slope, as determined from a least squares analysis, equal to 2090 cm^2 gm^{-1}. This is the approximate difference between the X-ray mass-absorption coefficients of iron (2502 cm^2 gm^{-1}) and silicon (710 cm^2 gm^{-1}) X-rays in the specimen, a value that is in agreement with that predicted by equation 4.

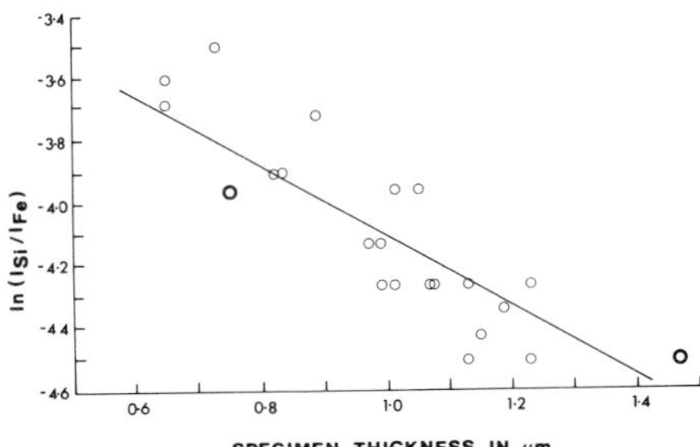

Fig.4. $\ln(I_{SiK}/I_{FeK\alpha})$ as a function of thickness for a 2wt% Si steel. From Al-Salman et al (1977).

4. Characteristic fluorescence correction

Fluorescence effects in thin foils are less than those observed in bulk specimens, but when strong fluorescence is predicted in the bulk, e.g. the fluorescence of Cr K$_\alpha$ by Fe K radiation, it is necessary to make a fluorescence correction. Correction procedures for thin films have been developed by Philibert and Tixier (1975) and Nockolds et al (1979). The former predicts a $(\rho t)^2$ dependence while the latter a ρt (0.9 − $\ln \rho t$) dependence for the correction (equation 5).

$$\frac{I_F^A}{I_A} = C_B \cdot \omega_{K_B} \cdot \frac{r_A - 1}{r_A} \cdot \frac{A_A}{A_B} \cdot \mu_B^A \cdot \frac{U_B \, \ln U_B}{U_A \, \ln U_A} \cdot \frac{\rho t}{2} \left| 0.923 - \ln(\mu_B^A \rho t) \right| \quad (5)$$

where U_A and U_B are the overvoltage ratios for elements A and B respectively, A_A and A_B are the relevant atomic weights, μ_B^A and μ_B are the mass absorption coefficients of X-radiation from element B in element A and the specimen, respectively, C_A is the weight fraction of element A, ω_{K_B} is the fluorescence yield of element B and r_A is the absorption-edge jump-ratio. (Note that if the sample is tilted through an angle α then expression (5) must be multiplied by sec α).

5. Beam broadening in the specimen

Beam spreading in thin foils is currently a lively topic of both experimental investigation and theoretical analysis, and the reader is referred to recently published articles by Goldstein and Williams (1981), Doig et al (1981), Stephenson et al (1981) and Cliff and Lorimer (1981) which summarize the present state-of-the-art.

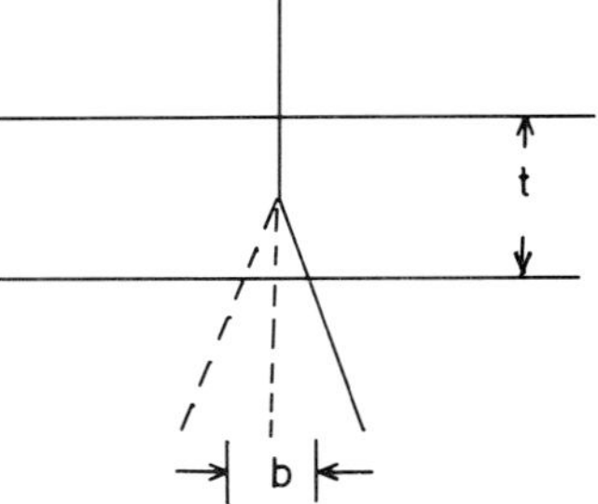

Fig.5. Simple model for beam spreading from Goldstein et al (1977).

A simple approach to the problem of beam spreading has been proposed by Goldstein et al (1977) who assumed that a single, elastic scattering event occurred at the centre of the foil and defined the X-ray source size as that volume in which 90% of the electron trajectories lie. The beam broadening b, from a point probe is given by:

$$b = 6.25 \times 10^5 \; \frac{Z}{E_o} \; \left[\frac{\rho}{A} \right]^{\frac{1}{2}} \; t^{3/2} \; cm \qquad (6)$$

where Z is the atomic number, E_o the accelerating voltage, ρ the density, A the atomic weight and t the sample thickness. Equation 6 was derived for a single scattering event, and hence is strictly valid for only very thin foils, but it agrees moderately well with the Monte Carlo calculations of Kyser (1979), Geiss and Kyser (1979) and Newbury and Myklebust (1979) and the few experimental measurements of beam broadening, Hutchings et al (1979) and Cliff and Lorimer (1981). Cliff and Lorimer (1981) have shown that the expression of Goldstein et al (1977) corresponds exactly to one limit predicted by plural scattering theory, that of a Gaussian distribution of scattering events,while the other limit, a single scatter- ing event, predicts less broadening and coincides with Monte Carlo calculations.

It is essential to know the electron intensity distribution in the incident probe if beam broadening in the specimen is to be calculated and the activated volume for X-ray microanalysis to be determined. The electron intensity distribution in the probe is often assumed to be Gaussian (Doig et al (1981)), but this is often not the true situation. Fig.6a

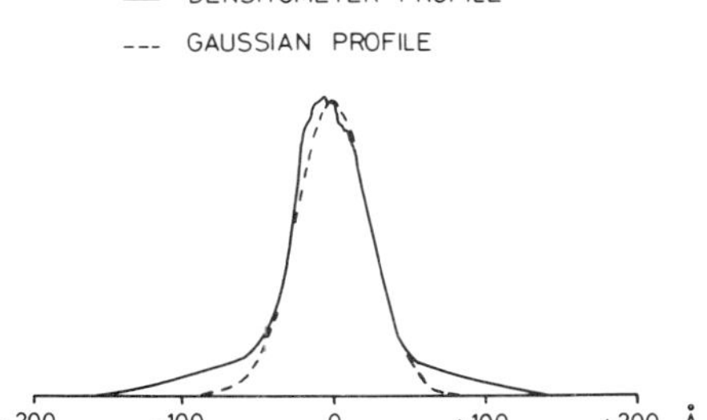

Fig.6a Microdensitometer profile of a focussed probe with a Gaussian profile superimposed.

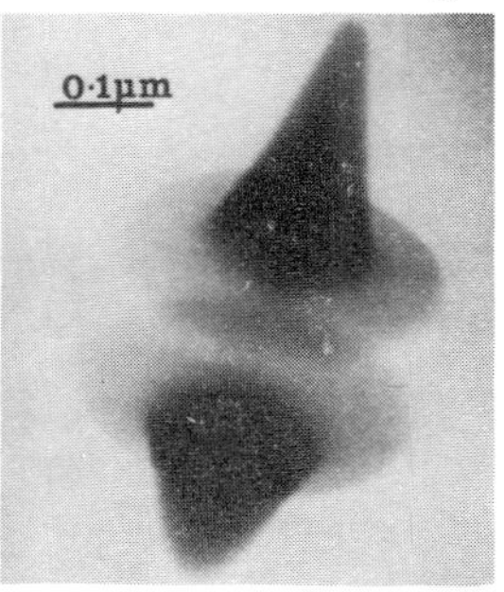

Fig.6b Micrograph of con- tamination spot. Nominal probe diameter 10nm, JEOL 100-CX. From Love et al (1981).

shows a microdensitometer trace of a micrograph of the analysis probe in a Philips EM400T operating in the TEM nanoprobe mode, 100kV, eucentric position, spot size 6, and a Gaussian distribution is superimposed on the experimentally measured profile. The experimental probe has a long, non- Gaussian, tail which, in the illustration, contains a significant pro- portion of the total current (the true 3-D ratio of 'tail' to 'probe' current is larger than the apparent ratio in the 2-D profile shown in Fig.6a). The ratio of the electron intensity distribution in the 'tail' to that in the 'probe' depends on the convergence angle of the beam. A highly convergent beam, which produces the smallest possible 'probe', for a given set of operating conditions, also produces a large 'tail'. With a less convergent beam the 'tail' can be dramatically reduced, although the 'probe' diameter is increased. The origin of the 'tail' is thought to be a combination of the high illumination angle and spherical aberration. Fig.6b, from the work of Rae et al (1981), shows a con-

tamination spot, formed with a probe of nominal diameter 10nm with a
'witch's hat' shape. We propose that the shape of this contamination spot
reflects an electron intensity distribution similar to that shown in Fig.
6a. It is possible to use a photographic technique to determine the
electron intensity distribution in the TEM mode of operation, but this
cannot be done as easily in STEM. The use of 'image resolution' in STEM
to define the probe diameter is not a valid measurement of the analysis
probe diameter; a probe with an electron intensity distribution similar to
that shown in Fig. 6a would produce an 'image resolution' of 6nm, but an
'X-ray resolution' of $\approx$30nm. As shown by Venables and Janssen (1980) an
estimation of the true electron intensity distribution in a STEM probe
can be obtained by monitoring an electron induced signal when the probe is
scanned across the edge of the specimen.

6. Acknowledgements

The author is indebted to Mr. G. Cliff and P. Kenway for helpful discussions and for permission to use the unpublished data concerning beam
broadening shown in Fig.6a. SRC and NERC provided funds for capital
equipment.

7. References

Cliff G and Lorimer G W 1975 J. Microsc. 103 203-207.
Cliff G and Lorimer G W 1981 Quantitative Microanalysis with High Spatial
    Resolution ed G W Lorimer, M Jacobs and P Doig (London : Metals
    Society) pp 47-52.
Doig P, Lonsdale D and Flewitt P E J ibid pp 41-46.
Geiss R H and Kyser D F 1979 Ultramicroscopy 3 397-400.
Goldstein J I, Costley J L, Lorimer G W 1977 Reed S J B Quantitative
    X-ray Analysis in the Electron Microscope SEM 1977,   Proc.
    Workshop on Analytical E M ed O Johari (Chicago, I I T R I) pp 315-324.
Goldstein J I and Williams D B 1981 Quantitative Microanalysis with
    High Spatial Resolution ed G W Lorimer, M Jacobs and P Doig
    (London : Metals Society) pp 5-14.
Green M and Cosslett V E 1961 Proc. Phys. Soc. 71 1206-1214.
Hutchings R, Loretto M H, Jones I P and Smallman R E 1979 Ultramicroscopy
    3 401-405.
Kyser D F 1979 Introduction to Analytical Electron Microscopy ed J J Hren,
    J I Goldstein and D C Joy (New York : Plenum) pp 199-219.
Lorimer G W, Al-Salman S A and Cliff G 1977 Developments in Electron
    Microscopy and Analysis ed D C Misell (London : Institute of Physics)
    pp 369-372.
McGill R J and Hubbard F H 1981 Quantitative Microanalysis with High
    Spatial Resolution ed G W Lorimer, M Jacobs and P Doig (London :
    Metals Society) pp 30-34.
Newbury D E and Myklebust 1979 Ultramicroscopy 3 391-396.
Nockholds C, Nasir M J, Cliff G and Lorimer G W 1979 Developments in
    Electron Microscopy and Analysis ed T Mulvey (London : Institute of
    Physics) pp 417-420.
Philibert J and Tixier R 1975 Physical Aspects of Electron Microscopy
    and Microbeam Analysis ed B M Siegel and D R Beaman (New York :
    York : Wiley) pp 333-354.
Rae D A, Scott V D and Love G 1981 Quantitative Microanalysis with High
    Spatial Resolution ed G W Lorimer, M Jacobs and P Doig (London :
    Metals Society) pp 57-62.
Stenton N, Notis M R, Goldstein J I and Williams D B ibid pp 35-40.
Wood J, Wiliams D B and Goldstein J I ibid pp 24-29.

X-ray elemental analysis with a TEM

S. Isakozawa, M. Shinohara, M. Kubozoe
Naka Works, Hitachi, Ltd., 882 Ichige, Katsuta, Japan 312

Analytical transmission electron microscope (ATEM) which
combines energy dispersive x-ray spectrometer and transmission
electron microscope has been greatly improved in its analytical
performance. We have incorporated a Kevex 7000Q EDX system
on Hitachi H-600 (100 kV TEM) and tested quantitative accuracy
in the x-ray microanalysis mode using thin and bulk specimens.
We have obtained satisfactory results.

Fig. 1 shows x-ray take-off geometry that is used on the H-600.
The electron probe is demagnified by the pre-field of objective
lens and it becomes 1 through 100 nm in diameter arriving at
the specimen. X-rays are taken through the magnetic field of
the objective lens at a high take-off angle of 68 degrees with
the specimen held normal to the electron beam (or at right
angles to the beam). This high take-off angle is effective
for better x-ray sensitivity as well as peak to background
ratio.

Fig. 2 is an analysis result of a thin mineral specimen called
Kaersutite which is by courtesy of Prof. Aoki, Tohoku Univ- [1]
ersity, Japan. The analysis was done using Cliff-Lorimer's
technique in which K-value was obtained through computation
without a use of standard specimens. The result is in good
agreement with a quoted chemical analysis. This analysis
shows a good quantitative accuracy for Na (sodium) which has
been regarded difficult upto this time. The analysis was
done at 100 kV, probe current 1×10^{-9} A, and data acquisition
time 200 sec.

Fig. 3 is a result of analysis with a bulk specimen of a
metallic compound called SRM-348 which is distributed by
National Bureau of Standards (NBS) in the U.S.A. Both the
x-ray spectrum and the background noise which needs to be
subtracted are shown. This background noise is a computed
value in the Kevex 7000Q EDX system. However, it is in good
agreement with the background noise which is actually
generated in the microscope column. Analytical conditions are
accelerating voltage 25 kV, probe current 2×10^{-11} A, and
data acquisition time 200 sec. The analysis was done with
the H-600.

Fig. 4 is a comparison of ZAF corrected result using standard
specimens and a semi-quantiative analysis taken in approximation or APP mode in the Kevex 7000Q system. The APP mode does
not require any standard specimen. It is clear that the APP
mode is fairly good for such elements as Si (Z=14) or heavier.

We have introduced some of the quantitative analysis results
run on Hitachi's analytical electron microscopes. These
results indicate that the ATEMs are no longer limited to
qualitative analysis. The ATEMs have now come to a level that
permits quantitative analysis of various specimens with the
results being equal to conventional chemical analysis.

Reference: 1) Cliff, G. and Lorimer, G. W., J. Microscopy,
                103 (1975)

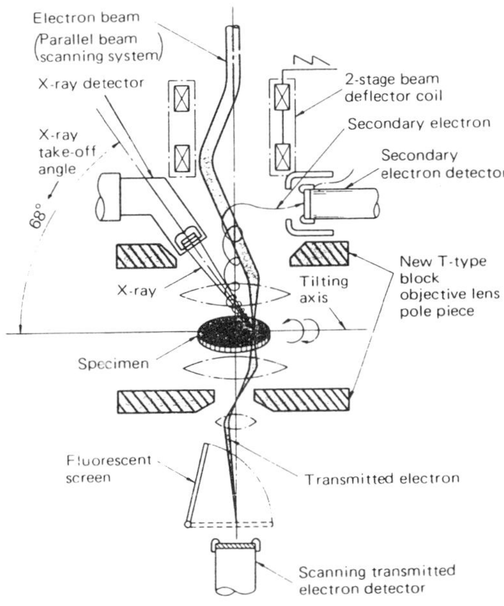

Fig. 1   X-ray Take-off Geometry

Element	FOIL Analysis	Chemical Analysis
Na	2.5	2.77
Mg	13.3	13.22
Al	13.0	12.73
Si	42.3	41.31
K	0.5	0.45
Ca	11.8	11.36
Ti	4.0	3.99
Mn	0.3	0.21
Fe	12.3	12.20

Fig. 2   Elemental Analysis
         Result of KAERSUTITE
         (oxide per cent)

Fig. 3   X-ray Spectrum of
         SRM-348

Element	ZAF. Analysis	APP. Analysis	Chemical Analysis
Si	0.6wt%	0.6wt%	0.54wt%
Ti	2.2	2.7	2.24
Cr	14.3	15.3	14.55
Mn	1.5	1.5	1.48
Fe	53.5	53.8	53.80
Ni	25.7	24.1	25.80
Mo	1.3	1.2	1.30

Fig. 4   Elemental Analysis Result
         of SRM-348

The correlation between specimen current, beam current and X-ray yield measurements over the primary energy range 75–200 keV

J.P. Broomfield and E. Metcalfe

CERL, Kelvin Avenue, Leatherhead, U.K.

1. Introduction

The Hitachi H700H STEM is a 200 kV TEM with a scanning attachment for STEM operation. The EDS detector which is fitted to the column has a high positive take-off angle (68°). This, combined with a low background kit fitted to the column to reduce spurious X-rays, allows a high X-ray performance to be obtained without sacrificing the high resolution imaging capability. This investigation of the correlation between X-ray yields and primary energy was carried out as part of a programme to define the best operational parameters of the microscope in order to ensure the optimum analytical capability.

2. Experimental Procedure

A Faraday cup based on the design of Nicholson (1981) was made using a Pt 10 micrometer TEM aperture set into a brass mount (Fig. 1). This was inserted into the low background specimen holder and examined in the SEM mode in the microscope. The low background holder was isolated from earth and a coaxial socket in the handle of the holder allowed currents to be measured (this was done using a Keithley 616 picoammeter with a range down to 10^{-15}A). Beam current and specimen current measurements were carried out on the Pt Faraday cup for the primary beam energies available in the H700H. X-ray spectra were also simultaneously collected. Dual element bulk standards of Pt and Al, Fe, Co or Au were then inserted into the microscope and the specimen currents for Al, Fe and Au were measured against the Pt specimen current. This was used to derive the true beam current, and hence the absorption coefficient (specimen current/beam current). X-ray measurements were then made on thin films of these elements.

3. Results

Table 1 shows the results of a large number of absorption coefficient measurements on the Pt Faraday cup. The measurements were taken on three different occasions with beam currents in the range 10^{-13} to 10^{-9} amps. It was found that the absorption coefficient was very reproducible. The specimen current measurements from the bulk elements are given in Table 2. The variation in total X-ray yield, normalised X-ray yield (c.s^{-1}.nA^{-1}) and peak to background ratio for Al, Fe and Au thin films (all less than 80 nm) are shown in Figs. 2–4 respectively.

4. Discussion

The absorption coefficients for Pt and Au are approximately constant and

equal over the primary energy range 75 to 200 keV. This is to be expected
of two neighbours in the periodic table with very similar large electron
scattering cross sections. The absorption coefficients for Al, Fe and Co
at primary energies less than 125 keV are higher than for Pt and Au and
increase with atomic number. The decrease in absorption coefficient with
primary energy for Fe leads to a cross-over with Pt at about 125 keV and
Fe has the lowest absorption coefficient above this value. These results
agree with those of Nicholson (1981) for Pt and Al at 60 and 100 keV, and
his results for Cu and Ni are in reasonable agreement with our Co results.
The low results for iron may be due to an oxide film on the surface of the
iron specimen affecting the number of secondary electrons escaping from
the surface. No attempt was made to bias the specimen to suppress the
generation of secondary electrons since reproducible results were obtained
without it, and the requirement was for reproducible results for
calculating beam currents.

The total X-ray yield increases approximately linearly with primary energy
for Al, Fe and Au (Figs. 2, 3 and 4 respectively). This is a function of
gun brightness which increases with primary energy for a given gun
geometry and is far greater than the decline in ionisation cross section
with increasing primary energy. This is a major advantage of carrying out
microanalysis in the STEM at higher primary beam energies, since under
these conditions the necessary counting time for a given statistical
accuracy is minimised. Short counting times also minimize the effects of
specimen drift and contamination. The effect of gun brightness has
recently been discussed by Clarke (1979).

When the effect of increasing gun brightness with primary energy is
eliminated by normalising the X-ray yield with respect to beam current,
the yield can be seen to decline with primary energy. This is due to the
declining ionisation cross section which is proportional to $(\ln E)/E$
(Worthington and Tomlin, 1956). However, since the energies are well above
the critical ionisation energy the decline is not strongly marked.

The peak to background ratio is also seen to be broadly invariant with
increasing primary energy, except in the case of Al where a definite
decline is observed. Theoretical estimates indicate that the peak/back-
ground ratio should increase with keV (Zaluzec, 1978). However the effect
can be difficult to demonstrate since additional scattering in the column
with increasing primary energy can contribute to the background and mask
the effect. This problem has probably been minimised in this equipment,
for hole counts of less than 1% have been reported (Metcalfe and Owen,
1981). Other experimental measurements of peak/background ratio show a
decline with keV. For example Faulkner and Norgaard (1978) took measurements from Ni in Inconel 600 at 50, 100 and 200 kV and found little
variation in peak/background ratio for thin films.

In conclusion, these measurements show a considerable increase in X-ray
yield due to increasing gun brightness as the primary energy is increased
from 75 to 200 keV. A slight decrease in normalised X-ray yield and peak/
background is also seen. However the large increase in total counts will
outweigh the small decline in the other two parameters which is
important when considering the minimum mass fraction limit during analysis.
This observation agrees with our current experience.

References

Clarke D R   1979 in Electron and Positron Spectroscopies in Materials
     Science and Engineering (Academic Press) p.315
Faulkner R G and Norrgard K 1978 X-Ray Spectrometry, 7, 184
Metcalfe E and Owen H T   1981 Conf. Proc. Quantitative Microanalysis
     with High Spatial Resolution (Manchester: Metals Soc.), Paper 13
Worthington C R and Tomlin S C   1956 Proc. Phys. Soc. A69, 401
Nicholson W A P 1981 J. Microscopy 121, 141
Zaluzec N J   1978 Ninth Int. Cong. on Electron Microscopy (Toronto:
     Micro. Soc. Canada) p. 548

Acknowledgement

This paper is published by permission of the Central Electricity
Generating Board.

Table 1

Platinum Absorption Coefficient

Primary Energy (keV)	Absorption Coefficient (specimen/beam current)	Standard Deviation	Bulk Pt Normalised Yield (counts/pA)
200	0.48	0.03	8441
175	0.49	0.01	8388
150	0.49	0.02	7651
100	0.49	0.02	4876
75	0.49	0.02	4088

Table 2

Specimen Current (SC) and Absorption Coefficient*
Measurements for Bulk Elements

Primary Energy (keV)	Al $\frac{SC_{Al}}{SC_{Pt}}$	Al Absorp. coeff.	Fe $\frac{SC_{Fe}}{SC_{Pt}}$	Fe Absorp. coeff.	Co $\frac{SC_{Co}}{SC_{Pt}}$	Co Absorp. coeff.	Au $\frac{SC_{Au}}{SC_{Pt}}$	Au Absorp. coeff.
200	1.25	0.60	0.42	0.22	1.31	0.63	1.00	0.48
175	1.27	0.62	0.52	0.27	1.31	0.64	1.00	0.49
150	1.33	0.65	0.68	0.35	1.31	0.64	1.00	0.49
100	1.45	0.71	1.18	0.57	1.31	0.64	0.98	0.48
75	1.45	0.71	1.27	0.63	1.32	0.65	0.99	0.49

$$* \quad \text{absorption coefficent} \ = \ \frac{SC_{Element}}{SC_{Pt}} \ \times \ \frac{SC_{Pt}}{\text{Beam Current}}$$

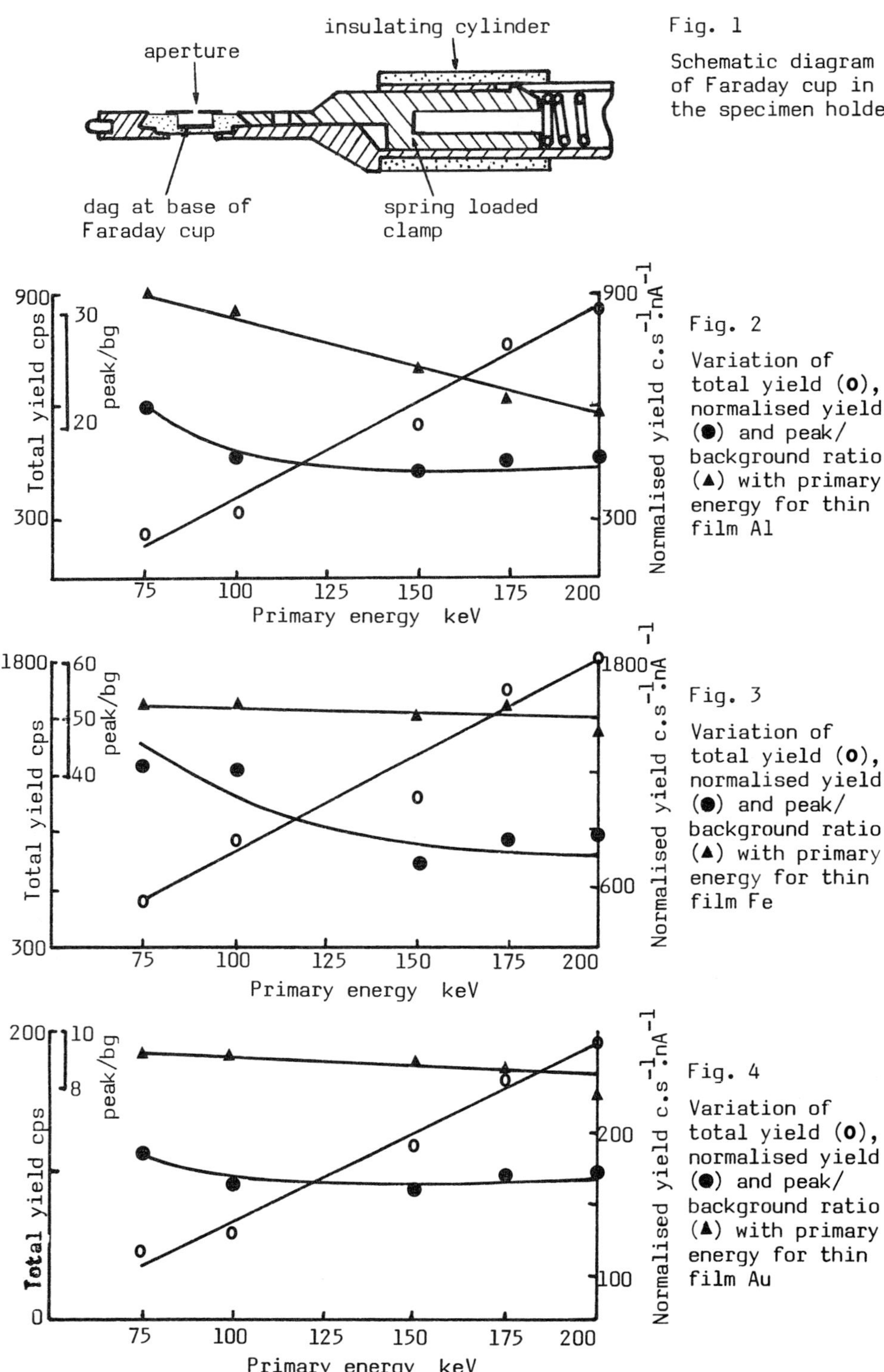

Fig. 1

Schematic diagram
of Faraday cup in
the specimen holder

Fig. 2

Variation of
total yield (○),
normalised yield
(●) and peak/
background ratio
(▲) with primary
energy for thin
film Al

Fig. 3

Variation of
total yield (○),
normalised yield
(●) and peak/
background ratio
(▲) with primary
energy for thin
film Fe

Fig. 4

Variation of
total yield (○),
normalised yield
(●) and peak/
background ratio
(▲) with primary
energy for thin
film Au

Rare earth element determinations by energy-dispersive electron microprobe techniques

D.G.W. Smith

Geology Dept., University of Alberta, Edmonton, Alberta T6G 2E3, Canada

S.J.B. Reed

Dept. of Earth Sciences, University of Cambridge, Downing St.,
Cambridge CB2 3EQ

1. Introduction

Over the last decade quantitative ED analysis has become routine, at least for elements in the atomic no. range 11–35, for which K lines are used. The main limitations are the relatively poor peak-to-background ratio, and interferences between lines, both of which are more serious for heavier elements, where L or M lines are used.

The rare-earth elements (REEs) pose particularly acute spectral interference problems, especially in natural mineral phases where all the REEs commonly co-exist. Some success has been achieved with WD analysis of REEs (e.g. Amli & Griffin, 1975; Drake & Weill, 1975; Exley, 1980) and interferences are certainly less severe than for ED analysis, owing to the better resolution. However, WD analysis entails laborious sequential acquisition of peak and background intensities for each element. The substantial time required for a complete analysis, as well as being inconvenient, may give rise to errors due to instrumental drift and sample contamination or degradation. It is therefore appropriate to explore the potential of the ED technique for REE analysis.

The REEs extend from atomic no. 57 (La) to 71 (Lu) and the L lines are used for analysis, the K series being of too high energy and the M series too low. The L spectrum for each element includes at least 10 lines with intensity $> 1\%$ of the $L\alpha$ line and for the REEs as a whole these cover the range 4–11 keV. Obviously the potential for interferences is great and fully quantitative ED analysis will require accurate overlap corrections. In naturally occurring compounds it is common for high concentrations of some REEs to be associated with relatively low concentrations of others, thereby exacerbating the interference problem.

2. Instrumental

An ARL EMX microprobe (X-ray take-off angle 52.5º), fitted with an Ortec ED system (FWHM – 154 eV at 5.9 keV; Be window thickness – 8 μm; distance from specimen – 27 cm) was used. The accelerating voltage was 15 kV and the probe current was adjusted to maintain an appsroximately constant count-rate of $\sim$ 3000 cps for the whole spectrum (typical current $\sim$ 30 nA).

3. Standards

Well-characterized REE mineral standards are very rare.   Some pure REE
metals are satisfactory, but others oxidise rapidly.   Oxide glasses tend
to be hygroscopic (Amli & Griffin, 1975).   Stable synthetic REE compounds
are available, however, including ferrites, Ga-garnets and Al alloys,
various of which have been used here.   These generally contain only one
REE each, thereby avoiding interference problems, but their compositions
do not resemble natural phases, hence ZAF corrections may be large.
Silicate glasses prepared by Drake & Weill (1972) are preferable in this
respect, but each contain several REEs and are subject to interferences.

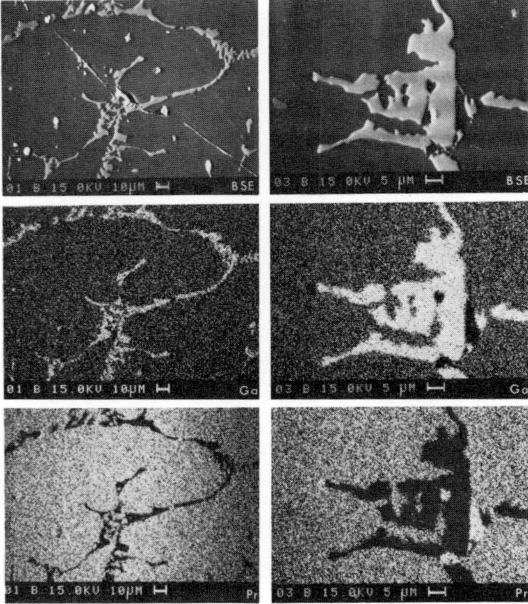

During examination of a suppos-
edly homogeneous Pr–Ga-garnet
it became apparent that it contained Pr and Ga oxide phases
(see Fig. 1), which detract from
its usefulness as a standard.
It is clearly important that the
homogeneity of potential standards on a micro-scale should be
examined carefully.   ED analysis combined with scanning
imagery is well-suited to such
investigations, which are also
relevant to the practical application of such materials.

Fig.1   Scanning images of
Pr–Ga-garnet, showing intergrowth with Ga_2O_3 and Pr_2O_3

4. Sensitivity factors

The conversion of observed line intensities into concentrations requires
knowledge of the sensitivity factors (cps per %) for the elements concerned.   Fig.2 shows relative
sensitivity factors determined
experimentally and corrected
for ZAF effects, for those REEs
for which there were suitable
standards.   The smooth Z-dependence suggests that interpolated values can be used
where standards are lacking.
The possibility of 'standard-
less' analysis is one advantage of ED compared to WD analysis, where such interpolation
is less practicable.

Fig.2   Relative sensitivity
factors for REEs (L_α line)

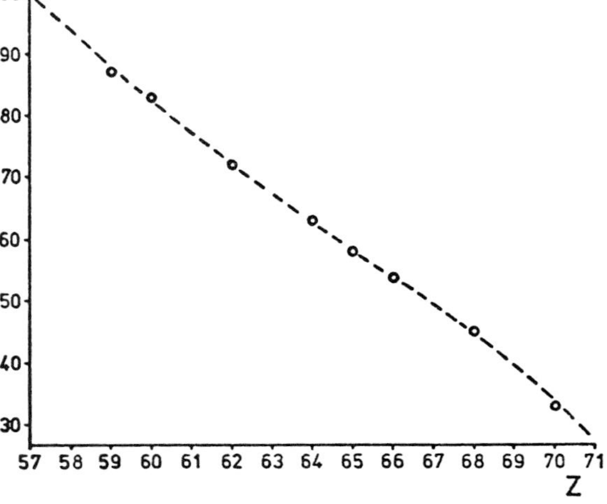

5. Overlap coefficients

Quantitative ED analysis can be carried out by integrating the counts contained in groups of channels corresponding to the appropriate X-ray lines and correcting the intensities obtained for interferences (Reed & Ware, 1973; Smith, 1976). 'Overlap coefficients' for REE L_α lines have been determined experimentally from pure element and oxide spectra, using the EDATA 2 program (Smith et al., 1979) for background corrections etc.

The results given in table 1 are expressed as the percentage of the measured L_α intensity (integrated over 1.2 x FWHM) of the interfering element appearing in the integrated region of the element interfered with. The overlapping element is identified by the atomic no. difference, ΔZ. These coefficients are strictly applicable only to this particular ED system, since they depend on resolution and peak shape; however, table 1 provides a useful guide to the amount of overlap between REEs for a typical system.

Table 1   Overlap coefficients (%) for L_α lines of REEs

overlapped el. (Z)	-8	-7	-6	-5	-4	-3	-2	-1	+1	+2	+3	+4
La 57									8	1	5	
Ce 58							13		8	1		1
Pr 59							58	10	8		4	1
Nd 60						18	42	10		1	4	2
Sm 62				2	17	25	30		6	1	4	2
Eu 63			7	2	19	33		8	6	1	3	3
Gd 64		3	7	2	18		10	8	5	1	3	3
Tb 65	1	3	6	4		43	6	8	5	0	2	4
Dy 66	1	3	4		12	49	2	8	5	0	1	4
Ho 67	2	5		8	12	47	1	8	4	0	1	5
Er 68	3		0	12	12	41	1	8	4	0	0	
Tm 69		7	0	15	14	31	0	7	4	0		
Yb 70	2	6	0	17	20	21	0	7	5			
Lu 71	3	3	0	18	29	10	0	6				

The most serious interference is that caused by the L_β lines for atomic no. differences between 2 and 5, the largest overlap coefficient being 58% for La interfering with Pr. The vertical columns in table 1 exhibit a smooth Z-dependence, hence interpolation of overlap coefficients as well as sensitivity factors may be feasible when standards are lacking.
(Note that the requirement for an overlap standard is somewhat different: it is merely necessary that the elemental spectrum should be 'pure' - the composition need not be known with high accuracy.)

In REE analysis using L_α lines there are clearly cases where overlap is so severe as to make accurate results unattainable. It may then be advantageous to use the L_{β_1} line instead, though the intensity will be lower by a factor of approximately 2. Overlap coefficients have also been determined for the L_{β_1} lines and in order to make these directly comparable with the L_α data they were multiplied by the L_β/L_α intensity ratio for the overlapped element, giving a figure which takes into account the lower intensity of the L_{β_1} line relative to the interfering line.

The results are given in table 2, in which asterisks denote cases where L_α overlap is significant (> 10%) and the L_β overlap is less than half that for L_α, suggesting that the L_β line is preferable. However, in complex samples all possible interferences must be considered and it may be that changing from L_α to L_β makes another interference worse. Tables 1 & 2 enable the optimum choice to be made in such cases.

Table 2   Overlap coefficients (%) for L_{β_1} lines, adjusted for L_{β_1} intensity

overlapped el. (Z)	-6	-5	-4	-3	-2	-1	+1	+2	+3	+4	+5	+6
La 57							19	225	17		8	3
Ce 58						32	11	221		1	7	5
Pr 59					28*	25	6		44	1	5	7
Nd 60				11	24	27		153	81	1	3	9
Sm 62		3	7*	6*	25		2	77	156	2	1	9
Eu 63	0	3	9*	4*		23	2	42	195	3	1	6
Gd 64	0	5	12		20	25	2	20	230	8	0	3
Tb 65	0	5		2*	19	26	2	8	214	22	0	0
Dy 66	0		13	2*	17	27	2	3	205	43	0	
Ho 67		6	16	1*	14	30	1	1	140	112		
Er 68	1	6	16	1*	12	34	1	0	81			
Tm 69	1	7*	17	1*	11	34	0	2				
Yb 70	1	7*	17	1*	7	45	2					
Lu 71	1	7*	17	0*	5	44						

6. Results

To examine the potentialities of ED analysis for REEs, various REE-bearing materials were analysed and the results are given in Table 3.   The agreement with the 'true' compositions is encouraging, especially for REEs present as major constituents. Somewhat less close agreement is found when the concentration is too low and/or interferences are particularly severe (e.g. Pr in monazite, which suffers interference from La and Ce).

Clearly ED analysis will be unsuitable for minerals such as apatite and zircon in which the REE concentrations are relatively low, and in such cases WD analysis is more appropriate. (Ion microprobe analysis is capable of much greater sensitivity but is more difficult to quantify.)   Within its limitations, quantitative ED analysis for REEs has a useful role in view of its convenience and the shorter analysis time.

Table 3   ED analyses of REE compounds

1. Pr–Ga–garnet

Pr	44.12	(43.83)
Ga	35.98	(36.19)

2. Nd–Ga garnet

Nd	44.78	(44.46)
Ga	35.54	(35.81)

3. Gd–Ga–garnet

Gd	46.15	(46.60)
Ga	34.82	(34.44)

4. Dy–Gd–Ga–garnet

Dy	4.00	(3.85)
Gd	42.10	(42.83)
Ga	34.88	(34.38)

5. Sm–ferrite

Sm	57.78	(59.15)
Fe	21.68	(21.97)

6. $TbAl_2$

Tb	74.96	(74.65)
Al	25.04	(25.35)

7. REE glass

La	3.54	(3.65)
Ce	3.50	(3.42)
Pr	3.74	(3.79)
Y	2.92	(3.21)
Ca	17.56	(18.10)
Al	12.23	(12.67)
Si	17.27	(16.37)

8. Monazite

La	14.20	(12.87)
Ce	29.87	(30.86)
Pr	2.85	(4.19)
Th	2.78	(3.78)
P	12.70	(12.13)

Note: results expressed in wt.%, with 'true' values in brackets, based either on stoichiometry (nos. 1–6) or independent analyses (nos. 7,8).   All samples except no. 6 contained O, which was calculated by difference in the ZAF procedure.

Acknowledgements

We are grateful to Drs Amli & Griffin for the oxide glasses, R. Jordan for REE–Al_2 alloys, P. Suddaby for REE ferrites and J.M. Page for REE–Ga–garnets.   Also we thank D.A. Tomlinson & S. Launspach for microprobe measurements and D. Wynne for data processing.

References

Amli R & Griffin W L 1975 Amer. Mineral. 60 599
Drake M J & Weill D F 1972 Chem. Geol. 10 179
Drake M J & Weill D F 1975 Geochim. Cosmochim. Acta 39 689
Exley R A 1980 Earth Planet. Sci. Lett. 48 97
Reed S J B & Ware N G 1973 X-ray Spectrom. 2 69
Smith D G W 1976 Short Course Handbook in Microbeam Techniques
    ed D G W Smith (Mineral. Assoc. Canada) pp 63–106
Smith D G W & Gold C M 1979 Microbeam Analysis – 1979 ed D E Newbury
    (San Francisco: San Francisco Press) pp 273–278

The STEM-EDS X-ray microanalysis of rod-shaped precipitates in an Al–4wt%Fe alloy

P. Doig, D. Lonsdale and P.E.J. Flewitt, CEGB South Eastern Region, Scientific Services Department, Gravesend, Kent.

1. Introduction

The authors have presented a theoretical analysis describing the dispersion of a fine electron probe within a thin foil (Doig et al, 1980) based on a Gaussian distribution of electron intensity within the incident probe. This analysis has been modified (Doig et al, 1981) to include an electron beam broadening versus foil thickness, t, dependence of $t^{3/2}$ in accord with the predictions of simple single electron scattering models (Goldstein et al, 1977) and Monte Carlo analyses (Hutchings et al, 1979; Kyser, 1979). It provides a quantitative description of the electron intensity distribution within the foil which, when combined with a given elemental distribution, yields a value for the relative X-ray intensity generated from any microstructural feature. A similar approach has recently been reported for the X-ray microanalysis of grain boundary segregation profiles (Hall et al, 1981). The present work investigates the application of this analysis by measuring the X-ray intensity from and thus the composition of, cylindrical rod precipitates in thin foils of an Al-4wt%Fe eutectic alloy.

2. X-ray Microanalysis of a Cylindrical Rod Precipitate

The characteristic X-ray intensity, I, emitted from any region within a thin metal foil during microanalysis is given by a combination of the local electron beam intensity, I_v, and the elemental composition distribution, C_v, such that:

$$I = K \int_V \{C_v . I_v\} \, dV \tag{1}$$

where V is the foil volume and K is a constant defining the efficiency of X-ray generation, emission and detection. The electron flux, I_v, has been given analytically by Doig et al (1981):

$$I_v = I(r,t) = I_e \{\pi(2\sigma^2 + \beta t^3)\}^{-1} . \exp \{-r^2/(2\sigma^2 + \beta t^3)\} \tag{2}$$

where I_e is the total incident electron current, r defines a position in the plane normal to the incident electron probe whose centre defines the origin and t is a co-ordinate defining the depth within the foil in the direction of the electron probe. β is a coefficient with dimensions of length^{-1} defining the dispersion of the electron beam within the foil due to random elastic scattering and σ is the standard deviation of the incident electron probe intensity distribution and defines its diameter, B, from the full width at half maximum peak intensity (FWHM) $\simeq 2.35\sigma$. The dispersion coefficient, β, depends on the foil material and electron accelerating voltage such that:

$$\beta = (\frac{4Z}{E_o})^2 . \frac{\rho}{A} . \frac{10^3}{2} \tag{3}$$

where E_o is the electron accelerating voltage (V), t is the foil thickness in nm, ρ is the specific gravity, Z is the atomic number and A is the atomic

weight (g). The X-ray intensity, I, generated from a rod precipitate of
radius r, and composition C_1 in a matrix of composition C_o, Figure 1, is
given from equations 1 and 2:

$$I = KI_e/t \int_0^{t_1} \{ (C_1-C_o)(1-\exp(-r_1^2/(2\sigma^2 + \beta t^3))) + C_o \} \, dt. \qquad (4)$$

This equation may be evaluated numerically for any precipitate radius, r_1,
foil thickness, t_1, incident probe diameter, σ, and electron dispersion
coefficient β. For the Al-Fe alloy considered, the composition of the
Al_6Fe rod precipitates, C_1, is given from stoichiometry as 25.6 wt%Fe and
the low solubility of iron in aluminium means C_o is negligible. Results
are shown in Figure 2(a) for 100 kV and Figure 2(b) for 200 kV electron
energies and a fixed incident probe diameter, B, of 5 nm. The calculated
concentration of iron measured from these rod precipitates decreases with
increasing foil thickness and decreasing precipitate diameter i.e. as the
proportion of the incident electron probe scattered outside the precipitate
increases. The effect is more significant for lower electron energies and
larger incident beam diameters.

3. Experimental

3 mm disc specimens were machined from a rod sample of a directionally
solidified $Al-Al_6Fe$ eutectic composite with overall composition Al-4wt%Fe.
These were thinned for transmission electron microscopy by electrochemical
jetting using a 33% nitric acid - 67% methanol by volume electrolyte maintained at 233K with an applied voltage of 35-40V. Specimens were examined
in both a JEOL 100 CX and a JEOL 200 CX* transmission electron microscope
each fitted with an ASID 4D scanning transmission attachment together with
an energy dispersive X-ray spectrometer. The X-ray microanalysis procedure
has been described elsewhere (Doig et al, 1980).

X-ray energy spectra, obtained from thin regions containing the large Al_6Fe
(25.6 wt%Fe) grain boundary precipitates, were recorded for calibration of
the Link RTS2 FLS thin film energy dispersive X-ray analysis program used
in this work. Spectra were recorded from a number of rod precipitates,
remote from grain boundaries, with axes parallel to the incident electron
beam at various foil thicknesses in the range 100-600 nm at 100 kV and 100-
1500 nm at 200 kV. Foil thicknesses and rod precipitate diameters were
measured from transmission electron micrographs recorded after tilting the
foil through a known angle. Since the precipitates extend through the foil,
their projected length gives an unambiguous and direct measure of foil thickness.

4. Results and Discussion

The microstructure of the alloy, Figure 3, consists of a distribution of
cylindrical rod precipitates formed by a eutectic decomposition and confirmed
by electron diffraction to be Al_6Fe in an aluminium rich matrix. The precipitate rods were generally of uniform section with diameters in the range
65-90 nm and extended through the foil thickness. Rod axes were linear over
the range of foil thicknesses considered and their direction varied in
different regions of the foil. Large Al_6Fe precipitates $\sim$ 1 µm in diameter
were observed on the grain boundaries together with an associated precipitate
free zone $\sim$ 0.5 µm wide.

Measured count rates were $\sim$ 300 counts s^{-1} per 100 nm foil thickness for 100k'
and $\sim$ 40 counts s^{-1} per 100 nm foil thickness for 200 kv. Consequently, for

---

* Instrument facility kindly provided by JEOL (UK) Ltd, Colindale, London

the fixed time recording the statistical accuracy of spectrum analysis
improved with increasing foil thickness. Measured X-ray intensities were
corrected for absorption using a numerical procedure which considered the
local X-ray intensity generation distribution within the foil in conjunction with the phase distribution and X-ray path to the detector. Typically,
the statistical recording and processing error (standard deviation) in iron
analysis varied from ± 0.2 wt% (100 kV) or ± 0.6 wt% (200 kV) for 100 nm
foil thickness to ± 0.1 wt% for 500 nm foil thickness at 100 kV to ± 0.2 wt%
for 1000 nm foil thickness at 200 kV. Precipitate diameters and lengths
were measured to an accuracy of ∿ 2% resulting in an error in the calculated
value of iron composition which varies from ∿ zero for thin foils up to a
maximum of ∿ 0.2 wt%Fe for the smallest measured precipitate diameters and
largest foil thicknesses.

Results of the experimentally measured iron compositions within the rod
precipitates are shown in Figure 4 with respect to those predicted by the
present analysis (Section 2) for (a) 100 kV and (b) 200 kV accelerating
voltages. These measurements were made on rods of various diameters and
foils of different thickness. The lines with slope = 1 represent the
condition for agreement between theory and experiment. In general, the
trend of the results confirm such an agreement for the range of precipitate
dimensions considered.

The present work has examined the X-ray emission from a simple cylindrical
Al_6Fe precipitate in an almost pure Al matrix in order to test the application of the previously described theoretical analysis (Doig et al, 1981).
The very good agreement between the measured and predicted iron concentrations (Figure 4) lends support to the accuracy and validity of this
analysis. The experimental work has allowed the effects of precipitate
diameter, foil thickness and electron accelerating voltage to be examined.
The shape of the predicted analysis curves clearly shows that attempts to
linearly extrapolate measured compositions to a zero foil thickness in
order to eliminate the effect of electron probe spreading may be in error.
The results show that the simple analytical description of electron
scattering within thin Al-4wt%Fe foils adequately predicts the influence of
electron probe spread on the measured composition of the rod precipitates.

5. Acknowledgements

This paper is published by permission of the Director-General, Central
Electricity Generating Board, South Eastern Region.

6. References

Doig, P., Lonsdale, D., and Flewitt, P.E.J. (1980) Phil. Mag. <u>A41</u>, 761.
Doig, P., Lonsdale, D., and Flewitt, P.E.J. (1981) "Quantitative Micro-
analysis with High Spatial Resolution", Metals Society, U.M.I.S.T.,
Manchester. (Papers 22 and 23), London
Goldstein, J.I., Costley, J.L., Lorimer, G.W., and Reed, S.J.B. (1977)
Scanning Electron Microscopy, (SEM inc.) <u>1</u>, p.315.
Hall, E.L., Imeson, D., and Vander Sande (1981) Phil. Mag., <u>A43</u>, 1569.
Hutchings, R, Loretto, M.H., Jones, I.P., and Smallman, R.E., (1979)
Ultramicroscopy, <u>3</u>, 401.
Kyser, D.F., (1979) "Introduction to Analytical Electron Microscopy"
Ed. J.J.Hren, J.I.Goldstein and D.C.Joy, Pub: Plenum Pub. Corp., New York.

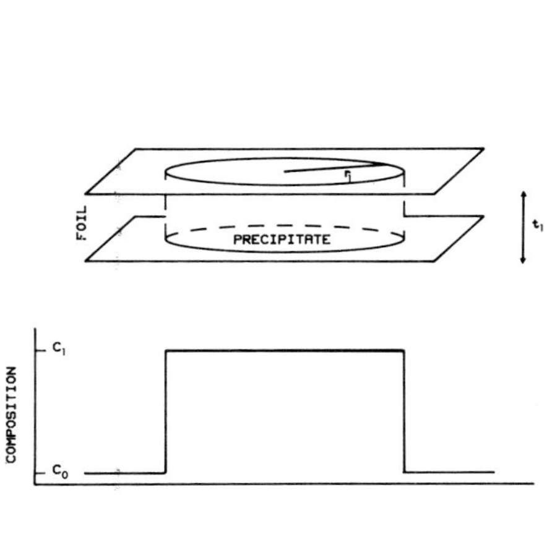

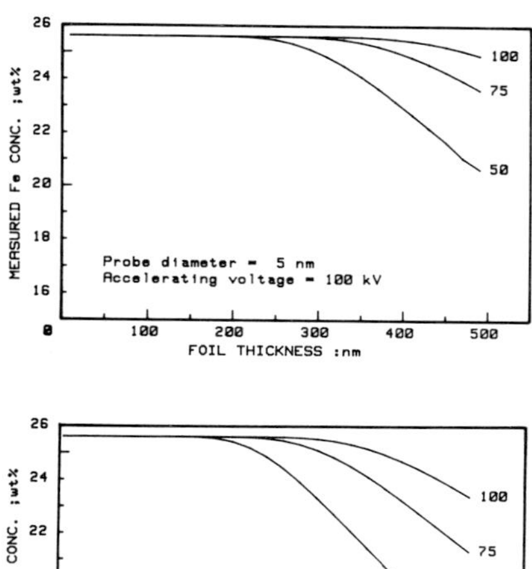

Fig 1. Schematic diagram showing geometry and composition of rod precipitate.

Fig 2. Calculated plots of measured iron concentration as a function of foil thickness for a range of rod diameters (nm).

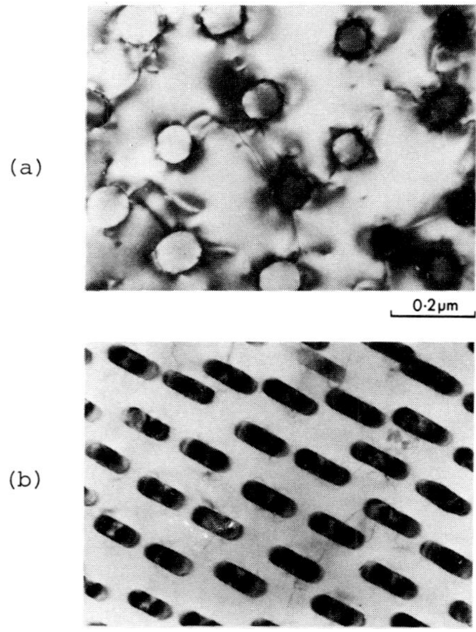

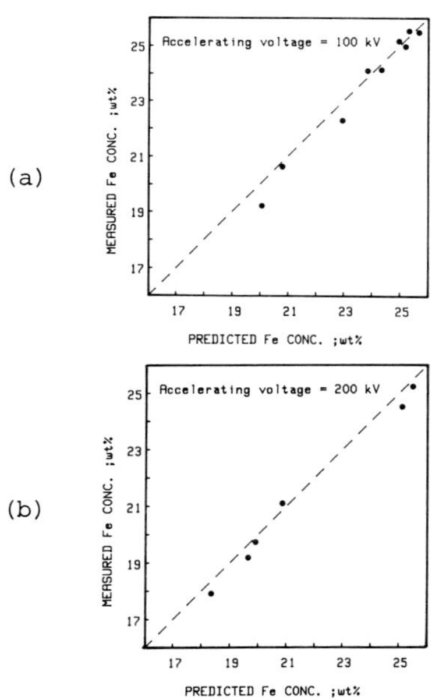

Fig 3. TEM micrographs of rod precipitates (a) parallel and (b) tilted w.r.t. the electron beam.

Fig 4. Experimentally measured versus predicted Fe concentrations in Al_6Fe rod precipitates.

The detection of monolayer grain boundary segregations in steels using STEM-EDS X-ray microanalysis

P. Doig and P.E.J. Flewitt
Central Electricity Generating Board, Scientific Services Department,
Gravesend, Kent, U.K.

1. Introduction

The equilibrium segregation of Groups IV to VI impurity elements to the
prior austenite grain boundaries of ferritic steels during heat treatment
in the temperature range 700 to 850K can significantly modify their mechanical and chemical properties (Hondros and Seah, 1977). Recently, the authors
have shown that embrittling segregations of elements (atomic number >11) can
be detected in conventional transmission electron microscope thin foil
specimens using the scanning transmission microscope combined with an energy
dispersive X-ray spectrometer (STEM-EDS) (Doig and Flewitt 1978; Doig et al
1978). Subsequent theoretical analysis (Doig et al, 1980, 1981(a), 1981(b))
predict that the proportion of X-ray signal obtained from the grain boundary
region is improved by reducing foil thickness and incident electron probe
size and increasing electron accelerating voltage. Such observations, however, must be considered in conjunction with the problem of maximising X-ray
counting statistics. Decreasing incident electron probe size and foil
thickness and increasing the accelerating voltage all serve to reduce the
X-ray intensity emitted for a given electron source brightness. Thus any
benefit of improved signal from the segregant at the grain boundary region
will be offset by decreased statistical significance of the X-ray data.

The purpose of the present paper is to investigate the relative effects of
electron probe size, accelerating voltage and foil thickness on the
detection of grain boundary segregations typical of those associated with
the impurity element embrittlement of ferritic steels.

2. STEM-EDS X-ray Microanalysis of Grain Boundary Segregations

The electron intensity distribution in a STEM electron probe passing through
a thin foil may be approximated by (Doig et al 1981(a)):

$$I(r,t) = I_e \{\pi(2\sigma^2 + \beta t^3)\}^{-1}.\exp\{-r^2/(2\sigma^2 + \beta t^3)\} \tag{1}$$

for a total incident electron intensity of I_e where $I(r,t)$ is the electron
intensity at a distance r from the centre of the probe and depth t in the
foil, σ is a measure of the incident electron probe size with the probe
diameter, $B(FWHM) \simeq 2.35\sigma$, and $\beta(nm^{-1})$ is a parameter defining the electron
scattering characteristic of the foil material given by:

$$\beta = (4Z/E_o)^2 .\rho/A . 1000/2 . \tag{2}$$

Here Z, A and ρ are the atomic number, the atomic weight and density of
the foil material and E_o is the electron accelerating voltage in eV.

We shall assume a grain boundary segregation profile which is described by:

$$C_x = C_o \exp(-|x|/w) \tag{3}$$

where C_x is the concentration of the segregating species at distance x from the grain boundary, C_o is the concentration on the grain boundary and w is a length parameter defining the spatial extent of the segregation. For the present work we shall consider a segregation described by w = O.lnm. Combining the electron intensity distribution with this assumed segregation profile gives the X-ray signal from the grain boundary I*(O,t):

$$I^*(O,t) = K\int_V \{C_x \cdot I(r,t)\}\, dV \tag{4}$$

where K is a constant relating the local electron flux to the detected X-ray intensity and $r^2 = x^2 + y^2$. Thus from equations 1, 3 and 4 we may evaluate I*(O,t) in iron base foils as a function of foil thickness, t_1, electron accelerating voltage and electron probe diameter.

The accuracy of any X-ray measurement is limited by statistical consideration of the detected X-ray quanta and is related to the counts in the characteristic X-ray energy peak, N_p, compared with those in the background, N_b. Confidence in any measured peak intensity is dependent on both the peak and background such that the net error (standard deviation) in the measured total intensity to background intensity ratio is given by:

$$\{(N_p + N_b)/N_b\}\{(N_p + N_b)^{-\frac{1}{2}} + N_b^{-\frac{1}{2}}\} \tag{5}$$

For any given electron source the total incident electron intensity increases with the electron probe diameter, B, formed on the specimen by:

$$I_e \propto \lambda B^{8/3} \tag{6}$$

where λ is a measure of the electron source brightness (electron flux). The background intensity increases linearly with foil thickness, t_1, total incident electron intensity, I_e, and spectrum recording time, T, such that:

$$N_b \propto KI_e T t_1 \text{ which from equation (6) gives } N_b = K'B^{8/3} t_1 \tag{7}$$

where K' is a constant characterising the electron accelerating voltage, source brightness, and the spectrum recording time, T. Combining equations (4), (5) and (7) allows evaluation of the accuracy of any X-ray measurement of grain boundary segregation as a function of t_1, K', B and R* (i.e. C_o). Figure 1 shows typical predicted X-ray intensity ratios $(N_p + N_b)/N_b$ together with the expected error in measurement defined by equation (5). The error in peak to background measurement decreases with increasing t_1, K' and B. Where the error in measurement is large such that the error bar crosses the line $(N_p + N_b)/N_b = 1$ then the accuracy of the measurement is not sufficient to detect a peak of value N_p. Comparing Figures 1(a) and (b) show that increasing B from 2 to 10 nm improves the counting statistics sufficiently for detection of a grain boundary segregation over the foil thickness range 20-500 nm. Increasing the parameter K' (Figures 1(c) and (d)) further improves the statistics to allow more accurate measurement. For any given set of these parameters there is a finite range of foil thickness over which the segregation is detectable. This is shown in the plots of Figure 2 where the foil thicknesses required for detection of grain boundary segregation are plotted as a function of the parameter K' for a range of R_o. For any given value of R_o, boundary segregation can only be detected with those combinations of K' and foil thickness to the right of the specified peak to background line. These results show that for a given effective electron intensity the sensitivity of detecting grain boundary

---

*R_o is the X-ray peak to background ratio characteristic of C_o

segregation improves with increasing electron probe diameter and, to a lesser extent, accelerating voltage. In general, the optimum sensitivity does not occur at the minimum foil thickness; for lower values of peak to background ratio there exists an optimum foil thickness of $\sim$ 100nm for detection of segregation. Calculated results demonstrate the sensitivity to detection improves slowly with increasing brightness such that a tenfold increase in brightness improves the sensitivity by a factor of $\sim$ 2.5.

3. Discussion

The analysis described in Section 2 combines the influence of electron scattering in a thin foil with a consideration of the statistical confidence in the measured data for the technically important case of embrittling segregation to grain boundaries. In the present work we have utilised a parameter K' which defines the background count, N_b, per unit time. In practice the recording time is limited by mechanical stability and specimen contamination. For a given electron flux, the X-ray generation efficiency is related to the accelerating voltage by the factor $(\ell nE)/E$ resulting in a net decrease in K' by a factor of $\sim$ 1.75 when changing the electron accelerating voltage from 100 to 200 kV. However, reduced electron scattering at the higher kV localises the electron probe to the grain boundary thereby increasing the proportion of X-ray intensity generated from the segregated species. For the range of foil thicknesses and electron probe diameters considered in this work however, the relative increase is small and less than the contribution due to the changing X-ray counting statistics. For an electron probe with brightness $> 10^{-4}$ and diameter of 10 nm typical of that in a conventional (S)TEM the minimum detectable grain boundary composition at 200 kV is increased to $\sim$ 1.2 of that at 100 kV. The background X-ray intensity and peak to background intensity ratio are both complex functions of the foil material and instrumental variables. It is clear that any increase in the value of the peak to background ratio, R_o, per unit concentration being measured will result in an improved composition resolution. R_o is approximately independent of electron accelerating voltage for the range currently used in STEM-EDS microanalysis. The present analysis predicts detection limits for C_o of < 2 at%Sn and < 20 at%P for our JEOL JEM 100CX TEMSCAN/LINK model 860 EDS system for an incident electron probe diameter of 5 nm at 100 kV using a conventional thermal hairpin filament.

4. Conclusions

(a) Monolayer segregations to grain boundaries in ferritic steels can be detected and quantitatively measured using STEM-EDS X-ray microanalysis on conventional thin foils.

(b) Increasing electron accelerating voltage from 100 to 200 kV decreases the detectability of any segregation by a factor of about 1.2.

5. Acknowledgement

This paper is published by permission of the Director-General, South Eastern Region, Central Electricity Generating Board.

6. References

P. Doig, P.E.J. Flewitt (1978) J. Microsc., 112, 257.
P. Doig, P.E.J. Flewitt and R.K. Wild (1978) Phil. Mag., 37, 759.
P Doig, D Lonsdale and P.E.J. Flewitt (1980) Phil. Mag., 41, 761.
P Doig, D Lonsdale and P.E.J. Flewitt (1981(a)) Proc. of "Quantitative Microanalysis with High Spatial Resolution", Manchester, Inst. Met., London.
P. Doig, D Lonsdale and P.E.J. Flewitt (1981(b)) Met. Trans. 12A, 1277.
E. D. Hondros and M.P. Seah (1977) Int. Met. Rev., 222, 262.

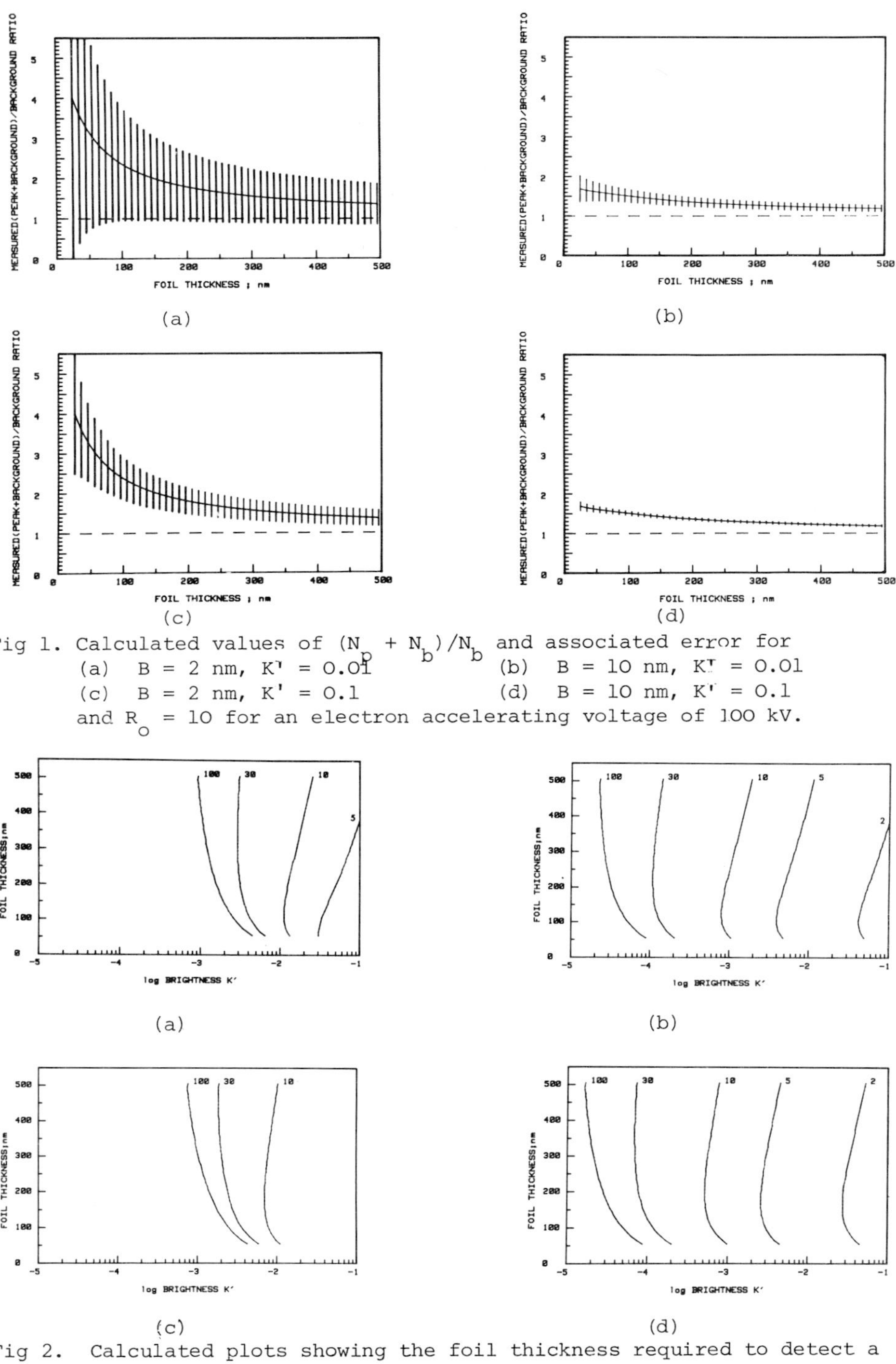

Fig 1. Calculated values of $(N_p + N_b)/N_b$ and associated error for
(a)   B = 2 nm, K' = 0.01   (b)   B = 10 nm, K' = 0.01
(c)   B = 2 nm, K' = 0.1   (d)   B = 10 nm, K' = 0.1
and R_O = 10 for an electron accelerating voltage of 100 kV.

Fig 2.   Calculated plots showing the foil thickness required to detect a
g.b. segregation characterised by a range of R_O as a function of K'
(a)   B = 2 nm, 100 kV   (b)   B = 10 nm, 100 kV
(c)   B = 2 nm, 200 kV   (d)   B = 10 nm, 200 kV.

Sensitivity limits in X-ray emission spectroscopy of catalysts

H Dexpert, J P Lynch and E Freund

Institut Français du Pétrole, BP311, 92506 Rueil Malmaison, France

1. Introduction

A VG HB5 STEM equipped with a Kevex energy dispersive X-ray spectrometer has
been used to study the chemical composition of several industrial catalyst
systems at high spatial resolution. The catalysts studied consist of a
highly divided aluminium oxide support which has been impregnated with
heavy metal compounds such as H_2PtCl_6. Calcination and reduction are then
carried out, to fix the metal, in conditions chosen to maximise the dispersion (surface area) (Moss 1978). Sizes of the resulting particles are
typically less than 10 nm. The systems therefore offer a useful test of
sensitivity for such analyses. The form of the light oxide support, platelets up to 100 nm wide and a few tens of nm thick, allows absorption and
secondary fluorescence effects to be ignored.

Some care is needed, however, to minimise spurious counts which might mask
the very low signals from the metal particles. A graphite nosed cartridge
and often nylon support grids are used to minimise stray X-rays from material surrounding the specimen. The probe is defined by a virtual objective
aperture, not visible to the detector, so that photons produced at the
aperture edge will not be detected. It is found that the remaining hole
count is significantly reduced by the use of a selected area diffraction
aperture, which presumably serves to limit X-rays produced in the column
reaching the specimen and causing secondary fluorescence. The 'real"
objective aperture is not used in analysis but is equipped with an electro-
meter so that, by blocking the beam, the probe current may be measured.

2. Specimen Limitations

Specimen stability is the major limiting factor for analysis under such
conditions. Contamination and drift may be reduced by baking and allowing
the stage to cool before analysis so that these are not usually problems.
Beam induced specimen changes may however reduce the usefullness of the
analysis. An example is shown in Figure 1 of two consecutive analyses of a
Pd/Al_2O_3 catalyst precursor (i.e. not yet reduced to the metallic state).
An area apparently void of particles was analysed in scan mode for approximately 100 secs and a small peak due to Pd is just visible above background,
accompanied by Al and Ca due to the support and Cu from the grid. Image 2,
taken after analysis, shows that some change in the specimen has occured—
the analysed area is now dark. A subsequent analysis of the same area
reveals a much higher Pd count, indicating that Pd is segregating to the
area under the effect of the beam. In this case it cannot be said that
the first analysis is representative of area shown in image 1.

Problems encountered more commonly are destruction of the support and displacement or melting of the particle. Except in the case of very stable specimens these effects limit particle lifetimes under the beam to below 200 secs.

3. Particle Analyses

An important contribution can be made by the STEM in the study of bimetallic catalysts, where alloying is expected to enhance selectivity (Sachtler and van Santen 1977). Figure 2 shows a $Pt-Pd/Al_2O_3$ catalyst with particles ranging in size from 5 nm to below 1 nm. The particles remain stable for analysis times well over 100 secs so that several particles on the same support platelet may be analysed to determine concentration ratios within $\pm 10\%$ using atomic number corrections (Reed 1975). In this specimen approximately one third of the particles are pure Pt – the remainder ranging from $Pt_{0.3}Pd_{0.7}$ to $Pt_{0.5}Pd_{0.5}$.

Insufficient counts are obtained in 100 secs to determine the accurate composition of very small particles. If the analysis time is increased to 900 secs significant signals may be obtained from particles down to 1.25 nm at the expenses of destroying the particle. Figure 4 shows a typical case, from the same $Pt-Pd/Al_2O_3$ catalyst. The support and particle shown in (a) were analysed in succession. The support spectrum, (b), shows only Al and a small, spurious, Fe peak above background. The particle spectrum, (c), shows in addition significant Pt, Pd and Ca peaks. Although interpretation is difficult because of severe specimen damage, (d), the comparison of (b) and (c) shows these three elements to have been localised near the particle. The most likely explanation is thought to be nucleation of the Pt-Pd particle at an area of the support locally rich in calcium, presumably remaining from the calcination during preparation.

References

Moss R L 1978 in Experimental Methods in Catalytic Research, Anderson R B and Dawson P T (eds) (New York : Academic Press)
Reed S J B 1975 Electron Microprobe Analysis (Cambridge : Cambridge University Press)
Sachtler W M H and van Santen R A 1977 Adv. Catal. <u>26</u> 69

Acknowledgements

J P Lynch would like to thank the Royal Society for an Exchange Fellowship.

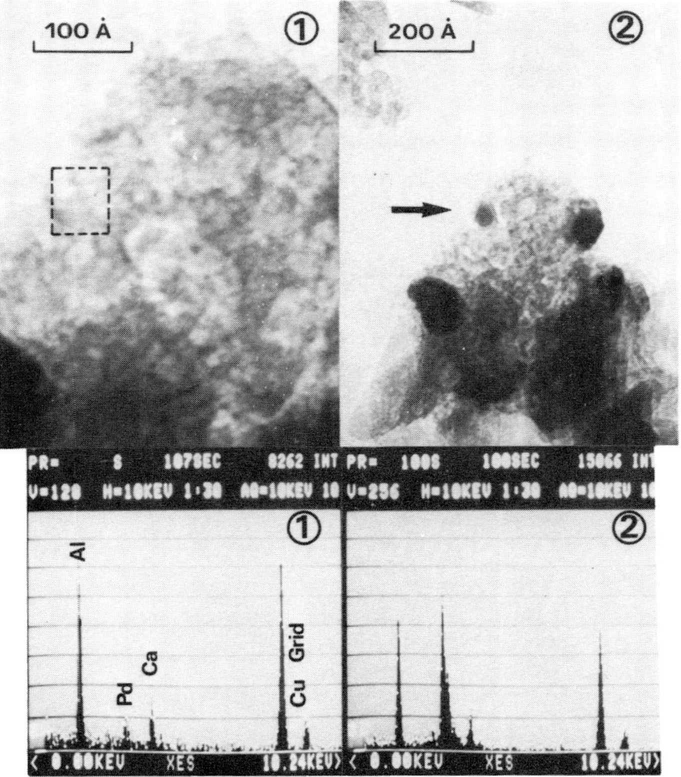

Figure 1

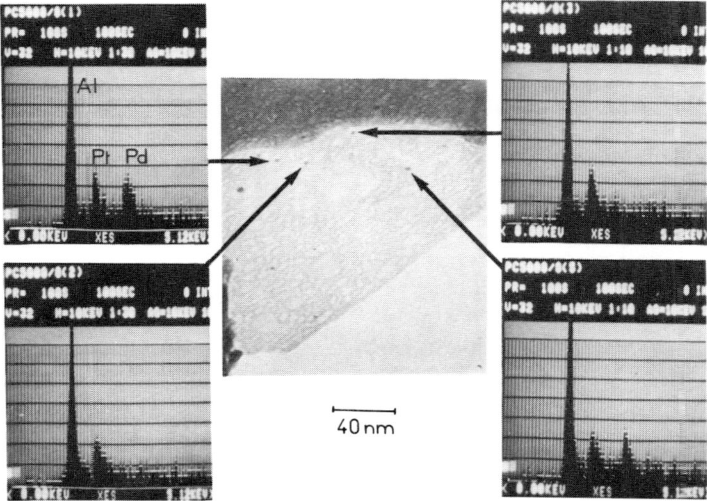

Figure 2 : beam current = 1.2 nA

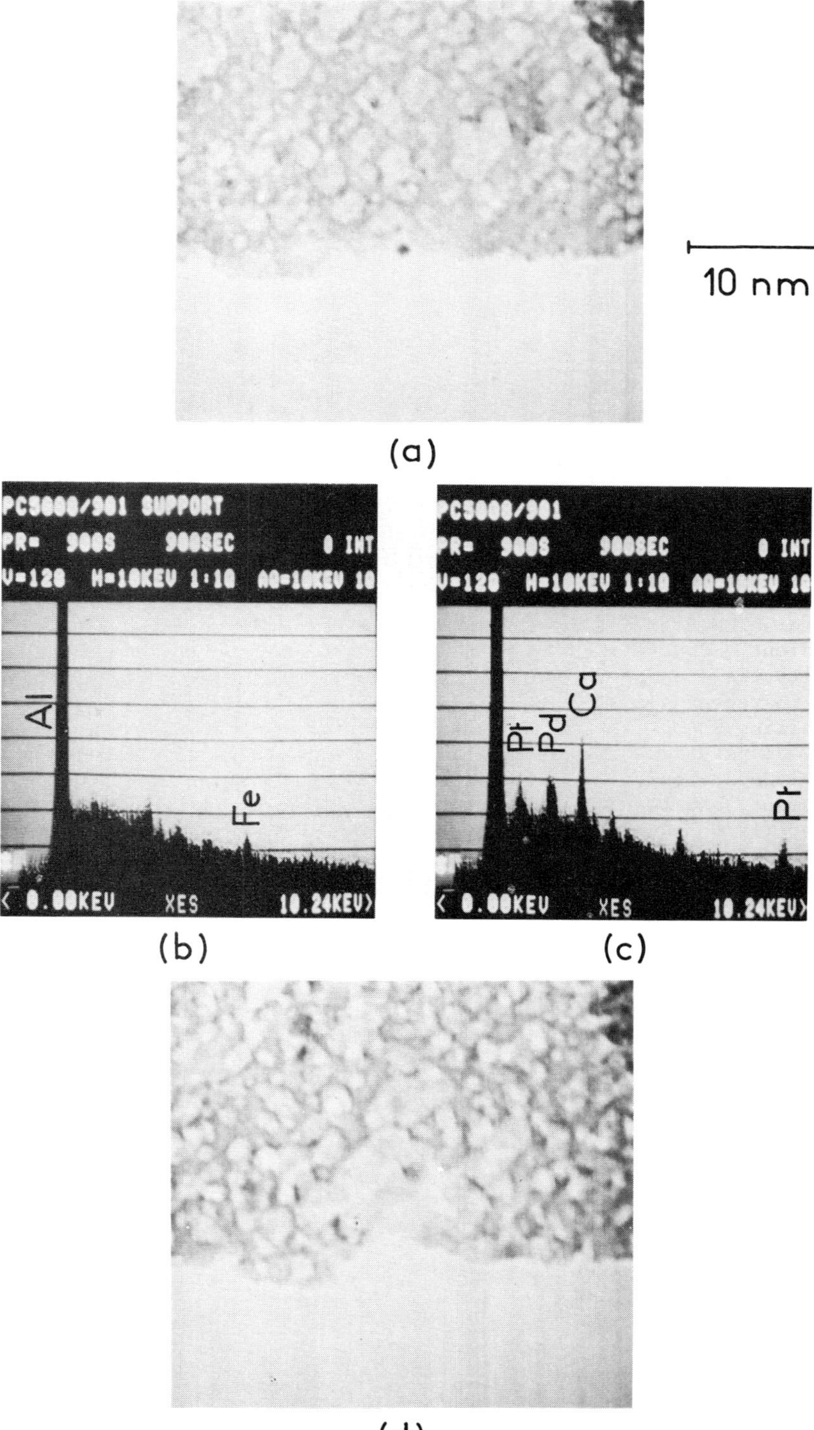

Figure 3 : beam current = 0.7 nA

On the role of an FEG combined TEM/STEM system in catalysis studies

M D Shannon, J M Bell and S J Pennycook[*]

ICI Corporate Laboratory, PO Box 11, The Heath, Runcorn.
*Cavendish Laboratory, Cambridge.

1. Introduction

Characterisation of catalysts by electron microscopy can give information
about particle size distributions, relative positions of particles,
particle structures and compositions. The aim is to correlate this with
catalyst manufacture and with performance. A wide range of materials is
encountered (eg supported metals, alloys, oxides, zeolites) and particle
sizes can range from several microns down to perhaps 1nm. For many
studies a TEM/STEM with conventional tungsten hairpin filament is
sufficient. However, for very small particles (< 3nm) the dedicated STEM
with its field emission source and alternative imaging modes has been
shown in recent years to be invaluable (Pennycook 1981, Treacy et al 1980).
The Philips EM400T fitted with field emission gun and STEM attachment plus
energy dispersive X-ray micro-analysis facility provides a machine which
not only operates fairly well in the regimes covered by the two types of
instrument, but also adds one or two new possibilities. Carpenter and
Bentley (1979) have described the performance of a prototype model with
regard to metals and ceramics. Here we outline the scope and performance
of a production instrument in catalyst characterisation.

The gun is of the thermally assisted type. The maximum brightness of the
present tip is estimated to be 2×10^8 A/cm^2/Sr for 20µA total current at
100kV. The gun will operate at up to 120kV. Due to the thermal
stabilization there is some flicker in the illumination intensity. The
exposure meter displays $\pm$ 10% fluctuations for the present tip. An
earlier tip gave only $\pm$ 2% fluctuations. The mean exposure time is
constant over several hours for a fixed operating temperature and
extraction voltage ($\sim$3kV). The present tip has run for some 200 hours.

2. Imaging

In the TEM mode exposure times of a fraction of a second with beam
divergence less than 1 mrad are possible. The resolution should be
around 0.4nm point to point. However the maximum focussed probe size is
around 15nm and overfocussed illumination is therefore generally used but
can lead to problems of varying astigmatism within the illuminated area.
Nonetheless $\sim$1nm Ru particles can be recognized in Fig 1.

In the STEM mode of operation the illumination flicker currently restricts
operation to the DF/BF signal. Edge to edge resolution of 0.5nm was
demonstrated on installation. Fig 2 is a DF/BF image from another Ru on

SiO_2 catalyst. 1.0-2.0nm particles can be discerned in much higher
contrast than in Fig 1. The angular limitations on the dark field
detector were 16 to 140mrad approximately. The image is of similar
quality to the nearest equivalent (ADF/ΔE) from Pennycook (1981) taken on
a VG-HB5 collecting between 13 and 50mrad. Treacy et al (1980) describe
a high-angle annular detector spanning 100 to 400mrad and its application
to imaging Pt on a crystalline supportγ-Al_2O_3. The EM400T can achieve 65
to 150mrad which, whilst it does not go very far into the region dominated
by incoherent thermal diffuse scattering, should only contain very weak
Bragg reflections from γ-Al_2O_3. The imaging of Ru on γ-Al_2O_3 has still to
be attempted. The resolution in the secondary electron mode has not been
measured. However, 3nm diameter silver particles on α-Al_2O_3 have been
observed (Fig 3).

3. <u>Microanalysis</u>

In X-ray analysis there are three possible modes of operation -
"microprobe", "nanoprobe" (these are two modes of TEM operation in the
EM400T - see Van der Mast (1980) for details) and STEM. The Table
summarizes the probe diameters (nm) and beam divergence (mrad) for given
C_2 aperture sizes and together with the brightness enables probe currents
to be estimated. In the "microprobe" mode the probe has an electron tail
which extends about 20nm, and is only used where spatial resolution is not
important. In the STEM mode X-ray mapping can prove useful when
investigating the distribution of two phases as in Fig 4. Good spatial
resolution is apparent - a band 15nm wide and about 30nm thick is imaged
as 18nm. All results presented were obtained from a detector at 90° to
the incident beam subtending 0.04srad.

However it is the analysis of sub 5nm particles which is of greatest
interest. For a given brightness and beam divergence there is clearly
more current in the "nanoprobe" than a STEM probe of 0.5-1nm diameter, by
roughly an order of magnitude. Of course the brightness and/or beam
divergence could be increased in a STEM, at the cost of increased current
density, to compensate. Fig 5 shows a spectrum obtained in the "nanoprobe"
mode from a 2.5-3.0nm Ru particle on SiO_2. The probe current was about
0.12nA and analysis time 30 seconds. It is possible to obtain 2nA probe
currents in this mode but even at 0.12nA the Ru particle was twisting
probably because of damage to the SiO_2 support. Pennycook (1981) took
very much longer times (c. 1500 secs) in the HB5 STEM to obtain spectra
with similar Ru counts, but probably used much lower current densities in
order to record microdiffraction patterns simultaneously. Spatial
resolution is good in the "nanoprobe" mode. Fig 6 shows a spectrum from
a 4nm Ru particle obtained under the same conditions as that in Fig 5, and
one taken 5nm from the particle. Only a trace of the Ru signal was
detected in the latter. The background signal from Cu grid bars is higher
than in the HB5 and is enhanced when scattering of the primary beam by Ru
metal takes place. There is some drift of the focussed probe of $\stackrel{\sim}{\sim}$ 1.5nm
per minute, which may be a problem where longer analysis times are needed
for smaller particles or alloy systems. Contamination is generally not
observed in the "nanoprobe" mode but can be a slight problem in the STEM
mode (see Fig 4). Where possible the "nanoprobe" is preferred in the
EM400T, and has even been applied to "surface" analyses of the edge of
particles in the 10-50nm range.

The FEG version of the EM400T offers the same excellent microdiffraction
facilities as the basic instrument but from smaller areas (see Table). The

great advantage over most dedicated STEMs is the direct recording of the diffraction pattern on photographic film. In fact the smaller probe sizes can be a problem when dealing with beam sensitive materials. So although convergent-beam diffraction was possible on some zeolites (Fig 7) before the FEG was fitted, it is now more difficult.

The FEG version of the EM400T can be used on a wide range of catalyst problems including those usually assigned to a dedicated STEM.

References

Carpenter, R W and Bentley, J 1980, SEM/1979/I, 153
Pennycook, S J 1981 J Microscopy to be published
Treacy, M M J , Howie, A and Pennycook, S J 1980, Electron Microscopy and Analysis 1979, Inst Phys Conf Series, 52, 261
Van der Mast, K D 1980, Electron Microscopy 1980 Volume I, 72

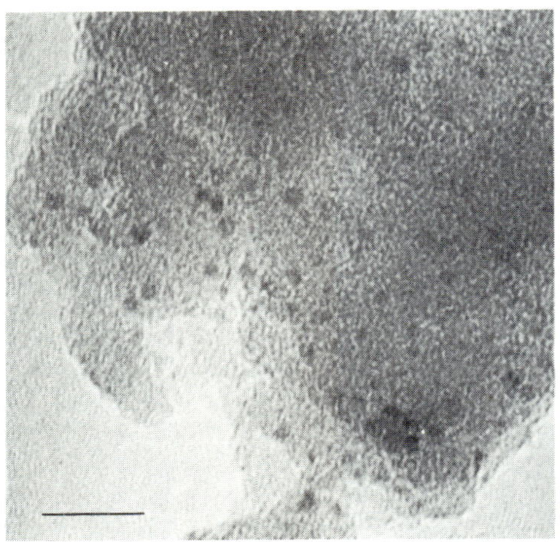

Fig 1.   Bright field TEM image of Ru on SiO_2. (Marker is 10nm)

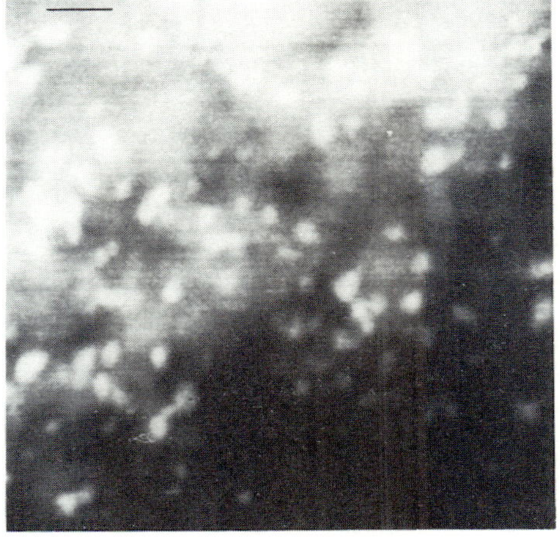

Fig 2· Dark field/bright field STEM image of Ru on SiO_2 (Marker is 10nm)

		Table		
Mode	Probe(nm)	100µ	50µ	15µ
Micro	10	4.3	2.4	0.9
Nano	2.5		8.6	3.1
STEM	0.5		11.2	4.0

Fig 3.   Secondary electron image of silver on $\alpha\text{-}Al_2O_3$ (Marker is 10nm)

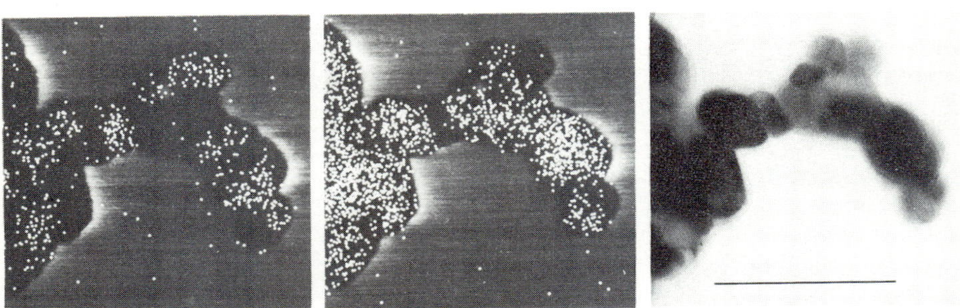

Fig 4 .   X-ray mapping of sample containing oxides of A and B
(a) element A, (b) element B, (c) TEM bright field.   (Marker 100 nm).

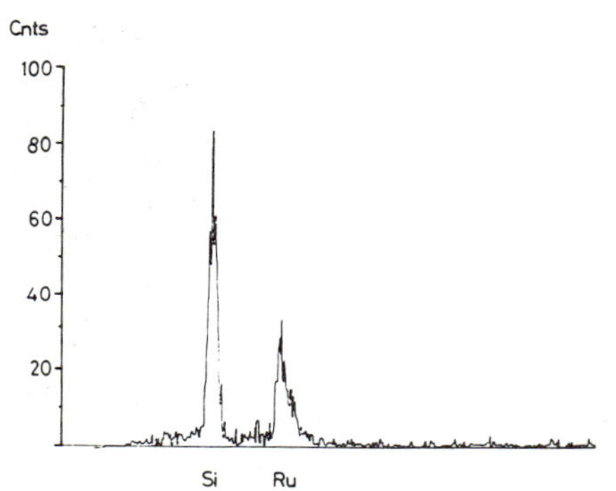

Fig 5.   Spectrum from 2.5-3.0nm Ru
particle on SiO_2

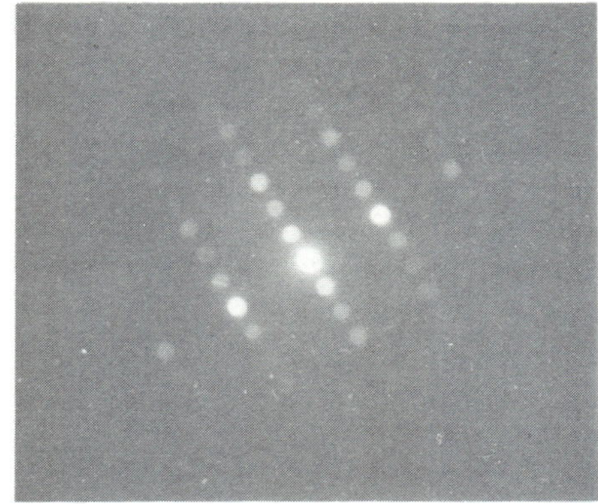

Fig 7.   Convergent-beam diffraction
pattern from a zeolite.

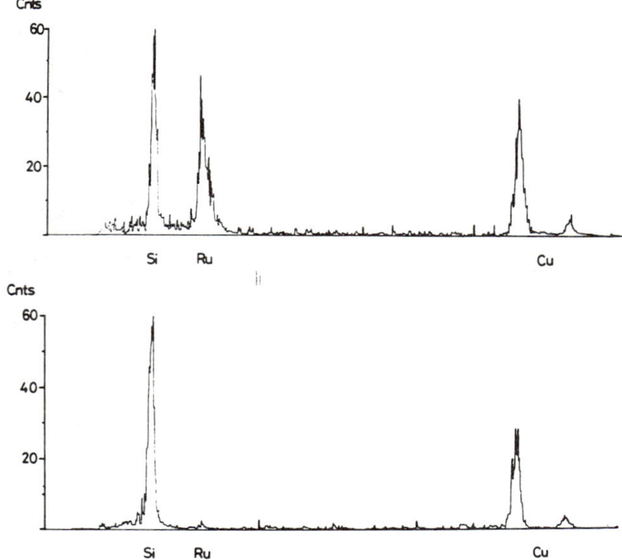

Fig 6.   Spectra from (a) 4nm Ru particle on SiO_2. (b) 5nm from (a)

Ultramicrotomy as a specimen preparation technique for thin foil microanalysis

M.D. Ball, R.C. Furneaux

Alcan International Limited:Banbury Laboratories,
Southam Road, Banbury, Oxon OX16 7SP

The value of thin foil microanalysis is already well established, providing good quantitative information at high spatial resolution. However, as the technique has developed, a growing number of examples has been found where surface compositional variations have significantly distorted the results (e.g. P.L.Morris et al 1977). These variations in composition can usually be attributed to specimen preparation procedures. Chemical or electro-chemical polishing is frequently complicated by the selective removal or deposition of certain elements, and similarly ion beam thinning can also result in enriched or depleted surface regions.

By preparing thin sections of the sample of interest using an ultra-microtome, (e.g. Furneaux et al 1978) these problems can generally be overcome (Figure 1). In addition, the ultramicrotomy offers a number of other advantages where microanalytical information is important.

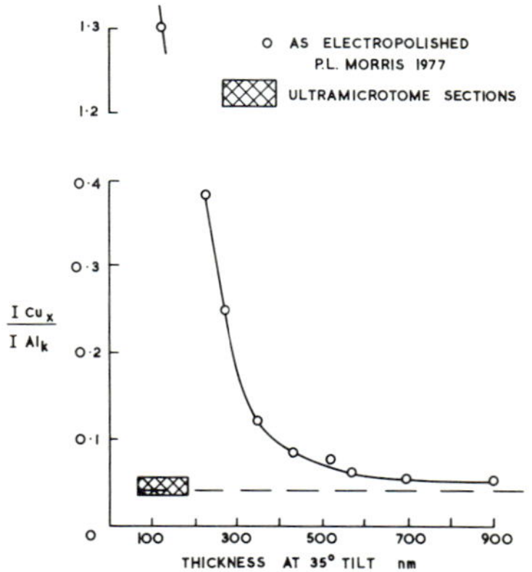

FIGURE 1.

VARIATION OF X-RAY INTENSITY RATIO WITH THICKNESS IN AN Al-4% Cu ALLOY
Microanalysis of electropolished specimens of Al-4% Cu alloy indicates the presence of a surface enrichment of Cu which dramatically distorts the measured compositions in thin areas. The compositions which were obtained from ultramicrotomed sections for thicknesses between about 50 and 200nm (cross hatched) do not exhibit this anomalous behaviour and the measured compositions correspond to the bulk analysis (dotted line).

The advantages of samples prepared by ultramicrotomy can be summarised:

1. Compositional artefacts such as surface enrichment or depletion effects do not occur (Figure 1).

2. The sections have a comparatively uniform controllable thickness. This facilitates the microanalysis of samples where absorption corrections are important, since it obviates the need to make repeated thickness measurements at each analysis point.

3. It is possible to select the area from which the section is taken.

4. Sections can be prepared from difficult samples such as bimetallic junctions, welds, porous and multiphase structures, surface films, etc.

5. Thin sections of large area can be prepared (a typical section from an aluminium alloy would be about 200 μm x 200 μm x 50 nm).

6. Ultramicrotome sections generally exhibit a lower contamination rate in the TEM than comparable electropolished specimens.

7. Ultramicrotomy offers the possibility of 3-D investigations by using serial sectioning.

   The main disadvantages of ultramicrotomy are:-

1. The microstructure is sometimes severely distorted and damaged.

2. Buckling and folding of the sections can lead to difficulties in determining the orientation of the area of interest (this is important for calculations of microanalysis corrections involving X-ray mass path length).

To summarise, ultramicrotomy can be a valuable addition to the range of techniques available to the analytical electron microscopist, complementing the more conventional specimen preparation methods.

References

R.C. Furneaux, G.E. Thompson, G.C. Wood 1978. Corrosion Science 18, pp 853-881.

P.L. Morris, N.C. Davies, J.A. Treverton 1977. Developments in Electron Microscopy and Analysis, EMAG 77, Ed.D.L. Misell, pp 377-380.

Errors in dead-time correction with EDS analysis

D.J.Bloomfield, G.Love and V.D.Scott

School of Materials Science, University of Bath, U.K.

In quantitative electron-probe microanalysis, dead-time corrections must be applied to take account of the finite time required for processing individual x-ray signals in the x-ray detector and associated electronics. In EDS systems, dead-time corrections are handled by the electronic circuitry and, although they may be substantial, the operator normally assumes that the process is satisfactorily carried out. Here experiments on two EDS systems are described and it is shown that one of them may give rise to significant errors in the dead-time correction.

1. The method of Russ et al (1973)

In EDS systems an amplifier with a short time constant ($\sim$ 1µs) is placed in parallel with the main amplifier in order to deal with the pulse pile-up problem (Williams 1968), i.e. when the main amplifier cannot separate two pulses arriving close together and records them as a single event of higher energy. However, the fast amplifier fulfils a second purpose in the method of Russ et al, where it is assumed that it recognizes all the input pulses and thus gives the true count rate. The main amplifier is recording at a lower rate due to dead-time and so the system periodically stops the 'live-time' clock to allow the main amplifier to acquire extra counts and match the count rate indicated by the fast amplifier.

The method was evaluated using an EDAX 711 analyser and an ECON detector. Figure 1 shows data of counting rate and corrected count rate for pure specimens of silicon, iron and zinc measured at different keV with the beryllium detector window in position. It may be seen that a smooth curve can be drawn through the silicon and iron results but that the zinc measurements appear to have been undercorrected by the system. The error is worse the lower the kV and the higher the count rate. Figure 2 shows the amount of undercorrection for different specimens. Inspection of the

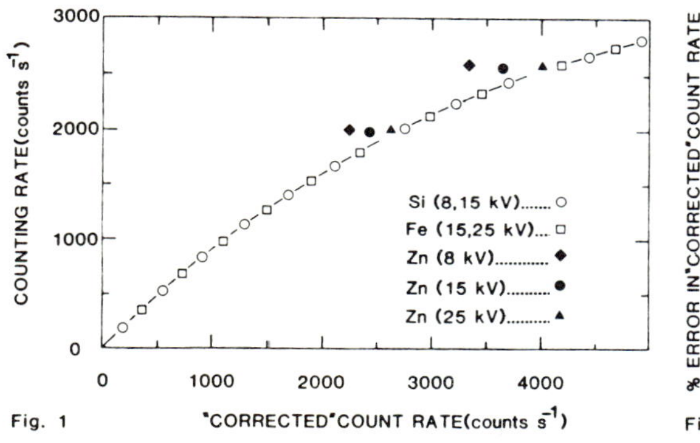

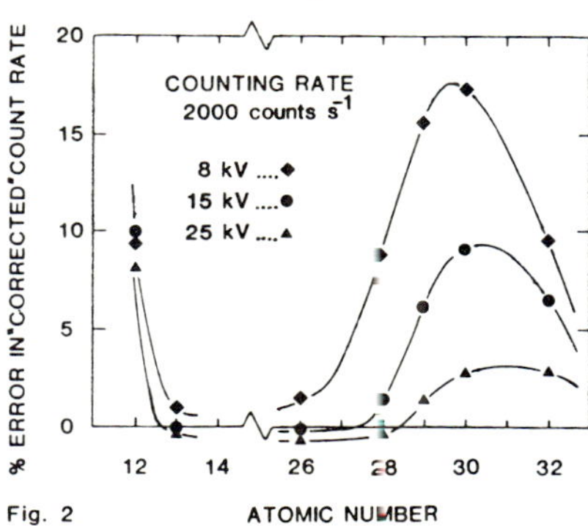

data indicates that the errors appear to be associated with the presence
of a low energy (<1500eV) peak in the recorded emission spectrum. For
specimens of atomic number Z<14, there are K peaks present while for
26<Z<32 the L emissions have energies below 1500eV. Although elements
between the two ranges may emit L radiations, these are absorbed in the
beryllium window and hence not detected.

The effects may be explained by considering the characteristics of the
fast amplifier where, in order to achieve speed, resolution has been
sacrificed ($\sim$ 700eV compared with the main amplifier resolution of $\sim$150eV).
Consequently, to avoid triggering of the fast amplifier by noise, the low
energy threshold has to be maintained at a high value, $\sim$ 800 eV. This
means that pulses with energies below the sum of these two energies
($\sim$1500eV) may not be detected, although most would be recorded by the
main amplifier (threshold at $\sim$150eV). The fact that pulses are missed
by the fast amplifier causes underestimation of the true count rate and
insufficient dead-time correction is applied to the main amplifier.
The undercorrection will be worse, the greater the fraction of pulses
below 1500eV, which explains why errors are worse at lower electron
accelerating voltages. The progressively larger errors as the specimen
atomic number is increased from Z = 26 to Z = 30 are associated mainly
with a progressive increase in transmission of L radiations through the
beryllium window. For Z>30, however, the fast amplifier now starts to
'see' more of the L x-ray pulses since they are more energetic, and the
error decreases accordingly. The observed decrease in the error between
$12 \leq Z \leq 14$ is for similar reasons, although now it is the K radiation
from these elements which is responsible for the undercorrection.

Replacing the beryllium window with the plastic window increased the
errors due to a greater fraction of lower energy x-ray photons being
transmitted. For lighter elements ($Z \leq 11$), no dead-time correction
was applied by the system and this resulted in errors of $\sim$ 25% for
counting rates $\sim$ 2000 sec^{-1}. In addition, significant errors were now
found when analysing iron, manganese and chromium.

2.   The method of Covell et al (1960)

In this method the clock is stopped by the main amplifier when a pulse
is being processed and so the output should be given directly in terms
of the live-time of the system. The task of the fast discriminator is
merely to deal with pulse pile-up. Hence, it would not be expected that
this method of correcting for dead-time would suffer from the major
problem experienced with that of Russ et al. Indeed, when the Covell
method was checked using a LINK systems 2010/1057, this was found to
be the case.

Acknowledgments to the SRC and MOD, Aldermaston, for support.

References

Covell D.F., Sandomire M.M. and Eichen M.S. 1960 Anal.Chem. <u>32</u> 1086.
Russ J.C., Sandborg A.O., Barnhart M.W., Soderquist, C.E., Lichtinger R.W.,
  and Walsh C.J., 1973 Adv. X-ray Anal. <u>16</u> 284.
Williams C.W., 1968 IEEE Trans. Nucl. Sci. NS <u>15</u> (1) 297.

Electron energy loss spectroscopy in the electron microscope: a review of recent progress

C Colliex, O L Krivanek* and P Trebbia

Laboratoire de Physique des Solides associé au CNRS, Bâtiment 510
Université Paris-Sud, 91405 Orsay, France

*Perm. address: Center for Solid State Science, Arizona State University,
Tempe, AZ 85287, USA

1. Introduction

Since 1978 (Colliex and Trebbia) significant advances in several aspects of
electron energy loss spectroscopy (EELS) have been made. Important developments in the design of energy loss spectrometers for EM applications
have taken place. Second-order aberrations both parallel and perpendicular
to the median plane of the spectrometer have been corrected (see, for
instance, Parker et al. 1978, Shuman 1980, Egerton 1980, Krivanek and
Swann 1981), electromagnetic focusing and aberration correction have been
introduced, detection systems capable of fast and reliable single electron
counting have been developed (Krivanek 1979, Engel 1981), and progress has
been made towards efficient parallel recording systems (Johnson 1979,
Egerton 1981, Shuman 1981). These developments have now led to many new
experimental advances, several of which are reviewed here.

As an example of the work now routinely possible we show spectra of some
'standard' specimens obtained with a Gatan 607 electron spectrometer
mounted on a VG HB5 STEM. Fig. 1a shows the zero loss peak, the surface
plasmon and the bulk plasmon in Si. The zero loss half-height width is
about 0.4 eV. Fig. 1b shows the Si $L_{2,3}$ edge in the same specimen revealing the characteristic pre-ionization peak at the edge threshold.
Figs. 1c and 1d show the well-known fine structure of the carbon K-edge in
amorphous and graphitic carbon respectively. All spectra were acquired
with a 20 Å probe in typically 100 msec per channel. The collection half-
angle (measured at the specimen) was 2 mrad for 1a and 20 mrad for 1b-d.
No post-specimen lenses were used to compress the angular range except the
post-specimen field of the VG high excitation objective lens pole piece.
This gave an angular compression of about 6x with minimum aberrations, so
that a 20 mrad collection half-angle was achieved with a spectrometer
entrance half-angle of 3.5 mrad. The determining aperture was the hole in
the dark-field detector of the HB5, especially enlarged to the point where
the spectrometer 3rd order and remaining 2nd order aberrations were just
beginning to have an effect.

2. Towards single atom identification

Isaacson and Johnson (1975) predicted that single atom identification by
EELS may be possible and Colliex and Trebbia (1979) developed some rough

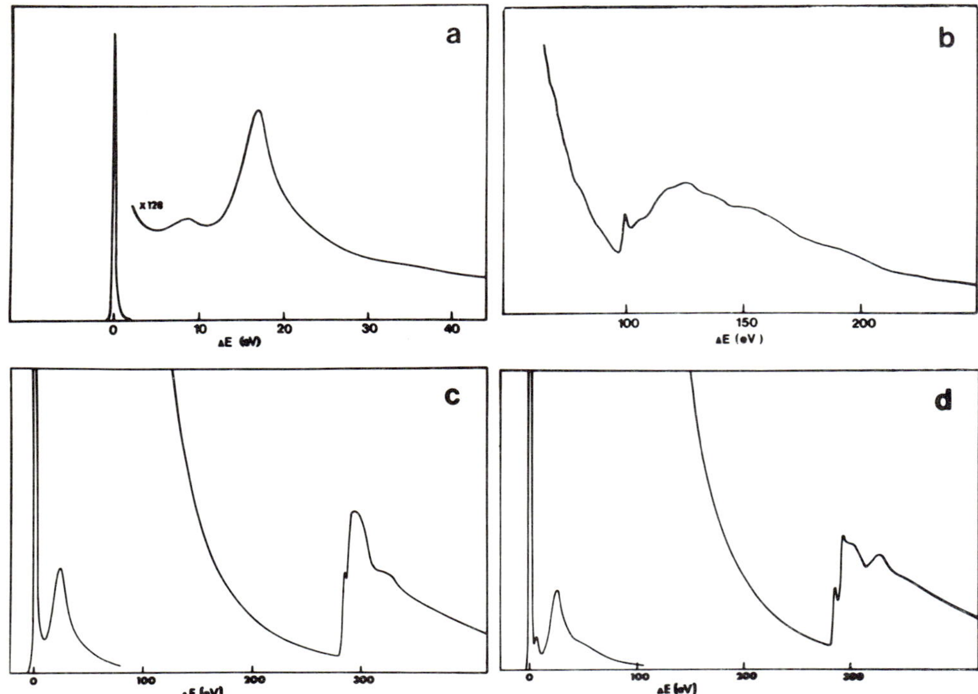

Fig.1   Electron energy loss spectra recorded with a 20 Å probe from
a) Si (low losses), b) Si (Si $L_{2,3}$), c) amorphous carbon,
d) graphitised carbon.

calculations about that feasibility in practical situations. The instrumentation advances mentioned in the previous section are now bringing us
to a point where this goal no longer seems experimentally unattainable.

Fig. 2a shows an elastic dark field image of a folded RNA strand heavily
stained with uranium and supported on an ultra-thin carbon foil, obtained
with a 5 Å probe. The small dots in the image were seen to move between
successive frame scans without changing intensity and were therefore
interpreted as images of either single uranium atoms or of very small
(2 to 3 atom) U clusters (see Mory, Colliex and Krivanek, these proceedings). The larger clusters attached to the RNA chain contained a few
hundred U atoms. A spectrum from one such cluster, obtained with a 20 Å
probe, is shown in Fig. 2d. The U $O_{4,5}$ edge with its characteristic
double peak structure is clearly visible. Figs. 2b and 2c are energy-
filtered bright field images formed with a 5 Å probe and electrons of
energies just before and just after the $O_{4,5}$ edge respectively. 2b shows
weak contrast corresponding to mass-thickness effects. 2c shows contrast
similar to 2a. The smallest U clusters are not visible, but the large
clusters and even clusters down to about 10 Å in size that probably contained 10 to 20 uranium atoms show up clearly. Both the energy-filtered
images were acquired as single electron dot maps (each dot = one electron
detected) in 50 secs per image and were not processed in any way. It is a
typical case for which the image processing suggested by Jeanguillaume et
al.(1978) should be highly favourable.

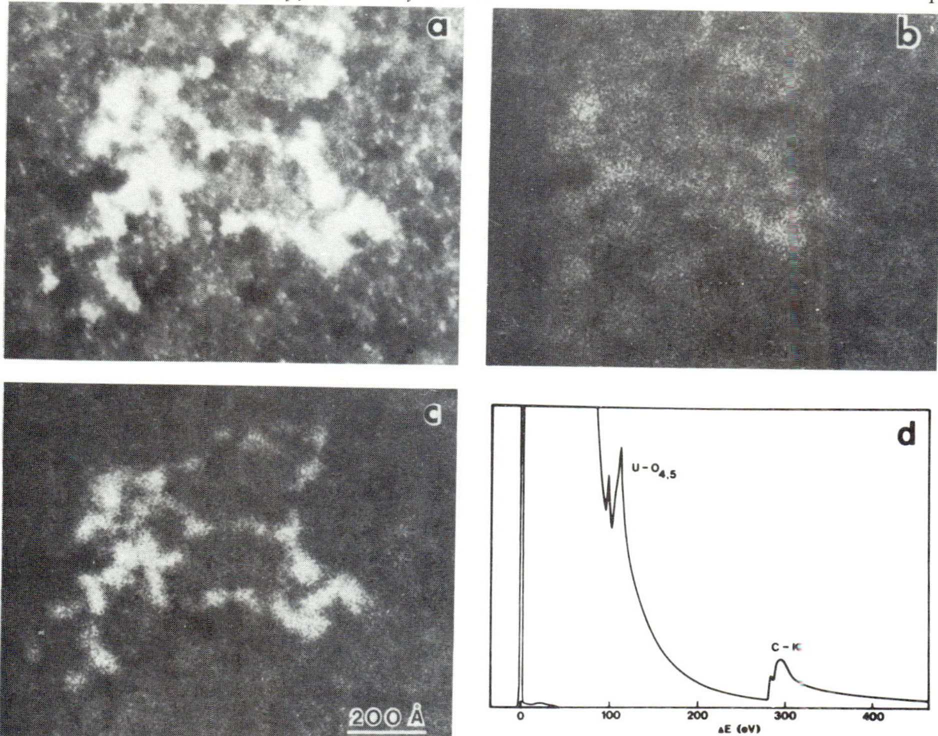

Fig.2   Images of a twisted RNA strand supported on an ultra thin
C film and heavily stained with U. a) annular dark field, b) energy
filtered bright field, 10 eV wide energy window just in front of the
U $O_{4,5}$ edge, c) as b) on top of the edge, d) energy loss spectrum
from a U cluster.

The fact that the U $O_{4,5}$ image 2c is relatively noise-free and yet its
resolution is much worse than 5 Å (between 20 and 50 Å) shows that the
U $O_{4,5}$ ionization process is delocalized, i.e. that it can be excited by
a fast electron passing several tens of Å away from the atom. The localization of the inelastic scattering process depends on the magnitude of
the energy loss roughly as $2\lambda E_0/\Delta E$ (see Howie 1981). For a 100 kV primary
beam energy this gives 60 Å as the width of the image of an uranium atom
formed with U $O_{4,5}$ electrons. In this characteristic image, the occurrence of multiple scattering processes is of very low probability, so
that the white contrast is mainly due to single core loss processes. The
resolution is therefore mostly chemical resolution, contrary to the
origin of the visibility of lattice fringes in the Dy   characteristic
images of Dy_2O_3 crystals previously reported by Craven and Colliex (1977).
In the present study, fixed probe spectra have been recorded at various
places on the clusters and on the background between them. In this
latter case, the visibility of weak edges seems to fix the present detection limit at about 10 uranium atoms. This lack of localization makes
low losses ($\lesssim$ 500 eV) unsuitable for single atom imaging and spectroscopy,
but at higher losses the goal of single atom identification by EELS may
be within reach.

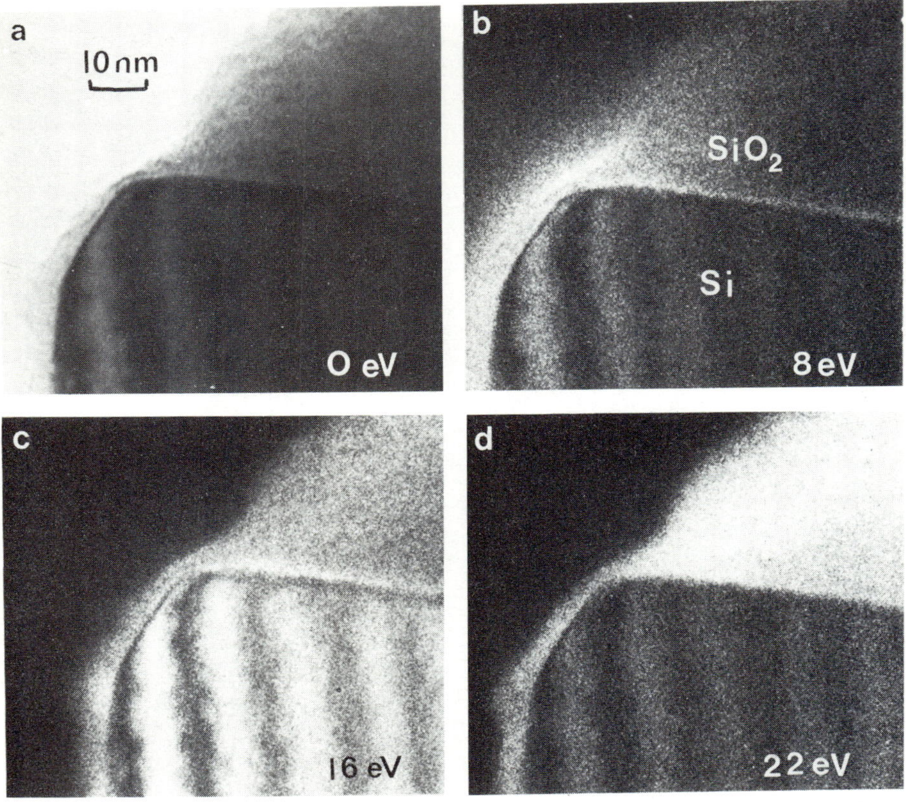

Fig.3   Low energy loss filtered images of a Si-SiO$_2$ interface viewed
        edge-on. a) elastic bright field ($\Delta E = 0 \pm 2$ eV), b) surface
        and interface plasmon ($\Delta E = 8 \pm 2$ eV), c) Si bulk plasmon
        ($\Delta E = 16 \pm 2$ eV), d) SiO$_2$ bulk plasmon ($\Delta E = 22 \pm 2$ eV). The
        width of the interface image is less than 10 Å.

3. Towards full utilization of EELS data

The variety of physical phenomena which contribute to EEL spectra is very
large, and the potential for extracting useful information from the spectra
other than simple elemental concentrations is considerable. We point out
and illustrate some interesting recent developments.

Surface and interface plasmons have recently been observed with a 5 Å
probe FEG STEM, and the region where they are excited has been precisely
localized around small Aℓ spheres covered with a thin oxide layer (Batson
and Treacy 1980). We show a similar example in Fig. 3, where a Si-SiO$_2$
interface is imaged edge-on in a thin foil. Note how the surface and
interface plasmons are imaged in the 8 eV loss image, how the diffraction
contrast (thickness fringes) of the Si crystal is preserved in all the
energy-filtered images, and that the interface plasmon is most strongly
excited on the oxide side of the interface.

The surface plasmon has also been recently observed in glancing angle
reflection from bulk Si and NiO, and it has been shown that the depth

sensitivity of glancing reflection EELS can be varied from less than 30 Å
to over 50 Å simply by changing the magnitude of the reflection angle
(Krivanek 1981). Fig. 4 shows an energy loss spectrum obtained with the
(666) Bragg-reflected beam from a (111) NiO surface. The Ni $M_{2,3}$ edge is
clearly visible. This raised the possibility that EELS may be used as a
surface analytical technique complementary to high resolution glancing-
angle reflection imaging.

Chemical shift and chemistry-sensitive changes in the near-edge fine structure have been known for some time. For instance, the sharp pre-ionization
peak at the Si $L_{2,3}$ edge threshold appears at 99 eV in elemental Si, 103 eV
in SiO, and 106 eV in SiO_2. Fig. 5 shows a spectrum obtained with a 20 Å
probe from the $Si-SiO_2$ interface of Fig. 3. Two peaks at 99 eV and 106 eV,
due to the Si and the SiO_2, are well resolved. An interesting question
was whether EELS could also pick up a third peak at 103 eV due to the monolayer of SiO though to exist at the interface. The answer appears to be
negative for now, but even so it is clear that EELS studies of interfaces
may provide a valuable alternative to Auger and other interface analysis
techniques.

Electron channelling through a crystal is a consequence of dynamical diffraction. It can be controlled (for instance by changing the direction of
the incident electron beam) so that the electron density peaks at different
atomic sites in a crystal. Energy loss spectra obtained in the channelling
configuration then show which type of atoms occupied the highlighted site
(Taftø and Krivanek 1981). The technique can be used to identify the
lattice sites of impurities in complex crystals, and also to study the
localization of the inelastic collision.

Extended energy loss fine structure studies (EXELFS) promise to provide
accurate data on the near-neighbour atomic environment of a selected
atomic species in sample regions as small as $(50 \text{ Å})^3$. An interesting
recent development of the technique is the demonstration that by careful
control of the momentum exchange q, EXELFS can be made to pick out nearest
neighbours in just one particular direction (Disko et al.1981), in a
manner analogous to polarized EXAFS experiments.

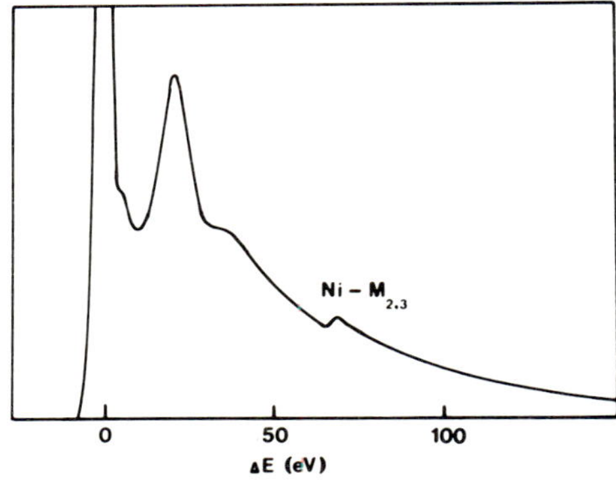

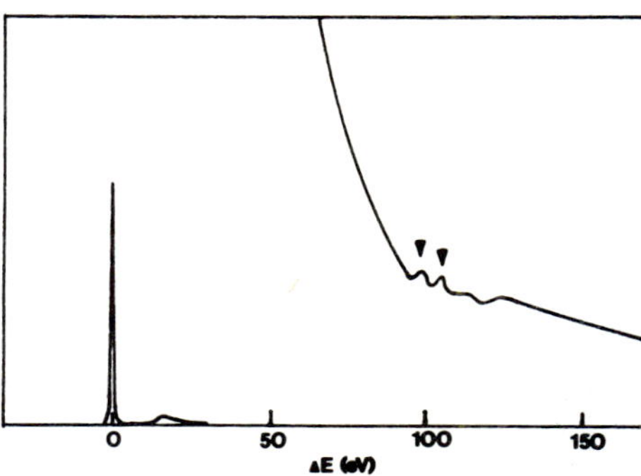

Fig.4 Spectrum from the surface of
bulk NiO obtained with 120 kV
electrons in glancing reflection.

Fig.5 Spectrum from the $Si-SiO_2$
interface (cf.fig.3). The split
Si $L_{2,3}$ pre-ionization peak
(arrowed) is due to the 7 eV
chemical shift in the oxide.

4. Other developments

Progress has also been rapid in the optimization of computer-controlled
EELS acquisition and processing systems. It is now possible to scan
slowly over the spectral regions of interest and fast inbetween; change
the detector gain under software command; automatically identify edges,
perform background subtraction and work out concentrations. Following the
general principles established by Egerton (1978) and thoroughly discussed
by Maher (1979), some examples can be found in Trebbia and Colliex (1979),
and among others, one versatile and convenient computer based acquisition
system has been described by Statham and Watson (1981). Progress has also
been made recently to recover single-scattering profiles by deconvolution of
multiple scattering features, to perform EXELFS or standard radial distribution function analysis (from an energy filtered diffraction pattern)
almost on-line, and to acquire elemental concentration maps properly corrected for thickness variation by energy filtered imaging, though not all
in one system as yet (see, for examples of recent developments, Leapman
and Swyt 1981 or Rez and Ahn 1981).

Acknowledgments

The STEM project in Orsay has been supported by DRET 72/465 and DGRST 73/
3/0489 grants. Thanks are due to Gatan Inc. for the loan of a spectrometer.

References

Batson P E and Treacy M M J 1980 Proc. 38th EMSA Meeting (Reno), 126
Colliex C and Trebbia P 1978 Proc. 9th Int. E.M.Congress (Toronto) III 268
Colliex C and Trebbia P 1979 Microbeam analysis in biology, 65, Ed.
  C Lechene and R Warner, Acad. Press. NY
Craven A J and Colliex C 1977 J. Micros. Spectros. Electron 1, 511
Disko M M, Krivanek O L and Rez P 1981 submitted to Phys. Rev. Letters
Egerton R F 1978 SEM 1 133
Egerton R F 1980 Proc. 38th EMSA Meeting (Reno) 278
Egerton R F 1981 Proc. 39th EMSA Meeting (Atlanta)
Engel A, Christen F and Michel B 1981 Ultramicroscopy, in press
Howie A 1981 Proc. 39th EMSA Meeting (Atlanta)
Isaacson M S and Johnson D E 1975 Ultramicroscopy 1 33
Jeanguillaume C, Trebbia P and Colliex C 1978 Ultramicroscopy 3 237
Krivanek O L 1979 Proc. 37th EMSA Meeting (San Antonio) 530
Krivanek O L 1981 Proc. 39th EMSA Meeting (Atlanta) 216
Krivanek O L and Swann P R 1981 Proc. Quant. Micr. Meeting (Manchester)
  to be published
Leapman R D and Swyt G 1981 Proc. 39th EMSA Meeting (Atlanta) 196
Maher D M 1979 Introduction to analytical electron microscopy, 259,
  Ed. T T Hren, T I Goldstein and D C Joy, Plenum Press
Parker N W, Utlaut M and Isaacson M S 1978 Optik 51 333
Rez P and Ahn C 1981 Proc. 39th EMSA Meeting (Atlanta) 262
Shuman H 1980 Ultramicroscopy 5 45, and in press
Statham P J and Watson T M 1981 Proc. Quant. Micr. Meeting (Manchester)
  to be published
Taftø T and Krivanek O L 1981 Proc. 39th EMSA Meeting (Atlanta), 190
Trebbia P and Colliex C 1979 Inst. Phys. Conf. Series 52 333
Johnson D E 1979 Ultramicroscopy 3 361

Optimum design and use of homogeneous magnetic field spectrometers in EELS

C. Jeanguillaume, O. Krivanek[*] and C. Colliex

Laboratoire de Physique des Solides, bât. 510, Université Paris-Sud, 91405 Orsay (France).

[*] Permanent address : Center for Solid State Science, Arizona State University, Tempe, Az 85281, U.S.A.

The optical properties of dipole magnet and of their coupling with the microscope were studied to characterize and improve the performance of an energy loss spectrometer attached with a dedicated STEM.

1. Theoretical approach

Computation methods were applied to the problem of the transfer of a given distribution of electrons (in energy and solid angle) from the specimen level to the detection slit after the spectrometer. A Monte-Carlo program is used to simulate many different trajectories between the object plane $(x_o, y_o, t_o, u_o, \gamma)$ and the image plane $(x_i, y_i, t_i, u_i, \gamma)$ of the spectrometer ($\gamma = \Delta p/p$ characterizes the energy loss). Consequently the intensity distribution $I_1(x_i, y_i, t_i, u_i, \gamma)$ at the detector level is computed from the knowledge of the distribution $I_o(x_o \ldots \gamma)$ at the specimen. This approach leads to the calculation of several important properties of the spectrometer. An integration over the y_i, t_i, u_i coordinates defines the image function $I_1(x_i)$ following the direction of energy dispersion of the electrons.

- The transmission efficiency of the spectrometer is determined by the integral of the image function over the slit width, divided by the intensity entering the spectrometer.

- The energy resolving power is defined with different criteria when one superposes the spectrometer answers for two different energies. For very general image intensity profiles, it is most convenient to define it as the standard deviation of the statistical distribution of x_i. A simplified formula introduced by Egerton and Lyman (1976) is currently used for the discussion of the practical properties of a spectrometer :

$$D\delta E = \{(D.\delta E_o)^2 + (M\, r_m)^2 + (C\theta_m^2)^2\}^{1/2}$$

The object function is assumed to be gaussian with the standard deviations r_m and θ_m and only one first order and one second order term are considered in the development in transfer coefficients at successive orders (see for instance Enge (1967)). The energy resolving power depends not only on the spectrometer properties but also on the object intensity distribution. To claim a 0.6 eV resolution does not mean anything without reference to the angular and spatial conditions in which it has been obtained. These conditions can be modified before the entrance in the spectrometer by the introduction of post specimen optics. Several authors (Crewe (1977), Egerton (1980), Johnson (1980), Craven (1981)) studied the optimization of such an electron optical coupling between the microscope and the spectrometer.

An illustrative description of it can be obtained in terms of emittance diagrams : Figure 1 shows isoresolution curves as functions of R_m and θ_m. If the δE_o term is neglected, the analytical expression of these curves can be easily derived :

$$\theta_M = \frac{1}{\sqrt{C}} \{ (D\delta E - MR_m)(D\delta E + MR_m) \}^{1/4}$$

When EELS Analysis is performed, the illuminating and collecting conditions are often imposed by experimental parameters such as : selected area on the sample, microscope working mode, needed efficiency. The point A (fig.1) can be an example of such conditions. In Image STEM mode, R_M is usually very small and θ_M as large as possible for getting the best efficiency. Only rather poor resolution can be obtained by spectrometers without transfer devices. In the case of insertion of magnetic lenses, the beam characteristics (R_M, θ_M) move on an hyperbola (dashed lines in fig.1 or 2) and the best conditions are obtained when this hyperbola intersects the best isoresolution curve (point B).

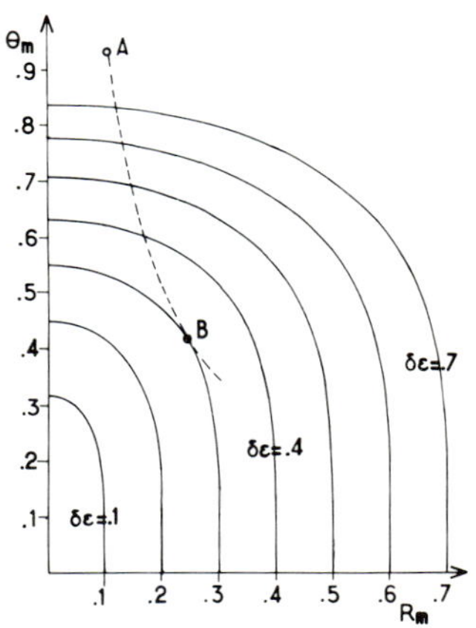

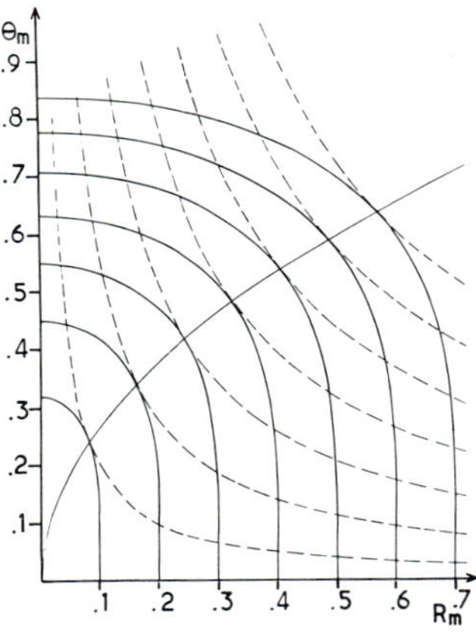

Fig. 1   Fig. 2

Depending on the initial beam characteristics, a set of hyperbolae must be considered, and the resulting optimal conditions are located on a parabola branch (see fig.2), the equation of which has been already given by Egerton (1980) :

$$\theta_M = \frac{1}{2^{1/4}} \sqrt{\frac{M R_M}{C}} \quad \text{where } C = (\frac{x_i}{t_o^2}) + (\frac{x_i}{u_o^2})$$

The two factors in coefficient C are equally treated by magnetic round lenses, but can be selectively adjusted with a quadrupole doublet. Such a design is successfully used in the Gatan Spectrometer (see thereafter).

2. Experimental study of homogenous magnetic field spectrometers

The observations are relevant to the specific case of a VG-HB-501 microscope, either fitted with the original spectrometer or with a Gatan one (Krivanek and **Swann** (1981)). The focusing and aberration properties are determined by the first order x_i/t_o and second order x_i/t_o^2 and x_i/u_o^2 coefficients. If the spectrometer is not focused, $x_i = F_r t_o$. If it is focused ($F_r = 0$) and :

$$x_i = At_o^2 + Bu_o^2 + Ht_o u_o$$

where H = 0 for symmetry consideration. It is easy in the S.A. diffraction mode to record the map x_i as a function of (t_o, u_o) - it has been called an alignment figure by Egerton (1981). The slits correspond to the curves

x_i = Constant = $\pm \ell/2$ in the plane defined by the (t_o, u_o) set of axis. If the spectrometer is non focused, the image of the slit consists in two straight edges F_r $t_o = \pm$ Ct. If it is focused and corrected at all orders the intensity $I_i(x_i)$ must be constant for any angle (t_o, u_o). The screen is uniformly bright or dark owing to the fact that this angle is intercepted or not by the slits.

Examples of such alignment figures are shown in figures 3 and 6. In one case, the observed hyperbolae $At_o^2 + Bu_o^2 = \pm$ Cte correspond to second order aberrations. It is confirmed by the law $t_o = g(x_i)$ which is shown in fig.5. In the other case, efficient second order correction of aberrations due to improved design of the pole-pieces and better optical coupling, is responsible for a good transmissivity of the spectrometer even with large solid angles (see figures 6 and 7). With a probe as small as 0.5 nm, it has been possible to get good signal/noise and a resolution of 1 eV on core loss spectra such as the calcium L_{2-3} edge shown in figure 8 or the other examples in the contribution of Colliex et al. (these proceedings).

Fig.3 – Alignment figure for the spectrometer supplied with the VG-HB 5 STEM. t and u refer to the angular coordinates in the specimen plane the scale being calibrated from a diffraction pattern and from the aluminium plasmon. In this case the size of the analyzed area diffraction aperture (that is typically between 1 and 5 μm at the specimen level) and the solid angle of collection is governed by the collection aperture between typically 5×10^{-4} and 5×10^{-3} rad.

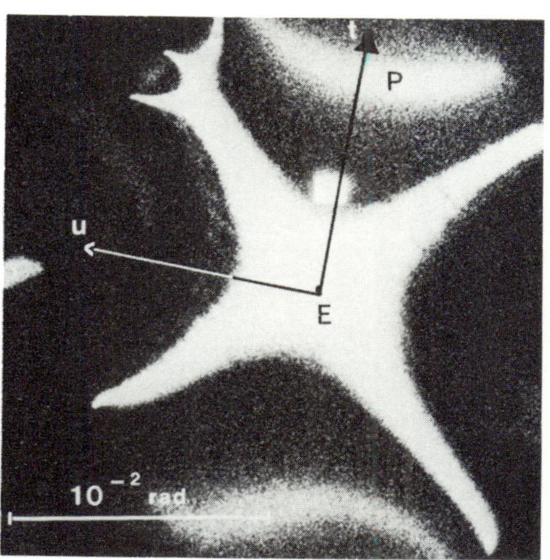

Fig. 4 – FWHM of the elastic peak corresponding to the alignment figure for the VG spectrometer, in the selected area diffraction mode.

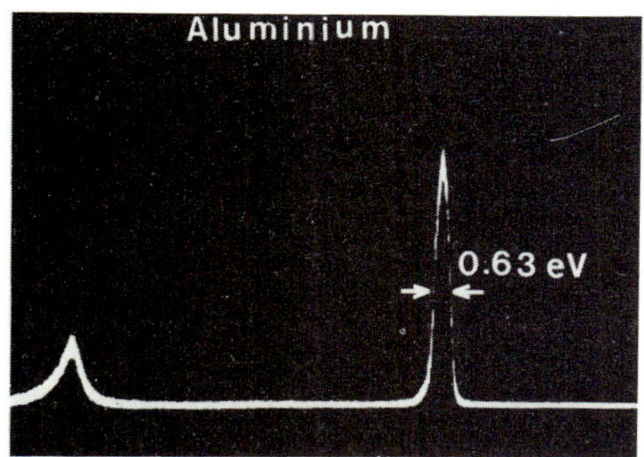

Fig. 5 – Curve showing the relationship between the energy loss scale and the angular coordinates in the alignment figure of the VG spectrometer. This parabolic curve is a confirmation of the dependence $\Delta E \propto t^2$.

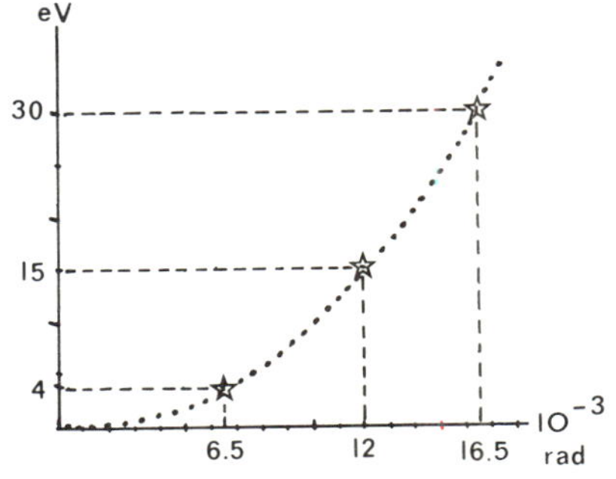

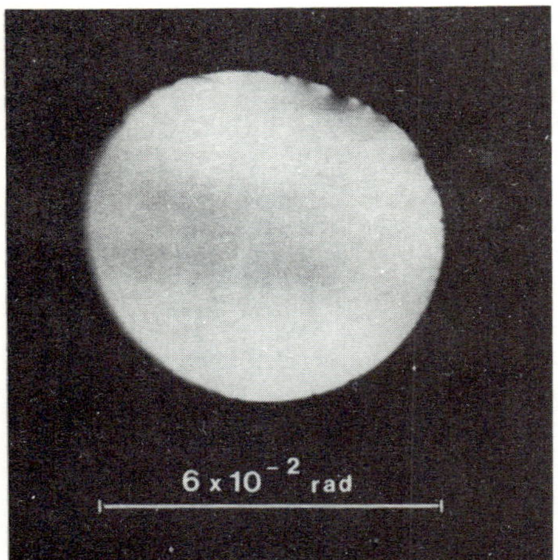

Fig.7 - FWHM of the elastic peak
corresponding to the alignment
figure for the Gatan spectrometer,
in the image mode. By reducing the
divergence of the primary beam in the
low Mag mode, a FWHM of about 0.35 eV
can be obtained (see the figure 1 of
Colliex et al., these proceedings).

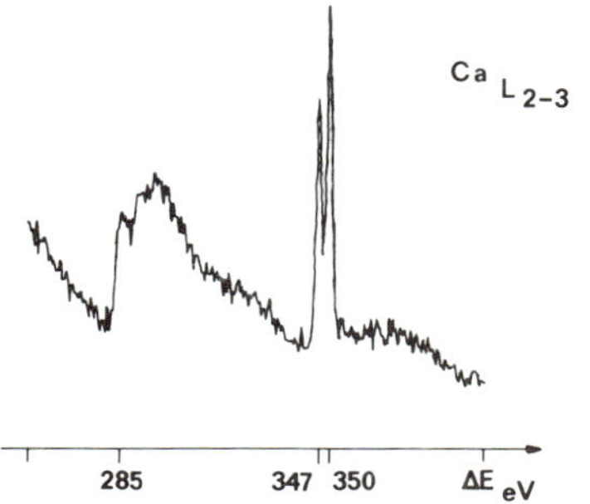

Fig. 6 - Alignment figure for the Gatan
spectrometer fitted on the VG HB 5 STEM.
Due to the good correction of the second order aberrations and to its efficient coupling to the microscope, a
solid angle as large as 6×10^{-2} rad can
be fully illuminated, even with the
slits at the detector closed to achieve less than 1 eV resolution. This
spectrometer is very efficient for recording energy loss spectra in the image
illumination mode, that is with a probe
fixed on the specimen. Practically both
the size of the analysed area and the
solid angle of collection are determined by the objective aperture. For
a 50 μm aperture the probe size is of
the order of 0.5 nm and the solid
angle $2\alpha_o$ of 15 mrad.

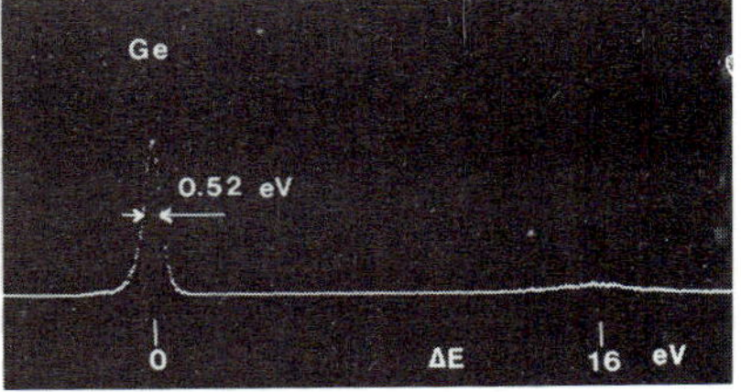

Fig. 8 - Example of energy resolution
achieved on a core-loss line with the
Gatan spectrometer in the image mode.
A probe of about 2 nm size is focused
on a section of human bone, consisting
mainly of calcium phosphate (hydroxy-
apatite). The two L$_2$ and L$_3$ lines the
spin orbit splitting of which is about
3 eV, are clearly resolved, exhibiting
a typical 1 eV resolution.

*The STEM project in Orsay has been supported by DRET 79/465 and DGRST/79/
7/0389 grants. Thanks are due to Gatan Inc. for the loan of a model 607
electron energy loss spectrometer.*

Brown K.L., Belbeoch R., Bounin P., Rev. Sci. Instr. 35, 418 (1964).
Craven A.J., UMIST Manchester - 25/27 March 1981.
Crewe A.V., Optik, 47, n° 3, 299-311 (1977).
Egerton R.F., Lyman C.E., Ed. J.A. Venables, Acad. Press., p. 35 (1975).
Egerton R.F., Optik 57, n° 2, 229-242 (1980).
Egerton R.F., Ultramicroscopy, 6, 93-96 (1981).
Enge H.A., A. Septier Ed. Vol. 2, ch. 4.2, Acad. Press, New-York (1967).
Johnson D.E., Ultramicroscopy, 5, 163-174 (1980).
Krivanek O.L., and Swann P.R., to be published (1981).

The design of an electron spectrometer for use in STEM free of second- and third-order aberrations

T Tang

Cavendish Laboratory, Madingley Road, Cambridge CB3 OHE

For the study of the fine structure of absorption edges, good energy resolution of 1 – 3 eV is necessary. Not only light elements, but elements with medium atomic number (Z = 11 – 15), such as Na, Si and P, also are of interest. This means a maximum energy spectrum range of $\sim$ 2 keV is desirable. The characteristic scattering angle θ_E is 5.36 mrad for Na, 9.2 mrad for Si and 10.73 mrad for P (K losses). The collection efficiency for inelastic electrons can be near 50% only if the acceptance angle γ is 4 – 6 times θ_E (Isaacson 1978). With respect to signal-noise ratio, the best condition for γ varies for different elements, but it is of the same order as θ_E and it is larger than 10 mrad for some elements (Joy and Maher 1978). Therefore, for the development of EELS technique, an advanced spectrometer with energy resolution of $\sim$ 1 – 3 eV, spectrum range of $\sim$ 2 keV and maximum acceptance angle of 10 – 20 mrad is required. The design of such a spectrometer for the STEM in the Cavendish Laboratory is reported here.

A matching lens between the spectrometer and the specimen can be used to compress the beam angle (Egerton 1980a, Crewe 1977). Such a lens must be close to the specimen if its aberrations are not to have a serious effect on performance. Alternatively the post-specimen field of the objective lens can provide a useful amount of beam compression with low aberrations. However, in the case of the Cavendish STEM there is no room for a lens to be positioned close to the specimen and the existing objective lens cannot be excited strongly enough and its post-specimen field can only compress the beam angle about 10 – 20 %. Therefore the only way to achieve the design target is to correct the aberrations of the spectrometer.

There have been many designs in recent literature (Shuman 1980, Egerton 1980b, Field 1977, Parker et al.1978) where second order aberrations are corrected by the use of contoured magnet edges or by hexapole lenses, assuming third order aberrations are not important. However, up to now good energy resolution can only be obtained when the acceptance angle is relatively small. To the author's knowledge it seems that there are no experimental results where $\sim$ 1 eV resolution can be achieved with an acceptance angle larger than 5 mrad.

In the designs mentioned above, two circular edges, one convex and the other concave, are used to correct second order aberrations. Circular edges have hexapole field components. The shape of an oblique circular edge can be expressed in terms of ρ, the curvature radius and ε, the oblique angle (fig. 1) by the equation

0305-2346/82/0061-0193$01.50   © 1982 The Institute of Physics

$$z = \mathrm{tg}\,\varepsilon\, x + \frac{x^2}{2\,\rho\,\cos^3\varepsilon} + \frac{\mathrm{tg}\,\varepsilon\, x^3}{2\rho^2\cos^4\varepsilon} + \ldots \tag{1}$$

As pointed out by Tang (to be published) approximately a quadrupole strength – $\mathrm{tg}\varepsilon$, hexapole strength – $1/2\rho\cos^3\varepsilon$, octopole strength – $\mathrm{tg}\varepsilon/2\rho^2\cos^4\varepsilon$ and so on are created by the edge. At the same time when the hexapole component corrects the second order aberrations, the hexapole and octopole components will create extra third order aberrations. It has been found from the calculation that usually a simple spectrometer with straight line edges possesses only moderate third order aberrations, of the order of R_O, the radius of the central trajectory or perhaps up to 10 times greater. But some special designs with no second order aberrations do have much larger third order aberrations. They can be several hundred times R_O. In the author's opinion, it is the large third order aberrations that set an obstacle to getting high resolution together with large acceptance angles.

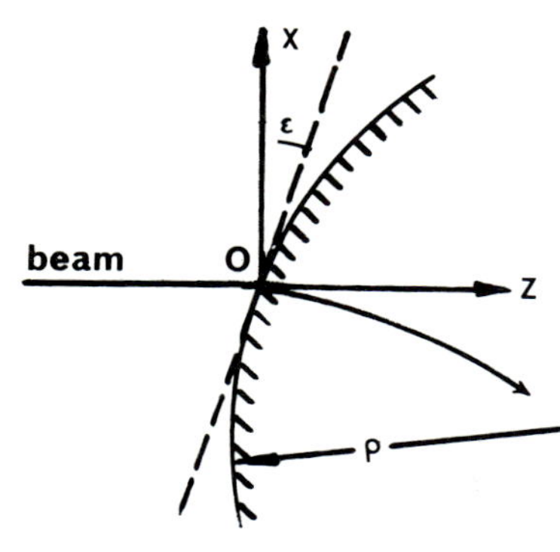

Fig.1   The circular edge

In this design, the third order aberrations were calculated and an attempt was made to keep the third order aberration to a moderate level while second order aberrations are completely corrected. As the octopole strength is approximately – $\mathrm{tg}\varepsilon/2\rho^2\cos^2\varepsilon$, the total octopole strength of two edges can cancel each other if the two oblique angles have different signs. By experience and from simple theory (Tang to be published), good results can be obtained when the magnifications in the radial and axial directions M_x and M_y are significantly different from each other.

Third order particle trajectories in a uniform magnetic field have been calculated by Ludwig (1967) and by Matsuo and Matsuda (1971) in the form of a transfer matrix. The third order transfer matrix of the fringing field has been developed by Matsuda and Wollnik (1970). In this work these transfer matrix theories have been connected and applied to the design of the spectrometer. The fringing field has been calculated by the finite element method. The integrals I_1 – I_4 required in the transfer matrix of the fringing field have been calculated by a specially written computer program and a computer optimization program has been written to minimize the third order aberration $C_3 = [(x/x'^3)^2 + (x/x'y'^2)]^{\frac{1}{2}}$ (Brown 1967) subject to the following conditions:

(1) first order double focusing;
(2) correction of second order aberrations;
(3) suitable first order parameters including M_x, M_y and D/M_x, where
    $D = \frac{1}{2}\,\Delta E/E_O$ is the dispersion and D/M_x is proportional to the
    first order resolution;

(4) some practical considerations.

Some design results are listed in Table 1, where γ_{max} is the largest γ limited by third order aberration for a resolution of 1 eV in 80 keV. It is possible to obtain smaller third order aberrations if a larger ℓ_2/ℓ_1 is used, but this will make the spectrometer very large (ℓ_1 is fixed in this STEM) and lead to bad first order parameters.

Table 1   Computer Results of the Spectrometer

Design	ϕ (deg)	ε_1 (deg)	ε_2	ℓ_1/R_o	ℓ_2/R_o	ρ_1/R_o	ρ_2/R_o	G/R_o	$D/R_o M_x$	C_3/R_o	γ_{max} (mrad)
A	59.2	-28.4	45.4	1.58	3.53	21.8	-1.66	0.28	1.45	4.46	19.6
B	81.5	-11.6	45.8	1.50	2.03	1.48	-1.29	0.17	2.13	7.28	15.8
C	71.7	-13.1	45.7	2.00	2.00	2.02	-1.28	0.16	2.27	8.14	14.5
D	66.6	-15.0	45.8	2.25	2.06	2.34	-1.30	0.18	2.31	8.63	14.0
E	54.7	-22.1	45.8	2.50	2.50	3.56	-1.47	0.19	2.04	8.14	14.0
F	53.0	-22.2	45.8	2.70	2.50	3.38	-1.46	0.20	2.13	8.63	13.5

The designs in Table 1 give a resolution of 1 eV in 80 keV together with maximum γ larger than 13.5 mrad. However, in order to further correct the residual third order aberrations and compensate aberrations caused by different practical factors such as inhomogeneity of material, inaccuracy of machining and assembling, multipole lens assemblies before and after the magnet are used. Besides the alignment coil, there are one hexapole and one octopole lens in an assembly. Electrical multipole lenses were chosen because they are easier to construct and more compact than magnetic lenses. If the residual aberrations (x/x'^2, x/y'^2, x/x'^3, $x/x'y'^2$) are A_2, B_2, A_3 and B_3 respectively, the required hexapole and octopole voltages $\Delta U_{61}/U_0$, $\Delta U_{62}/U_0$, $\Delta U_{81}/U_0$ and $\Delta U_{82}/U_0$ are (Tang to be published)

$$\frac{\Delta U_{61}}{U_0} = \frac{8a^3}{3K_6 p_1 M_x}\left[A_2 + \frac{M_y^2}{M_x^2 - M_y^2}(A_2 + B_2)\right] \tag{2}$$

$$\frac{\Delta U_{62}}{U_0} = \frac{8a^3 M_x^2}{3K_6 p_2}\frac{M_y^2}{M_x^2 - M_y^2}(A_2 + B_2) \tag{3}$$

$$\frac{\Delta U_{81}}{U_0} = \frac{5a^4}{2K_8 q_1 M_x}\left[A_3 + \frac{M_y^2}{M_x^2 - M_y^2}\left(A_3 + \frac{B_3}{3}\right)\right] \tag{4}$$

$$\frac{\Delta U_{82}}{U_0} = \frac{-5a^4 M_x^3}{2q_2 k_8}\frac{M_y^2}{M_x^2 - M_y^2}\left(A_3 + \frac{B_3}{3}\right) \tag{5}$$

where a is the radius of the circle tangent to the multipole electrodes and

$$p_1 = (\ell_{61} + t_{61})^4 - \ell_{61}^4 , \qquad p_2 = (\ell_{62} + t_{62})^4 - \ell_{62}^4$$
$$q_1 = (\ell_{81} + t_{81})^5 - \ell_{61}^5 , \qquad q_2 = (\ell_{82} + t_{82})^5 - \ell_{82}^5$$

all the lengths in (2-5) are in the unit of R_o; t_{61}, t_{62}, t_{81}, t_{82} are

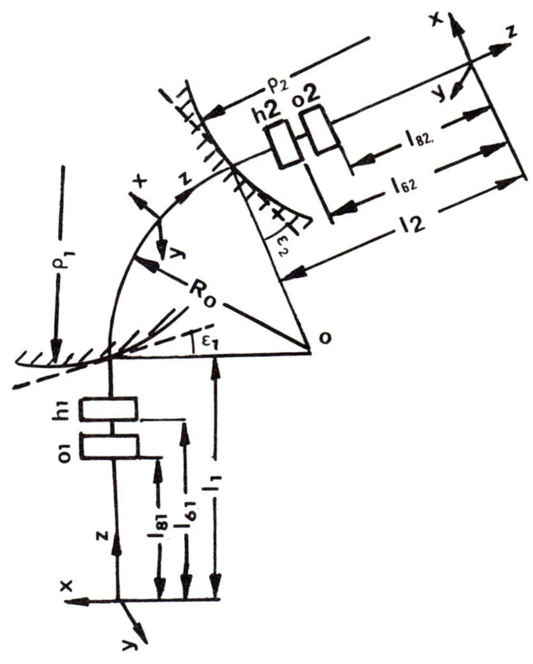

Fig.2   The spectrometer system

the equivalent lengths of the hexapole and octopole lenses; $\ell_{61},\ell_{62},\ell_{81},\ell_{82}$ are distances between those lenses and the objective and image planes. For a well designed spectrometer with moderate third order aberrations and small residual second order aberrations, $U_{61}/U_0, U_{62}/U_0, U_{81}/U_0, U_{82}/U_0$ are very small. The multipole lenses can be compact and this makes final adjustment after assembling possible.

For both high resolution and wide acceptance angle, good homogeneity of the magnetic field is required over a large area in the gap. A special alloy, having high permeability at low field is adopted for the pole piece and a window type magnet is used. The homogeneity of the field has been calculated numerically and the geometry of magnet and a position for the coil obtained which give good homogeneity. Fringing field clamps are used to shorten the extent of the fringing field and this makes the fringing field effect adjustable.

ACKNOWLEDGMENT

The author wishes to thank Dr V E Cosslett, Dr A Howie, Dr L M Brown and Dr D McMullan for helpful and stimulating discussions.

REFERENCES

Brown K L   1967   Advances in Particle Physics Vol. 1 77
Crewe A V et al   1971   Rev. Sci. Instrum. 42 411
Crewe A V   1977   Optik 47 299
Egerton R F   1980a   Optik 56 363
Egerton R F   1980b   Optik 57 229
Field J R   1977   Ultramicroscopy 2 311
Isaacson M S   1978   SEM Vol 1 763
Joy D and Maher D M   1978   Ultramicroscopy 3 69
Ludwig R   1967   Z. Naturforsch. 22a 553
Matsuo T and Matsuda H   1971   Int. J. Mass. Spectr. Ion Phys. 6 361
Matsuda H and Wollnik H   1970 Nucl. Instrum. Meth. 77 283
Parker N W et al   1978 Optik 51 335
Shuman H   1980   Ultramicroscopy 5   45
Tang T   To be published

Optimization of post-specimen lenses for use in STEM

T W Buggy and A J Craven

Department of Natural Philosophy, University of Glasgow, Glasgow G12 8QQ

1. Introduction

The primary function of post-specimen lenses in a scanning transmission
electron microscope (STEM) is to match selected portions of the characteristic distribution of scattered electrons leaving the specimen to detectors
which have a fixed angular acceptance or physical size. In the case of
electron energy loss spectroscopy (EELS), the half-angle of acceptance of
the spectrometer, β_S, is determined by the spectrometer aberrations and the
energy resolution required. The post-specimen lenses allow the desired
angular range of scattered electrons to be matched to this acceptance
angle. In the case of diffraction patterns, the post-specimen lenses allow
a wide range of camera lengths to be obtained at a diffraction screen of
fixed diameter; this is situated in the far-field in our VG Microscopes HB5.

In order to understand the EELS spectra and diffraction patterns obtained
using the post-specimen lenses, it is necessary to take account of the
relevant aberrations they introduce and how these vary with the excitations
of the lenses. Detailed theoretical calculations have been reported elsewhere (Craven and Buggy 1981), based on the programs of Munro 1973).
However, since the detectors lie at various planes in the far-field, new
aberration integrals had to be derived. This paper describes some
experimental measurements made to check the validity of our calculations
and emphasises the importance of taking the effects of these aberrations
into account when analysing data.

2. Electron Energy Loss Spectroscopy

In addition to selecting the angular range of scattering entering the
electron spectrometer, the post-specimen lenses also provide an object for
the spectrometer at a fixed plane conjugate with the slit of the spectrometer. In our case the spectrometer object is usually an image of the
small illuminated area on the specimen. When electrons lose energy in the
specimen, their trajectories in subsequent optical components differ from
those of electrons elastically scattered through the same angles. Such
chromatic effects cause a change in the axial position of the spectrometer
object and a change in the angular magnification. As a consequence both
the collection efficiency and the defocus of the spectrum vary with energy
loss. These effects are illustrated schematically in fig. 1 where the
dashed lines represent the trajectories of electrons which have lost energy.
Two modes of operation (I,II) are indicated showing that it is possible to
have an increase (mode I) or a decrease (mode II) in collection efficiency
with energy loss.

This dependence of collection efficiency on energy loss can be expressed as
a compression factor $C(\Delta E)$ which converts the geometric acceptance angle of
the spectrometer, β_S, to the angle, $\beta(\Delta E)$, subtended at the specimen for
electrons which have lost energy, ΔE. For elastically scattered electrons
$\beta(O) = C(O)\beta_S$ where $C(O)$ is the angular compression for electrons which
have lost no energy. This is equal to the linear magnification from the
specimen to the spectrometer object plane. Fig. 2 shows the variation of
$C(\Delta E)$ as a function of $C(O)$ for energy losses corresponding to the K
ionization edges of nitrogen and silicon for the modes defined in fig. 1.
The accuracy of these calculations has been checked by comparing the
relative intensity in the silicon K-edge at various values of $C(O)$ with
that expected theoretically (Craven, Buggy and Ferrier 1981). The agreement is reasonable but the experimental errors are appreciable for the
reasons discussed in this earlier paper.

It is worth noting the errors which can occur if no account is taken of
this effect. In calculating a composition from an EELS spectrum use is
made of the ratio of the partial ionization cross-sections. These are
calculated by integrating the differential cross-section out to the angle
of scattering accepted by the spectrometer. If a system containing nitrogen and silicon is analysed using a spectrometer with $\beta_S = 1$ mrad, $C(O) = 10$
and optics corresponding to fig. 2, then the ratio of partial K-shell
ionization cross-sections of nitrogen and silicon should be 18 if mode I
is used and 43 if mode II is used. (The partial cross-sections were taken
from Egerton 1978). If $C(O)$ is used rather than $C(\Delta E)$, ie no chromatic
correction is considered in determining the collection half-angle, then a
value of 23 is obtained. Hence very significant errors can occur if the
correction is not made.

As a further check on the calculations, we have measured the change in
excitation parameter of a lens required to refocus the spectrum for a
given energy loss. The spectrometer and lenses were set up to give a
focussed zero loss peak. The spectrometer magnet excitation was adjusted
to bring the nickel $L_{2,3}$ edge into the centre of the spectrum. In this
way the spectrometer itself had been refocussed to correct for the change
in energy. Thus the change of the lens excitation required to focus the
spectrum was that needed to put the spectrometer object back onto the
correct plane. Fig. 3 shows the **defocus** expected for the $L_{2,3}$ ionization
edge of nickel as a function of $C(O)$. The points marked are calculated
from the measured changes of excitation parameter. The error bars result
from the range over which the spectrum can be considered to be in focus.
These calculations can be used in the design of a dynamic correction to
the spectrum focus.

3. Diffraction Patterns

When recording diffraction patterns, the level of radial and spiral distortion introduced by the post-specimen lenses must be acceptable.
Detailed calculations of these distortions for a wide range of conditions
show that acceptable distortions (< 2% at the edge of a 20mm diameter
screen) can be obtained for camera lengths greater than 100mm (Craven and
Buggy 1981). To check these calculations, measurements have been taken
from [111] zone axis patterns of silicon recorded for a range of conditions. To simplify the comparison of experiment with theory, the camera
length was varied by changing only the excitation of the final post-
specimen lens, P III, with the excitation of the objective lens and first
post-specimen lens, P I, held constant. At one value of the excitation of

P III, the zero loss peak of the electron spectrum was in focus providing
a known experimental configuration which could be related easily to the
calculations. Figures 4 and 5 show the percentage distortion at the edge
of a 20mm diameter screen as a function of the camera length, for two
values of the excitation of P I. The agreement between theory and
experiment is good and we can now select with some confidence lens
excitations to optimise the performance.

4. Conclusions

A detailed understanding of the electron optical performance of the post-
specimen lenses is necessary to optimise their performance for a particular
application and to interpret correctly data recorded using them. The
experimental results obtained to date suggest that our calculations are
correct and that they can be used with some confidence.

Acknowledgements

The authors would like to thank the Science Research Council for providing
the equipment and a research studentship to TWB in collaboration with
VG Microscopes.

Craven A J and Buggy T W (1981) Ultramicroscopy. In press.
Craven A J, Buggy T W, Ferrier R P (1981). To be published in Conference
    Proceedings "Quantitative Microanalysis with High Spatial Resolution"
    (UMIST Manchester 25-27 April 1981), Metals Society, London.
Egerton R F (1978) Scanning Electron Microscopy, SEM Inc., AMF O'Hare, 133.
Munro E (1973) Ph D Thesis, Cambridge University.

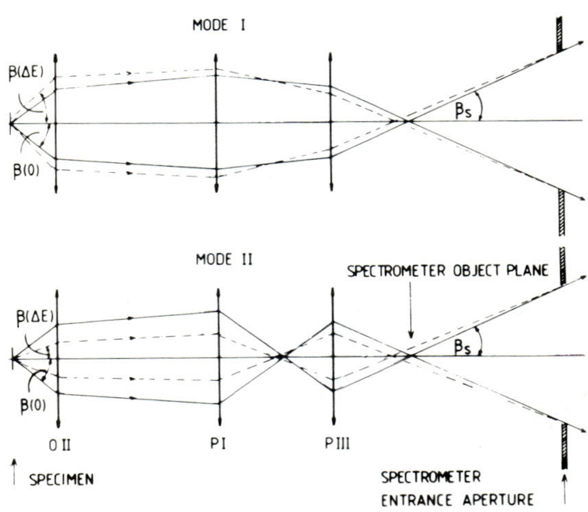

Figure 1: Schematic illustration of post-specimen optics together with
the change in trajectory which results when electrons lose energy (dashed
lines). The two modes of operation shown indicate that it is possible to
have an increase (I) or decrease (II) in collection angle, β(ΔE), with
energy loss ΔE.

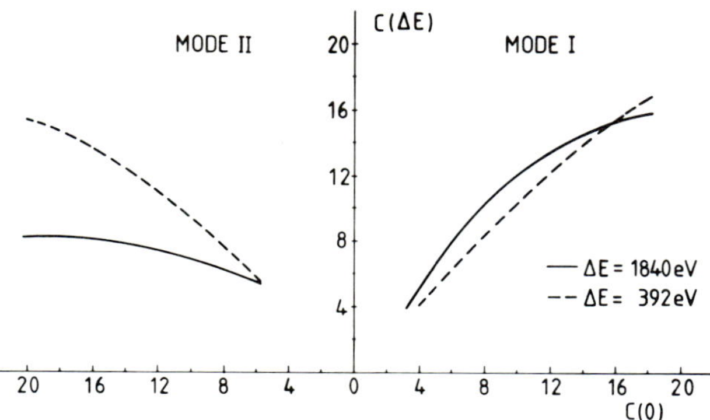

Figure 2: Plot of calculated compression factor, $C(\Delta E)$, against zero loss compression, $C(O)$, for N(ΔE = 392 eV) and Si (ΔE = 1840 eV) K-edges

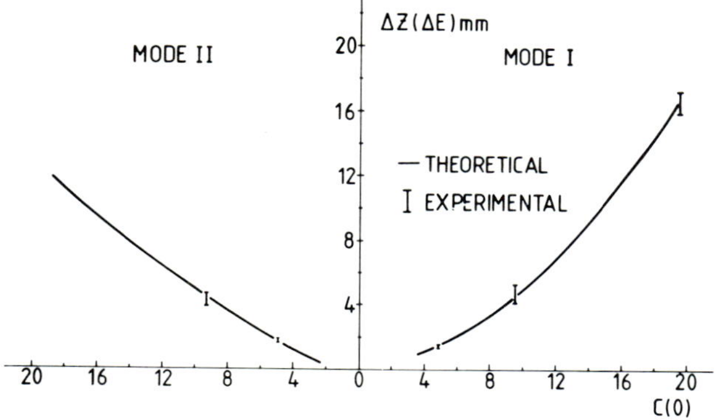

Figure 3: Comparison of theoretical and experimental defocus values ΔZ, for Ni $L_{2,3}$ ionization edge (ΔE = 852 eV)

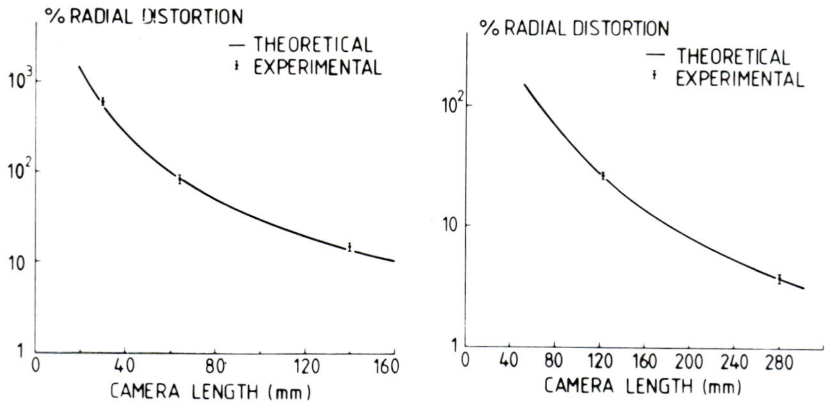

Figures 4,5: Comparison of theoretical and experimental radial distortions (see text).

Minimum detectable atomic fractions of carbon and oxygen in iron

D-R Liu and L M Brown

Cavendish Laboratory, Madingley Road, Cambridge CB3 OHE

Introduction

Electron energy loss spectroscopy (EELS) by STEM is a convenient method of measuring the contents of light elements in a thin specimen. If element (1) is present in a matrix of element (2), then the atomic ratio of (1) to (2) is given by:

$$\frac{N_1}{N_2} = \frac{\sigma_2(\Delta,\beta)}{\sigma_1(\Delta,\beta)} \cdot \frac{I_1(\Delta,\beta)}{I_2(\Delta,\beta)} \tag{1}$$

Here $\sigma(\Delta,\beta)$ is the partial cross-section corresponding to an energy window Δ and a maximum scattering angle β; and $I(\Delta,\beta)$ is the corresponding count in the spectrometer (see fig.2a and Egerton (1978) for clarification).

Many papers have estimated theoretically the minimum amount of an element in a matrix (Isaacson and Johnson 1975, Joy and Maher 1980). This paper is intended to demonstrate some practical aspects. Theoretically speaking, the detection limits are set only by the requirement of statistics:

$$S \geq A\sqrt{B} \tag{2}$$

where S and B are integrated signals for a characteristic edge and for the background under the edge respectively (fig.2a). Taking the coefficient A = 1 corresponds to 63% accuracy of the detection limit, no matter how many collecting channels one has used for S and B; A = 3 gives 98% accuracy. However, it must be borne in mind that expression (2) does not take account of the extrapolation error which may also be important.

Carbon and oxygen are two common elements in industrial steels: it is of considerable interest to perform the detection experiments for those elements in thin iron films.

Sample preparation

Thin iron film was evaporated at 5×10^{-5} torr onto a piece of rock salt, then floated off in distilled water and picked up on a grid. To get a variation of the amount of carbon over the iron, a razor blade was used as a mask to cover half a grid on which was mounted a piece of 'pure' iron film and then a thin carbon film was evaporated onto this grid. Fig.1 shows the grid and the position of the boundary of the carbon film. The experiment depends upon the assumption that the energy-loss spectrum of carbon dissolved in iron will be identical with the spectrum from the same number of carbon atoms in a surface of a thin film. Because the

characteristic energy losses are determined mainly by atomic structure
and not by solid-state effects, this assumption should be valid.

To get a minimum oxidisation for the O-Fe series of experiments, an iron
specimen was immediately put into the microscope after being picked up
from the surface of the distilled water, even while the specimen was wet.
The specimen later was oxidised at 250°C and 650°C.

Carbon detection
────────────────

The cross-section in equation (1) can be calculated theoretically, however,
since in this experiment the carbon loss is a K-loss while the iron loss
is an L-loss, it seemed better to calibrate the C to Fe ratio by some
direct experimental measurements. A special C-Fe specimen was made by
picking up a portion of Fe film and then a portion of C film so that on
one grid there were three regions: (1) only Fe, (2) only C and (3) C-Fe
overlapping. Under these circumstances two independent thickness
measurements are possible.

The first one utilises the plasmon intensity. The ratio of the counts of
the first plasmon loss at energy E_p to that in the zero-loss is given by
$P(E_p)/P(0) = t/\lambda$, where t is the foil thickness and λ the mean free path,
to a good approximation, $\lambda = A_0/\theta_E \cdot (\ln \theta_c/\theta_E)^{-1}$, where A_0 is 0.53 Å,
$\theta_E = E_p/2E_0$ and $\theta_c = \sqrt{E_p/E_0}$ Assuming that the densities of the C and Fe
films are 2 gcm^{-3} and 6.6 gcm^{-3}, one obtains the atomic ratios of the two
elements in our specimen. The second method is by optical interferometry,
using an 'Å-scope interferometer kindly lent to us by the PCS group in
our laboratory. This method gives directly the thickness of the film,
but it is necessary once again to assume their density to determine the
atomic ratio.

Such a group of measurements for a specimen is given below, along with the
result by EELS, where the cross-sections in eqn.(1) were computed with
Egerton's programs SIGMAK, SIGMAL:

Method		by plasmon	by interferometry	by EELS
Measured	C	408 Å	400 Å ± 10%	
thickness	Fe	441 Å	344 Å ± 10%	
Atomic ratio N_C/N_{Fe}		1.32	1.62 +20%	1.50 ± 8% (1.38 ∿ 1.62)

It can be seen that all three estimates of
the atomic ratio agree to within the experimental error so that it is possible to
determine C to Fe ratio confidently along
the 'concentration gradient' perpendicular
to the carbon film boundary shown in
fig.1. Such a profile is shown in fig.3.
the arrow shows the last point in our
experiment from which carbon can be
quantified with ± 10% accuracy. The
carbon-iron ratio at that point is 0.03.
The number of counts in a channel just
before the C-K edge is 1.5 x 10^5. Because
of the characteristic square-root

Fig.1 Arrangement for obtaining
carbon concentration profile on
a Fe foil

variation of uncertainty in the mean signal derived from N counts, it is
possible to generalise this result. If one can collect more counts N, a
lower detectable limit can be reached with similar accuracy: thus

$$k = \left(\frac{N_C}{N_{Fe}}\right)_{min.\,detectable} = 0.03 \ \sqrt{\frac{1.5 \times 10^5}{N}} \tag{3}$$

in this case.

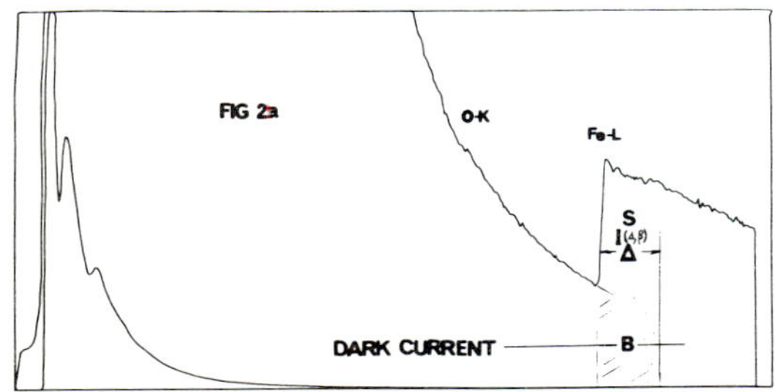

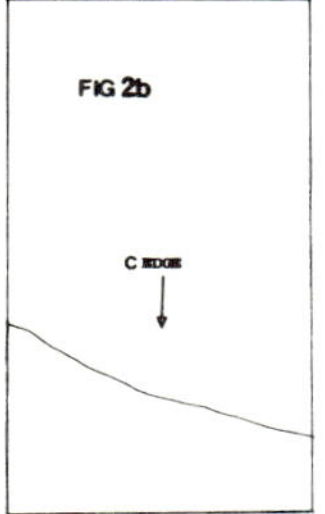

Fig.2a   A spectrum showing N_C/N_{Fe} = 0.06   Fig.2b   Appearance of the spec-
                                                        trum for N_C/N_{Fe} = 0.03 in fig.3

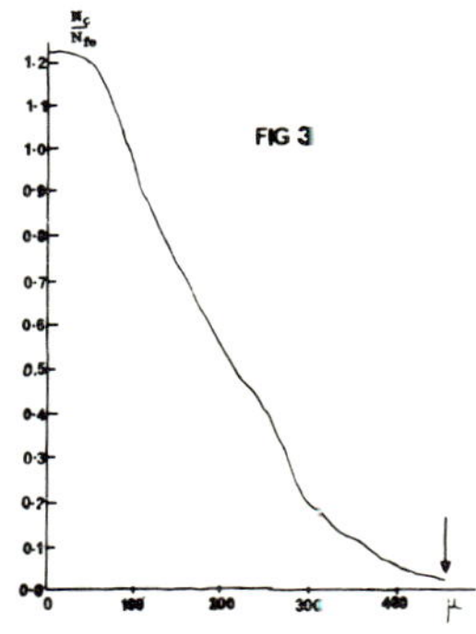

Fig.3   Carbon concen-
tration profile

Some comments should be made concerning eqn.(3).
Firstly, the minimum detectable ratio corresponds to the spectrum of fig.2b, in which the
edge signal defined by the number of counts
under the carbon edge is about ten times the
square root of the background counts defined
by the number under the extrapolated background;
thus the constant A in expression (2) is
approximately 10. Clearly it might be possible
to reduce the value of A but in practice
uncertainty in the extrapolation which is not
included in expression (2) will always make
the value A greater than 2 ∿ 3. If the accuracy
is reduced, the minimum detectable atomic
fraction would be correspondingly reduced. If
the uncertainty of the carbon content is allowed to increase to ± 50%, then assuming Gaussian
statistics, the value of A would change by
approximately a factor of 2, becoming 5 rather
than 10. Under these circumstances, the
minimum detectable atomic fraction is half that given by equation (3).
The second comment concerns the dependence of equation (3) upon instrumental factors. The equation does not depend upon the brightness of the
gun, the size of the probe, or the spectrometer efficiency; it depends
slightly on the apertures employed because the ratio of the cross-sections
depends only weakly on aperture size. Thus equation (3) should apply to
any instrument. Note, however, that in eqn.(3) N refers to the number
of counts in a single channel in an array of 20 channels. The last
comment is that it seems one cannot really define a 'detection limit'.
If one can afford the time to accumulate enough N, the k could be as
small as one wishes with well-defined accuracy.

The estimates given here differ from previous ones (Joy and Maher 1980,
Isaacson and Johnson 1975) in the following ways: we have included the uncertainty in the background extrapolation as it is encountered in practical
computations and our estimate is fundamentally a ratio; it cannot be reduced to a 'mass' of carbon without knowing instrumental details. For our
VG HB5, with a total three minutes' acquisition time with 5 eV per channel
and $\sim$ 1 nm probe size incident on an iron foil $\sim$ 30 nm thick, the number
of carbon atoms detectable is about 100. This is in agreement with experience and corresponds roughly to the earlier estimates based on different instrumental conditions.

Oxygen detection

The principle here is the same as for carbon, except that the oxygen combines with the iron to form oxides so no very convenient independent
measurement of oxygen could be found. However, the electron diffraction
patterns can be analysed assuming that the film consists of mixtures of
Fe, FeO, Fe_3O_4 and Fe_2O_3. If the X-ray ratios of diffracted rings in the
ASTM card index can be assumed to correspond roughly to the electron
ratios, the total oxygen concentration can be estimated from the diffraction patterns and compared with the EELS values. For example, after aging the
specimens   at 250°C and 650°C, diffraction patterns were obtained and
transferred to histograms, from which one got the oxygen/iron ratio of .35
and 1.37 respectively; the corresponding EELS ratios were .41 and 1.33.
The lowest oxygen concentration we have got is 0.06 (fig.2a) with an
accuracy ± 5% from a specimen 37 nm thick, for which the number of counts
in the channel just before the Fe-L edge is $\sim$ 2 x 10^4. If one could
collect more counts N, less oxygen should be detected:

$$\left(\frac{N_O}{N_{Fe}}\right)_{\min\ detectable} = .06\sqrt{2}\ x\ 10^4/\sqrt{N} \tag{4}$$

It seems to be impossible to prepare iron films with less oxygen, because
one usually cannot get rid of the exposure of the iron foil to atmosphere
(Pennycook, Batson and Fisher 1980) during the preparation. Equation (4)
is subject to the same comments as equation (3).

Acknowledgments

The authors are grateful to Dr R F Egerton for offering his programs, to
Drs D McMullan and S J Pennycook for many valuable discussions and to the
PCS group of the Laboratory for lending their interferometer.

References

Egerton R F 1978 Scanning Electron Microscopy 1 133
Isaacson M and Johnson D 1975 Ultramicroscopy 1 33
Joy D C and Maher D M 1980 Ultramicroscopy 1 333
Pennycook S J, Batson P E and Fisher R M 1980 Inst. Phys. Conf. Ser.
  52 337

Measurement of electron momentum densities using an electron microscope—a new approach to Compton scattering

B G Williams* and A J Bourdillon+

*Department of Physical Chemistry, University of Cambridge, Lensfield Road, Cambridge CB2 1EP. +Department of Metallurgy and Materials Science, University of Cambridge, Pembroke Street, Cambridge, CB2 3QZ

Abstract. Compton profiles of amorphous carbon and graphite have been measured in an electron microscope with an energy loss spectrometer. From these results the reciprocal form factors have been computed and are compared with the results of pseudo-potential and molecular orbital calculations.

In Compton scattering experiments a monochromatic beam of gamma-rays or electrons is scattered inelastically from the electrons in a sample. An energy analysis cf the particles scattered through a fixed angle reveals a Doppler broadening of the scattered beam so that the ground state momentum of the electrons in the sample can be deduced (Williams,1977).

Conventional gamma-ray experiments are limited by the weakness of the interaction so that large samples ($\sim$1 cc.) and long counting times ($\sim$days) are needed. The momentum resolution is limited to about 0.4 a.u. so that information concerning the valence electron profiles, which are typically about 2.0 a.u. wide, is significantly degraded. Electron scattering offers enormous potential advantages. The strength of the interaction and the intensity in the incident beam are both several orders of magnitude greater than in the gamma-ray case so that much higher count rates can be obtained from microscope samples. Furthermore, resolutions of about 0.05 a.u. have been obtained in experiments on gases (Barlas, et al., 1978).'

Provided the energy transferred in the scattering process is significantly greater than the binding energy of the electrons under investigation Bonham and Wellenstein (1977) have shown that the binary encounter impulse approximation is valid. Solving the equations for the conservation of energy and momentum one then obtains the following expression for the energy transferred to the scattering electron:

$$\Delta E = \Delta E_o + q\sqrt{2\Delta E_o} \qquad\qquad 1$$

$$\Delta E_o = E_1 \sin^2 (\phi) \qquad\qquad 2$$

where E_1 represents the incident energy, ϕ the scattering angle, q the component in the direction of the scattering vector of the ground state momentum of the scattering electron, ΔE_o the energy loss for scattering from stationary electrons and the second term on the RHS of equation 1 the Doppler broadening (in units m = c = $\hbar$ = 1).

From these two equations the energies, angles and resolution needed for these experiments can be determined. Using an incident energy of 120 kV

and an energy transfer of 1 kV gives a scattering angle of about 5 degrees
which is readily obtained in most microscopes. The energy resolution is
given in terms of the momentum resolution (from equations 1 and 2) as
(Williams et al., 1981)

$$\delta E = \delta q \sqrt{2\Delta E_o} \qquad\qquad 3$$

so that for a momentum resolution of 0.1 a.u. an easily accessible resolution of 23 eV is required. The angular resolution is given by

$$\delta \Phi = \delta q / \sqrt{2E_1} \qquad\qquad 4$$

which under these conditions gives 0.06 degrees or about 1% of the scattering angle.

Our first experiments have been carried out using samples of amorphous
carbon and graphite since thin samples may be readily prepared which are
highly resistant to radiation damage. Furthermore, since carbon is a light
element multiple scattering is kept to manageable proportions (Humphreys
and Hirsch, 1968).

The specimen was placed in a Philips EM 400 electron microscope equipped
with an energy loss spectrometer (Bourdillon, Jepps, Stobbs, and Krivanek,
1981) and a computer supplied by Link Systems. The incident beam was
focussed onto the sample so that an area of about 10 microns was illuminated.
A 40 micron condenser aperture was used which fixes the angular divergence
in the incident beam at 5 mrad. The incident beam was then tilted through
an angle of about 5 degrees and a 20 micron objective aperture used to
limit the angular divergence in the scattered beam to 5 mrad. Together
these divergences give a momentum resolution of 0.5 a.u.

Figure 1 shows the Compton profile of graphite measured using 120 kV incident electrons. The carbon K-edge occurs at 284 eV. The profile is
compared with a free atom calculation (Weiss, Harvey and Phillips, 1968).

The tail of the profile is dominated by the core electrons and beyond 1700
eV (2.5 a.u. of momentum) the experimental profile is in excellent agreement with the free atom theory and the experimental results reflect the
effect of bonding on the valence electrons. The data have been analysed in
the following way. The peak energy was found by fitting a parabola to the
peak of the Compton profile to find the zero of the momentum scale. The
spectrum at higher energy-losses was converted to momentum space (equations
1 and 2) and Fourier transformed to obtain the reciprocal form factor, $B(\underline{r})$.
This is a function in real space instead of momentum space and is the autocorrelation function of the ground state wavefunction. A bonding orbital
in carbon produces a negative dip at a distance related to the bond length.
This explains the change in sign of $B(\underline{r})$ at about 0.2 nm in figure 2. The
differences that occur between $B(\underline{r})$ for amorphous carbon and for graphite
refelct the differences in bonding, including directional differences. In
figure 3 a pseudo-potential calculation for graphite in two directions and
a molecular model averaged over all directions in the basal plane are given
for comparison (Reed et al., 1974). The change in sign in $B(\underline{r})$ correlates
with experimental data as does the strong minimum at about 4 A in the $[11\bar{2}0]$
direction. The effect of the finite momentum resolution is to attenuate
$B(\underline{r})$ and this must be taken into account before more accurate comparisons
can be made.

The attenuation of $B(\underline{r})$ can be reduced in principle by reducing the probe
and collection angle divergences provided new techniques can be introduced

to increase signal intensity. In the present data about 4×10^5 Compton events were detected in about 20 minutes. The statistical accuracy can be improved by using parallel detection in a linear diode array. We have found that the small bore of electron microscopes makes the use of annular apertures, large enough to define the angles involved, impractical. The scattered intensity can be increased by reducing the scattering angle but if it is not sufficiently large the impulse approximation used in the theory breaks down. For amorphous carbon and graphite, beam damage and multiple scattering do not appear to present problems. Our effective sample thickness is similar to that used by Barlas et al. (1978) in their experiments on gases while we are using higher incident beam energies for which the attentuation is reduced.

We have considered the dynamical effects which govern the incident beam electron distribution in a crystalline field. By orienting the foil to well defined angles relative to the incident beam the electron current may be directed predominantly on the atomic planes or between them. Figure 4 shows preliminary measurements taken on a graphite foil tilted to a Laue condition and to a Bragg condition. In the former the Compton profile is larger as the current density is then predominantly on the atomic planes where the ground state electron density is high. The small change in peak energy on moving the foil to the Bragg condition is apparently due to a small change in the scattering angle, which can be eliminated by arranging the Laue zone to fall normal to the scattering vector. The impact parameter is sufficiently small (Bourdillon, Self and Stobbs, 1981) to allow measurement of momentum distributions at different parts of the unit cell. With improved detection efficiency it will be possible to undertake an entirely new type of study of crystalline foils.

References
Barlas A D, Rueckner W H E and Wellenstein H F 1978 J. Phys. B 11 3381
Bonham R A and Wellenstein H F 1977 Compton Scattering ed B G Williams
   (McGraw-Hill: New York)
Bourdillon A J , Self P G and Stobbs W M 1981 Phil. Mag. to be published
Bourdillon A J , Jepps N W, Stobbs W M and Krivanek O L 1981 J. Micr. 124
   to be published.
Humphries C J and Hirsch P B 1968 Phil. Mag. 18 115.
Reed W A, Eisenberger P, Pandey K C and Snyder L C 1974 Phys. Rev. B 10
   1507.
Weiss R J, Harvey A and Phillips W C 1968 Phil. Mag. 17 241
Williams B G 1977 ed Compton Scattering (McGraw-Hill: New York)
Williams B G, Parkinson G M, Eckhardt C J, Thomas J M and Sparrow T 1981
   Chem. Phys. Lett. 78 434.

Acknowledgements
We are grateful to G M Parkinson for preparing specimens and to H F Wellenstein, W M Stobbs, J M Thomas and G M Parkinson for their comments and to Professor R W K Honeycombe for the provision of laboratory facilities.

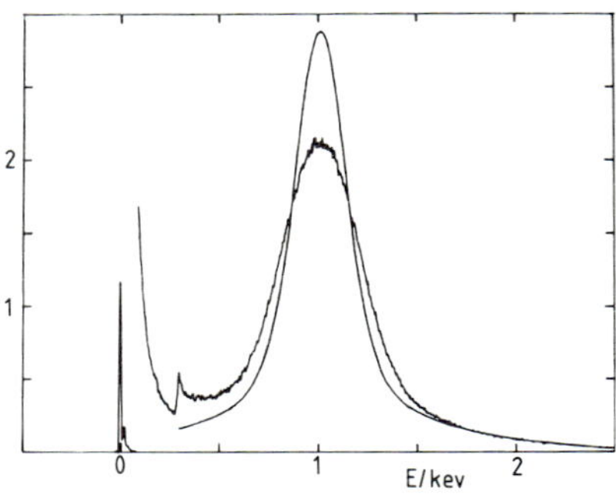

Fig.1. Compton profile of amorphous carbon measured with an electron energy-loss spectrometer using 120 kV electrons and a scattering angle of about 5 degrees (jagged line) and computed free atom Compton profile (smooth line). The zero-loss line has been scaled down by a factor of 1000.

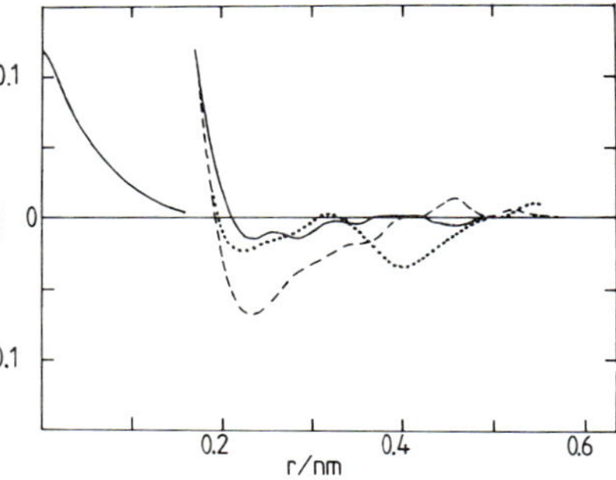

Fig.2. Reciprocal form factor $B(\underline{r})$ computed from measurements on amorphous carbon (full line) and graphite oriented with the scattering vector parallel to the $[10\bar{1}0]$ direction (dashed line) and to the $[11\bar{2}0]$ direction (dotted line).Data below 0.2 nm have been scaled down by a factor of 50.

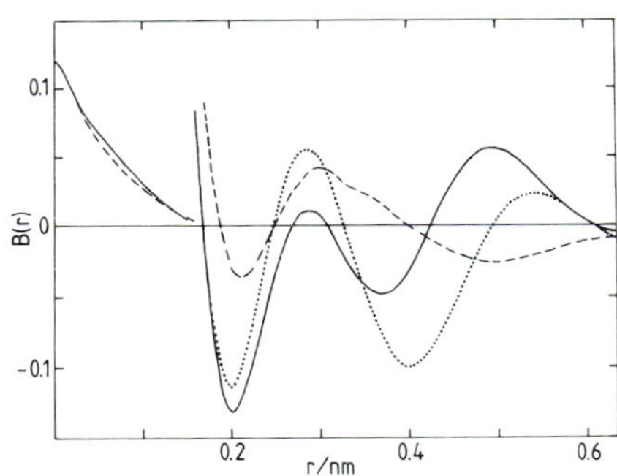

Fig.3. Reciprocal form factor $B(\underline{r})$ computed using a pseudo-potential model for graphite with the scattering vector in the $[10\bar{1}0]$ direction (full line) and the $[11\bar{2}0]$ direction (dotted line); and using a molecular model averaged in the basal plane (dashed line). Data below 0.2 nm have been scaled down by a factor of 50.

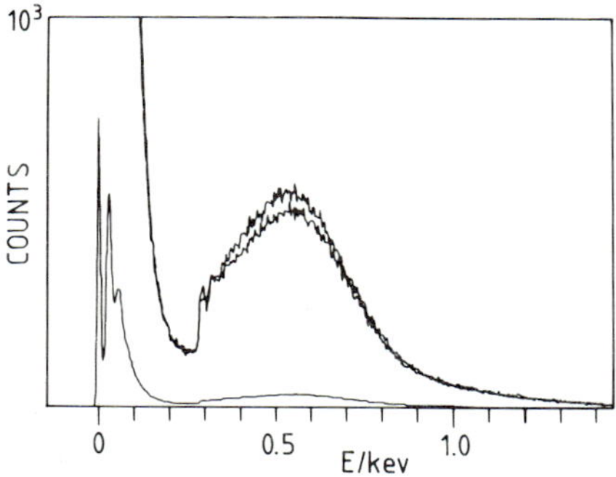

Fig.4. Orientation effects found in the Compton profile of graphite oriented to the Laue condition (upper trace) and the $[100]$ Bragg condition (middle trace) with c-axis almost vertical. Lower trace shows the zero-loss on a log scale.

Applications of STEM in solid state chemistry

H Dexpert[*], E Freund[*], J P Lynch[*] and S J Pennycook[**]

[*] Institut Français du Pétrole, B.P. 311, 92506 Rueil-Malmaison Cedex, France
[**] Cavendish Laboratory, Madingley Road, Cambridge CB3 OHE, England

1. <u>Introduction</u>

In various fields from materials science, solid state physico-chemistry
and geology to biology both basic research and engineering developments
have to correlate macroscopic observations with a detailed knowledge of
local composition and structure. Understanding sample behavior and
sensitivity to any crystallochemical change in order to improve their
properties relies strongly on these relations.

The micro- which tends to become nano- scopic aspect of the work has been
tremendously enhanced during the past decade with the birth of High Resolution Analytical Electron Microscopy (HRAEM). The need for such information is particularly imperative when one tackles questions posed by non
stoichiometry, ordered or disordered phenomena, nucleation processes,
epitaxial growth, highly divided solids and their influence on phase
transformations or crystallogenesis. Qualitative examples illustrating
HRAEM capabilities in some subjects with which we have been concerned
will be given here.

HRAEM has a tool, the Scanning Transmission Electron Microscope (STEM),
which provides images, diffraction and analytical facilities by several
channels from regions of the order of nanometers in diameter (see for
example Hren et al., 1979).

We will just recall two important features of the STEM. The first comes
from the brightness of the field emission tip which allows high current
densities in probes limited in size. This leads to reasonable signals
from very localised areas of thin specimens. The second point is that
signals can be treated both on line and off. First, their combination
in many different ways improves the information content of the image.
Second, after digitalization and storage in image memories or by computer,
treatment may be carried out to give required presentation.

The three main analytical outputs :
(i)   The spectroscopic analysis of the emitted X rays (XES or EDX) or
      light (cathodoluminescence: STEM-CL)
(ii) The angular distribution of the transmitted electrons (micro-
      diffraction μD)
(iii) The energy distribution of these electrons (EELS)

have been used and some successful applications in each case are
presented. The machine is a V.G. Microscopes HB5 equipped with a Kevex
energy dispersive X-ray spectrometer. The pole pieces compatible with
this spectrometer have a point to point resolution of about 0.45 nm and
the electron spectrometer used for EELS has a 1.8 eV resolution. The
commercial microdiffraction facilities have been modified for simultaneous
acquisition of the bright field image with its diffraction pattern.

2. Examples of XES and CL analysis

Industrial catalysts are of great interest to show the capabilities of
high resolution microanalysis, particularly with XES, because they consist
of heavy metal particles from 10 nm in diameter down to single atoms
dispersed on the surface of a highly divided light oxide which is in the
form of very thin platelets (less than 10 nm in thickness). The information needed to characterize these complex systems for fresh or aged,
polluted, catalysts comprises the local elemental composition, the microstructure, the nature of exposed faces and the chemical state of the
carrier and crystallite.

A limitation in XES studies of these small areas is the low specimen count
rate and possible spurious background radiation coming from stray electrons (hole count) interacting mainly with the objective aperture. The
use of a virtual objective aperture in a plane conjugate to the real
objective aperture, and not in line of sight of the Si(Li) detector, is
essential to analyse these very small crystallites.

Quantitative analysis has to take care of several classical difficulties
such as the possible absorption by the copper grid of the K emission lines
of the light elements constituting the carrier; or overlappings, for
example the well known case of L and K lines of molybdenum and sulphur for
selective hydrogenation catalysts.

With attention to these sources of errors, such materials are useful
tests for sensitivity evaluation in EDX analysis. This point is fully
discussed in another talk of the Conference (Dexpert et al., 1981).

So, using XES, we have determined the elemental composition and distribution of these particles and their extrinsic poisons (due to impurities
contained in the reactants: sulphur, lead, etc.) (Dexpert et al., 1981).
An illustration is presented in figure 1 of a Pt-Ru/γalumina reforming
catalyst. To improve the detection limit, the carrier has been in this case extracted by a replica technique and different Pt-Ru-S ratios are easily determined from small aggregates separated by a few nanometers. These
characterizations can be performed simultaneously with a sequential EELS
acquisition and intrinsic poisoning of aged catalysts by coke (carbonace-
ous deposit resulting from hydrocarbon conversion) has been located.
Carbon is recognized in a few seconds by its K edge at 285 eV and the
interesting point is that the same beam conditions, in particular the
size of the objective aperture, are allowing reasonably good signals from
the two techniques in a short time.

The XES ultimate detection limit has been ascertained with other bimetallic
systems, Pt-Pd and Pt-Rh .   Alloy formation is detectable for particles
ranging between 1 and 2 nm and various ratios have been found for the
first system where significant Pd-L and Pt-M lines appear in 100 secs.
The typical signal heights are 3 to 4 times greater than the background

level. In the Pt-Rh alloy, the criterion $R = \text{signal}/[2\ \text{background}]^{1/2} > 3$ (for
definite detection) is around 3 for Pt and half this for Rh (particle size:
1.5 nm- recording time : 400 secs). By increasing the counting time up to
15 minutes, we estimate that the minimum detectable numbers of atoms would
then be set at about 30.

Of the usual outputs, light has been less used mainly because of the low
luminescence efficiency of the materials to which the technique has been
applied. There is however a large class of solids, phosphors, for which
cathodoluminescence is an important mean of investigation. The STEM gives
the possibility of studying with a resolution of some hundreds of angstroems
 or less   their luminescence and their elemental composition either by
EELS or XES using a special cartridge (Pennycook et al., 1981).

Associated to STEM-CL experiments, XES has a slight advantage over EELS.
The excited regions must be thick enough to emit a sufficient number of
photons which fall on to the photomultiplier through the light guide of
the CL detector. In this situation EELS can suffer from multiple scattering effects which will reduce the precision of the characterization.
We have nevertheless been able to do CL+EELS observations on thin flakes
of mixed rare earth oxysulphides crystals where the activator concentration was high enough to produce an intense CL response, the rare earth
being identified by the strong N_{IV-V} excitation of its ten 4d electrons.
In these cases, the spatial resolution could be reduced to a few
nanometers. When using XES, the thickness can be low enough for absorption corrections to remain negligible so quantitative analysis is
simply done. Another difficulty is that the phosphors grains irradiated
must be in the same range of thickness to be sure that CL is comparable
from one grain to another. The relative thickness is estimated directly
from the rate of X-ray emission. The contrast difference in the annular
dark field image is also a good indication. This is shown on figure 2
where elemental maps from yttrium K line and lanthanum L line and the ADF
image suggest that regions (a) and (b) are of similar thickness. The
luminescence of grain (a) is lower than that of crystal (b), indicating
in this case the preferential substitution of europium with yttrium as
shown by the EDX spectra which give roughly $(La_{2/3} Y_{1/3} O)_2 S$ and
$(La_{1/3} Y_{2/3} O)_2 S$ compositions for crystals (a) and (b) respectively.

3. <u>Examples for angular distribution studies</u>

In stopping the focussed probe onto the crystal, a series of diverging
cones emerges from the exit side, each one corresponding to a particular
diffraction, the whole forming the µD pattern. We have developed a
parallel microdiffraction system, first described by Cowley et al. (1979)
which allows simultaneous display of µD pattern and image.

Particular zone axes may be easily distinguished, opening the possibility
of looking at particle-support orientations, the diffracting area being
about 2 to 3 nm in diameter. The technique and applications will be
discussed in detail in another talk of the Conference (J P Lynch et al.,
1981). I shall nevertheless show here a particular result illustrating
the usefullness of the structural characterization of transition solids
such as the numerous neighbouring forms of alumina. Figure 3 shows two
parallel µD patterns of the same alumina platelet (δ form) taken in about
2 secs. The larger area (b) which covers 100 times more unit surface
cells than the smaller, is a $[1\bar{1}0]$ zone axis pattern of the δ-structure

a spinel with the cube side multiplied by 3 in one direction.
In μD (a), the structural arrangements presented by this structure is not
averaged because of the small number of diffracting cells and the intensities differ. This form is yet unlabelled and can be considered as the
basic unit cell leading to δ alumina.

4. Examples of electron energy loss analysis

Despite its multiple scattering limitation with increasing thickness, EELS
is very powerful when one wants to combine both spatial and energy resolution. Carbon contamination formed during the oxydation of the metallic
rare earth thin film by beam exposure within the CTEM has been found to
depend on the two possible oxide structures, in this case monoclinic or
cubic at a 5 nm scale. The latter oxide was the more contaminated by
reason of the perfect epitaxy existing between the two networks in
contact, graphite and cubic Y_2O_3, the $[111]$ zone axis of the cubic

crystal being parallel to the $[001]$ of graphite layers at the surface of
the oxide film (Dexpert et al., 1979).

In an attempt to evaluate how sensitive EELS could be in high spatial
resolution applications, we investigated a rare earth doped form of beta-
alumina and the result is shown in figure 4 (Dexpert-Ghys et al., 1981).
This compound may be described as a succession of alumina spinel blocks
separated each eleven $\overset{\circ}{A}$ by a mirror plane where lanthanum and europium
have been substituted (for use in visible-UV spectroscopic experiments)for
the usual constitu nt sodium. When the probe is stopped onto a very thin
region of the spinel blocks, four features appear: a plasmon loss (a) at
22 eV corresponding to the excitation of aluminium atoms and three core
losses measured at 70 eV (1), 83 eV (2) and 98 eV(3) corresponding respectively to L_{II-III}, L_I edges of aluminium and N_{IV-V} of lanthanum. The latter
element is known to be normally absent from the spinel block, its presence
being explained by the fact that the alumina spinel blocks melt rapidly
under the stopped beam. Moving to a mirror plane, the EELS spectrum
becomes different. The low loss region now presents three bands, still
(a) plus (b) and (c) coming out in the region where the second rare earth
plasmon peaks, the high loss part of the spectrum showing only the N_{IV-V}

edge of the rare earth. The counting time was about ten minutes for each
spectrum and special care was taken to avoid any mechanical drift or
electronic shift. The specimen was free of detectable contamination.
Attempts have to be made in filtered imaging, particularly with the 22 eV
aluminium loss and also in parallel μD to compare their practical capabilities and complementarities.

5. Conclusion

From all these examples it is clear that the STEM is well suited for the
examination at a scale close to a few unit cells of a wide range of
solids. The large range of information it can provide is a unique
contribution to the knowledge of local variations in structure and
composition which are often decisive for the global properties of
the material. Many techniques are still in progress (window-less
X ray detector, parallel detection system in EELS, etc.) and the
future of these machines, despite the large investment they require
is undoubtly very promising. Its increasing use by specialists coming
from very different horizons make for the reality of a microscience,
very helpful to anyone interested in fundamental solid state work
as well as in applied domains.

<u>References</u>

Cowley JM and Spence JCH 1979 Ultramicroscopy <u>3</u> 433
Dexpert H, Freund E and Lynch JP 1981 This Conference
Dexpert H, Freund E and Lynch JP 1981 Metals Society Conference, In press
Dexpert H, Pennycook SJ and Brown LM 1979 Inst. of Physics Conference
    <u>52</u> 339
Dexpert-Ghys J and Dexpert H 1981 Unpublished results
Hren JJ, Goldstein JI and Joy DC 1979 Introduction to Analytical Electron
    Microscopy (New-York: Plenum Press)
Lynch JP, Lesage E, Dexpert H and Freund E 1981 This Conference
Pennycook SJ, Dexpert H and Charreire Y 1981 J. de Spec. et Mic. Elect.
    In press

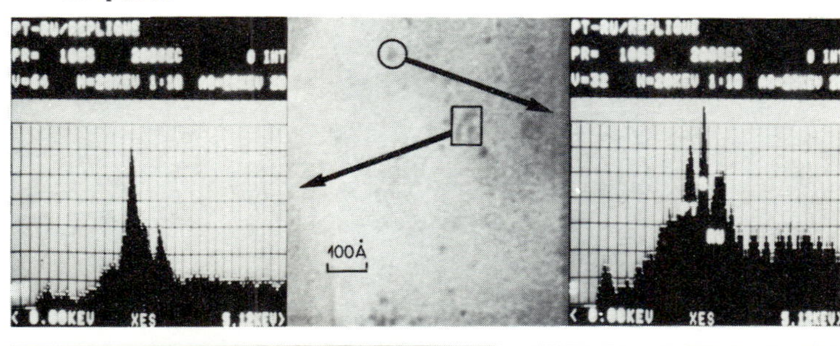

Fig. 1

Typical example of a XES catalyst characterization.

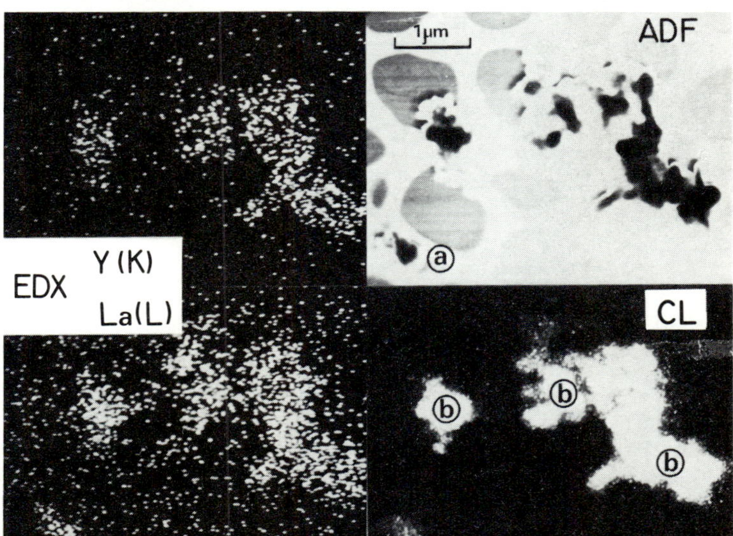

Fig.2

Example of STEM CL and EDX analysis with the (La-Y) oxysulphide system.

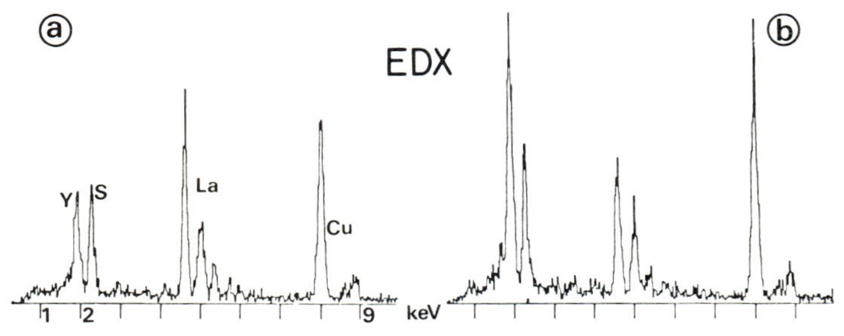

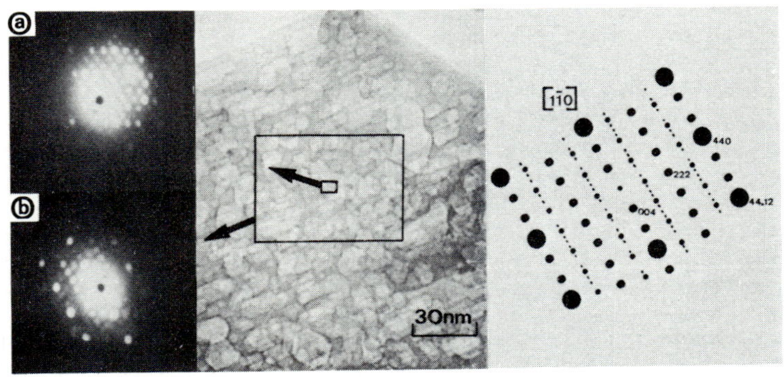

Fig. 3

Two alumina microdiffraction patterns : a) stopped beam and
b) scanned beam obtained in parallel with the corresponding
bright field image. Recording time around 2 secs.

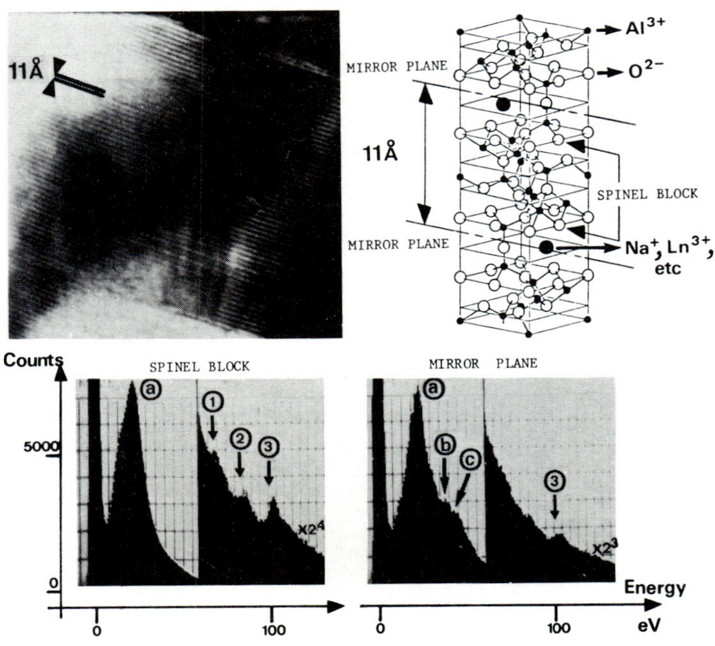

Fig. 4

High spatial resolution example of EELS on a rare earth (La, Eu)
doped beta alumina. Differences in the low loss region of the spectra
[(a), (b) and (c)]   and in the core loss region   [(1), (2) and (3)]
are clearly visible.

Electron microscopy and microanalysis of an industrial oxidation catalyst

P L Gai,[†] J C J Bart* and E D Boyes[†]

[†]Department of Metallurgy & Science of Materials,
 University of Oxford, Parks Road, Oxford OX1 3PH
*Montedison Chemical Company, Italy.

1. Introduction

Binary oxide catalysts of tellurium-molybdenum are used industrially in the
ammoxidation of propylene (C_3H_6) to acrylonitrile (C_3H_3N) and to acrolein
(C_3H_4O). In-situ high voltage electron microscopy (HVEM) for dynamic
observations of such heterogeneous catalytic reactions conducted in parallel with the chemical studies (Bart et al 1975) have confirmed α-Te_2MoO_7
as the active phase in the maximum conversion to C_3H_3N. Degradation in
C_3H_3N selectivity was also observed at 673K. Mechanisms of the redox reactions and nature of degradation of Te_2MoO_7 have therefore, been examined
using the in-situ HVEM, state of the art high resolution EM(HREM) with diffraction, electron probe X-ray microanalysis (EPMA) and analytical EM with
a field emission gun STEM (AEM-FEG STEM) as well as electron spin resonance
(ESR) methods.

2. Experimental

The Te-Mo oxides including α-Te_2MoO_7 were prepared by $(NH_4)_6Mo_7O_{24}.4H_2O$ and
H_6TeO_6. Crystals of Te_2MoO_7 were also grown by fusion in air. The samples
had MoO_3 as a coexisting phase. Direct observations of C_3H_3N reaction were
carried out using mixtures of $C_3H_6|NH_3|$ air and also of reduction indepen-
dently in C_3H_6 or H_2 to examine the sequence of events in a gas reaction
cell fitted to an AEI-EM7 HVEM at 1MeV with total gas pressures of 100torr.
Parallel experiments were conducted in a VG-Q7B mass spectrometer/Varian
3700 gas chromatograph. For HREM, a JEOL 200CX EM at 200kV was used with a
point resolution of $\leqslant 1.7$Å and a CTF first zero of 2.4Å at optimum defocus
(Boyes et al 1980), and structure images were recorded with the objective
aperture out to 0.6Å^{-1}. X-ray microanalysis of the samples and reduced
crystals ($\leqslant 1000$Å in size) was carried out in the Oxford FEG-STEM at 80keV
calibrated with the known standard compounds Te_2MoO_7 and also $TeMo_5O_{16}$
(Arnaud and Guidot, 1977). The results were complemented by EPMA of larger
crystals (a few microns) using an automated Cameca Camebax fitted with
wavelength dispersive spectrometers (WDS) as well as EDS. The L lines of
Te and Mo were used at 20kV and oxygen content analysed by difference-with
quantitative corrections using CORREX files in the standard MBXCOR programs.
Once the possible phases in the samples have been identified by the various
techniques including EPMA, the distribution was determined by an AEM finger
printing method in which the EDS data accumulation was allowed to proceed
only long enough to obtain a significant identification of the respective
phase from the cation ratio. The method also allowed much smaller
particles and mixed phases to be analysed reliably (Boyes and Gai, 1981).

Results

α-Te$_2$MoO$_7$ (a = 4.25, b = 8.6, c = 15.9Å, β = 95.6°, Te^{4+}, Mo^{6+}) consists of edge sharing Mo-O octahedra linked by Te-O polyhedra. The structure images at different defocus settings are shown in Fig.1(a), (b), (c) and (c) corresponds to the image near optimum defocus (ΔF = -600Å) estimated by minimum contrast at zero focus as a reference or by carbon contamination. A preliminary calculation is inset for a foil thickness (t) of 100Å. The image and calculations are consistent given the uncertainties in the experimental Δf and t. In the HVEM, in Te$_2$MoO$_7$ reduction, no microstructural changes were observed from room temperature to $\sim$ 573K although the ESR and mass spectrometry showed abstraction of oxygen from the catalyst. Fig.2 shows an example of high resolution ESR spectrum in H$_2$ with Mo(V) occupying different geometrical (g) sites characterised by the line intensities. At 630K some surface features (not crystallographic shear (CS) planes) were observed and at catalyst operating temperatures of $\sim$ 673K, the sample disintegrated into small crystals. The d.p. and the lattice image of the reduced crystals are shown in Fig.3(a) and (b) giving orth. cell structure of a $\sim$ 20Å, b $\sim$ 7.2, c $\sim$ 4Å. The quantitative analysis of the main reduced phase indicated Te/Mo/O atom ratio of 1:5:16 in the unit giving a composition of TeMo$_5$O$_{16}$ within the experimental accuracy of $\pm$ 2% in the corrected atomic concentrations. The results of typical quantitative analysis of the reduced material and the standard Te$_2$MoO$_7$ are shown in the table in Fig.4. The EDS results from the FEG STEM are shown in Fig.5 (a), (c) and (b) for the standard Te$_2$MoO$_7$, reduced material and the known TeMo$_5$O$_{16}$ (French standard) prepared independently using 5MoO$_3$ + $\frac{1}{2}$TeO$_2$ + $\frac{1}{2}$Te (Arnaud et al 1977). However, the unit cell dimensions of the reduced phase did not match the published TeMo$_5$O$_{16}$ structure (Arnaud et al 1977) although the EDS/WDS indicated the same chemical composition. To confirm this, HREM of independently prepared TeMo$_5$O$_{16}$ was carried out and the d.p. and the lattice image are shown in Fig.6(a) and (b) (with a $\sim$ 10Å, b $\sim$ 14.43, c $\sim$ 8.16, β $\sim$ 90.85°). On the basis of AEM/HREM observations we therefore, conclude that the reduced material is an allotropic modification of TeMo$_5$O$_{16}$ to which we have assigned the designation β-TeMo$_5$O$_{16}$ (Gai et al 1981). The proposed structure of β-TeMo$_5$O$_{16}$ with Mo in octahedral and Te in bipyramidal configurations is shown in (001) in Fig.7.

The observations have demonstrated that CS planes are not present in some industrial catalytic reactions at operating temperatures. The role of microanalysis has been central to this work in identifying various phases and indicates the importance of AEM chemical analysis in HREM interpretation.

We thank the SRC (UK) for financial support.

References

Arnaud Y and Guidot J 1977 Acta.Cryst. B33 2151
Bart J C J and Guiordano N 1975 Gazz.Chim.Ital. 109 73
Boyes et al 1980 Inst.Phys.Conf.Ser. 52 445
Boyes E D and Gai P L 1981 Proc. AEM Workshop MAS Conf. Vail (USA)
Gai P L, Bart J C J and Boyes E D, Phil.Mag. (in press)

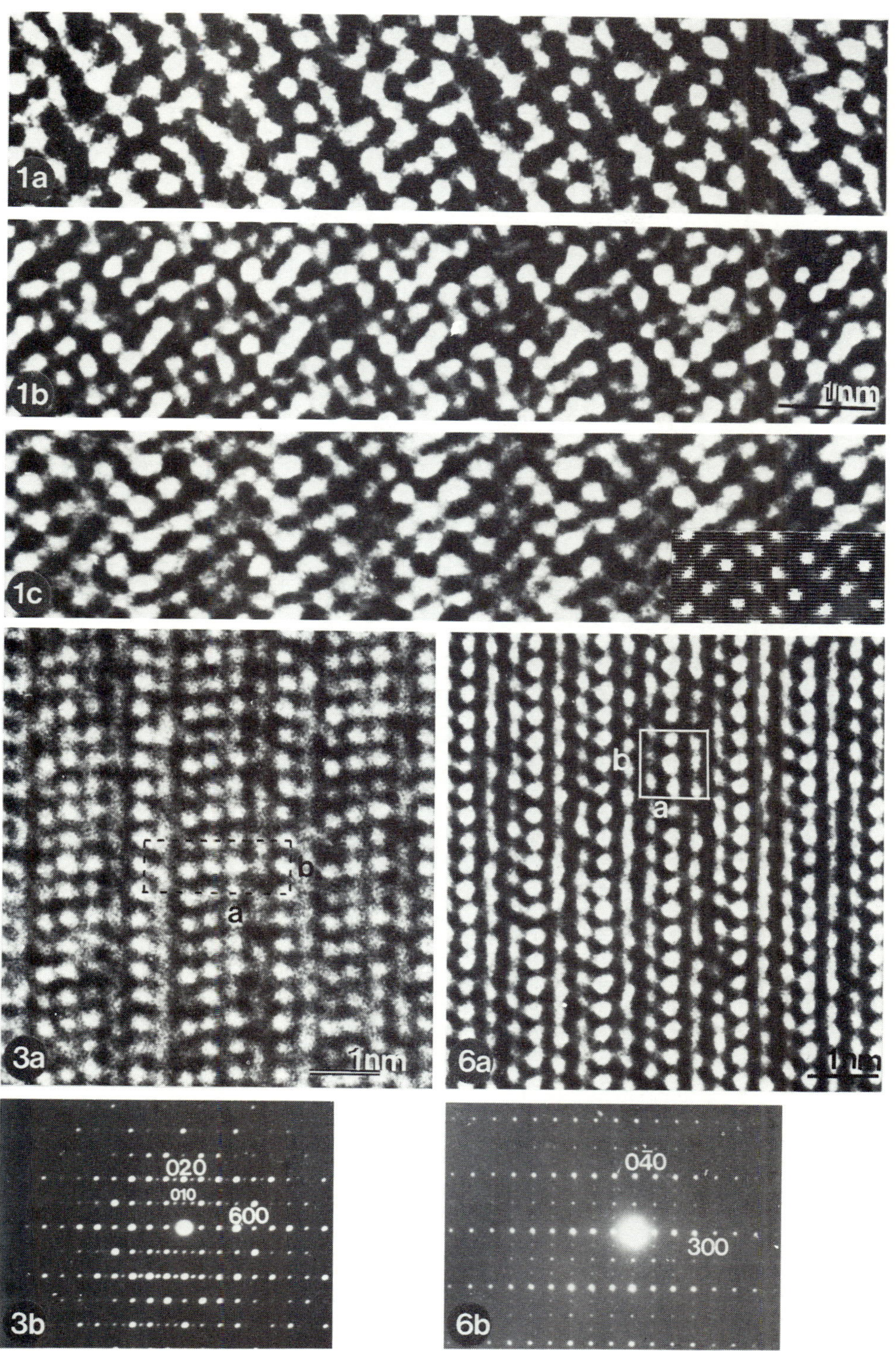
1a
1b
1nm
1c
3a
1nm
b
a
6a
b
a
1nm
020
010
600
3b
040
300
6b

Fig.4.   EXAMPLES OF QUANTITATIVE XR MICROANALYSIS

1. Standard α-Te$_2$MoO$_7$

ELEMENT	POSITION	PEAK	BACKGR.	K.RATIO
MO	61775	327.3	5.8	0.1800
TE	37537	2592.6	27.6	0.4592
TOTAL				0.6391

ELEMENT	K- I.X./I.STD.	K.RATIO	CONCEN.	ATOM C ▼
MO	1.0141	0.1800	0.2095	0.1009
TE	0.9952	0.4592	0.5481	0.1985
O			0.2425	0.7006 BY DIFFERENCE

2. Reduced Material

ELEMENT	POSITION	PEAK	BACKGR.	K.RATIO
MO	61775	909.0	5.9	0.5055
TE	37537	612.3	20.5	0.1057
TOTAL				0.6115

ELEMENT	K- I.X./I.STD.	K.RATIO	CONCEN.	ATOM C ▼
MO	2.8487	0.5055	0.5553	0.2274
TE	0.2296	0.1059	0.1488	0.0458
O			0.2959	0.7268 BY DIFFERENCE

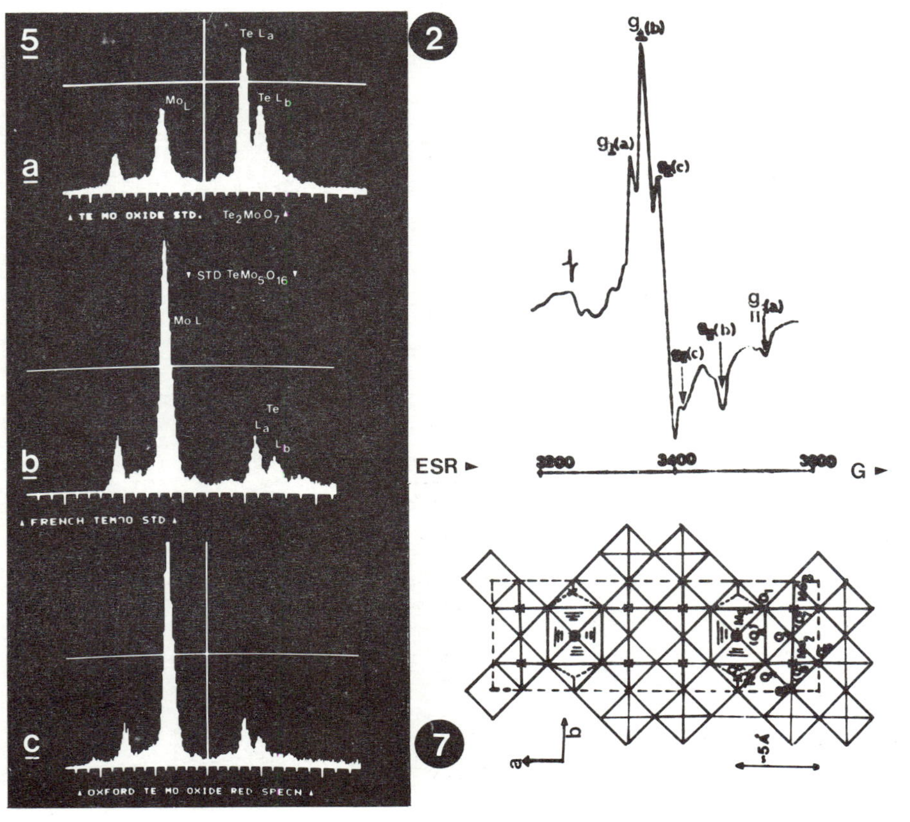

The study of hydrogenated amorphous silicon using STEM

A J Craven, T W Buggy and A Patterson

Department of Natural Philosophy, University of Glasgow, Glasgow G12 8QQ

Hydrogenated amorphous silicon is under investigation for use in semi-conductor devices, e.g. solar cells. It can be produced either by R.F. glow discharge of silane or by sputtering silicon in an atmosphere containing hydrogen. Studies of the material prepared by glow discharge show some correlation between the preparation conditions, the electrical properties and the structure observed in conventional transmission electron microscope (CTEM) images (Knights 1980). These studies suggest that the material is in the form of rods $\approx$ 10nm diameter which run through the film from top to bottom. The rods are thought to be continuous random networks of silicon held together by a less dense material thought to contain a higher concentration of hydrogen. The amount of this less dense material varies with the preparation conditions.

The electrical properties of sputtered material have been under investigation in our department for a number of years. We wish to see if similar structure occurs in this material and correlate it with the electrical properties. This paper reports our preliminary studies.

A film 30 ± 5nm thick was sputtered onto a KCl substrate at 260°C in an atmosphere of argon and hydrogen in the ratio 2:1. Microscope specimens were prepared by floating the film off on water and picking pieces up on grids. Fig. 1 shows an elastic bright field (EBF) image, a plasmon bright field (PBF) image and an annular dark field (ADF) image of the same area of film. Each image has had the contrast increased by use of the same amount of black level. Structure on a scale > 3nm is visible and this differs from that observed in the glow discharge material. The fluctuation in the ADF signal is ± 20%. Because of the non-linear variation of the ADF signal with thickness, this corresponds to mass thickness fluctuations of about ± 25%. The contrast in the EBF image is approximately the inverse of that in the ADF image as expected but that of the PBF image is much lower. In some areas this contrast follows the ADF image and in others it follows the EBF image.

If the contrast observed is the result of density rather than thickness fluctuations, a shift in the plasmon energy from point to point is expected (Raether 1980) provided that the plasmon is localised within the contrast feature. The localisation of the plasmon is determined by the angle of scattering accepted in the spectrum and for large angles is less than 5nm (Cundy et al 1968). Electron energy loss spectra with a large angle of acceptance ($\approx$ 30 mrad) were recorded from areas of differing signal level. There is no change of plasmon energy within a ± 2% experimental error implying density changes less than ± 4%. The ratio of the area under the

plasmon peak to that under the zero loss peak corresponds to t/λ_p the effective mean free path for plasmon excitation. This ratio varies from 0.17
to 0.27 with an average value of 0.22. This agrees well with the fluctuations in thickness inferred from the ADF signal. No accurate values are
available for λ_p but an estimate using the equation of Raether (1980) gives
a value of 135nm and hence a mean thickness of 30nm with an error of about
20% in reasonable agreement with the film thickness obtained from the deposition conditions. Small variations of the plasmon width were observed
but the errors were large and there was no correlation with t/λ_p.

The low contrast and anomalous effects in the PBF image can be explained by
thickness variations in a film of constant density if the thickness is
close to λ_T, the mean free path for all scattering events which remove intensity from the plasmon. This includes all the inelastic scattering and
the majority of the elastic scattering since the width of the elastic
scattering distribution is much greater than the angle of scattering accepted by the spectrometer, 8 mrad in this case. (See, for example, the
calculations of von Keil, Zeitler and Zinn (1960) for elastic scattering
from an assembly of incoherent scatterers.) For a thin amorphous film,
the PBF signal should be well represented by (t/λ_p) exp $(-t/\lambda_T)$ which is a
maximum when $t = \lambda_T$. In the region of this maximum, the change of the PBF
signal with thickness will be small and for $t < \lambda_T$ it will increase with
thickness as does the ADF signal while for $t > \lambda_T$ it will decrease with
thickness as does the EBF signal. No values are available for λ_T but it
can be estimated at 42nm with an error of $\pm$ 20%. This estimate uses a
film density of 2.1×10^3 kg/m^3. (Freeman and Paul 1979), the elastic
scattering cross-section of silicon from Langmore, Wall and Isaacson (1973)
and a total inelastic mean free path for silicon estimated from the data
for aluminium given by Ritchie and Howie (1977). This is sufficiently
close to the mean film thickness to explain the observed contrast.

Fig. 2 shows a secondary electron image of the film. This images gives a
strong impression of surface topography corresponding to granules of
material piled on top of each other. While stereo-microscopy is needed
to confirm this, a model in which the film is made of granules 3-10 nm in
size which coalesce or are held together by less dense material explains
the observations made to date. The thickness fluctuations correspond to a
varying number of granules through the film. The results of the plasmon
measurements correspond to an average over the granules and intergranular
material through the film and hence no shift in plasmon energy is likely
to be observed. Only by going to thinner specimens might it be possible
to establish whether there are density as well as thickness fluctuations
in this material.

Acknowledgments

The authors would like to thank the SERC for funds to purchase the equipment and for a studentship in collaboration with V.G. Microscopes for
TWB. The authors would also like to thank Dr. A.R. Long and Mr. I. Gibb
for help and advice.

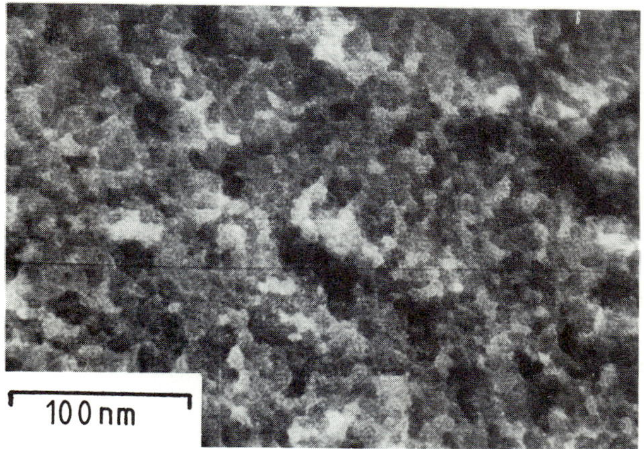

Fig. 1a.   Energy filtered bright field image using elastically scattered
electrons, (referred to as EBF image in the text).

Fig. 1b.   Energy filtered bright field image of the same area using
electrons which have excited a plasmon, (referred to as PBF
image in the text).

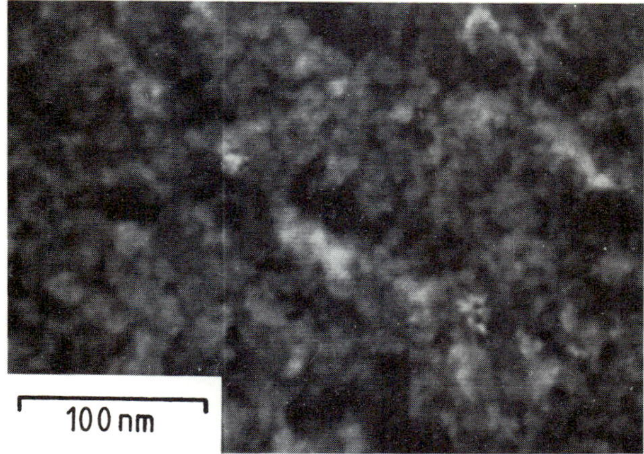

Fig. 1c.   Annular dark field image of the same area

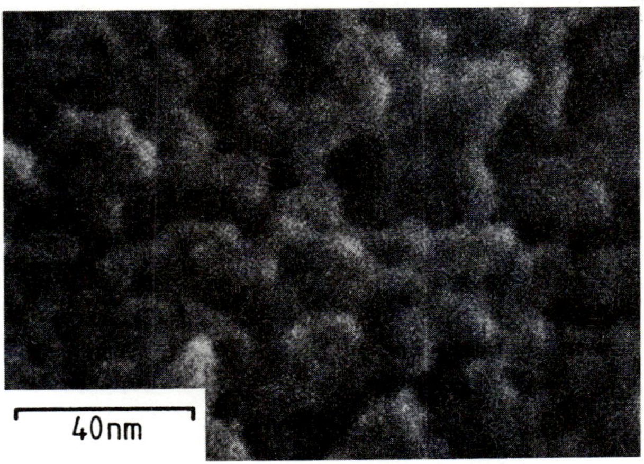

Fig. 2. Secondary electron image

References

S L Cundy, A J F Methrell, M.J Whelan, P N T Unwin, R B Nicholson
    (1968) Proc Roy Soc A30 7267
E C Freeman and W Paul (1979) Phys Rev B20 716
E von Keil, E Zeitler and W Zinn (1960) Z Naturforschg 15A 1031
J C Knights (1980) Journal of Non-Crystalline Solids 35 36 159
J P Langmore, J Wall and M S Isaacson (1973) Optik 38 335
H Raether (1980) **Excitation** of Plasmons and Interband Transitions by
    Electrons (Springer Verlag Berlin 1980)
R H Ritchie and A Howie (1977) Phil Mag 36 463

Cathodoluminescence of slip bands in plastically deformed MgO single crystals

B Sieber

Cavendish Laboratory, Madingley Road, Cambridge CB3 0HE, England and
LPM, CNRS, 1 Place A. Briand, 92190 Meudon, France.

1. Introduction

So far, the results of the cathodoluminescence studies performed in a scanning electron microscope (CL/SEM) of deformed MgO single crystals have been quite controversial. The crystals have been deformed either by indentation (Velednitskaya et al 1975, Pennycook and Brown 1979, Pennycook et al 1980) or by plastic deformation (Llopis et al 1978, Datta et al 1979) or by contact damage(Chaudhri et al 1980). The origin of the CL signal has been suggested to be due to interstitial point defects (Velednitskaya et al 1975) or to dislocations (Pennycook and Brown 1979, Pennycook et al 1980, Chaudhri et al 1980) or to point defects in association with dislocations (Llopis et al 1978). A detailed study has shown that plastic deformation in MgO gives rise to a blue CL, while a red band can be associated with iron impurities in various valence states (Datta et al 1979). In this paper the results are described of CL/SEM and photoluminescence (PL) experiments performed on the bulk material. Coupled with a correlation between the dislocation struc ture −observed in a high voltage electron microscope− and the CL/SEM image of a thin foil, they show that the role played by dislocations as radiative recombination centres is important, as suggested by Pennycook et al 1980.

2. Experimental

Fresh dislocations are introduced in as grown 3N purity crystals by mechani cal tests performed in compression ($\sigma///<100>$) at constant strain rate($\approx 10^{-5}$ s^{-1}) in air at room temperature.The strain is approximately 1%. Sheets(thickness$\approx 500\mu$m) are cleaved from the crystals, then mechanically and chemically (in hot H_3PO_4) polished to a thickness of $\approx 200\mu$m. CL/SEM studies are made at 300K at 30 keV, using a photomultiplier with S20 photocathode. PL studies are carried out also at 300K, using the excitation of the 482 nm line of a krypton-laser, and a photomultiplier with a GaAs photocathode[x]. To get the correspondence of the CL/SEM picture with the dislocation structure, dish shaped foils are chemically thinned in hot H_3PO_4 to produce suitable areas for observation at 450 kV.

3. CL/SEM spectra and images and comparison with PL results

The spectra shown in the following have been corrected for the response of the S20 photocathode. A typical CL spectrum of an undeformed specimen consists of a very intense red band peaking at λ=790 nm, and of a weak broad blue band (λ=450 to 550 nm) (Fig 1). Since the intensity of the red band is

[x]performed in Laboratoire de Physique des Solides, CNRS, Meudon.

not constant all over the specimen(Fig 1), the recombination centres giving
rise to it are certainly heterogeneous in distribution and/or density.
It has been pointed out,first by Llopis et al(1978) that in sheets coming
from deformed specimens, bright and dark bands can be observed, the former
ones being correlated with the slip bands. The CL spectrum of the bright
bands consists of a red band overlapping with a very broad blue band ranging from λ=600 to 400 nm (Fig 2). The spectrum of a dark band is usually
similar to that of the undeformed material (Fig 2). Monochromatic pictures
show that the slip bands are luminescent from the blue region to the red
one (Fig 3). It has to be noticed that a higher beam current is needed to
get a good image at 730nm than at 500nm. In earlier work, Datta et al(1979)
observed the bright bands only in the blue region and concluded that the
red band is due to iron in various valence states, which is the major impurity in MgO. Their observations were taken at constant beam current. However
one can imagine that impurities have a sufficient concentration to contribu
te to the luminescence. The blue CL comes directly from the plastic deformation process, and it seems to influence the red luminescence.
When photoexcited in the blue region, both the undeformed and deformed specimens exhibit a red PL with the same spectral structure (Fig 4). This means that plastic deformation does not influence the red luminescence. Thus,
why, in CL, are the slip bands imaged in the red region of the spectrum, and
not only in the blue region, as suggested by the PL results and by Datta et
al(1979)? This can be explained by a secondary red PL, excited by the blue
photons created by the defects introduced during the plastic deformation.
This is supported by the fact that the number of photons created in our CL
experiments is higher that the number of blue photons needed to produce the
observed red PL (see Appendix). Furthermore, Chen et al(1975), by exciting
deformed single crystals above the band gap, have observed only a blue luminescence. Thus, we can conclude, in agreement with Datta et al(1979) that
the defects introduced during the deformation are responsible for the blue
CL.

4. Correlation of the dislocation structure (HVEM) with the CL/SEM image

The same area of a thin foil coming from a 0.86% deformed specimen (the CL
spectrum of the foil is shown on Fig 2) is imaged in the CL mode (Fig 5)
and in transmission (Fig 6). The bright bands called A,B and C on the SEM
image are parallel to the $|100|$ direction, and have been reported in dashed
lines on the HVEM image; they correspond to the slip bands. They are composed of tangles of mixed dislocations lying in many various directions incli
ned with respect to the foil plane, and with mainly Burgers vectors $\vec{b}$ of
<110> types. In the dark bands, there are only long and straight screw dis
locations -$\vec{b}$=<110>- lying in the plane of the foil and which enter the slip
bands. They have perhaps been introduced during the mechanical polishing.
Cusped dislocations are also observed; they have been observed in a thin
foil coming from an undeformed specimen. Since the HVEM picture has been
taken in a 2-beam condition, a whole set of pure edge dislocations are out
of contrast; they lie parallel to the $|001|$ direction and have a $\vec{b}$ of the
<110> type. The presence of tangles which produce a lot of dislocation in-
tersections and thus probably point defects in the bands A, B and C could
explain the origin of the CL, as well as the high density of dislocations
themselves (4.10^9 cm^{-2} in the bands, 10^8 cm^{-2} outside). This assumption is
supported by the fact that the blue CL is very weak in undeformed specimen
and increases when the specimen is deformed. On the other hand, a density
of 10^{11} cm^{-2} around an indent has been shown to produce a very strong blue
signal (Pennycook and Brown 1979, Pennycook et al 1980, Velednitskaya et
al 1975).

5. Conclusion

Care has to be taken when interpreting the CL monochromatic images, because
of the occurence of a fluorescence effect in which photons from the blue
end of the spectrum excite red photons from impurity levels. Different kinds
of impurities act as radiative recombination centres in the red region of
the visible spectrum, while dislocations, associated or not with point defects, have a major contribution to the blue CL; their density is likely an
important factor, a high density being required to produce observable luminescence in a SEM.

Appendix: Comparison between laser excitation and the focussed electron
probe

The density of photons of frequency ν in the laser beam can be calculated
from the laser power P and the beam area A; it is $N_\nu \simeq P/(A.h.\nu.c)$
For P=8mW, A=3x10^{-7} m^{-2} and photons at 482nm, we find $N_\nu \simeq 10^{14}$ m^{-3};(h and c
have their usual meaning; c is the velocity in the crystal).
The corresponding density in the electron probe is somewhat more difficult
to estimate. The average energy required to create an electron-hole pair is
three times the band gap (Egap$\simeq$ 7.8 ev for MgO) and each electron-hole pair
creates one photon. The radius of the interaction volume associated with the
focussed probe R is about 6μm for 30 kV fast electrons (Kanaya and Okayama
1972). Each photon remains for a time R/c in the interaction volume. We
find $N'_\nu \simeq (E_i.I/e)(R/c)/(3.Egap.4\pi R^3/3)$. I here is the beam current normally
10nA. Insertion of the appropriate values yields $N'_\nu \simeq 2.10^{15}$ m^{-3}, about twenty
times larger than the value achieved for the laser. From a qualitative point
of view, the power into the electron probe is comparable with the laser power, but the illuminated volume is much smaller, so the level of excitation
is greater. It follows that the electron probe will excite levels below
the band gap energy by fluorescence more strongly than will the laser.

Chaudhri M M, Hagan J T, and Wells J K, 1980, J. Mat. Sci. 15 1189
Chen Y, Abraham M M, Turner T J, and Nelson C M, 1975, Phil.Mag. 32 99
Datta S, Boswarva I M, and Holt D B, 1979, J. Phys. Chem. Solids 40 567
Kanaya K, and Okayama S, 1972, J. Phys. D5 43
Llopis J, Piqueras J,and Bru L, 1978, J. Mat. Sci. Letts. 13 1361
Pennycook S J, and Brown L M, 1979, J. Luminescence 18/19 905
Pennycook S J, Brown L M, and McGovern S, 1980, Dev. Elect. Micr. and Anal.
52 161
Velednitskaya M A, Rozhanskii V N, Comolova L F, Saparin G V, Schreiber J,
and Brummer O, 1975, Phys. Stat. Sol. (a) 32 123

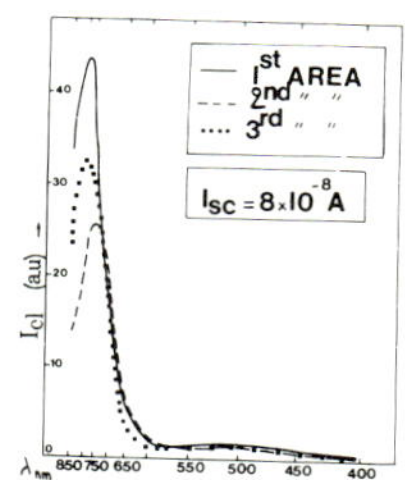

Fig 1: CL/SEM spec-
trum; undeformed
specimen

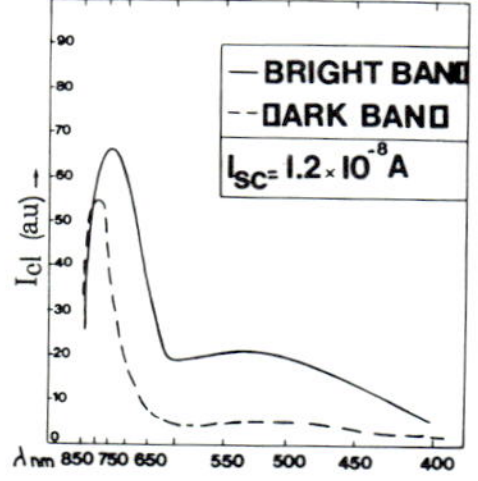

Fig 2: CL/SEM spec-
trum; deformed spe-
cimen

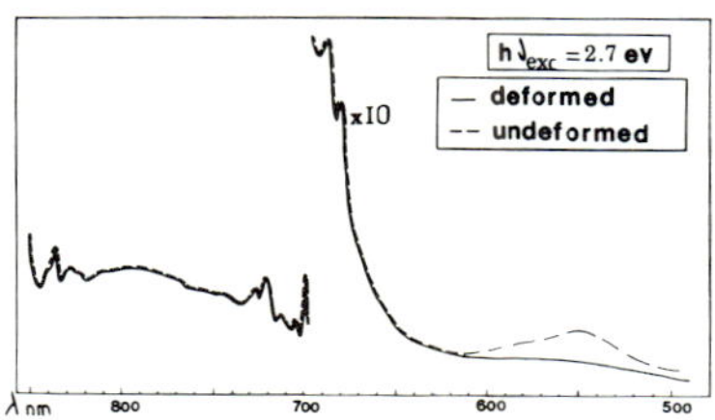

Fig 4: PL spectra

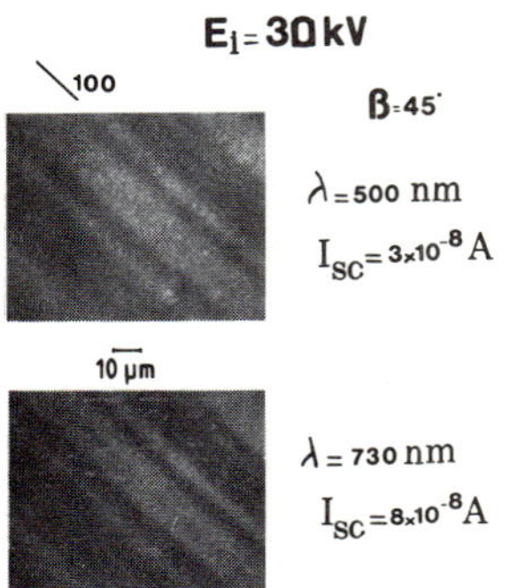

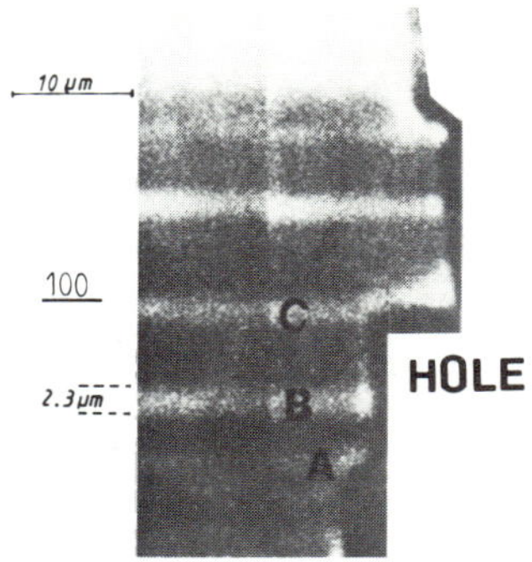

Fig 3: CL/SEM monochromatic images of a deformed specimen

Fig 5: CL/SEM panchromatic image of a 0.86% deformed thin foil. $E_i = 10$ kV

Fig 6: HVEM 450 kV of the same area as on Fig 5.

The author is deeply indebted to Dr L.M.Brown for helpful discussions. She also thanks Dr A.Howie for his critical suggestions, and R.Legros, G.Neu, M.Miloche and T.Sparrow for their assistance. The work has been carried out with the financial support of IBM England.

Applications of SEM-based CL analyses to high efficiency phosphor powders

B P Richards, A D Trigg

The General Electric Co Ltd Hirst Research Centre Wembley England

1   Introduction

The phenomenon of cathodoluminescence (CL) to investigate microscopic
properties of CL materials has frequently been used on solid samples,
usually to identify compositional inhomogeneities and to locate structural defects. The small amount of work reported on polycrystalline powders has concerned large area examinations of materials such as Y_2O_2S:Eu
(Robbins 1976) and Zn_S:Cu,Al (Kuboniwa et al 1980) used mainly for colour
TV tubes. Little (Ishikawa et al 1972) has been reported on microscopic
investigations on high brightness phosphors intended for large scale applications. However there is a need to assess the 'quality' of UV excited
phosphors for use in fluorescent lamps without expensive fabrication. In
the past this assessment has been in terms of simple particle sizes and
the overall performance of the lamp. Rogue particles, or poorly efficient
components, causing a degradation in total efficiency, therefore remained
unidentified. The application of dispersive CL on a microscopic scale to
high brightness phosphors is described as a means of providing an adequate,
relatively inexpensive basis on which to assess powder 'quality' prior to
use. The brightness levels permit use of low beam current (10^{-8} A) for
kV$\geq$20.

2   Experimental procedures

Two experimental approaches have been employed. The first is based on a
Cambridge S4 SEM, and suffers from poor collection efficiency ($\sim$5%). The
second, dedicated, system is based on an AEI electron probe microanalyser,
with a purpose-built specimen chamber and stage, incorporating an ellipsoidal mirror, having a theoretical collection efficiency >90%. The powders were examined in two ways: (a) deposited onto a cast film of synthetic rubber on a polished Al stub, or (b) in-situ on sections cut from
lamp tubes, coated with a thin conducting layer to minimise charging
effects in the glass tube. Problems of charging and CL decay in the powder were overcome using instrumental variables (e.g. $\sim$0.2s line time).
Particle sizes were large enough to minimise beam spreading effects. Vali-
dation of this CL technique as a meaningful tool for high brightness phosphors normally excited under UV (254 nm) was obtained from comparative CL
and UV spectra from the same materials. A close correspondence between
the spectra was found for the halophosphate and for the BAM (Ba aluminate:
Mn), CAT (Ce aluminate:Tb) and YOE (Y oxide:Eu) materials.

3   Applications

3.1 Single component phosphors

The phosphor routinely used in fluorescent lamps is a halophosphate phosphor $BCa_3(PO_4)_2.Ca(Cl,F)_2$ of apatite structure formed by synthesis from $CaHPO_4$, $CaCO_3$, CaF_2 and NH_4Cl; Sb and Mn are added as activators.

The CL spectrum of a typical halophosphate consists of two peaks; the main peak (~570 nm) corresponds to emission from the Mn activator and the shoulder (~480 nm) to emission from the Sb activator (see figure 1). One problem encountered in the fabrication of such fluorescent lamps is the cause of a slight loss in output (a few lumens/watt for a typical rating of 75 1/w). It is difficult to ascertain the cause of this problem from powder or full lamp-testing. The CL spectrum from powder from such a 'poor' lamp suggested a slight loss in Sb emission (figure 1). A comparison of CL images obtained at both wavelengths shows that the loss in emission was restricted to relatively few particles. A parallel EDX examination indicated that the Ca:P ratio was significantly lower in these 'anomalous' particles than in the bulk of the material. This difference in Ca:P ratio is consistent with these particles being calcium pyrophosphate (as confirmed using X-ray diffraction), their origin probably associated with inadequate firing procedures, resulting in only partial conversion to the halophosphate. In addition to providing a rapid solution to the problem, the use of CL has evidently provided an early microscopic QA/diagnostic capability for the phosphor powders.

3.2 Multi-component phosphors

To satisfy the requirements for better and different colour rendering, and higher efficiency a mixture of phosphors may be used. A typical multi-component phosphor contains a blue emitter (e.g. BAM) a yellow emitter (e.g. CAT) and a red emitter (e.g. YOE) and the resulting spectrum exhibits a series of band emissions from 440 nm to 630 nm (see figure 2). Maximising the output from such multi-phase phosphors depends not only on the thickness and packing density of the layer but also on the particle distributions of the individual components. The CL approach enables such information to be gathered thus leading to optimisation of the mixture and its distribution in the tube.

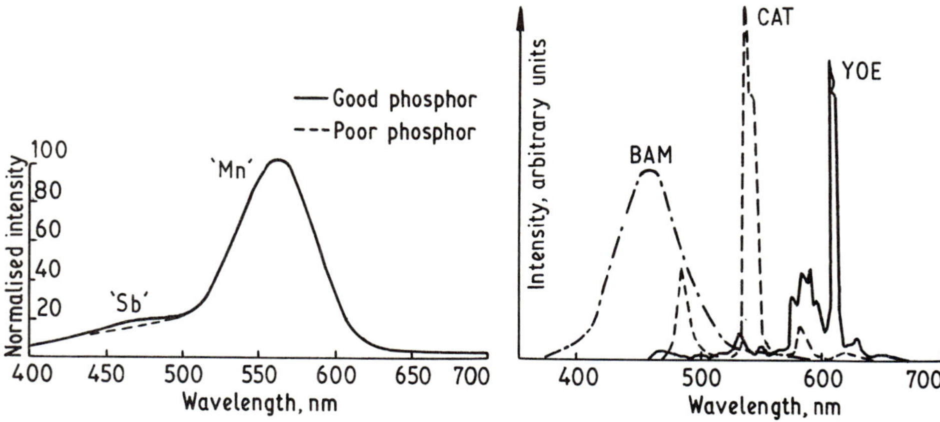

Bandwidth : 10 nm - uncorrected

Fig 1 CL spectra for halophosphate phosphors

Bandwidth: 1 nm - uncorrected

Fig 2 Spectrum of multi-component phosphor

An acceptable multi-phase coating is illustrated in figures 3-8 which show
a SE image of cross-section through a lamp, the corresponding CL images for
the BAM (450 nm), the CAT (544 nm) and the YOE particles (612 nm) and X-ray
distribution images for Al and Y.   In this case the constituents of the
phosphor were distributed evenly through the coating with little segregation across its thickness.   (The CL approach distinguishes readily between the BAM and CAT components whereas EDX cannot easily do so).   Some
difficulty was experienced with the BAM component since its CL intensity
fell during exposure to the electron beam implying that some degradation
had occurred.

As in the case of the Ca halophosphate, CL analysis can be used to examine
'poor' powder and thereby locate rogue particles and identify the cause of
any inhomogeneities.

The spectrum for an acceptable CAT (see figure 2) contains 3 lines (at
490, 540 and 590 nm), whilst that from similar material of poor quality exhibited distinct shoulders around the main 540 nm peak, with particular
emission (and shorter decay time) at 520 nm.   A 520 nm CL image was used
to locate the rogue particles, which appears to exhibit a slight yellowish
body colour instead of the usual white.   EDX analysis on individual rogue
particles gave a Ce:Tb ratio of 1:3 rather than the 2:1 expected from the
normal material.

Examination of the as-blended powder revealed the presence of a few small
lumps containing both yellow and white regions exhibiting CL spectra similar to those from particles in the 'poor' powder.   While the white regions
consisted of individual grains sintered together with some enclosed
porosity, the yellow regions were solid with little porosity.   X-ray diffraction examination of this (separated) lump suggested that the yellow
mass was a material similar to cubic $Tb_3Al_5O_3$ with a garnet-type structure.   EDX confirmed the presence of an extremely high level of Tb.SE, CL
and X-ray images confirmed the data obtained on the in-situ powder, particularly with respect to the distribution of the Tb, and clearly indicated that the cause of the inhomogeneity was in the blending of the powder, and this has a drastic effect on emission properties.

This type of sequential analysis − (a) spectroscopic CL of powders to de-
pict irregularities in the spectral outputs, (b) monochromatic CL imaging
to locate particular constituents, (c) dispersive CL and EDX analyses to
characterise these constituents, − has been successfully used on systems
containing up to 4 components.   One such system contained strontium phosphate, calcium halophosphate, strontium orthophosphate and magnesium fluo-
rogermanate emitting at 450, 570, 575 and 654 nm respectively.

3.3 Quantitative measurements of CRT phosphor efficiencies.

The comparative evaluation of a series of phosphors for potential CRT use
has necessitated the determination of relative overall efficiency.   In
choosing between phosphors for a given application it is necessary to
consider the complex variation in efficiency with several parameters, e.g.
whilst the efficiency of $YVO_4:Eu^{3+}$ and ZnS:Cu vary with beam voltage and
current respectively, those of $LaOCl:Tb^{3+}$ and $ZnSiO_4:Mn$ are independent
of these parameters.   However, these latter materials can be distinguished
in terms of the variation of efficiency with beam power.

Acknowledgement

We acknowledge Osram (GEC) Ltd for permission to publish this work.

References

Ishikawa A, Mizuno F, Uchikawa Y, Maruse S : 1972 Jap J Appl Phys 12 286
Kuboniwa S, Kawai H, Hoshina T : 1980 Jap J Appl Phys 19 1647
Robbins D J : 1976 J Electrochem Soc 123 1219.

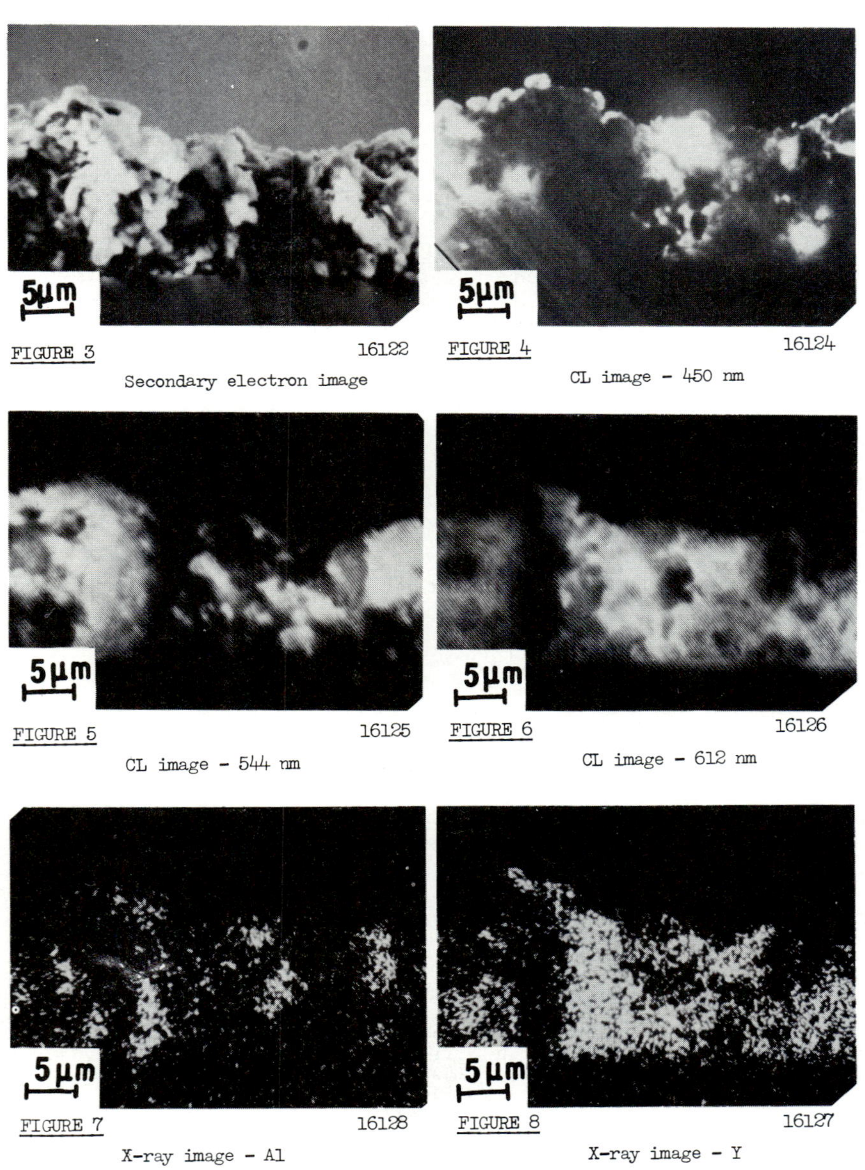

FIGURE 3   16122

Secondary electron image

FIGURE 4   16124

CL image — 450 nm

FIGURE 5   16125

CL image — 544 nm

FIGURE 6   16126

CL image — 612 nm

FIGURE 7   16128

X-ray image — Al

FIGURE 8   16127

X-ray image — Y

Electron microscopy of Li_2O, Al_2O_3 and SiO_2 glass ceramics

Q Q Chen and P L Gai

Department of Metallurgy & Science of Materials,
University of Oxford, Parks Road, Oxford OX1 3PH

1. Introduction

The superior properties of glass ceramics depend to a large extent on their
microstructure, morphology and the residual uncrystallised glass that may
be present (McMillan, 1974). In Li_2O Al_2O_3 $4SiO_2$ glass ceramic containing
TiO_2 as a nucleating agent (system A) some residual glass may be present
but was not detected by the conventional TEM methods (Barry, 1976; Bold &
Groves, 1978). Also, differences of opinions exist regarding the relationship between the glass phase and the final stage grain growth in glass
ceramics, e.g. (a) grain growth obeys a cube root of time law and that the
residual glass is the rate limiting process for grain boundary migration,
based on the observations by replica methods (Chyung, 1969), but on the
contrary, (b) grain growth accelerated by the presence of the glass phase
(Barry, 1976). The object of the present experiments therefore, is to
investigate (a) crystallographic properties of A, the presence of residual
glass and compare with the observations in the commercial glass ceramic
system $0.68Li_2O_3$, Al_2O_3, $6.1SiO_2$, $0.13ZnO$, $0.03Na_2O$, $0.01k_2O$, $0.11TiO_2$,
$0.077ZrO_2$ (system D) and (b) characteristics of grain growth and microstructure using a combination of high resolution EM (HREM) for direct
imaging of the glass ceramics with a point resolution of $\sim 1.7\overset{\circ}{A}$, STEM, X-
ray microanalysis in an analytical electron microscope (AEM), X-ray/
electron diffraction (d.p.) and dark-field techniques as the results would
be complementary and more reliable than the replica EM methods.

2. Experimental

Preparations of A and D are described in Bold & Groves, 1978; Barry, 1976.
Powdered samples and thin foils ($< 500\overset{\circ}{A}$) were used for EM and the foil
thickness (t) in the latter was estimated by stereo microscopy. For STEM
and AEM, thin foils were prepared by ion thinning and for HREM, the surfaces were carbon coated to minimise charging of these very beam sensitive
samples. For the studies of grain growth, STEM and replica methods were
used and in the replica, samples were prepared by mechanical polishing/
etching in an EDTA solution. The grain size was measured by intercept
method, obtaining the average size from ~ 100 grains. The volume fraction
of the glass phase was measured by point counting. For the growth studies,
samples were heated from R.T. (room temperature) to $1150°C$ for 5, 10, 50
and 100 hours. For STEM/AEM, a JEOL 100C/100CX Temscan at 100kV (with
probe size ~ 100-$300\overset{\circ}{A}$) as well as Oxford FEG STEM for high resolution
(with probe $20\overset{\circ}{A}$) were used. In AEM, the composition of precipitates and
compositional changes near the grain boundaries (oriented parallel to the

beam) were examined by point analyses and the X-ray data were collected for
$\sim$ 100secs. Elemental maps were also taken to examine Si/Al distribution.
For HREM, a JEOL 200CX EM at 200keV was used (Boyes et al 1980).

3. Results

The d.p.s. and lattice images of A showed the main phase to be β-spodumene
($LiAlSi_2O_6$-II, tetra, a = 7.541, c = 9.156$\AA$) as shown by Fig.1(a) and (b)
respectively for (010). The structural model showing aluminosilicate network and computed image (at defocus Δf = -700$\AA$, estimated experimentally
with minimum contrast at zero focus as reference, t = 75$\AA$) are inset. The
microstructure of A in TEM and STEM are shown by Fig.2(a) and (b) respectively, with TiO_2 (anatase) as particles analysed by AEM/HREM. Direct
imaging of intergranluar films was attempted by orienting the boundary
plane parallel to the beam and also using displaced aperture d.f. and
through focus b.f. methods (Clarke, 1979). Only a few boundaries exhibited
what appeared to be an amorphous film extending to a few lattice dimensions
with the d.p.s. showing a weak amorphous ring. Fig.3(a) shows the direct
image, and (b) shows the glass pocket in displaced d.f. mode. Small
amounts of γ-spodumene ($LiAlSi_2O_6$-III, hex., a = 5.21, c = 5.46A) was also
present in A and Fig.4   for example shows the   lattice image in
(0001). The structure and computed models are inset (Δf = -800$\AA$, t=100$\AA$).γ
spodumene is the main component of low expansion Li-based glass ceramics.
Experiments conducted on D (main phase β-spodumene solid solution) showed
that amorphous glass exists in larger amounts, as pockets or intergranular
films. The volume fraction of glass in D was $\sim$ 14-16%. Fig.5(a) illustrates microstructure of D (in TEM) showing TiO_2, either as anatase/rutile
particles (at P) or TiO_2-ZrO_2 solid solution as needles (T) and gahnite (Zn
Al_2O_4) crystals (S) confirmed by AEM; 5(b) shows residual glass pocket at
g. Through focus b.f. images of the boundary are shown in Fig.6 with (a)
underfocus (b) near focus and (c) overfocus conditions at 200kV.

4. X-ray microanalysis

In the compositional microanalysis of the grain boundaries the basic ratio
method (Cliff & Lorimer, 1975) was used for the thin foils (< 500$\AA$) reported here. Characteristic X-ray line intensity data (I) was converted to
the relative atomic concentrations as following: $C_{Si}/C_{Al} = k_{Si/Al}(I_{Si}/I_{Al})$,
where C_{Si} and C_{Al} are the elemental atomic concentrations of Si and Al and
$k_{Si/Al}$ is a constant for relative detection efficiency and was determined
using a standard jadeite specimen. More than 30 boundaries were investigated in D and the composition profiles normal to the boundary examined.
Consistent results were obtained when the experiment was repeated in different areas. Fig.7 shows an example and is plotted as the grain boundary
composition profile. Fig.8(a) and (b) show examples of elemental maps of
the sample with the boundary region (at triple point) arrowed, (a) Si map
(b) Al map. From these preliminary observations it seems likely therefore,
that the intergranular regions are Al rich with $C_{Si}/C_{Al} \sim 3$ and
not $5Al_2O_3$ $95SiO_2$ as has been suggested previously (Chyung, 1969).
(Results for thicker regions incorporating absorption corrections will be
discussed elsewhere).

5. Grain growth

Average grain sizes were measured by STEM (presented here) as well as by
replica method. Fig.9(a) and (b) show examples of grain in D at R.T. and
at 1150°C for 50h. Very little residual glass is present in A and

therefore, the grain growth limiting process in A can be considered to be
due to impurity drag effect. In D however, decreasing of the glass phase
and increase in the grain size occurred simultaneously, the glass phase
partially penetrating along the grains and rutile as well as gahnite
crystals were observed at the boundary. Apart from the drag effect the
presence of grain boundary liquid/glass phase may accelerate or decelerate
the rate of growth. Such liquid at the boundary may also facilitate
accumulation of ions at the boundaries thus increasing the ion diffusion
rate. Fig.10 shows isothermal grain growth data for samples D at 1150°C
with log D_g (average grain diameter) plotted vs log t (time). It fits the
general rate of growth formula, $(dD_g/dt) = k(1/D_g)$ where k = rate constant.
The rate of growth in the system D was found to be greater than in A thus
establishing that in this system Barry's suggestion (1976) may be more
acceptable.

We thank the SRC for support.

References

Barry T I 1976 Ceram. Micr. 479
Bold S E and Groves G W 1978 J.Mat.Sci. 13 611
Boyes et al 1980 Inst.Phys.Conf.Ser. 52 445
Clarke D R 1979 Ultra.Micr. 4 33
Cliff G and Lorimer G W 1975 J.Micr. 103 203
Chyung C K 1969 J.Amer.Ceram.Soc. 52 242
McMillan P W 1974 Glass.Tech. 15 5

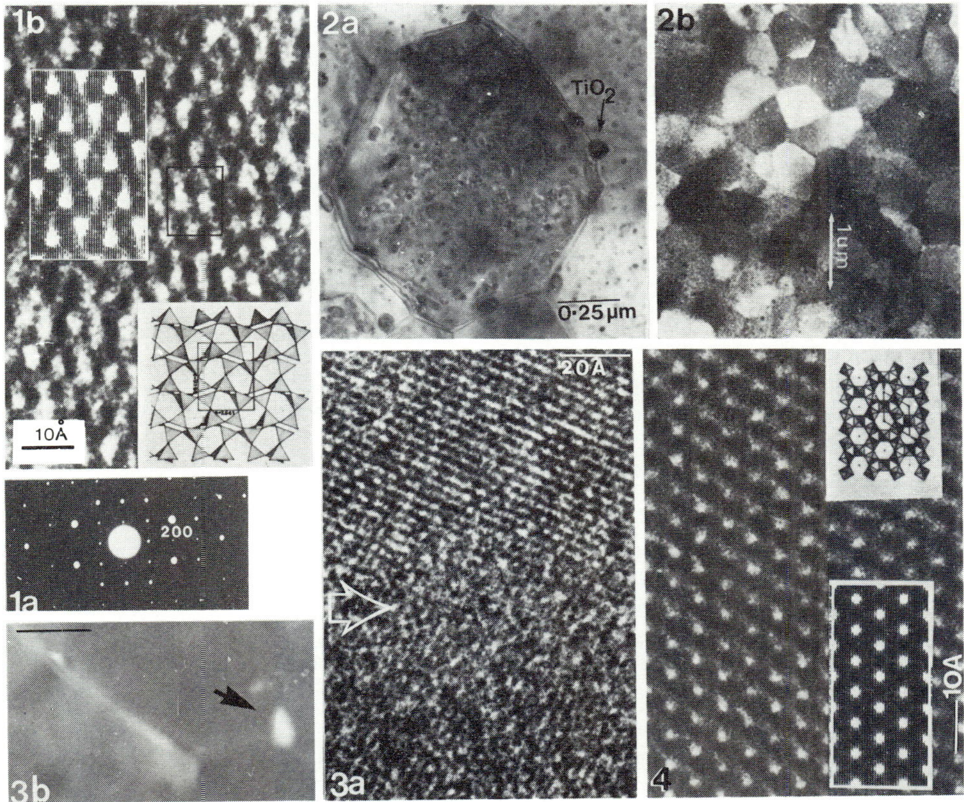

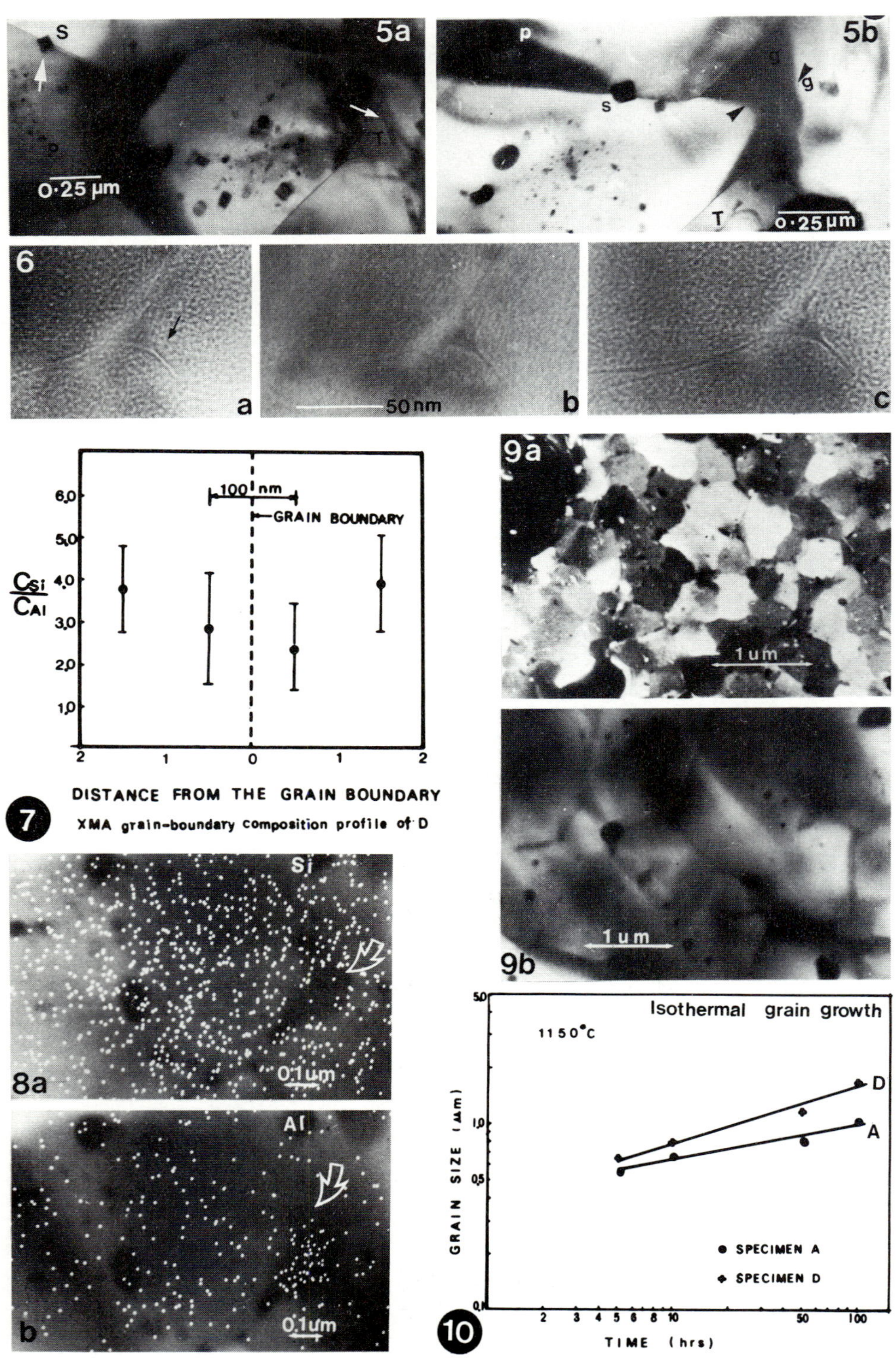

S
5a
0·25 μm
T
p
5b
s
g
g
T
o·25 μm
6
a
50 nm
b
c
6,0
5,0
4,0
3,0
2,0
1,0
100 nm
GRAIN BOUNDARY
C_Si / C_Al
2   1   0   1   2
DISTANCE FROM THE GRAIN BOUNDARY
7
XMA grain-boundary composition profile of D
Si
8a
0·1 um
Al
b
0·1 um
9a
1 um
9b
1 um
Isothermal grain growth
1150°C
D
A
GRAIN SIZE (μm)
1,0
0,5
0,1
2 3 4 5 6 8 10   50 100
SPECIMEN A
SPECIMEN D
10
TIME (hrs)

Weld metal microstructures in duplex austenite–ferrite alloys

P.R. Howell[1], S. Paetke[2] and R.A. Ricks[3]

1. Dept. of Materials Science and Engineering, Pennsylvania State Univ.
2. Shell Research Ltd. Thornton Research Centre, Chester.
3. Dept. Metallurgy and Materials Science, University of Cambridge.

1. Introduction

Adjustment of the nickel and chromium contents in stainless steels can promote the development of duplex austenite/ferrite microstructures. These duplex structures can offer a good combination of properties including resistance to crevice and stress corrosion (e.g. Cunningham (1972)), a reduced susceptibility to hot cracking when compared to fully austenitic structures (e.g. Astrom et al. (1976)) and excellent tensile strength.

In the present investigation, the effect of stress relieving treatments on one particular duplex stainless steel weldment has been examined. Both conventional TEM and STEM/EDS techniques have been employed to provide structural and chemical information on both the as–welded alloy and on weldments which have been subsequently heat treated at 1050°C. The results detailed below relate mainly to the heat treated weldment.

2. Experimental

The experimental weld had the following composition (wt%)

0.051%C, 2.0%Mn, 0.68%Si, 3.20%Mo, 8.51%Ni, 20.0%Cr, balance Fe.

Specimens for transmission electron microscopy were prepared in a twin jet electropolisher employing an electrolyte consisting of 10% perchloric acid, 15% glycerol and 75% ethanol. Previous studies (Ricks et al. (1981)) have shown that, using this technique, no solute rich surface layer is formed on the specimens. The thin foils were examined in a Philips EM400 TEM/STEM operating at 120KV. The compositions of the various features were determined using a LINK E.D.S. system employing standard thin foil correction programmes.

3. Results

Figures 1a,b are light optical micrographs of the as welded structure (figure 1a) and of the weld which had received an anneal at 1050°C for 1 hour (figure 1b). The cooling cycle of the weld metal has resulted in a fine network of ferrite both at dendritic and interdendritic areas (figure 1a) together with a number of refractory weld metal inclusions. Heat treating the weld for 1 hour leads to some coarsening of the duplex aggregate (figure 1b) and it is noticeable that some of the ferritic areas show dissimilar etching characteristics. Typical transmission electron micrographs of the heat treated material are shown in figures 2-4. In

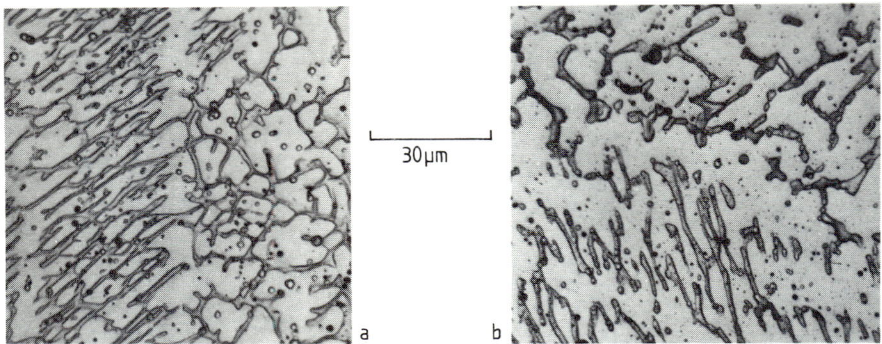

<u>Fig.1</u>   The as-welded microstructure (a) and the effect of annealing at
1050°C for 1 hour (b).

general, the ferrite was observed to be located at austenite/austenite
grain boundaries (e.g. figures 2 and 3a);   these boundaries often being
low angle related.   It was also found that the austenite and ferrite were
frequently related by either the Kurdjumov-Sachs (1930) or the Nishiyama
(1934)-Wasserman (1933) orientation relationships as has been noted in
previous studies of duplex weld metal microstructures (e.g. Astrom et al.
(1976)).   Figures 3b,c are selected area diffraction patterns of γ_1 and
a_1(figure 3a) respectively;   in this instance the two phases are related
by the N-W relationship.   It is also interesting to note that $M_{23}C_6$
has nucleated on the γ_1/a_1 interface and that these precipitates are
inhibiting the growth of the ferrite grain.   Crystallographic analysis of
the three phases showed that the $M_{23}C_6$ was cube/cube related to γ_1 and
NW related to a_1.   This type of three phase crystallography has been
documented previously in other duplex steels and for interphase
precipitation of $M_{23}C_6$ in Fe-Cr-C alloys (e.g. Howell et al. (1979)).

Perhaps the most significant observation, however, is that the ferrite had
partially transformed to the σ phase (figure 4 and see figure 1b).   This
transformation is somewhat unexpected considering the composition of the
weld metal and the ageing temperature employed (1050°C).   Long term
ageing experiments revealed that the σ phase was unstable to prolonged
exposure at 1050°C, reversion to ferrite ultimately occuring.

In order to investigate the possible reasons for σ phase formation, the
compositions of the various microstructural features illustrated in
figures 2-4 were obtained using STEM/EDS.   The results of these analyses
are presented in table 1.

Several points can be made concerning these analyses:
    (i) partitioning of the alloying elements between austenite and ferrite
has occurred.   Hence the austenite is enriched with respect to ferrite, in
Mn and Ni, and depleted with respect to ferrite in Si, Cr and Mo;
    (ii) the σ phase composition is very similar to that of the ferrite,
the only major difference being the partial replacement of Fe by Cr and
Mo;
  (iii) both the ferrite and the austenite contain more Cr than would be
expected from the bulk composition of the weldment.

Table 1

EDS analyses of the weld metal aged at 1050°C for 1 hour

	Si	Mn	Cr	Fe	Ni	Mo
Austenite	1.2	2.5	23.0	63.6	9.3	2.3
Ferrite	1.5	1.4	27.6	60.5	5.5	4.0
Sigma	1.7	1.3	31.5	53.6	5.2	7.6
$M_{23}C_6$			64.8	23.3	1.7	10.2
Inclusions	29.4	56.3	6.1	5.4	0.9	0.9

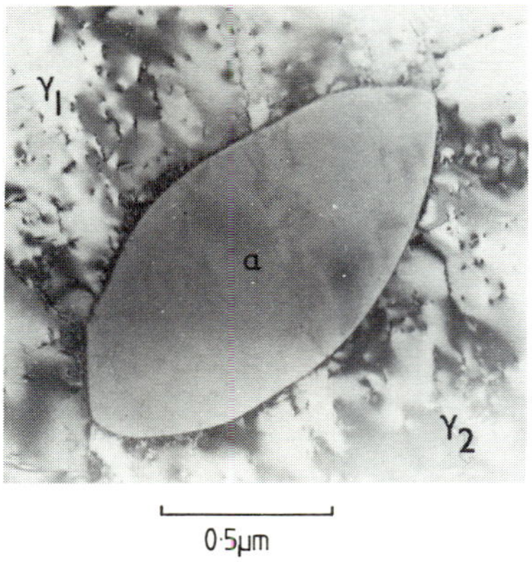

2

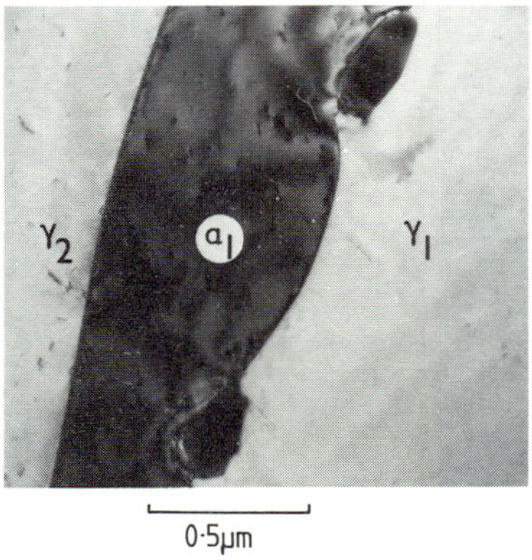

3a

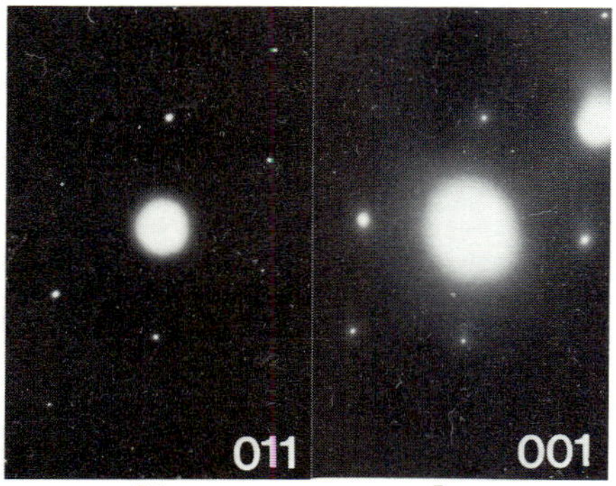

3b 3c

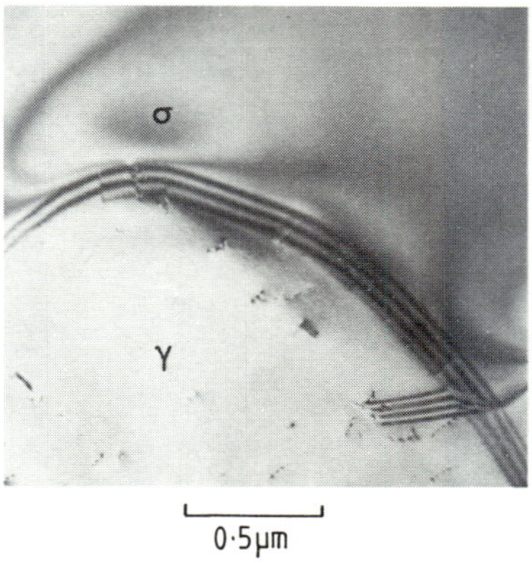

4

Fig.2 Ferrite at a low angle austenite grain boundary.
Fig.3 Precipitation of $M_{23}C_6$ at an austenite/ferrite interface.
Fig.4 σ phase in the annealed weld metal.

The increased Cr levels observed in austenite and ferrite areas containing σ phase particles may be explained in terms of the segregation pattern which obtains during the solidification of the weld. Hence, it is likely that the initial solidification product (i.e. δ-ferrite) will be enriched in Cr and Mo and that this enrichment will not be wholly dissipated during the subsequent solid state transformation to austenite. This increased Cr level then promotes the σ phase transformation at $1050^{\circ}C$ since reference to table 1 shows that there is little compositional difference between ferrite and σ. The subsequent dissolution of σ can then be rationalised in terms of the removal of the macrosegreation profile during prolonged exposure at $1050^{\circ}C$. Further work is now in progress to substantiate this hypothesis.

Finally it is worth noting that a high Mo level is observed in the $M_{23}C_6$ precipitates and that the inclusions are most probably Mn silicates.

4. <u>Summary</u>

Both conventional TEM and STEM/EDS techniques have been employed to analyse microstructural development in a duplex stainless steel weldment. The preliminary results suggest that σ phase formation occurs as a result of macrosegregation during solidification of the weld pool.

<u>Acknowledgements</u>

The authors are grateful to Professor R.W.K.Honeycombe FRS for the provision of laboratory facilities and for the constant advice and encouragement. The authors are also grateful to G.S.Barritte and G. Honeyman for their helpful advice and comments. Financial support was received from the SRC.

<u>References</u>

Astrom H, Loberg B, Bengtsson B and Easterling K E (1976) Met Sci 225.
Cunningham E R (1972) Met Prog <u>57</u>.
Howell P.R, Bee J V and Honeycombe R W K (1979) Met Trans <u>A</u> <u>10A</u> 1213.
Kurdjumov G.and Sachs G (1930) Z Phys <u>64</u> 325.
Nishiyama Z.(1934) Scient Rep Tohuku Univ <u>23</u> 637.
Ricks R.A, Southwick P D and Howell P R (1981) J Mic (in press).
Wasserman G (1933) Arch Eisenhuttwen <u>16</u> 647.

Microanalysis of carbides in ferritic steels

T.N. Baker, A.J. Craven[+],S.P. Duckworth, F. Glas[+]

Department of Metallurgy, University of Strathclyde, Glasgow G1 LXN, U.K.

+ Department of Natural Philosophy, University of Glasgow, Glasgow G12 8QQ.

1. Introduction

A realistic assessment of particle volume fraction (f) in an alloy system
is crucial to accurate predictions of strengthening due to particles which
are frequently <10 nm in diameter.   Because direct measurement of (f) is
difficult, values have frequently been calculated using bulk chemical composition and assuming a stoichiometric composition for the particles.

Microalloyed ferritic steels containing a fine dispersion of carbides,
nitrides or carbonitrides of the transition metals niobium, titanium or
vanadium (M), singly or in combination, are known to benefit from precipitation strengthening (Honeycombe 1981).   Agreement between theoretical
predictions of particle strengthening and values calculated from microstructural measurements is only moderate.

Recently, analytical electron microscopy with STEM/EDA systems has shown
that many of the small particles in the above systems contain significant
quantities of iron, which could influence volume fraction of the precipitation hardening particles (Baker 1980, Hirsch and Parker 1980, Baker and
Duckworth 1981).

The present work studied a possible relationship between particle size and
Fe/M ratio and extended the investigation to light element analysis.

2. Experimental Method

Since carbon replicas would affect the analysis of carbon in extracted
particles, aluminium replicas (Garrett-Reed, 1981) were prepared. Aluminium was evaporated, by resistance heating a tungsten basket containing
aluminium powder, in a Balzer Micro BA3 evaporator. Particular attention
was paid to the thickness of the resultant film which was controlled by
the change in the colour of an adjacently placed gold film, previously
evaporated onto a thin sheet of aluminium.   Films ∿15 nm thick fragmented
on separation from the specimen while films ∿30 nm thick consisted of a
very fine grain structure which obscured the smaller particles.
A thickness of ∿20 to 25 nm proved to be satisfactory.
Carbon replicas were also prepared as control samples.

Electron microscopy and analysis of the particles on the extraction
replicas was carried out (a) on a Philips EM400T instrument fitted with a
STEM unit and with an EDAX9100 energy dispersive analyser, (University of

Strathclyde) and (b) a V.G. Microscopes HB5 STEM instrument (University of Glasgow). The HB5 was still awaiting data acquisition and processing facilities.

An experimental vanadium steel of composition (in weight percent) 0.12C 0.48V 0.002N **0.003 oxygen** was examined in both instruments. For instrument (a), preset counting times were 50, 60 and 100s live time, and an EDAX semi quantitative analysis programme which included ZAF corrections was used, whilst for (b) EELS spectra were collected with 100s scans. A drift of $\sim$3 nm was noted during this scan time.

3. Results

The steel was solution treated at 1473K, directly quenched to 873K or 955K and isothermally aged for times between 55s and 35 min followed by water quenching.
(a) Between 20 and 50 EDAX analyses were recorded for each specimen. The Fe/V ratio, (R) for ageing at 873K was an average of 0.1 after 55s ageing, and an average of 0.3 after 230s, whilst after 955K ageing 155s, R $\leqslant$ 0.2. Fig. 1 compares the frequency vs Fe/V ratio for 873 and 955K ageing temperatures after 100s. Both nanoprobe and STEM modes were employed and the reproducibility in R for successive 60s counts on the same specimen was $\pm$ 5%. Fig. 2 shows an EDAX spectrum from a particle with a low R value.
(b) Samples aged isothermally for times of 115s and 1080s at 955K were studied in the HB5 to obtain EELS spectra.
(i) <u>115s Ageing time</u> Typical spectra are shown in Figs. 3 and 4 for high and low M/C ratios. 14 particles containing vanadium and carbon in the range 5 to 50 nm dia. were examined. Iron was detected in 13 of the particles. Comparing the height of the edge above the background gives an approximate Fe_L/V_L estimate. This was $\leqslant$0.08. Corresponding $(Fe_L + V_L)/C_K$ height ratios (N) varied from 0.02 to 20. No peaks associated with nitrogen and oxygen were found.
(ii) <u>1080s Ageing time</u> Particles coarsened after the longer ageing time, and the particle density decreased; elongated particles 100 x 10 nm, R = .1 to .4 for 9 particles; round particles, 12-20 nm dia., R = .01 to 10 for 5 particles - 1 round particle 6 nm dia., R = .5. Fig. 5 shows a typical distribution containing both elongated and round particles. In several instances, the larger particles, both round and elongated, appeared to be growing from a nucleus. The R and N values of the nucleus were higher than the outer regions, and towards the end of the range. N values varied from .06 to 0.4. No correlation was found between particle size and R and N values.

4. Discussion

The composition of vanadium carbide determined from particles present after long term ageing of Fe-C-V alloys has been found to be about $VC_{0.84}$, (Hollox et al 1971, Dunlop and Porter 1977, Dunlop and Honeycombe 1978). Several of the 115s aged particles have the correct order of N to fit vanadium carbide, frequently with a few percent of iron.

However, a number of the particles with N values of <<1 do not fit any compound known to precipitate in microalloyed steels. Oxygen and nitrogen peaks were not found in any spectra.

When analysing very small particles for carbon, contamination is always a problem. In some experiments, contamination was definitely occurring as

illustrated in Fig. 6 which shows the increasing contamination on three
successive 100s scans. The degradation of the edges is very marked.   In
other experiments contamination was virtually undetectable after pre-baking
the specimen and flooding it with the electron beam.   This may explain
some of the unexpectedly low metal to carbon ratios.

While the ratios of the heights of the V_L to C_K ionisation edges (N) does
not give an accurate atomic ratio, the error involved is relatively small
because the effect of the sharp spike at the L edge of vanadium approximately compensates for the difference in partial-cross section estimated
from the data of Leapman, Rez and Mayers (1980).

The variation in R, (Fe/V ratio) implies that equilibrium is reached at
shorter times in the 955K samples.   At lower ageing temperatures and/or
shorter ageing times, a significant departure from equilibrium could exist
with a concomitent increase in the iron content of the particles.   Neither
the results of technique (a) nor (b) showed any relationship between R
values and particle size as found by Hirsch and Parker (1980).

In these preliminary experiments, the wide range of N values is a matter
for concern and more data are necessary to give confident figures for particle composition.   On the other hand, the R values are in reasonable agreement for both EDAX and EELS techniques.   Furthermore, once the data
acquisition facilities are available on the HB5 instrument, data collection times will be shorter and drift reduced.

References

Baker T N 1980 Electron Microscopy 1980 3 44
Baker T N and Duckworth S P 1981 Quantitative Microanalysis with High
   Spatial Resolution (to be published by Metals Society)
Dunlop G L and Porter D A 1977 Scand. J. Met. 6 19
Dunlop G L and Honeycombe R W K 1978 Metal Sci. 12 367
Garrett-Reed A J 1981 Quantitative Microanalysis with High Spatial
   Resolution (to be published by Metals Society)

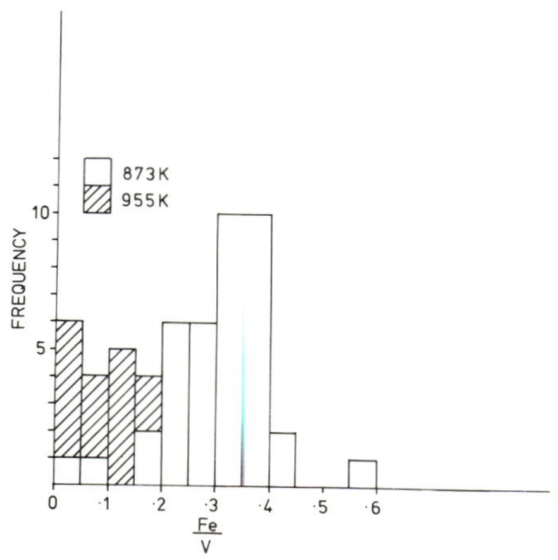

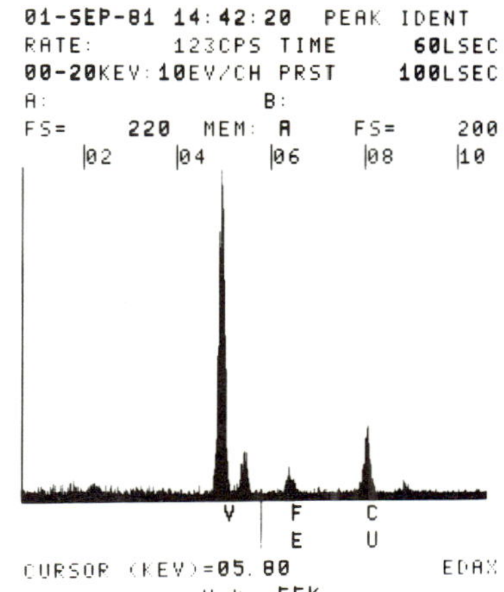

Fig. 1. Histogram: Fe/V ratio vs
particle frequency.   Steel aged at
873K and 955K for 100s.

Fig. 2. EDAX spectrum from a
particle showing a low Fe/V
ratio.

Hirsch Y C and Parker B A 1980 Micron <u>11</u> 317
Hollox G E, Edington J W and Scarlin R B 1971 J. Iron Steel Inst. <u>209</u> 839
Honeycombe R W K 1981 Steels - Microstructure and Properties (London:
    Arnold) pp.166-176
Leapman, R D, Rez P and Mayers D F 1980 J. Chem. Phys. <u>72</u> 1232

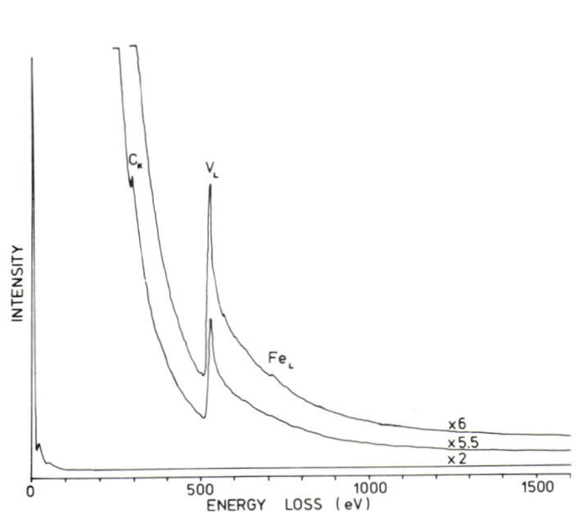

Fig. 3.   EELS Spectra.   High N
ratio with K edge of carbon (284eV),
L edges of vanadium (513ev) and
iron (710ev).   Gains x1(x2),
x2500(x5.5) and x5000(x6).

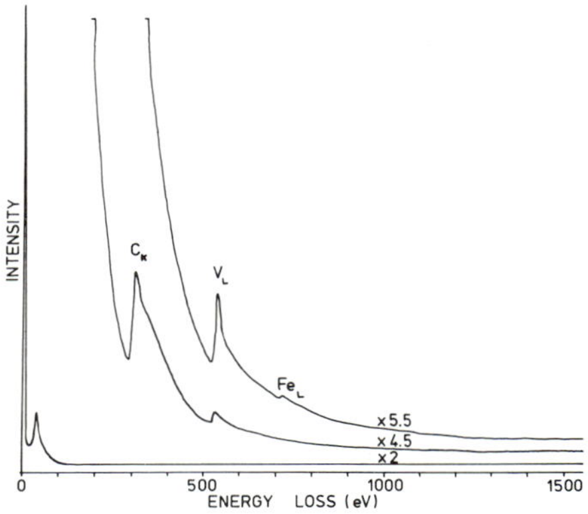

Fig. 4.   EELS spectra.   Low N ratio
with gains of x1(x2), x540(x4.5)
and x2500(x5.5).

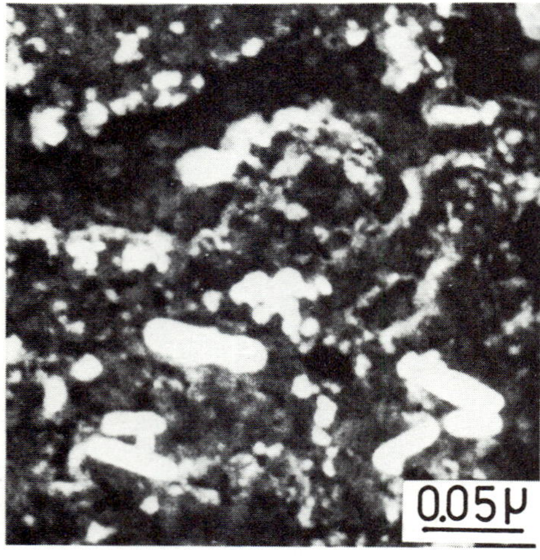

Fig. 5.   Dark-field image of typical
particle distribution analysed on
the HB5 instrument.

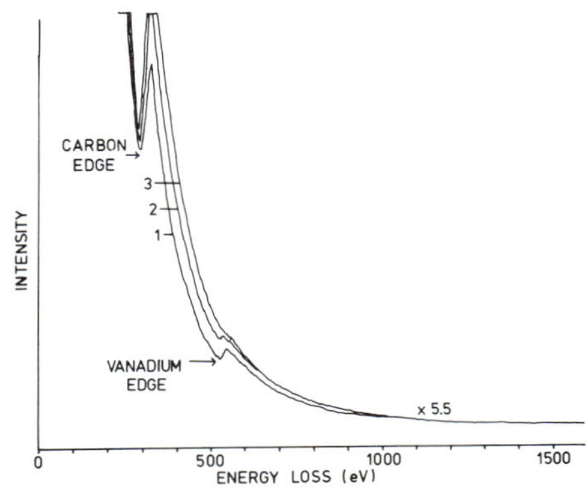

Fig. 6.   EELS spectra showing the
effect of carbon contamination on
three successive 100s scans on the
same particle.

Conditions for all EELS spectra: acceptance angle of spectrometer ∿28m
Rad;   primary energy 100keV;   Spectrometer resolution ∿10ev.

X-ray microanalysis of carbide precipitates in an embrittled commercial alloy steel

A B Muhammad and Z C Szkopiak

Department of Metallurgy and Materials Technology, University of Surrey, Guildford, Surrey.

The present work has been concerned with the application of STEM and X-ray microanalysis to the study of the composition and morphology of carbide precipitates in an embrittled steel using the carbon extraction replica technique. The material used was a commercial $2\frac{1}{4}$Cr-1Mo steel in as-tempered and embrittled (at 500°C) conditions. Carbon extraction replicas were prepared using a standard technique. The replicas were examined in JEOL-200 TEM fitted with a scanning transmission attachment and an energy dispersive spectrometer. The spectra of carbides were recorded using a stationary beam, size of 100Å, at accelerating voltage of 100 KV.

In Fig. 1 are shown carbon replica electron micrographs illustrating the morphology of carbides in an as-tempered (a) and tempered and embrittled (b) condition. Figs. 2 and 3 show typical energy dispersive X-ray spectra of carbide precipitates from which the types of carbides can be determined and the relative proportions of metallic constituents in them estimated after different heat treatments. From these spectra and electron diff-raction patterns (Titchmarsh 1979) the following types of carbides have been identified : Mo-rich (M_2C, M_6C), Cr-rich (M_7C_3) and Fe-rich (M_3C, $M_{23}C_6$). In most cases the component M of the carbides depends on heat treatments, e.g. in a sample tempered at 600°C for 10 hours X-ray microanalysis showed that the M_3C carbide contained predominantly Fe, some Cr and possibly some Mn, but Mo was not detected. The Fe($K\alpha$):Cr($K\alpha$) peak ratio varied from 4:1 to 2:1. But after embrittling treatments the Fe($K\alpha$) : Cr($K\alpha$) X-ray peak ratio varied from 2:1 to 1:1, and a definite Mo($K\alpha$) peak at 17.441 keV was also seen. Similar observation was made by Kula and Anctil (1969) and Titchmarsh (1979) who reported that the Cr content of M_3C increased to saturation before M_3C was replaced by Cr-rich M_7C_3.

The main constituents of M_2C of the present steel are Mo, Cr and Fe, with the Mo(L) : Cr($K\alpha$) X-ray peak ratio varying from the range of 2:1 to 3:1 for the as-tempered condition to the range of 2:1 to 5:1 after embrittling treatments. Furthermore, after embrittling treatments the spectra for M_2C and M_6C also contained a small peak at 2.01 keV (Fig. 2 (c) and 3). It is believed that this peak is due to P($K\alpha$) X-ray emission. (The presence of P in these carbides is subject to further investigations). A similar peak was not observed in the spectrum of M_2C carbides either from within the grains of embrittled specimens or from grain boundaries of the as-tempered specimens. Furthermore, it is apparent that this peak is present in spectra of Mo-rich carbides only. This is in agreement with the work of Lonsdale and Flewitt (1979) who, when studying creep life of $2\frac{1}{4}$Cr-1Mo steel, observed significant concentrations of P in Mo-rich $M_{23}C_6$ carbides and not in Cr-rich ones.

The implication of the present results therefore is that the X-ray microanalysis technique may provide additional information concerning the
segregation of P to grain boundaries during embrittling treatments, and
its association with the segregation of major alloying elements (Yu and
McMahon 1980).

The replica technique has the advantage over the thin foil technique in
X-ray microanalysis in so far that, because of the very small carbide
particle size ($\sim$300 Å), any observed P-peak would be due to the P
associated with carbide particles and not grain boundaries. The difficulty in differentiating between the P present at the surface and that
within the carbide particles remains still to be resolved.

In summary, the results of the present work show that the replica technique is useful for rapidly characterising carbide particles into various
types by means of X-ray microanalysis. Also, if the presence of P at the
surface of and/or within the M_2C and M_6C carbides was confirmed, the
replica technique will provide a convenient way for studying the segregation of P during embrittling treatments.

This work is supported by AERE, Harwell; the authors gratefully acknowledge the assistance of Dr. J.A. Hudson and his colleagues.

References

Kula E.B. and Anctil A.A., 1969 J. Mat. 4, 817
Lonsdale D. and Flewitt P.E.J., 1979 Mat. Sci. Eng. 41, 127
Titchmarsh J.M.,1979 Harwell Report AERE-R9661
Yu J. and McMahon Jr. C.J., 1980 Met. Trans. 11A, 277

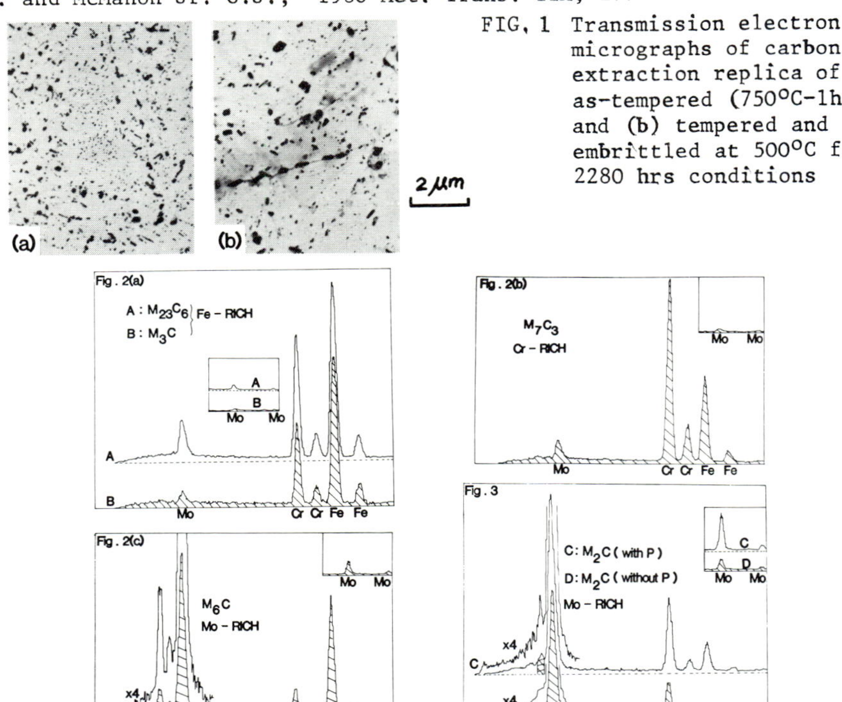

FIG. 1 Transmission electron
micrographs of carbon
extraction replica of (a)
as-tempered (750°C-1hr)
and (b) tempered and
embrittled at 500°C for
2280 hrs conditions

2 μm

FIGS 2 & 3.   EDXA Spectra for different types of carbides.   The insets
represent the high energy Mo(Kα) and Mo(Kβ).

Some aspects of microanalysis in dedicated STEM

D C Houghton, McMaster University, University of Toronto
Joint STEM Facility

The unique microanalytical capabilities of a Vacuum Generator's HB5 STEM have been utilized in conjunction with some simple techniques to investigate the evolution of precipitate morphologies induced by certain rather complex phase transformations.

A new technique for parallel recording of diffraction patterns has been developed for multisignal micro-analysis and was used throughout this work, Houghton (1981). The dummy specimen cartridge in fig 1 (a) can be used in conjurction with the HB5 airlock infra-red heater to generate specimen thermal cycles of the form seen in Fig 1(b). The low thermal mass of this device permits thermal cycling experiments, decontamination and pulse heating in addition to protracted anneals at temperatures up to 250^C and at pressures $< 5.10^{-9}$ torr, within the microscope's 'clean' environment.

The annealing behaviour of a metastable Cu–6% Cr thin film formed by cosputtering was studied using the infra-red airlock heater described above. Microdiffraction, EDX and simultaneous STEM-SEM imaging were utilized to characterize the morphological changes observed after isothermal reheating. The as-sputtered film depicted in fig 2(a) shows typical mottled contrast in bright field and a diffuse FCC ring pattern in SADP from ∼100A region, insert. The morphology coarsens on heating for 10.5h at 250°C in UHV see fig 2(b) but in addition many large facetted particles were observed with major axes out of the plane of the film fig 2 (c). Microdiffraction and EDX analysis of these large facetted particles revealed that they are single crystals considerably copper enriched fig 2(d). No local microchemical gradients were detected.

The metastable needle-like precipitates shown in figs 3 and 4 occur in Al-6Zn-2Mg alloys after step quenching and ageing operations, Ryum (1971) and lie along <001> and <011>$_\alpha$. Considerable uncertainty as to the exact chemical composition and crystal structure of this transient phase persist, since microanalysis is impeded by a strong matrix contribution to both diffracted intensities and EDX Spectra. The central micrograph in fig 3 is a bright field image of a needle precipitate for which the $(200)_\alpha$ reflection was closely parallel to the long axis of the precipitate and moire fringes are visible ∼ 3.8nm apart. The SAPD in fig 3(a) indicates a foil normal close to $(013)_\alpha$ in a 2 region around the precipitate. The micro-diffraction patterns in figs 3(b) (c) and (d) were taken with decreasing angle of convergenceThe diffraction pattern in fig 3(d) at a half angle of ∼ 1.7m rods readily reveals the contribution of the precipitate to diffracted intensity. Hexagonal symmetry for these precipitates is strongly suggested. The EDX Spectra presented in Fig 4 indicate that the needle-like precipitates (2) contain a very low concentration of magnesium relative to the adjacent equilibrium phase (1) η Mg2n$_2$, suggesting the possibility of zinc-rich zone formation.

The composite particle in fig 5 is a titanium–niobium carbonitride extracted from a control-rolled HSLA steel by conventional carbon replication. The complex precipitation sequence in these alloys is characterized by high temperature core-nitride cuboids acting as nucleation sites at lower temperatures for a niobium rich carbonitride layer Houghton et al (1981). The particle displayed in fig 5 is typical and consists of a central cuboid of the form $Nb_{40} Ti_{60}N$ with top and bottom roughly hemispherical caps revealed by EDX analysis to be $Nb_{90}Ti_{10}$ (C,N). These carbonitride compounds are completely isomorphous with B$_1$ structure lattice parameter ∼ 0.43 nm, Goldshmidt (1967). Microdiffraction analysis reveals no significant difference in orientation along the vertical axis with in the composite particle which was close to (110) indicating epitaxial growth of thenobium-rich envelopes.

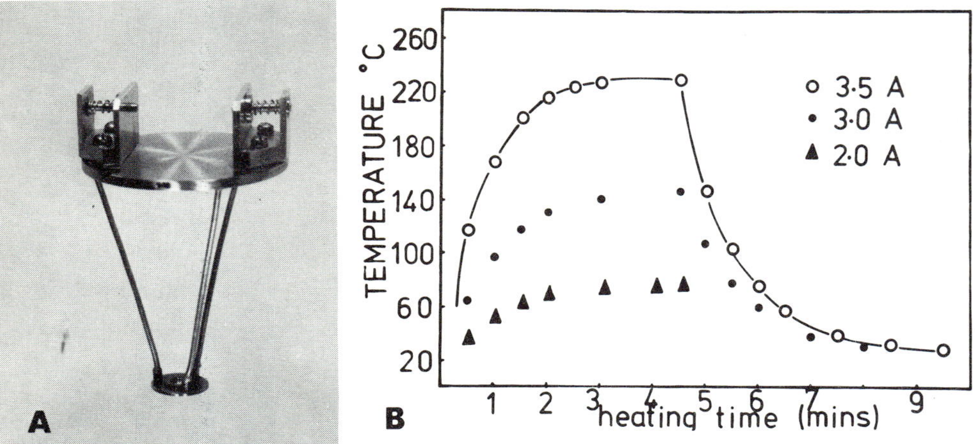

Fig 1A.   Dummy specimen cartridge of low thermal mass for airlock thermal cycling. (B) Time-temperature characteristics as a function of I.R. heater current.

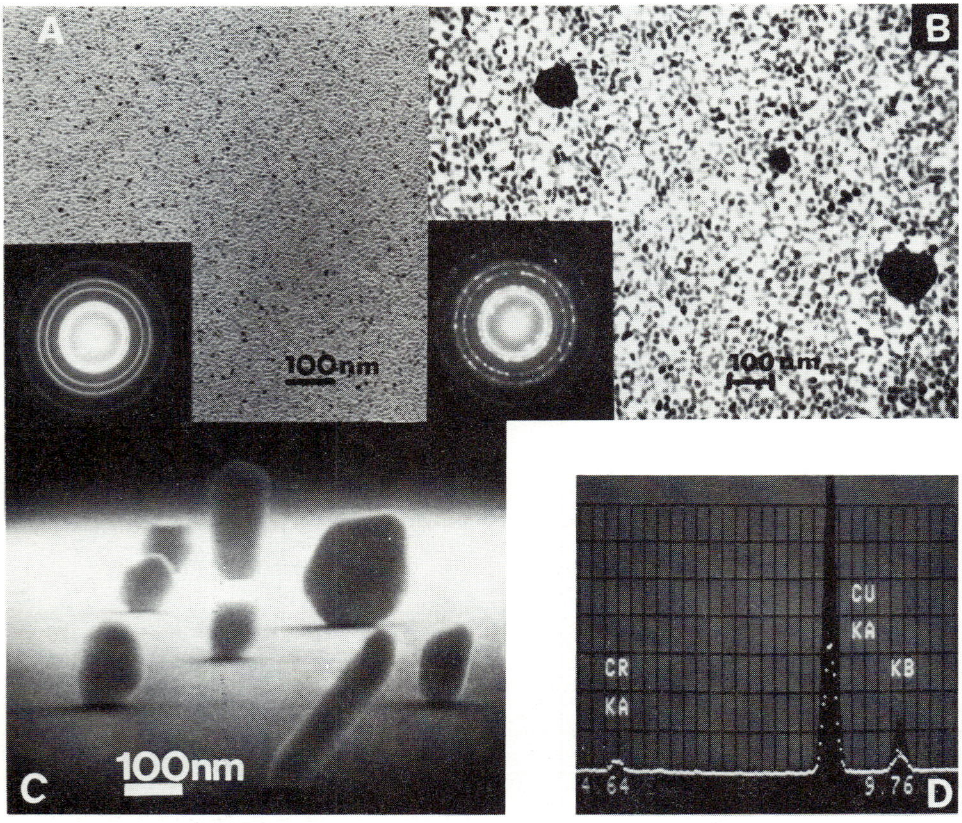

Fig. 2A.   As-sputtered metastable Cu-6% thin film formed by cosputtering, as seen in STEM B.F. with inset SADP from ∿100Å region. (B) STEM B.F. micrograph of sputtered film annealed 10.5 h, 250°C at 5.10^{-9} torr in airlock, inset SADP from 100A region showing spotty rings. (C) HB5 SEM Micrograph of copper-rich single crystals seen in (B). Comparative EDX Spectra normalised to Cr peak showing that the facetted particles are Cu enriched.

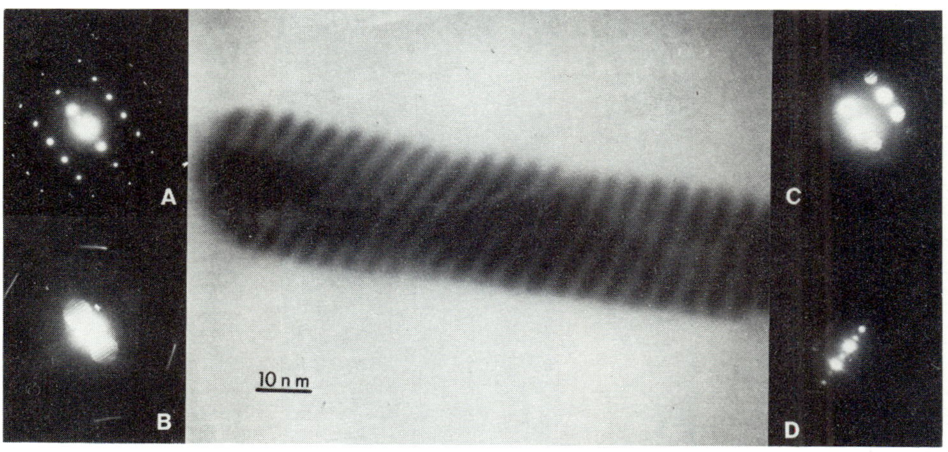

Fig 3. B.F. STEM micrograph of metastable precipitate formed by step quenching A1-6Zn-2Mg. Needles form along$\langle 100\rangle$ $\langle 110\rangle$ A1, g =(200)A1. (A) SADP from a 2 micron region around the precipitate. (B), (C) and (D) are microdiffraction patterns from the precipitate at 8, 4 and 1.7m rads respectively.

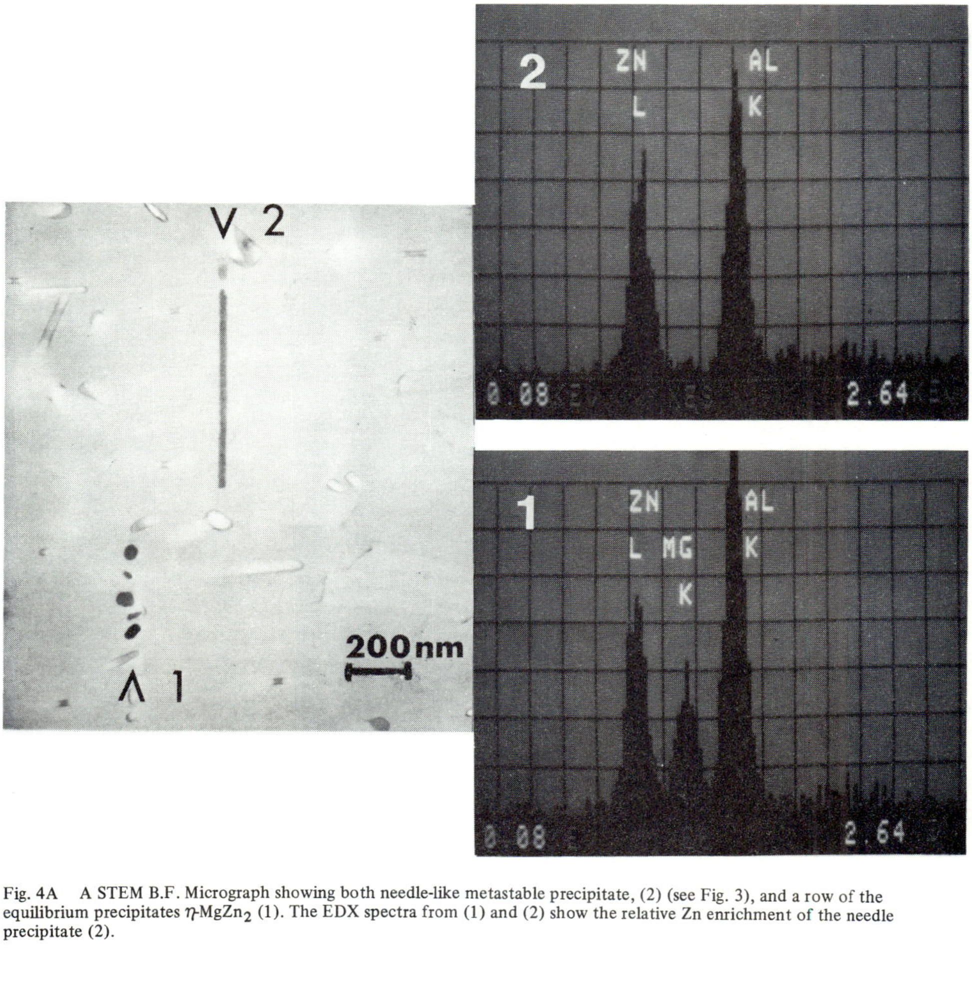

Fig. 4A   A STEM B.F. Micrograph showing both needle-like metastable precipitate, (2) (see Fig. 3), and a row of the equilibrium precipitates η-MgZn$_2$ (1). The EDX spectra from (1) and (2) show the relative Zn enrichment of the needle precipitate (2).

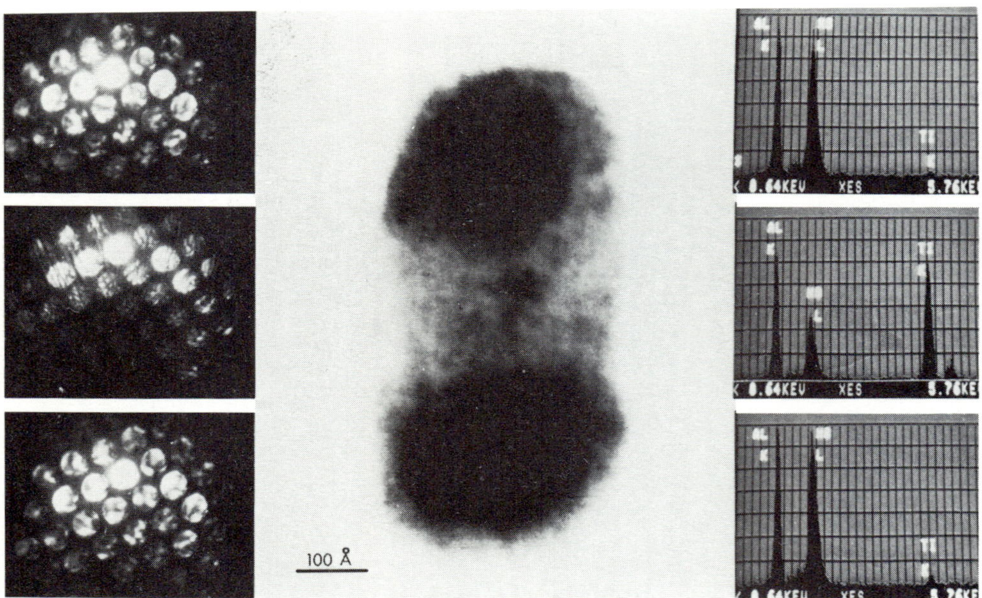

Fig. 5. STEM Micrograph of a composite carbonitride particle extracted for HSLA steel specimen. The EDX and Micro-diffraction data indicate complex microchemistry and growth epitaxy.

The particle shown in figure 6 was extracted from a metallurgical coke specimen and may be seen to contain many heterogeneously distributed impurities. The region ringed was found to be tin-rich by EDX analysis. The microdiffraction data indicated that the area in dark contrast was crystalline of hexagonal symmetry with a_o=0.29 nm, c_o=0.48 nm, which suggests that a tin-rich layered distorted intercalation compound may be formed.

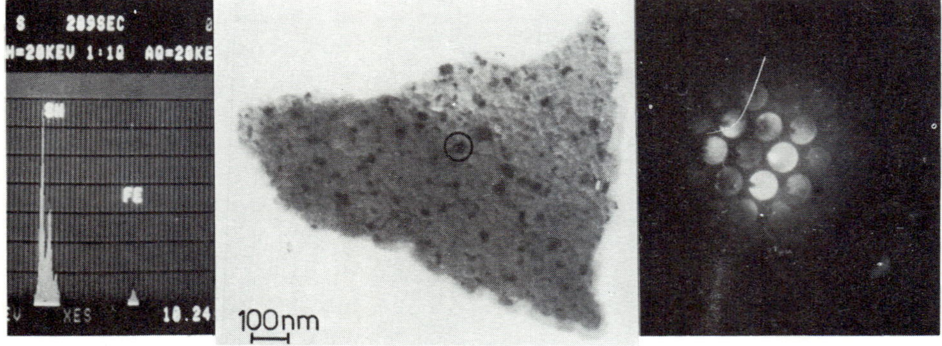

Fig. 6. Heterogeneously distributed impurities in a metallurgical coke particle. The regions exhibiting dark contrast in the central bright field STEM micrograph are seen to be tin-rich from EDX analysis and crystalline with hexagonal symmetry by microdiffraction investigation.

Goldshmidt H.J. (1967) 'Interstitial alloys' Plenum NY. 1967.

Houghton D C (1981) Microdiffraction in dedicated STEM EMSA 1981

Houghton D C, Embury J D and Weatherly G C (1981) Thermomechanical processing of microalloyed austerite, AIME.

Ryum N (1971) PhD Thesis Trondheim Norway P. 23

Systematic errors in thickness measurement by Grigson microdiffraction

J P Lynch [*] and LM Brown

Cavendish Laboratory, Madingly Road, Cambridge CB3 OHE

1. Introduction

Kelly et al.(1975) have recently demonstrated a simple method for obtaining
values of thickness, t, and extinction distance, ξ_g , using the fringe
structure present in convergent beam electron diffraction (CBED) patterns
of cristals oriented in the two beam condition. The fringe structure results from variations in the diffracted itensity I_g with deviation parameter s (Hirsch at al. 1965) :

$$I_g \propto \sin^2(\pi t s_{eff}) / s_{eff}^2 \qquad \ldots 1$$

where
$$s_{eff}^2 = s^2 + 1/\xi_g^2 \qquad \ldots 2$$

so that minima are expected in I_g when ts_{eff} is an integer, n. For a given
thickness the minima occur at values of s :

$$s_n^2 = (n/t)^2 - 1/\xi_g^2 \qquad \ldots 3$$

Values for s_n may be calculated from the diffraction pattern using the
separation, R , of disc centres since the angular deviation from the Bragg
condition is

$$ds_n = x_n/L = x_n 2\theta_B/R = \lambda x_n/dR \qquad \ldots 4$$

x_n is the separation of the nth minimum from the Bragg position, θ_B the
Bragg angle for an interplanar spacing d, L the camera length and λ the
electron wavelength. Equation 3 has no solution for $n < t/\xi_g$ so that these
minima are not present. If the minima are correctly numbered a graph of
s_n^2 vs n^2 should yield a straight line of slope $1/t^2$ and intercept $1/\xi_g^2$.

The dedicated STEM is capable of producing CBED patterns from regions less
than 1 nm in diameter and so might be expected to yield similar information with very high spatial resolution. Although conventional CBED patterns
are recorded on photographic plate, in a STEM the pattern is scanned past
a collector aperture which defines the angular resolution (Grigson 1965)
and is termed Grigson microdiffraction.

2. Results

A VG HB5 STEM operating at 80kV was used to produce microdiffraction
patterns from aluminium oriented near the (200) Bragg condition. An objective aperture of semi-angle α_o = 8.5 mrads, just small enough to avoid disc

[*]current address : IFP, BP311, 92506 Rueil Malmaison, France

overlap, and collector aperture of semi-angle α_c = 1 mrad were used. Intensity distributions, equivalent to rocking curves, were then recorded during a line scan through the disc centres parallel to g_{200}.

Figure 1 (a) to (c) shows three traces for increasing specimen thickness. It will be seen that, when fringes are present (in (a) for example), the diffracted intensity does not fall to zero between maxima, as would be expected from equation 1. This is partly due to absorbtion effects (also evident in the assymetry of the (000) disc) and partly to the averaging effect of the collector aperture.

Allen (1981) has calculated that absorbtion limits the maximum measurable thickness in aluminium to about 600 nm. The minimum measurable thickness, determined by the requirement of at least one diffraction minimum being below the limiting angle is about 20 nm. The effect of the collector aperture may be modelled by convolution of I_g with a "circular" aperture function A in the diffraction plane :

$$A(x) = \begin{cases} 1 - (x/\alpha_c L)^2 & : x/L < \alpha_c \\ 0 & : x/L > \alpha_c \end{cases} \qquad \dots 5$$

Typical results are shown in Figure 2 (a) to (c), again for increasing thickness. Integral values of ts_{eff} are marked for (a) and (c). For t=100nm little displacement of fringes occurs, the major effect being to remove the zeroes of I_g. As expected, when averaging over one period of the fringes, the structure is washed out. Since the limiting periodicity in s for large s is 1/t, this occurs at an angular periodicity d/t about equal to $2\alpha_c$. Because a circular aperture, and not a square one, is used the fringe destruction in fact occurs at t=2d/3α_c. For greater thickness, Figure 2(c), averaging takes place over more than one period so that oscillations reappear weaker and with reversed contrast. Evidently, thickness measurement will not be possible for certain values (2d/3α_c=130 nm for the conditions used here) and will suffer from errors, induced by displacement of minima, for greater thicknesses. The variation of apparent thickness from microdiffraction may be calculated from I_g*A, by the method of Kelly et al. (1975), using all the minima withinα_o, by fixing α_c and varying t.

In order to measure the errors a reference is required for t. This was calculated from the plasmon spectrum, at the same specimen point, by fitting to the probability distribution

$$P_j = (t/\lambda_p)^j \frac{e^{-t/\lambda_p}}{j!} \qquad \dots 6$$

where P_j is the probability of exciting the jth plasmon (proportional to the area under the jth peak) and λ_p is the mean free path for plasmon excitation. This measurement of t relies on accurate knowledge of λ_p or may be used to give thickness in units of λ_p.

Figure 3 showns the measured variation of thickness derived from microdiffraction(t') and the plasmon spectrum (t/λ_p) together with the expected variation, calculated from the model. Below 100 nm the two methods give values varying linearly, from which λ_p was calculated to be (105$\pm$4) nm. This is comparable to the theoretical value under the conditions used (Ferrel 1956) of 113 nm. Using this value of λ_p reasonable agreement is found between the measured and calculated variations of t' with thickness.

The average value of ξ_{200} from the reliable data was (70$\pm$4) nm, the theoreti-

cal value (Hirsch et al. 1965) being 67.3 nm. In thicker regions the apparent value was (48$\pm$8) nm. Grigson microdiffraction thus gives reliable
thickness and extinction distance measurements, and may be used to calibrate other methods, when not limited by angular resolution. Fringe structures
apparent beyond this limit cannot be regarded as useful.

References

Allen S M 1981 Phil. Mag. 43 325
Ferrel R A 1956 Phys. Rev. 101 554
Grigson C W B 1965 Rev. Sci. Inst. 36 1587
Hirsch P B, Howie A, Nicholson R B, Pashley D W and Whelan M J 1965
    Electron Microscopy of Thin Crystals (London : Butterworths)
Kelly P M, Jostcn A, Blake R G and Napier J C 1975
    Phys. Stat. Sol. (a) 31 771

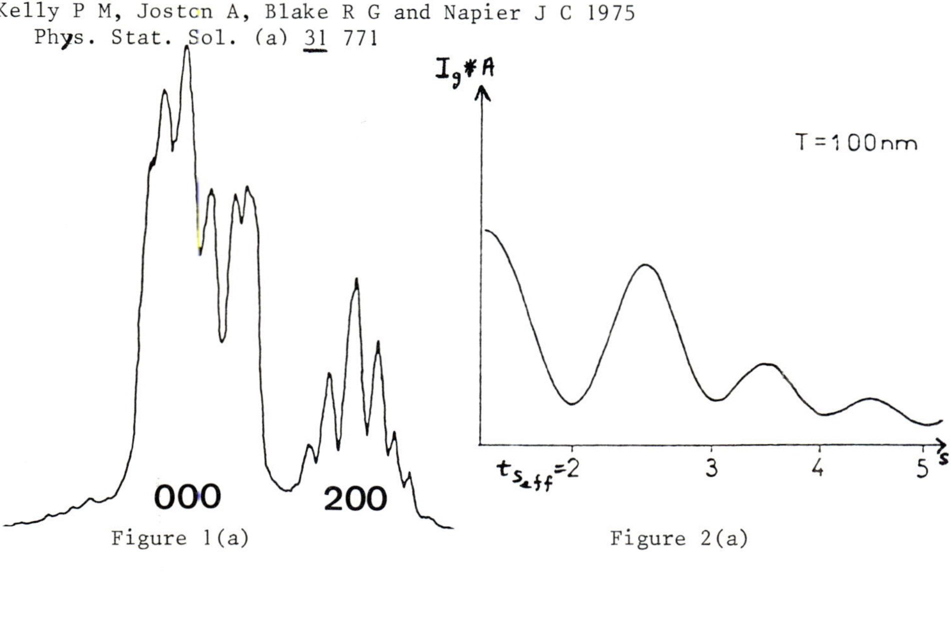

Figure 1(a)   Figure 2(a)

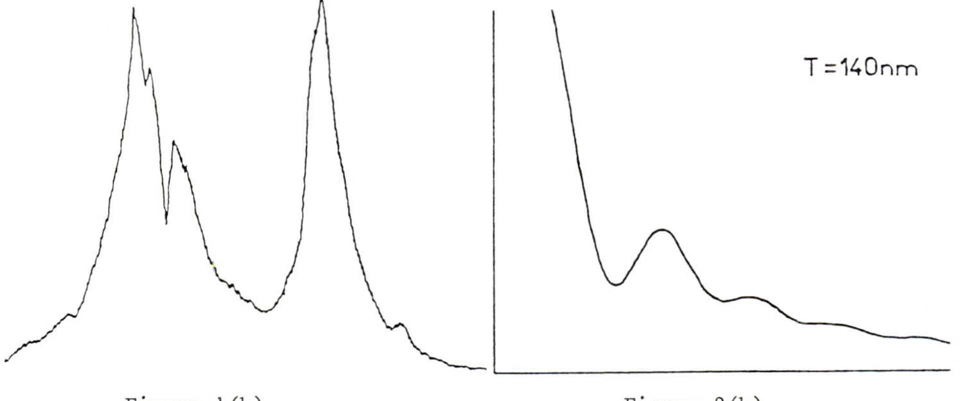

Figure 1(b)   Figure 2(b)

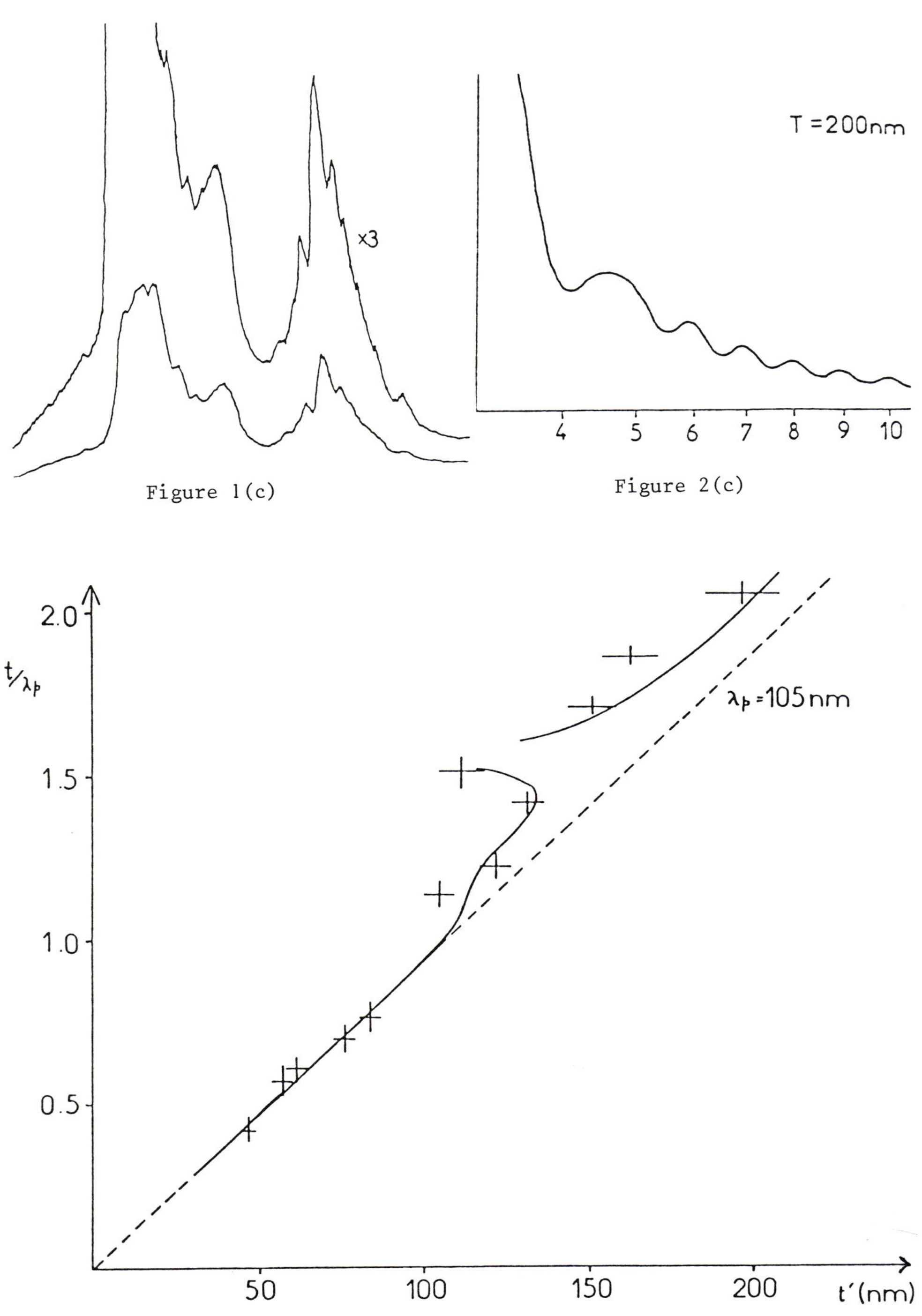

Figure 1(c)

Figure 2(c)

Figure 3

The crystallographic information in electron energy loss spectra

J. Spence, O. Krivanek, J. Taftø and M. Disko
Department of Physics, Arizona State University
Tempe, Arizona 85287 U.S.A.

Apart from the work done at very high energy resolution and using the dielectric response function, (Geiger 1981), the interpretation of fast
electron energy loss spectra has mainly been based on the well established
theories for isolated atoms and for plasmon excitation. In neither case
are crystallographic effects important. Here we discuss the results of
three recent experiments in which crystallographic information may be extracted from fast electron transmission energy loss spectra.

1.   Directional Effects in EXELFS Spectra

The first attempt to apply the EXAFS theory to fast electron energy loss
spectra appears to be the work here in Cambridge by Leapman and Cosslett
(1976). We have attempted to extend their work by investigating the influence of crystallographic orientation on EXELFS spectra. The differential cross section for the inelastic scattering of a fast electron in the
Born approximation is

$$\frac{d^2\sigma}{d\Omega dE} = \frac{4}{q^2 a_o^2} \mid <f\mid \hat{q}\cdot r \mid i> \mid^2 \qquad\qquad 1$$

with q the inelastic scattering vector and a_o the Bohr radius. In addition, conservation of momentum and energy requires that

$$q^2 = \frac{2mE_o}{\hbar^2} \ [(\Delta E/2E_o)^2 + \theta^2] \qquad\qquad 2$$

for a small scattering angle θ and energy loss ΔE. The total transition
rate describing the atomic absorption of a photon beam is, however,

$$W = \frac{2\Pi e^2}{\hbar^2} \sum_f \mid <f\mid \hat{\epsilon} \cdot r \mid i> \mid^2 \delta(E_i + h\omega - E_f) \qquad\qquad 3$$

Where $\hat{\epsilon}$ is a unit vector in the direction of the photon beam polarization.
Since the analysis of EXAFS data (based on equation 3) has allowed inter-
atomic distances in the experimentally predetermined direction of $\hat{\epsilon}$ to be
measured (Heald and Stern 1978), we have attempted to observe a similar
directional effect in EXELFS (extended electron energy loss fine structure)
spectra, based on equation 1. Directional EXAFS results from the diffraction of the ejected photo-electron and subsequent modification of the
final state $<f\mid$ which can be written as the sum of an outgoing and reflected wave. It is heavily weighted in the direction of $\hat{\epsilon}$. In EXELFS both the
direction and magnitude of hq, the momentum imparted to the "photo
electron", can be controlled experimentally by choice of the fast electron
detector position as shown in figure 1. Thus different directions in the
crystal may be probed by varying the scattering angle. Both EXAFS and

EXELFS can be thought of as "LEED with the source and detector at the same point" (the target atom). By mounting a graphite crystal with its c axis at 45° to the electron beam (similar to the experiments of Leapman and Silcox (1979), off-axis detectors for the transmitted electrons could be placed in directions such that $\underset{\sim}{q}$ lay either parallel or normal to the crystal c axis. For a small detector equation 1 then suggests that the dominant contribution to the EXELFS will come from crystallographic directions which are roughly aligned with $\underset{\sim}{\hat{q}}$.

Radial distribution functions derived from energy loss spectra recorded successively from the two scattering directions indicated in figure 1 are shown in figure 2. The algorithm used (similar to that used in EXAFS) has been discussed extensively elsewhere (Teo and Joy 1980). Briefly this consisted of background removal (using a function of form AE^{-r}, r a constant), scaling the data against $\underset{\sim}{k}$, the photoelectron wavevector, and removal of the non-oscillatory background using a fifth degree polynomial, followed by Fourier analysis. The k=o energy loss corresponding to the onset of the EXELFS oscillations was chosen by requiring that $|F(R)|$ and $Im|F(R)|$ both peak at the same radial distance R (Lee and Beni 1977), where F(R) is the experimentally derived radial distribution function. Spectra were recorded using a GATAN model 607 magnetic sector energy analyser fitted to a Philips EM400 Electron microscope (LaB_6 filament). These were obtained at 120 kV from 300 Å thick regions (thickness estimated using CBED patterns) using 0.3 mrad illumination and collection semiangles. The count rate at the carbon K-edge was 600 kH_z for a 3eV energy window. Details of the data analysis method used, including background subtraction, the effects of finite detector size, Debye-Waller factor effects, the choice of filtering window and EXELFS onset energy, and the theoretical backscattered phase-shifts used (Teo and Lee 1979) are given elsewhere (Disko, Krivanek and Rez 1981). The large peaks in the spectra shown in figure 2 agree closely with the known 3.64Å second-nearest neighbour interplane distance in 2 (a) (12 or 6 atoms), and with the 1.42Å in-plane nearest neighbour distance in 2 (b) (3 atoms). These dominant peaks are rather insensitive to changes in the parameters used in the Fourier analysis procedure.

2. Channelling (Borrmann) Effects On Characteristic Energy Loss Spectra

The height of core-loss peaks in transmission electron energy loss spectra from crystals has been found to depend strongly on the incident beam direction (diffraction conditions) and degree of localization. The possible uses of this effect for identifying the sites of light atom impurities in crystals has therefore been investigated, using natural spinel ($M_gAl_2O_4$) as a test sample. By choosing a systematics orientation and placing the detector at various positions along a line normal to the systematics line, *the diffraction conditions and localization could be varied independently.*

Figure 3 shows the structure of spinel projected in two directions normal to [hoo]. Spinel contains light atoms (M_g and Al) in distinguishable octahedral (Al) and tetrahedral (M_g) sites. Also shown are the results of eleven beam dynamical calculations for the elastic fast electron intensity over the unit cell for the two orientations used to collect energy loss spectra. For the dotted curve (orientation 1), the excitation error $S(400) = \lambda U_{400}$, while for the continuous curve (orientation 2) $S(400) = -\lambda U_{400}$. Thus we expect an enhancement (reduction) of the oxygen and Aluminium K-loss peaks relative to the M_g K-loss peak in orientation 2 (1) by analogy with the corresponding Borrman effect in X-ray

microanalysis techniques (Taftø 1979; Cherns, Howie and Jacobs 1973). A
crucial difference, however is that the elastic diffraction of the inelastically scattered electron must be considered (using, for example the
method described in 3 below) and this effect (for which there is no corresponding effect in X-ray microanalysis) causes the orientation effect
on ELS spectra to be much stronger than that on characteristic X-ray production.

The results shown in figures 4 and 5 were obtained using the instrumentation described in (1) above. Detectors and illumination semiangles were
set to one quarter of the (400) Bragg angle. The inset to the figures
shows the direct beam (continuous circle) and detector position (dotted
circle) relative to the Brillouin Zone as revealed by the (400) Kikuchi
lines (vertical lines), which run normal to the systematics line. In
figure 4, using in-line source and detector, the change from positive
(figure 4(a)) to negative(figure 4(b))(400) excitation error - i.e. from
orientation 1 to 2 of figure 3 - causes little change in the core-loss
peak intensity ratios. However the use of a larger scattering angle and
correspondingly smaller impact parameter shown in figure 5 causes a large
change in peak intensity ratios for the same change in diffraction conditions. From the observed ratio trends we readily confirm that the
cation distribution corresponds to normal spinel (Al in octahedal
positions) and not inverse spinel. A theoretical analysis of this effect,
and a comparison of the channelling effect on X-ray production and core
losses in ZnS is given elsewhere (Taftø, Krivanek and Spence 1981). Table
1 summarizes the results of this work on ZnS. Here the ratio of the
intensity of the zinc L shell ELS peak to that of the sulphur K shell ELS
peak for negative excitation error (written $(I(L,Zn)/I(K,S))S^-$) has been
divided by the same ratio for positive excitation error. Two scattering
angles (normal to the systematic line) are shown for the electron energy
loss case, and these are compared with the results for the same ratio recorded at the same time on the same sample from peaks in the X-ray
fluorescent spectra. The orientation effect on the magnitude of core
loss peaks in transmission energy loss spectra is seen to be larger than
that on the X-ray emission peaks if large scattering angles are used. As
with orientation effects in X-ray microanalysis, it would appear necessary
to take these channelling effects into account when attempting microanalysis by transmitted electron energy loss spectroscopy. Finally we
note the possibility of using the channelling effect to enhance the EXELFS
(or EXAFS) signal, and so to distinguish non-equivalent crystallographic
sites containing the same species.

3. Core-loss Energy Filtered Images and Diffraction Patterns

Experimental $O_{4,5}$ shell excitation energy filtered images from small groups
of Uranium atoms have recently been obtained using the GATAN 607 electron
spectrometer fitted to the HB5 STEM instrument (Colliex, Krivanek and
Trebia 1981). Dynamical calculations for these images and diffraction
patterns, including multiple elastic scattering of the inelastically scattered electrons reveal the following: (i) assuming negligible electron
optical aberrations, core loss filtered HREM images of one species may include images of any other species which lie within a distance L or W
(whichever is the larger) of the selected species. Here $L \approx 2\lambda E_0/\Delta E$ is the
localization (ΔE the energy loss) and W is a "column width" as indicated
in figure 6. This conclusion is confirmed by the dynamical calculations
shown in figure 7 in which an energy filter tuned to the Oxygen K loss was

assumed and despite which the neighboring copper atom also appears in the image (Spence 1981). (ii) Core loss filtered diffraction patterns (plane wave illumination, not CBED patterns) may not express the crystal symmetry but rather that of the local atomic coordination around the selected species. Thus, in figure 6 the filtered diffraction pattern from atom A (with localization L as shown) may show six-fold symmetry (two incoherently superimposed three-fold patterns related by the crystal two-fold axis). Thus the symmetry of these filtered inelastic patterns may provide additional information on the sites of atoms within co-ordination polyhedra (Spence 1981).

References

Cherns D, Howie A and Jacobs M H 1973 Z. Naturforsch <u>28a</u>, 565
Colliex C, Krivanek O L and Trebia P 1981 These proceedings
Disko M M, Krivanek O L and Rez P 1981 Phys. Rev. Letts. Submitted
Geiger J 981 Proc. Thirty-ninth Annual EMSA meeting ed. G. W. Bailey
    (Claitor's Baton Rouge) pp. 182-185
Heald S M and Stern E A 1978 Phys. Rev. <u>B17</u>, 4069
Leapman R and Cosslett V 1976 J. Phys. <u>D9</u>, L29
Leapman R and Silcox J 1979 Phys. Rev. Letts. <u>42</u>, 1361
Spence J C H 1981 Ultramicroscopy, in press
Taftø J 1979 Z Naturforsch <u>34a</u>, 452
Taftø J, Krivanek O L and Spence J 1981 Ultramicroscopy, submitted
Teo B K and Joy D C 1980 EXAFS Spectroscopy (New York: Plenum)
Teo B K and Lee P A 1979 J. Am. Chem. Soc. <u>101</u>, 2815

Acknowledgement

This work was supported by NSF grant #DMR8002108, by ARO contract #DAAG-29-80-C-0080 and by the services of the NSF HREM facility at Arizona State University.

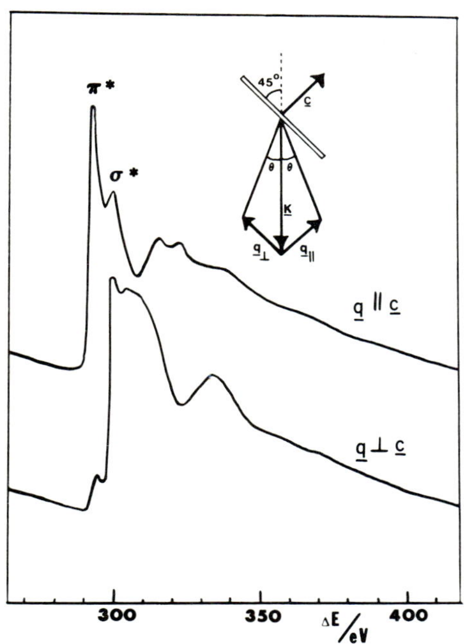

Figure 1.   Transmission energy loss
spectra recorded at 120 kV. from an
inclined graphite crystal using two
different detector positions.   For
a given energy loss the length of q
is fixed by equation 2; the direction of q is then fixed by θ.

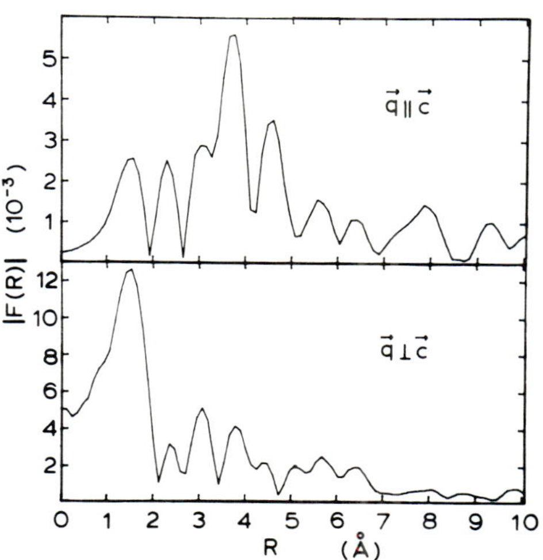

Figure 2.   Radial distributions obtained from the two spectra of figure
1 using the EXAFS algorithm.   The
major peak in the lower plot corresponds to the 1.42Å in-plane nearest
neighbor distance, while that in the
upper plot represents the unresolved
nearest (3.35Å) and second nearest
(3.64Å) neighbors in the approximate
c axis direction.

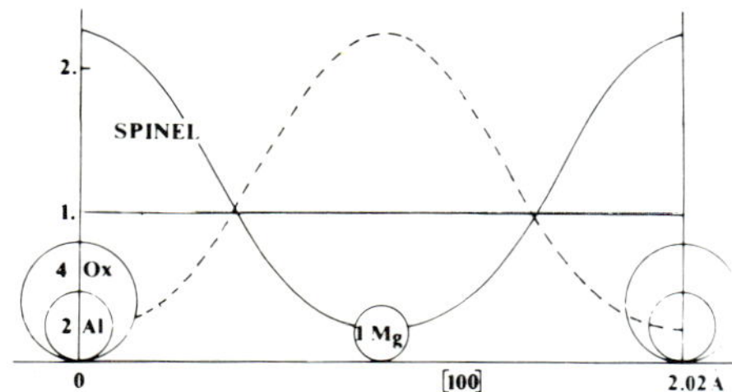

Figure 3.   Eleven beam dynamical
calculations for the electron intensity across the spinel unit cell
for two incident beam orientations.
Approximate 2-beam conditions with
positive (400) excitation error
(dotted curve) and negative (400)
excitation error ((400) outside
Ewald sphere, continuous curve).

scattering angle ╲ Ratio	$\dfrac{(I(L,Zn)/I(K,S))s^-}{(I(L,Zn)/I(K,S))s^+}$
0	1.8
1/2(200)	3.3
X-ray emission	1.4

Table 1.   Ratio of Zinc L shell to
sulphur K shell intensities compared
for two orientations (S⁻ and S⁺) in
Zinc sulphide.   The scattering angle
is measured in the plane which contains the beam and is normal to the
systematics line.

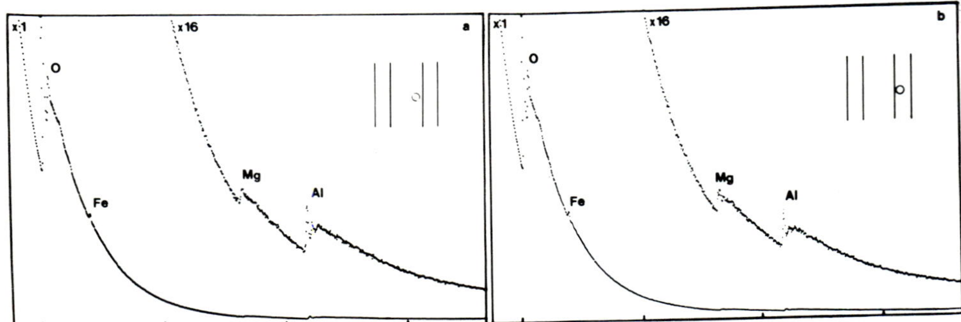

Figure 4.   ELS spectra from Spinel ($MgAl_2O_4$) for two approximate two-beam orientations.   Positive excitation error a), Negative excitation error b) Incident beam direction intersects detector.   Ordinate scaling divided by 16 for clarity at 1 keV loss.

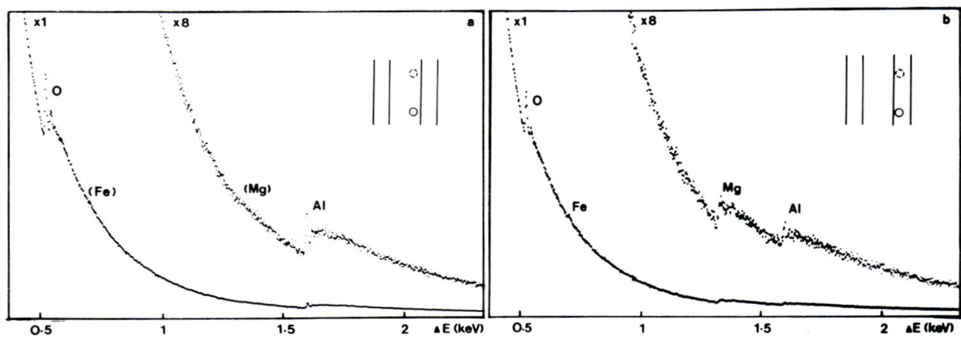

Figure 5.   Similar to figure 4 but with off-axis detector, displaced normal to the systematics line and the beam, thus collecting electrons scattered through a larger angle which have passed closer to the nucleus. Same diffraction conditions as 4(a) and 4(b).   Note change in size of Mg core-loss peak.

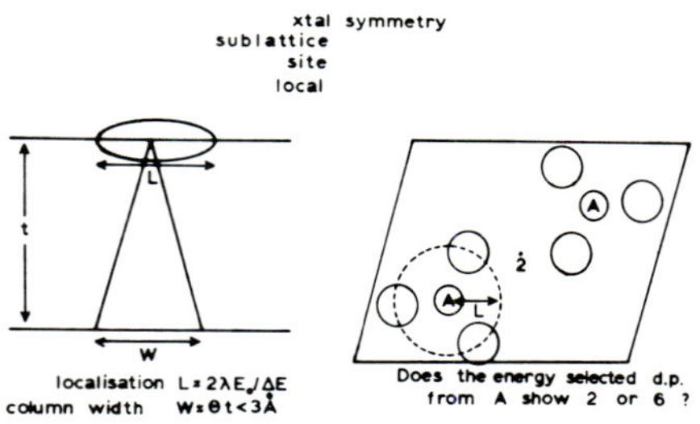

Figure 6.   Localized inelastic scattering coherently generated within a volume L^3 and spreading by a "column width" w to the specimen exit face.   Also shown are two projected tetrahedra with "local" three fold symmetry.

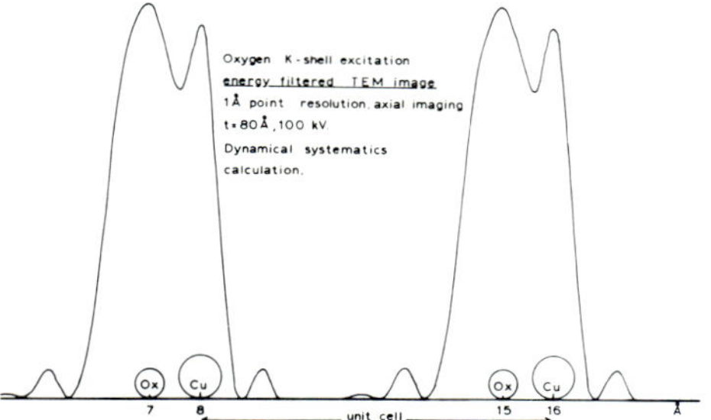

Figure 7.   Computed Oxygen K-shell energy filtered TEM image at 1Å resolution from the artical crystal shown.   Thickness 80Å.   Despite filtering, the copper atom also appears.

Imaging low energy losses from MgO smoke particles

L D Marks

Cavendish Laboratory, Madingley Road, Cambridge CB3 OHE

1. Introduction

When fast electrons travel near to a surface, surface plasmons are excited
by the interaction with the image potential. The most favourable configuration for examining these effects is at glancing or parallel incidence,
when the excitation probability for a semi-infinite surface in a classical
dielectric takes the form (Ritchie, private communication):

$$\frac{dP}{d\omega dz} = \frac{2e^2}{\pi\hbar^2 v^2}\; \mathrm{Im}\left(\frac{-2}{1+\varepsilon}\right)\, K_o\left(\frac{2\omega x}{\hbar v}\right) \qquad\qquad x > 0$$

$$= \frac{2e^2}{\pi\hbar^2 v^2}\left[\mathrm{Im}\left(\frac{-1}{\varepsilon}\right)\left\{\ell n\left(\frac{k_c\hbar v}{\omega}\right) - K_o\left(\frac{2\omega x}{\hbar v}\right)\right\}\right.$$

$$\left. +\; \mathrm{Im}\, -\left(\frac{2}{1+\varepsilon}\right)\, K_o\left(\frac{2\omega x}{\hbar v}\right)\right] \qquad\qquad x > 0 \qquad\qquad (1)$$

where the particle is taken to lie in the region $x < 0$, v is the electron
velocity, ω is the energy loss in eV.

An ideal specimen for the study of surface inelastic events is MgO smoke
particles, which are known to possess atomically flat (100) surfaces.
Furthermore, the dielectric function is known (Roessler and Walker 1967,
Venghaus 1971). For reference, values of the bulk and surface (Ritchie
1957) energy loss functions ($\mathrm{Im}(-1/\varepsilon)$ and $\mathrm{Im}(-2/1+\varepsilon)$ respectively) are
shown in fig.1. It should be noted that both functions display interband
excitations at 8.0, 11.8, 14, 15.8 and 24, to go with the bulk plasma at
22.5 eV and a surface plasmon at $\sim$ 20 eV.

2. Experimental Procedure

Burning magnesium smoke was collected directly onto freshly cauterised
copper grids and loaded immediately into an HB5 STEM equipped with an
electron energy loss detector of $\sim$ 3 eV resolution. Energy loss spectra
with a stationary probe of $\sim$ 2 mm diameter located on or near the particles, as well as energy filtered images, were obtained from particles
with faces oriented within a few degrees of parallel to the optic axis.

3. Results

The bulk EELS obtained by transmission through the particles (see fig.2)
showed the same general features as described by Venghaus (1971). When

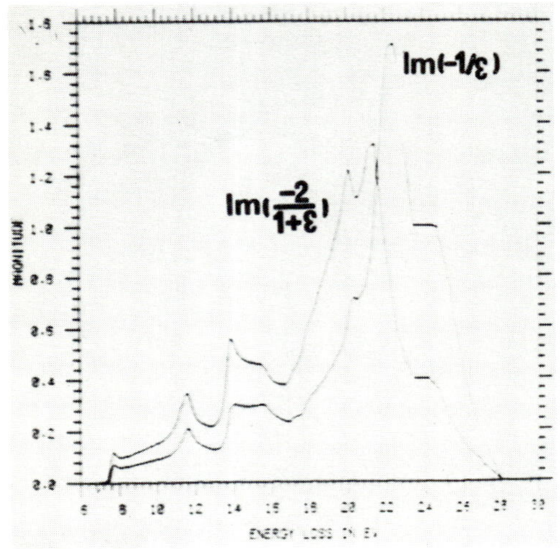

Fig.1 Plots of the bulk (Im(−1/ε)) and surface (Im(−2/1+ε)) loss functions for MgO, taken from the optical work of Roessler and Walker (1967).

Fig.2 EELS taken from different positions with a stationary probe: B, through the centre of a particle; F, down a face and E, by an edge. (The latter two are magnified × 4 relative to the bulk spectrum). The particle used is arrowed on fig.2

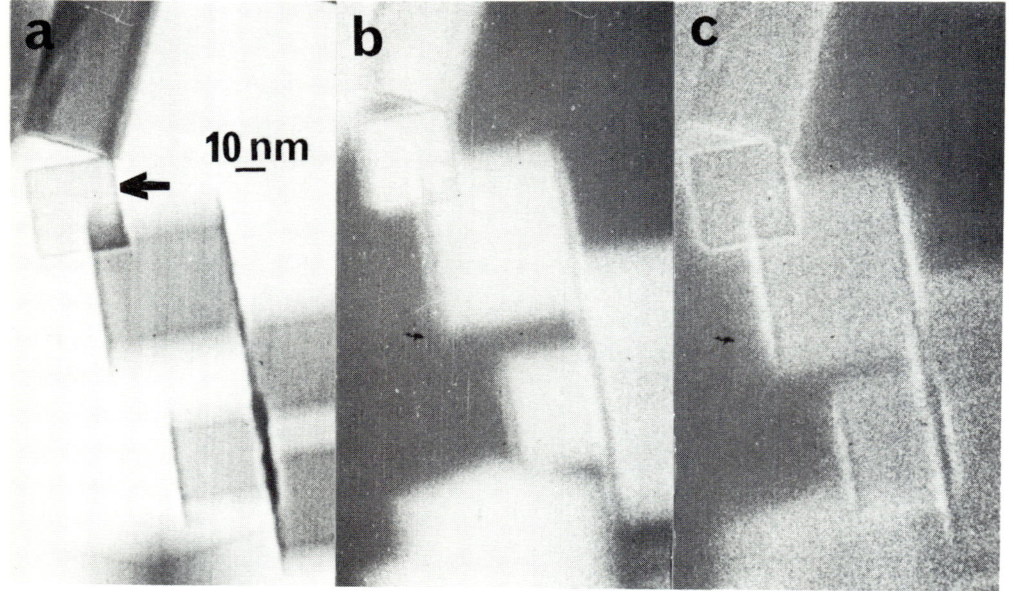

Fig.3 Energy filter images of MgO smoke particles taken with a) zero, b) 22 eV and c) 18 eV energy loss electrons.   The particle arrowed was used for the spectra shown in fig.3

the probe was moved slightly outside the particles, and then to positions further from the surface, a pronounced shift towards lower frequencies was observed (as well as a substantial drop in intensity). This can be explained as the influence of the distance weighting ($K_o(2\omega x/\hbar v)$ in eqn.(1).

When imaged with the energy filter, the bulk plasmon signal was clearly located primarily within the particles, the surface plasmon outside, as shown in fig.3. The strongest surface signal was observed at 18 eV, although surface localised losses were observed at lower frequencies as well (due to the extended range of $Im(-2/1+\varepsilon)$ in MgO). In practice, cleaner signals could generally be obtained at 15 or 13 eV (when the bulk signal was substantially smaller).

When examined in detail, it was observed that the maximum intensities for the surface excitations occurred well outside the bright field particle surface, as shown in fig.4. Typical distances ranged from 1 - 5 nm, generally increasing with the particle size. At large distances the excitation showed an exponential decay, in agreement with the long range behaviour predicted by eqn.(1). Inside the particles the bulk plasmons also showed exponential behaviour consistent with the edge correction term in eqn.(1).

One particularly strong feature observed with the surface plasmons was strong phase contrast, as shown in fig.5. To first order these effects are consistent with a small angle reflection component to the inelastic scattering. (A more rigorous image inversion is currently in progress.)

4. Discussion

It is apparent that substantial quantitative data about surface inelastic scattering can be obtained by energy filtered imaging at glancing incidence. Good agreement has been reached with the theoretically predicted behaviour of a classical dielectric, both quantitatively for the exponential decay and qualitatively for the surface loss function $Im(-2/1+\varepsilon)$. Furthermore, the coherent behaviour of these losses promises that detailed phase as well as amplitude information can be obtained.

Perhaps the most striking feature of the results presented herein is the presence of a clear maximum in the intensity of the surface plasmons outside the particles. The source of this is probably a combination of various effects. For example, preliminary calculations have shown that multiple losses cannot be neglected close to the surface. However, other effects, such as diffraction, microscope aberrations (C_S = 7 mm) or the size of the probe cannot be ignored. Some preliminary experiments (not included here) have also suggested a link between surface diffraction (Turner and Cowley 1981) and surface inelastic processes.

Acknowledgments

The author would like to thank the SRC for financial support, A Howie and R J Ritchie for their extensive advice and S J Pennycook for assistance in the use of the HB5 STEM

References

R H Ritchie 1957 Phys. Rev. <u>106</u> 874
D M Roessler and W C Walker 1967 Phys. Rev. <u>159</u> 733

P S Turner and J M Cowley   1981   Ultramicroscopy 6 125
H Venghaus   1971   Optics Comm. 2 447

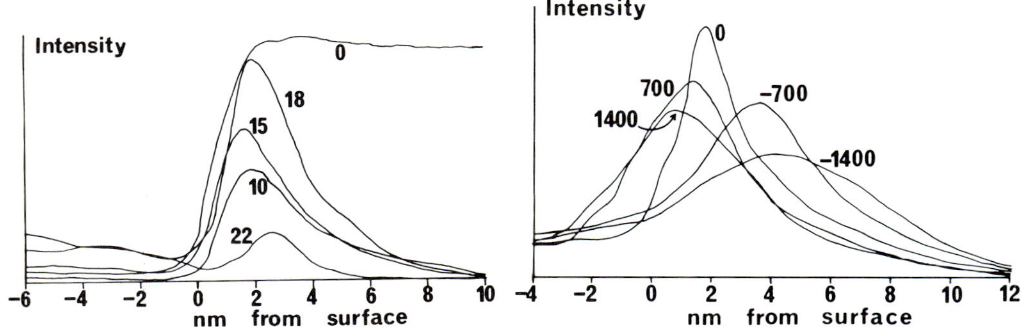

<table>
<tr><td>

Fig.4 Line traces from the sur-
face of a 100 nm particle oriented
near a (100) pole, with the energy
losses used marked on the figure.
The zero loss scan has been re-
duced in intensity by a factor
of 30.

</td><td>

Fig.5 Line traces from a 80 nm
particle with an energy window
at 13 eV showing phase contrast
effects. Defocus values in nm
are indicated on the figure.

</td></tr>
</table>

On the detection of point defects in crystals using high-angle diffuse scattering in the STEM

M.M.J. Treacy[*] and J.M. Gibson[**]

[*] IBM, T.J. Watson Research Center, Yorktown Hts., NY 10598

[**] Bell Labs, 600 Mountain Ave., Murray Hill, NJ 07974

One of the principal attractions of the annular detector signal in the scanning transmission electron microscope (STEM) is its strong dependence on the atomic number Z of the specimen (Cosslett 1965). For very thin specimens where the component atoms can be regarded as independent scatterers the signal from each atom type is proportional to the local mass thickness and the screened Rutherford elastic scattering cross section ($\propto Z^{4/3}$ on the Thomas-Fermi atomic model)(Langmore et al 1973). This simple model forms the basis of the Z-contrast technique which has been successfuly used for imaging single heavy atoms supported on thin low Z substrates (Isaacson 1979, Mory et al 1981).

For studying complex crystalline materials however, it is advantageous to use an annular detector collecting at large scattering angles only(>100mrad) (Treacy et al 1980, Butler 1981). At these angles Bragg reflected intensities are weak due to Debye-Waller attenuation, which thereby minimises the deleterious effects of diffraction contrast in the annular detector signal. The detected scattering now arises predominantly from near head-on collisions with atomic nuclei and follows the full Z^2 dependence of the unscreened Rutherford formula (Humphreys et al 1973).

However, for thick (> 5 nm) crystalline specimens, channeling can be an important source of contrast in the high angle detector (HAD) signal, and thus the lattice site location of each atom type becomes an important consideration. This arises because the incident fast electrons form Bloch waves within the crystal and the resultant current density varies across the unit cell. The current density profile is strongly dependent on the exact diffraction conditions and is depicted in figure 1 for a two-beam case. Since the impact parameter for high angle scattering is small compared with inter-atomic dimensions, each current density profile in fig 1 will result in a different high angle scattered intensity. This effect has been studied in detail by Hall et al (1967) for the case of a uniform solution of point defects in crystals. It appears therefore,by judiciously controlling the specimen diffraction conditions it may be possible to use the STEM HAD signal to distinguish between individual interstitial and substitutional high-Z impurity atoms in a low-Z crystal matrix.

To investigate this possibility we need to treat the impurity atom and host crystal together as a single system and not as an assembly of independent scatterers.

* Present address : CNET 196 rue de Paris, 92220 BAGNEUX. (FRANCE).

Following the treatment due to Cherns et al (1973) for the analogous case
of X-ray production, we represent the HAD scattering by a probability $P(\underline{r})$
which peaks on the atomic sites. The HAD signal can be written as :

$$I_{HAD}(\underline{r}) = \int_0^t |\psi(\underline{r},z)|^2 \, P(\underline{r})dz \qquad (1)$$

The integral over z represents the contribution from a column at image pos-
ition $\underline{r}$, and ψ is the total wavefunction. $|\psi|^2$ is directly proportional to
the current density plotted in fig 1. The suitability of this expression
for the HAD case can be checked by comparing computed and experimental contrasts for the case of a stacking fault imaged under two-beam conditions
(fig 2). It is clear from fig 2d that for a good fit with experiment it is
necessary to include the spread in deviation parameter w, introduced by
the finite angular width of the objective aperture and which tends to wash
out the effects of channeling.

For the case of a single impurity atom of atomic number Z_a we wish to consider the contrast C = I(atom + crystal)/I(crystal) as a function of crystal
orientation and for a probe of diameter δ. From eqn.1 we can derive the
simplified expression

$$C = 1 + (4\Omega/\pi\delta^2).(Z_a/Z_x)^2 . \frac{\langle |\psi_a|^2 \rangle}{\langle \int_0^t |\psi|^2 dz \rangle} \qquad (2)$$

Z_x and Ω are the atomic number and atomic volume of the host crystal atoms,
and ψ_a is the wave function incident on the impurity atom. The angular brackets represent averaging over the angular spread of the incident probe. ψ
should include the effects of lattice distortion around the impurity atom.
Inelastic scattering can be crudely accounted for in this treatment by
replacing Z^2 with $Z(Z+1)$. Probe spreading in the crystal has been ignored.

Fig 3 shows calculated contrasts for the case of Au (Z=79) impurity atoms
in Si (Z=14) one extinction distance thick, for two different depths in
the Si lattice (corresponding to X and Y in fig 1). The full line shows the
computed contrast if we neglect the averaging effect of probe convergence
for the interstitial case. The dashed line and the dashed-dotted line show
the contrasts for interstitial and substitutional Au atoms respectively,
with probe convergence included. Both interstitial and substitutional defects show an increase in contrast for w > 0, although this effect is much
weaker for the substitutional case. Although for w >0 the current density
peaks between the lattice planes, away from the substitutional atoms, the
contrast still appears to increase for the substitutional case because the
signal from the host crystal has fallen as well.

It appears from fig 3 that to simply detect a Au atom in the crystal we
need to be able to detect contrast changes of   about 5 % in the final image. Taking into account the effects of shot noise from both the atom and
the crystal, this will require images containing at least 30 grey levels
of information or about 10^3 detected electrons per picture point. For a HAD
collecting 1 % of the incident current ($\sim 10^{-10}$A) this will require exposure times of about 30 seconds for images containing 1000 x 1000 picture
points. To distinguish between substitutional and interstitial sites we need
to be capable of detecting contrast changes of around 1 %, and this can lead
to unrealistic exposure times in excess of 10 minutes. For lower Z impurity
atoms such as Cu(Z=29) even longer exposures would be required and higher
probe currents would be desirable.

A possible alternative to the channeling method would be to reduce the
probe size to below that of the lattice spacing and to resolve directly
the impurity atom position with respect to the host lattice. For this, wide
probe convergence angles and low values of objective lens spherical aberration are needed. Due to the $1/\delta^2$ dependence of the contrast (eqn. 2) the
smaller probe size would increase the atom contrast. Furthermore, the wider
probe angles would wash out the channeling effects and would increase the
total current in the probe.

Preliminary HAD channeling studies for the cases of Au and Cu in Si have
met with no success, probably because the specimens studied have been too
thick ($\sim$ 70 nm), and in the case of Cu, because there is not sufficient
disparity in Z between the Cu and Si. It is anticipated that graphite intercalated with heavy atoms may provide more favourable specimens for these
studies.

The authors would like to thank Drs P.E. Batson, K.N. Tu, T.Y. Tan,
D.A. Smith and A. Howie for valuable discussions.

Butler J H, Proc. of 39th Ann. EMSA Meeting. 1981
Cherns D, Howie A and Jacobs M H Z. Naturforsch 28a 565 1973
Mory C, Colliex C, Revet B and Delain E 1981 to be published
Cosslett V E, in 'Quantitative Electron Microscopy' ed G F Bahr and E H
  Zeitler (Baltimore, Maryland: The Williams and Wilkins Co) 271 1965
Hall C R, Hirsch P B and Booker G R, Phil. Mag. 20 979 1967
Humphreys C J, Sandstrom and Spenser J P IITRI Proceedings vol 2 233 1973
Isaacson M S in 'Introduction to Analytical Electron Microscopy' ed Hren
  Goldstein and Joy, Plenum Press 1979
Langmore J P, Wall J and Isaacson M S Optik 38 335 1973
Treacy M M J, Howie A and Pennycook S J Inst. Phys. Conf. Ser. 52 251 1980

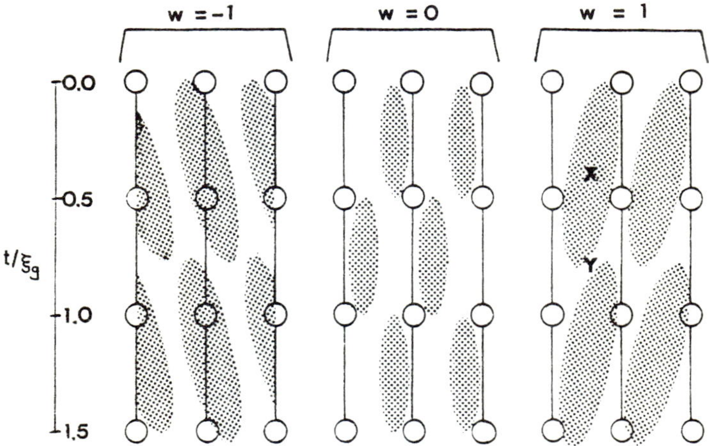

FIGURE 1: Distribution of the electron beam intensity for a crystal in a
two-beam diffracting condition. The vertical scale represents depth within
the crystal as a multiple of the extinction distance ξ_g and the shaded
regions represent, schematically, regions of high beam intensity. The
diffraction conditions are varied across the page as a function of the
deviation parameter $w = |g|\xi_g\Delta\theta$ where $\Delta\theta$ is the angular deviation from
the exact Bragg condition.

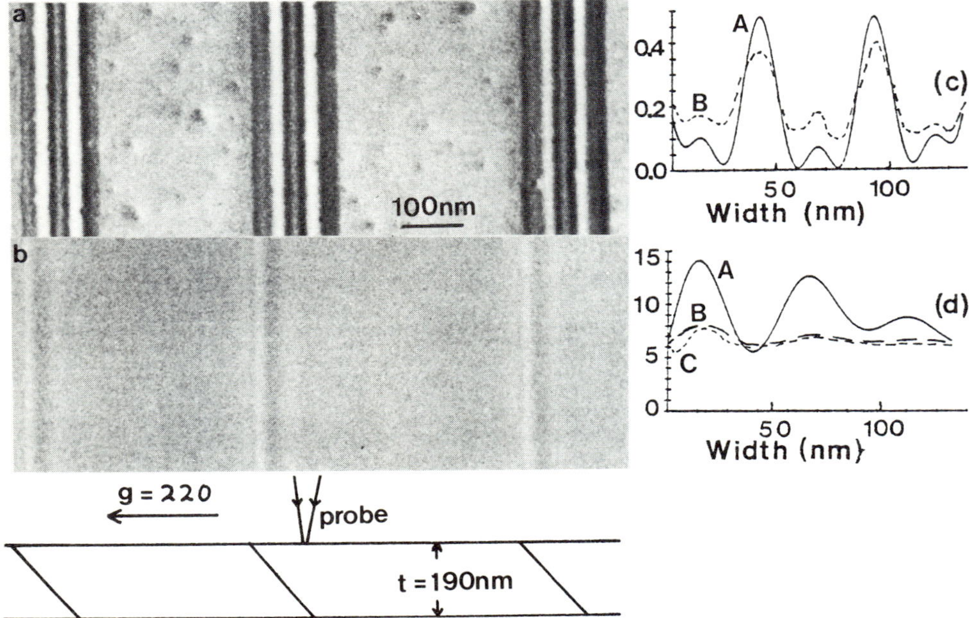

FIGURE 2: STEM images of parallel intrinsic stacking faults in Si (g=220 w=0.1 and thickness t=190nm). (a) STEM bright field image. (b) STEM high angle detector (HAD) image. (c) Comparison of computed and experimental bright field intensity profiles. A....computed; B....experimental. (d) Comparison of computed and experimental HAD intensity profiles. A....computed from eqn 1 ignoring beam convergence. B ...computed from eqn 1 with beam convergence ($\Delta w = \pm 3$). C....experimental.

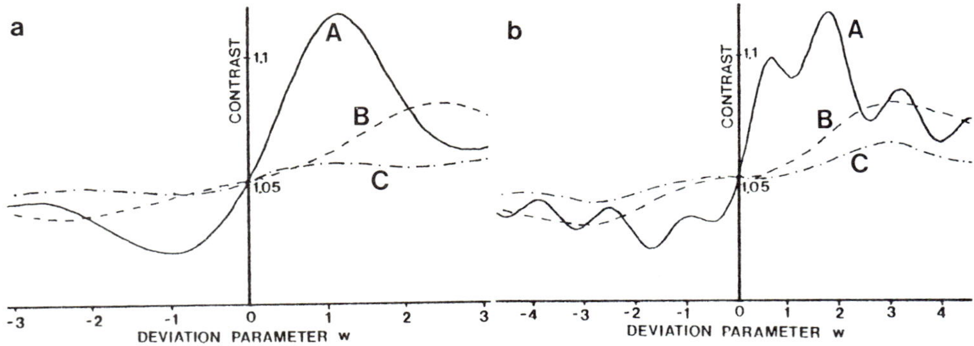

FIGURE 3: Computed contrast in the STEM high angle annular detector for a Au impurity atom in Si as a function of deviation parameter w for the two beam diffracting case. Si thickness is t = ξ_g and the probe size δ = 5nm. (a) Au atom at a depth z = $0.35\xi_g$ (corresponding to X in fig 1). (b) Au atom at a depth z = $0.71\xi_g$ (corresponding to Y in fig 1).
                A....interstitial site, ignoring beam convergence.
                B....interstitial site with beam convergence included.
                C....substitutional site with beam convergence included.

The theory of the emission of channelling radiation by fast electrons

D M Bird

TCM, Cavendish Laboratory, Madingley Road, Cambridge CB3 OHE

1. Introduction

There has been considerable interest recently in the radiation emitted
when highly relativistic particles are channelled through crystals (for
a detailed discussion see the review by Wedell (1980)). In a classical
picture the radiation is due to the periodic transverse motion of the
channelled particles about the atomic strings (axial channelling) or
atomic planes (planar channelling). The observed high frequencies and
forward peaking are due to a Doppler shift and relativistic searchlight
effect. Such a classical theory works well for planar channelled positrons
because the potential they see is approximately harmonic and the frequency
of the transverse motion is therefore independent of its amplitude,
leading to a single strong peak in the frequency spectrum of the observed
radiation (Alguard et al 1979). This classical picture breaks down for
channelled electrons. Swent et al (1979) and Anderson et al (1981)
observed several well defined peaks in the frequency spectrum of the
radiation emitted by highly relativistic planar and axial channelled
electrons. A quantum picture is essential because there are only a few
widely separated tranverse levels bound in the wells of the one or two
dimensional channelling potential. The observed peaks are then due to
"transitions" between these levels (eg Kumakhov and Wedell 1977, Humphreys
1980).

2. Theory

We shall derive the cross section $\dfrac{d^2\sigma}{d\Omega_p\,dE_p}$ for the emission of a photon of
energy between E_p and $E_p + dE_p$ into a solid angle $d\Omega_p$ for the case of a
crystal with negligible absorption. In the following sections we will use
this result to discuss various aspects of the radiation. For simplicity
we shall concentrate on planar channelling, in the language of HEED
this means that we will be considering systematic diffraction. Channelled
electrons tend to be localised on the atomic planes where the potential
wells are deepest.

In general we should determine the initial and final state spinors from
the Dirac equation. However, we have shown (for further details of the
calculation see Bird and Buxton (1981)) that, just as in HEED, we can use
the standard non-relativistic formulae for the transition rate provided
we use relativistically corrected masses and wavevectors. Also, provided
the energy of the emitted photon is considerably less than the electron
rest mass energy then the spin term is negligible, as is usually the
case in allowed atomic transitions (eg Messiah 1961).

If we sum over the photon polarisations the result for the cross section becomes

$$d\sigma = \frac{E_p}{2\pi c^3} \frac{e^2}{4\pi\varepsilon_0} \frac{1}{\gamma^2 m_0^2 v} \; |\hat{q}{\times}\underline{M}|^2 \; d\Omega_p \; \frac{d^3k'}{(2\pi)^3} \qquad\qquad 2.1$$

where v is the electron velocity, γ is the Lorentz factor, k' labels the possible final electron states below the crystal and $\hat{q}$ is a unit vector in the direction of the photon wavevector $\underline{q}$. The matrix element $\underline{M}$ is given by

$$\underline{M} = \int d^3\underline{r} \; e^{-i\underline{q}\cdot\underline{r}} \; \psi_f^* \; \underline{\nabla}\psi_i . \qquad\qquad 2.2$$

where the Schrödinger wavefunctions ψ_i and ψ_f^* match to plane waves $e^{i\underline{k}\cdot\underline{r}}$ and $e^{-i\underline{k}'\cdot\underline{r}}$ above and below the crystal respectively. Using the notation of Berry (1971) these can be written as

$$\psi_i = \sum_j \varepsilon_j \tau_j(x) \; e^{ik_z^j z} \; e^{ik_y y}$$
$$\psi_f^* = \sum_{j'} \varepsilon_{j'}^* \tau_{j'}^*(x) \; e^{-ik_z^{j'}(z-t)} \; e^{-ik_y' y} \qquad\qquad 2.3$$

where the τ's are one dimensional Bloch functions with transverse energy s_j. The most important part of the matrix element 2.2 is the z integral which, assuming for the moment an infinite specimen, gives just $\delta(k_z^{j'}+q_z-k_z^j)$. In a finite specimen this should be a function with spread of order 1/t but provided the crystal is sufficiently thick for there to be little overlap between these functions the Bloch waves can be taken to be independent. This will be a good approximation provided 1/t is less than a typical separation $|k_z^j-k_z^{j'}|$, that is, if the crystal is thicker than the extinction distance.

The delta function above fixes the energy of the photons. If the photon is emitted at an angle θ with respect to the z axis then we find

$$\delta(k_z^{j'}+q_z-k_z^j) = \frac{\hbar c\beta}{1-\beta\cos\theta} \quad \delta\left(E_p - \frac{\hbar^2(s_j-s_{j'})}{2\gamma m_0(1-\beta\cos\theta)}\right) \qquad 2.4$$

where $\beta=v/c$ and we have assumed that the photon energy is much less than the incident electron energy so that both s_j and $s_{j'}$ are evaluated at the same energy. After integrating out the final electron directions the cross section is finally given by

$$d\sigma = \sum_{jj'} \frac{E_p}{2\pi c^3} \frac{e^2}{4\pi\varepsilon_0} \frac{1}{\gamma^2 m_0^2 v} \frac{At}{a} \frac{|\varepsilon_j|^2}{a} \frac{\beta}{1-\beta\cos\theta} \delta\left(E_p \frac{\hbar^2(s_j-s_{j'})}{2\gamma m_0(1-\beta\cos\theta)}\right)$$

$$x \; |M_x^{j'j}|^2 \; \left(\sin^2\phi + \frac{(\cos\theta-\beta)^2\cos^2\phi}{(1-\beta\cos\theta)^2}\right) \sin\theta \, d\theta \, d\phi \, dE_p \qquad 2.5$$

where A is the area and t the thickness of the crystal, a is the size of a unit cell and M_x is given by

$$M_x^{j'j} = \int_{-a/2}^{a/2} dx \; \tau_{j'}^* \; \frac{d\tau_j}{dx} \; e^{-iq_x x} \qquad\qquad 2.6.$$

3. The Spectrum and Angular Distribution

It can be seen from 2.5 and 2.6 that in the dipole approximation (when $e^{-iq_x x}=1$ in 2.6) the radiation is exactly the same as that from a classical dipole, pointing in the x direction, oscillating with a frequency $\hbar(s_j - s_{j'})\dfrac{1}{2\gamma m_0}$ and moving at velocity $v=\beta c$ in the z direction. The $(1-\beta\cos\theta)^{-1}$ factor in the delta function which determines the photon energy is just the usual Doppler shift and the other angular factors can be obtained by applying the relativistic aberration formulae to the angular distribution of dipole radiation in the rest frame (eg Landau and Lifshitz 1975). For highly relativistic electrons the photon distribution is strongly peaked at $\theta=0$ and most of the radiation is emitted into a cone of semi-angle $1/\gamma$ about the z axis. When γ is large the photon energy is also a sensitive function of θ. The maximum energy E_{max} occurs for $\theta=0$;

$$E_{max} = \frac{\hbar^2 \gamma}{m_0}(s_j(k_x)-s_{j'}(k_x-q_x)) \qquad\qquad 3.1$$

and it can be seen that the photon energy is simply related to the difference in the transverse energy levels. This leads to the viewpoint that the radiation is due to "transitions" between these transverse levels and that the radiation is effectively from one dimensional "atoms" moving relativistically.

As well as the explicit dependence on γ in 3.1 the scaling of the peak energies with incident electron energy is also affected by the variation of s_j as the electron mass increases. The effective strength of the systematic potential scales as γ and if we take the potential to be roughly linear near the atomic planes then the transverse energies s scale like $\gamma^{2/3}$. Thus the channelling radiation energies will scale roughly like $\gamma^{5/3}$.

4. The Spectral Linewidth

In the example above of a crystal which is thinner than a typical absorption length there are three main contributions to the observed linewidth. Two of these are extrinsic and could be reduced by improving the experimental conditions. They are the effect of the finite size of the photon collector aperture and the effect of a finite bandwidth of the tranverse levels due to the overlap of the wavefunction in adjacent wells. If γ is large, the photon energy is a sensitive function of angle and if the detector is placed on the z axis and subtends a small angle $2\alpha_c$ at the crystal, the resulting linewidth is given by

$$\Delta E_p(\text{collector aperture}) = \frac{\gamma^2\alpha_c^2}{1+\gamma^2\alpha_c^2}E_{max} \qquad\qquad 4.1$$

which can obviously be reduced by making α_c smaller. If the collimation of the incident beam is perfect and the photon detector aperture is small then equation 3.1 fixes the energies of the peaks and <u>all</u> the transverse levels produce sharp lines, even if the transitions involve unbound states. However, in a real experiment at high energies it will be difficult to avoid a sufficiently large angular spread in the incident beam such that the possible k_x's will fill a Brillouin zone. Thus the whole bandwidth will be seen in the resulting radiation and we find

$$\Delta E_p(\text{bandstructure}) = \frac{h^2\gamma}{m_0} (W_j + W_{j'}) \qquad\qquad 4.2$$

where W_j and $W_{j'}$ are the widths of the bands j and j'. Thus sharp spectral lines will be seen only if both levels involved in a transition are tightly bound. Transitions involving unbound states **will** just contribute to the broad background bremsstrahlung.

In a thin crystal the intrinsic linewidth mechanism is that the delta function which determines the photon energy in 2.5 should really take the form $\frac{\sin^2 x}{x^2}$ with a width of order $1/t$

For a thicker crystal absorption must be considered. What is meant by this is that thermal diffuse scattering (TDS) can cause <u>non</u>- radiative trans-itions between the transverse levels and will also lead to dechannelling by transitions to unbound levels. Thus in 2.6 the excitation coefficients willbecome depth dependent as the TDS (which is a considerably stronger scattering mechanism than channelling radiation) causes a "diffusion" of electrons over the entire dispersion surface. The linewidth mechanisms are also rather different. The bandstructure effect is now intrinsic because the TDS will cause redistribution throughout the Brillouin zone on each level j.For transitions involving only the tightly bound states the linewidth is mainly due to the finite coherence length caused by the TDS. We can imagine that the levels in the one dimensional atoms become broadened into Lorentzians of width $2k\lambda^j$ where λ^j is the absorbative part of the wavevector k_z^j. This leads to a Lorentzian profile for the channell-ing radiation peaks, and for large γ we find

$$\Delta E_p(\text{TDS}) = 2\hbar c\beta\gamma^2 (\lambda^j + \lambda^{j'}) \qquad\qquad 4.3$$

This width will be largest for transitions involving the most tightly bound states because the more localised the state the faster the scattering rate. We might also expect that as the channelling states become more localised as γ is increased, the dechannelling rate will increase and so, in a thick crystal, the ratio of the peaks to the background in the channelling radiation spectrum will decrease.

<u>References</u>

Alguard M J et al 1979 Phys. Rev. Lett. <u>42</u> 1148
Anderson J U et al 1981 Institute of Physics, University of Aarhus preprint
Berry M V 1971 J. Phys. C<u>4</u> 697
Bird D M and Buxton B F 1981 to be published in Proc. Roy. Soc.
Humphreys C J 1980 in Electron Microscopy 1980 Vol<u>4</u> (ed P Brederoo and
  J Van Landuyt) 68
Kumakhov M A and Wedell R 1977 Phys. Stat. Sol. <u>b84</u> 581
Landau L D and Lifshitz E M 1975 The Classical Theory of Fields
  (Pergamon Press: Oxford)
Messiah A 1961 Quantum Mechanics Vol II (North Holland: Amsterdam)
Swent R L et al 1979 Phys. Rev. Lett. <u>43</u> 1723
Wedell R 1980 Phys. Stat. Sol. <u>b99</u> 11

Combined data methods for three-dimensional structures

P. Goodman, A.F. Moodie, H.J. Whitfield, A.J. Morton and C.J. Rossouw

CSIRO Division of Chemical Physics,
P.O. Box 160, Clayton, Victoria, Australia 3168

In this paper we are concerned to show, from present studies, how certain aspects of moderately large unit-cell structures can be studied by combining convergent beam diffraction with lattice imaging, and furthermore to show how future incorporation of accurate beam-rocking facilities might overcome some of the present difficulties.

Until fairly recently, selected area diffraction (SAED) has been the main adjunct to high resolution lattice imaging. This can be understood since this latter technique was established initially with the study of defect oxide structures having one short (3.8 Å) axis, giving readily interpretable projections, and SAED provides all the necessary diffraction data.

In the meantime, convergent beam diffraction was developed as a separate technique, and was found to be most powerful in long-axis projections from otherwise simple structures. In other words, the unit cell requirements for these two techniques were inversely related. This was because on the one hand we can obtain high resolution images (HRI) in x-y space from a projection with a slight angular tolerance on an ideally parallel (plane wave) illumination, and on the other we are looking with convergent (partially spherical wave) illumination at a θ-ϕ view of a structure, recorded on an x-y plane.

Recently we have been asked by chemists to examine 3-dimensional structures with no short axis. In these cases we have found it extremely valuable to examine the HRI and CBED data jointly, so that very frequently the deficiency of the one are overcome by the strengths of the other. Ideally, these two sets of data should be taken in the one instrument, but where this is not possible, sequential examination has been used.

At first the structure $BaNd_2Ti_3O_6$, which actually does have interpretable projections as well as the ready-cleavage long-axis projection (7.6 Å x 28.2 Å x 7.6 Å), was examined. Here an X-ray determined average structure was refined in stages, first with HRI/SAED data from different projections (Olsen and Roth, 1981); it was then space-group refined using CBED. Finally the HR image from the long-axis projection could be examined and interpreted in terms of deviations from the exact zone-axis projection which arise over the micrograph field of a slightly curved crystal.

The next example was a truly 3-dimensional (non-layer) structure with a substituted wurzite lattice: CuAsSe (11.75 Å x 6.79 Å x 19.21 Å)

(Whitfield, 1981). Here many problems typical of 3-D studies were encountered. Two 'structural variants' were found to be different cleavages/projections from a single structure. The slight curvatures which occur even in a relatively hard substance and which cause difficulties initially in both HRI and CBED could finally be exploited to provide a 'stereoscopic' view of the 3-D structure. The key step in this analysis lay in the identification of a glide plane by means of the extinction (Gjønnes-Moodie) band in CBED, and observing its effect in the HRI. Here also the problem of overlapping CBED discs was encountered. By good fortune in this case the space group-forbidden reflexions ($00\ell = 2n + 1$) could be examined in some detail since the intervening 002, 004, reflexions were allowed but extremely weak. The zone axis pattern and curved crystal image are shown in Figure 1; in this case the projection axis is 11.75 $\overset{\circ}{A}$.

As a third example, a particular anti-phase-domain structure of γ-brass was examined in order to determine some of its symmetry elements, and in particular the question of centro-symmetry. For the CB work principle projections not containing the long (70 $\overset{\circ}{A}$) superlattice axis were chosen. After obtaining a [111] axis projection (Figure 3) a centre of symmetry for the structure was confirmed by obtaining equivalent $hkl/\overline{hkl}$ settings within that zone pattern, and exploring over an angular region by using the dark-field tilt controls provided with the 200CX microscope. Alternatively, a wider angle could be explored using an overlapping disc pattern in which the $hkl/\overline{hkl}$ pair were simultaneous excited, and in which there was sufficiently sharp detail to be visible through the overlap (Goodman and Morton, 1981). The obvious improvement over both manual magnetic scanning and overlapping discs is to use the technique of beam-rocking electron diffraction (BRED). Our experience indicates that such a system would overcome many of our present problems if used in conjunction with, for example, a desaturated LaB_6 source, which would allow a sufficiently small region to be illuminated to avoid problems of crystal curvature.

Finally, some of the possibilities opened up by the beam-rocking technique of Eades (1980) have been examined. In the immediate context, it permits an alternative means for direct recording of convergent beam data, with apparently less upheaval to the optical requirements of a high resolution microscope. In addition, the coupling of BRED with ELS (electron less spectroscopy) appears to allow the possibility of analytical CBED. Previously existing theory was found to be too primitive to be predictive but recent advances (Rossouw, 1981) have largely changed that situation. Because of the fact that, in this technique, we are able to adhere to approximately zero-angle scattering, we can overcome much of the averaging involved in Kikuchi line analysis, and approach the ideal situation (for a limited region of the periodic table) where we can use the inelastic process to provide a localized wave-field, and the following N-beam interaction to provide structural phase information. At the present there are some simple tests to be made to see how practical this will be.

<u>Ordering References</u>

1.   Eades, J.A. (1980) Inst. Phys. Conf. No. 52, 9-12.
2.   Goodman, P. and Morton, A.J. (1981) Submitted to Acta Cryst.
3.   Olsen, A. and Roth, R.S. (1981) J. Solid State Chem. (in press).
4.   Rossouw, C.J. (1981) Ultramicroscopy (in press).
5.   Whitfield, H.J. (1981) J. Solid State Chem. (in press).

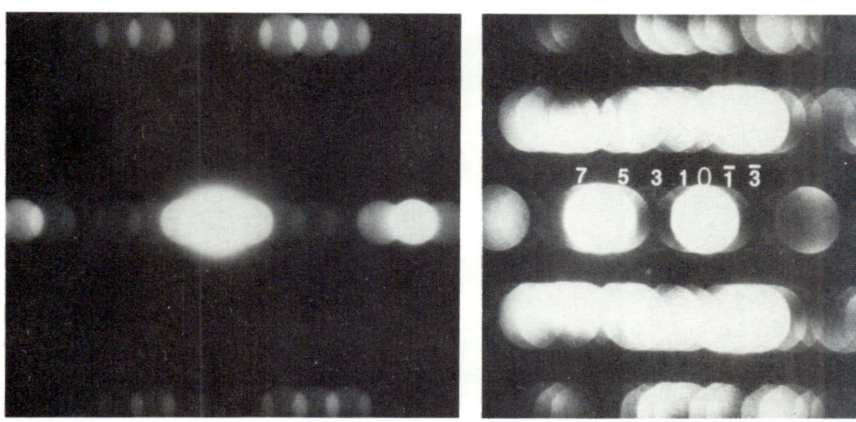

Figure 1 (a). SAED pattern from the [100] zone axis, showing strong 001, 005, 007 reflexions, contributed to by crystal curvature.

   (b). CBED pattern from near the same setting, but exactly exciting the 006 reflexion. Here the extinction bands are visible in the 00ℓ, ℓ = 2n + 1 reflexions, and the angular tolerance required to maintain this extinction is displayed.

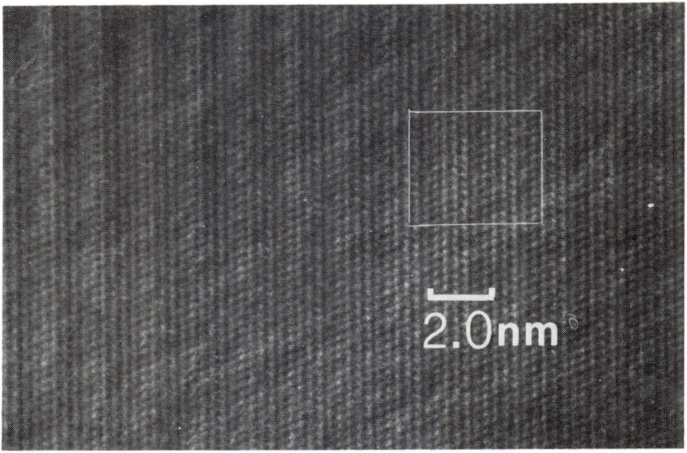

Figure 2. Micrograph from near the same setting as above. Boxed area shows the actual symmetry of the zone; variations over the field are due to crystal curvatures.

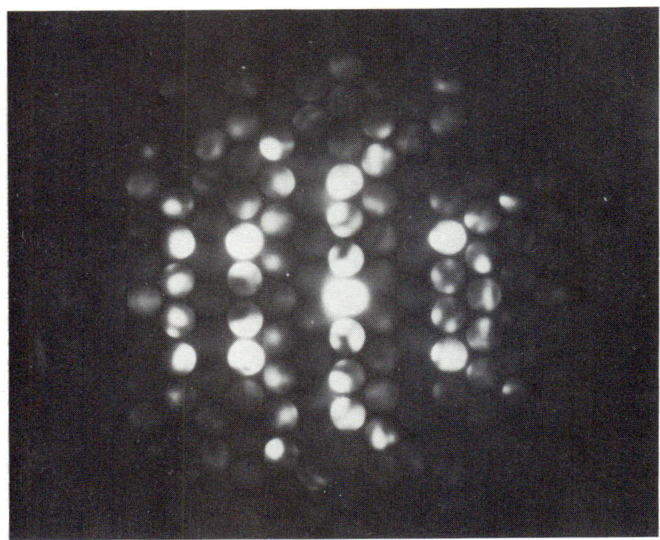

Figure 3. Zone axis CBED pattern taken at a [111]-type zone of γ-brass (in-situ: 200CX) showing departure from hexagonal symmetry.

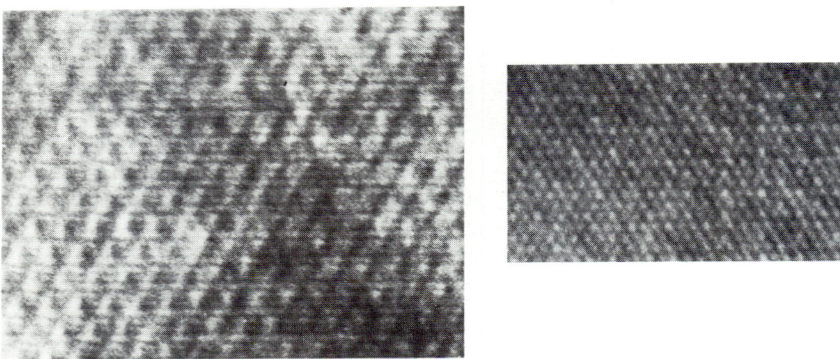

Figure 4. (a) Micrograph taken from the same projection as above. By using a large incidence convergence and a thick region the unit cell dimension is seen in many parts of the field, with non-hexagonal symmetry. Also, pseudo superlattice effects appear in some areas.

(b) A micrograph taken with less incident convergence and from a thinner region shows mainly the simple sub-lattice.

Diffraction symmetries for elastic scattering

R Portier and D Gratias

C.E.C.M./C.N.R.S. 15 rue Georges Urbain, 94400 VITRY, France

Scattering experiments consist in measuring the transition probability
for an initial state, say $|i>$, of the incident wave at time zero to reach
a final state, say $|f>$ at time t after interaction with a specified scat-
tering potential. This transition probability $\omega_{i \to f}$ is formally given by:

$$\omega_{i \to f} = |\ <f|\ U\ (t)\ |i>\ |^2 \tag{1}$$

where $U(t)$ is the evolution operator of the system defined by

$$U(t) = e^{-i \frac{H}{\hbar} t} \tag{2}$$

for a time independent system with Hamiltonian H.
Certain symmetries for the diffraction arise from the fact that the scalar
product defined in eq.(1) is invariant through unitary and antiunitary
transformations. Such invarieance is written, for a unitary transformation S as:

$$<f|\ U(t)\ |i> \ = <Sf|\ S\ U(t)\ S^+\ |Si> \tag{3}$$

and for an antiunitary transformation I as :

$$<f|\ U(t)\ |i> \ = <Ii|\ I\ U(t)^+ I^+\ |If> \tag{4}$$

where $^+$ designates the hermitian adjoint of the operator.

Special cases of (3) and (4) are, respectively, the space isotropy and
the time reversal symmetry usually encountered in quantum mechanics.

For the case of a monochromatic incident beam, both initial and asymptotic
final states are plane waves

$$|i> \, = e^{i\ \underset{\sim}{k}_i \cdot \underset{\sim}{r}} = |k_i> \quad (5) \qquad\qquad |f> \ = e^{i\ \underset{\sim}{k}_f \cdot \underset{\sim}{r}} = |k_f> \quad (6)$$

and the application of any $S = (\alpha/\underset{\sim}{\tau})$ space isometry transforms eq.(3) into

$$<k_f|\ U(t)\ |k_i> \ = <\alpha^{-1}\ k_f|S\ U(t)\ S^+|\ \alpha^{-1}k_i> \ e^{i(\underset{\sim}{k}_i - \underset{\sim}{k}_f) \cdot \underset{\sim}{\tau}} \tag{7}$$

while the application of the time reversal operator I transforms eq.(4) into

$$<k_f|\ U(t)\ |k_i> \ = <-\ k_i|\ IU(t)^+\ I^+\ |-k_f> \tag{8}$$

These two invariants (7) and (8) are sufficient for determining the symmetry properties of diffraction in crystals. For instance, if both S and I
commute with the Hamiltonian (i.e. belong to the symmetry group of the
crystal), then:

$$\overline{U}(t) \equiv S\ U(t)\ S^+ \equiv I\ U(t)^+\ I^+ \tag{9}$$

showing the equivalence between the three diffraction situations sketched:

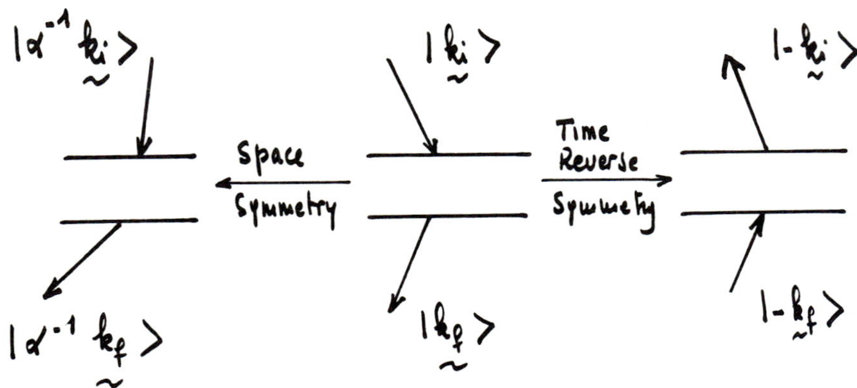

1. The extinction conditions

If we suppose now that there exists a $(\alpha|\tilde{\tau})$ symmetry operation of the crystal such that :

$$\begin{cases} \tilde{k_i} = \alpha^{-1}\tilde{k_i} \\ \tilde{k_f} = \alpha^{-1}\tilde{k_f} \end{cases} \quad \text{or} \quad \begin{cases} -\tilde{k_i} = \alpha^{-1}\tilde{k_f} \\ -\tilde{k_f} = \alpha^{-1}\tilde{k_i} \end{cases}$$

then from (7) and (8) we obtain the following relation :

$$\langle k_f | U(t) | k_i \rangle = \langle k_f | U(t) | k_i \rangle \, e^{i(\tilde{k_i}-\tilde{k_f})\tilde{\tau}} \tag{10}$$

by designating by $\tilde{h}$ the diffracting vector

$$\tilde{h} = \tilde{k_f} - \tilde{k_i} \tag{11}$$

and $\tilde{K}$ the vector,

$$\tilde{K} = \tilde{k_f} + \tilde{k_i} \tag{12}$$

we reach the conclusion that the transition probability from $|i\rangle$ to $|f\rangle$ is zero if there exists a space symmetric element $(\alpha|\tilde{\tau})$ such that

$$\text{(a)} \begin{cases} \alpha\tilde{h} = \tilde{h} \\ \tilde{h}\cdot\tilde{\tau} \neq 2\pi n \ (n \text{ integer}) \end{cases} \tag{13}$$

$$\text{(b)} \quad \alpha\tilde{K} = \pm\tilde{K}$$

Conditions (13a) are the usual kinemanical conditions for the h-reflexion to be zero. The dynamical absences require the additional relation (13b). The only possible α point operations which may satisfy conditions (13) are :

  i) the mirror plane containing $\tilde{K}$ and $\tilde{h}$

 ii) a binary axis containing $\tilde{h}$

iii) a mirror passing through $\tilde{h}$ and perpendicular to $\tilde{K}$

Case i) is satisfied if the projection of the Ewald sphere on the diffracting plane is along the $\tilde{h}$ direction (A-locus).
Case ii) is satisfied if the projection of the Ewald sphere is along the perpendicular bisector of $\tilde{h}$ (B-locus) and case iii) only at Γ point, intersection of A and B.

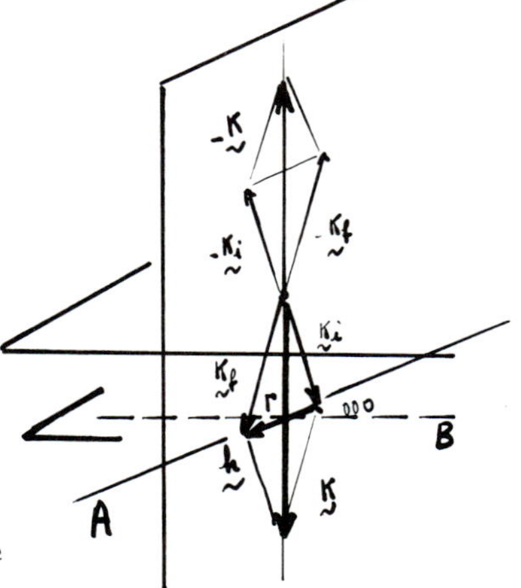

2. The reciprocity theorem for small angle scattering

Relation (8) is the exact application of the time reversal symmetry (T.R.S.) for a quantum mechanical system. However the so-called reciprocity theorem used in electron microscopy is not a direct consequence of the T.R.S.: assuming that relation (9) is satisfied (the Hamiltonian is real) we see that the T.R.S. involves comparisons between $-k_i$ and $\alpha^{-1}k_f$ (also $-k_f$ and $\alpha^{-1}k_i$. In practice, the symmetry information we can obtain is due to the fact that the vectors $\alpha^{-1}k_f$ and $\alpha^{-1}k_i$ may be superimposed, after some geometric manipulation with either k_f and k_i (space symmetry) or $-k_i$ and $-k_f$ (T.R.S.). In this latter case, the length of $\alpha^{-1}k_f$ and $-k_i$ have to be identical: the condition $\alpha K = -K$ implies the h-reflexion to be at Bragg reflexion (microreversibility principle). T.R.S. principle never relates reflexions with non zero excitation error. The additional hypothesis which allows us to relate such reflexions is the small angle scattering approximation which essentially consists in transforming the 3-dim stationary collision problem into a 2-dim "time" dependent problem where the "time parameter" is the z direction of propagation of the electron flux (D Gratias and R Portier 1981). The T.R.S. consists in inverting the z parameter into $-z$. The new involved symmetry group of the crystal is no longer the actual 3-dim space group but a 2-dim two colour space group (B F Buxton et al. 1976): the S operators are the symmetry operators of the 3-dim space group which leave the z direction invariant whereas the I antiunitary operators are the operators of the space group which transform z into $-z$. For relation (9) to be satisfied, the condition that the potential has to be real is not sufficient: we must further have in the 3-dim space group at least one antiunitary operator. The new S and I operators may be written as

$$S \;=\; (\alpha|\underset{\sim}{\tau}) \quad ; \quad I \;=\; (\alpha'|\underset{\sim}{\tau'}) \tag{14}$$

where α, α' are 2-dim point symmetry operators and τ, τ' their associated translations in the (x,y) plane (the z translation symmetry of the infinite crystal has been lost). Formulae (3) and (4) transform into

$$\langle\underset{\sim}{\jmath}_f|U(z)|\underset{\sim}{\jmath}_i\rangle \;=\; \langle\alpha^{-1}\underset{\sim}{\jmath}_f|SU(z)\,S^+|\alpha^{-1}\underset{\sim}{\jmath}_i\rangle\, e^{\,i(\underset{\sim}{\jmath}_i-\underset{\sim}{\jmath}_f)\underset{\sim}{\tau}} \tag{15}$$

$$\langle\underset{\sim}{\jmath}_f|U(z)|\underset{\sim}{\jmath}_i\rangle \;=\; \langle-\alpha'^{-1}\underset{\sim}{\jmath}_i|\,IU^+(z)\,I^+|-\alpha'^{-1}\underset{\sim}{\jmath}_f\rangle\, e^{\,-i(\underset{\sim}{\jmath}_i-\underset{\sim}{\jmath}_f)\underset{\sim}{\tau'}} \tag{16}$$

where $\underset{\sim}{\jmath}_f$ and $\underset{\sim}{\jmath}_i$ are the projections on the (xy) plane of respectively $\underset{\sim}{k}_f$ and $\underset{\sim}{k}_i$. The extinction conditions obtained for small angle scattering may be derived in the same way as in the previous section ; for unitary operations one obtains

$$\alpha\underset{\sim}{h} \;=\; \underset{\sim}{h} \qquad \alpha\underset{\sim}{\jmath} = \underset{\sim}{\jmath} \qquad \underset{\vee\sim}{h\tau} \;\neq\; 2\pi n \tag{17}$$

and for antiunitary operations

$$\alpha'\underset{\sim}{h} \;=\; \underset{\sim}{h} \qquad \alpha'\underset{\sim}{\jmath} = -\underset{\sim}{\jmath} \qquad \underset{\sim}{h\tau'} \;\neq\; 2\pi n \tag{18}$$

where

$$\underset{\sim}{h} = \underset{\sim}{\jmath}_f - \underset{\sim}{\jmath}_i \qquad \text{and} \qquad \underset{\sim}{\jmath} = \underset{\sim}{\jmath}_f + \underset{\sim}{\jmath}_i$$

These conditions lead to the same geometrical results as in § 1 and has been previously derived by Gjønnes and Moodie (1965) by analysing the possible invariants of the general perturbation term in small angle scattering diffraction (Optical formulation. Cowley J. and Moodie A.F. 1957). The reciprocity theorem also applies for reflexion out of the Bragg position. Let us assume that the structure shows a pure mirror perpendicular to z. With respect to x and y the I operator is simply the unity and (16) becomes

$$\langle\underset{\sim}{\jmath}_f|U(z)|\underset{\sim}{\jmath}_i\rangle \;=\; \langle-\underset{\sim}{\jmath}_i|U(z)|-\underset{\sim}{\jmath}_f\rangle$$

Putting $\underset{\sim}{\jmath}_i = \underset{\sim}{Q} - \tfrac{1}{2}\underset{\sim}{h}$

$$\underset{\sim}{\jmath}_f = \underset{\sim}{Q} - \tfrac{1}{2}\underset{\sim}{h} \tag{19}$$

we obtain :

$$\langle \underset{\sim}{Q} + \tfrac{1}{2}\,\underset{\sim}{h}|U(z)|\,\underset{\sim}{Q} - \tfrac{1}{2}\,\underset{\sim}{h}\rangle = \langle -\underset{\sim}{Q} + \tfrac{1}{2}\,\underset{\sim}{h}\,|U(z)|\,-\underset{\sim}{Q} - \tfrac{1}{2}\,\underset{\sim}{h}\rangle \qquad (20)$$

showing that the +Q situation is equivalent to the -Q situation
(Pogany and Turner 1968).

A further approximation consists in averaging the z-dependence of the potential by the projected potential (zero layer approximation). Such an
assumption results in adding to any type of structure a mirror perpendicular to z in such a way that the previous equivalence +Q, -Q is always
satisfied. The extinction conditions then become

$$\alpha\underset{\sim}{h} = \underset{\sim}{h} \qquad \alpha\underset{\sim}{k} = \pm\,\underset{\sim}{k} \qquad \underset{\wedge\sim}{h\tau} \neq 2\pi n \qquad (21)$$

where $(\alpha|\tau)$ are unitary or antiunitary operations. As a result, a binary
axis along h or a mirror containing h and z are not differentiated (same
projection!). Therefore, the geometrical loci of dynamical absences are
both A and B. For the mirror perpendicular to z. the dynamical absence
condition reduces to the kinematical condition (no condition on   ).

3. <u>Conclusion</u>

It is shown through general symmetry arguments that for the case of diffraction of a monochromatic plane wave, simple geometrical rules lead to
the sufficient conditions (13) for a transition $|i\rangle \to |f\rangle$ to be forbidden.
These general rules are:
 i) invariance of the h reflexion through a space symmetry operator α such
that its associated translation τ introduces a phase change different from
$2\pi n$. (Usual condition for the structure factor to be zero.)
ii) the sum of the incident and the diffracted wave vectors is invariant
(space symmetry) or inverted (T.R.S.) for the case of a real Hamiltonian
by the space operation considered in i).

The condition ii) is always satisfied for reflexions in exact Bragg position and α point operator of order 2 ($\alpha^2 = 1$). This condition is therefore
fulfilled in X-ray diffraction whereas it imposes precise orientations of
the crystal to be satisfied in electron diffraction.

In a second part, the physical signification of the reciprocity theorem,
as used in electron microscopy has been discussed: the application of the
reciprocity theorem for reflexions out of the Bragg position is valid as
far as the small angle scattering approximation is correct. *Sensu stricto*
this theorem applies on the z coordinate and not on time.

The derivation presented here was limited to the case of extinction of
reflexions. However the general formulation (7) and (8) have other applications and are of a very convenient use for any type of problem involving
symmetry considerations. This formulation does not require any explicit
development of the diffraction theory. It only requires the determination
of the symmetry group of the evolution operator U(t).

<u>References</u>

Buxton B.F., Eades J.A., Steeds J.W. and Rackham G.M. (1976) Phil. Trans.
   Roy. Soc. <u>281</u> 171
Cowley J.M. and Moodie A.F. (1957) Proc. Phys. Soc. B<u>70</u>. 486, 497, 505
Gjønnes J. and Moodie A.F. (1965) Acta Cryst. <u>19</u> 65
Gratias D. and Portier R. (1981) in this volume
Pogany A.P. and Turner P.S. (1968) Acta Cryst. A <u>24</u> 103

The orientation dependence of anomalous absorption in electron diffraction

G R Anstis

Department of Metallurgy and Science of Materials, University of Oxford,
Parks Road, Oxford OX1 3PH

1.   Introduction

The importance of thermal vibrations in a crystal lattice in explaining
anomalous absorption has been demonstrated by Yoshioka and Kainuma (1962)
and by Whelan (1965). These authors noted that absorption should depend
strongly on orientation and this has been demonstrated by Goodman (1971)
who showed by convergent beam methods that when the 200, 020 and 220
reflections of MgO are simultaneously excited at a temperature of 450°C
absorption parameters are two to three times the value for single 200
excitation.

It is the purpose of this note to show how such an increase can be calculated. Goodman points out that the concentration of thermal diffuse
scattering around Bragg reflections, as predicted by Debye's model, is
important. An additional factor is the range of correlations of atomic
vibrations in relation to the region of coherence of the electron wave.

A modification to the calculation of Yoshioka et al. which allows for the
singularity introduced by Debye's model is first presented and it is shown
that the dependence of absorption on orientation is small. It is then
argued that for long wavelength phonons it is appropriate to consider how
the Bragg intensities are modified by a single phonon. Such a model cannot
be treated easily analytically but preliminary calculations point to it
providing the desired dependence on the direction of the incident beam.

2.   Scattering by many phonons

With a deformable ion model a phonon of wavevector q leads to a Fourier
component (Whelan, 1965).

$$V_{g+q} = 2\pi (MN)^{-\frac{1}{2}} \sum_{p=1}^{3} a_{qp} \, \ell_{qp} \cdot g \, V_g \tag{1}$$

where ℓ_{qp} is a unit vector in the pth direction of polarization, M is the
mass of an atom and N is the number of atoms. In Debye's model at high
temperatures

$$a_{qp} = (K_B T/2)^{\frac{1}{2}} \frac{1}{vq} \tag{2}$$

where v is the speed of sound. When there are many phonons, each with an
associated random phase factor, in the region of coherence of the electron
beam the dynamical equations for Bragg reflections are

$$\frac{d\phi_g}{dz} = 2\pi i \ S_g \phi_g(z) + i\sigma \sum_h V_h \phi_{g-h}(z)$$

$$- \sigma^2 \sum_h \left(\int_0^z dz' \ \sum_{k,q} e^{2\pi i S_{k+q}(z-z')} V_{h-k-q} V_{k+q} \right) \phi_{g-h}(z') \qquad (3)$$

where S_k is the excitation error and σ is the interaction constant. If $V_{h-k-q} \ V_{k+q}$ varies slowly with q the usual expression for the absorption potential results

$$V_h' = \tfrac{1}{2} \ \sigma \sum_{k,Q} V_{h-k-Q} V_{k+Q} \qquad (4)$$

where the sum is over wavevectors Q for which $S_{k+Q} = 0$.

If the Ewald sphere passes close to a Bragg reflection this analysis is not applicable. Eq.(3) can be analysed by substituting into the last term an expression for $\phi_{g-h}(z')$ obtained from the equations for a perfect crystal, evaluating the integral and expressing the result in terms of $\phi_g(z)$. This leads to real and imaginary corrections to the Fourier coefficients V_h. Consider the contribution from $V_{h-k-q} \ V_{k+q}$ when $S_{k+q} \simeq 0$. Writing q in terms of Q, its component parallel to the surface of the Ewald sphere, and q' we must evaluate

$$\tfrac{1}{2} \ \sigma \ \overline{u^2} \ (4\pi^2/q_m) \left| h-k \right| V_{h-k} \left| k \right| V_k \ \cos \phi$$

$$\times \int_0^z dz' 2\pi \int_0^{q_m} dQ \ Q \int_{-q_m}^{q_m} dq' \ e^{2\pi i (S_{k+q'})} \frac{1}{Q^2+q'^2} \sum_i a_i e^{i\lambda_i(z-z')} \qquad (5)$$

where $\overline{u^2} = K_B T/Mv^2 q_m^2$, ϕ is the angle between k and h-k and $\phi_{g-h}(z-z')$ is expressed in terms of the eigenvalues λ_i obtained from the equations for a perfect crystal. Evaluating the integrals we obtain contributions to V_h' of the form

$$\tfrac{1}{2} \ \sigma \ \overline{u^2}(4\pi^2/q_m) \left| h-k \right| V_{h-k} \left| k \right| V_k \ \cos \phi (a_i/2\pi) \ln[1+4\pi^2 q_m^2/(2\pi S_k - \lambda_i)^2] \qquad (6)$$

which are of the same order of magnitude as contributions calculated by Einstein's model of a vibrating lattice. This lack of dependency on the distribution of diffuse scattering is due to the large number of independent phonons scattering the coherent incident wave. A similar conclusion was reached by Doyle (1969) in calculations of the distribution of diffuse scattering.

There is also a real correction to the potential

$$\tfrac{1}{2} \ \sigma \ \overline{u^2}(4\pi^2/q_m) \left| h-k \right| V_{h-k} \left| k \right| V_k \ \cos \phi \ (a_i/\pi) \ \tan^{-1} [2\pi q_m/(2\pi S_k - \lambda_i)] \qquad (7)$$

which is of the order of 1% of the coefficients V_h.

3.   A single phonon model

The lateral coherence of the electron beam is determined by the angle subtended by the condenser and in a convergent beam experiment may be of the order of 100Å while the degree of coherence in the direction of propagation may be of the order of 1μm (e.g. Cowley, 1975). Thus in calculating the effects of phonons of long wavelength it may be appropriate to assume there are few phonons effective in the region of coherence. Consider the limiting case of one phonon. To model the situation of Goodman's experiment consider reflections g = 200, h = 020 and the diffuse reflection g + q. Asymmetry of the 000 convergent beam disc, which provides a measure of $V_h{}'$, arises from terms such as $V_{g+q}V_{h-g-q}V_{-h}$. The intensity variation in the disc is obtained by summing <u>intensities</u> for all values of q rather than amplitudes as in the theory described previously.

A many-beam calculation should be performed to calculate the variation in intensity but some indication of the amount of asymmetry can be obtained by taking S_g and S_h large, in which case there is a contribution to the asymmetry in the central disc of

$$\frac{1}{\pi S_h} \sigma^3 V_{g+q} V_{h-g-q} V_{-h} z \frac{\sin[2(\pi^2 S_{g+q}^2 + \sigma^2 V_{g+q}^2)^{\frac{1}{2}} z]}{\pi^2 S_{g+q}^2 + \sigma^2 V_{g+q}^2} \qquad (8)$$

provided $S_{g+q} \ll 1$. After summing over all phonons the resultant expression can be compared with that obtained from two beam theory with absorption (Cowley, 1975)

$$(2/\pi S_h) \sigma^3 V_{-h} z \sum_Q V_{g+Q} V_{h-g-Q} \qquad (9)$$

The modified theory appears to predict the greater amount of absorption although confirmation by numerical calculations is required.

<u>References</u>

Cowley J M 1975 "Diffraction Physics" North Holland
Doyle P A 1969 Acta Cryst. <u>A25</u> 569
Goodman P 1971 Acta Cryst. <u>A27</u> 140
Whelan M J 1965 J.Appl.Phys. <u>36</u> 2103
Yoshioka H and Kainuma Y 1962 J.Phys.Soc.Japan <u>17</u> Suppl.B-II, 134

Structure factor information from HOLZ beam intensities in convergent-beam HEED

J R Baker, S McKernan
H H Wills Physics Laboratory, University of Bristol, Royal Fort,
Tyndall Avenue, Bristol BS8 1TL

1. Introduction

Structure determination by electron diffraction requires the use of thin
crystals to reduce the effect of the strong interaction between the elec-
-tron and the crystal. For crystal thicknesses over a few hundred
Angstrom, diffraction patterns show the usual dynamic behaviour which
breaks the linear relation between intensity and thickness. A useful
tool in analysing dynamical effects has been the presence of higher-
order Laue zone (HOLZ) reflections. It has been shown by Jones et al
(1977) that the position of the HOLZ lines in a zone-axis pattern is essen-
-tially a map of the zero-layer (zl) dispersion surface generated in the
projection approximation. It has also been noted that the intensities of
the HOLZ reflections seem to bear a direct relationship to the structure
factor for that beam, for example, in the layer structures studied by
Fung (1979). It is this last point we look at.

2. Theory

The theory applicable to the behaviour of the HOLZ beams is given by
Buxton (1976). This allows a set of HOLZ beams to interact with an unper-
-turbed wave state which is made up of zero-layer reflections only
(projection approximation). The eigenvalues of the zero-layer state and
HOLZ state may be equal at certain points, so that degenerate perturbation
theory is required. The interaction between the HOLZ and zero-layer is
only significant in small hybridizing regions, where it behaves locally
as a quasi-two-beam interaction (Jones et al 1977). In contrast to the
usual two-beam theory, the unperturbed waves involved are no longer plane
waves. In all other respects, the analysis is that for the two-beam
theory. At the Bragg position,

$$\zeta_\beta = 2\pi / \left| \beta_j \right| \quad \text{where} \quad \beta_j = \left\langle \tau_H(\mathbf{R}) e^{2\pi i g_z z} \left| U_H \right| \tau_{jo} \right\rangle \tag{1}$$

The intensity in the HOLZ beam is

$$\frac{\left| \epsilon_j \right|^2 \left| C_g \right|^2 \left| \beta_j \right|^2 \sin^2 \left(\tfrac{1}{2} t \sqrt{k^2 + \left| \beta_j \right|^2} \right)}{\kappa^2 + \left| \beta_j \right|^2} \tag{2}$$

which is given by Buxton (1976). This does not include absorption.

For thicknesses much smaller than the extinction distance, we have the
kinematic result integrated over the intersection region.

0305-2346/82/0061-0283$01.50   © 1982 The Institute of Physics

$$I_g^j \quad \alpha |\beta_j| (|\varepsilon_j|^2 |C_g|^2) t \tag{3}$$

For zone-axes in which g_z is large, the interactions between the layers are weak, and substantial effects only occur in very narrow regions (HOLZ lines). The extinction distance is typically over 5000Å, so for many samples, we have a chance of getting kinematic information as described by Eqn 3.

The matrix element β_j is expanded in real space below

$$\beta_j = \int_{\underline{R} \text{ in cell}} \tau_{jo}(\underline{R}) \tau_H^*(\underline{R}) \left[\int_z U_H(\underline{r}) e^{-2\pi i\, g_z z} dz \right] d^2R \tag{4}$$

We define

$$U_H^{pp}(\underline{R}) = \int_z U_H(\underline{r}) e^{-2\pi i g_z z} dz \tag{5}$$

which is the projected potential of the g_z Fourier coefficient of the full crystal potential. In contrast to the usual projected potential ($g_z=0$), this projected potential may contain phase information, and some wells may not project at all.

τ_H can be close to a plane wave $e^{2\pi i\, \underline{G}.\underline{R}}$ in cases where there are weak interactions between the HOLZ beams. The integral in Eqn.4 will then simply pick out the $\underline{G}$ Fourier component of the product $\tau_{jo}U_H^{pp}$. If τ_{jo} is much smoother than the parts of the projected potential which produce the large HOLZ Fourier component $\underline{G}$ (which comes from the atomic cores), there the integral is proportional to U_g. To illustrate the character of τ_{jo} in real space, it is helpful to consider the restricted case of a string potential (studied by Buxton et al 1978). The zero-layer Bloch waves τ_{jo} may then be characterized in terms of atomic and molecular wavefunctions. τ_{jo} may vary significantly near the atomic cores, so the matrix element β_j will depend on the type of state of the branch j. Below we suggest how β_j may differ from U_g. Strongly bound states are not very orientation dependent, and sit in the deepest wells in the projected potential (and are strongly absorbed for this reason). Less bound states change character and excitation with orientation. We expect the first few branches to reflect a structure in which the heavier atoms are picked out preferentially. The higher branches give information about the nature of the wave states τ_{jo} as well as the potential.

In summary, the intensity of our HOLZ beam is kinematically proportional to $|\beta_j|\,|\varepsilon_j|^2$. By voltage variation, we should be able to map this quan--tity out in the Brillouin Zone for each branch. In cases where $|\varepsilon_j|$ is not strongly varying, we can obtain $|\beta_j|$ around the HOLZ ring, and hence have information about the function $\tau_{jo}\, U_H^{pp}$.

3. Experiment and calculation

Some experimental examples and many-beam computed data have been interpreted using this theory. Fig. 1 shows a portion of the first HOLZ ring in Si $<111>$. A computed section of the dispersion surface with relevant excitations is given in Fig.2 (following examples in Steeds 1980), which matches well to the picture.

Fig 1 (opposite)
Section of first HOLZ ring
in Si $<111>$ at 100 keV
(a) interaction with adjacent
    spots
(b) loss at Kikuchi lines
(c) enhanced diffuse scatter

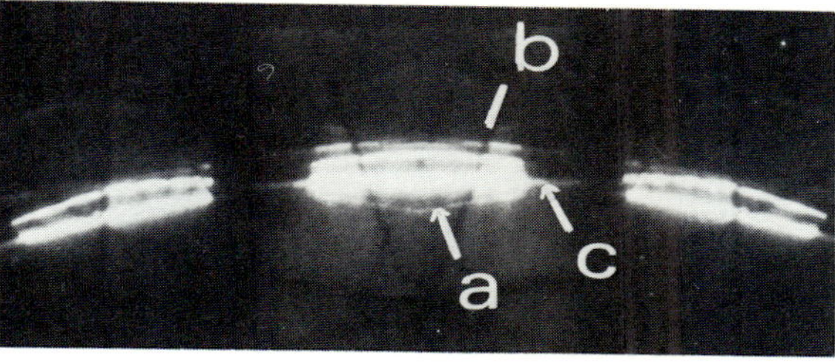

$\bar{3}\bar{7}11$ $\bar{5}\bar{5}11$ $\bar{7}\bar{3}11$

Fig 2.(opposite) Si $<111>$ 100 keV
computed dispersion surface section perpen-
-dicular to the $\bar{5}\bar{5}11$ beam. It is plotted
on the same plan as Fig 1. The thickness
of a branch is proportion to its excitation.
Dotted lines are essentially unexcited.

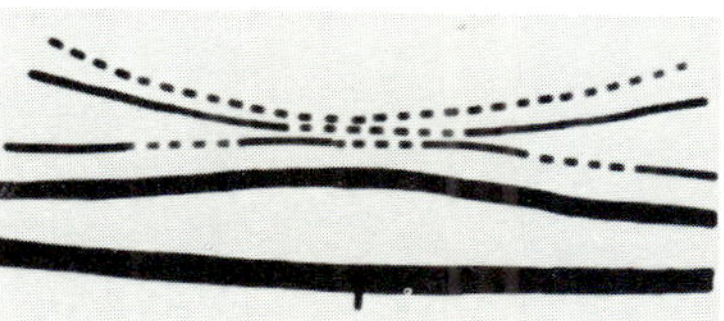

Extinction distances have been calculated in Si across the zero-layer
branch (1) and branch (2) intersections. Two beam curves have been fitted
to these, and show excellent matching. Intensity profiles also fit well
to the expected two-beam form.

The extinction distance calculations show that $|\beta_1|/|U_g|=3.28$ and
$|\beta_2|/|U_g|=2.67$. Explicit evaluation of β_j assuming τ_H to be a plane wave
support this and also shows that β varies significantly with orientation
for the higher branches.

In MgO, zone-axis patterns on $<100>$ and $<110>$ have been studied.
On $<100>$, we have a simple square array of potential. The HOLZ lines
show little variation around the ring. This is expected, as the excited
Bloch waves will be simple s-states occupying the wells equally. On
$<110>$, the reflections are expected to alternate in strength, as there
are two different strength wells in the projected potential. Observations
show that the branch (1)HOLZ line does not vary, but that the higher lines
alternate. Calculations show that the most strongly bound state is an S1
state sitting in the deeper well only, so that the lack of alternation
in this line is expected. The higher states occupy both wells, so that
alternation in intensity is anticipated for these.

4.   Further Analysis

Complications to the simple approach presented are apparent from the pictu-
-res. The HOLZ states do interact with each other to produce hybridized
states. These interactions have been described by Jones et al (1977).
Extreme cases have been seen by Fung 1979 in which the HOLZ lines become
HOLZ rings. β_j now depends on the G component of $\tau_{jo}\ \tau^*_H\ U^{pp}_H$. When
τ_H is not strongly perturbed from its plane wave state, the position
of the HOLZ line will be affected, but not its intensity. In Fig.1
(at a) the effect of the interaction of the spots adjacent to the $\bar{5}\bar{5}11$
beam is shown by the faint lines in front of the branch (1) line, which
are due to the adjacent HOLZ beams. Another manifestation of interaction
with other HOLZ beams occurs at the loss of intensity in the same
position as the Kikuchi lines in Fig.1 (at b). This is due to hybridiza-
-tion with beams in other HOLZ layers.

We have used non-degenerate perturbation theory to deal with absorption in
the calculations. When two branches become degenerate, Buxton et al (1977)
have shown that branch points in the complex dispersion surface may invali-
-date this, and that full complex diagonalisation is necessary. However,
observable   effects were not found. The rigorous calculations will be
required if the difference between the absorption lengths of the two
unperturbed states is shorter than the extinction distance for the inter-
-action. In mitigation, waves which are strongly absorbed also give
shorter extinction distances.

A further effect of inelastic scattering lies in the enhancement of the
diffuse scatter at the HOLZ line positions (Fig.1 at c). These arise
because thermally scattered beams may be built from Bloch waves at a
slightly lower energy, but with the same general band structure. These
wave states may undergo elastic scatter to the HOLZ beams. Indeed, the
distribution of the diffuse scatter in the HOLZ beams may offer a sensitive
probe of the inelastic wave states.

5. Conclusion

The results presented suggest that we can extract structure factor infor-
-mation round a ring of HOLZ beams in a crystal several thousand Angstroms
thick. The complications noted may make accurate potential reconstruction
less straightforward, but the positions of the atoms should not be distor-
-ted. The difficulties may also provide an interesting guide to the
nature of the zero-layer wave states.

Acknowledgements

We wish to thank Dr.G M Rackham for providing pictures of Si $\langle 111 \rangle$
one of which is exhibited in Fig.1., and Dr.J W Steeds for directing
the work.

6. References

Buxton B F 1976 Proc.Roy.Soc.A350 pp 335-361
Buxton B F, Loveluck J E 1977 J.Phys.C. 10 3941
Buxton B F, Loveluck J E, Steeds J W 1978 Phil.Mag.38 pp 254-278
Fung K K 1979 Ph.D. thesis, University of Bristol
Jones P M, Rackham G M, Steeds J W 1977 Proc.Roy.Soc. A354 pp 197
Steeds J W 1980 Electron Microscopy, Antwerp Vol.4 pp 96-103

Optimum detection of backscattered electron channeling contrast

S M Payne

University of Oxford, Dept of Metallurgy & Science of Materials,
Parks Road, Oxford OX1 3PH

Introduction

Electron channeling patterns are being increasingly used for obtaining
orientation information from the surfaces of bulk crystalline specimens.
The channeling contrast signal is a maximum of 5% of the total backscattered electron (BSE) yield and the contrast of higher order lines necessary
for accurate orientation determination is much lower than this, so a high
level of signal processing is necessary. This being so, it is important
that the quality of the initially detected signal is as high as possible.
This paper is concerned with aspects of improving this signal.

Distinguishing channeling contrast carrying electrons

Electrons are backscattered from all parts of the specimen at less than
the penetration depth from the surface, but the channeling contrast carrying electrons (CCEs) arise only from the top $1.5/\mu_o$ (μ_o is the mean absorption coefficient) of this. The CCEs will have a higher average energy
than the background BSEs and they will be emitted with the spatial distribution characteristic of a thin film $1.5/\mu_o$ deep, this signal will be
superimposed on a background of the total backscattered yield from the
whole activated volume.

Choice of detector

The problem of optimising the detected signal reduces to (a) maximising
the overall signal, and (b) increasing the signal to noise ratio. The
specimen current satisfies (a), giving the equivalent of 2π detection of
BSEs, but will have a very high background and the signal due to BSEs
emitted at different angles are indistinguishable. Low loss imageing
(Wells 1972) increases the proportion of CCEs which escape from the
surface but the high tilt on the specimen complicates the contrast produced. Solid state backscattered detectors automatically enhance the CCE
contribution to the BSE signal since they have an amplification factor
proportional to the electron energy. They are also geometrically flexible,
so it is possible to exploit the difference in spatial distribution of
the CCEs and other BSEs.

Maximising overall signal

Murata (1974) calculates the exact spatial distribution of the total BSE
yield but this varies approximately as $\cos\theta$ where θ is the angle to the
incident beam. Where maximising the total BSE signal is the chief concern

eg when using atomic number contrast, a detector of fixed area should be positioned to maximise the solid angle close to the incident beam. For a 1.5/u thick film the angular distribution of BSEs may be approximated by

$$\frac{dN}{d\Omega} \propto K + (1 - K)\sin\frac{\theta\pi}{2\theta_{max}} \qquad 0 < \theta < \theta_{max}$$

$$\text{and } \frac{dN}{d\Omega} \propto \cos(\theta - \theta_{max}) \qquad \theta_{max} < \theta < \frac{\pi}{2}$$

where K is a constant and θ_{max} the angle of maximum BSE yield. For low Z, $K \approx 0.4$ and $\theta_{max} \approx 70°$; for high Z, $K \approx 0.8$ and $\theta_{max} \approx 60°$. Detailed calculations of the optimum detector position including the effects of varying Z are given elsewhere (Payne 1981). To maximise the signal from CCEs for ECP observation or channeling contrast imageing a detector should be annular and positioned so as to maximise the integral of this expression over the solid angle subtended at the specimen.

Increasing the signal to noise ratio

By careful choice of working distance it is possible to selectively collect BSEs from angles where the bulk BSE yield is low and the CCE yield high. The ratio of CCEs to other BSEs increases with θ but at high collection angles the CCE yield is small and where ECP recording is amplification limited, the best detector position is that to maximise the CCE yield.

Application to ISI100B and comparison with observation

The ISI100B at Cambridge has a 30mm dia annular Schottky barrier detector with an 8mm dia hole, fixed 6mm below the final lens. For 40kV electrons incident on Co-Fe, $1.5/\mu_o$ =60nm, K=0.5 and θ_{max} =67°. Fig 1 shows that although the solid detector angle subtends at the specimen is maximum for a working distance of approx. 11mm, the working distance for maximum BSE collection (ie the optimum position for Z contrast) is at a working distance of 14mm and the optimum position for the collection of CCEs is at a working distance of 10.5mm. This agrees with the observed conditions for best recording of both ECPs and channeling contrast.

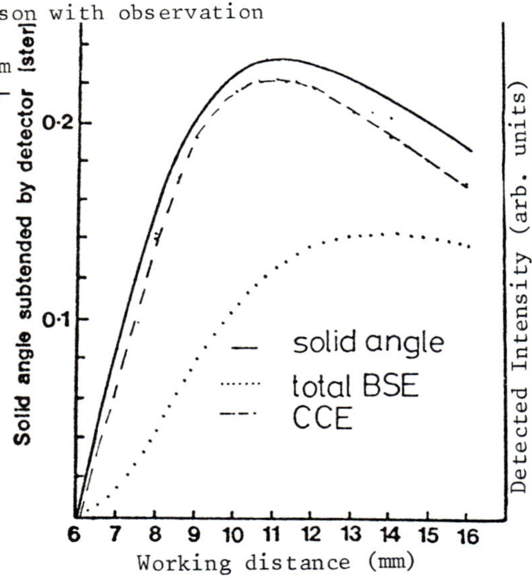

Fig 1 Variation of solid angle of collection to collected backscattered signals with specimen height.

Conclusion

By careful positioning of an annular solid state BSE detector it is possible to enhance either Z or channeling contrast in the BSE signal.

References
Murata 1974 J. Appl. Phys. 45, 4110
Payne, S.M. 1981 In preparation
Wells, O.C. 1972 Proc. V An SEM Symposium IIT Research Inst Chicago
Yamamoto, T. 1977 phys. stat. sol. (a) 44 137.

A unified description of the dynamical theory in the small angle scattering approximation

D Gratias and R Portier

C.E.C.M./C.N.R.S. 15 rue Georges Urbain, 94400 VITRY, France

Theoretical approaches of the dynamical theory of fast electrons may be roughly divided into two categories:
i. those for which the scattering object is assumed to be an infinite 3-dim medium (Bethe 1928).
ii. those for which the scattering object is a sequence of 2-dim objects of infinitesimal thickness (Cowley and Moodie 1957).

In both categories, the fact that the propagation of fast electron waves is close to the forward direction (small angle approximation) is taken into account in seemingly different although equivalent manners.

The direct introduction of the small angle scattering approximation into the general 3-dim Schrödinger equation for stationary collision

$$(\nabla^2 + k_o^2)\,\Psi = V\,\Psi \tag{1}$$

leads to a time dependent-like Schrödinger equation (Leontovich and Fock 1946).

$$i\frac{\partial}{\partial z}\,\Psi(z) = \frac{1}{2K}\left(-\nabla^2_{xy} - \lambda^2 + V(z)\,\right)\Psi(z) = H(z)\,\Psi(z) \tag{2}$$

where K and   are the components of ko incident wave vector respectively parallel and perpendicular to the z direction. (z) is the slowly oscillating part of (z).

$$\Psi(z) = \Psi(z)\,e^{i\,Kz} \tag{3}$$

Instead of solving eq (2) using standard quantum mechanical techniques of time dependent perturbation, it is more convenient to search for the evolution operator U (z,z_o) which is governed by equation (see for instance Messiah 1959):

$$i\frac{\partial}{\partial z}\,U\,(z,z_o) = H(z)\,U\,(z,z_o) \tag{4}$$

The central problem of electron microscopy is to calculate, once a certain initial state, say $|i\!>$, is given, the transition probabilities to a specified final state, say $|f\!>$. Such a transition probability is formally given by

$$\omega_{i\,\to\,f} = \left|\,<\!f|\;U(z,z_o)|\;i\!>\,\right|^2 \tag{5}$$

the final state $|f\!>$ is an eigenfunction either of the momentum q (Diffraction pattern) or of the position ρ (Image). Both initial and final states are defined in the 2-dim x.y configurational space. The remaining z-coordinate is formally a time-dependent-like parameter. It will be seen here that the different ways of solving eq (4) correspond to the different usual theoretical approaches of dynamical diffraction.

1. z-independent potential: the projected potential approximation

Eq (4) may be directly integrated if the potential $\overline{V}$ is independent of z, choosing $z_o = 0$ we obtain

$$U(z) \;=\; e^{-iHz} \tag{6}$$

Let $|j\rangle$ and γ_i be the eigenvectors and eigenvalues of the Hamiltonian

$$\frac{1}{2K}\left(-\nabla^2 - \chi^2 + \overline{V}\right)|j\rangle \;=\; \gamma_j \; |j\rangle \tag{7}$$

The evolution operator is then written

$$U(z) \;=\; \sum_j e^{-i\gamma_j z}\, |j\rangle\langle j| \tag{8}$$

By designating by $|0\rangle = e^{i\rho}$ the initial state at $z = 0$, we obtain the amplitude of the transition probability to a final state $|q\rangle$:

$$\langle q|\,U(z)\,|0\rangle \;=\; \sum_j e^{-i\gamma_j z}\,\langle q|j\rangle\langle j|0\rangle \tag{9}$$

where $\langle j|0\rangle$ is the component of the initial state $|0\rangle$ on the eigenstate $|j\rangle$ and $\langle q|j\rangle$ the component of the eigenstate $|j\rangle$ on the final state $|q\rangle$. For a periodic potential the eigenstates $|j\rangle$ are the usual 2-dim Bloch waves and the components $\langle q|j\rangle$ are designated by C_q^j. C_q^j coefficients and γ_j eigenvalues are determined by projecting the eigenstate eq (7) on the q basis

$$\langle q|\frac{1}{2K}\left(-\nabla^2 - \chi^2 + \overline{V}\right)|j\rangle \;=\; \gamma_j \,\langle q|j\rangle \tag{10}$$

Inserting the closure relation $\sum_q |q\rangle\langle q| = 1$ into (10) we finally obtain the usual secular equation:

$$\frac{q^2 - \chi^2}{2K}\,\langle q|j\rangle + \sum_{q'}\langle q|\frac{\overline{V}}{2K}|q'\rangle\langle q'|j\rangle \;=\; \gamma_j \,\langle q|j\rangle \tag{11}$$

where $q^2-\chi^2/2K$ are the excitation parameter $\zeta(q)$ (times 2π) and $\langle q|\frac{\overline{V}}{2K}|q'\rangle$ the 2dim Fourier transforms $V_{q'-q}$ of the potential $\overline{V}/2K$.

2. z-dependent potential

For the general case of a z-dependent potential, no direct analytical integration may be obtained and perturbation theory has to be used. From standard quantum mechanic techniques, the evolution operator U(z) of a perturbated system may be expanded in a "time" perturbation series.

$$U(z,0) \;=\; U^{(0)}(z,0) + \sum_{n=1}^{\infty} U^{(n)}(z,0) \tag{12}$$

$$\text{with } U^{(n)}(z,0) = (-i)^n \int_0^z d\tau_n \int_0^{\tau_n} d\tau_{n-1} \cdots \int_0^{\tau_2} d\tau_1$$

$$z > \tau_n > \tau_{n-1} > \ldots > \tau_1 > 0 \tag{13}$$

$$U^{(0)}(z,\tau_n)V(\tau_n)\,U^{(0)}(\tau_n,\tau_{n-1})\cdots U^{(0)}(\tau_2,\tau_1)V(\tau_1)\,U^{(0)}(\tau_1,0)$$

where $U^{(0)}$ is the evolution operator corresponding to the unperturbated Hamiltonian. Two different theories have been developed in electron microscopy which are implicitly based on expansion (13). The Cowley-Moodie (1957) formulation takes as unperturbated Hamiltonian the Hamiltonian of the free electron

$$H^\circ_{CM} \;=\; \frac{1}{2K}\left(-\nabla^2 - \chi^2\right) \tag{14}$$

The more recent approach of Buxton (1976) considers the Hamiltonian corresponding of the projected potential (Eq (7)).

$$H^\circ_{B} \;=\; \frac{1}{2K}\left(-\nabla^2 - \chi^2 + \overline{V}\right) \tag{15}$$

Both of these approximations may simultaneously be treated by eq (13) :

Let $|j>$ and γ_j be the eigenvectors and eigenfunctions of the unperturba-
ted Hamiltonian $(H^{(0)}{}_{CM}$ or $H^{(0)}{}_B$; the n-th order perturbation term is then,
according to eq (8):

$$U^{(n)}(z,0) = (-i)^n \int_0^{z} d\,\tau_n \cdots \int_0^{\tau_t} d\tau_1 \sum_{j_n} \cdots \sum_{j_0} <j_n|V(\tau_n)|j_{n-1}> \quad (16)$$

$$<j_{n-1}|V(\tau_{n-1})|j_{n-2}> \cdots <j_1|V(\tau_1)|j_0>$$

$$e^{-i[\gamma_{j_n}(z-\tau_n)+\gamma_{j_{n-1}}(\tau_n-\tau_{n-1})+\ldots+\gamma_{j_1}(\tau_2-\tau_1)+\gamma_{j_0}\tau_1]}|j_n><j_0|$$

In the C.M. formulation, $|j>$ are plane waves $|q>$ and the eigenvalues are
the excitation parameters (q) (apart from the 2π scaling factor)

$$|j>_{C.M.} = |q> = e^{i \cdot \underset{\sim}{q} \cdot \underset{\sim}{\rho}} \quad (17)$$

$$\gamma_j{}_{C.M.} = \frac{q^2 - \chi^2}{2\,K} = \zeta(q) \quad (18)$$

the elements of the V matrix are the 2-dim Fourier coefficients of the
potential. We obtain the n-th multiple scattering term by calculating
$<q|U^{(n)}{}_{(z,o)}|0>$

In the B. formulation, $|j>$ are the Bloch waves (§ 1) and γ_j the "anpassung"
parameters. The matrix elements of the potential may be expressed with
respect to the Fourier coefficients:

$$<j|V(\tau)|j'> = \sum_q \sum_{q'} <j|q> <q|V(\tau)|q'> <q'|j'>$$

$$= \sum_q \sum_{q'} C^{*j}_q V_{q-q'}(\tau) C^{j'}_{q'} \quad (19)$$

If only the first term of the perturbation series is retained, we obtain
the single transition approximation which corresponds respectively to the
kinematical theory with $H_{C.M.}$ and to the single interband transition with
H_B. The single transition probability from $|j>$ to $|j'>$ is

$$<j'|U^{(1)}(z,0)|j> = -i \int_0^{z} d\tau <j'|V(\tau)|j> e^{-i(\gamma_{j'}(z-\tau)+\gamma_j\tau)} \quad (20)$$

If the potential is periodic along z

$$V(\tau) = \sum_{\ell} V_\ell\, e^{i\ell\omega z} \quad (21)$$

we obtain :

$$<j'|U^{(1)}(z,0)|j> = -i \sum_{\ell} <j'|V_\ell|j> e^{-i(\gamma_{j'}+\gamma_j+\ell\omega)z/2}$$

$$\frac{\sin(\gamma_{j'}-\gamma_j+\ell\omega)z/2}{(\gamma_{j'}-\gamma_j+\ell\omega)\,z/2} \quad (22)$$

showing that the only significant term for sufficiently thick crystals
comes for the ℓ-th V_ℓ Fourier component such that

$$|\gamma_{j'}-\gamma_j+\ell\omega| \lesssim 1/Z \quad (23)$$

Hence the 3-dim nature of the reciprocal lattice is essentially recovered:
the periodicity of V along z induces resonance-like peaks along z, the
centre of which corresponds to the Bragg positions.

3. The other derivations

Attention has been focused here on two major approaches of electron diffraction. However the Tournarie semi-reciprocal approach (1962) or Sturkey's

scattering matrix formulation (1962) might be derived as well. The connections between these theories are summarized in the Table. A detailed discussion is given elsewhere (Gratias and Portier 1981).

When the crystal is thin enough, the sudden perturbation approximation may be used, consisting in replacing the evolution operator by unity

$$U^o(z,o) \simeq 1 \tag{24}$$

In order to take into account the potential V of the crystal, we apply the perturbation technique:

$$U(z,o) = 1 - i \int_o^z d\tau_1\, V(\tau_1) + (-i)^2 \int_o^z d\tau_2 \int_o^{\tau_2} d\tau_1\, V(\tau_2)\, V(\tau_1) + \cdots \tag{25}$$

leading to (after a few algebraic manipulations)

$$U(z,o) = e^{-i \int_o^z V(\tau)\, d\tau} \tag{26}$$

The evolution operator (26) corresponds to the phase grating approximation. It essentially means that the interaction time is sufficiently short, so that the propagation of the particle between any two successive interactions is negligible. If the perturbation series (25) is limited to the first order term, we obtain the weak phase object. The kinematic theory and the weak phase object approximation are in essence very similar, both resulting from the single transition assumption.

On the other hand, neglecting the potential of the crystal, the evolution operator corresponds to Fresnel propagation. This operator and operator (26) are the basic ingredients of the so-called multi slice method (Goodman and Moodie 1974).

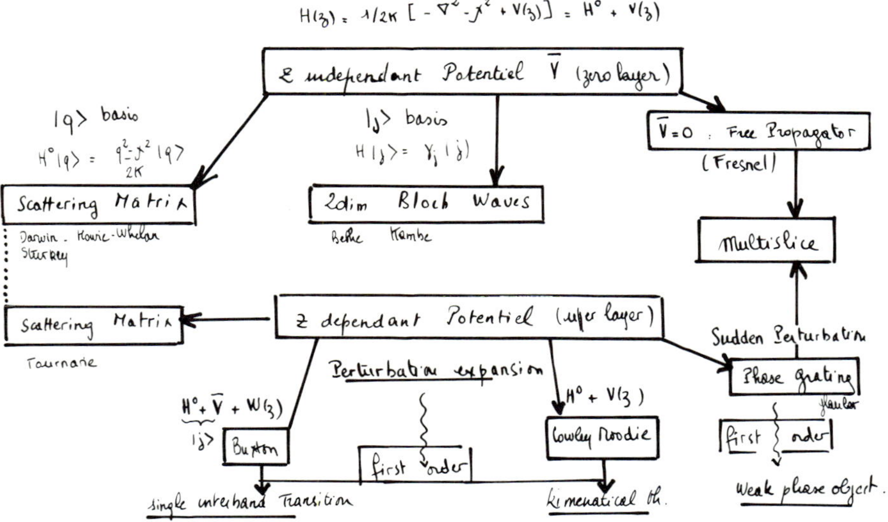

Bethe H.A., Ann. Physik (1928) 87, 55
Buxton B.F., Proc. Roy. Soc. Lond. (1976) A 350, 335
Cowley J.M. & Moodie A.F. (1957) Proc. Phys. Soc. B 70, 486 & 497 & 505
Goodmann P. & Moodie A.F. (1974) Acta Cryst. A 30, 280
Gratias D. & Portier R. (1981) Acta Cryst. submitted
Leontovich M. & Fock V., J. of Phys. (1946) 10, 1, 13
Messiah A. (1959) Mécanique Quantique, Dunod, Paris
Sturkey L. (1962) Proc. Roy. Soc. 80, 321
Tournarie M., J. of Phys. Soc. Jap. (1962) 17, B II, 98

Critical voltage effects in backscattered electron channeling patterns

S M Payne

University of Oxford, Department of Metallurgy and Science of Materials,
Parks Road, Oxford OX1 3PH

1. Introduction

The critical voltage effect (Thomas et al. 1974) which arises from the
accidental degeneracy of 2 Bloch waves at an accelerating voltage V_c, is
usually characterised by the disappearance of a particular line in a
Kikuchi pattern. In general, V_c decreases as the atomic number of the
scattering material increases and as the magnitude of the scattering vector
decreases. The low order critical voltages for heavy b.c.c. materials
such as tungsten lie in the voltage range of SEMs.

2. Observation of critical voltage effects in the SEM

Farrow and Joy (1980a), studying electron channeling patterns (ECPs) of
tungsten over a range of accelerating voltages, found a minimum in the
measured 220 line width at an accelerating voltage of 10.5 ± 0.05 kV and
a reversal of the normal contrast in the 220 line at accelerating
voltages above this. The constrast of the other lines was unaffected.
They attributed these changes to a second order critical voltage effect
and following Schulson's (1971) treatment of the widths of ECP lines
interpreted the line width changes in terms of the interchange of the wave
vectors of Bloch waves 2 and 3 at the critical voltage. To explain the
reversal of line contrast they involved the dynamical diffraction
expression for the intensity of the transmitted second order diffracted
beam (Humphreys 1979). This will not however apply to the intensity of an
ECP line. In this paper the theory of Spencer et al. (1972) for backscattered intensities is used to account for the observed contrast
reversal.

3. Theory of ECP formation

For near normal incidence, the backscattering coefficient $p^{(j)}$ of the jth
Bloch wave is given by

$$p^{(j)} = \frac{\pi}{\Omega N_c} \left(\frac{e^2 m Z \lambda^2}{h^2} \right) \sum_g \sum_h C_g^{(j)} C_h^{(j)} \exp(-M_{h-g}) \sum_J \exp 2\pi i (\underline{h}-\underline{g}) \cdot \underline{r}_J$$

The total backscattering coefficient $p^{(o)} = \sum \dfrac{p^{(j)}}{n}$ where n is the number
of Bloch waves considered.
For a perfect crystal the total backscattered intensity $I_B(o)$ is given by

$$I_B(o) = \frac{1}{1+p^{(o)}t} \left[p^{(o)}t + \sum_J I_o^{(j)} \frac{(p^{(j)} - p^{(o)})}{\mu^{(j)}} (1 - \exp(-\mu^{(j)}t)) \right]$$

0305-2346/82/0061-0293$01.50   © 1982 The Institute of Physics

where t is the effective thickness for backscattered electrons. The total
ECP contrast is less than 5% giving $p^{(o)}$ approximately independent of
orientation so the 1st term in the brackets is a background term and the
second gives rise to an ECP. $(1 - \exp(-\mu^{(j)}t))$ is slowly varying and for
near normal incidence $I^{(j)}(o) = |C_o^{(j)}|^2$. In this case the contrast is
mainly governed by

$$\sum_j \frac{|C_o^{(j)}|^2 p^{(j)}}{\mu^{(j)}}$$

4.   Interpretation of observations

$C_o^{(j)}$ can be calculated by standard dynamical matrix calculations
(Hirsch et al. 1965). The variation of $C_o^{(j)}$ with orientation for 110
systematic reflections just above V_c is shown in fig.1. The excitation of
all Bloch waves higher than the fifth is small in the orientation range
shown. $\mu^{(j)}$ may be calculated approximately from the absorption parameters
given by Humphreys and Hirsch (1968) and the variation of $\mu^{(j)}$ with
orientation just above V_c is shown in fig.2. Fig.3 shows the variation of
$p^{(j)}$ with orientation at the same accelerating voltage. Below V_c the
variation of $C_o^{(j)}$ in the orientation range considered is little changed
from that shown in fig.1 except that $C^{(2)}$ drops even more sharply just
outside 220. The values of $\mu^{(2)}$ and $\mu^{(3)}$ and $p^{(2)}$ and $p^{(3)}$ interchange at
220, 440 etc as the accelerating voltage changes through V_c. Figure 4
shows the variation of $p^{(j)}$ with orientation below V_c. Figure 2 shows the
effect on $\mu^{(2)}$ and $\mu^{(3)}$ on passing through V_c is small compared to the
change in $p^{(2)}$ and $p^{(3)}$ shown in figs.3 and 4. ECPs are recorded at
accelerating voltages below V_c for most reflections and materials so
'normal' SEM contrast is that below V_c, with the lines showing bright-dark
contrast going out from the centre. $p^{(j)}$ and $\mu^{(j)}$ are almost symmetrical
about the Brillouin zone boundaries so the asymmetry of the lines is
introduced by the Bloch wave excitations. Fig.1 shows that around the 220
line Bloch waves 1, 2 and 4 are more strongly excited within the zone and
wave 3 outside, so waves 1, 2 and 4 will act to give normal contrast
whereas wave 3 will act to give reverse contrast. Figure 4 shows that
below V_c the contribution of wave 3 to the backscattered intensity around
220 is small. Just above V_c $p^{(2)}$ and $p^{(3)}$ are interchanged so wave 2
contributes little to the observed contrast whilst the large asymmetry of
the wave 3 contribution dominates the contribution from waves 1 and 4 and
the line contrast is reversed. Figure 5 shows the calculated contrast
across the {110} lines for tungsten above V_c showing reversed contrast in
agreement with the observation of Farrow and Joy.

5.   Application to lattice potential determination

Accurate measurement of V_c allows refinement of the value for the low
order Fourier components of potential (Thomas et al. 1974). Observation of
contrast reversal of ECP lines at V_c is hampered by the low contrast of
ECPs and the broadness of the low order lines away from V_c. There is
greater contrast in the line when exhibiting normal contrast than when
contrast is reversed so the range of accelerating voltages over which the
line is invisible will not be centred on the true V_c. However by using
signal averaging and gating techniques (Farrow and Joy 1980b) very
accurate 110 V_c determinations for Ta, Mo and Nb could be performed in
addition to the measurements already made for W (Farrow and Joy 1980a).

5. Conclusion

The contrast reversal of certain lines in channeling patterns at accelerating voltage above their V_C is attributable to a critical voltage effect and can be explained by application of the theory of Spencer et al. (1972).

References

Farrow R C and Joy D C 1980a Phys.Rev.Lett. <u>44</u> 1590
Farrow R C and Joy D C 1980b Scanning (in press)
Hirsch P B, Howie A, Nicholson R B, Pashley D W and Whelan M J 1965
  Electron Microscopy of Thin Crystals (New York : Krieger)
Humphreys C J 1979 Rep.Prog.Phys. <u>42</u> 1825
Humphreys C J and Hirsch P B 1968 Phil.Mag. <u>18</u> 115
Schulson E M 1971 Phys.stat.sol.(b) <u>46</u> 95
Spencer J P, Humphreys C.J. and Hirsch P B 1972 Phil.Mag. <u>26</u> 193
Thomas L E, Shirley C G, Lally J S and Fisher R M 1974 Proc.III Int.Conf.
  on HVEM ed. Swann et al (New York : Academic Press) p.38

Acknowledgements

The author would like to thank Professor Sir Peter Hirsch FRS for providing laboratory facilities and the Science Research Council for provision of a grant.

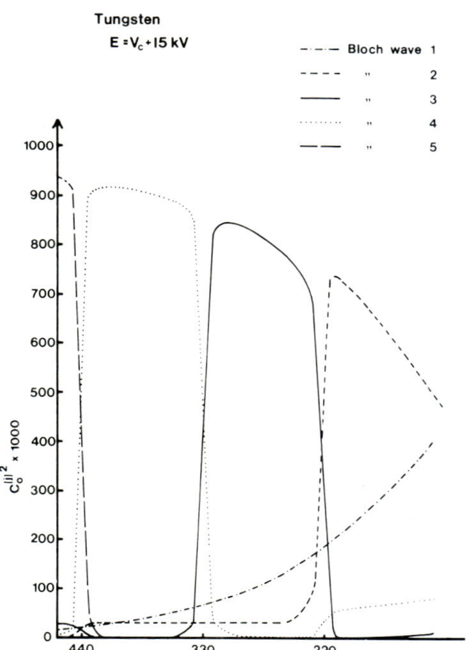

FIG.1. Variation of principal Bloch wave excitations with orientation just above V_c.

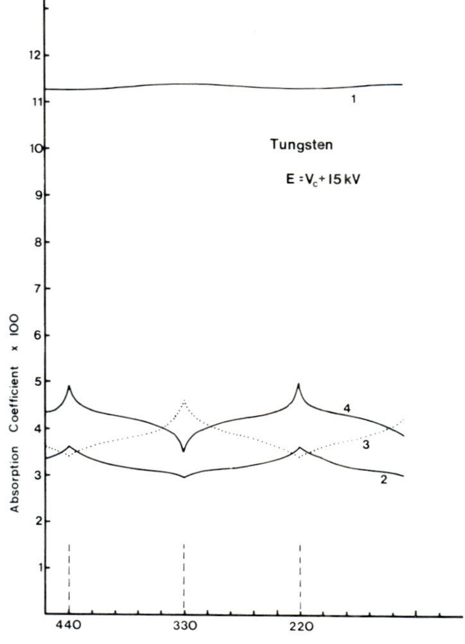

FIG.2. Variation of principal Bloch wave absorption with orientation just above V_c.

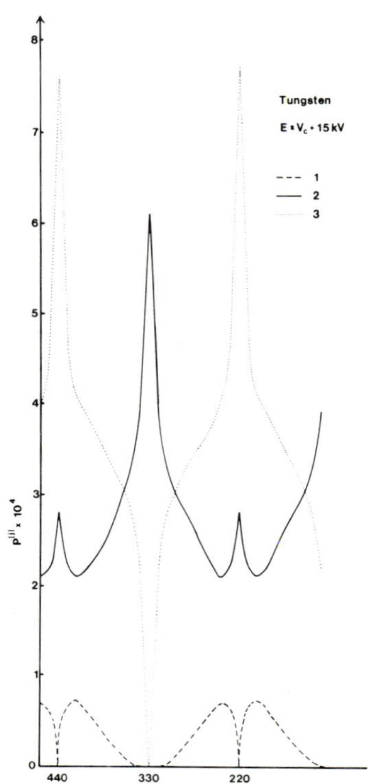

FIG.3. Variation of back scattering coefficients of principal Bloch
waves with orientation just above
V_c.

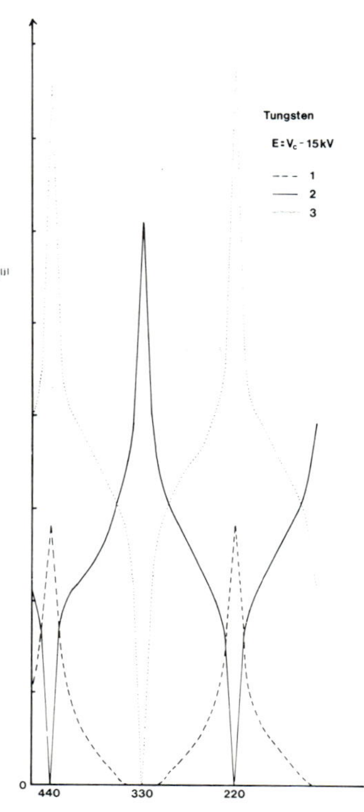

FIG.4. Variation of back scattering coefficient of principal Bloch
waves just below V_c.

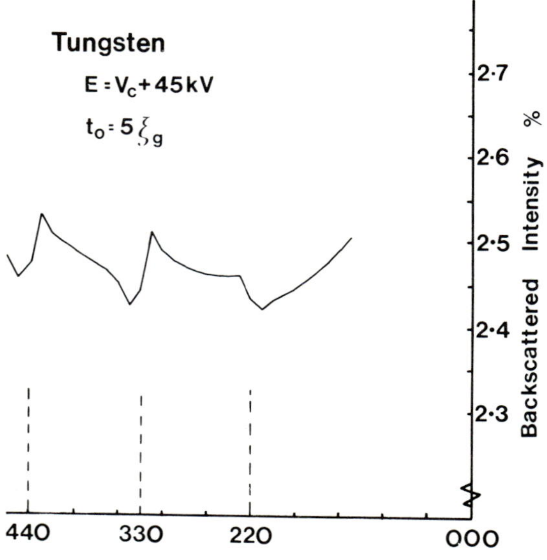

FIG.5. Calculated profile
across {110} lines of a
tungsten ECP above V_c.

Anomalous contrast from stacking faults

M.P. Shaw, P.G. Self and W.M. Stobbs

Department of Metallurgy and Materials Science,
University of Cambridge, Cambridge CB2 3QZ, U.K.

Introduction

Weak beam images of extrinsic faulted dislocation loops exhibit anomalous contrast. The observations described here arose from a study of interstitial (extrinsic) faults produced by radiation damage in a commercial nickel-base superalloy, Nimonic PE16 (Shaw, Ralph and Stobbs 1981). This alloy has an f.c.c. matrix containing both a dispersion of ordered (Ll_2) γ' precipitates and various carbides.

Results and Discussion

Figure 1 is a bright field image showing faulted loops on all four $\{111\}$ planes. Complementary images taken in a variety of different two beam diffracting conditions showed the faults to be $^a/3 \langle 111 \rangle$ interstitial faults surrounded by $^a/3 \langle 111 \rangle$ pure edge (Frank) partial dislocations. When images were taken under weak beam conditions, contrast from the stacking faults was found to be critically dependent on the sign of g.R (where g is the diffracting vector and R the displacement vector of the fault). Figure 2 shows weak beam images taken using + and −g with varying degrees of deviation from the Bragg condition. Strong contrast is produced from the faults when g.R is positive but contrast falling below the background level when g.R is negative. The effect becomes stronger as the deviation parameter and the fault angle are increased, but is evident for small values of both.
The effect was first observed by Cullis and Booker (1972). At first sight the general character of the contrast must appear to be the result of the fact that a stacking fault locally destroys the centrosymmetric nature of an f.c.c. lattice. Therefore, some asymmetry is expected in images of a fault from complementary beams under complementary diffracting conditions. However, the size of the contrast asymmetry is inexplicable within the normal computational framework of dynamical electron diffraction theory. Cockayne (1980) has suggested that the effect may be accounted for by allowing for the finite thickness of the fault, although the anomaly would then be expected to be critically dependent upon the deviation parameter. Foll, Carter and Wilkens (1980) have investigated this model for the contrast asymmetry from stacking faults in silicon, using a kinematical approach. Their results neither explain the magnitude of the effect they observed for 220 silicon images nor our observations in Nimonic PE16. Critically, the model gives no reasonable explanation for either the lack of asymmetry of 111 weak beam images in silicon or the fact that the asymmetry is strong for two intrinsic faults overlapping at a high angle (so that the spacing between them varies rapidly) in silicon whatever the g used. Further analysis of possible origins for the effect would appear

to be required.

The standard computational approach to fault imaging is to assume that the fault lies at a certain depth in the foil and that there is perfect crystal above and below the fault. These two pieces of crystal are shifted relative to each other by an amount consistent with the $g.R$ value of the fault. This approach is a valid one, for a stacking fault, because the fault causes little distortion to the lattice at distances further than a couple of atomic layers away from it. As the Bloch wave solutions are well known in perfect crystals, the contrast problem is one of matching the electron current from the upper perfect crystal to the lower perfect crystal in a way which is consistent with the nature of the fault. That is, the problem is one of correctly determining the excitation coefficients of the Bloch waves (the Bloch waves themselves being well defined) in the lower half of the crystal.

The obvious places to check for any inconsistency in a computational problem are the approximations used in solving that problem. Several of the more widely used approximations of electron diffraction theory can be readily dismissed. As the asymmetry can be seen in orientations where the fault is nearly flat with respect to the imaging beam direction, it is valid to use the column approximation. For the same reason, the neglect of fault inclination in the boundary conditions for the calculation cannot be important. Furthermore, the treatment of absorption by perturbation methods (rather than fully) has equally little effect. However, it must be emphasised that the anomaly itself shows changes from beam to beam at a given orientation which show the correct symmetries for the expected Bloch waves. This implies that it is only the magnitudes of the current transfer which are anomalous and that it is the matching procedures which are inadequate.

Bloch wave calculations, using the finite fault thickness concept used by Foll, Carter and Wilkens (1980), give an asymmetry in intensity corresponding to the observed dependence on $g.R$ but the relative magnitudes of the intensity difference are nowhere near large enough (see fig.3) and contrast below background level is never observed. Perhaps then, the Bloch potentials are inadequately defined at the fault. This suggests that a better approach should be to consider the whole extrinsic fault on a projected potential (phase grating) basis. Image intensity profiles from $a/3\langle 111\rangle$ extrinsic faults using this approach are shown in figure 4. This method provides an upperlimiting approximation for the diffracted intensity at the fault. Despite this, the contrast asymmetry predicted is neither bigger than that suggested by the full Bloch wave treatment nor big enough to account for experimental observations. Larger contrast asymmetries could be produced and over emphasised by the above approach by substantially increasing the fault thickness, but this is also unrealistic.

The calculations show that the effect cannot be adequately described within the normal framework of the dynamical theory of electron diffraction. Furthermore, we have no explanation for the lack of asymmetry in silicon 111 images or the strong thickness independent effect for overlapping intrinsic faults. The calculations have neglected segregation effects which have been shown to be present (Shaw, Ralph and Stobbs 1981) as a local increase in γ' density at the fault. However, it is unlikely that significant modifications in the appropriate local atomic scattering factors could be responsible for both the effect and the

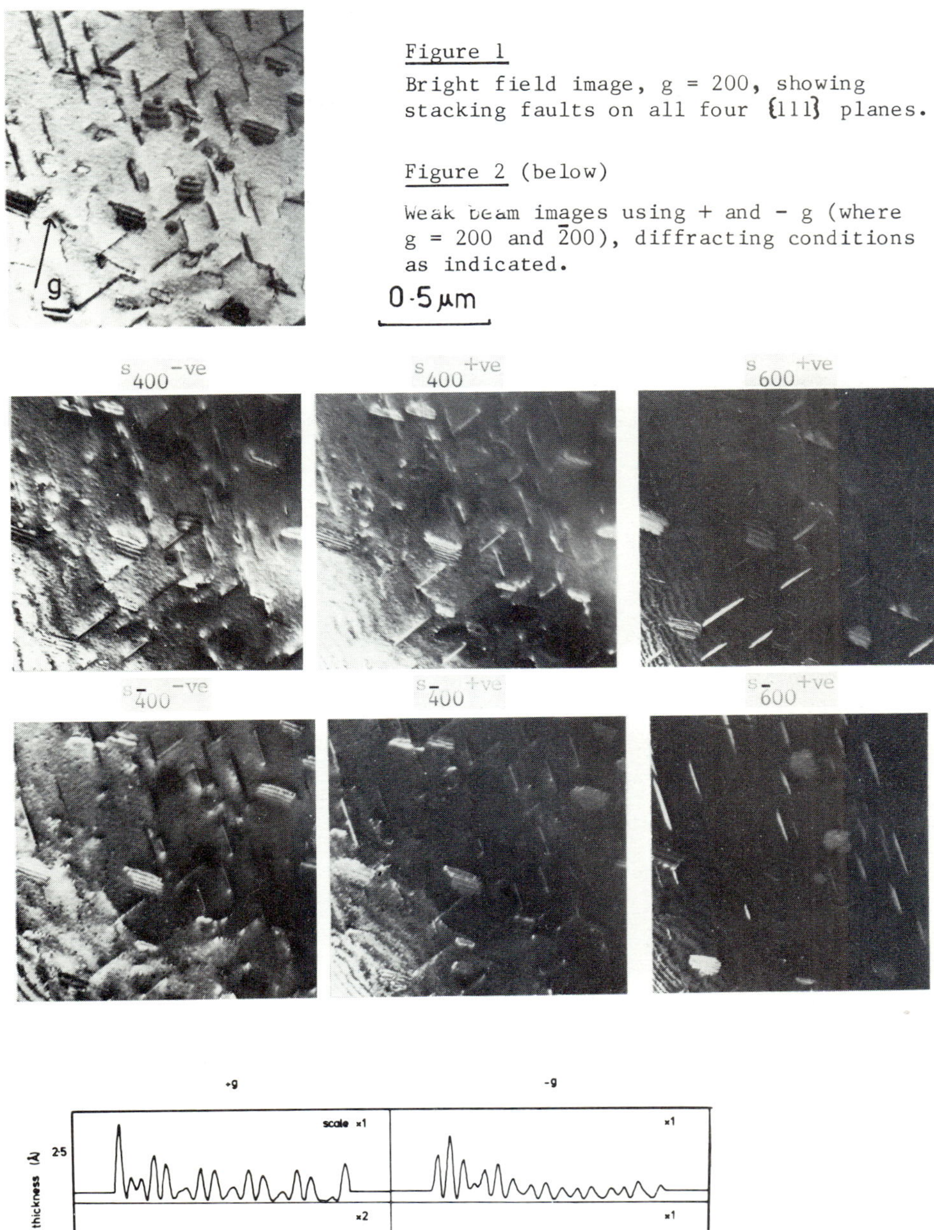

Figure 1

Bright field image, g = 200, showing stacking faults on all four {111} planes.

Figure 2 (below)

Weak beam images using + and − g (where g = 200 and $\bar{2}$00), diffracting conditions as indicated.

0·5 μm

Figure 4. As figure 3, but using the phase grating approximation to describe the fault layer (see text).

differences between the silicon and Nimonic PE16 observations. Perhaps a
more fruitful approach would be to include contributions from upper and
lower Laue zones. This can give relatively high intensities in forbidden
reflections (Tu and Howie 1976) but the contributions are much more
strongly dependent upon orientation than the contrast effects observed.

The most significant point arising from the above discussion is that we
are convinced that it is essentially the matching procedure which is at
fault rather than the need for any qualitatively new effect to be
included. If this is in fact the case, any calculations which seek to
determine small physical displacements at defects by the observation of
small changes in generally strong contrast would appear to be doomed.

Acknowledgements

Thanks are due to U K A E A (Harwell) and the S R C for financial support.
We are grateful to Prof. R W K Honeycombe FRS for provision of laboratory
facilities.

References

Cockayne, D. J. H, 1980 Micron, 11, Suppl. No.1, 33.
Cullis, A.G. and Looker, G.R. 1972, Proc. Fifth European Congress on Electron
    Microscopy, p.532.
Foll, H. Carter, C. B. and Wilkens, M. 1980, Phys Stat Sol. (a) 58 393.
Shaw. M.P, Ralph. B. and Stobbs W.M. 1981, J. Nucl Mater. in press.
Tu K. N. and Howie A. 1976 Proc. Sixth European Congress on Electron
    Microscopy p288.

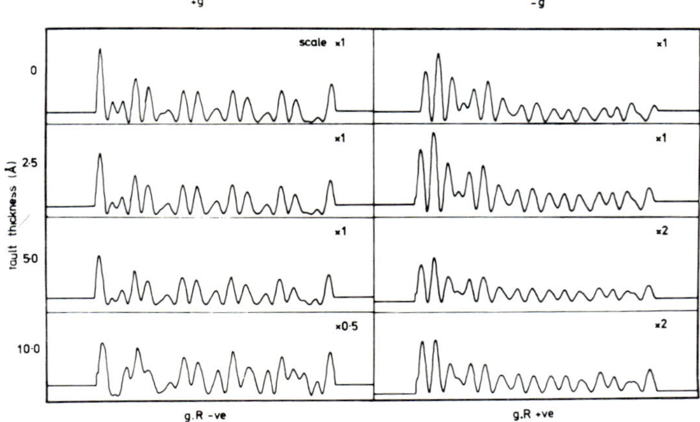

Figure 3. Computed image intensity profile for stacking fault imaged in
200 type beam. Weak beam conditions (i.e. + and −3g, just + ve). Full
Bloch wave calculation using finite fault thickness concept.

'Spike' electron diffraction patterns from platelet-containing diamonds

N. Sumida and A.R. Lang

H.H. Wills Physics Laboratory, Tyndall Avenue, Bristol BS8 1TL, U.K.

Abstract   The commonest variety of natural diamonds (classified as Type Ia) are noted for their diffuse X-ray reflections which extend as spikes along some or all $<100>$ directions surrounding reciprocal lattice points. The spikes arise because of platelet defects lying randomly on $\{100\}$ planes of the diamond matrix. Analogous diffuse reflections have now been studied in electron diffraction, and are shown to correspond in essential respects with the X-ray spikes, as expected from diffraction theory. Similarities and differences between electron and X-ray spike patterns are described.

1. Introduction

Just 40 years ago Lonsdale and Smith (1941) reported that the diffuse X-ray reflections from diamond discovered by Raman and Nilakantan (1940) fell into two categories: (a) the thermal diffuse reflections, qualitatively similar to those exhibited by other cubic crystals (but relatively weak because of the high Debye temperature of diamond) and (b) 'spike-like' diffuse reflections which were substantially temperature independent but specimen dependent. The spikes are (in Lonsdale and Smith's words) 'regions of very sharp intense diffraction along some or all of the reciprocal $[100]$, $[010]$, $[001]$ axes. There are six such 'horns' extending in reciprocal space from the 111 points, more than half-way across the Brillouin zone in each case'. Lonsdale and Smith found that spikes were absent in the X-ray diffraction patterns of those rarer, ultraviolet transparent diamonds classified by Robertson, Fox and Martin (1934) as Type II, but were present (in strength varying from specimen to specimen (Lonsdale 1942)) in the commoner diamonds classified nowadays as Type Ia (Dyer, Raal, Du Preez and Loubser 1965) and known to contain nitrogen impurity (Kaiser and Bond 1959) and (as a general rule) electron microscopically visible platelet defects on $\{100\}$ (Evans and Phaal 1962). We now report observations on electron diffraction spike patterns from Type Ia diamonds. These patterns can be obtained with remarkable sharpness and contrast. Study of them usefully complements the older, X-ray techniques, and gives indications of some features at variance with the accepted findings of X-ray experiments.

2. Observations

Slices of natural diamonds were sawn and polished parallel to (001) or (110) and ion-beam thinned for transmission electron microscopy. Diffraction patterns taken with the electron beam parallel to symmetry axes such as [001], [110] and [111] display well the symmetry of the spike reflections, and provide panoramas of spike phenomena that are impossible to obtain by X-ray diffraction with the wavelengths commonly used in X-ray crystallography. For example, with electrons of 120 kV energy, whose wavelength is 0.00335 nm, the radius of the Ewald sphere is 46 times greater than when CuKα X-radiation, wavelength 0.154 nm, is used. As a consequence of the flatness of the Ewald sphere in the electron case, it follows that if the primary beam is directed parallel to [001], then spikes parallel to both [100] and [010], and centred on low-order reciprocal lattice points (and also on the reciprocal lattice origin) tangentially graze the Ewald sphere and generate an extensive, fourfold symmetric pattern of sharp streaks. The [110] zone axis spike diffraction pattern has, correspondingly, twofold symmetry since only the [001] spikes are then tangential to the Ewald sphere. When the electron beam is along, or near to [111], the spikes produce triplets of sharp spots surrounding weakly excited Bragg reflections in cases where the Ewald sphere passes close to but not precisely through the reciprocal lattice points. Fig. 1 shows part of the [001] zone axis electron diffraction pattern of a Type Ia, platelet-rich natural diamond. Though well visible on the original film, the spike pattern is most difficult to reproduce in photographic printing because the fine spikes are superimposed on a strongly varying background. The area shown includes two strong 220-type reflections, from which spikes issue both downwards and also inwards towards the bright dot in the centre of the field. The bright dot is the kinematically forbidden 200 reflection which is generated through multiple scattering involving higher-order reflections not contained within the hk0 reciprocal lattice layer (Tu and Howie 1978).

In contrast to Fig. 1, Fig. 2 illustrates how electron diffraction spike patterns very similar to the known X-ray patterns (e.g. plate 34(b) of Lonsdale (1942)) can be produced. The X-ray pattern (Fig. 2a) was obtained from a specimen that exhibited exceptionally strong and sharp spikes, and it was photographed under conditions such that the size of the spike image was dominantly controlled by the X-ray beam cross-section and monochromaticity, and by the obliquity of the spikes to the Ewald sphere (Suzuki and Lang 1976). Beam divergence and spike obliquity correspondingly largely determine the size of the electron diffraction spike images in Fig. 2b. In the X-ray case the white-radiation, Laue reflection appears shifted well away from the centroid of the spike triplet because of the specimen rotation (typically $\sim 1^{\circ}$) performed in order to remove the Ewald sphere of the monochromatic component in the radiation sufficiently away from exact satisfaction of the Bragg condition. In the electron case, all diffraction angles are small. Then, provided that the beam direction is not far from the axis of the kinematical diffraction spike normal to the foil that necessarily exists even in the case of defect-free crystals, the weak, off-peak, perfect-crystal reflection remains not far from the centre of the spike triplet. This difference in diffraction geometry, plus (in this instance) relatively stronger thermal diffuse scattering in the electron diffraction pattern, simply explain the difference of patterns between Fig. 2a and b.

3. Discussion

We have confirmed that the electron diffraction spikes of platelet-containing

Fig. 1. Part of electron diffraction pattern of a Type Ia diamond taken with electron beam direction close to [001]. Electron energy 120 kV. The diffraction pattern centre (i.e. reciprocal lattice origin) lies below the field shown. The two strong Bragg reflections included in the field are indexed 220 (right) and 2$\bar{2}$0(left). The bright dot in the field centre is the forbidden 200 reflection. Spikes parallel to $\pm$[010] issue from the 220 and 2$\bar{2}$0 reflections in horizontal directions, nearly meeting at the point 200. Spikes parallel to $\pm$[100] are vertical.

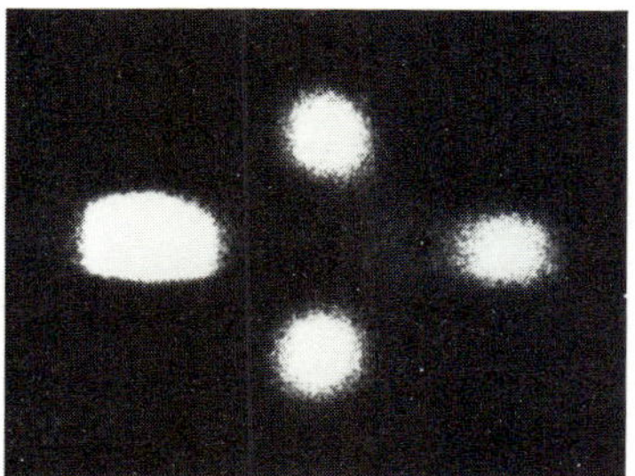 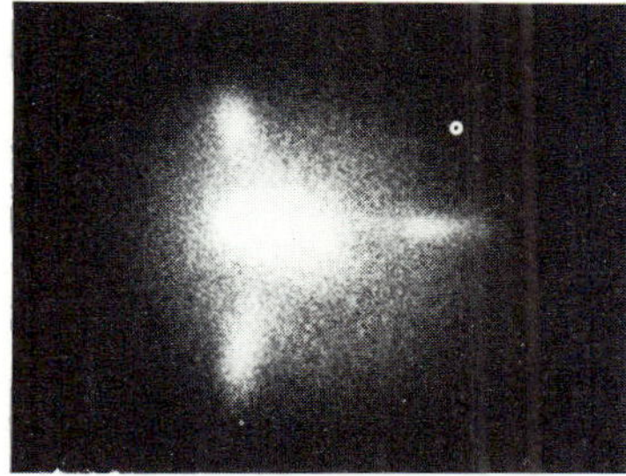

Fig. 2(a, Left). X-ray spike pattern associated with the 111 reciprocal lattice point of a Type Ia diamond. CuKα radiation, wavelength 0.154 nm. The diffraction geometry is identical with that shown in Fig. 1a of Suzuki and Lang (1976), and the mis-setting off the exact Bragg angle is 1.2°. The intense white-radiation, Laue reflection from (111) planes is on the left, the <001> spike triplet is on the right. **(b, Right).** Electron diffraction spike pattern associated with a 2$\bar{2}$4-type reciprocal lattice point of a Type Ia diamond. Electron energy 120 kV, wavelength 0.00335 nm. The beam direction is very close to [111] and the specimen foil is parallel to (110). The triangle of <001> spikes encloses the weak, off-peak 2$\bar{2}$4 reflection and the thermal diffuse scattering that surrounds the reciprocal lattice point. Figs. 2a and 2b are printed to give roughly equal sizes of spike triplet triangles.

Type Ia diamonds are parallel to <100> directions, that they are absent in Type II specimens (which are platelet-free) and are seen most strongly in specimens very rich in platelets. We have evidence that spikes from diamonds containing platelets of large diameter (say 100 nm) are sharper than those from diamonds containing small-diameter platelets ($\sim$ 20 nm). (Since the platelets can be imaged directly in the electron microscope, there is no dependence upon spike sharpness measurements to provide estimates of mean platelet size as is the case in X-ray diffraction experiments (Suzuki and Lang 1976, Moore and Lang 1972, 1977). The prevalence of multiple scattering in the electron case makes it difficult to establish a relation between spike strength and order of reflection as has been done in the case of X-ray spikes (Hoerni and Wooster 1955). Our finding of non-vanishing spike strength in the third order is in accord with our finding from electron microscopic moiré patterns that the diamond matrix displacement by platelets is appreciably greater than one third of the f.c.c. cell edge of diamond(Bursill, Hutchison, Sumida and Lang 1981). However, another observation, that there are quite strong spikes extending from the origin of the reciprocal lattice, is at variance with the reported non-detectability of such spikes in the X-ray case (Hoerni and Wooster 1955). The general similarity evident between X-ray and electron spike patterns satisfyingly illustrates the wide applicability of the common underlying diffraction theory. Some experiments can be done in the electron case that are impossible with X-rays. For example, using the selected area diffraction technique, and including only one platelet in the field illuminated, the spike pattern reduces to one containing only spikes normal to the platelet concerned. On the other hand, in the electron case it is more difficult to set up an experiment analogous to the spike X-ray topograph (Takagi and Lang 1964). Such an analogue would be a dark field electron micrograph using one spike reflection to form the image. Some success towards achieving this has been obtained (Sumida and Lang, submitted to J. Appl. Crystallogr.)

Acknowledgements

The authors thank the Science and Engineering Research Council and De Beers Industrial Diamond Divn. (Pty) Ltd., for financial support.

References

Bursill L A, Hutchison J L, Sumida N and Lang A R 1981 Nature, Lond. __292__ 518-520

Dyer H B, Raal F A, Du Preez L and Loubser J H N 1965 Phil.Mag.__11__ 763-774

Evans T and Phaal C 1962 Proc.R.Soc.Lond. A__270__ 535-552

Hoerni J A and Wooster W A 1955 Acta Crystallogr. __8__ 187-194

Kaiser W and Bond W L 1959 Phys. Rev. __115__ 857-863

Lonsdale K 1942 Proc.R.Soc.Lond. A__237__ 315-320

Lonsdale K and Smith H 1941 Nature, Lond. __148__ 112-113

Moore M and Lang A R 1972 Phil. Mag. __25__ 219-227

Moore M and Lang A R 1977 J.Appl. Crystallogr. __10__ 422-425

Raman C V and Nilakantan P 1940 Proc. Indian Acad. Sci. A__11__ 389-397

Robertson R, Fox J J and Martin A E 1934 Phil.Trans.R.Soc.Lond. A__232__ 463-535

Suzuki S and Lang A R 1976 J. Appl. Crystallogr. __9__ 95-97

Takagi M and Lang A R 1964 Proc.R.Soc.Lond. A__281__ 310-322

Tu K N and Howie A 1978 Phil. Mag. B__37__ 73-81

An electron diffraction study of twist boundaries in 2H layer structures

G.J. Tatlock

Department of Metallurgy and Materials Science,
University of Liverpool, P.O. Box 147, Liverpool L69 3BX

1. Introduction

Diffraction patterns from layer structures of the CdI_2 type often exhibit
complex superlattices due to periodic distortions of the crystals, related
to charge density waves. However, as pointed out previously (Tatlock
1978), these can sometimes be confused with other superlattices caused by
double diffraction between rotated crystallites. The nature of these
patterns suggests that the crystallites have non-random misorientations.
Hence it was decided to carry out a more systematic study of the twist
boundaries between pairs of crystallites and compare the observations with
a simple coincidence site lattice (CSL) model. This paper reports
preliminary results for the 2H polytypes of TaS_2, MoS_2 and $MoSe_2$.

2. Boundary Structure

Projections of the unit cells of 2H TaS_2 and 2H MoS_2 are shown in Fig. 1.
$MoSe_2$ is isotypic with MoS_2 with minor changes in lattice parameters. The
structures consist of close packed layers of sulphur or selenium atoms
with metal atoms sandwiched between. This sulphur-metal-sulphur sandwich
is only held to the next one by weak Van der Waals forces, hence the
frequent occurrence of twist boundaries about the $[0001]$ axis during growth
from the vapour. Since it is most likely that the boundary plane lies
between two hexagonal close packed layers of sulphur atoms, a simple 2-D
CSL model is the most appropriate starting point with which to investigate
this behaviour. Comparison with the full 3-D CSL theory (e.g. Bollmann
1970) shows that the five boundaries with the highest density of
coincidence sites have Σ values of 7, 13, 19, 31 and 37, corresponding to
misorientations of 21.8°, 27.8°, 13.2°, 17.9° and 9.4° respectively.
Alternatively a simple geometrical method similar to that of Ranganathan
(1966) can be used to generate the Σ values as a set within the series
$\Sigma = \frac{1}{4}(3n^2 + m^2)$ where n and m are integers with n + m even and n > m > 0.

(hk.0) diffraction patterns from pairs of crystallites at these special
relative orientations exhibit commensurate superlattices. Double
diffraction effects ensure that not only spots corresponding to the two
sets of matrix reflections are present in hexagonal networks, but that
combinations of the scattering vectors also give rise to extra reflections.
Hence, for example, in Fig. 2a the pattern from a $\Sigma7$ boundary has 6
additional spots between the matrix reflections along the $[a_1^* + 2a_2^*]$
directions and the pattern from a $\Sigma13$ boundary (Fig. 2b) shows 12
additional spots along $[a_1^* + 3a_2^*]$ etc. Small relative misorientations
give characteristic patterns of satelite spots (Fig. 2c), while complex

patterns are obtained at higher angles near low Σ valued orientations
such as the 16.5° pattern shown in Fig. 2d. However, the relative
misorientations may still be determined in each case.

3. <u>Results and Discussion</u>

At least 80 diffraction patterns were taken from regions containing
rotated crystallites for each batch of crystals. The histograms shown in
Fig. 3 were then constructed from the measured misorientations which were
plotted at intervals of 0.75°. Only the angular range 0° - 30° is needed
due to the crystal symmetry. Special care was taken to avoid regions of
the samples where twist boundaries may have been generated by accidental
overlap. Also identical boundaries were not counted more than once in
adjacent regions since they may have been remnants of one larger boundary
which broke up during sample cleavage. (This may, of course, have
artificially reduced the peak heights in the distributions if several
individual but identical boundaries were present in the one sample.)

The most obvious feature of the histograms in Fig. 3 is the lack of small
angle misorientations in the range 0° - 4°. This effect has been noted
previously in other work on twist boundaries in MgO and CdO by
Chaudhari and Matthews (1971) and was ascribed to rotation of the newly
formed crystallites with these misorientations into the perfect 0°
orientation. In the present case, however, this may have been
systematically overemphasised since very small misorientations can easily
be missed when examining diffraction patterns in the TEM. The other main
features of the histograms are the preferred occurrence of certain
orientations. Although there is a rather high background noise level in
the results, certain orientations are particularly favoured. The widths
of the distributions round these maxima are not just due to experimental
errors in the measurements since these should be $\sim \pm \frac{1}{2}^{\circ}$.

2H TaS$_2$ was the first material to be examined and Fig. 3a shows the
results from a batch of good quality crystals grown with no iodine
transporting agent. Most of the low Σ valued orientations were favoured,
although the distribution around the $\Sigma 13$ orientation is very broad.
Indeed, other results from another batch of poorer quality 2H TaS$_2$
crystals showed little preference for orientations near $\Sigma 13$ and this
effect may be due to impurities or the presence of other polytypes of
TaS$_2$. However, it does emphasise the caution needed in interpreting the
results. In order to investigate the influence of second nearest
neighbours on the boundary orientations 2H MoS$_2$ was also examined. Fig. 1
shows the difference in structure between TaS$_2$ and MoS$_2$ and in particular
the change in site of the metal atoms relative to the two close packed
sulphur layers. The MoS$_2$ results are shown in Fig. 3b and comparison with
Fig. 3a reveals a much sharper distribution round the $\Sigma 13$ orientation and
the absence of the $\Sigma 31$ orientation. The results of changing from sulphur
to selenium in the MoS$_2$/MoSe$_2$ structure are illustrated in Fig. 3b and 3c.
One of the main changes appears at low angles where the $\Sigma 37$ boundary seems
to be much more favoured in the selenide; and the possible reasons for all
these changes will be discussed more fully at the conference.

In conclusion, these preliminary results suggest that certain twist
boundary orientations, often with low Σ values, are preferred in 2H layer
structures such as TaS$_2$, MoS$_2$ and MoSe$_2$. However, comparison with
previous work on CdO and MgO by Chaudhari and Matthews (1971) and Mykura
et al (1980) shows that there is a greater random element in the layer

Fig. 1.   (11$\bar{2}$0) projections of the unit cells of   (a)   2H TaS$_2$
          (b)   2H MoS$_2$

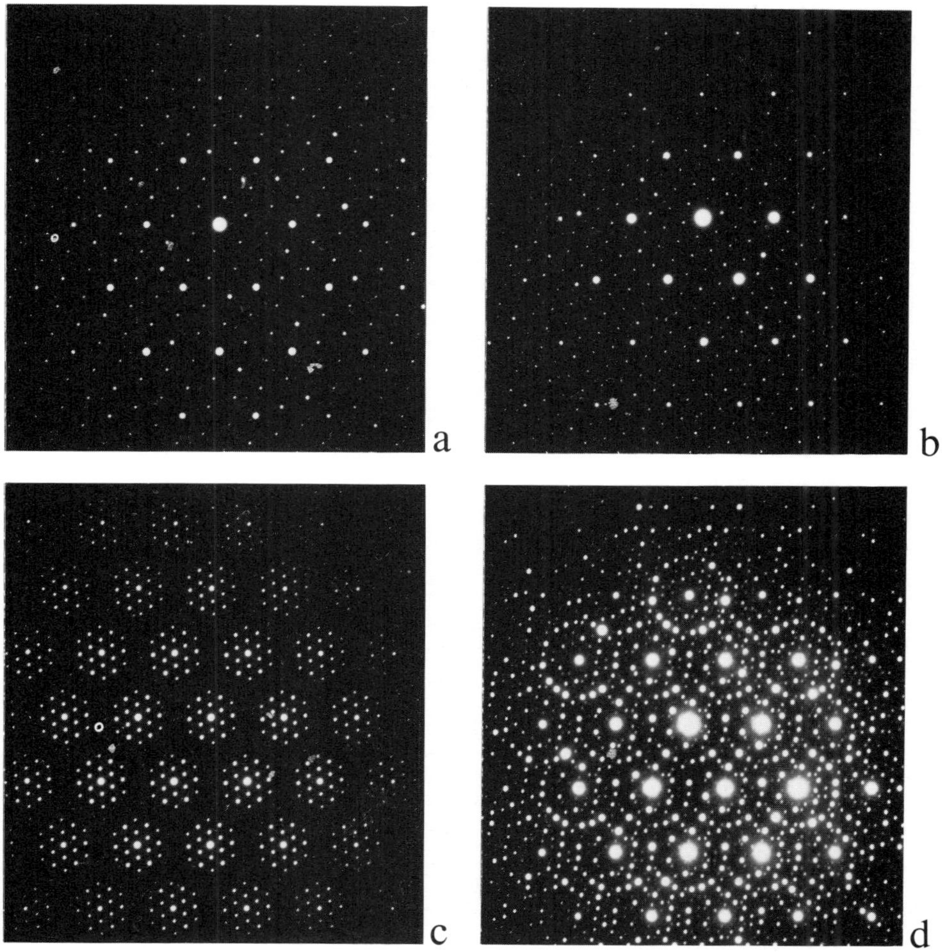

Fig. 2.   (hk.0) diffraction patterns from [0001] twist boundaries in
          (a)   2H MoSe$_2$ (Σ7)   (b)   2H TaS$_2$ (Σ13)   (c)   2H MoSe$_2$ (10°)
          (d)   2H TaS$_2$ (16.5°)

structure results, possibly due to the growth conditions employed. As
expected it also appears that the boundaries are very sensitive to crystal
perfection, impurities etc. Changing the second nearest neighbour sites
or the type of atom at the interface may also have more subtle effects on
the distributions, although care is needed in interpreting these results
in view of the high noise level in the histograms.

4. Acknowledgements

The author is indebted to Dr. R.C. Pond for useful discussions and
Mr. M.E. Lewis for some of the $MoSe_2$ measurements. He is also grateful
to Dr. A.D. Yoffee and his group for the supply of crystals.

5. References

Bollmann, W. 1970 Crystal Defects and Crystalline Interfaces (Berlin:
                    Springer Verlag)
Chaudhari, P. and Matthews, J.W. 1971 J.Appl. Phys. <u>42</u> 3063
Mykura, H., Bansel, P.S. and Lewis, M.H. 1980 Phil. Mag. <u>A42</u> 225
Ranganathan, S. 1966 Acta. Cryst. <u>21</u> 197
Tatlock, G.J. 1978 Proc. 9th Int. Cong. on EM ed J.M. Sturgess
                    (Toronto: Microscopical Soc. Canada) 558

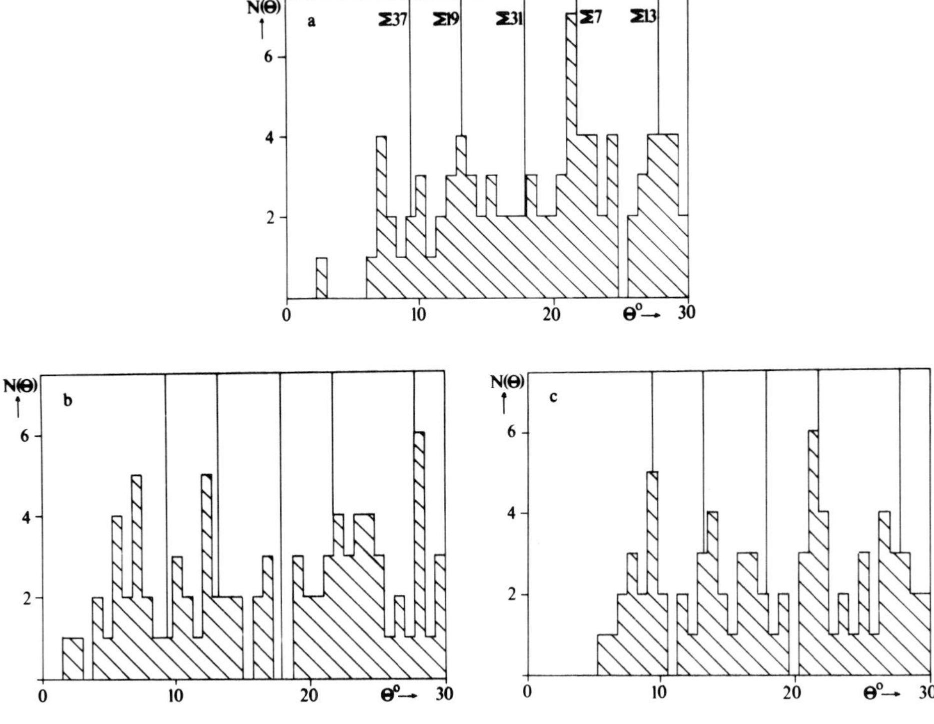

Fig. 3. Histogram of the number of $[000\overline{1}]$ twist boundaries ($N(\theta)$) at
         angle θ, against θ, for (a) 2H TaS_2 (b) 2H MoS_2
         (c) 2H $MoSe_2$

Low temperature phase changes in layered compounds

S McKernan, K K Fung[*] and J W Steeds

Physics Department, University of Bristol, Bristol BS8 1TL

Introduction

Many layer-structured transition metal dichalcogenides are known to
undergo charge density wave (CDW) transitions on cooling (e.g. Wilson
et al (1975)). These transitions are accompanied by periodic structural
distortions of the host lattice, forming a superlattice which is readily
observable in the diffraction patterns of these materials. In the majority of such materials the periodicity of the distortion (and hence the
superlattice) at the transition temperature, T_O, is incommensurate with
the underlying lattice, but becomes commensurate on cooling to the "lock-
in" temperature T_d. The CDW formation is very largely fermi surface
dependent, and consequently affects many of the materials' physical
properties. This accounts for the large theoretical and experimental
interest shown, within many fields, in these compounds.

The structure of these materials takes the form of three-layer sand-
wiches: a hexagonal sheet of metal atoms between two similar sheets of
chalcogen atoms. The chalcogens may have either octachedral or trigonal
prismatic coordination with respect to the metal atoms; and this coordination may vary in a regular way, giving rise to a variety of stacking
sequences.

The material reported here - 2H TaSe$_2$ - has a two layer stacking sequence
of trigonal prismatically coordinated chalcogens. It forms an incommensurate superlattice at $T_O \sim 120$ K, and a commensurate $3a_O$ x $3a_O$ one
below $T_d \sim 90$ K. The symmetry of the commensurate phase has been previously reported to be hexagonal, the same as the undistorted crystal
(P6$_3$/mmc) (Brouwer 1980, Moncton 1977). This symmetry for the low temperature phase is shown to be the result of averaging over several
microtwinned orthorhombic domains.

Samples of a single crystal 2H TaSe$_2$ were prepared for use in a Philips
EM400 by repeated cleavage to thicknesses of 50 - 100 nm. The samples
were cooled to below 70 K in a double-tilt liquid helium cooled stage
designed by J A Eades and built in this laboratory.

Satellite diffraction spots - arising from the CDW are normally rather
weak, but suitable tilting of the specimen makes one or more of these
spots "flare out" strongly. Dark field images from these strongly
excited spots reveal the domain structure of the CDW. We have used this

*Present address: Institute of Physics, Academica Sinica,Beijing, China.

technique to study the domain structure in other CDW-bearing materials
(Fung and Steeds 1980a and 1980b).

<u>Results</u>

Convergent beam patterns (CBPs) from single domains show marked depart-
ures from hexagonal symmetry. In many patterns a varying degree of
trigonality is shown, which is also evident above the transition temperature. The amount of this trigonal distortion is associated with faint
wavy dislocations visible in bright field. These dislocations may be the
product of beam damage, as they move readily under the influence of the
electron beam and the number of the dislocations increases with prolonged
exposure. They lie in the basal plane and, because of their faintness,
may involve displacements of the selenium atoms only, disturbing their
coordination or the overall stacking sequence. In either case a stacking
fault parallel to the specimen surfaces is produced. Tatlock (1973) has
shown how such faulting can reduce the observed symmetry properties of a
crystal, in particular from hexagonal to trigonal.

Apart from trigonal distortions, the low temperature CBPs contain some
discs which do not show six fold symmetry particularly the $\{40\bar{4}0\}$ and
$\{11\ 12\ \bar{23}\ 1\}$ reflections (see figure 1). In an ideal crystal (i.e. no
trigonality) this would reduce the whole pattern symmetry from 6 mm to
2 mm producing an orthorhombic rather than hexagonal diffraction pattern.
Using the hexagonal-to-orthorhombic conversion $\bar{1}100 = 020$, $\bar{1}\bar{1}20 = 200$ we
find that the first order Laue zone reflections $\pm\ 22\ 0\ 1$ appear to have
zero intensity, despite neighbouring reflections ($\pm\ 22$, $\pm\ 2$, 1 and $\pm\ 23$,
$\pm\ 1$, 1) being strongly excited, implying the existence of a glide plane
perpendicular to the b axis. Combining this with the zero layer geometry
of two centred faces, indicates that the space group for the low temperature phase is either number 36 ($Cmc2_1, C_{2v}{}^{12}$) or number 63 ($Cmcm, D_{2h}{}^{17}$).
These groups differ only by an inversion centre and could easily be distinguished in a perfect specimen. Present evidence seems to favour the
latter (Cmcm) space group, containing an inversion centre.

Dark field images formed using strongly excited hki0 satellite reflections reveal many domains, of average size 500 µm, which are not visible
in bright field images. $h0\bar{h}0$ images show the domain interiors in bright
or dark contrast, but reveal only a portion of the boundaries (see
figure 2). Images formed from the weaker $hh\bar{2}h0$ satellites reveal all of
the boundaries in strong contrast, but give only weak interior contrast.
Dark field images in the symmetry related satellites $h0\bar{h}0$, $0\bar{h}h0$ and $\bar{h}h00$
show that any particular region appears bright (i.e. strongly diffract-
ing) in two of the images and dark (i.e. weakly diffracting) in the third.
The sign of the reflection appears to have little effect on the domain
contrast. Similarly any particular boundary is visible in two of the
images and not in the third. CBPs obtained from different sides of a
boundary show a mutual rotation of the orthorhombic axes of $+2\pi/3$
(or $\pm\ 2\pi/6$). This is consistent with the orthorhombic CBP where the 0k0
reflections are essentially absent for $k \neq 6n$. Domains of similar
orientation appear to be connected together, forming necklace-like chains
which often form together in groups of three.

<u>Discussion</u>

Using satellite dark field images to locate individual domains, and small
probe convergent beam techniques to explore their symmetry, it is clear

that the apparent hexagonal symmetry of low temperature diffraction patterns is due to the averaged effect of many twinned orthorhombic domains. The contrast of these domains is then easily interpreted in terms of the intensity of the orthorhombic reflections. The morphology of the domains, however, is the result of processes occurring in the incommensurate phase between 120 K and 90 K. Details of this phase are to be published elsewhere.

Acknowledgements

Support for this work from the Science Research Council is gratefully acknowledged. We are indebted to Drs J A Wilson and J A Eades for much helpful discussion and Mr J H Burrow for the supply of the crystals.

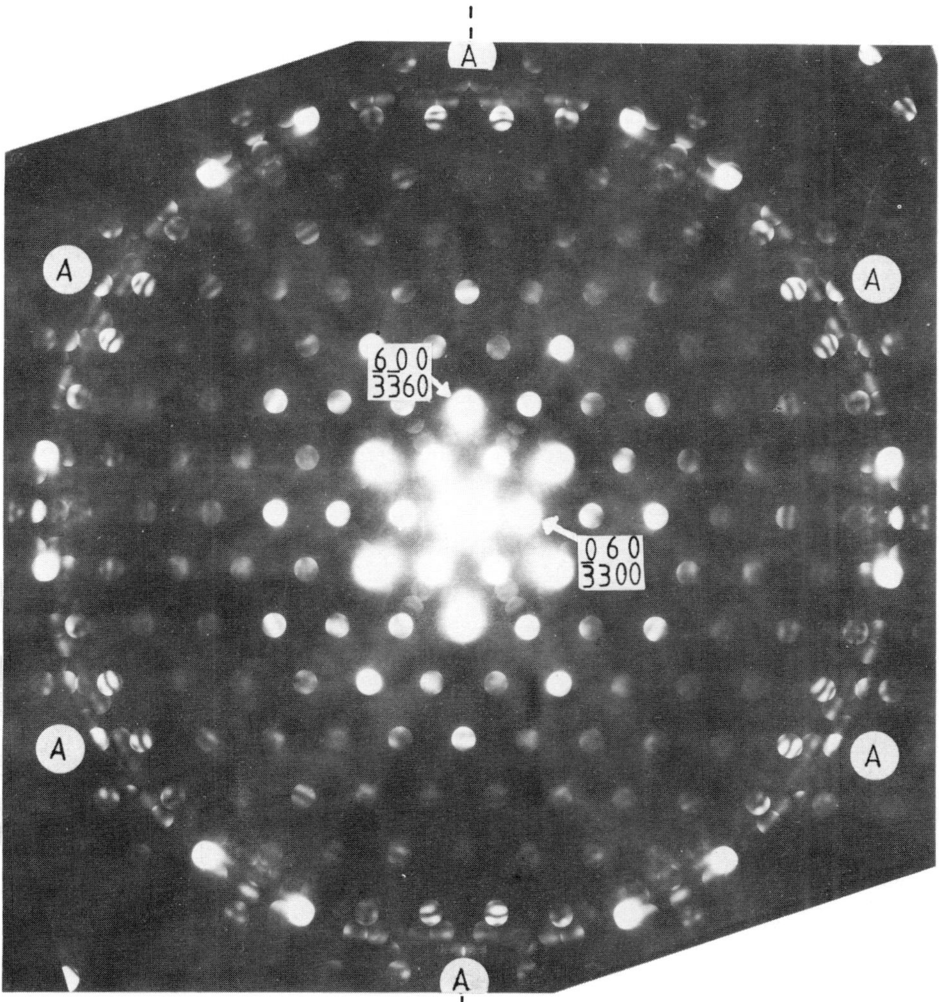

Fig 1. Convergent beam pattern from a single orthorhombic domain in 2H TaSe$_2$ at $\sim$ 70 K, 120 kV. The 2 mm symmetry is most easily seen at points A. Mirror lines shown dashed.

References

Brouwer R and Jellinek F (1980) Physica B+C 99 51 .
Fung K K and Steeds (1980a) Phys Rev Lett 45 1696.
Fung K K and Steeds (1980b)Proceedings EMSA 338.
Moncton D E,Axe J D,and DiSalvo F J (1977) Phys Rev B 16 801.
Tatlock G J (1973) PhD Thesis, University of Bristol.
Wilson J A, DiSalvo F J and Mahajan S (1975) Adv Phys 24 117.

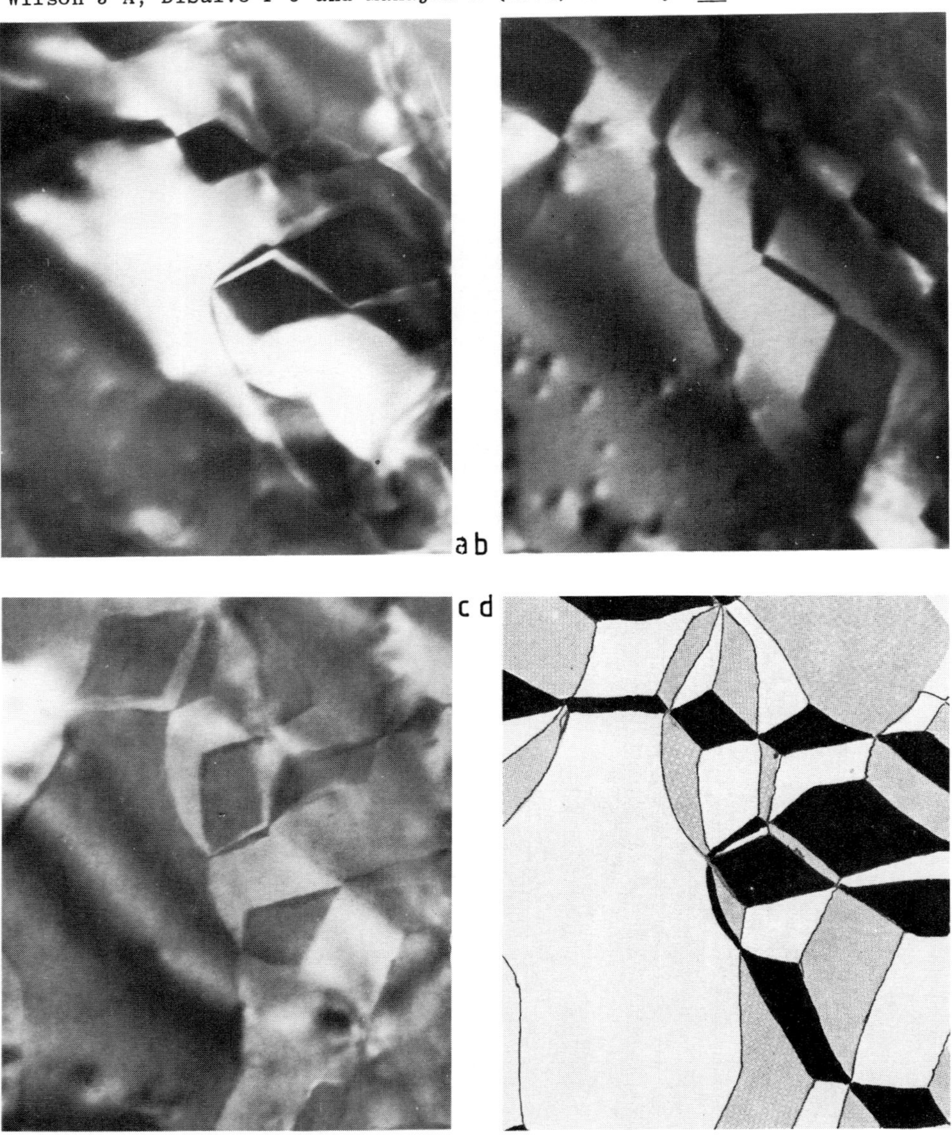

Fig 2.   Satellite dark field images taken using a) 1010, b) 1100 and
c) 0110 reflections. d) is a composite diagram showing how the domains
fill all space and the interlocking nature of the domains.

Low temperature structural studies of intercalated organometallics by electron microscopy

M. Audier, J.M.Thomas, R.Schlögl*, G.M. Parkinson and W. Jones

Department of Physical Chemistry, University of Cambridge, U.K.
*Institut fur Anorganische Chemie der Universität Munchen, F.D.R.

The nature of the intercalation of organic and organometallic molecules
in layer compounds is being investigated by electron diffraction. The
intercalated samples are studied in a modified Siemens Elmiskop 1A in the
temperature range, 300 to 15 K using a double-tilting liquid helium cooled
stage. Information complementary to the diffraction data is obtainable
by the simultaneous measurement of mass-spectrometric analysis and cathodoluminescence spectra and the effect of the direct excitation of intercalated molecules may be studied using the facility of in-situ UV irradiation.
The low temperatures have been used to prevent de-intercalation in vacuo,
to study the effect of molecular ordering on cooling and to decrease the
rate of electron-induced radiation damage of the organic molecules.

Samples of $Sn(CH_3)_4$ intercalated in flakes of natural graphite have been
studied in the above manner and it has been possible to derive a detailed
structural model for this material at low temperatures. The intercalation
was carried out by using a variation on the photochemical technique of
Schlogl and Boehm (1980). The material produced is a second-stage intercalate, the c-axis repeat distance determined by X-ray diffraction to be
approximately 12.9Å at room temperature.

Fig. 1 shows the electron diffraction pattern of a sample of $Sn(CH_3)_4$ in
graphite cooled to 80° K and oriented so that the electron beam is parallel
to the [0001] graphite zone axis. Whereas the graphite reflections are
sharp those arising from the intercalate are arced by 10° indicating some
degree of rotational disorder about the [0001] intercalate zone axis (which
is coincident with the same zone axis of graphite). The other prominent
features of the diffraction pattern, the short arcs and "triangles"
arise from multiple scattering. No coincidences between reflections from
the graphite lattice and the intercalate lattice are seen in Fig.1, indicating that the intercalate matrix is incommensurate with that of the
graphite. In order to determine what degree of long range order would be
required for coincidence to occur, the corresponding distances in real
space along the [1100] and [1120] directions were compared in increasing
multiples until very nearly coincident distances were obtained. Good
correspondence was obtained between $19xd_{[1\bar{1}00]}$ graphite and $13xd_{[11\bar{2}0]}$
intercalate. Using this distance to construct an intercalate lattice
it was found that an exactly commensurate structure could be produced
by rotating the graphite and intercalate lattices with respect to each
other by $\pm 5.2^{\circ}$. On this basis it is possible to construct a model of the
intercalated layer consisting of domains, separated by grain boundaries
or vacancies, rotated by $\pm 5^{\circ}$ with respect to the graphite. Investigation
of this model by optical diffractometry confirms that the observed arcing

of the intercalate reflections may be accounted for in this way (Harburn et al., 1975).   The commensurate structure may be described as $\sqrt{28}$ x $\sqrt{28}$ R 16.77°.

That a commensurate structure results from rotation of the intercalate layer by 5.2° with respect to the graphite layer suggests the possibility of observing a rotation of one graphite layer with respect to another: one graphite layer below the $Sn(CH_3)_4$ rotated by +5.2°, the layer above rotated by -5.2°. Fig. 2(a) shows the electron diffraction pattern from a different area of sample where this rotation is observed, giving rise to a doubling of the graphite and intercalate reflections and the multiple diffraction "triangles". Fig. 2(b), shows a diffraction pattern from a sample titlted so that the electron beam is no longer perpendicular to the basal layer and (000ℓ) reflections are seen. The c-axis repeat distance is determined to be 13.2±0.1Å.

Using the measured crystallographic parameters and the known Van der Waals radii and covalent bond lengths, a three dimensional structure for the intercalate was constructed (Fig.3). From the commensurate relationship between the two lattices it is calculated that the distance projected onto (0001) separating the tin atoms is 2.078Å and thus in order to avoid interpenetration of Van der Waals volumes it is necessary for molecules to be staggered along the c-axis, in total four-layers of $Sn(CH_3)_4$ occurring between two graphite layers. Assuming second stage intercalation, the proposed model gives a c-axis repeat distance of 13.2Å in good agreement with the experimentally observed value. The discrepancy between the room temperature value for the c-axis (12.9Å measured by X-ray diffraction) and the value at 80 K may be understood in terms of the intercalate forming two-dimensional liquid-like layers at room temperature which, upon cooling undergo some lateral contraction to produce close packed layers. As a result of this the now fixed orientations of the molecules in the four intercalate layers make them distinct and, to avoid interpenetration, slight expansion along the c-axis is required. During the cooling and "crystallisation" of the intercalate some of the observed rotations of graphite layers with respect to each other may also occur.

References
R Schlögl and H P Boehm 1980 Carbon Conference, Baden-Baden, preprints.
G Harburn, C A Taylor and T R Welberry 1975 Atlas of Optical Transforms
   (London: Bell), p.16.

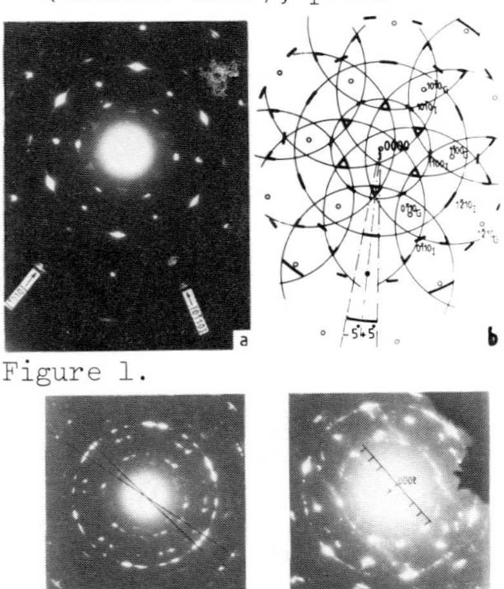

Figure 1.

Figure 2.

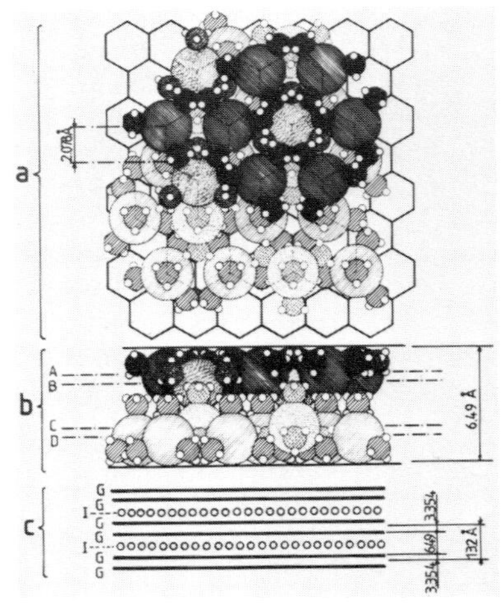

Figure 3.

STEM multiple images, the Karhunen–Loeve transform and data compression

R.E. Burge and A.F. Clark
Department of Physics, Queen Elizabeth College, University of London,
Campden Hill Road, LONDON W8 7AH

1.   Introduction

The geometry of the STEM detector plane lends itself to the simultaneous
collection of multiple signals in both bright and dark field (Burge and
van Toorn, 1980), and the sequential formation of the images makes the
instrument ideal for on-line digitisation.

The multi-channel recording system in use on our prototype Vacuum
Generators HB5 STEM (Browne, Charalambous and Burge, 1981) can digitise
the signals from up to eight scintillator/photomultiplier combinations
simultaneously, using dual slope integration.   Eight channels may be
recorded at 256 x 256 pixels per image, using a 64 µs sample time, or four
channels of 512 x 512 pixels, using a 32 µs sample time.   Eight bit
conversions are used.   The digitised images are stored in video memories
and may be replayed on TV monitors or transferred to a Data General Eclipse
mini-computer, where they are stored on disk.   Up to eight "scans", each
of eight 256 x 256 images, or four scans of four 512 x 512 images may be
recorded before the images must be transferred to magnetic tape for long-
term storage or processing.   The titling of images, selection of scan size
and transfer of images from video memories to disk and from disk to tape,
are all under computer control.

The detector configuration used for this work consists of (Ward et al 1980)
four quadrant detectors, a small, central bright-field detector and an
annular dark-field detector.   The separate recording of signals in
inelastic scattering has not yet been implemented.

The signals recorded from the different detectors correspond to the scattering of incident electrons into different solid angles;   this is analogous to multi-spectral optical images, for example LandSat images, where
different channels correspond to optical reflectance in different wave-
bands.

It is usually found that there is some degree of redundancy in the
information content of the images from the various channels, for a given
object;   the optimum method of analysing this redundancy is with the
Karhunen-Loeve {K-L} or eigenvector transform {Hall 1979}, which re-
distributes the signal power of the multi-channel images into "principal
components" of decreasing power, which are orthonormal {uncorrelated}.

We shall consider here the application of the K-L transform to data compression from both multiple images taken simultaneously in six parallel
channels and single images.   Because there is no fast algorithm for
computing the K-L transform, for compression of single images the discrete
cosine transform (DCT) is considered.   The DCT closely approximates the
K-L transform and is a suitable alternative for data sets greater than

128 x 128 pixels.

2.   Model calculations

The action of the K-L transform on multiple images will be illustrated by computations using the full theory of image formation under conditions of probe partial spatial coherence (Burge & Dainty, 1976) of the noise-free images, as seen by the six detectors noted above, of an object comprising five uranium atoms distributed on a 10Å thick carbon film (Fig.1). Transforming six channels, each of 64 x 64 pixels, takes 25 sec. using the DEC VAX 11/780 computer with double-precision arithmetic.

The images from the opposite quadrant detectors (a&c or b&d in Fig.1) are similar, but exhibit reversed contrast with respect to each other.   It is therefore reasonable to expect that at least two of the channels carry redundant information.   As the K-L transform is a determination of basis vectors in N-space (where N is the number of channels), calculated by maximising the variance along the basis vectors, any images which are redundant have reduced contrast in the transform.   Calculation of the transform of these images confirms this.   The transforms of the images of Fig.1 are shown in Fig.2. in the form of six principal components images. To allow inspection of detail each transformed image has been printed with similar contrast;   in reality the effective image power is proportional to the variance and the relative intensities of the images decrease rapidly with decreasing eigenvalue.   The variances of the images before trans-formation (arbitrary units) were 5569, 6073, 5571, 5006 (quadrant detect-ors), 43 (dark field) and 1 (point central aperture).   The variations in the variance for the four quadrants is due to the problem of the central off-set when representing a circular aperture by a square sampling raster. The variances of the principal components images following transformation are 10333, 10133, 1720, 47, 29, 1;   the sum of the power spectra of the images before and after transformation is constant.

The three images with the lowest variances (Fig.2) contribute only 0.3% to the total variance and may therefore be discarded;   indeed, even if the four lowest principal components are discarded, over 90% of the signal power is still present.   This reduction in the order of the data greatly simplifies attempts at feature extraction, classification, etc.

It is of interest to observe the relationship between the original images and their principal components.   Empirically the following identification can be made.

The first principal component is very similar to the difference images from a vertically split Dekkers/de Lang (half-plane) detector (a+b-c-d in Fig.1), while the second component is similar to the detector split horizontally (a+d-b-c in Fig.1).   The third component corresponds to the sum of opposing quadrants (a+c or b+d).   Identification of the fourth and sixth components is unclear, while the fifth component resembles the image from the middle annulus of a Rose centrosymmetrical detector (central circular collector and two annuli filling the bright field cone). Apportionment of significance to the features of the principal components images is discussed by Lowitz (1978) in connection with multispectral images.   He points out that the homogeneous zones of a scene (large scale structure) are found in the principal components images of low order and the texture information is found in the higher order images.

If the six initial images are made up differently, e.g. by using for four of the signals the sums and differences of signals in opposite quadrant detectors, the results are a little different; although the transformed

images are similar, the power is contained almost entirely within the
first two components.   For an annular (Rose-type) system of detectors,
the individual images are reduced to effectively one component,which
correlates highly with the image from the middle annulus of the Rose
detector.

3.   Application to Experimental Images

We have applied the K-L transform to six images of each of two specimens.
The first (Fig.3) is the globular iron storage protein ferritin in which
each particle has a central core of about 50Å diameter of hydrated iron
oxide, and the second (Fig.4) is the cilia of <u>Tetrahymena pyriformis</u>
splayed by biochemical treatment and negatively stained with uranyl-
acetate.   Both specimens are supported on thin films of evaporated
carbon.   The principal components images are shown in Figs.5 and 6 for
ferritin and Fig.7 for cilia.   In Fig.5 an attempt to represent the
relative image powers has been made to allow the assessment of the
relative significance of the images.   In Fig.6 the higher principal components images have been artifically enhanced to show their structure.

For ferritin, several of the principal components images may be discarded;
the variances of the components are 6, 11, 204, 818, 1122 and 3457
respectively.   Examination of the component images shows that the object
has become separated from the background.   This observation is confirmed
by the calculation of the cross-correlation coefficient; the correlation
between the background of the image from the central bright-field detector
in which the background is most easily visible, and the third component of
the transform is over 80%.   Furthermore, the scan-line distortion of the
images has been largely concentrated in the fourth transform component.

This separation is due entirely to the nature of the images; the means of
the carbon film and the ferritin differ widely.   As there are more pixels
in the background than in the image and the background has a higher
variance than the object, the mean of the image as a whole is close to
that of the background.   Since the K-L transform maximises the variance
of the first principal component relative to the mean, the object becomes
separated from the background.   A similar effect explains the separation
of the scan-line distortion.   This significantly enhances the contrast
of the first principal component.   These results are consistent with
those of Lowitz (1978).

Several of the ferritin images also exhibit similar relationships to those
noted for the simulated images and their transforms.   The first principal
component is similar to the sum of signals from opposing quadrants, while
the second and fifth components correspond to opposite quadrant difference
images.

For the cilia specimen, the small contribution from the dark field and the
similarity of the quadrant images effectively reduce the data to one
principal component (Fig.7).   Comparison of the original and component
images shows that the first image is similar to a quadrant sum image,
while the second and fourth components are similar to quadrant difference
images, the same effect as was found for the simulated and ferritin images.

4.   Data Compression

One of the problems encountered in processing the large amounts of data
produced by a multi-channel STEM is that of storage.   By reducing the
variation in the higher component images, the quantity of data has been
compressed.   All that need be stored is the kernel of the transform
(N x N, where N is the number of channels), the highest variance images

and the means of the discarded images.   The original images may be reconstructed by an inverse transform, which takes about 5 seconds on a VAX-
11/780 (double-precision).   The degree of fidelity between original and
compressed images is obviously important, and may be described by the
cross-correlation coefficient.   The results obtained for the samples of
Figs. 6 & 7 are given in Table 1.

No. Images Retained	Ave. correlation (Ferritin)	Av. correlation (Cilia)
1	0.786	0.900
2	0.854	0.910
3	0.913	0.931
4	0.943	0.983
5	0.975	0.996
6	1.000	1.000

Table 1:   Average correlation between original and compressed
images for Ferritin and Cilia images

For ferritin, the signal power is spread predominantly across the first
three components;   compression to three components gives reconstructed
images with over 90% correlation.   For the cilia specimen, the power is
concentrated in the first component and the compression is even greater.
The data may be reduced to 1 image for storage while the reconstructed
image still has 90% fidelity.   In general, the reconstructed images show
large features well;   it is the details of the images (due to their small
variance) which are degraded.

It is also possible, by partitioning an image into "blocks", to compute
the K-L transform of a single image, providing that the size of the
blocks is greater than the correlation length of points in the image.
The background of the ferritin images consists of amorphous carbon, so its
correlation length is small;   blocks of 8 x 8 pixels are sufficiently
large, while allowing reasonably quick computation.   The single-image
transform reduces the central bright-field image from 64 x 64 pixels to
about 12 x 64 pixels.   The reduction for the first principal component,
which has a smoother background, is even greater, to about 8 x 64 pixels.
In both cases, the correlation between original and reconstruction is
better than 90%.   Again, it is the detail in the image which is degraded.

Due to the uniformity of fairly large areas of the cilia specimen, the
amount of redundancy in each block is greater and the compression is
better;   for the central bright-field detector the data is reduced to
6 x 64 pixels and for the first component image the reduction is to
3 x 64 pixels, both with 90% fidelity.   This latter reduction of the data
is by a factor of better than 21:1

The K-L transform is computationally inefficient for the usual 512 x 512
pixel images normally considered for storage and the adaptive cosine
transform (Burge & Wu, 1981) has been modified and used for the compression of single electron images.   This fast transform also lends itself
to the implementation of pattern recognition and texture classification
measures (Wu & Burge, 1982) which increase the tolerable degree of image
compression for constant image fidelity.   Compression ratios of 15 to 20
are reasonable with most electron images of a distributed nature, and
up to 50 or more may be achieved for specimens with distributions of
particles.

5. Discussion

We have shown how some approaches to image analysis developed for other fields of research, notably for satellite image processing, may be adapted for electron microscopy. The K-L transform provides an efficient means of bringing together the information available for collection across the STEM detector plane. One motivation for the work is the minimising of electron beam damage for a given electron exposure and another is the reduction in redundant information and the consequent efficient storage and display of the large quantities of digital image data that may be rapidly accumulated with STEM.

6. Acknowledgments

We thank the Science & Engineering Research Council and the Royal Society for their support of the work.

7. References

Browne, M.T., Charalambous, P., and Burge, R.E. 1981 In preparation.
Burge, R.E. and Dainty, J.C. 1976 Optik 46, 229-40.

Burge, R.E. and van Toorn, P. 1980 Scanning Electron Microscopy
Vol.I, SEM Inc., AMF O'Hare, 81-91.

Burge, R.E. and Wu S.K. 1981 Ultramicroscopy (In press).

Hall, E. 1979 Computer Image Processing & Recognition, London,
Academic Press, 115.

Lowitz, G.E. 1978 Pattern Recognition 10, 3591363.

Ward, J.F.L.,Browne, M.T. and Burge, R.E. 1980
Inst. Phys. Conf. Ser. 52, 85-8.

Wu, J.K. and Burge R.E. 1982
Computer graphics and Image processing (in press).

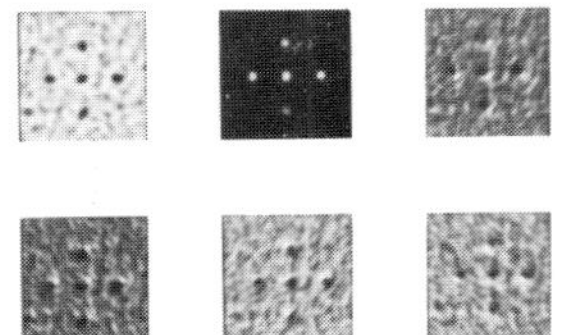

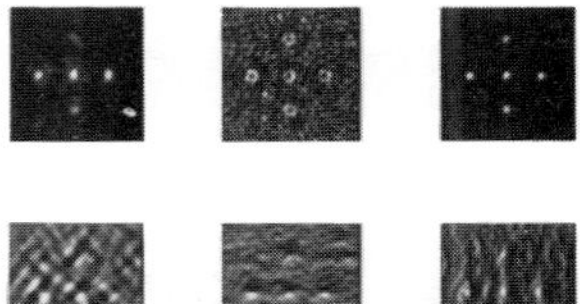

Fig.1 Calculated images of U atoms on C field, 80 Kev STEM. Images (R to L) are: top line, quadrant (d), dark field and axial point; lower line, quadrants (a) to (c).

Fig.2 Principal components derived from Fig.1. Images (R to L) are: 4th to 6th components, lower line 1st to 3rd components. Contrast of components higher than first artifically enhanced.

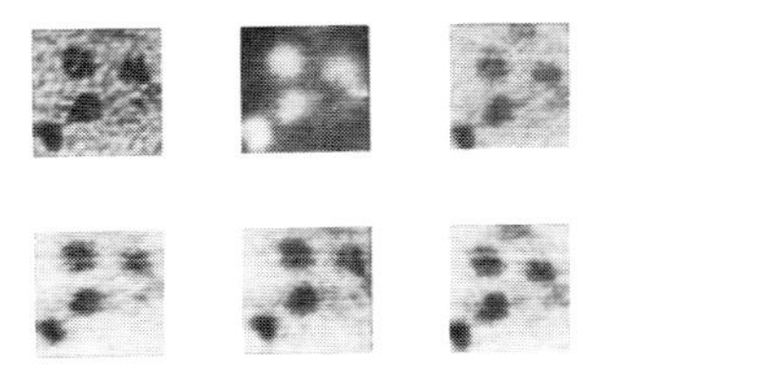

Fig.3   Experimental STEM images
of ferritin cores.diameters
50Å

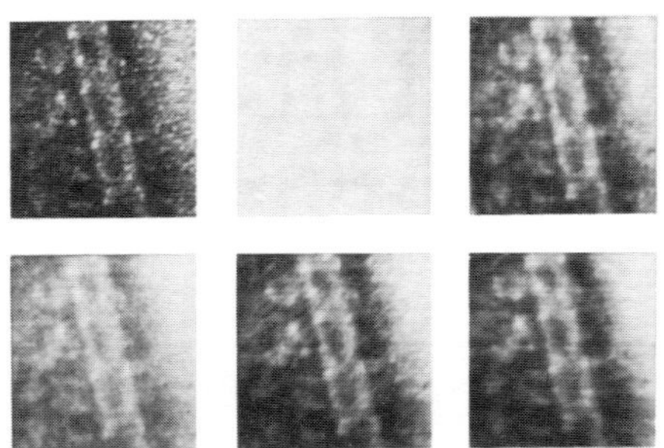

Fig.4   Experimental STEM images
of cilia, frame width 3000Å
Images as in Fig.1

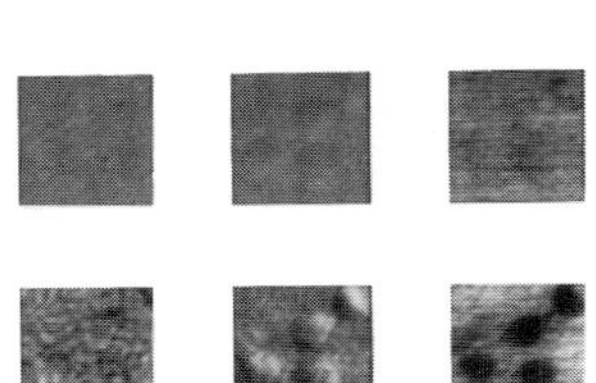

Fig.5 Principal components
derived from Fig.3.   Images
as Fig.2; without
enhancement

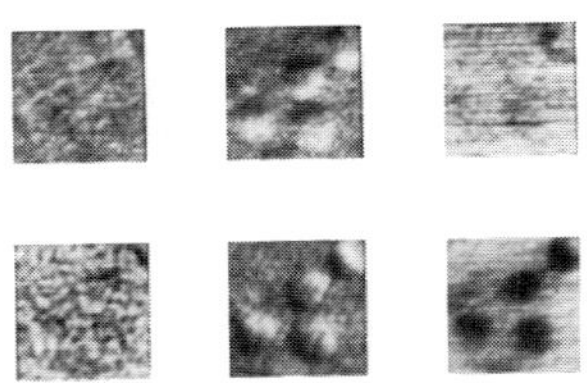

Fig. 6   Principal components
as in Fig.5 Contrast enhanced

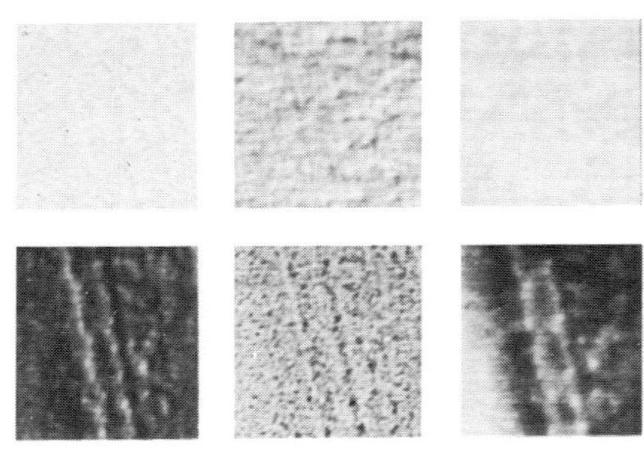

Fig.7   Principal components for
cilia images.   Images as in
Fig.2.   Contrast enhanced.

Aspects of high resolution imaging in STEM

C. Mory, C. Colliex, O. Krivanek

Laboratoire de Physique des Solides, Bât. 510, Université Paris-Sud, 91405 Orsay, France.

Since the first report by Crewe et al. (1970) that they have visualized single atoms, STEM instruments have been shown to offer several unique features for the recording of information below the nanometer scale. For a good insight into experimental state of STEM atomic imaging, see Crewe (1978), Wall (1978) and Ohtsuki (1980). In this contribution the present capabilities of a VG-HB 5 microscope fitted with two condenser lenses and a high excitation objective (f $\simeq$ 3mm, C_s $\simeq$ 3.2 mm) have been evaluated in terms of high resolution imaging in an analytical environment. It deals specifically with the imaging of metallic clusters and of single heavy atoms, and with the direct observation of the dynamic behaviour of atoms in clusters.

The microscope is operated at 100 kV primary voltage and 10 µA total field emitted current. Illumination conditions are set so that a current of 5 x 10^{-10} A is focused into a probe of 0.4 – 0.5 nm diameter (50 µm aperture in the back focal plane of the objective) on the specimen. The contribution of the source size is small (< 0.2 nm) and the angle of convergence (α = 7.7 x 10^{-3} rad) close to the optimum value :

$$\alpha_{opt} = 1.4 \ C_s^{-\frac{1}{4}}.\lambda^{\frac{1}{4}} \simeq 7,5 \ x \ 10^{-3} \ rad$$

Vibrations and stray fields effects remain below 0.1 nm. Two detectors – the axial bright field (BF) and the annular dark field (ADF) – simultaneously collect the electrons scattered from the specimen within solid angles defined in fig.1. Magnification factors above 1 M are used, so that each image pixel is oversampled, and micrographs are typically recorded in 20 ms per line and 20 s per frame. The currents available at the output of the PMT can be permanently displayed on the scopes, offering the possibility of quantitative contrast measurements along line traces or through Y modulated pictures. The sample is a 2 nm thick amorphous carbon film on which DNA or RNA strands slightly stained with uracyl acetate have been deposited (Mory et al. (1981)).

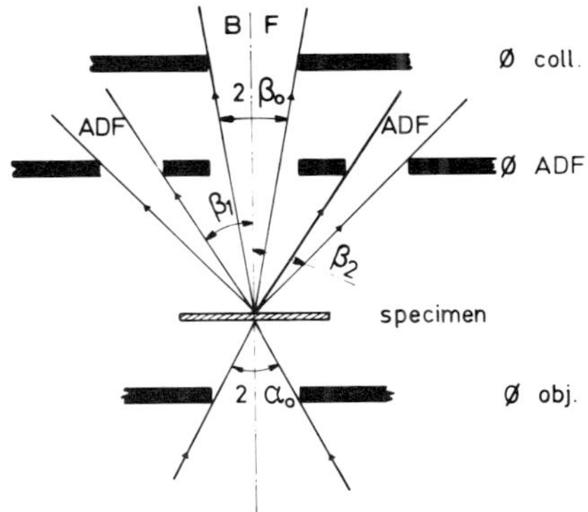

Fig. 1 Typical angular conditions for high resolution imaging :
$2\alpha_0$ = 1.5 x 10^{-2} rad
$2\beta_0$ = 6 x 10^{-3} rad
$2\beta_1$ = 2.8 x 10^{-2} rad
$2\beta_2$ = 5 x 10^{-1} rad

The annular dark field mode is best suited for the visualization of the heavy staining atoms because it is very efficient to detect the electrons elastically scattered at large angles and because it averages the phase contrast   of the carbon foil . Uranium atoms are visible, either gathered in clusters, containing typically $\sim$ 20 atoms, regularly spaced along the biological molecules, or as smaller spots scattered on the support film between them (fig.2). These spots correspond to a distribution of clusters of variable size, the smallest ones being 0.5 nm wide in accordance with the beam diameter on the specimen. In certain favourable circumstances, these discrete spots move under the beam, coagulating into clusters or escaping from them. They correspond to single uranium atoms and following the criterion used by Isaacson et al. (1976)

$$\frac{I_A}{I_S} = (\frac{Z_A}{Z_S})^{3/2} \frac{0.94}{N_S T_S}\delta^2 = (\frac{92}{6})^{3/2} \frac{0.94}{0.11 \times 20 \times 25} \simeq 1$$

their contrast is sufficient with such a thin layer of carbon, to be visualized even with a probe of 0.5 nm size. In the above formula $N_S T_S$ is the mass thickness of the substrate and δ is the probe size.

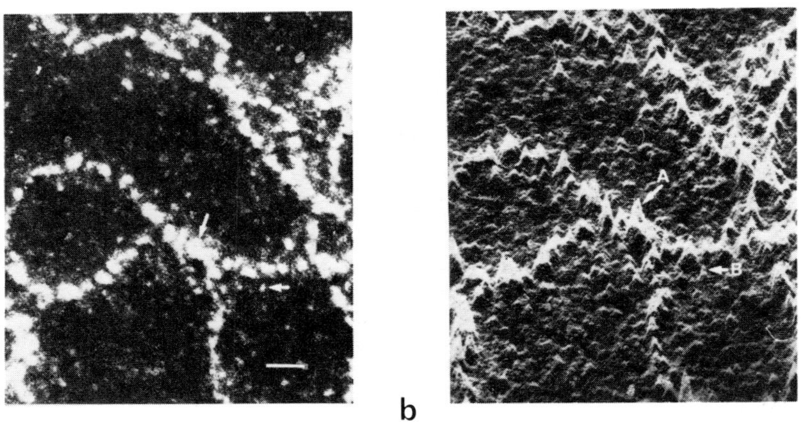

a   b

Fig. 2 Annular dark field (a) and its corresponding Y-modulated image (b) of a segment of twisted ds PM_2 DNA molecule. Each one of the strongest staining peaks (see A) along the molecule contains roughly 10 to 20 uranium atoms. The smaller dots visible on the background (see B) are responsible of peaks of intensity about 10 times smaller and of width of the order of 0.5 nm. They likely correspond to the smallest uranium "clusters" (1 to 2 atoms), for which the contrast can be discriminated over the mean intensity due to the support carbon film. The bar represents 5 nm.

Fig. 3 shows two successive pictures of the same area where the displacement of uranium atoms can be seen. Their dynamic behaviour has been observed at TV rates.

The contrast of clusters made of a few atoms has been measured in different imaging modes(fig. 4). The annular dark field exhibits white spots superposed on a rather homogeneous dark substrate. The axial bright field picture is governed by phase contrast effects, depending on the focusing

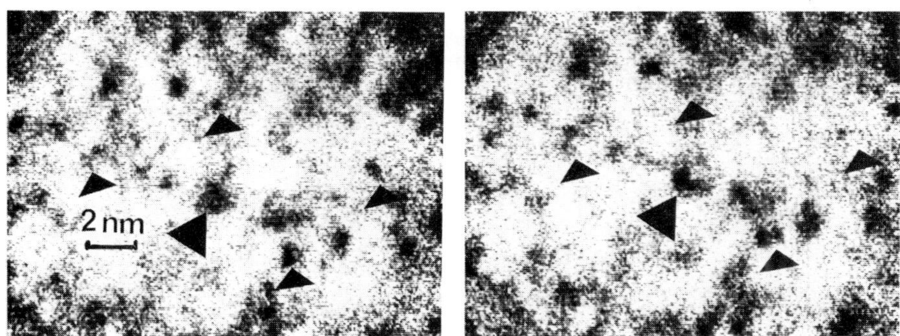

Fig. 3   Two micrographs of the same area recorded successively to
demonstrate the movement of individual uranium atoms between larger
clusters. See the change of the shape of the big cluster indicated
by the larger triangle or the other slight modifications. The pictures are printed negatively for a better visual perception of the
atoms.

conditions and the collection aperture. At Scherzer focus, a cluster
appears as a small dark spot immersed in a randomly fluctuating background.
The real time analog processing of simultaneous signals enables various
combinations, such as the ratio

$$I_{API} = \frac{aBF - bDF}{aBF + bDF}$$

as suggested by Jones and Leonard (1978). The coefficients a and b are
chosen such as to make the contrast nul in the denominator. It is easy to
show that the contrast of the ratio signal is then about twice the bright
field one, while averaging the background fluctuations. Consequently, by
varying the zero level, it is possible to visualize atoms as tiny dark
spots (fig. 4c).

*The STEM project in Orsay has been supported by DRET 79/465 and DGRST/79/
7/0489 grants and the pluridisciplinary aspect of this work is part of a
CNRS RCP 647. Thanks are due to B. Revet and E. Delain for providing the
specimen. One of us (O.K.), whose permanent adress is : Center for Solid
State Science, Arizona State University, Tempe, Az. 85281,U.S.A.has obtained a short time CNRS grant.*

Crewe A V, Wall J and Langmore J 1970 Science 168 1338
Crewe A V 1978-79 Chemica Scripta 14 17
Isaacson M S, Langmore J, Parker N W, Kopf D and Utlaut M 1976 Ultra-
    microscopy 1 359
Jones A V and Leonard K R 1978 Nature 271 659
Mory C, Colliex C, Revet B and Delain E 1981 to be published
Ohtsuki M 1980 Ultramicroscopy 5 325
Wall J S 1978-79 Chemica Scripta 14 271

Fig.4 (next page). Study of the contrast of uranium clusters deposited on a thin support carbon film.(a) Schematic representation of
the model and of the various imaging signals. (b) and (c) DF and
API micrographs localizing five clusters and a few individual atoms
indicated by triangles in (c). (d) to (f) Intensity profiles recorded in the various modes along the line shown in (b).

U cluster
a
≅ 2 nm
API = (aBF − bDF)/(aBF + bDF) = N/D
BF
DF
D
N
API
b
1 2 3 4 5
DF
c
20 nm
API
d
1 2 3 4 5
DF
e
BF
f
N
g
API

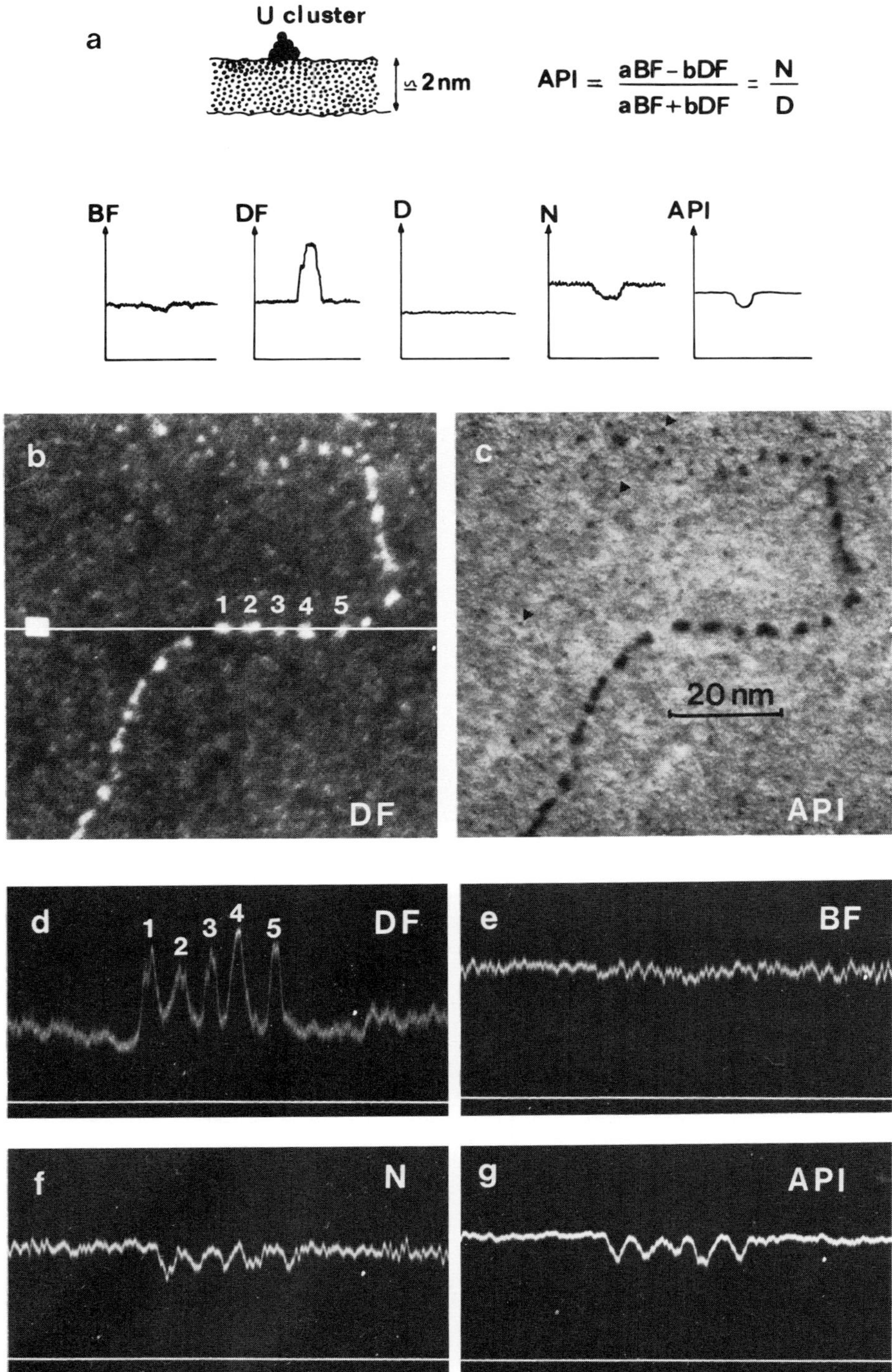

Figure 4

Recursive methods with a transfer function for image processing

P.W. Hawkes

Laboratoire d'Optique Electronique du C.N.R.S., B.P. 4347, F-31055 Toulouse
Cedex.

1. Introduction

Among the various methods invoked for restoring degraded images, those
using one or other of a variety of two-dimensional transforms have been
most extensively used and studied. In some situations, however, such methods seem inappropriate or become impractical. Thus if the image is generated by a scanning microscope (electron or optical), the possibility of
beginning image processing as the picture is collected is attractive ; if
the image contains some gross defect but is otherwise of good quality and
represents some structure difficult to record, it is inconvenient to be
restricted to a square picture area ; finally, it may be desirable to study a larger number of pixels than the computer available can tolerate, if
transforms have to be performed. For such reasons as these, the class of
methods that are known generically as "recursive" is beginning to be extended to two dimensions ; such methods have been explored in some detail in
the field of one dimensional signal processing.

The underlying idea is simple : information about the neighbourhood of each
point is to be used to calculate an improved estimate of the picture value
at this point. The neighbourhood is chosen to suit the particular case.
Thus in the case of a scanning image, where the improvement is to be effec-
ted as the image is formed, only points already scanned can of course be
included ; alternatively, whole lines might be improved using all the lines
already completed. If, however, the method is to be applied to an image
already recorded, for which transform methods are inappropriate, we should
probably use the nearest neighbours of each point, as in relaxation for
example.

We have of course to choose some criterion for "improvement". Here we have
followed current practice and adopted the least mean square error constraint.

The case of image restoration in the presence of noise has been studied
thoroughly (Rosenfeld and Kak, 1976) but most of the literature on recursive methods is heavily impregnated with the notions and priorities of
electrical signal processing. Stability is important whatever the field
of application but the synthesis of recursive filters is perhaps not so
vital a preoccupation in digital picture processing. For lengthy discussion
of these topics, we refer to Ramamoorthy and Bruton (1981), O'Connor and
Huang (1981) and Woods (1981), and for a readable survey putting these
methods in perspective, tc Jain (1981).

2. <u>Derivation of a Recursive Operator in Direct space, Including the Effect</u>
   <u>of a Transfer Function</u>

We set out from the assumption that the image contrast C is related to
some specimen transparency function t by a discrete convolution, of the
form :

$$C(i,j) = \sum_{p,q} K(i - p, j - q)\, t(p,q) + n\,(i,j) \tag{1}$$

in which K is a point-spread function and n denotes noise. In order to
explain the principle clearly, we restrict the present discussion to a single-component (typically pure phase) specimen. The two-component case, for
which problems analogous to those encountered with the Schiske filter
(Schiske, 1973) arise, is discussed in a separate paper ; we anticipate that
a way round these, not unlike that given by Saxton (1978) and Kirkland et
al. (1980), can be found.

We now seek an estimate $\hat{t}$ (i,j) of the true value of t at (i,j) such
that $<| t - \hat{t} |^2>$ is minimized and $\hat{t}$ is a weighted linear combination of
contrast values already measured :

$$\hat{t}\,(i,j) = \sum_{k,l} d_{i,j}\,(k,l)\, C\,(k,l) \tag{2}$$

Here, k and l are permitted to take values corresponding only to measured
points ; in Fig. 1, we show a typical situation in which the image is scanned from top to bottom and hence k and l in eq. (2) are confined to the
region A(i,j).

The angle brackets in the quantity to be minimized can be interpreted in
two ways, not necessarily identical. If we had a large number of versions
of the same image, we should interpret them as the ensemble average over
this set of images but this does not correspond to reality, in microscopy
at least. The alternative view is to regard them as signifying an average
over the image itself. These averages will only be the same if the ergodic
hypothesis is valid (up to second-order statistics) ; we shall say no more
about this thorny question here.

Differentiating eq. (2) with respect to the weights $d_{i,j}$ (k,l), we find
that $<|t - t|^2>$ passes through an extremum if the following set of equa-
tions is satisfied for every value of k,l in A(i,j) :

$$< \{t(i,j) - \hat{t}(i,j)\}\; C\,(k,l)> \,= 0 \tag{3}$$

In practice, however, it is unrealistic to require that eq. (3) should be
satisfied for all values of k,l lying in A(i,j), for why should information
about points distant from the current point improve our estimate of the
true value at the latter ? Furthermore, the computation would become highly
complicated and cumbersome if more than a few points were retained.
We therefore introduce the restricted domain D of nearest measured neighbours of the current point (Fig. 1). This domain consists only of the set
of points {i - k, j - l}, for which { (k,l)} = { (1,0) ; (0,1) ; and (1,1)}.

Returning to eq. (2), it clearly seems preferable to use not the raw data
C but the estimates already available, and we therefore write :

$$\hat{t}\,(i,j) = \sum_{k,l} d_{i,j}\,(k,l)\, \hat{t}\,(i - k, j - l) + d_{i,j}^{(0)}\, C\,(i,j) \tag{4}$$

where k and l are confined to D. We now seek weights $d_{i,j}(k,l)$ and $d_{i,j}^{(0)}$ such that eq. (2) is satisfied as well as possible for all values of k and l in $A(i,j)$. (The change from eq. 2 to eq. 4 recalls the introduction of successive displacement in relaxation calculations).

Space does not permit us to set out the full calculation here and we therefore give only the main steps. First we substitute the expression for $\hat{t}$ of eq. (4) into condition (3) and then attempt to derive the values of $d_{i,j}(k,l)$ and $d_{i,j}^{(0)}$. After some manipulation, we find that :

$$< \sum_{r,s} K(m-r, n-s) \, t(r,s) \, t(i,j)>$$

$$- \; d_{i,j}^{(0)} < \sum_{p,q} K(i-p, j-q) \, t(p,q) \; \sum_{r,s} K(m-r, n-s) \, t(r,s)>$$

$$- \; \sum_{k,l} d_{i,j}(k,l) < \sum_{r,s} K(m-r, n-s) \, t(i-k, j-l) \, t(r,s)> \; = 0 \qquad (5)$$

Into this we set (m,n) equal to $(i-1,j)$, $(i-1, j-1)$ and $(i, j-1)$ in turn, which yields three complicated equations. A fourth equation is obtained by including the current point (i,j). Interpreting the angle brackets as averages over the image, we obtain a set of four linear equations for the four unknown weights $d_{i,j}(k,l)$ and $d_{i,j}^{(0)}$. These are then solved to give the weights required in the recursion relation.

3. Discussion

Although the procedure described above is always capable of providing a formal solution, and will indeed do so however many points we retain in the recursion relation, its practical implementation is not straightforward owing to the fact that some assumption must be made about the object statistics. The simultaneous equations that yield the weights contain not only the autocorrelation function (over the image) of the point-spread function K but also that of the specimen transparency, t. This is analogous to the problem of choosing the denominator correctly when applying a Wiener filter to an image, and can, we hope, be managed in the same way.

Two extensions of this procedure are of considerable interest : to two-component specimens and to line estimation. The formal treatment of the two-component specimen is very similar to that of the single-component case but leads to inaccessible correlations of the two specimen transparency functions - a method of preventing these correlations from appearing, analogous to that used by Saxton (1978) and Kirkland et al. (1980) can in principle be devised. Nevertheless, the two-component case manifestly needs a separate treatment, for the fact that two images (with different point-spread functions) are required means that we need no longer confine k and l to "past" values, and hence to D (or A).

The question of line-by-line recursive restoration has been considered in some detail by Murphy and Silverman (1978), who include both noise and "blur", and by Jain (see Jain, 1981). Considerable investigation is needed to establish what performance can be expected from these procedures. The work of Murphy and Silverman was originally intended to be used to correct blurring due to motion of the image (Aboutalib et al., 1977) but the results given by Murphy and Silverman (1978) are applicable to more general types of restoration.

<u>References</u>

Aboutalib A O, Murphy M S and Silverman L M 1977 IEEE Trans. <u>AC-22</u> 294-302.
Jain A K 1981 Proc. IEEE <u>69</u> 502-528.
Kirkland E J, Siegel B M, Uyeda N and Fujiyoshi Y 1980 Ultramicroscopy <u>5</u>
    479-503.
Murphy M S and Silverman L M 1978 IEEE Trans. <u>AC-23</u> 809-816.
O'Connor B T and Huang T S 1981 In Two-dimensional Digital Signal Proces-
    sing ed by T S Huang (Berlin and New York : Springer) vol. I, pp 85-154.
Ramamoorthy P A and Bruton L T 1981 In Two-dimensional Digital Signal Pro-
    cessing ed by T S Huang (Berlin and New York : Springer) vol. I, pp 41-83.
Rosenfeld A and Kak A C 1976 Digital Picture Processing (New York and
    London : Academic).
Saxton W O 1978 Computer Techniques for Image Processing in Electron Micros-
    copy (New York and London : Academic) p 243.
Schiske P 1973 In Image Processing and Computer-aided Design in Electron
    Optics ed P W Hawkes (New York and London : Academic) pp 82-90.
Woods J W 1981 In Two-dimensional Digital Signal Processing ed by T S Huang
(Berlin and New York : Springer) vol. I, pp 155-205.

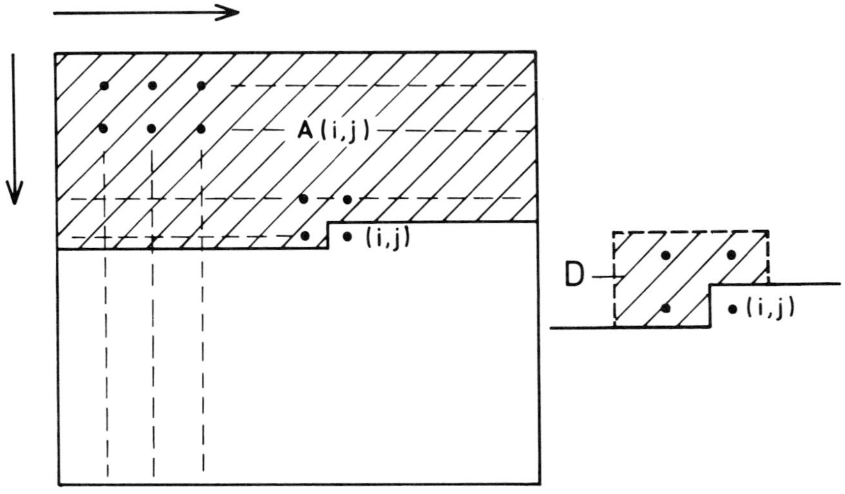

Fig. 1. The domains D and A relative
to the current point (i,j).
Scanning from left to right and
from top to bottom.

STEM imaging with a quadrant detector

G R Morrison and J N Chapman

Department of Natural Philosophy, University of Glasgow, Glasgow G12 8QQ

1.   Introduction

A number of imaging conditions not readily attainable in a CTEM become
possible in a STEM if a configured detector is used. A particularly
interesting case results when signal combinations are chosen so that the
bright-field detector response is an odd function, for then an image is
obtained which relates to one component of the phase gradient of the
specimen transmittance. By dissecting the usual bright-field detector
into four quadrants, four different directions of differentiation may be
realised; the difference signal from adjacent semicircles, which simulates
the original split detector of Dekkers and de Lang (1974), or the difference signal from diagonally opposite quadrants (e.g. Morrison et al (1980)).
A significant advantage of such differential phase contrast (DPC) systems
is their ability to tolerate relatively large phase excursions at low
spatial frequencies and still image linearly (Waddell and Chapman (1979)),
and this makes them particularly well suited to Lorentz electron microscopy (Chapman et al (1978)).

2.   The Quadrant Detector System

A quadrant detector system based on a windowless quadrant photodiode
detector (a Centronic QD-100) has been attached to an extended Vacuum
Generators HB501 FEGSTEM fitted with 3 post-specimen lenses. This detector
is mounted on a retractable carriage within the detector chamber, and when
in use lies just below the annular dark-field detector. In order to fully
exploit the potential of this detector it is important that the criterion
$\alpha_d > \alpha_o$ is always satisfied, where α_d and α_o are the half-angles subtended
by the detector and probe-forming aperture respectively. Angle α_o can
vary over more than an order of magnitude, depending on whether or not the
objective lens is used as one of the probe-forming lenses, so that the
post-specimen lenses are required to provide a range of different effective
camera lengths to match the angular scattering distribution to the 11mm
diameter active area of the detector. The operation of the post-specimen
lenses has been considered in detail by Craven and Buggy (1981), and probe-
forming conditions suitable for Lorentz microscopy have been discussed by
Chapman et al (1980). The one other important requirement when using a
detector which is sensitive to an asymmetric intensity distribution in the
detector plane is the ability to remove any systematic motion across the
detector as the probe is scanned across the specimen. This motion arises
because the detector lies in the far-field with respect to the specimen,
rather than in a true Fraunhofer plane, and is corrected by feeding an
appropriate signal to the existing Grigson coils located between the

specimen and the first post-specimen lens. Accurate descanning of the beam proves to be a particularly stringent requirement under the operating conditions appropriate for Lorentz microscopy, when the objective lens is off or only weakly excited and camera lengths up to 10m are used.

3. <u>Interpretation of Magnetic Domain Images</u>

In a ferromagnetic thin film the phase gradient of the specimen transmittance can be written as $\nabla\phi = -2\pi et(\underset{\sim}{B} \times \underset{\sim}{n})/h$ when only the magnetic interaction is considered. B denotes the magnetic induction, $\underset{\sim}{n}$ is a unit vector parallel to the incident beam direction, t is the thickness of the foil, and e and h have their usual meanings. In a real specimen, however, it is unlikely that the in-plane flux density will be the only contribution to the phase gradient. Variations in t can also result in differential contrast, through a term of the form $(\pi V_O/\lambda E)\nabla t$, where V_O is the mean inner potential and E is the energy of the electron beam. The magnitude of a DPC signal is also affected by the intensity of the beam falling on the detector, so that scattering contrast is also evident on a DPC image, although this effect can be considerably diminished if the DPC signal is normalised by the sum signal from all 4 quadrants.

Figure 1c shows the incoherent bright-field image of a crystalline iron foil obtained from the sum signal, and it can be seen that the area of foil under examination is of relatively non-uniform thickness. The DPC images of the same area shown in figures 1a,b,d and e were all normalised by the sum signal, and at low resolution scattering contrast is largely absent from these images. When examining relatively large areas of foil at low resolution a qualitative interpretation of the image in terms of $B_O t$ is possible because any variations occur only slowly within any one domain, and it is not usually necessary to take into account the spatial frequency dependence of the transfer function. It should be possible, at least in principle, to obtain a complete 2-dimensional map of the product $B_O t$, either with the pair of images taken with the two split detector combinations (figures 1a and 1b), or with the pair taken using the opposite quadrant detector configurations (figures 1d and 1e). However it is found that a visual interpretation of the domain pattern is considerably facilitated by the ability to rotate the direction of differentiation in steps of 45° by making use of both split and opposite quadrant configurations. A schematic view of the magnetisation distribution found in this area of foil is shown in figure 1f, and from this it is clear that 90°, 180°, and intermediate angle domain walls are present.

At a higher resolution the variation of the phase gradient across a domain wall can be studied. The DPC image of the boundaries between some anti-parallel domains is shown in figure 2. The direction of differentiation is perpendicular to the domain wall, and a Y-modulation image of the small inset region is shown in figure 3a, but an immediate interpretation in terms of the domain wall profile is precluded by the following considerations. 1) The traces were obtained with a probe size estimated to be $\sim$ 20nm. Since the variation across the domain wall takes place over $\sim$ 70nm the resolution available is clearly insufficient. 2) As the phase gradient varies quite rapidly across a domain wall, proper account must be taken of the contrast transfer function before a quantitative interpretation is possible. 3) The thickness variations which are clearly evident on the bright-field Y-modulation image of figure 3b indicate that there may be a significant contribution to the phase gradient from the term proportional to ∇t.

Despite these difficulties the ease with which the traces of figure 3a
were obtained, and the absence of noise on them, clearly indicates the
possibility of a much more direct and accurate determination of the wall
profile than is possible using the Fresnel mode of Lorentz microscopy. In
this latter technique the insensitivity of the defocussed image to the
exact form of the domain wall profile presents a formidable obstacle to
the accurate experimental determination of it. The problems that still
exist with DPC Lorentz microscopy should be overcome by examining more
uniform foils, so that $\tilde{\nabla}t \simeq 0$, by using a smaller probe size , and by recording the signals from each of the four quadrants digitally, so that corrections for the transfer function and other processing can be carried out
more easily.

Chapman J N, Batson P E, Waddell E M, Ferrier R P 1978 Ultramicroscopy <u>3</u>
   203-214
Chapman J N, Morrison G R, Ferrier R P 1980 "Electron Microscopy 1980" Vol 1
   pp90-91 Eds P Brederoo, G Boom (7th Eur Congr on EM Foundation: Leiden)
Craven A J, Buggy T W 1981 Ultramicroscopy. In press
Dekkers N H, de Lang H 1974 Optik <u>41</u> 452-456
Morrison G R, Chapman J N, Craven A J 1980 "Electron Microscopy and
   Analysis 1979" pp257-260 Ed T Mulvey (IOP: Bristol)
Waddell E M, Chapman J N 1979 Optik <u>54</u> 83-96.

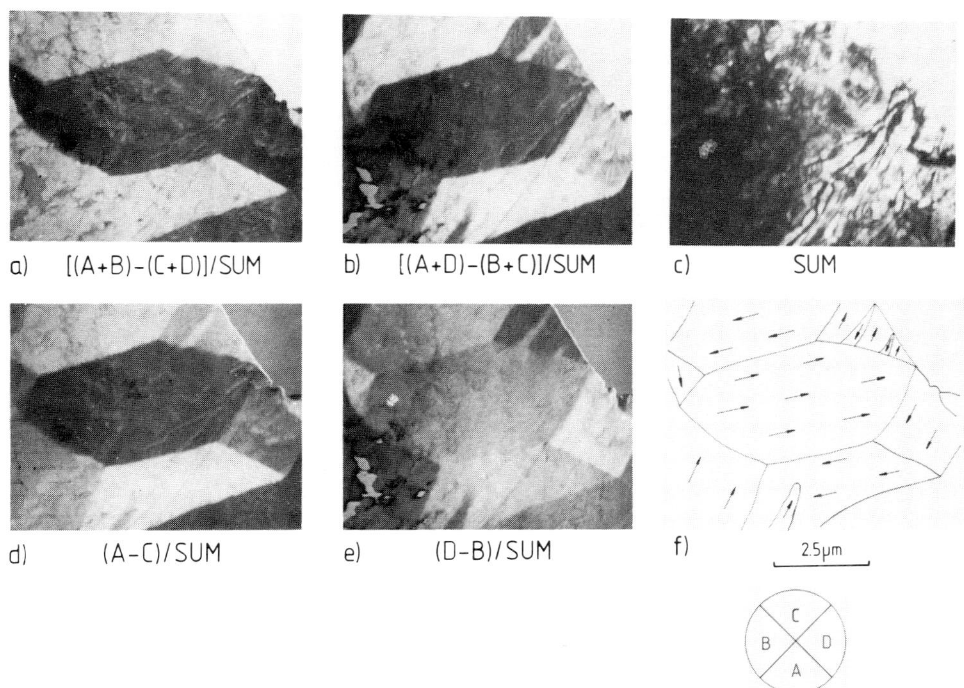

Figure 1: Five images of a crystalline iron film taken with the
quadrant detector. A schematic view of the in-plane magnetisation
distribtuion is also shown.

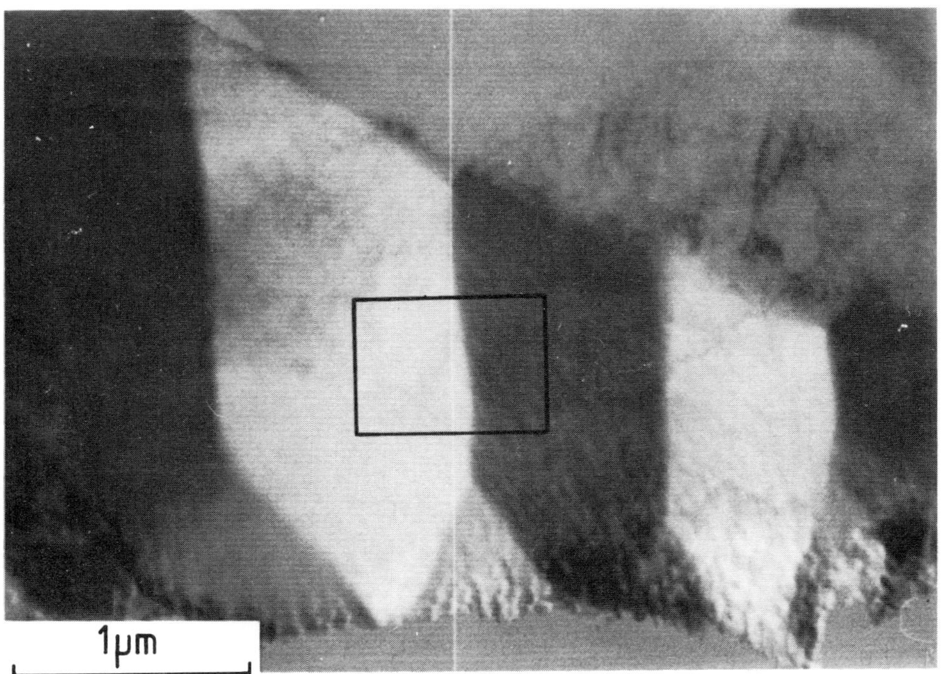

Figure 2: An array of anti-parallel domains in a crystalline iron foil

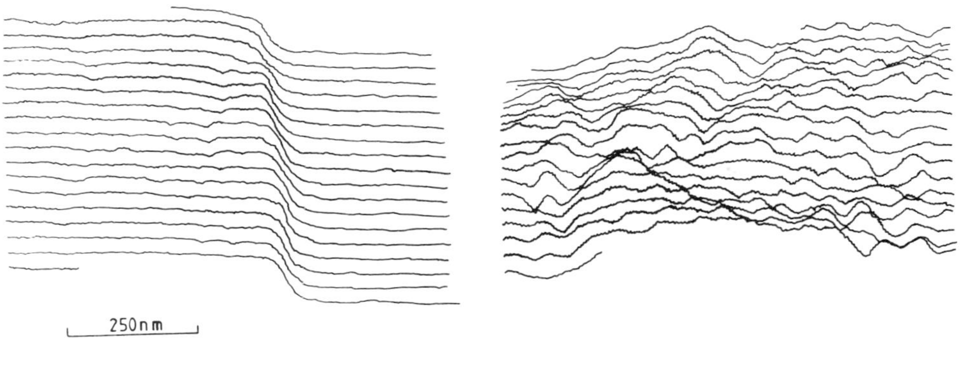

a)   b)

Figure 3: Y-modulation images of the small inset area shown in Figure 2
   a)   DPC image with the direction of differentiation
        perpendicular to the domain wall.
   b)   Incoherent bright-field signal obtained from the sum of
        all four quadrants.

Image averaging for biological specimens: the limits imposed by imperfect crystallinity

W.O. Saxton and W. Baumeister

High Resolution Electron Microscope, Free School Lane, Cambridge, CB2 3RQ.
and
Institut fur Elektronenmikroskopie und Biophysik, Moorenstr. 5,
D-4000 Dusseldorf, W. Germany.

1. Introduction

The averaging of the images of many identical molecules or molecular
complexes arranged in 2-D crystalline sheets has proved a most fruitful
way of overcoming the radiation sensitivity of biological specimens; the
poor statistical definition (i.e. noisy appearance) of electron micrographs recorded at magnifications and doses low enough to avoid serious
structural disruption is circumvented by a posteriori image averaging
(carried out either digitally or optically, by real-space superposition or
Fourier-space filtering) in which the noise is greatly reduced while the
signal survives.

The methods conventionally used to achieve the averaging however rely on
perfect specimen crystallinity, and the long and short range distortions
actually present in most specimens impose a fundamental limit on the
resolution they can achieve.   Random molecular displacements, normally
distributed with an rms r_0 for example, would lead to attentuation of a
specimen periodicity r by a factor exp $(-\pi^2 r_0^2/r^2)$, thus setting a
(grating) resolution limit of 2-2.5r_0 for cutoffs set at 10-20%.   We
present here some results obtained with the newer method of correlation
averaging, which overcomes this limit by direct identification of, and
compensation for, the distortions; averages are shown which derive from
well and less well ordered areas of the same specimen, together with
further averages involving a comparable number of molecules selected from
a larger specimen area on such bases as low local distortion levels.

2. The correlation averaging method

Correlation averaging involves direct digital superposition of individual
molecular images after their positions have been determined via their high
correlation level with a common reference image.   In a typical application to periodic specimens (Saxton, 1980; Frank and Goldfarb, 1980;
Crepeau and Fram, 1981), a small patch of the image to be averaged is
selected visually as a reference, and cross correlated with the whole
area.   The cross-correlation function is filtered to remove very low
and very high frequencies.   Rough molecular positions indicated by local
correlation peaks are refined by local centre-of-mass determinations, and
the average is then produced by superposing regions centred on the final
peak positions.   The peaks may be sorted into descending mass order and

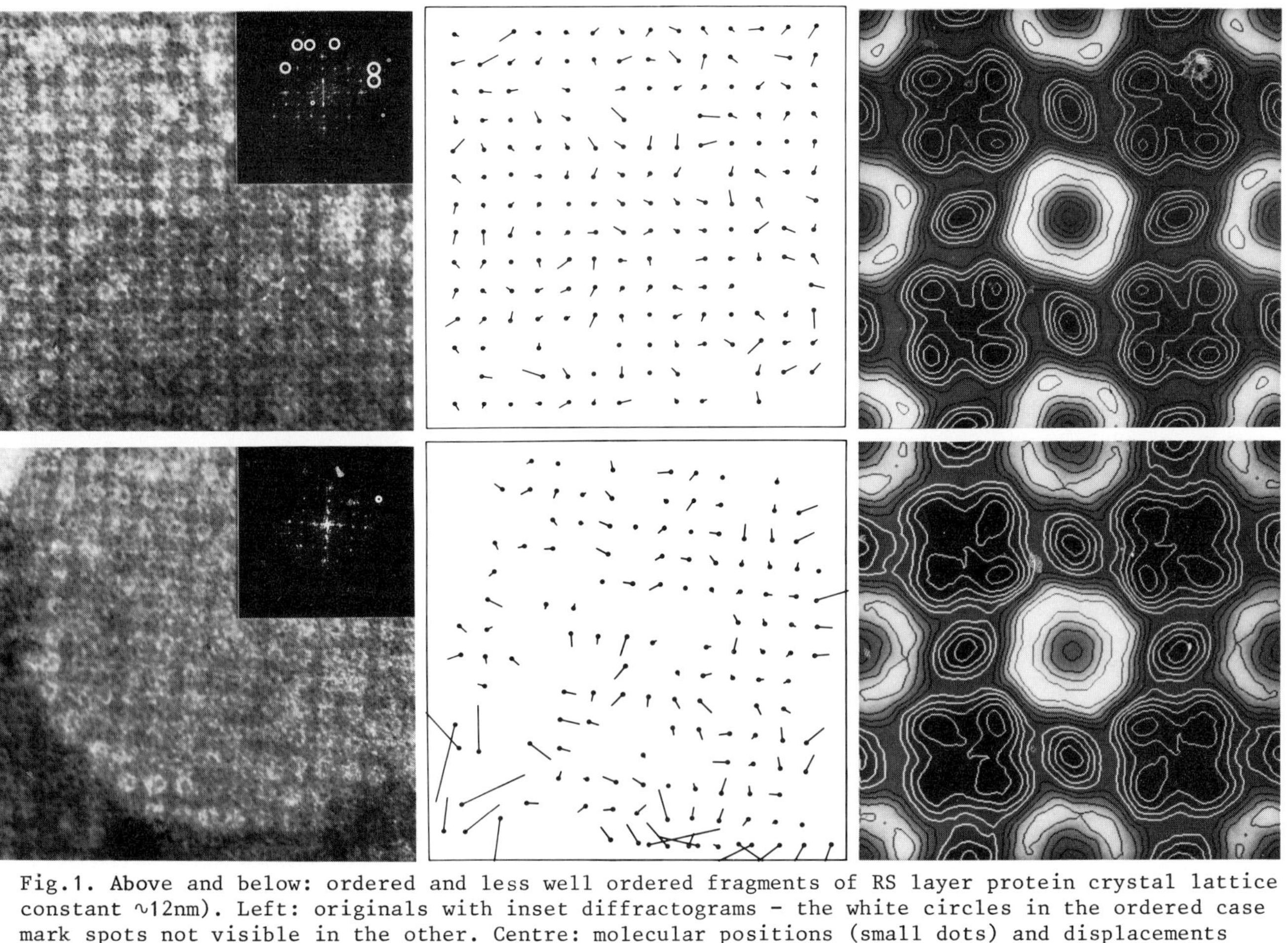

Fig.1. Above and below: ordered and less well ordered fragments of RS layer protein crystal lattice constant ∿12nm). Left: originals with inset diffractograms – the white circles in the ordered case mark spots not visible in the other. Centre: molecular positions (small dots) and displacements (magnified five times) – rms displacements 0.80nm and 1.78nm. Right: contoured averages after 4-fold rotational averaging.

an average prepared using the strongest peaks only: The whole process can be iterated using an initial average as a reference in its turn, either to allow the use of a smaller reference (Saxton and Frank, 1977), or to eliminate bias towards the particular reference used originally.   Such a procedure allows averages such as those in fig. 1 to be obtained almost completely automatically - certainly more conveniently than is possible with the conventional approach - and obviously compensates adequately for long-range crystal distortions.   The distortions actually present in the crystal can be established by assuming that the undistorted structure is perfectly crystalline: rough estimates of the lattice vectors, sufficient to allow correct indexation of the molecular positions, are obtained manually and refined by a least-squares fitting procedure (linear regression), after which the displacement field is identified via the deviations of the molecular positions from the corresponding lattice sites (cf. fig. 1).

While long-range distortions do not necessarily imply any serious dis- tortions of the individual molecules, the same is not true of short-range distortions, and an average may remain degraded by variations in molecular shape even after correction of the overall displacements.   With this in mind, we have pursued the possibility of using the displacement fields now established to derive strain fields on the basis of which an average can be prepared including only the least strained molecules.   As a simple measure of the strain at a given site, we have used the rms displacement change between the site and its immediate neighbours, normalised by the lattice vector length - this seems preferable for various reasons to a full regression fit of the four strain tensor components at each site.

3. Application to the RS layer

The cell envelope of sporosarcina ureae contains a 2-D crystalline protein layer called the RS layer; this can be isolated in sheets which, although of considerable size, develop fissures and local distortions in flattening down on to a support film.   Electron micrographs of negatively stained preparations of the RS layer have been recorded at a primary magnification of about 70 000 x under low (but not minimum) dose conditions.   Averages prepared from two different regions, 512 pixels/180nm square, of a single crystal fragment are compared in fig. 1; the two regions differ markedly in how well ordered they are (compare the inset diffraction patterns and the displacement maps), but similar averages are obtained from both.   The slightly higher resolution evident in the central molecule of each reflects the fact that it is for this molecule only that exact registration of all repeats is achieved.

Figure 2 shows a larger area, 1024 pixels/360nm square, of the fragment (including the area shown in fig. 1), and averages formed from a quarter only of the nearly 600 molecules present, selecting in one case molecules with high correlation peak levels, and in the other those with low local strain.

The structure obtained shows many striking similarities to other cell envelope protein crystals, e.g. the HPI layer of micrococcus radiodurans (Saxton and Baumeister, 1981), and will be discussed more fully elsewhere. The support of the Deutsche Forschungsgemeinschaft (Sonderforschungsbereich 160), the UK Science Research Council and the European Molecular Biology Organisation is acknowledged.

References

Crepeau, R.H., and Fram, E.K., 1981 Ultramicr. 6 7
Frank, J. and Goldfarb, W., 1980 Electron Microscopy at Molecular
    Dimensions eds. W. Baumeister and W. Vogell (Berlin: Springer) pp261-9.
Saxton, W.O., 1980 Electron Microscopy at Molecular Dimensions
    eds. W. Baumeister and W. Vogell (Berlin: Springer) pp 245-55.
Saxton, W.O. and Frank, J., 1977 Ultramicr. 2 219.
Saxton, W.O. and Baumeister, W., 1981 submitted to Proc. Nat. Acad. Sci.

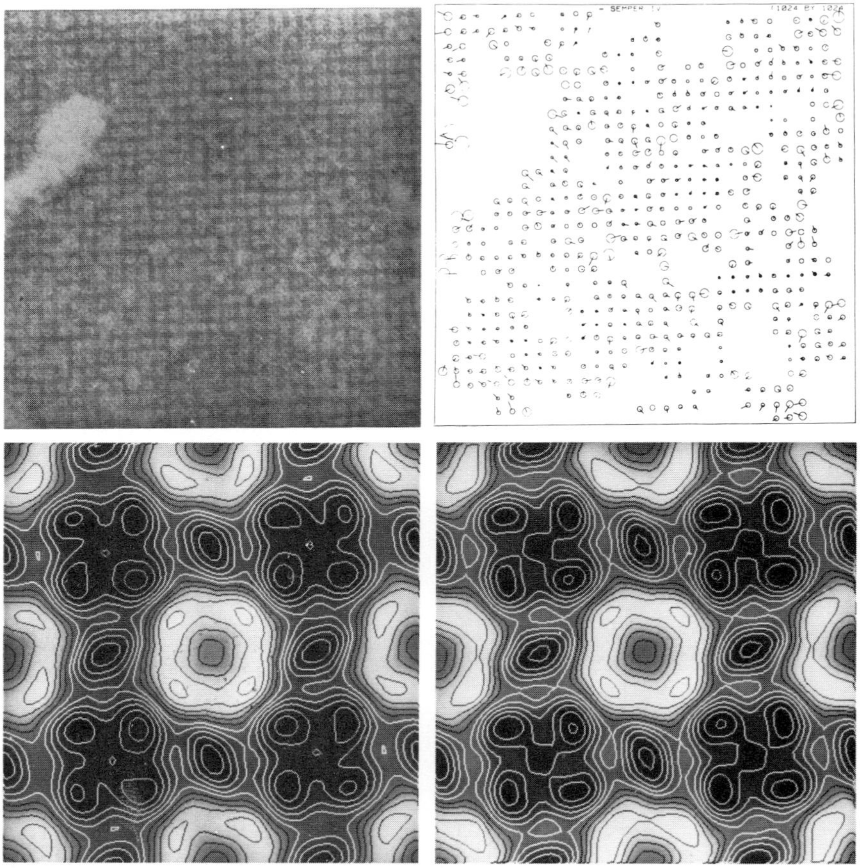

Fig.2. Above left: larger area of RS layer crystal; above right: molecular
displacements magnified 5 times (short lines) and rms strains (circles
with radius proportional to strain) - the rms displacement was 0.93nm, and
the strains ranged from 0.02 to 0.28 with a mode around 0.09nm. Below left
and right; averages prepared using respectively the molecules correlating
most strongly with the reference (a preliminary average), and those with
lowest strains.   No great improvement due to selection by strain level
can be claimed in the present case but at least the necessary techniques
are now established.

Image interpretation for aromatic hydrocarbons by computer simulation

M. A. O'Keefe, J.R. Fryer[*] and David J. Smith.

High Resolution Electron Microscope, University of Cambridge, Cambridge.
*Chemistry Department, University of Glasgow, Glasgow.

1. Introduction

It is well-known that high-resolution image interpretation is validated
by the criterion of agreement between experimental and computed images.
From a knowledge of the crystal structure of the material it is possible
to compute the image appearance as a function of orientation, defocus,
thickness and experimental resolution. This is of particular value for
beam-sensitive materials since "minimum-exposure" techniques (Williams and
Fisher, 1970; Fryer, 1978) prohibit the recording of focal series and often
result in images at non-optimum defocus.

Previous image simulation studies, which have generally concentrated on
materials of high atomic number (O'Keefe et al, 1978), have shown that
qualitative interpretation in terms of the Weak-Phase-Object (WPO) approximation is possible only for very thin crystals and over severely-limited
ranges of defocus. It might be expected that the low-Z polynuclear
aromatic hydrocarbons of the present investigation would produce less
dynamic scattering thereby making images interpretable to greater thickness. Indeed, Uyeda and Ishizuka (1974) found a critical thickness of
20 to 30 cells (75 to 110Å) for images of an organometallic compound
computed at 500kV, and Jap and Glaeser (1980) found a value of 168Å for
cytosine, also at 500kV. The present results, for materials of even
lower mean atomic number, confirm this trend and indicate a critical
value of around 250Å at 500kV. In addition, they display the effects
of standard parameters and enable determination of the resolution levels
required firstly to resolve the molecule as a discrete shape, then the
approximate molecular shape, and finally, the carbon hexagons of which the
molecules are composed.

2. Computation

Images of the aromatic hydrocarbons quaterrylene ($C_{40}H_{20}$), coronene
($C_{24}H_{12}$) and ovalene ($C_{32}H_{16}$) have been computed using the SHRLI package
(O'Keefe and Buseck, 1979) of Fortran programs installed on the Cambridge
University IBM 370/165. The atom positions for the simulations were
taken from published structural data (quaterrylene; Kerr et al, 1975;
ovalene: Donaldson and Robertson, 1954; coronene: Robertson and White,
1945) and the scattering factors from the International Tables of Crystallography, Vol. IV, table 2.2.B.

A typical simulation (for coronene) commenced with the computation of the

3685 structure factors for the h0l zone out to 4Å^{-1} (corresponding to h = ±64 and l = ±41). These were used to produce an atlas of all WPO images (O'Keefe and Pitt, 1980) in the resolution range from 7.5Å (corresponding to 5 contributing reflections) to 1Å (249 reflections). Images were stored on magnetic tape and later written directly onto film using an Optronics Photowrite System P-1500 with 256 grey levels and 256 pixels of 50µm size across each image. To incorporate the effects of dynamical diffraction and microscope lens aberrations, full multi-slice programs were used. Dynamical scattering was calculated in the thickness steps of 4.695Å and the results stored after 5 slices, 10 slices and then every 10 up to 100 slices (469.5Å) for later imaging calculations. All 3685 reflections were carried through the computation; their normalised intensity sum was 0.99915 after 100 slices. Images were primarily calculated at a voltage of 500kV, with C_s = 3.5mm, α_i = 0.3mrad and focal-spread half-width, Δ, of 250Å. The objective aperture cut-off was chosen to correspond to 0.7Å^{-1}. Output images were again stored on tape for later Photowrite processing.

3. Results

WPO images – Selected images from the WPO series for quaterrylene, ovalene and coronene are shown in Figs. 1(a), (b) and (c) together with maps of their respective projected potentials, In each series, the molecular positions first appear at resolutions of around 6-7Å. Improved resolution of around 4Å provides good molecular shape definition; by 2Å individual rings are clearly resolved, with the individual carbon atoms appearing near 1Å WPO-resolution.

Focal series – Images of coronene crystals were simulated for thickness of 24Å (5 slices) and 94Å (20 slices), from zero (Gaussian) defocus to – 3000Å underfocus, in steps of 100Å (see Fig. 2). These indicate that the basic molecular-shape WPO-like image persists from about – 400Å through to –1000Å defocus. This range is not very sensitive to thickness and suggests a reasonable probability of obtaining this type of image when using a minimal-exposure technique. Furthermore, whereas the typical block-oxide image might reverse contrast several times in a defocus series covering 3000Å, no such reversals are apparent here, although comparison with the WPO-series indicates a progressive degradation with defocus in the resolution of detail which can be related to the structure. Finally, note that contrast, defined as maximum intensity minus minimum, over their sum, slowly increases with defocus. For example, for 5 slices (23.5Å) thickness, the increase is from 0.08 at –400Å defocus to 0.18 at –1000Å and 0.28 at –2000Å; at 20 slices (94Å), the corresponding values are 0.32, 0.65 and 0.79 respectively.

Thickness series – Image variations with thickness were computed in 10-slice steps up to 100 slices, for five defocus values in the range –450Å to –1050Å, As shown in Fig. 3, the molecular shape persists to 100 slices (470Å), although the amount of false (non-structural) detail increases with thickness, becoming quite severe after 50 slices (235Å); i.e. dynamical scattering has produced phase changes in the diffracted beams which are sufficiently large to cause some pseudo-reversal in the image making the molecules appear white with black borders.

4. Discussion

The full simulations presented above demonstrate that interpretation of
500kV images of aromatic hydrocarbons in terms of WPO-images is plausible
over quite wide ranges of thickness (up to 250Å and defocus (-400Å to
-1000Å for best resolution, and to -2000Å for progressively degraded
resolution)). Minimum exposure techniques thus have a reasonable probability of success at least for resolutions in the approximate range 3-6Å.
Indeed, such a high-resolution image of quaterrylene has been recently
obtained (Smith and Fryer, 1981). Matching with image simulations shows
that, whilst the experimental image is slightly tilted, in addition to
containing the inevitable noise introduced by the low-dose technique, it
can be readily identified as having a resolution of between 3.8Å and
3.2Å, despite the radiation sensitivity of quaterrylene.

It is expected that the same benign imaging results will apply to other
hydrocarbons and organic molecules where the crystal structure produces
structure factors falling off monotonically with increasing diffracted
orders. However, once a high-order structure factor of larger amplitude
is included in the image (as will occur when the higher resolution detail,
corresponding to individual ring definition, contributes to the image),
then the high-resolution directly-interpretable image may only appear over
a very narrow range of defocus and thickness. For example, at 600kV,
with C_s=2.0mm, and Δ=150Å, our calculations show that images with better
than 2Å resolution appear only over the very restricted defocus range of
-500 to -600Å. ie. any improvement in radiation sensitivity, for example
by cooling to liquid helium temperature, may not necessarily lead to
better intra-molecular definition than currently achieved unless a very
precise defocus position is fortuitously obtained.

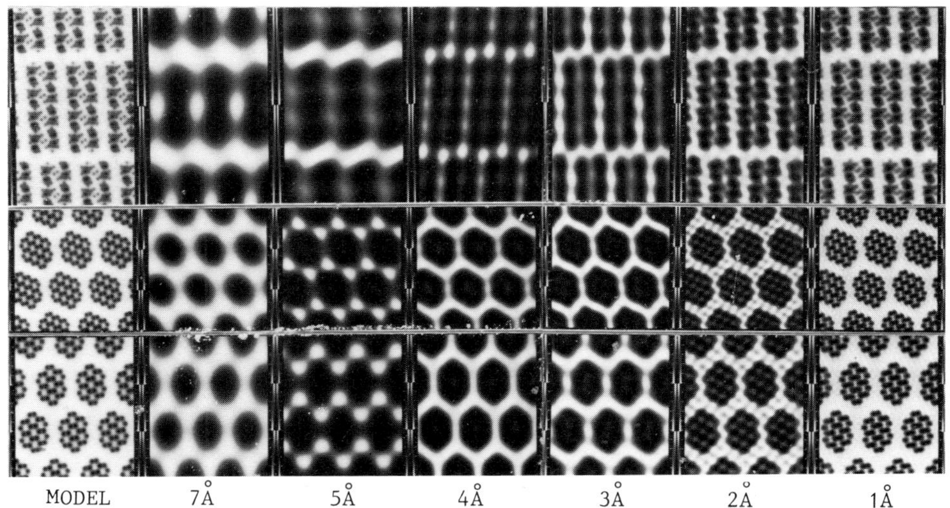

Figure 1: Projected potential (left column) and WPO images (resolutions
indicated) for crystals of quaterrylene (upper row) and ovalene (center
row) in 110 orientation, and for coronene (lower row) in 010 orientation;
the three unit cells across each image correspond to 22Å, 29Å, and 24Å
respectively.

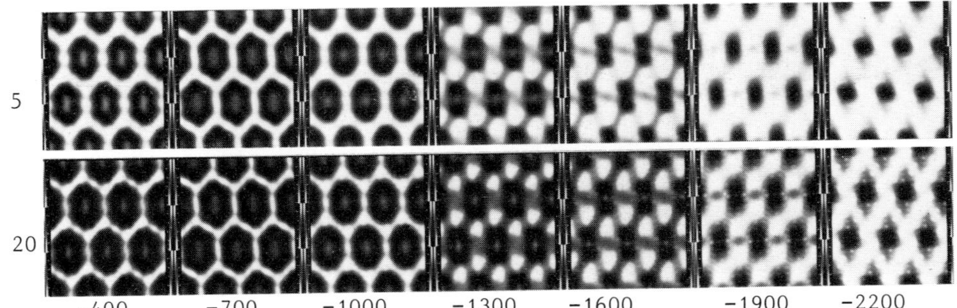

Figure 2: Focal series of coronene images simulated at crystal thicknesses of 5 slices (23Å) and 20 slices (94Å).   Defocus values (Å) are marked.

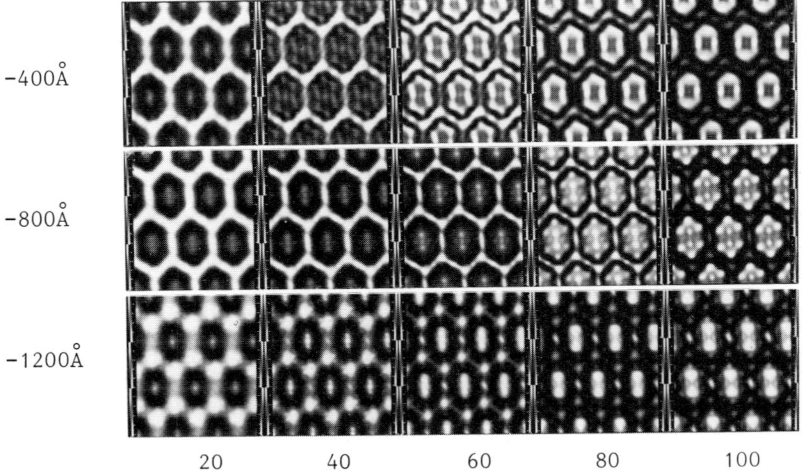

Figure 3: Thickness series of coronene images from 20 slices to 100 slices at three defocus values (marked at left); one slice is equal to 4.695Å

Financial support from the Science and Engineering Research Council is gratefully acknowledged.

References

Donaldson D M and Robertson J M 1954 Proc. R. Soc. A220 157
Fryer J R Acta Cryst 1978 A34 603
Jap B K and Glaeser R M 1980 Acta Cryst. A36 57
Kerr K A, Ashmore J P and Speakman J C 1975 Proc. R. Soc. A344 199
O'Keefe M A and Buseck P R 1979 Trans ACA 15 27
O'Keefe M A, Buseck P R and Iijima S 1978 Nature 274 322
O'Keefe M A and Pitt A J 1980 Electron Microscopy 1 122-3
Robertson J M and White J G 1945 J. Chem. Soc. 607
Smith D J and Fryer J R 1981 Nature 291 481-2
Uyeda N and Ishizuka K J. 1974 Electron Mic. 23 179
Williams R C and Fisher H W 1970 J. Mol. Biol. 52 121

Applications of the $2\frac{1}{2}$D TEM technique

R. Sinclair
Department of Materials Science and Engineering
Stanford University
Stanford, California 94305 USA

The purpose of this article is to draw attention to the $2\frac{1}{2}$D technique of transmission electron microscopy. When studying solids with complex micro-structures, it is common to find that many reflections occur close together in the diffraction pattern. This often precludes taking dark-field images with a particular spot, which is the usual way one associates a feature with its diffraction effects. In a similar vein, micrographs obtained with very small objective apertures are often blurred, lack clarity of de-tail and require long exposure times. The $2\frac{1}{2}$D method circumvents these problems. Dark field pictures are taken using a reasonably sized objective aperture incorporating a number of reflections of interest. Thus several diffracting regions appear in the image simultaneously. They are associ-ated with their particular reflections by the following procedure.

A pair of dark field images is taken, at different objective focus settings (usually 1–10μ apart). Features corresponding to two different reflections will move relative to one another as the focus is changed (Δf) according to the relation $\Delta \tilde{y}_{12} = M \; \Delta f \; \Delta \tilde{g}_{12}$ i.e. proportional to the separation ($\Delta \tilde{g}_{12}$) of their reflections in reciprocal space (Bell 1976). When one views the two images as a stereo pair, this relative image shift is converted to differences in level of height (Bell 1976; Sinclair, Michal and Yamashita 1981). The level of a feature is determined by the position of its dif-raction spot in the objective aperture. Michal and Sinclair (1980) have suggested the following criterion for image interpretation. With the diffraction pattern and images appropriately oriented, with the more over-focus micrograph of the pair on the right, the highest feature (closest to the observer) corresponds to the reflection on the extreme left in the objective aperture. A scheme can then be devised for sorting out which features correspond to particular diffraction spots.

In our experience the method is especially useful for characterizing non-simple microstructures, which often occur in contemporary alloys, ceramics and various materials. A number of examples have been presented elsewhere (Sinclair et al. 1981), ranging from application to a dual phase steel to zircalloy, thin film cobalt, martensitic nitinol and partially ordered Ni_4Mo. Only a few other applications have appeared in the literature. (Mitchell and Bell 1976; Rao 1975; Guy and Butler 1980; Moine, Michal and Sinclair 1981).

The technique is easy to apply and extends the usefulness of dark field microscopy far beyond conventional practice. Any image blurring due to defocus can also be reduced by defocussing the second condenser lens

(Michal and Sinclair 1980) and by using the in-focus picture as one of the images (Moine et al. 1981). In addition, one's perception of height is so sensitive that relative height levels can be distinguished to much finer levels than the normal lateral resolution of the eye (about tenfold, in our experience), allowing identification of features even with closely overlapping diffraction spots (Michal and Sinclair 1980).

It is our conclusion that the 2½D method is much more useful than is commonly appreciated, especially in the area of microstructural characterization and that the electron microscopist should be encouraged to apply this technique in appropriate situations.

Funding for this work was provided by the NSF-MRL program through CMR at Stanford University, the article being written while the author was on sabbatical leave at the Centre d'Etudes Nucleaires, Grenoble. Several colleagues contributed to my appreciation of the technique, including P. Moine, T. Yamashita and J. G. Malcor, and most particularly G. M. Michal, Dr. W. L. Bell, the originator of the method, has always been helpful with comments on our work.

References

Bell W L 1976 J. Appl. Phys. <u>47</u> 1676
Guy K B and Butler E P 1980 Scr. Met. <u>14</u> 607
Michal G M and Sinclair R 1980 Phil. Mag. A<u>42</u> 691
Mitchell J B and Bell W L 1976 Acta Met. <u>24</u> 147
Moine P, Michal G M and Sinclair R 1981 Acta Met. in press
Rao P 1975 Phil. Mag. <u>32</u> 755
Sinclair R, Michal G M and Yamashita T 1981 Met. Trans. <u>12A</u> 1503

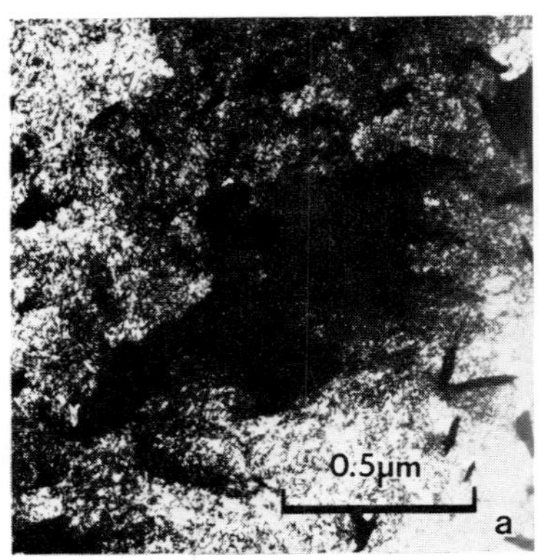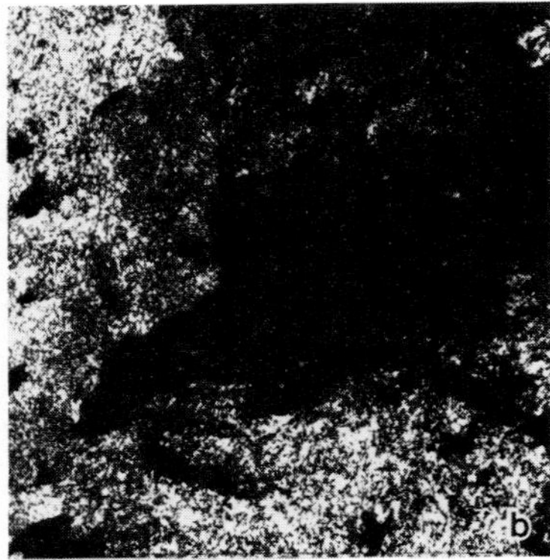

A 2½D image pair of subgrains in a cobalt thin film. The various subgrains appear at different levels of height in the fused image whereas they cannot be discerned in the individual dark field pictures (Sinclair et al. 1981).

High resolution—in spite of misalignment?

W O Saxton and M A O'Keefe

High Resolution Electron Microscope, University of Cambridge, Cambridge

High resolution electron micrographs of crystal with known symmetries frequently do not exhibit the expected symmetry very convincingly. In many cases this is of course because the region of crystal imaged is tilted slightly away from the zone axis being imaged; thin crystals are rarely perfectly planar. However similar image degradation can also be produced by misalignments of the electron beam with respect to the optical axis, i.e. by illumination tilt, quite independently of crystal tilt, and the amount of misalignment necessary is substantially smaller than the crystal tilt required for comparable levels of degradation. Moreover, a posteriori measurement of such beam tilt (magnitude and direction) is extremely difficult. Here we hope to illustrate the problem with calculated contrast transfer functions and simulated images for thin and somewhat thicker specimens.

Theory

The image formation equations in the presence of beam tilt have been given previously (O'Keefe 1979; Saxton 1980) and are not repeated here. However it illustrates our theme to cite here the expression for the contribution made to the image intensity by the mutual interference of a pair of beams f_i and f_j intersecting the back focal plane of the microscope at $\underline{k}_i$ and $\underline{k}_j$ to the Fourier component $\underline{k}_i-\underline{k}_j$ of the intensity namely,

$$f_i f_j^* \, e^{-[\gamma(k_i)-\gamma(k_j)]} \cdot e^{-\pi^2 s^2 \left| (k_i^2-D)k_i - (k_j^2-D)k_j \right|^2} \cdot e^{-\frac{1}{2}\pi^2 d^2 (k_i^2-k_j^2)^2} \tag{1}$$

γ, D, s and δ are the wave aberration function, underfocus, rms beam divergence and rms focus spread, and reduced units are used for brevity (Sch=$C_s^2\lambda^2$, Gl = $C_s^{1/4}\lambda^{3/4}$; see Hawkes, 1980).

The first factor of this expression describes lateral shifting of image intensity components, while the remaining two describe their attenuation as a result of finite beam divergence and focus spread respectively. It will be obvious that a slight movement of the primary beam, which will add a small constant vector to all $\underline{k}_i$ and $\underline{k}_j$, will destroy any original symmetry in all three terms; however it is enough for the present to look at the last term only. This shows that beams equidistant from the optical axis are passed without attenuation by focus spread, but for high order reflections small changes in $\underline{k}_i$, $\underline{k}_j$ due to slight beam tilt will produce large values of $k_i^2-k_j^2$ when either k lies along the tilt direction, resulting in complete removal of the corresponding image intensity component.

As an extreme example, beams at $\underline{k}$ and $-\underline{k}$, whose mutual interference would be strong (divergence permitting) under perfectly axial conditions, will disappear completely for a misalignment of $1/2\pi k\delta$ i.e. 1.3mrad for a pair of 2Å reflections in a 500keV/3mmC$_s$ microscope with δ=25nm, and 1.9mrad in a 100keV/1.8mmC$_s$ microscope with δ=17nm.

The expression (1) above is of course the <u>mutual</u> transfer function applicable to non-linear second order (scattered-scattered) interferences as well as to the simpler linear first order ones (primary-scattered). The latter are those to which the term contrast transfer function (CTF) is most commonly applied, and the form it takes for axial imaging when one of k_i and k_j is zero in each term is well known, namely the product of sin γ(k) and exponential factors like those in (1); its squared modulus is observable in diffractograms of micrographs of thin amorphous specimens, and these are accordingly widely used for confirming the imaging conditions after the event.   Now the introduction of slight primary beam tilt modifies the CTF substantially, (Wade and Jenkins, 1978; Saxton, 1978) particularly in the higher frequency range, introducing an antisymmetric imaginary part and so allowing some specimen frequency components to be imaged displaced and not simply attenuated or reversed as in axial imaging; yet the modulus of the CTF, and hence the vital diagnostic information provided by a diffractogram, does not change, making the tilt incapable of a posteriori detection by the usual methods.

Figs.1-2 show the linear CTF for axial and slightly tilted conditions, confirming the dramatic changes the misalignment can cause.   Fig.3 shows white noise filtered by axial and tilted CTFs, with inset diffractograms,i.e. simulated images of a fully amorphous material, and it can be seen that while there are differences between the two images, it is impossible to detect the tilt either from the diffractogram or the image texture.   Figs.4-6 present simulated 500kV images of a block-structured oxide, GeNb$_9$O$_{25}$ under various conditions of tilt.   These show clearly that beam tilt degrades the image much more rapidly than crystal tilt, and that the effects are already significant at tilts which are undetectable by the usual means. Similar conclusions can be drawn from the paper by O'Keefe and Sanders (1975).

<u>Experimental implications</u>

Unfortunately it is much harder to find solutions to the problem than to delineate it.   It is our impression that conventional current or voltage centre alignment may well leave residual beam tilts sufficient to cause difficulty, and the directionality of image texture produced by rather larger tilts can in any case be partially removed by introducing astigmatism (McFarlane 1975), disguising the tilt still further.

The procedure described by Zemlin et al (1978) is certainly capable of detecting and correcting any misalignment, but it is extremely laborious: a simpler expedient suggested to us by D.J. Smith may prove adequate. As in Zemlin et al's approach an amorphous specimen area is imaged, and the beam is deflected by known amounts in four orthogonal directions from the central position in turn, but instead of recording an image in each case for diffractogram analysis, the rms image contrast levels are simply compared visually and the central position of the beam is adjusted until the same contrast is observed for each direction of deflection.   The optimum deflection is obviously that for which the contrast varies most rapidly with tilt,and our calculations indicate that this occurs for a tilt of about 0.3G1^{-1} for both D=1 (Scherzer focus) and D=1.8 (first broad band condition). The contrast variation is not very great in any case however, and it may be that the on-line digital image acquisition/processing systems now becoming available (e.g. Catto et al, 1981) will find another important application in allowing its accurate measurement.

References

Catto CJD, Smith KCA, Nixon WC and Erasmus SJ 1981 this volume
Hawkes PW 1980 Ultramicrosc. $\underline{5}$ 67
McFarlane SC 1975   J. Phys. $\underline{C8}$ 2819
O'Keefe MA 1979 37th Ann. Proc. EMSA ed. GW Bailey $\underline{1}$ 556
O'Keefe MA and Sanders JV 1975 Acta Crystallogr. $\underline{A31}$ 307
Saxton WO 1978 Computer Techniques for Image Processing in Electron
    Microscopy (New York: Academic) pp9-19
Saxton WO 1980 J. Micr. Spectr. Electr. $\underline{5}$ 661
Wade RH and Jenkins WK 1978 Optik $\underline{50}$ 1
Zemlin F, Weiss K, Schiske P, Kunath W and Herrmann KH 1978
    Ultramicrosc. $\underline{3}$ 49

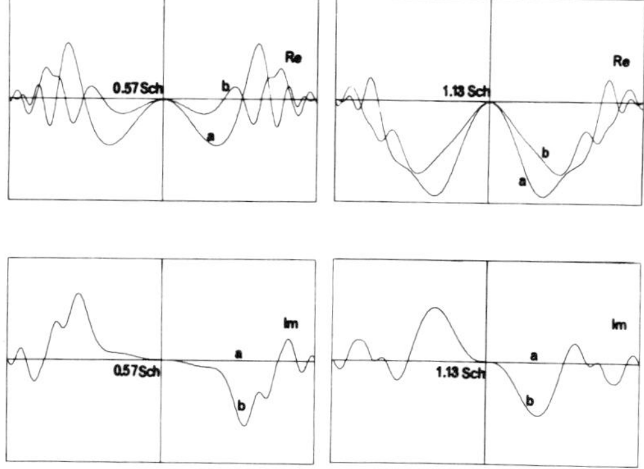

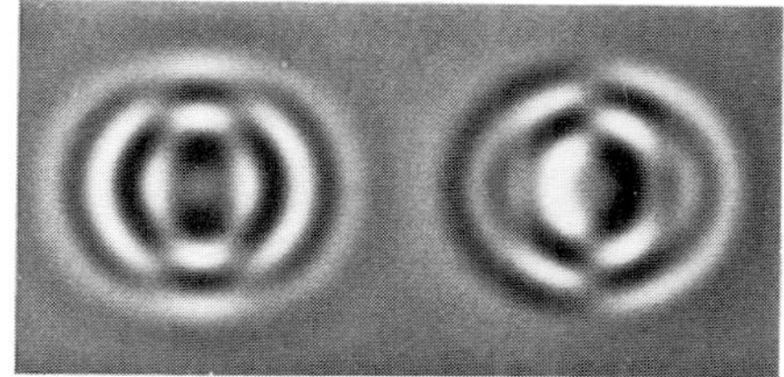

Fig.2. Full CTF for the conditions of fig.1, but at higher underfocus D=3 Sch. Left: real part; right: imaginary part. White represents positive contrast and black normal (i.e. negative) contrast. Edge of field 2.1 $\overline{GI}$. The tilt corresponds to only 1.3 mrad at 500kV, C_s=3.5mm.

Fig.1. CTF sections for axial (a) and slightly tilted (b) conditions; tilt 0.29 $\overline{GI}^{-1}$, s=0.052m =0.2. Upper row: real part; lower: imaginary. Underfocus left D=0.57 Sch, right D=1.13 Sch. These conditions correspond to those of figs.4-6 with a lower focus spread.

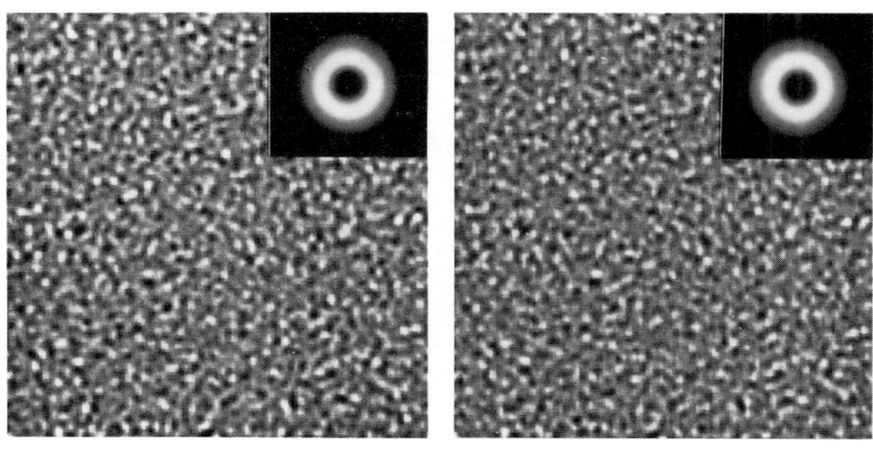

Fig.3. Simulated images of perfectly amorphous specimen for the conditions of fig.1 (D=1.13 Sch). Left: axial illumination; right: tilted.

Fig.4. Simulated 500kV images of $GeNb_9O_{25}$ 1.9 nm thick. Left: crystal and beam untilted; centre: crystal tilt of 1.3 mrad; right: beam tilt of 1.3mrad Upper row: underfocus 40nm; lower: 80nm (Scherzer focus). Rms beam divergence 0.23mrad, rms focus spread 25nm, field of view 4.6nm across.

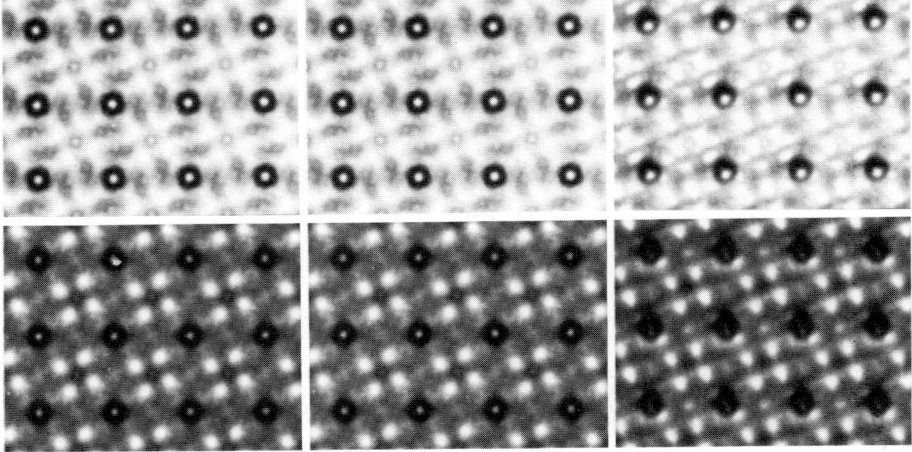

Fig.5. Simulated images as in fig.4, but for 7.6nm crystal thickness.

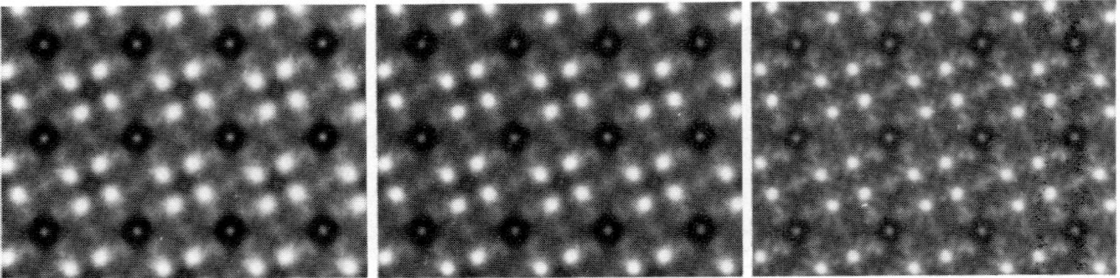

Fig.6. Simulated images for 7.6nm thickness and no beam tilt; crystal tilt left to right 0, 2mrad and 8mrad.

The support of the SRC is gratefully acknowledged.

Multi-slice calculations using an array processor

A J Skarnulis[†], D L Wild, G R Anstis, C J Humphreys and J C H Spence
Department of Metallurgy and Science of Materials, University of Oxford,
Parks Road, Oxford OX1 3PH
† Laboratory of Chemical Crystallography, Parks Road, Oxford

1. Introduction

The recent development of array processors, which use dedicated
programmeable hardware to speed up various array operations, such as fast
Fourier transforms (FFT's), has made dynamical electron diffraction multislice calculations, using large numbers of beams and slices, on small
computers a realistic prospect (Rez, 1980). Applications which require
such calculations include the simulation of high resolution images of
defects by the method of periodic continuation using a large unit cell and
the interpretation of lattice images of thick crystals.

An array processor is a fast processor, attached to a host computer, and
its parallel structure allows the overheads of array indexing, loop
counting and data fetching from the memory to be performed simultaneously
with arithmetic operations on the data. This results in a much faster
execution time than in a typical computer, where each of the above
operations occurs sequentially.

2. Method

The multislice method was formulated by Cowley and Moodie (1957), who
represented the transmission of electrons through a crystal by transmission through a set of two-dimensional phase objects separated by distances,
Δz. The amplitude of the wave function after $n + 1$ slices is given by:

$$\psi_{n+1}(x,y) = [\psi_n(x,y).q(x,y)]*p(x,y) . \tag{1}$$

Here * denotes convolution, p is a propagation function and q is the phase
grating given by $\exp(i\sigma v(r)\Delta z)$, where $v(r)$ is the projected potential and
σ is an interaction constant related to the energy of the electrons.
Goodman and Moodie (1974) have described a computational method which uses
the equivalent reciprocal space formulation:

$$\phi_{n+1}(h,k) = \sum_{h'}\sum_{k'} \phi_n(h',k')P(h',k')Q(h-h',k-k') \tag{2}$$

Here ϕ, P and Q denote the Fourier transforms of ψ, p and q respectively.
$P(h,k)$ is given by $\exp(2\pi i\, s_{h,k}\Delta z)$ where $s_{h,k}$ is the excitation error. Δz
must be chosen so that $2\pi i\, s_{h,k}\Delta z \ll 1$. Anstis (1977) has shown that by
this method, in the limit that the slice thickness approaches zero, the
wave function is the solution to the Howie-Whelan equations for a finite

number of beams.

Direct evaluation of the convolution in equation 2 is relatively slow.
Use of fast Fourier transforms and the convolution theorem, which for
finite transforms states that the Fourier transform of the product of two
functions is the cyclic convolution of their Fourier transforms, is more
efficient. (Ishizuka and Uyeda, 1977; Figure 1).

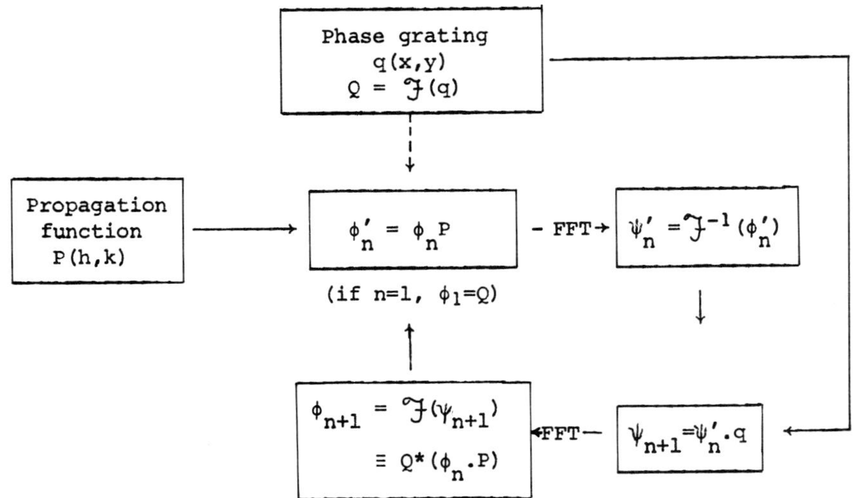

where 𝒥 denotes the Fourier transform operator.

<u>FIG. 1</u>

By this method one obtains, using an N × N transform, the solution to the
equations

$$\frac{d}{dz}\,\phi(h,k) = 2\pi i\,s_{h,k}\phi(h,k) + i\sigma \sum_{h'}\sum_{k'} V([h-h',N],[k-k',N])\phi(h',k')$$

where all indices ℓ satisfy $-\frac{1}{2}N < \ell \leqslant \frac{1}{2}N$ and $[\ell,N]$ is a number, obtained
from ℓ by adding multiples of N, which lies between $-\frac{1}{2}N$ and $\frac{1}{2}N$. As an
example of how these equations differ from the Howie-Whelan equations,
consider the equation for $h = \frac{1}{4}N + 1$. Then terms such as $V(-\frac{1}{2}N+1,[k-k',N])$
$\phi(-\frac{1}{4}N,k')$ arise. These describe scattering between beams with index
$h = -\frac{1}{4}N$ to those with index $h = \frac{1}{4}N + 1$ whereas the correct strength of the
interaction between such beams is proportional to $V(\frac{1}{2}N + 1, [k-k',N])$.

To avoid such errors only beams with indices in the range -N/4 to N/4
should be included. Thus in using the fast Fourier transform algorithm,
transforms of order 2N × 2N must be used in problems involving N × N beams.
The computing time, for N × N beams, using the Fourier transforms is
proportional to

$$2(2N)^2 \log_2 (2N)^2 + 2(2N)^2 = 8N^2 \log_2 8N^2$$

whilst the time based on the direct convolution method (equation 2) is
proportional to $(N^2)^2 + N^2 \simeq (N^2)^2$.

3. Results

An interactive multislice programme (Skarnulis, 1979) running on a
Prime 750 computer, has been modified to utilize an attached Floating Point
Systems AP120B Array Processor with 333ns memory, which offers Fortran
callable subroutines. Arrays are created in the host programme for the
phase grating, q, the propagation function, P, and the beam amplitudes and
phases for the starting slice. These arrays are then passed over to the
array processor, which performs the multislice cycle for the required
number of slices, and the beam amplitudes are then returned to the host.
The programme uses a 64×64 point complex FFT, which takes 55m sec on the
AP120B, as opposed to about 1.5 secs for a software FFT on the Prime 750.
Table 1 shows typical times for one multislice cycle for 1000 beams,
together with the set-up time which includes the overhead of transferring
the arrays to the Array Processor. This is of the order of 1 sec. for the
AP120B. Times obtained by Rez (1980,1981) using an FFT algorithm on a
CDC 7600 and on a PDP 11/04 with attached array processor (Analogic AP400)
are shown for comparison.

TABLE 1

Method	Mean time/slice for multislice calculation using 1000 beams (based on 64×64 point complex FFT) (secs)	SETUP TIME (secs.)
a) Direct convolution algorithm on Prime 750	460	0.0
b) FFT algorithm using Software FFT on Prime 750 (estimated)	3	0.0
c) FFT algorithm on Prime 750 + AP120B	0.2	∿1.0
d) FFT algorithm on CDC 7600 (Rez)	0.3	0.0
e) FFT algorithm on PDP 11/04 + AP400 (Rez)	0.1	∿50.0

4. Applications

4.1 Thick perfect crystal

As one example of a possible application the programme has been used to
investigate the variation of diffracted intensities with thickness for a
perfect crystal of silicon in a [110] symmetrical orientation. The
calculation, which was for 200KV electrons, included all reflections to
3.5Å^{-1} resolution (805 beams including dynamically forbidden ones), the
slice thickness used was 1.92Å, and no absorption was included. The sum of
intensities after 1600 slices was 0.99. Comparison with a similar calculation using a slice thickness of 3.84Å showed no significant differences
for low order reflections, indicating that the accuracy of the calculation
was adequate. Figure 2 shows the variation in the intensities of the (000),
(111) and kinematically forbidden (002) reflections with thickness.

Because absorption was not included, the calculation should not be
considered realistic for thick crystals, but it is nevertheless
interesting to observe that the periodic nature of the solution persists.

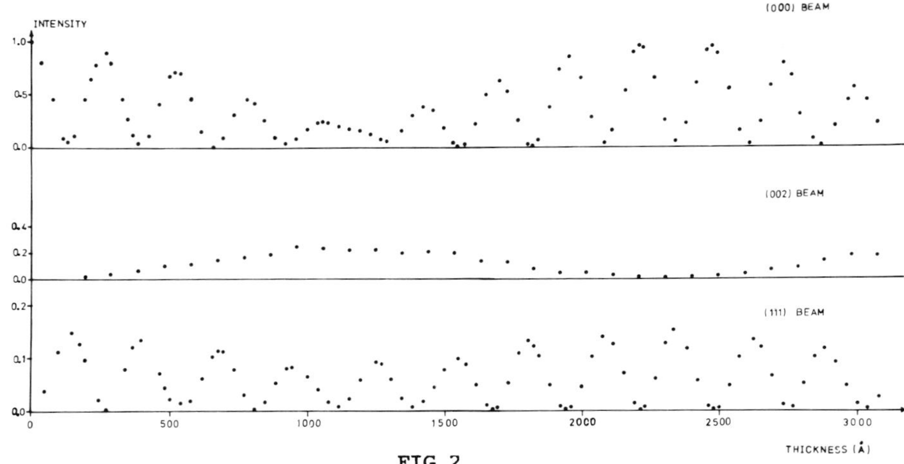

FIG.2

4.2 Large unit cell

The programme has also been used to simulate images of a Si/NiS_2 interface, in conjunction with experimental work being carried out in this laboratory. The calculation is based on a supercell of dimensions 3.3Å × 37.4Å × 3.8Å. 397 beams were included, which corresponds to a resolution of $1Å^{-1}$. Typical images are shown in Figure 3.

The above examples demonstrate the potential of the array processor for large multislice calculations. Other possible applications include calculations involving upper layer line interactions, microdiffraction and lattice images of complex oxides.

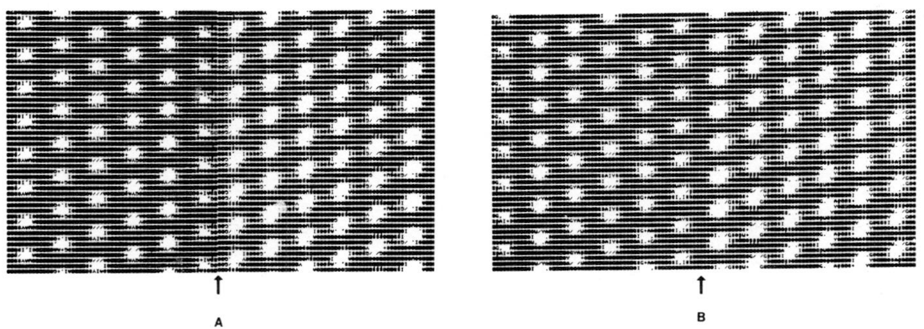

FIG.3 Images of two models of Si/NiS_2 interface at thickness of 38Å. Aperture = $0.4Å^{-1}$, defocus = -650Å, C_S = 1.2mm, 200KV.

References

Anstis G R 1977 Acta Cryst. A33, 844
Cowley J M and Moodie A F 1957 Acta Cryst. 10, 609
Goodman P and Moodie A F 1974 Acta Cryst. A30, 280
Ishizuka K and Uyeda N 1977 Acta Cryst. A33, 740
Rez P 1980 38th Ann.Proc.Electron Microscopy Soc.Amer. ed: G.W. Bailey
   (San Francisco) pp 180-181
Rez P 1981 Personal communication
Skarnulis A J 1979 J.Appl.Cryst. 12, 636

Contrast from amorphous hollow fibres

P. J. Goodhew and D. Chescoe

Department of Metallurgy & Materials Technology, University of Surrey,
Guildford GU2 5XH, U.K.

1. Introduction

Many inorganic materials may grow as tubes or scrolls on a very fine
scale. In the particular case of hydrated Ordinary Portland Cement (OPC)
it is a matter of current controversy whether certain fibrillar features
grow by extrusion of liquid up the centre of a hollow tube (Bailey and
Chescoe, 1979; Barnes et al, 1980). The alternative is the precipitation
from solution of a solid fibre and it is therefore important to be able to
distinguish, in the electron microscope, between solid and hollow rod-like
features with diameters generally less than 100nm. In ideal
circumstances it would be possible to resolve the point using high
resolution scanning microscopy on the ends of the rods. However this
fails to provide clear-cut evidence since the ends of a hollow tube could
well be blocked by a plug of solidified extrudate: it is also possible
that a fine tube which contained liquid "in vivo" will, after preparation
for microscopy, contain a core of solid of different composition from the
wall. For these reasons it is necessary to investigate the limits of
hole size, density difference between core and wall, and defocus which do
enable us to distinguish between solid and hollow tubes.

2. Method

Since electron diffraction has revealed that the majority of rod-like
features in hydrated OPC are essentially amorphous the scattering model
can be very simple and we have used a mass absorption coefficient which
gives an absorption coefficient proportional to density. In this simple
method the size of the aperture is not significant. The cross-section of
the fibre is shown in Figure 1, which also illustrates schematically one
column for the absorption calculation. In practice 60 steps were found
to be sufficient per column. The major variables in the absorption
calculation were the hole size ratio r_H/r_F and the hole absorption
parameter ratio, which under our assumptions is equal to the hole density
ratio ρ_H/ρ_F.

The emergent intensity distribution from a matrix of 128 x 128 columns
was defocussed using fast Fourier transform routines and a transfer
function appropriate to our JEM 200 CX microscope. The images were output
to microfilm using 32 grey levels.

3. Results and Discussion

If the fibre is truly hollow, with the hole contributing zero scattering,

then a hole only 10% of the diameter of the fibre is readily visible.
Figure 2 shows simulated micrographs of fibre containing four different
hole sizes. The fibre containing the smallest hole shown here, 5% of the
fibre diameter, is indistinguishable from a solid rod. We would expect
defocussing to reduce the visibility of the hole but although this does
eventually happen (Figure 3) the amount of defocus needed is larger than
would be likely in practical microscopy of such specimens. The third
effect, variation of the density and hence absorption coefficient of the
material in the hole, is shown in Figure 4. For core densities up to 50%
of the wall density the effect is only to reduce the apparent size of the
"hole" (Figure 4 a, b, c). However if the core density exceeds 60% of
that of the fibre wall the fibre once more appears to be a solid rod
(Figure 4d).

The computed images compare well with experimental micrographs of cement
fibres. For instance Figure 5a shows a region containing both hollow (H)
and solid (S) fibres. Microdensitometer traces across these fibres at a
variety of focus settings (Figure 5b) show that it is quite straightforward to differentiate between solid and hollow fibres without the need
to focus precisely. From the simulated image results it is possible to
estimate that fibre H contains a hole of about 30% of the fibre diameter,
containing material of density less than half that of the fibre wall.

It might be thought that the image of the crossover of two hollow fibres
would be more sensitive to hole size than the image of a single fibre.
However computed images show that the intensity minimum in the centre of
the diamond image visible in several places in Figure 5a is not a useful
parameter. Among other problems it is necessary to find two similar
fibres crossing, and to assume that no small defect is present in either,
in the superimposed regions.

4.   References

J.E. Bailey and D. Chescoe Proc.Brit.Ceram.Soc. 1979 **28** 165
P. Barnes, A. Ghose and A.L. Mackay Cem. Concrete Res. 1980 **10** 639

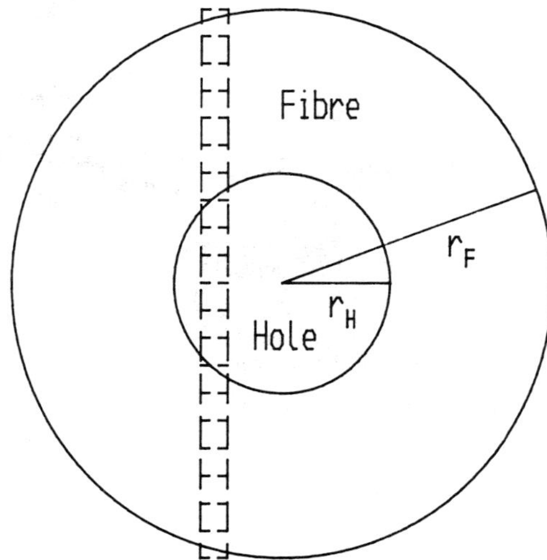

Figure 1. The computation scheme.

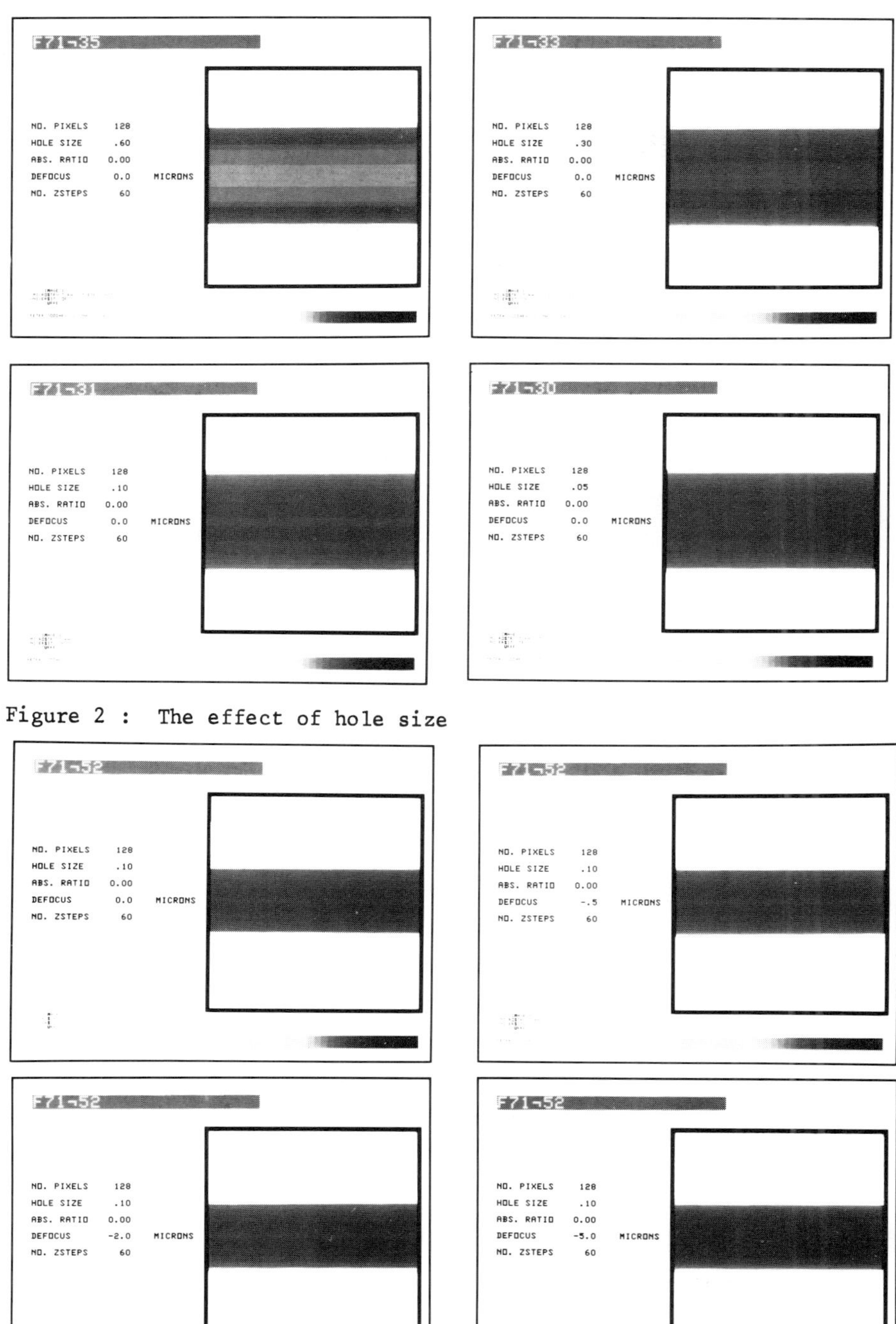

Figure 2 : The effect of hole size

Figure 3 : The effect of defocussing

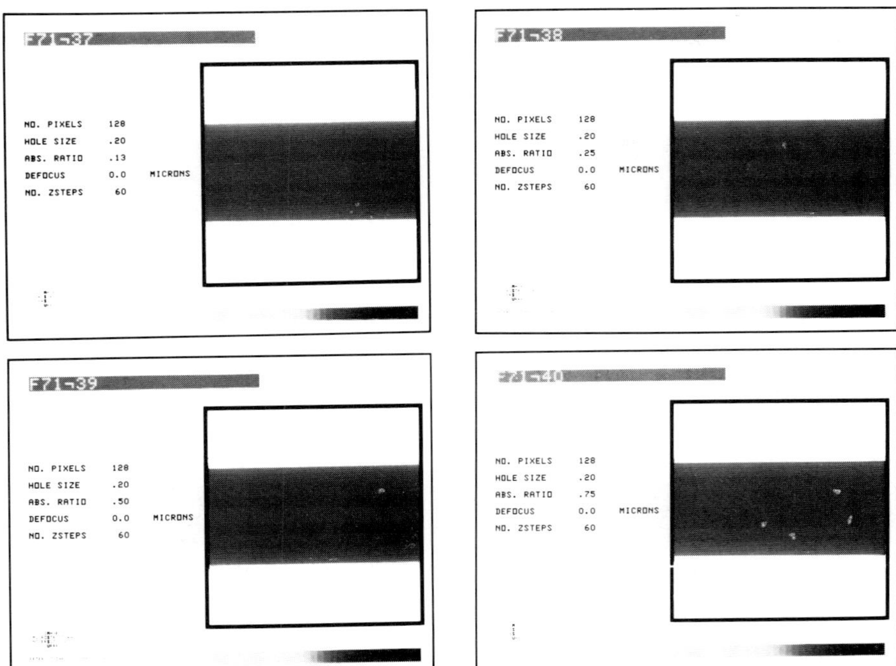

Figure 4: The effect of the density of the core material

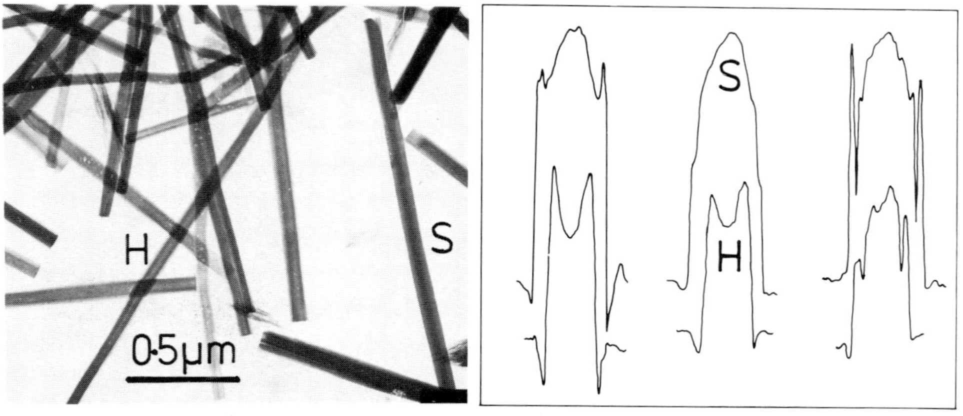

Figure 5: Hollow and solid cement fibres and microdensitometer traces across fibres under- in- and over-focussed.

Contrast from pressurized gas bubbles in metals

B. Cochrane, S.K. Tyler and P.J. Goodhew

Department of Metallurgy & Materials Technology, University of Surrey,
Guildford, GU2 5XH.

Voids and gas bubbles of nanometre dimensions are frequently created as
a result of irradiation damage by neutrons and/or alpha particles. Of
particular interest in the use of metals in fast or fusion reactors is
the gas pressure within the bubbles. Strain contrast in the transmission
electron microscope offers a way of detecting the strain field of a
bubble but quantitative deductions are complicated by the fact that in
the majority of cases such bubbles are strongly faceted : They cannot
therefore be assumed to be spherical and the strain field which surrounds
them is unlikely to be spherically symmetrical. It is generally
considered that use of a large $\underline{g}$ but a small deviation parameter, $\underline{w}$,
provides optimum imaging of strain fields, whereas use of a large $\underline{w}$ and
a considerable underfocus is preferred for revealing bubble shapes. We
have therefore considered the contrast from a bubble arising
simultaneously from mass thickness effects, from diffraction strain
contrast and from defocussing.

Two-beam calculations have been used and in these early stages of the
work the bubble has been assumed to facet perfectly on (100) planes
whereas the strain field has retained spherical symmetry. Diffraction
contrast and mass thickness effects have been calculated for a matrix of
128x128 columns using an array processor. Defocussing of the contrast
distribution at the exit face of the foil was carried out using FFT
routines and a transfer function appropriate to our JEM 200CX microscope.

Figure 1 illustrates the effect of $\underline{w}$ and defocussing on the image of a
40nm bubble containing a gas pressure of 49x its equilibrium value. The
upper computed image shows an in-focus bubble at $\underline{w}$ = o, where the
contrast is dominated by the expected dark lobes. For large $\underline{w}$ (centre
image) the projected shape of the bubble dominates whereas even at $\underline{w}$ = o
(lower image) a moderate defocus (0.3µm under) causes the shape to
dominate and all edges to become visible. Figure 2 shows the effect of
increasing the gas pressure within the bubble on the in-focus image at
low $\underline{w}$ (0.5). For small overpressure (1.07, upper image) the bubble is
almost invisible but as the pressure increases (9, centre and 20, lower)
approximately symmetrical lobes of strain contrast appear. Figure 3
illustrates, however, that the experimental image does not yet correlate
in every detail with the computed image under closely similar diffraction
conditions. This is probably because of the unrealistic assumption of a
spherical strain field surrounding a faceted bubble. Improved strain
fields are currently under development.

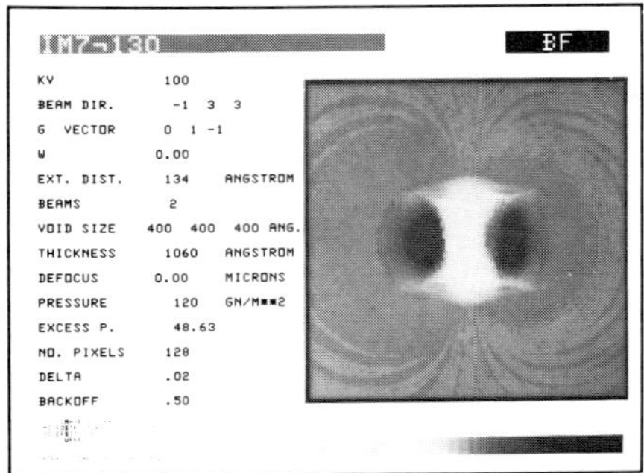

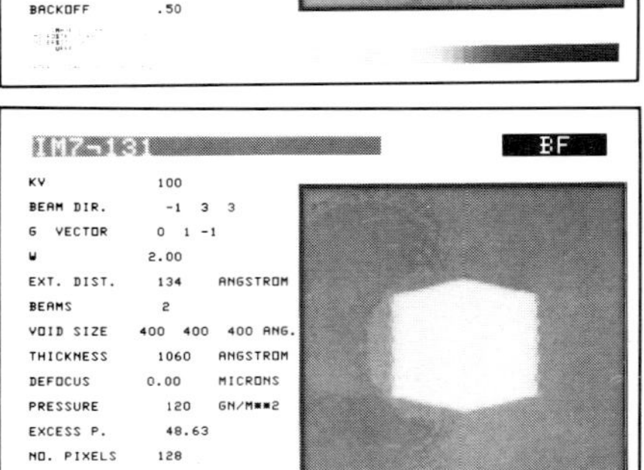

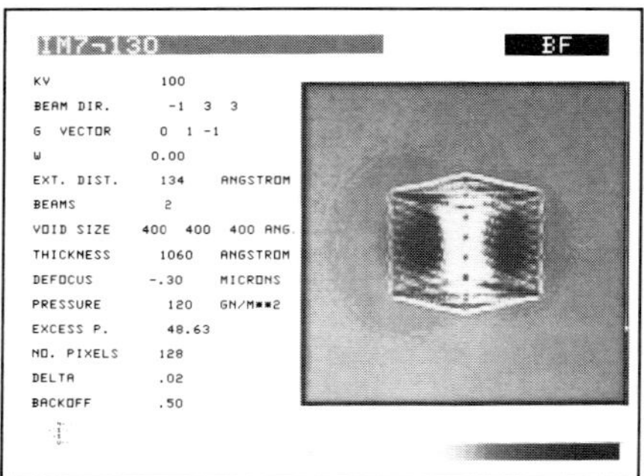

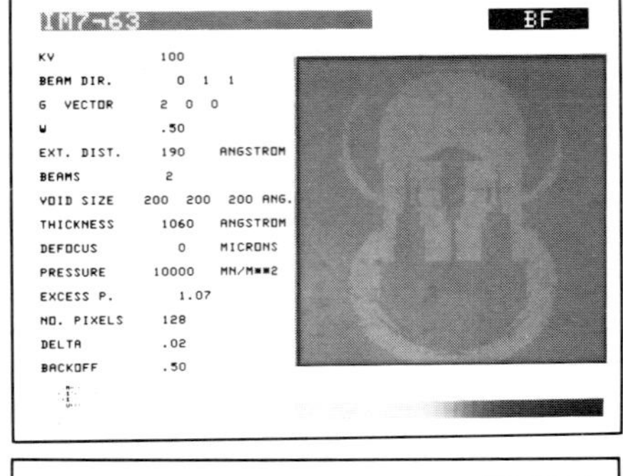

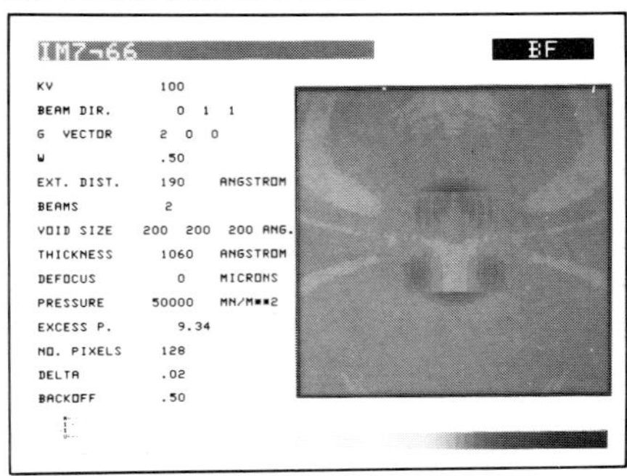

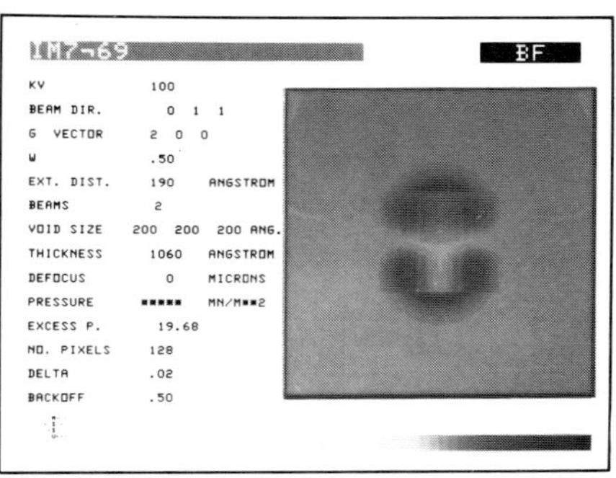

Figure 1   Figure 2

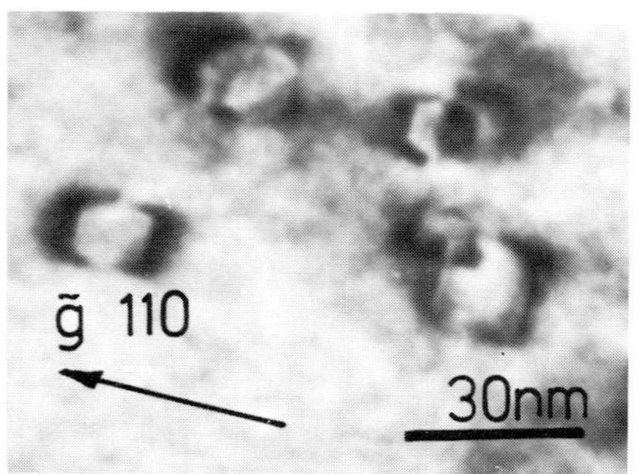

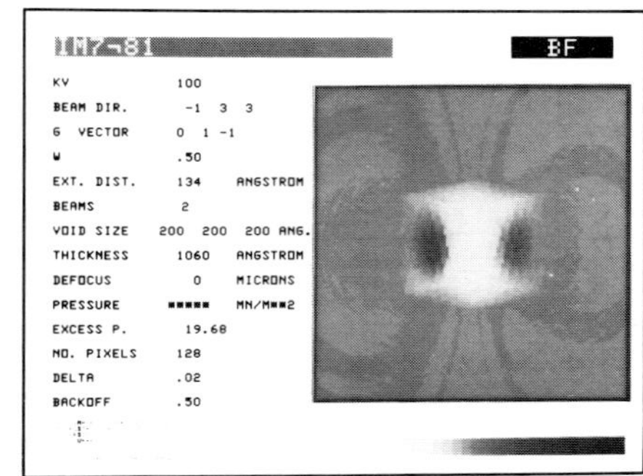

Figure 3

Atomic images of silicon and related materials: fact or artefact?

J L Hutchison, G R Anstis, C J Humphreys and A Ourmazd

Department of Metallurgy and Science of Materials, University of Oxford,
Parks Road, Oxford OX1 3PH, U.K.

1. Introduction

The study of defect structures in silicon and other semiconductors with the
diamond cubic or sphalerite type structures e.g. Ge, CdTe, GaAs, etc. is of
considerable interest in the context of microelectronics, and high
resolution imaging of these materials is currently being carried out in a
number of laboratories.

A <110> projection of the diamond cubic type lattice is shown in Fig 1. In
silicon, for which a = 5.43Å, nearest neighbour atoms (e.g. at 0 0 0 and
$\frac{1}{4}$ $\frac{1}{4}$ $\frac{1}{4}$) have a separation of 2.35Å;in the <110> projection this appears as
1.36Å (i.e. $\frac{a}{4}$) and in the majority of published <110> images these pairs of
atoms are imaged as composite blobs.   Clearly, if the individual atomic
columns could be resolved, the interpretation of image contrast at
dislocation cores would be greatly facilitated.

A well-known 100kV image of perfect silicon, obtained by Izui et al (1979)
shows pairs of white dots or 'dumbells' in what appeared to be correct
atomic positions, 1.36Å apart.   Recently, similar images have been
produced elsewhere, of Si (Hutchison et al 1980(a)), as shown in Fig 2,
GaAs (Hashimoto et al 1978-79) and CdTe (Yamashita, 1980; Krivanek & Rez,
1980).   However, it is now recognised that the splitting in the dumbells
or spot pairs is significantly larger than $\frac{a}{4}$; i.e. the images do not
reveal the correct atomic positions.   In this paper we present a
quantitative analysis of such images, indicating experimental criteria
which must be met if "correct" image contrast is to be obtained for axial
images of these structures.

2. Geometrical interpretation of images

In an earlier communication (Hutchison et al. 1980(b)), we described on a
purely geometrical basis how a lattice of intersecting planes could be
constructed corresponding to the hkl   reflections which contribute to
these images, and showed that certain intersections formed "nodes" about
1.8Å apart, near the positions of atom pairs in the projected crystal
structure: image calculations have now made the analysis quantitative.

3. Image Computation

The multi-slice method was used to calculate amplitudes and phases of
diffracted beams as a function of crystal thickness.   An array processor
was used to enable large calculations to be carried out quickly (Skarnulis

et al. 1981).   The calculations included all beams out to 1.4Å^{-1} with a
slice thickness of 3.84Å.   This size of calculation was found to be
adequate by comparison with results of much larger calculations.   Parameters included in the image intensity calculations were the radius (R) of
the objective aperture in reciprocal space, the spherical aberration (C_s)
and defocus (Δf) of the objective lens.   The half-width of the focal
spread (Δ) and the incident beam divergence semi-angle (α) were also
included, with non-linear contributions from the last two parameters, as
described by O'Keefe(1979) and Krivanek and Rez (1980).   For the images
shown here, however, these contributions were very small.   Table 1
summarizes the instrumental parameters used in the image calculations.

kV	C_s mm	ΔÅ	α mrad
100	0.7	50	1
200	1.2	30	0.2
400	1.0	50	0.1
600	3.3	50	0.3
1000	1.5	50	0.1

Table 1.   Instrumental parameters used in image calculation.

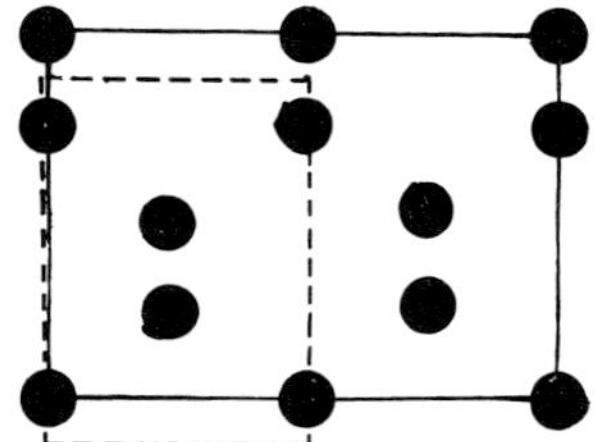

Fig 1.   <110> projection of silicon,
showing area in computed images
(dashed line).

Fig 3 shows images of thin crystals at "optimum defocus" ($\Delta f = -1.2(C_s\lambda)^{\frac{1}{2}}$),
the condition normally sought in conventional high resolution microscopy.
Atomic positions are resolved only with the 1 MeV instrument which has a
resolution of 1.2Å with the parameters indicated in the table.   Note,
however, that the focal spread is unrealistically small for both 600 and
1000 kV images.

Systematic calculations of images using parameters corresponding to the
Oxford JEOL 200CX microscope as used to obtain fig 2 showed that "dumbell"
images occur for foil thicknesses in the range ξ_{000} to 1.5 ξ_{000}.   Similar
images formed at the other voltages which are shown in fig 4 are also for
foil thicknesses in this range.   Note that the Δf values are not the
optimum values for thin crystals.   For the 1 MeV case, Δf is positive.
The spacing of the dots is not the atomic spacing, apart from at 1 MeV.
Furthermore in 100 and 400 kV images the white dots are in the positions of
tunnels rather than of atom pairs.

The separation of the spots depends on the ratio of the (002) and (111)
intensities.   This is demonstrated in fig 5 which shows images for
different thicknesses.   In this series the ratio of the (002) and (111)
intensities varies from 11 to 0.7.   The relative phases of these
reflections are also important as can be seen in the through focal series
in fig 6.

4. <u>Conclusions</u>

The above series of simulated images clearly demonstrate that the
appearance of pairs of dots in [110] images of silicon does not correspond
in any direct way to the 'resolution' of atomic sites; indeed pairs of dots
may be situated in tunnel positions in the lattice, as in the 100 and 400
kV images of fig 4.

It has been pointed out that images may be extremely sensitive to small

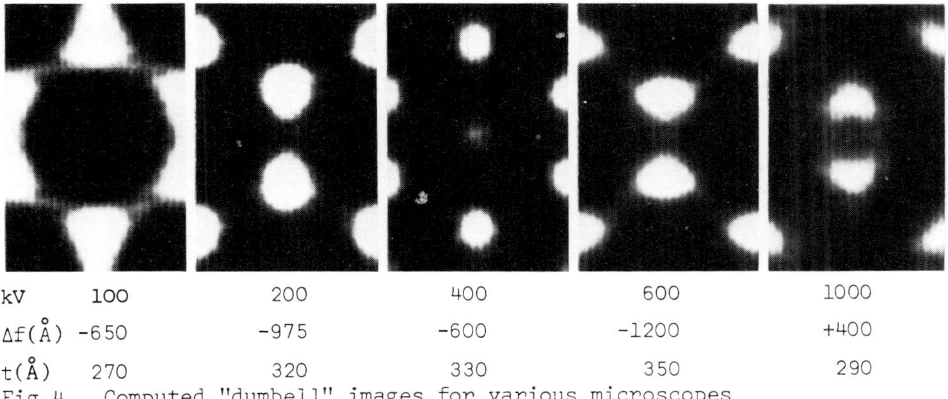

Fig 4. Computed "dumbell" images for various microscopes

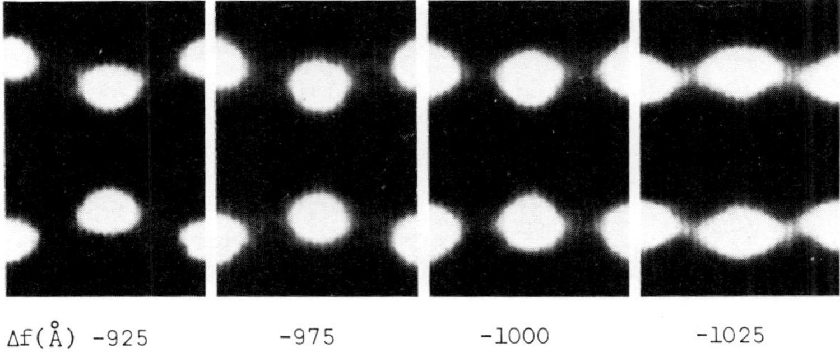

Fig 5. Through thickness series of computed images for 200kV microscope,
Δf = -975Å.

Fig 6. Through focal series of computed images for 200kV microscope,
thickness = 290Å

changes of orientation (Izui et al. 1979).   Small tilts may produce pairs
of dots of different intensities which exacerbates the problem of
identifying different atomic species in sphalerite structure compounds such
as CdTe.

For "atomic images" of silicon and related materials to become fact rather
than an electron optical artefact electron microscopes operating at high
voltage (1 MeV) with lens characteristics and high voltage stability
similar to today's best 100 and 200 kV machines will be required.   The
problems of irradiation damage will then have to be considered.

Hashimoto H, Endoh H, Takai Y, Tomioka H and Yokota Y, 1979, Chemica Scripta
     14, 23.
Hutchison J L, Humphreys C J and Ourmazd A, 1980(b) Inst.Physics Meeting on
     HREM, London.
Hutchison J L, Humphreys C J, Ourmazd A and Hirsch P B, 1980(a) Electron
     Microscopy, 1, 304.
Izui K, Furono A, Nishida T and Otsu H, 1979, Chemica Scripta 14, 99.
Krivanek O L and Rez P, 1980, 38th Ann.Proc. EMSA 180.
O'Keefe M A, 1979, 37th Ann.Proc. EMSA 556.
Skarnulis A J, Wild D L, Anstis G R, Humphreys C J and Spence J C H, these
     proceedings, p.
Yamashita T 1980 38th Ann. Proc. EMSA p168

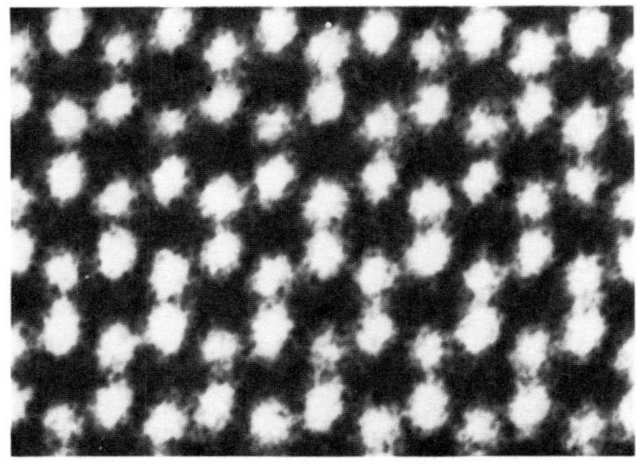

Fig 2.   High resolution <110> image of silicon, showing spot pairs
                          ∿ 1.9Å apart

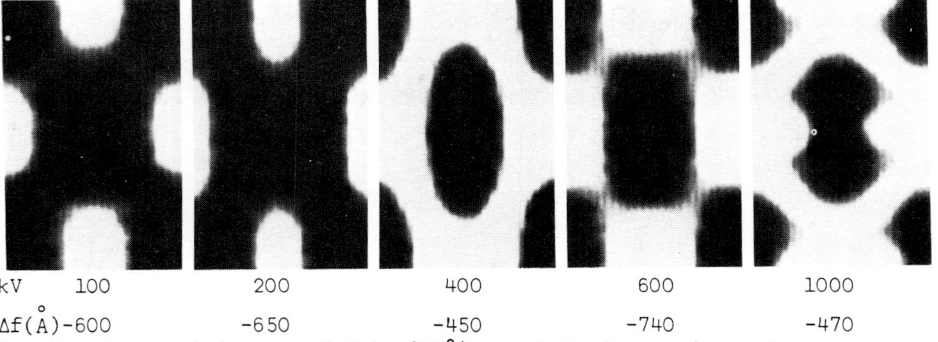

Fig 3.   Computed images of thin (38Å) crystals for various electron
         microscopes, at Scherzer defocus.

Petal and horseshoe shaped contrasts of the electron microscopic images of atoms on 011 gold crystal in deviated Bragg condition

Y Takai, H Hashimoto, H Endoh and N Ajika

Department of applied Physics, Osaka University, 2-1 Yamadaoka, Suita, Osaka 565, Japan

1. Introduction

The present authors (1977) observed doughnut shape contrast in the high resolution electron microscopic images of atoms in a gold crystal in 001 symmetry position and discussed the contrast in terms of flux flow of electron waves in the atom strings in the crystal and aberration free focus condition for recording the images. It was also pointed out that the inner structure of the image of atoms could be formed only when the second order Bragg reflections such as 400, 420 and 440 were included in the image formation (Hashimoto et al., 1978).

It is reported in the present paper that the fine structure with petal and horseshoe shape contrasts have been observed in the image of atoms in a gold crystal in 011 orientation deviated from the symmetry Bragg condition. The observed contrasts are discussed by the many beam dynamical theory of electron diffraction (Howie and Whelan, 1961) and the image formation theory considering the effects of the spacial and temporal coherence of electron wave (O'Keefe, 1979; Ishizuka, 1980).

2. Experimentals

Specimen gold crystals in 011 orientation were prepared by the deposition of gold in a vacuum on 011 single silver crystal at about 300°C and dissolving the substrate silver as was reported by Pashley (1959). Images of atoms were photographed by using a 100kV JEM-120C electron microscope with beam intensity of about 100 A/cm^2 emitted from a LaB_6 electron gun, the beam divergence was measured from the size of the diffraction spot to be 1×10^{-3} rad. The chromatic aberration coefficient of the objective lens was given by the JEM catalogue to be Cc=1.05mm. The spherical aberration coefficient was measured from optical diffractograms of the recorded images of a thin carbon film to be Cs=0.67±0.05mm (Krivanek, 1976).

The contrast of the images changed with the change of defocussing and Bragg reflecting condition. Fig. 1(a) shows one of the images which was photographed at the defocussing of about $\Delta f=2000\text{Å}$ with direct magnification of 2,000,000 times. The image was formed with axial illumination by using only 000, four 111 and two 200 reflected waves as shown in the diffraction pattern right below in the figure. Since the thin film of the specimen had small bending, images in various Bragg reflecting conditions could be photographed at the same time. The bright region in the left of the photograph which is indicated by a letter S is in the $0\bar{1}\bar{1}$ symmetry

position. The regions indicated by A,B,C and D are in the deviated Bragg
conditions which are shown in Fig. 1(b) as the projection of wave points
to the diffraction plane. As is seen in Fig. 1(a), the structure with the
spacings of 1.18Å and 1.44Å becomes predominant near the regions A,B,C and
D which correspond to the spacings of the 222 and 022 lattice planes. Since
222 and 022 Bragg reflections are cut off by the aperture and do not
contribute to the image, it can be concluded that such fine structures are
due to the interference between the waves scattered from {111}. The image
contrast with fine structures did not change very much by the change of
focus of ±30 Å, which is one unit of focus change of the microscope.

3. Comparison of observations with calculations

Figs. 2(a), (b), (c) and (d) show the magnified images of the regions
indicated by A, B, C and D in Fig. 1(a). Figs.2(a')-(d') show the
processed images of Figs. 2(a)-(d) respectively, which were obtained by
printing the image five times with shifting one atomic distance. The
petal and horseshoe shape contrasts can be seen clearly in Fig. 2(c) and

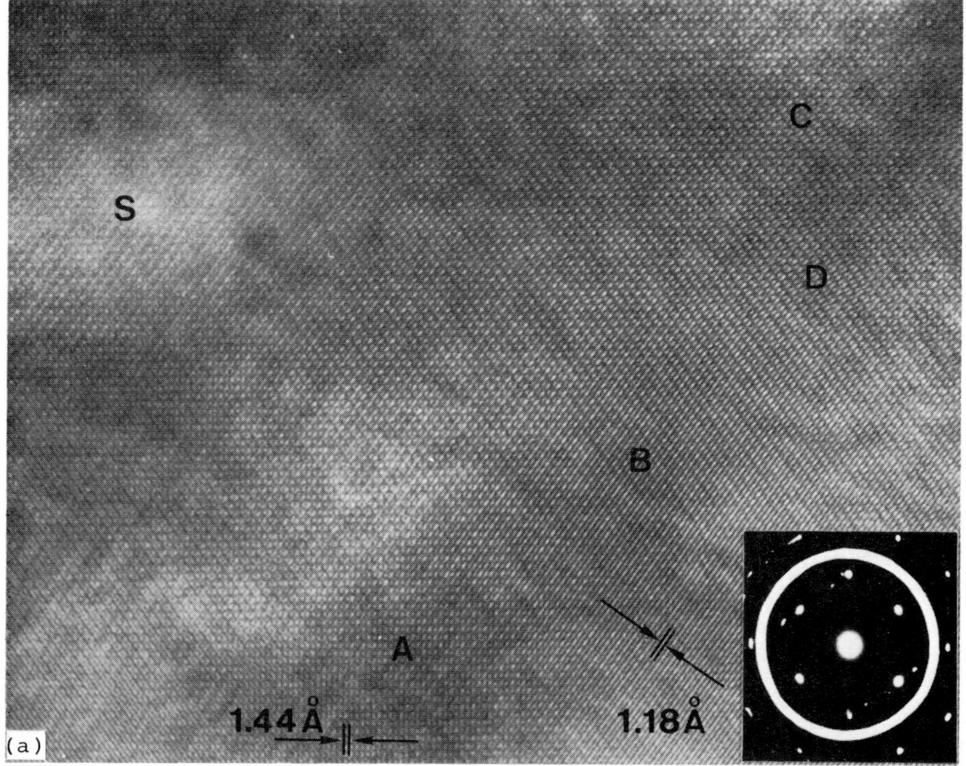

Fig. 1 (a) Electron microscopic image of a gold thin
film about 100 Å in 0ĪĪ orientation and corresponding
diffraction pattern with aperture size.
(b) Schematic illustration of diffraction pattern
with the projection of wave points corresponding to
the regions of A, B, C and D in (a).

(d) respectively. Figs. 2(a")-(d") show the contrast maps obtained by the calculation assuming the Bragg conditions shown in Fig. 1(b). Each figure shows the contrast corresponding to the crystal region shown in Fig. 2(e). The agreement of the observed and calculated contrast seems to be fairly good. According to the calculations, contrast of the images does not change very much in the range of crystal thickness $t=121.1\pm5$ Å, and defocus $\Delta f=2000\pm30$ Å, which agree with the observation. The calculation also suggests that the chromatic defocus value Δ which is one of the parameters representing partial coherence of the electron wave must be assumed to be more than 150 Å for obtaining these contrasts. The beam divergence has little influence on the image contrast at the defocus of 2000 Å in the range of $\alpha=0\sim10^{-3}$rad.

4. Discussion

Calculation suggests that the intensity distribution observed in the present experiment is not same with the one appearing at the bottom face of the crystal formed by the waves contributing the image contrast.
Fig. 3 shows the calculated images in the region indicated by C in Fig. 1(a) with different chromatic defocus value and beam divergence. As can be seen in Fig. 3, only the chromatic defocus value has much effect on the image contrast at the defocus of 2000Å. In order to see the order of interference between individual waves, the effect of interference of some selected pairs of waves is illustrated in Fig. 4. Fig.4 shows the value of $I(g",g')$ $=T(g",g')\phi_{g"}\phi_{g'}{}^{*}$, in term of chromatic defocus value Δ, where $T(g',g')$ is transmission cross coefficient and ϕ_g is the amplitude of a diffracted wave. As can be seen in Fig. 4, the contribution $I(g",g')$ of the interference between the transmitted 000 wave and the diffracted $1\bar{1}1$ wave decreases rapidly with increasing the chromatic value but the one between the diffracted waves themselves does not change very much. For example,$I(000,1\bar{1}1)$ is reduced to one sixtieth when the chromatic defocus value is 200 Å, but $I(1\bar{1}1,\bar{1}1\bar{1})$ does not change. The phase difference between the waves which give the same spacing, for example $(200,1\bar{1}1)$ and $(\bar{2}00,\bar{1}1\bar{1})$ will play important role to the image contrast. The phase difference due to the difference of Bragg reflecting condition gives the contrast change shown in Fig. 2. The phase difference due to the order of partial coherence of the illumination gives the contrast change shown in Fig. 3. As shown already, at the defocus of $\Delta f=2000$ Å, high contrast of the image is obtained and the contrast has little influence of beam divergence and defocus change. This defocus corresponds to the $m=3+3/4$ order of Fourier image both for 111 and 200 reflections, in which half spacing of their lattice plane can be seen, whereas AFF condition is satisfied at m=integer. The contrast transfer function at $\Delta f=2000$ Å have a flat wide zone in which the 111 and 200 reflections appear and hence no phase difference due to spherical aberration occurs between them. As shown already, the fine structure of the image is chiefly due to the interference of the waves in nearly achromatic condition.

References

Hashimoto H, Endoh H, Tanji T, Ono A and Watanabe E 1977 J. Phys. Soc. Japan 42 1073
Hashimoto H, and Endoh H 1978 Inst. Phys. Conf. Ser. No.41 London 64
Howie A and Whelan M J 1961 Proc. Roy. Soc. A263 217
Ishizuka K 1980 Ultramicroscopy 5 55
Krivanek O L 1976 Optik 45 97
O'Keefe M A 1979 37th Proc. EMSA vol. 1 556
Pashley D W 1959 Phil. Mag. 4 324

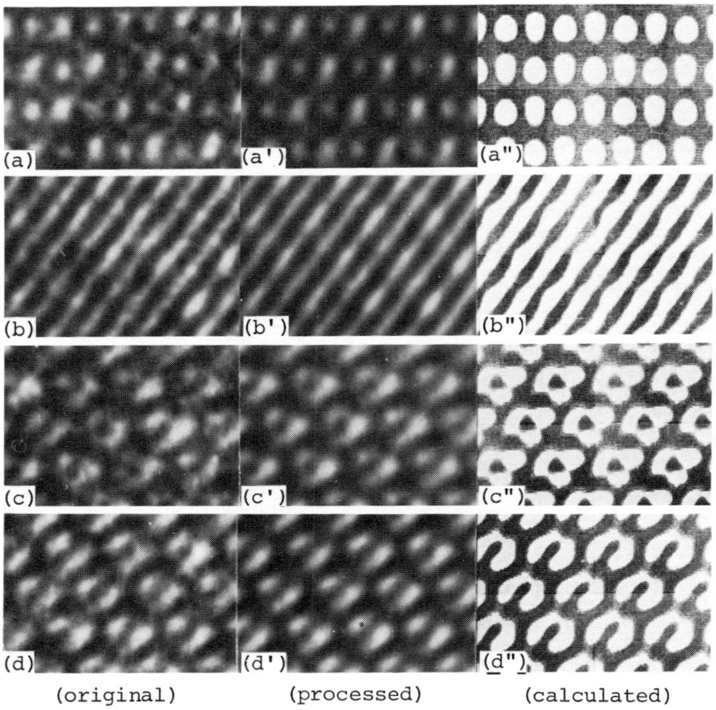

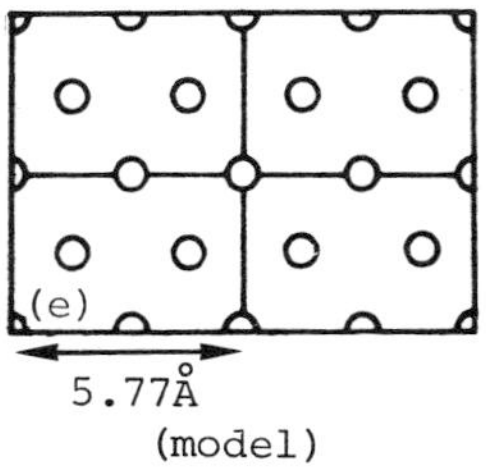

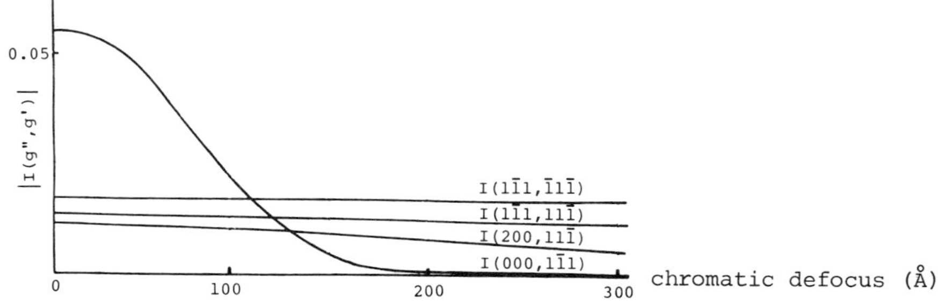

5.77Å

(model)

Fig.2(a),(b),(c),
(d) Enlargement of
A,B,C and D in Fig.
1(a). (e) Position
of Au atoms in 0$\bar{1}\bar{1}$
projection corre-
sponding to the
areas of (a),(b),
(c) and (d).

Calculation condi-
tion: t=121.2Å,
129 excited waves,
7 contributing
waves, Cs=0.7mm,
Δf=2000Å, α=
1x10^{-3}rad. Δ=200Å.

(original)   (processed)   (calculated)

(a) α=
1x10^{-4}rad.

(b) α=
1x10^{-3}rad.

Δ=0Å   Δ=100Å   Δ=150Å   Δ=200Å   Δ=300Å

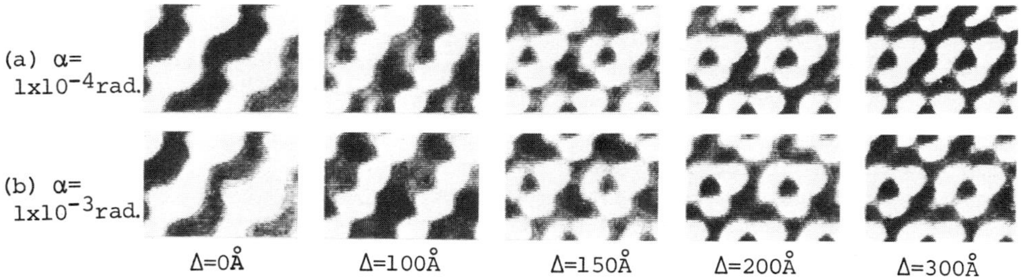

Fig. 3 Calculated contrast showing the dependence of chromatic defocus
value and beam divergence angle for the region C in Fig. 1(a).
(Cs=0.7mm, Δf=2000Å)

Fig. 4 Effect of chromatic defocus value to I(g",g') for the case (b)
shown in Fig. 3. (Cs=0.7mm, Δf=2000Å)

Analysis of atomic level resolution images of surfaces and inclined planar defects in (111)Au films

William Krakow

IBM Corporation, T.J. Watson Research Center, Yorktown Heights, New York 10598 USA

In this paper the imaging of the atomic surface lattice structure of thin vapor deposited (111) Au films is demonstrated which requires higher lateral resolution of surfaces than heretofore has been achieved in a conventional transmission microscope (CTEM). Images of incoherent double positioning boundaries (DPB's) which are inclined to the (111) surface have also been obtained and analyzed. It will be demonstrated that there is a relationship between the surface structure and that of a DPB which strongly influences the type of image contrast attained in a bright-field electron microscope image.

There have been many studies of crystalline surfaces where atomic resolution perpendicular to surface layers has been achieved. Dark-field images of the forbidden reflections from (111) Au films were employed to image lateral areas ~500Å in diameter.[1] However, aside from the inordinately long exposure times for dark-field imaging (40s), the lateral resolution is degraded by the diffraction limit of a small objective aperture. It is impossible to distinguish between different types of surface layer terminations because of the lack of resolution and lack of localized contrast variations. The first atomic level resolution pictures of the Au of (001) films were reported for bright-field imaging by Krakow and Ast.[2] Here optical diffractograms revealed the presence of forbidden surface reflections due to partially filled bulk lattice unit cells corresponding to surface lattice periodicies of ~2.86Å. Subsequently, improved electron micrographs were obtained at incident beam energies of 100keV and 1MeV where in the latter case the microscope resolution was better than 2Å.[3] In this study models of a "single" (001) Au surface were constructed and computer calculations performed to demonstrate the types of images attainable at different resolution levels and under different microscope conditions. It was also demonstrated that top and bottom surface periodicies could be separated. However, the question of the combined effect of observing these top and bottom surfaces were not answered. To evaluate this effect multislice dynamical diffraction computer programs were developed for "generalized" objects for $256 \times 256 = 65,536$ beams or reciprocal space lattice points.[4] The analysis of the (001) Au surfaces with either one or both top and bottom surfaces roughened was considered in another publication.[5] Here the important conclusion to empha-size was that images from different surfaces are shifted by $\frac{1}{2}$ the surface lattice constant of 2.86Å and requires a microscope resolution of ~2.86Å to resolve the surface atoms. General-ly, a granular image results and appears similar to an amorphous film structure if these two surfaces are statistically rough and are superimposed in projection.

Surface imaging of (111) Au films with a ABCABC.... stacking sequence requires a higher level of resolution than (001) films since each close packed surface plane for (111) Au has periodicies of ~2.48Å which appear as a hexagonal arrangement in a electron diffraction pattern. In this case, a JEOL 200CX microscope equipped with a high resolution objective lens (C_s=1.1mm at 200kV) was suitable for bright-field axial illumination. The first crossover in the optimum phase contrast transfer function is ~2.3Å. Therefore, the point resolution of the microscope is better than the separations of single atoms of a given Au (111) surface layer. Fig. 1 shows a result of a bright-field microscope imaging experiment of a vapor deposited 30Å thick Au film. Here the image suffers from some residual astigmatism; however, the fundamental surface periodicies of 2.48Å are present. The low contrast and granular appearance of this image is expected since the contrast from surface layer atoms is ~10-15% of the bright-field background intensity which produced a mean photographic density of ~1.5 on an electron microscopic plate.

Inspecting the experimental image of Fig. 1 it can be seen that both black and white atomic details are present and often there are merged details. In order to explain the observed image contrast features, multislice dynamical diffraction calculations and subsequent image simulations were performed. Statistically rough atomic position models of top and bottom surface layers were constructed via the computer as well as individual layers of the bulk lattice. Some results of the analysis of these (111) Au film models are displayed in Fig. 2. Images (a) to (c) are for one rough surface with additional statistical noise included to produce a signal-to-noise ratio of ~7 for the modulated component of image intensity, i.e. no background. Image (a) is under optimum microscope imaging conditions (Scherzer focus, $\Delta f = 650\text{Å}$, and no astigmatism), (b) is at a different defocus value ($\Delta f = 400\text{Å}$) while (c) was adjusted to attain reasonable agreement with the real micrograph of Fig. 1 by varying defocus and astigmatism values. These images generally show the same contrast features observed in the experimental micrograph of Fig. 1. Image (d), however, was obtained for both surfaces being statistically rough and shows almost a completely random granular structure which is not the case for the experimental micrograph of Fig. 1.

In the real and computed micrographs of (111) Au, it is apparent that surface lattice periodicities shift by small amounts along their lengths which occurs because of different surface termination layers. For the ABC.... stacking sequence, there could be beginning and terminating layers of the type A-A, B-B, or C-C corresponding to extra A, B, or C layers respectively which would be shifted by $(2.48/3)\text{Å}$ from each other. There can, of course, be extra AB, BC, CA layers. Using a vector summation rule such that three consecutive layers produce no diffraction amplitude for the surface reflections such that $A+B+C=0$, we can see the equivalences $AB \rightarrow -C$, $BC \rightarrow -A$ and $CA \rightarrow -B$. Therefore, surface lattice images of the types A, B, C, -A, -B, -C can occur which would be shifted in spatial position by amounts as small as $(2.48/6)\text{Å}$ or .41Å. Examples of well defined fringe shifts are given in Fig. 3 for excess layers where idealized model surfaces have been considered. The surfaces are atomically flat and divided into two regions corresponding to the upper and lower half of each image. The shifts can be seen by looking along the fringe directions which cross the boundary where the image results of Fig. 3 are (a) $\frac{C}{A}$ which is a $\frac{1}{3}$ fringe shift, (b) $\frac{-A}{A}$ which is a $\frac{1}{2}$ fringe shift, and (c) $\frac{A}{-2C}$ where an intrinsic stacking fault has been introduced for the bottom image part.

Inclined boundaries in Au films occur because the (111) planes of the deposited crystal stack parallel to the substrate surface layer in either an ABC.... or the reverse CBA.... stacking sequences. The boundary between these two regions is often referred to as an incoherent double positioning boundary (DPB). Experimental images and computer simulations of these inclined planar defects show that they exist on {111} planes inclined to the surface. It is possible to obtain images that have black, white, mixed or no atomic level contrast from a DPB. Two examples of black and white contrast are given in Fig. 4 for a resolution level which just resolves the surface periodicities of 2.48Å. This contrast can be explained by the choice of top layer termination (A, B or C), bottom layer termination and the choice of the every third coherent plane (A, B or C). Each of these gives three choices and hence $3 \times 3 \times 3 = 27$ total choices are possible. All of these cases have been investigated using multislice dynamical diffraction computations which include the effects of translational shifts of an inclined boundary as a function of layer number, i.e. depth in the crystal. These calculations have also been performed for DPB's with jogs due to coherent twinning planes. This leads to at least another factor of three in the total number of boundary images possible, i.e. 81 cases.

For a higher level of resolution which includes the bulk lattice spacings of 1.43Å it was possible to visualize a DPB as shown in Fig. 5a. This micrograph shows that partial coherence, i.e. convergent beam illumination, plays an important role in weighing the relative contrast of the surface and bulk lattice periodicities. The power spectrum of this type of micrograph in Fig. 5b reveals the bulk lattice periods which are visible because their lattice diffraction spots are several orders of magnitude more intense than the weak scattering from a disordered surface. A detailed analysis has been carried out for partially coherent images of (111) Au surfaces and DPB's.

1.   D. Cherns, Phil. Mag. **30**, 549 (1974).
2.   W. Krakow and D. G. Ast, Sur. Sci. **58**, 485 (1976).
3.   W. Krakow, Ultramicrosc. **4**, 55 (1979).
4.   W. Krakow, IBM J. Res. and Devel. **25**, 58 (1981).
5.   W. Krakow, Sur. Sci. (1981) in press.

Fig. 1.   Bright-field axial illumination micrograph of a 30Å thick (111) Au film. E=200kV and original microscope magnification =500,000X.

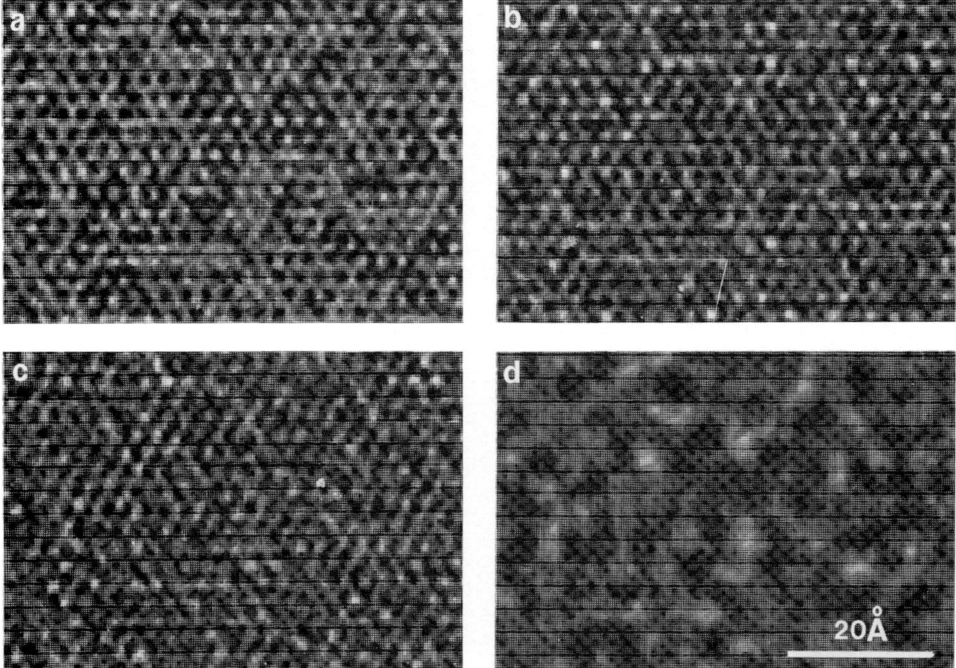

Fig. 2.   Computed images of (111) Au with one rough surface and: (a) Δf=650Å, (b) Δf=400Å, (c) Δf=300Å, Δf_a=400Å; (d) has two rough surfaces and Δf=650Å. All micrographs computed for E=200kV.

Fig. 3.   Boundary shifts for stacking sequences with different excess planes for the upper and lower part of each displayed image. (a) $\frac{C}{A}$, (b) $\frac{-A}{A}$ and (c) stacking fault with $\frac{A}{-2C}$.

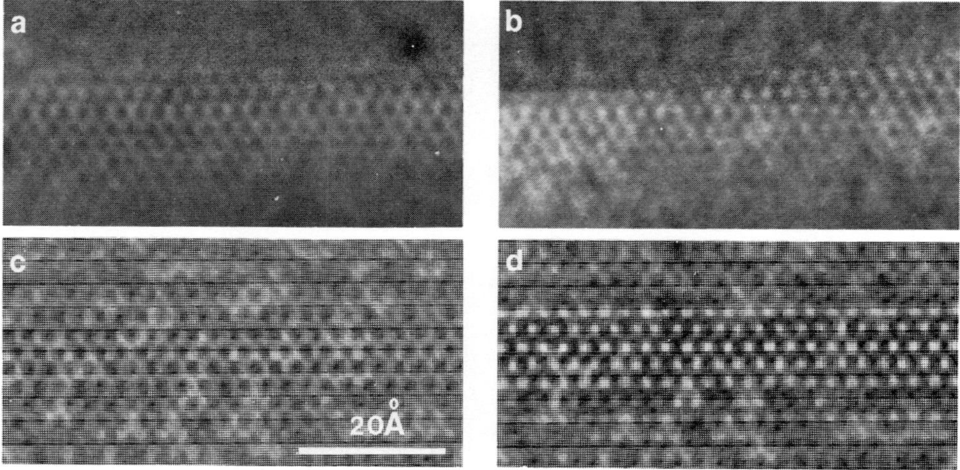

Fig. 4.   Images of an incoherent double positioning boundary: (a) and (b) are real micrographs and (c) and (d) were simulated.

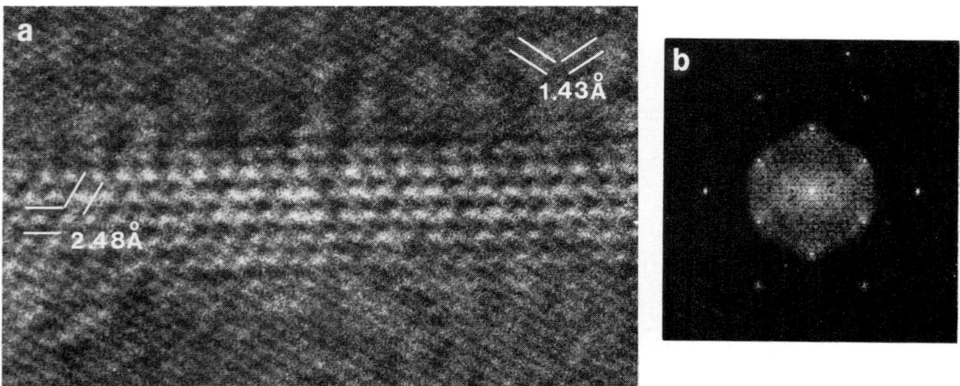

Fig. 5.   (a) Image of double positioning boundaries to the 1.43Å bulk lattice resolution level and (b) power spectrum of a simulated image showing the bulk lattice periods.

Structure of {100} platelet defects in natural diamonds

J L Hutchison, L A Bursill* and J C Barry*

Department of Metallurgy and Science of Materials, University of Oxford,
Parks Road, Oxford OX1 3PH, U.K.
*   School of Physics, University of Melbourne, Parkville, Victoria 3052,
Australia.

1.   Introduction

The composition, structure and origin of {100} platelet defects in Type Ia
diamond is a long-standing and controversial problem (see e.g. Raman and
Nilakatan 1940; Lang, 1979).   Recently two new approaches have significantly advanced our understanding: (i) Evans, Qi and Maquire (1981) showed
that synthetic diamonds containing high concentrations of dispersed nitrogen developed {100} platelets when heated at 2450°C under stabilizing
pressures of 9.5 GPa, proving that platelets contain nitrogen. (ii) we
here present high-resolution two-dimensional images which have demanded the
derivation of a new atomic structural model for the distribution of the
nitrogen in the platelets.

2.   High-Resolution Images

A type Ia diamond was selected by Professor A. Lang (Bristol) and a (110)
section prepared for electron microscopy by ion-thinning.   This was examined in a JEOL-200CX, using a LaB_6 source and ±10° goniometer in a high-
resolution (C_S = 1.2mm) objective lens.   The objective aperture included
000; 111(x4); 002(x2); 220(x2) and 222(x4) beams, having spacings of
2.06Å, 1.78Å, 1.26Å and 1.03Å respectively.

Fig la shows one of a through focal series of images of part of a 210Å
diameter platelet lying in a 300Å thick area, recorded with $\Delta f \approx$ -2025Å.
Note the appearance of an odd number (3) of rows of very intense white dots.
Asymmetry of the diamond matrix image across the defect indicates some
misalignment with respect to the optic axis, possibly including some local
strain.   Fig.2a shows a second type of platelet image in which the contrast
is relatively weak and an even number (4) of rows of white dots is present
at the platelet.

3.   Computer Simulation

Fig 3a gives a [110] projection of a two-layer N-platelet model which presupposes non-paramagnetic pairs of N atoms with bond lengths and angles
consistent with known crystal chemical principles (Lang, 1964).   Also it
predicts an extrinsic displacement vector $\underline{R}$ = <foo> (f = 0.35) in agreement
with earlier indirect X-ray diffraction and electron diffraction contrast
and also recent Moire and direct lattice fringe measurements. (Bursill,
Hutchison, Sumida and Lang, 1981; Barry, Bursill and Hutchison, 1981).

Carbon-interstitial platelets have also been proposed and an alternative model is shown in Fig 3b.   It contains pentagonal rings of C and has an effective f = 0.4, as suggested by computer simulation of diffraction contrast observations (Humble, 1981).

Lattice image simulations for a large number of variants of these and other structural models have now been completed (Barry 1981; Barry, Bursill and Hutchison 1981).   Periodic continuation and multislice technique programs due to Wilson (1981) were used.   Results for the Lang and Humble models, for 315Å thick crystal and Δf = -2250Å are given in Figs 1b, c.   Neither model is consistent with experiment: the Lang model predicts an even (4 for Δf = -2250Å) number of rows of white dots, while the Humble model predicts 5.04Å periodicity in image detail along the defect.   (A modified Humble model, obtained by staggering, rather than eclipsing the pentagonal rings along [110], is also ruled out by its computed images, Barry 1981).

The presence of an odd number (3 in Fig 1a) of rows of white dots suggested we try a model where the nitrogen was distributed over three rather than two atomic planes.   This was done by a simple modification of the Lang model (Fig 3c) whereby N-N pairs are placed in a zig-zag arrangement giving a central pure N plane with the two adjacent planes containing ordered half-occupancy by N and C atoms.   The corresponding computer simulations (Fig 1d) are in remarkably good agreement with Fig 1a.   This model also predicts anisotropy of projected potential with respect of [110] and [1$\bar{1}$0] projections.   The corresponding image calculation for [1$\bar{1}$0] shows an even number (4) of relatively low contrast rows of dots (Fig 2b) in reasonable agreement with Fig 2a.   We therefore interpret images as shown in Fig 1a and 2a as [110] and [1$\bar{1}$0] projections respectively.

4. Conclusion

Two-dimensional high-resolution images of diamond are shown to be sensitive to details of the defect structure for platelets lying in 300Å thick crystals and recorded using Δf = -2250Å.   Good agreement with experiment is obtained for a zig-zag arrangement of N-N pairs staggered about a (100) plane.   Note that in thin crystals, and at "optimum" defocus, the images obtained are not sufficiently model-sensitive to favour any one model. Full details of the structural models and image-matching are given elsewhere (Barry 1981; Barry, Bursill and Hutchison, 1981) as is a discussion of the consequences of the structural model for other properties of defects in diamond, and of mechanisms of formation and decomposition of the platelets to produce voidites.

5. Acknowledgements

We thank Professor A.Lang FRS for the specimen, Dr. N. Sumida for the ion-thinning, and S.E.R.C. for financial support.

6. References

1. Barry J.C., Ph.D. thesis, University of Melbourne, 1981
2. Barry J.C., Bursill L.A. and Hutchison J.L.(1981) in preparation
3. Bursill L.A., Hutchison J.L., Sumida N. and Lang A.R. 1981, Nature
        292, 518
4. Evans T.B., Qi Z. and Maquire J. 1981, J.Phys.C Solid State Physics,
        14, 1379.

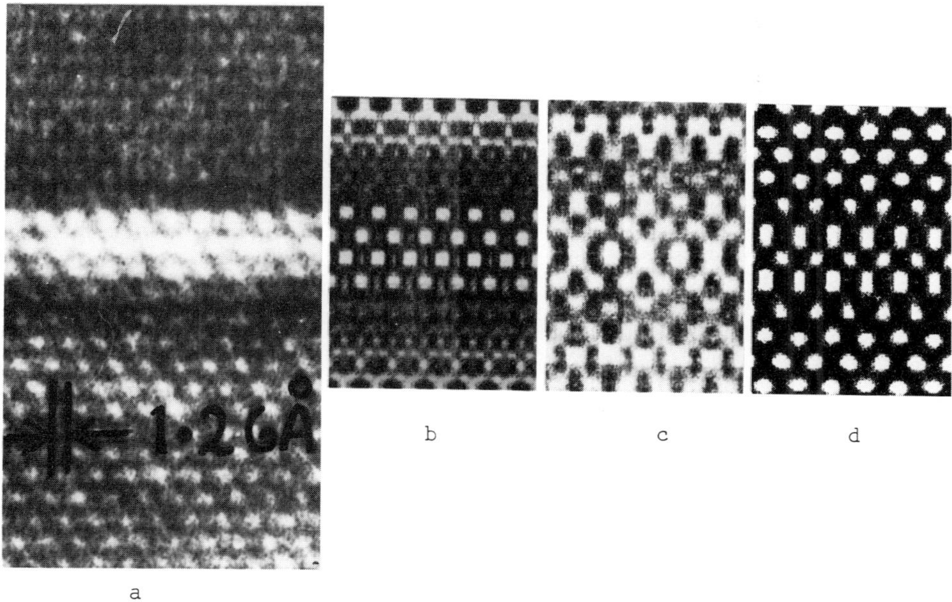

Fig 1.   (a) [110] lattice image of end-on platelet. (b,c,d) computed
         images for Δf = 2250Å, thickness = 315Å, for Lang, Humble and
         'zig-zag' models respectively.

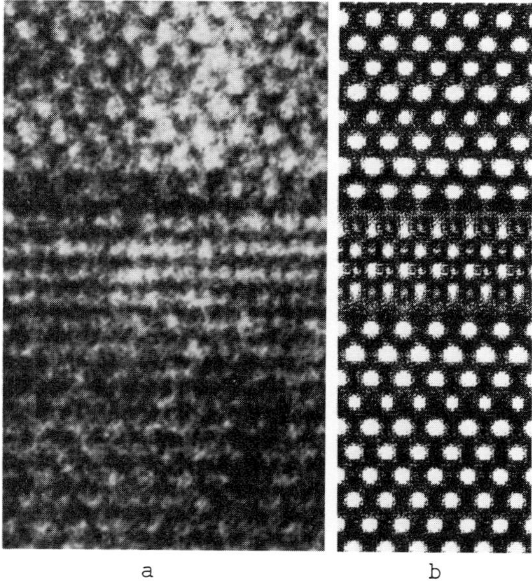

Fig 2.   (a) [1$\bar{1}$0] lattice image of end-on platelet with (b) computed image
         for zig-zag model.   The other models (Fig 1 b,c) are identical
         in both projections.

5.   Humble P. 1981, submitted
6.   Lang A.R., 1964, Proc.Phys.Soc. <u>84</u>, 871.
7.   Lang A.R. "The Properties of Diamond" J.E. Field (ed) Academic Press,
         N.Y., 1979, pp.425-469.
8.   Raman C.V. and Nilakatan P. 1940, Proc.Indian Acad.Sci.A <u>11</u>, 389
9.   Wilson A.R., Ph.D. thesis, University of Melbourne, 1981

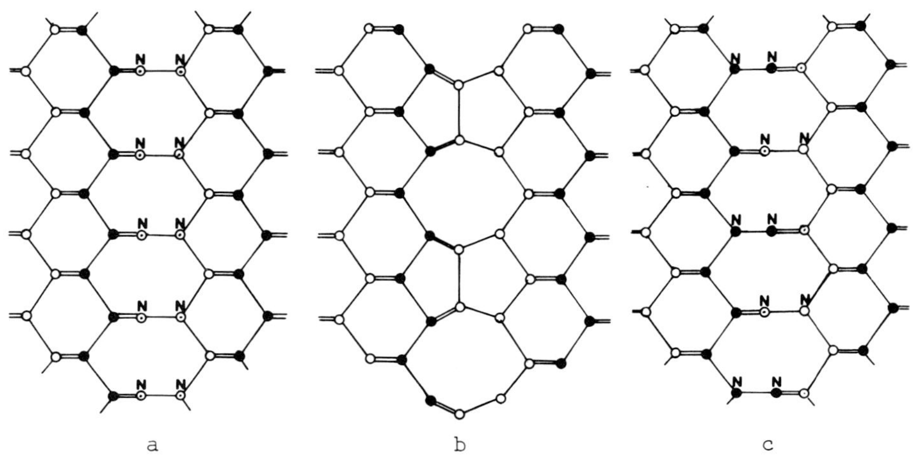

a   b   c

Fig 3.   [110] projection of three models for {100} platelets. (a) Nitrogen substitution model (Lang 1964) (b) Carbon interstitial model (Humble 1981) (c) "Zig-zag" nitrogen model derived from image matching .

A high resolution study of twin boundary structure in copper

W. M. Stobbs and David J. Smith[*]

Department of Metallurgy and Materials Science
*High Resolution Electron Microscope, University of Cambridge, Cambridge

1. Introduction

Many properties of a material, ranging from its creep resistance to its propensity to embrittle, are a function of its grain boundary structure. Rather little has, however, been done on the local and average displacements at given types of boundary and it is these which control the boundary's behaviour.   Carter and Sass (1981) have recently reviewed some of the techniques which may be used to study a grain boundary.   Direct HREM experimental results have generally been confined to the more open structures like germanium (Krivanek et al, 1977) though some high resolution axial images of low Σ boundaries in gold have recently been reported (Ichionose and Ishida, 1981).   While boundary modelling has generally been confined to descriptions based on hard spheres (e.g. Frost et al,1980) some effort has recently been made to determine the form of the relaxed boundary structures using plausible interatomic potentials.   Smith et al (1977) have, for example, completed simulations of some Al boundaries and Crocker and Faridi (1980) have made an interesting comparison of the relative energies and relaxation displacements to be expected for copper as a function of the way the appropriate Al and Cu potentials have rather different forms.   Indirect TEM methods for determining the average boundary displacements are currently confined either to the analysis of boundary fringe contrast (e.g. Pond, 1979) or the measurement of Moiré fringe displacements (Matthews and Stobbs, 1977).   We report here some of the preliminary results of an HREM study of both Σ=3 (11$\bar{1}$) twin boundaries in electrolytically-deposited copper and the way Σ=3 (12$\bar{1}$) twin boundaries relax in a thin foil.

2. Results and discussions

An axial ($\bar{1}$01) image of a Σ=3 symmetric twin in copper is shown in Fig. 1 as obtained with the Cambridge University 600kV HREM (Nixon et al, 1977; Cosslett, 1980).   The image was recorded near Scherzer defocus where the first zero in the contrast transfer function is a little beyond $1/a_{111}$. The area is also arguably thin enough to allow cursory interpretation on the basis of the weak-phase-object approximation.   Despite this, direct measurements of any (111) displacement, and variable atomic shifts, would even under these circumstances require $\underline{a}$ considerably greater point-to-point resolution.   Even at the exact ($\bar{1}$01) normal, some Fresnel effects are detectable in thicker regions and this suggests a small average displacement.   Careful measurement of 111 lattice fringe spacings across the boundary indicate that such a displacement must be less than 0.02nm.

0305-2346/82/0061-0373$01.50   © 1982 The Institute of Physics

Uncertainties in this figure are associated with variations with focus, themselves a function of either imperfect specimen tilt or slight non-axial alignment: asymmetric reflex images can have either origin. Incidentally, not only is the reflex image an insensitive test of crystal tilt alignment in very thin regions, but it is also affected by the asymmetries in the transfer function.

A number of imperfect $\Sigma=3$ boundaries were observed: that shown in Fig. 2 is rotated by one (111) plane about ($\bar{1}$01) between the two dislocations. While the region is far too thick for direct interpretation, greater Fresnel effects than for a perfect $\Sigma=3$ (111) boundary, and generally increased irregularity for the lattice fringe image reasonably distant from the defects, suggest that such a boundary rotation results in increased boundary displacements. Figs. 3a, b show another imperfect boundary at two different defoci. The appearance of the lattice image suggests some rotation about [111], and the variation in Fresnel fringe contrast is probably more indicative of a variation in the inclination of the boundary than of any variation in displacement across it. In principle, elastic side-band images can be used to determine the "thickness" of a boundary though this is difficult because of the need to correlate specimen tilt with beam tilt on sampling the boundary streak in reciprocal space. Qualitatively, such images are, however, useful in indicating irregularities in a boundary (see Fig. 4)

In view of the potentially much more drastic rearrangements likely at $\Sigma=3$ (1$\bar{2}$1) boundaries (Crocker and Faridi, 1980), and a previous measurement of the $\Sigma=3$ (1$\bar{2}$1) displacement in epitaxially-deposited Au islands, using the Moiré technique, as 0.056nm (Matthews and Stobbs, 1977), such boundaries were sought in copper. When interconnecting (111) twins they were invariably rotated about [111] from the (1$\bar{2}$1) plane perhaps because the shears and plane coalescences suggested by Crocker and Faridi, and which reduce the boundary energy, are then precluded by the (111) twin interconnections. An interesting interconnecting twin boundary showing apparent wedge-faceting (as indicated by the Moirés between the twins) is shown in Fig. 5. The geometry of the boundary has not been fully analysed, but the rather unusual faceting indicated is probably made possible as a method of reducing the energy of a (1$\bar{2}$1) twin only by the presence of the thin foil free surfaces

Finally, it should be noted that during examination with high beam currents at 500kV, in regions of <20nm thickness, stacking fault tetrahedra were observed to develop; presumably, in such thin regions, interstitials preferentially escaped to the foil surfaces reducing the probability for vacancy/interstitial annihilation. An example is shown in Fig. 6. Rather surprisingly, no particular association or otherwise of the tetrahedra was noted, with either (111) or other types of twin boundaries. The implications of these observations for the use of high voltage high resolution electron microscopes in the study of point and line defects at close to atomic resolution are thus demonstrated to be serious.

We currently feel that locally-variable displacements at grain boundaries will only be analysable quantitatively by HREM in thicker foils by comparison of image series with simulations of models for which at least the mean boundary displacement has been previously determined by an indirect method.

The Cambridge University 600kV High Resolution Electron Microscope was built as a joint project between the Cavendish Laboratory and the Department of Engineering with major financial support from the Science Research Council: continuing support is gratefully acknowledged.

<u>References</u>

Carter C B and Sass S L 1981 J. Am. Ceram. Soc. <u>64</u> 335-45
Cosslett V E 1980 Proc. R. Soc. (London) A370 1-18
Crocker A G and Faridi 1980 Acta Met. <u>28</u> 549-55
Frost M J, Ashby M R and Spaepen F 1980 Scripta Met. <u>14</u> 1051-56
Ichinose H and Ishida Y 1981 Phil. Mag. <u>43</u> 1253-64
Krivanek O L, Isoda S and Kobayashi K 1877 Phil. Mag. 36 931-40
Matthews J W and Stobbs W M 1977 Phil. Mag. <u>36</u> 373-83
Nixon W C, Ahmed H, Catto C J D, Cleaver J R A, Smith K C A, Timbs A E,
    Turner P W and Ross P M EMAG 1977 pp. 13-6
Pond R C 1979 J. Micros. <u>116</u> 105-14

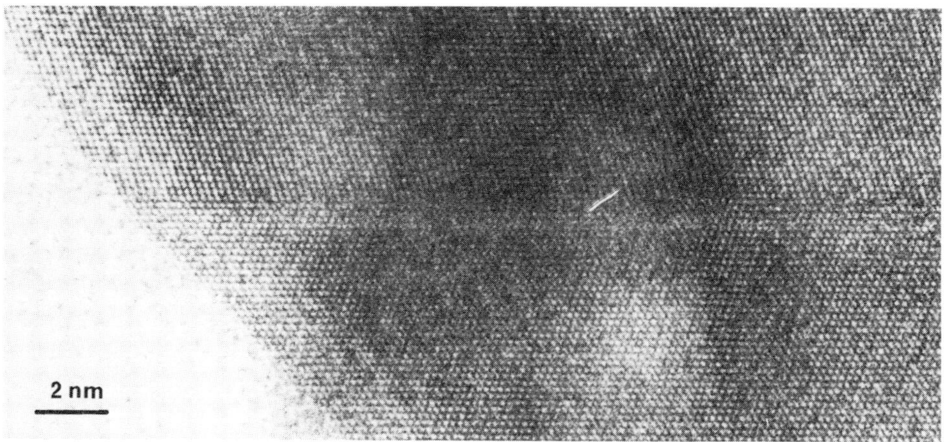

Fig.1. Σ3 (111) twin. ($\bar{1}$01) projection in this area; axial illumination.

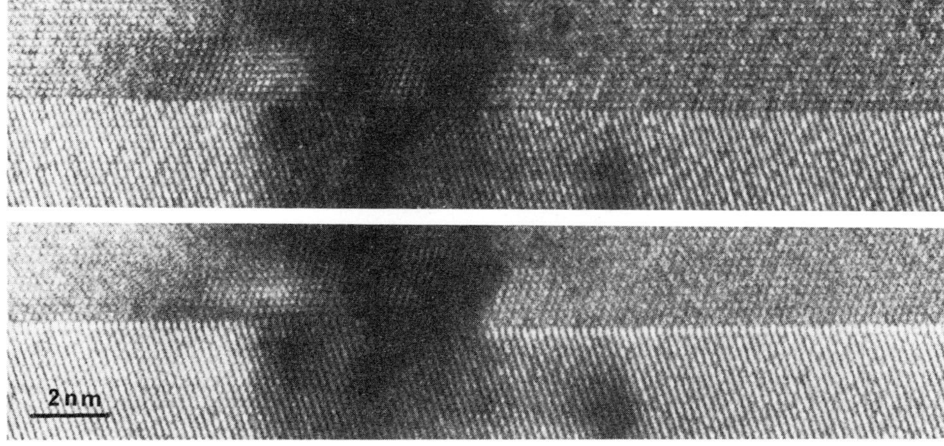

Fig.2. Imperfect Σ3 (111) twin. a) and b) at two different defocii. Fresnel contrast is stronger than for a perfect twin boundary.

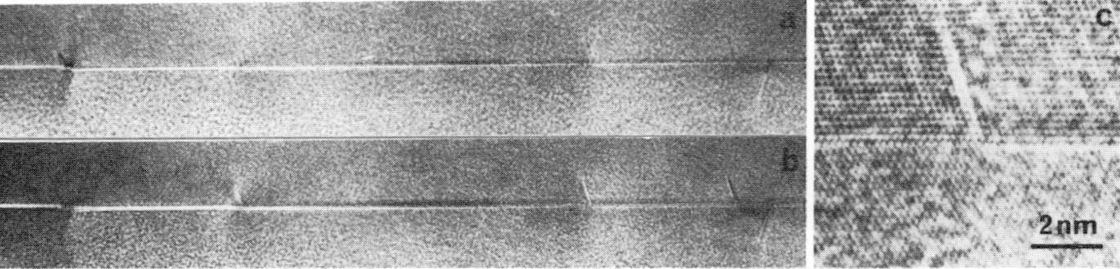

Fig.3. Imperfect Σ3 (111) twin. a) and b) at two different defocii;
    c) enlargement of part of a). Variation of Fresnel contrast along
    boundary indicative of variation of tilt angle.

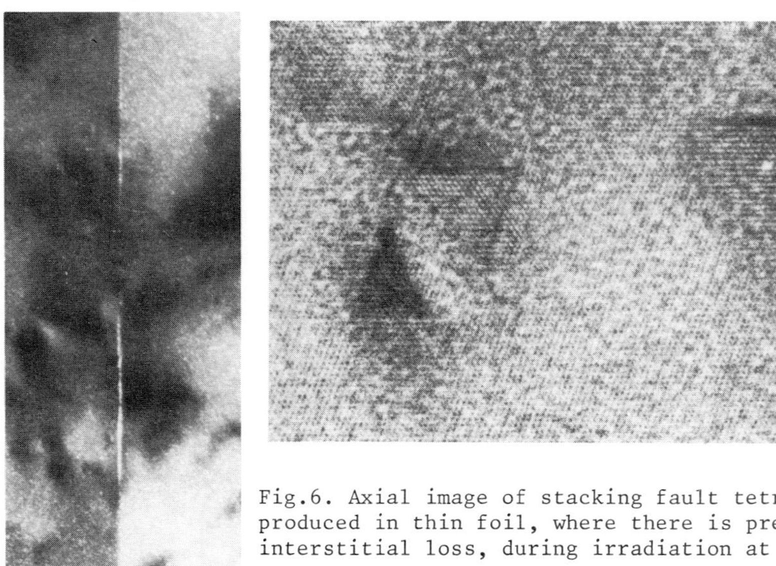

Fig.6. Axial image of stacking fault tetrahedra
produced in thin foil, where there is preferential
interstitial loss, during irradiation at 500kV.

Fig.4. Elastic side-band image of imperfect Σ3 (111)
twin boundary.

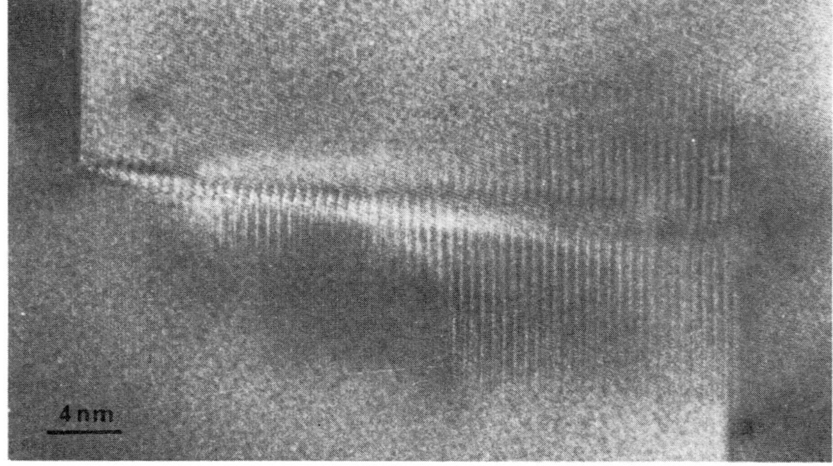

Fig.5. Moirés indicating wedge-faceted interconnecting boundary between
two Σ3 (111) twin boundaries. No (1$\bar{2}$1) boundaries were seen.

Computed high-resolution Guinier–Preston zone images in an Al–Cu alloy

M.J. Casanove-Lahana, D. Dorignac and B. Jouffrey,

Laboratoire d'Optique Electronique du C.N.R.S., 31055 Toulouse, France.

1. Introduction

Imaging of defects in crystalline materials having relatively small unit cells should enable us to distinguish between different structural models, whereas X-ray results give statistically averaged information. But, even with instrumental resolutions extending beyond the first diffraction peak the interpretation of high-resolution images remains difficult and needs the aid of the computer.

In this paper, we report on calculated images of Guinier-Preston (GP) zones, in an Al–4w%(1.7at%)Cu alloy, quenched from 540°C and aged at 100°C. In a previous note (Dorignac et al 1980), we presented an initial and not wholly convincing simulated image, roughly calculated with a simplified model, consisting in a pure substitution of Al atoms of the matrix by Cu atoms. Further studies have been made more recently (Casanove 1981), with several 2-dimensional geometrical models, derived from recent statistical X-ray ones (Fontaine et al 1979; Auvray et al 1981), including various strain-field deformations and Cu concentrations. They essentially showed that the models which seem to fit the experiment best agree (i) with zones constituted by $\sim$ $\cong 100\%$Cu (Auvray's results) and (ii) with relatively weak strain-field deformations ($\sim$3% of relative displacement at the centre of the zones for the second-nearest-neighbour Al plane). Our purpose here is only to illustrate this second point.

2. Experimental and computer procedures

In order to achieve any useful contrast at the required resolution of $\leqslant 0.2$nm (the Al crystal lattice distance is $a_{Al}/2 = 0.20248$ nm), the electron images were recorded using the Kyoto high-resolution microscope (500kV; $C_s = 1.06$mm), which can provide directly interpretable information in terms of the projected structure of the crystal to the Scherzer resolution limit of about $0.7\lambda^{3/4} C_s^{1/4} 0.17$nm. Figure 1 shows a typical GP zone image, with apparently two darker "Cu atom layers", about 14 atomic spacings long.

The computer simulations were carried out using the multi-slice method based on a modified version of a program developed earlier by Cowley's group at the Arizona State University. In particular, RHFS potentials (Carlson et al 1970), more exact than those usually employed, were included in the calculations. The coordinates for 144 atoms (128 Al and 16 Cu) were placed in superlattice cells (6x6x1 a_{Al}), not yet sufficiently extended entirely to include the zones (because of computer time and memory requirements) but adequate for analysing the structure of their borders.

Fig.1 Bright-field experimental image of a GP zone showing "two darker Cu atom layers". Beam voltage 500 kV; magnification 400k; Scherzer focus; axial-illumination with no objective aperture; convergence semi-angle 0.012 and theoretical rms spread of defocus 0.11 (in reduced units)

Fig.2 Model I. Projected potential (a, real part, in V) and focal series of simulated images for an artificially extended cell (100 x 100 pixels, including half a "biplane" zone, 6 atomic spacings long; zone centre at the bottom; focus indicated in reduced units; crystal thickness of 16 unit cells = 6.5 nm).

3. Results and discussion.

Comparison of the projected potentials with the corresponding images is shown in figures 2 and 3 for two particular 100%Cu structural models, with strong (I) or weak (II) negative strains. The corresponding interplane distances and relative displacements $(\Delta a/a_{Al})$ in the transverse direction at the zone centre, deduced from Auvray's diffuse X-ray scattering results, are given below:

Atoms	Cu	Cu	Al	Al	Al	Al	Al	Al
Planes (hkl)	1st	2nd	$\frac{1}{2}$00	100	$\frac{3}{2}$00	200	$\frac{5}{2}$00	300
Model I								
Distances /nm	0.165	0.166	0.171	0.202	0.230	0.209	0.203	
Displacements /%				16.8		9.75		7.94
Model II								
Distances /nm	0.165	0.185	0.201	0.208	0.207	0.204	0.203	
Displacements /%				3.24		0.47		0.14

The comparison shows that images will change rapidly with defocus, so that careful image matching would be required for structure analysis. However, it is clear that, for model II, it is possible, around the Scherzer focus, to correlate the individual atomic columns of the potential and of the simulated images directly, the dark blobs in the images corresponding to the potential peaks. On the contrary, this is quite impossible for model I, the image being in this case strongly blurred and deformed near the zone. Overall, the contrast and morphology of the computed images for model II do resemble those of the experimental image (except in the direction of the zone planes, probably because no strain was applied in this direction). Obviously, this does not completely solve the local structure problem of the zones, but it will provide a basis for further systematic studies. More sophisticated 3-dimensional models are being investigated.

References

Auvray X, Georgopoulos P and Cohen J.B, 1981, Acta Met., in press.
Carlson T.A, Lu C.C, Tucker T.C. and Nestor C.W, 1970 Oak Ridge Natl. Lab. Report 4614.
Casanove M J, 3rd Cycle Thesis, 1981, Toulouse.
Dorignac D, Casanove M J, Jagut R and Jouffrey B. 1980, EM80, 4, 174
Fontaine A, Lagarde P,, Naudon A, Raoux D and Spanjaard D, 1979, Phil.Mag. 40,17.

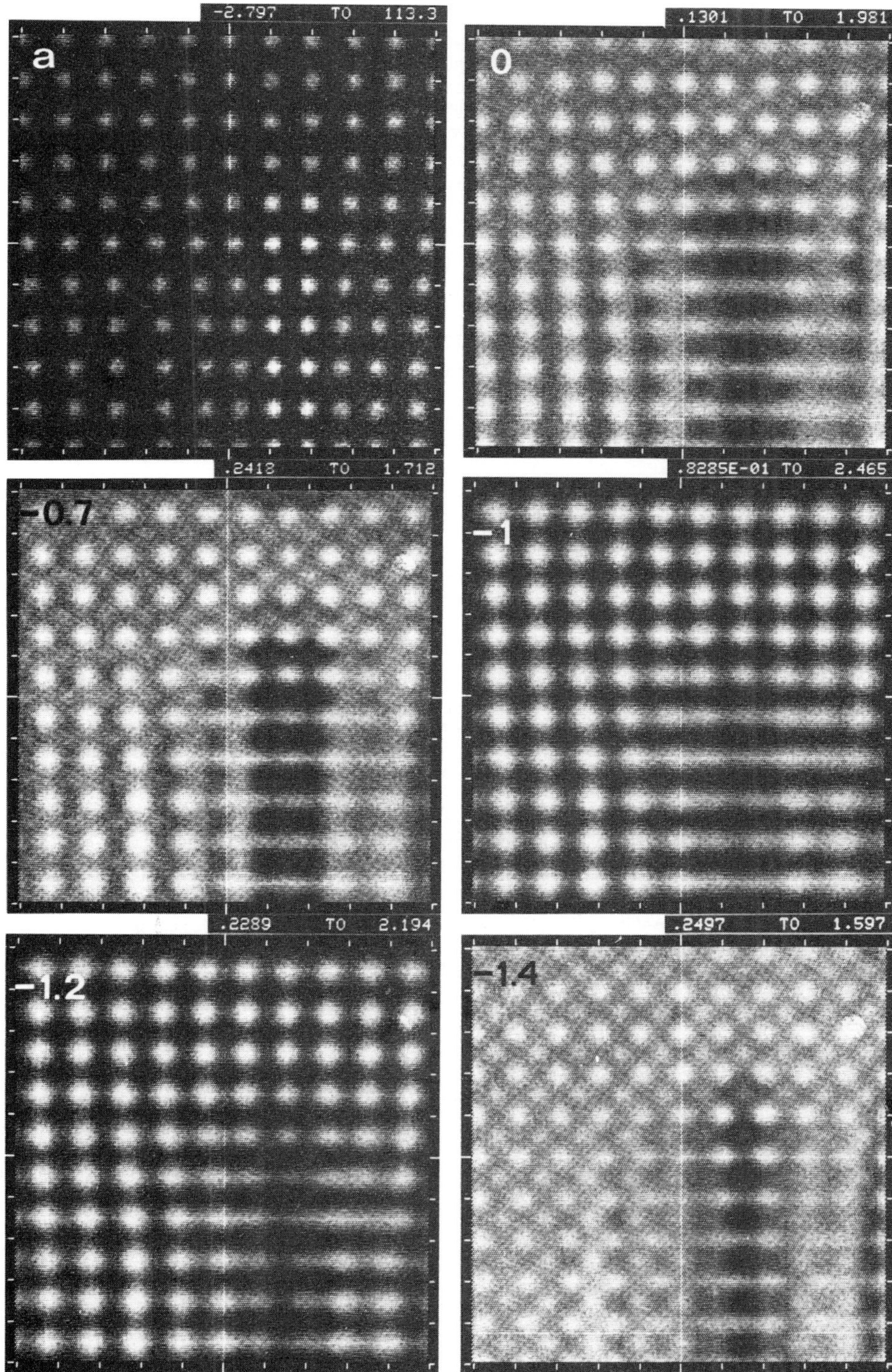
-2.797   TO   113.3
a
.1301   TO   1.981
0
.2418   TO   1.712
-0.7
.8285E-01 TO   2.465
-1
.2289   TO   2.194
-1.2
.2497   TO   1.597
-1.4

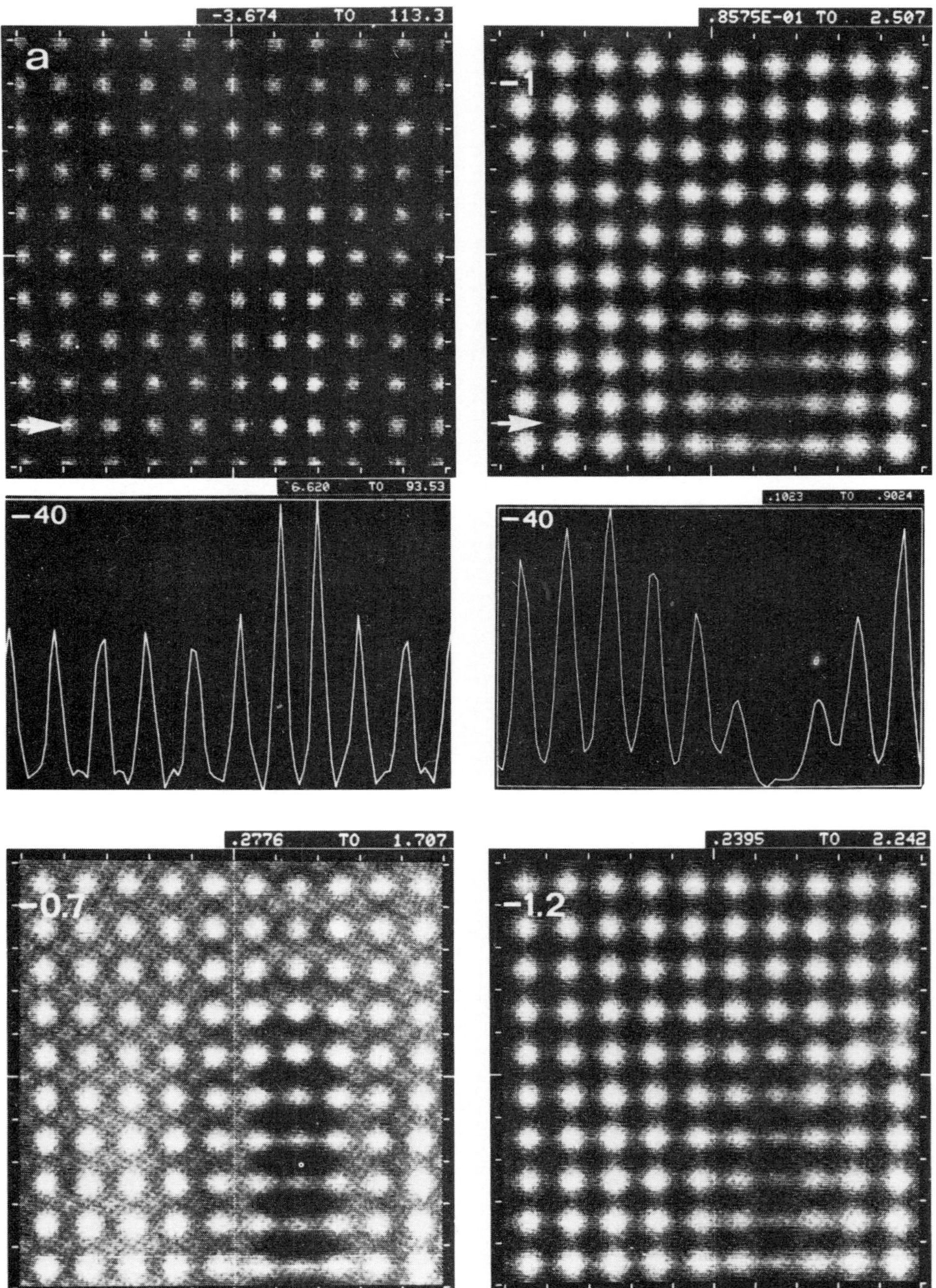

Figure 3 Model II. Projected potential (a) and focal series of simulated images around Scherzer focus (same conditions as in Figure 2). Comparison between image intensities and projected potential values is given for the line -40 (zone centre, arrowed).

Atomic resolution in the high voltage electron microscope

David J. Smith, R. A. Camps and L.A. Freeman

High Resolution Electron Microscope, University of Cambridge, Cambridge.

The potential of the high voltage electron microscope for resolving object details of atomic dimensions has been realised only comparatively recently in a small number of laboratories. This paper reviews the present limitations and future prospects of these microscopes as well as briefly describing applications where they are already being used to good effect. Some of the problems of high-resolution image interpretation are also discussed.

1. Introduction

Electron microscopes operating at 100kV are now very close to their theoretical resolution limits with little improvement due to mechanical, electrical and electron-optical design seemingly in prospect. Directly-interpretable image information is provided to resolutions of about 2.8Å with instrumental resolutions of better than 1.7Å obtainable under extreme operating conditions (Bursill and Wood, 1978) even though image interpretation is then far from straightforward without lengthy image simulation. Commercially-available 200kV instruments have also demonstrated structural resolutions extending to about 2.5Å (Boyes et al, 1979). In principle, the high voltage electron microscope (HVEM) offers considerable scope for improved image interpretability because of the extended resolution limit associated with its shorter-wavelength electron beam. In practice, perhaps only 4 HVEMs, namely those located at Kyoto (Kobayashi et al, 1974), Ibaraki (Horiuchi et al, 1977), Cambridge (Cosslett et al, 1979) and Sendai (Hirabayashi, 1980), have come close to realising these potentialities. All seem limited in performance by either mechanical or electrical stabilities and, unlike lower voltage instruments, it also appears that they remain far from optimised in terms of lens and stage design for top performance on a routine basis combined with ease of specimen manipulation. As discussed below, these various problems of instrumentation can, in principle, be overcome, but not without considerable expenditure of time, effort or money.

2. Image interpretation and resolution criteria

The primary objective of the electron microscopist is obviously to relate the details recorded in an electron micrograph to particular features of the specimen under observation. In high resolution studies, where the imaging process depends strongly on the contrast transfer characteristics of the objective lens, the final image is a very sensitive function of the objective lens defocus, as well as depending on specimen thickness. Furthermore, small alterations in either specimen orientation or incident beam alignment can have a marked influence on appearance (eg. Saxton and O'Keefe; Takai et al; these proceedings). For small-unit-cell materials, the situation is also complicated by the periodic recurrence with focus of Fourier images [Cowley and Moodie, 1957].   Intuitive interpretation of high resolution images is thus possible over very restricted ranges of thickness and focus, and normally only to resolutions corresponding to the first zero of the contrast transfer function at the optimum, or Scherzer, defocus.   In the high-resolution HVEMs mentioned above, these "structure image" resolution limits already correspond to 2Å or better and recent calculations suggest that there is still considerable scope for further improvements (Smith, 1980).   It should also be noted here that, as for imaging at 100kV, any detailed high-resolution HVEM investigation of defective structures needs to be accompanied by full image simulations, both to establish the precise imaging conditions (e.g. Wilson, 1980) and also to verify the image interpretation, particularly in situations where the high-resolution images may not, in fact, be very model-sensitive.   Indeed, the optimum defocus for distinguishing between alternative models may sometimes be well displaced from the Scherzer value (see Hutchison, Barry and Bursill, these proceedings).

It is now becoming common-place for both 100 and 200kV electron microscopes to produce micrographs with considerably finer details than their structural resolutions, primarily because their "instrumental" resolution is limited only by coherence effects due to beam energy spread and illumination angle rather than by more serious instrumental shortcomings such as mechanical or electrical instabilities.   Detail beyond the structural resolution limit does not often appear to occur at high voltage, even in most of the high resolution instruments.   This can be a blessing in that image interpretation is not unnecessarily complicated.   It also means that instrumental resolutions are generally better for the lower voltage instruments.   Note, however, that under these latter circumstances, image simulation is absolutely essential to ensure that erroneous image interpretation does not result.

The possible pitfalls are well-demonstrated by the recent widespread interest in <u>apparent</u> structure images of materials with zincblende or diamond-cubic structure, initiated by the observations of Si in (110) by Izui et al (1978) and discussed at length by Hutchison, Anstis et al in these proceedings.   Briefly, apparent atom pairs, or "dumbells",are resolved which seem to correctly represent the crystal structure.   However, closer inspection reveals that the images come from thicker crystal regions, the spot separations are considerably in error and the contrast is reversed (i.e. "atoms" are white).   In fact, the images are basically interference effects (Krivanek and Rez, 1980) which reflect the crystal structure (and the instrumental resolution) but which can not be easily related to the actual atomic positions and thus should not be called structure images. Fig.1. shows a similar type of image from a crystal of SnO_2 in the [100] projection recorded with the Cambridge University 600kV HREM (Cosslett, 1980; Nixon et al, 1977).   Detailed analysis will not be carried out but cursory examination of the image, including the adjacent thin crystal region, and comparison with the structural model (inset), shows that the "dumbells" here appear roughly at the positions of the pairs of oxygen atoms, but with a separation of 1.5Å rather than 1.8Å.   Moreover, there is no intensity at the positions of the much heavier Sn atoms! i.e. whilst reflecting the structure and symmetry of SnO_2 (and resembling the Si (110) images), this image is again clearly not a structure image.

3. <u>Practical realisation</u>

The major drawback of the high-resolution HVEM has previously been its inability to produce high quality images on a routine basis.   Various

instrumental factors, particularly electrical and mechanical, were close
to marginal and any slight deteriorations had an adverse effect on performance.   Recent improvements have minimised many former shortcomings but
electrical stability of the EHT supply, as well as overall mechanical stability in some cases, appears to remain limiting.   Notwithstanding these
limits, our experiences have demonstrated the value of providing for continuous monitoring of the most critical parameters, namely the EHT and
objective lens current stabilities, mechanical vibration levels and external
AC magnetic fields.   When any of these exceeds permissible levels, microscopy normally ceases until the exact cause (and a cure) is found; this
obviously eliminates the time and effort which might otherwise have been
wasted recording poor micrographs.   Moreover, there are a number of other
relatively simple steps, mostly based on improving the viewing conditions
so that image focussing and correction of image astigmatism can be carried
out directly at high magnifications (M>750,000), which assist in increasing
markedly the quantity of good micrographs (Smith, 1980a).   These include
choosing viewing binoculars to optimise light collection efficiency, using
a phosphor thickness and type appropriate for best resolution (we use
$\sim$10mgm/cm² of P22 for operation at 500kV) replacing the normal viewing
screen with a thin foil of high atomic number material, such as platinum,
which increases the yield of backscattered electrons and hence improves the
screen brightness.   Obviously, it will also help to provide a higher brightness electron gun, either with a pointed tungsten filament or, provided that
the accelerator vacuum is adequate (better than 10^{-6}mm), with a directly-
heated LaB_6 cathode.   These have the alternative advantage that more coherent illumination is also provided.

A longer term, though potentially more fruitful, means of realising improved
performance in the HVEM, is to re-design the objective lens to provide improved values of the lens aberration co-efficients.   With the singular exception of the Kyoto instrument (which has such a tight-fitting stage that
no tilting is possible) all other existing HVEMs appear, from published data,
to be far from optimised in this regard.   For example, the Ibaraki machine
has C_C=4mm; C_S=10mm (Horiuchi et al, 1978) and that at Sendai has C_C=10mm;
C_S=11mm (Hirabayashi, 1980).   By comparison, it is envisaged that the high
resolution HVEMs under construction (Gronsky, 1980; Honjo et al, 1980) will
have values for C_C and C_S of about $3\frac{1}{2}$ and $2\frac{1}{2}$mm respectively for operation
at 1MV, whilst still leaving reasonable space for specimen tilting and
handling facilities.   The immediate effect of lowering the value of C_C in
the other HVEMs would be to reduce the major influence of the high voltage
instabilities on the instrumental resolution.   The effects of improving C_S
are two-fold: first, some slight gain in the structural resolution is
obtained; and second, more subtle but more important, the specifications on
the incident illumination angle (spatial coherence) are relaxed considerably (see Smith, 1980) so that much larger incidence angles can be tolerated
without resolution being unduly impaired i.e. gun brightness requirements
are relaxed.

Finally, it is clear from our work that the high-resolution HVEM is capable
of regularly attaining the very high levels of instrumental stability sometimes reached by lower voltage machines.   Fig.2, for example, shows the
0.88Å (200) half-spacing lattice fringes recorded from a sample of nickel
at 575kV.   These fringes demonstrate both a very high level of mechanical
stability of both specimen stage and microscope, and well-aligned and highly-
coherent (i.e. small incidence angle) incident illumination.   Similar half-
spacing fringes are also often recorded from thicker regions of other
specimens observed with our microscope.

4. Applications

With the realisation of structural resolutions of atomic dimensions in the
HVEM, a vastly-increased range of defects in materials becomes amenable to
direct study; of particular interest are those in close-packed simple metals
and alloys as well as binary oxides such as rutile, TiO_2.   In these
materials, the projected atomic separations in major crystallographic directions are usually in the range 2.0 to 2.5Å. i.e. beyond the structural
resolution limit of lower voltage instruments.   Materials under active
investigation in high-resolution HVEMs include metglasses, small metal
particles (including model catalyst systems), twin boundaries and grain
boundaries generally, and various alloy systems, as well as ceramics,
minerals and complex inorganic oxides including perovskites.   These applications are described in detail elsewhere (see Electron Microscopy 1980,
Vol. 4, High Voltage and these proceedings).

An interesting aspect of our high resolution studies has been the differing
effects of radiation damage due to the electron beam on different materials.
Some beam-sensitive specimens, such as sulphides and carbonates, which give
poor quality lattice fringes at 100kV, if at all, have proved at least an
order of magnitude more stable at accelerating voltages of 400-500kV.
Conversely, we have found that moderate irradiation (jo $\sim$ 2-10A/cm²) of
thin Cu foils during a 500kV study of twin boundary structure was enough
to induce the formation of stacking fault tetrahedra (see Stobbs and Smith,
these proceedings).   This has serious implications for the use of high-
resolution HVEMs for studying atomic configurations in the vicinity of
lattice defects in low Z material.

5. Advantages and disadvantages of high voltage

In principle, direct image interpretability at high voltage extends to
thicker specimen regions; in practice, however, improved resolution reduces
the maximum thickness for a valid image.   Decreased diffraction angles,
because of the shorter electron wavelength, make precise crystal tilting
into zone axes orientations more difficult, as well as making slight local
variations in crystal orientation more noticeable in the image. It is also
appropriate here to point out the divergence between image simulation and
experimental micrograph with increasing thickness.   Accurate illumination
alignment, which is essential for analysis of images from thicker crystals,
becomes critical (see Saxton and O'Keefe, these proceedings).

There are other experimental factors which relate to the usefulness of the
HVEMs for high resolution studies.   A major problem lies in the uniqueness
of each individual high-resolution HVEM.   Conventional lower voltage
instruments are built in considerable numbers; experience with fault conditions etc. quickly accumulates, and overall instrumental down-time can
thus be minimised.   Each high-resolution HVEM is effectively a prototype
instrument and the "folk-lore" of fault conditions only develops slowly.
In this regard, it is very useful to retain skilled scientists on-site who
are highly familiar with all aspects of the instrument; this familiarity
also ensures that optimum use is made of the microscope's valuable operating time.

6. Future directions

The way forward will be far from simple.   Many existing high voltage
instruments are not easily upgraded (Smith, 1980a) and those few high-
resolution HVEMs under construction (e.g. Gronsky, 1980; Honjo et al, 1980)
are of almost prohibitive cost because of the problems associated with
providing adequate mechanical and electrical stabilities, as well as improved vacuum levels.   It should be possible, however, to modify some
existing machines to provide better specimen manipulation and viewing
conditions, perhaps even adding an image intensifier which could be linked
with an on-line, interactive, image processing system (e.g. Erasmus and
Smith, these proceedings) to ensure that only high quality micrographs are
recorded.   It should be regarded as essential that all possible steps are
taken to improve objective lens aberration co-efficients.   Better structural resolution will obviously result and, as pointed out above, the operating margins on gun and EHT performance, in particular, will also be
considerably relaxed.

It appears unlikely that high-resolution HVEMs will ever be built in large
numbers: those in existence, or under construction, should therefore
concentrate on applications which fully utilise their superior structural
resolution capabilities.

The micrographs included here were recorded with the Cambridge University
600kV High-Resolution Electron Microscope which was built as a joint project
between the Cavendish Laboratory and the Department of Engineering with
major financial support from SRC; continuing support from SERC is gratefully
acknowledged.

References

Boyes E D, Watanabe E, Skarnulis A J, Hutchison J L, Gai P L, Jenkins M L
    and Naruse M   EMAG 1979 pp. 445-48
Bursill L A and Wood G J 1978 Phil. Mag. A38 673-89
Cosslett V E 1980 Proc. R. Soc. (London) A$\overline{370}$ 1-18
Cosslett V E, Camps R A, Saxton W O, Smith D J, Nixon W C, Ahmed H,
    Catto C J D, Cleaver J R A, Smith K C A, Timbs A E, Turner P W and
    Ross P M 1979 Nature 281 49-51
Cowley J M and Moodie A F 1957 Proc. Phys. Soc. 70 486
Gronsky R 1980 Proc. 38th Ann. EMSA pp 2-5
Hirabayashi M Electron Microscopy 1980 4 142-49
Horiuchi S, Matsui Y, Bando Y, Katsuta T and Matsui I 1978 J. Electron
    Mic. 27 39-48
Honjo G, Yagi K, Takayanagi K, Nagakura S, Katagiri S, Kubozoe M and
    Matsui I Electron Microscopy 1980 4 22-25
Izui K, Furuno S, Nishida T, Otsu H and Kuwabara S 1978 J. Electron Micr.
    27 171-179
Kobayashi K, Suito E, Uyeda N, Watanabe M, Yanaka T, Etoh T, Watanabe H
    and Moriguchi M Electron Microscopy 1974 1 pp 30-31
Krivanek O L and Rez P 1980 Proc. 38th Ann. EMSA pp 170-71
Nixon W C, Ahmed H, Catto C J D, Cleaver J R A, Smith K C A, Timbs A E,
    Turner P W and Ross P M EMAG 1977 pp 13-16
Smith D J Electron Microscopy 1980 4 126-33
Smith D J 1980a Proc. 38th Ann. EMSA pp 822-5
Wilson A R 1980 Micron 11 281-83

Fig.1. Interference lattice image of tin dioxide, SnO_2, recorded at 500kV in [100] projection.   Inset shows structural model: o ≡ Sn; • ≡ O. Note that this is <u>not</u> a structure image – it was recorded at the first extinction condition for the (000) beam – despite its resemblance to the structure.

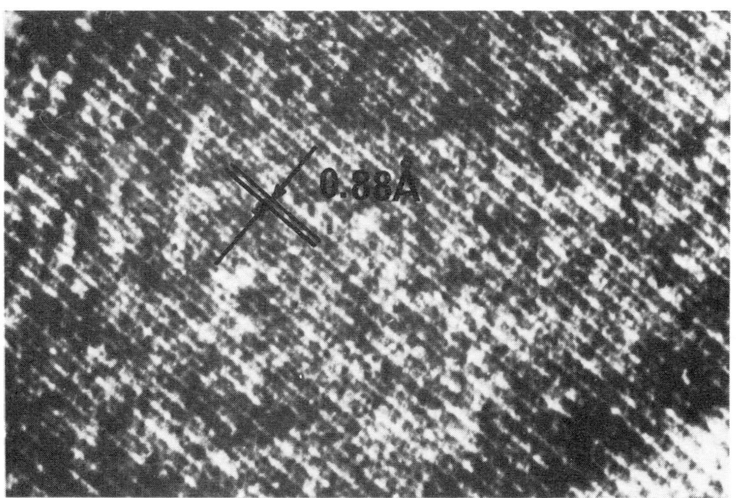

Fig.2. Nickel lattice image showing interference between ± (200) – beams to give fringes of spacing 0.88Å. Recorded at 575kV.

Incommensurate statistical structures among long period ordered alloys

M. Guymont[+], R. Portier[o], D. Gratias[o]

+ Laboratoire de Cristallographie, Université de Paris XI, 91405 Orsay,
France
o C.E.C.M., 15 rue Georges Urbain, 94400 Vitry sur Seine, France

The existence of two kinds of structures, both thermodynamically stable among long period ordered alloys is now firmly established (Guymont et al. 1979, Portier et al. 1980). In rational structures (e.g. Ag_3Mg, Cu_3Al α''), the mean long period varies discontinuously with composition, thus adapting the structure, a behaviour quite similar to that of long period oxides and polytypes. In incommensurate structures (e.g. AuCu II, $AuCu_3$ II, Cu_3Au II, AuCu-Zn II, both one and two-dimensional long period Cu_3Pd), the mean long period varies continuously with composition, thus taking irrational values (for instance, AuCu II 55 at.% Au gives a mean half long period $\bar{M}=5.13+0.02$, always measured in units of the disordered cell of same composition) (Guymont and Gratias 1978, Guymont et al. 1980).

These incommensurate structures actually are statistical long period structures, a feature which can be revealed only through high resolution microscopy imaging. We present here the result of some high resolution observations (see Table I, Fig. 1 and 2) on two typical statistical long period structures : gold-copper and copper-palladium alloys. The apparatus used was a Jeol 200CX; the samples were thinned by electro-gravity.

The common behaviour of these statistical structures can be easily pic-

Table I

alloys structures	compositions	thermodynamical conditions	$\bar{M}$
Cu_3Au II	32.2 at.%Au	annealed 10 days at 350°C	9.30+0.3
AuCu II	50 at.%Au	– 4 – – 400°C	5.00+0.01
Au_3Cu II	75 at.%Au	quenched from 910°C, then annealed 1 hour at 188°C	8.7+0.3
Cu_3Pd II	18.5 at.%Pd	annealed 4 days at 475°C	12.3+0.1
Cu_3Pd II	27 at.%Pd	– 4 – – 365°C	3.51+0.1
Cu_3Pd III	27 at.%Pd	– 4 – – 425°C	3.51+0.08 4.68+0.08

tured on the sketch of Fig. 3 : instead of being a quite regular stacking
of an integral number of units of disordered cell (which gives rise to
Ll_0 or Ll_2 Bravais cell when ordering), there are fluctuations of this
number by some units, resulting in a
mean half long period (as measured
o n the diffraction pattern. Thus, in
contrast to the images of rational
structures showing well-defined,
plane boundaries between antiphased
domains, the high resolution images
of statistical structures show fluctuating, badly-defined periodic antiphase boundaries.

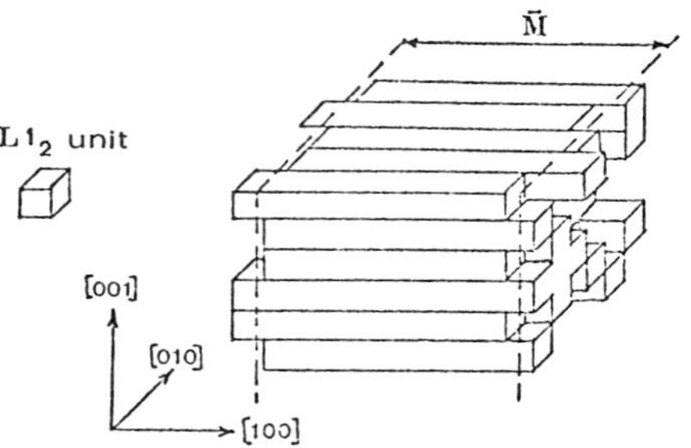

Fig.3 : Typical sequence of a
long period statistical structure

For Au-Cu alloys, statistical long
period structures can be obtained at
almost any composition greater than
25 at.%Au. Thus, for instance on
Cu_3Au II image (Fig.1a) each white point can be put into correspondence
with a Ll_2 unit : following the horizontal sequence of such points, we
reach the antiphase boundary; the mean length of the sequence is $\bar{M}=9.30$.
In Au_3Cu II, the fluctuating character of antiphase boundaries is much
more pronounced. For AuCu II the half mean long period is minimal ($\bar{M}=5$)
at the stoichiometry and increases on both sides of this composition. Two
different images are obtained according to the diffraction plane chosen
(Fig.1c) : on both the antiphase step is visible.

Cu_3Pd shows the same statistical behaviour with $\bar{M}$ varying between about
15 and 3 according to composition (Fig.2b and c). But a two-dimensional
statistical structure also exists (Cu_3Pd III) resulting in two distinct
mean long periods in two directions (see Fig.2a) where antiphase boundaries are clearly seen sinuous in both directions.

References

Guymont M and Gratias D 1978 J. Phys. Lettres 39 437
Guymont M, Gratias D and Portier R 1979 AIP Conf. Proc. Ser. 53 244
Guymont M, Portier R and Gratias D 1980 Acta Cryst. A36 792
Portier R, Gratias D, Guymont M and Stobbs W M Acta Cryst. A36 190

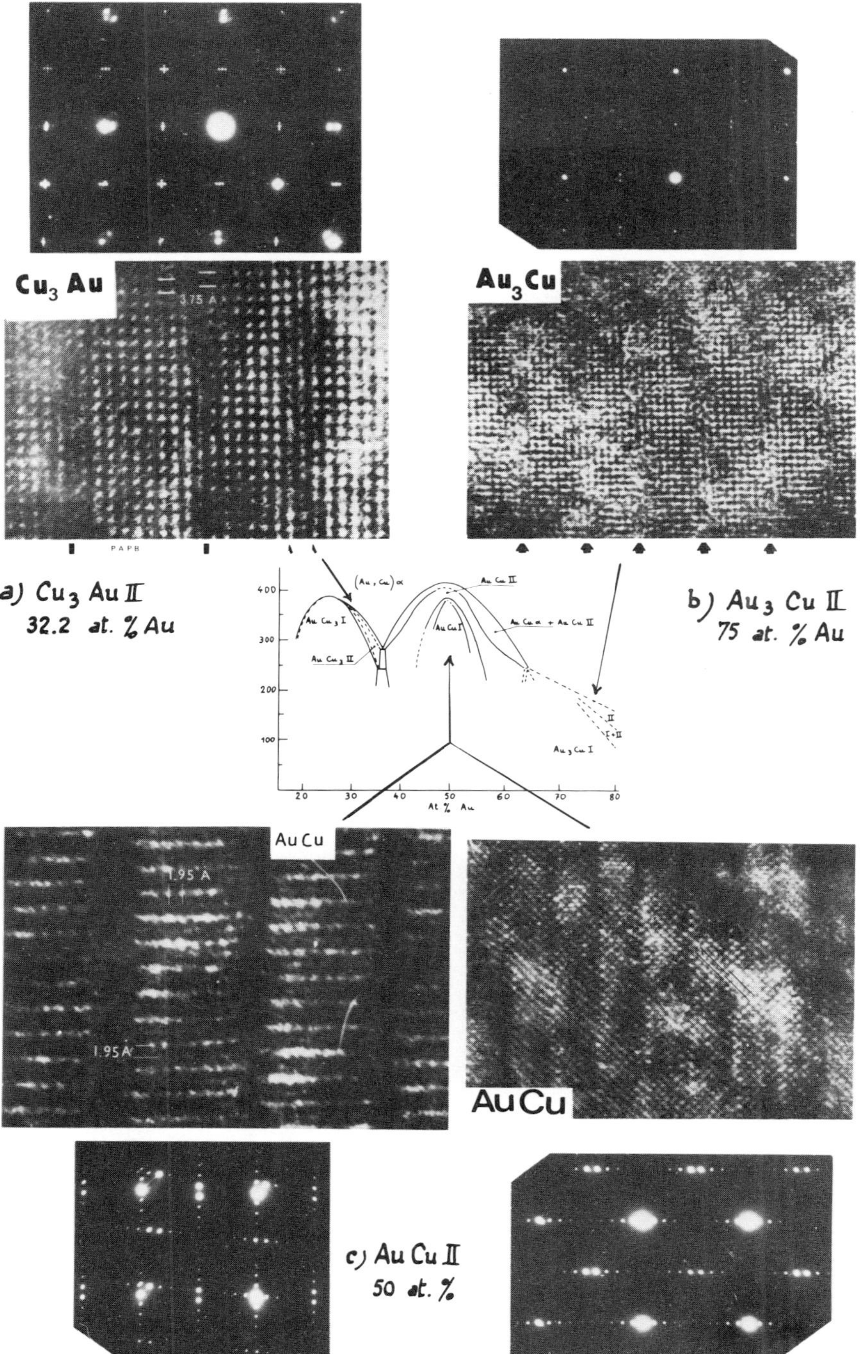
Fig. 1 : LONG PERIOD GOLD-COPPER ALLOYS
Cu₃ Au
3.75 A
PAPB
a) Cu₃ Au II
32.2 at. % Au
Au₃ Cu
b) Au₃ Cu II
75 at. % Au
(Au,Cu)α
Au Cu II
Au Cu₃ I
Au Cu I
Au Cu₃ II
Au Cu α + Au Cu II
Au₃ Cu I
I + II
II
At % Au
AuCu
1.95 A
1.95 A
AuCu
c) Au Cu II
50 at. %

Fig. 2 :   LONG PERIOD COPPER - PALLADIUM ALLOYS

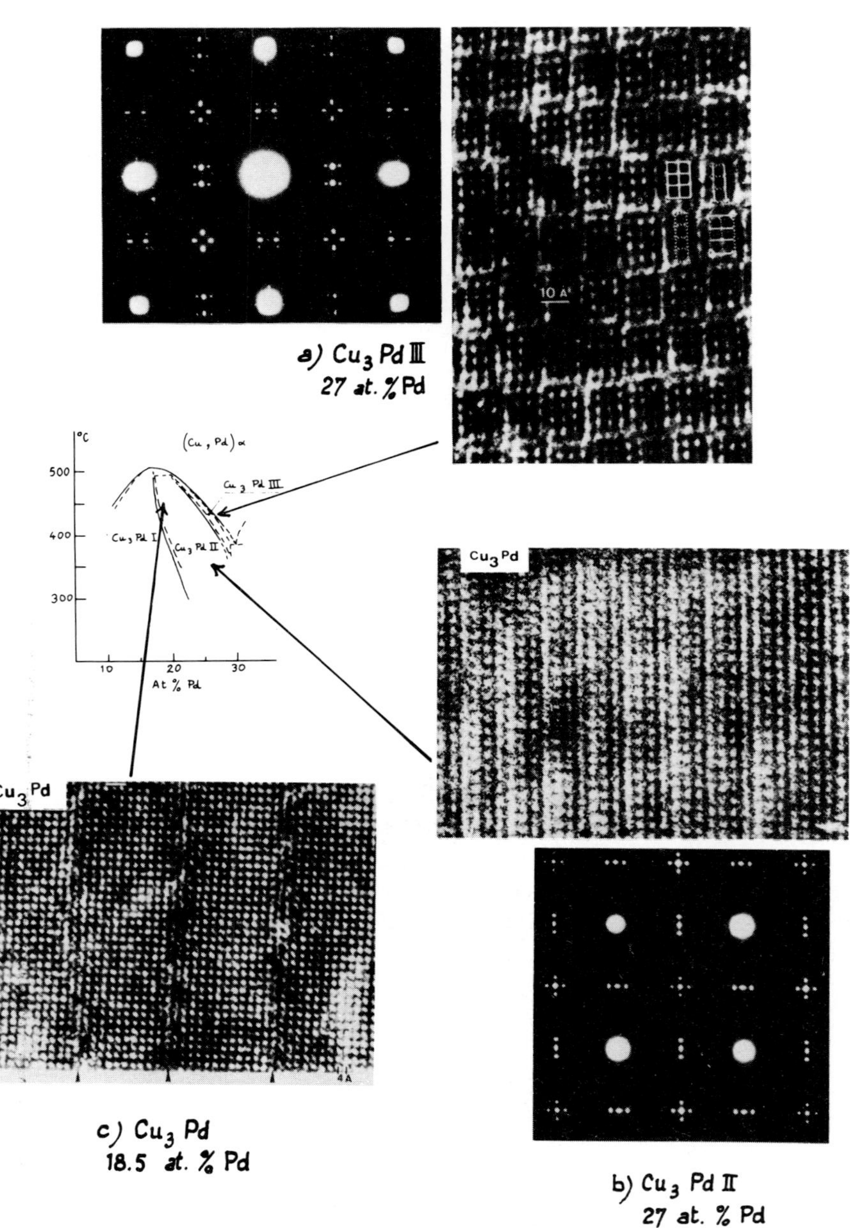

High resolution imaging of a poly-phased (microcrystalline-amorphous) $Cu_{60}Zr_{40}O_x$ alloy

J.P. Chevalier

C.E.C.M.–C.N.R.S. 15 rue G. Urbain, 94400 Vitry-sur-Seine, France

Amorphous $Cu_{60}Zr_{40}$ is sensitive to oxidation and specimens undergo oxidation during observation in the microscope, storage and heat treatment (Bigot et al, 1981). Initially, for room temperature oxidation in air, this leads to the appearance of extra diffuse rings in the diffraction pattern. Most of these can be indexed accurately as due to f.c.c. copper, but the strong small angle diffuse halo (close to the positions of the Debye rings of a cubic form of zirconia) appears to be due to an amorphous zirconia phase. In the thinnest regions of the foil there seems to be little of the parent amorphous phase left.

Dark field images taken with part of the amorphous zirconia halo (figure 1b) and with part of the (111) and (200) Debye rings (figure 1a) clearly show qualitative differences. That obtained with the zirconia halo consists of a speckle pattern typical of an amorphous material. The speckles are regularly sized with low contrast and their size is essentially governed by the objective aperture size (Lowenthal and Arsenault, 1970). Here the Airy disc for the aperture used is about 0.6 nm at half-height. In comparison, the image obtained using the copper rings shows strongly contrasting speckle with quite a large range of size (from about 0.6 nm to 10 nm), which deviates from the statistics expected in the case of a random image (Krivanek, 1976). Those speckles which are greater in size than about 1 nm can be identified with fair certainty as not being due to noise, but as images of copper microcrystals. If we use a larger objective aperture (figure 2b), with corresponding Airy disc diameter at half-height of about 0.3 nm it is still possible to correlate speckle with that of figure 2a down to about 0.8 nm in size. Strict comparison between these two images is not easy since the larger aperture enables a greater number of diffracting crystallites to be imaged, as well as including speckle due to the amorphous zirconia like phase. This imaging behaviour, together with the diffraction information, demonstrates that copper crystallites are present.

The use of change of focus effects on the dark-field image (so-called 2 1/2-D imaging) yields information concerning the localisation of the diffracted amplitude in reciprocal space (Gibson, 1980). Off-axis beams will produce a change in crystallite image position, on change of focus. This can be seen in figure 3a,b, by comparing the relative positions of groups of crystallites (some are arrowed). In this case the de-focus is 680 nm. This shows that the interpretation of these images as due to copper crystallites is consistent.

Given such a specimen, it is interesting to see whether axial bright

field imaging can indeed show lattice fringes from the copper crystallites, given the noise produced by the amorphous matrix, and effects of
randomly oriented crystallites. It is now generally considered (Howie,
1978) that these may only be visible for very thin films, and the films
used here are relatively thick (of order several tens of nanometers).
Figure 4 shows such a high resolution axial bright field image. (III) copper fringes (0.208 nm) are visible in several small regions. The smallest
region which can be observed is about 1 nm. However, the number of such
regions is very much less than would be expected given the number observed in dark-field from the same region. Imperfect correction of astigmatism can readily eliminate fringes and furthermore since the specimen
here is by no means a weak phase object, there is also no ideal defocus
value for all the crystallites.

Finally, Gibson et al (1977) have suggested that the use of incoherent
illumination in CTEM,or of an annular detector in STEM, reduces statistical noise with essentially random specimens. Here, the annular detector
in a VG HB5 picks up all scattering upwards from (and including) the
(III) copper Debye ring. Figure 5 shows such an image compared with a
STEM dark-field image. Surprisingly the annular detector image does not
show the copper crystallites and the little contrast there is appears to
be dominated by changes in the specimen thickness.

In conclusion, conventional high resolution dark-field techniques enable
copper microcrystals in an amorphous matrix to be imaged and identified
down to about 1 nm in size, when effects of aperture size and defocus
have been ascertained. High resolution axial bright field images also
show up these crystallites, but not as reliably as the dark field images.
Annular dark-field STEM images may not be as useful as originally
thought since the detector integrates over part of the diffuse scattering
from the amorphous matrix (in this case with large atomic scattering
factor) and the resulting contrast is thus very poor.

References

Bigot J.,Calvayrac Y., Harmelin M., Chevalier J.P. and Quivy A. (1981)
to be published Rapidly Quenched Metals IV

Gibson J.M., Howie A. and Stobbs W.M. (1977) EMAG 1977 Inst. Phys. Conf.
Ser.   36, 275

Gibson J.M. (1980) EMAG 1979 Inst. Phys. Conf. Ser.   52, 273

Howie A. (1978) J.   non-cryst. Sol. 31, 41

Krivanek O.L. (1976) Electron Microscopy, Jerusalem, p. 275

Lowenthal S. and Arsenault H. (1970) J. Opt. Soc. Am. 60, 1478

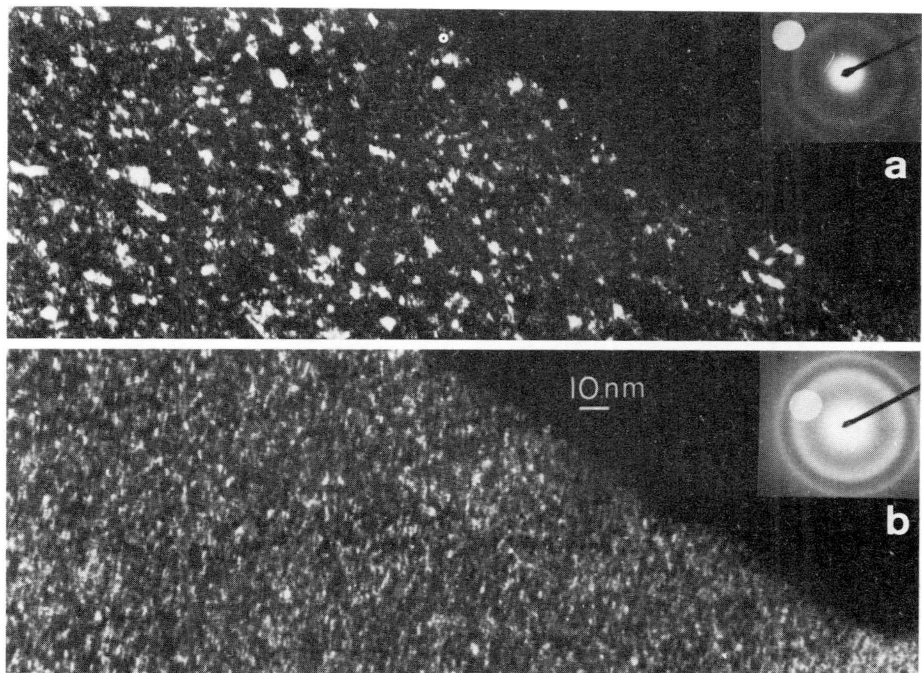

Figure 1.

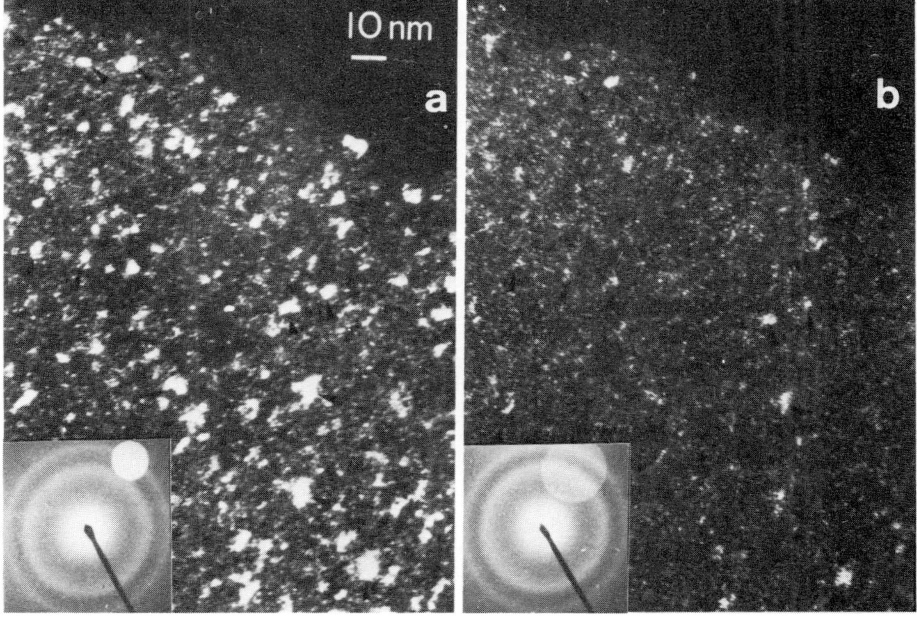

Figure 2.

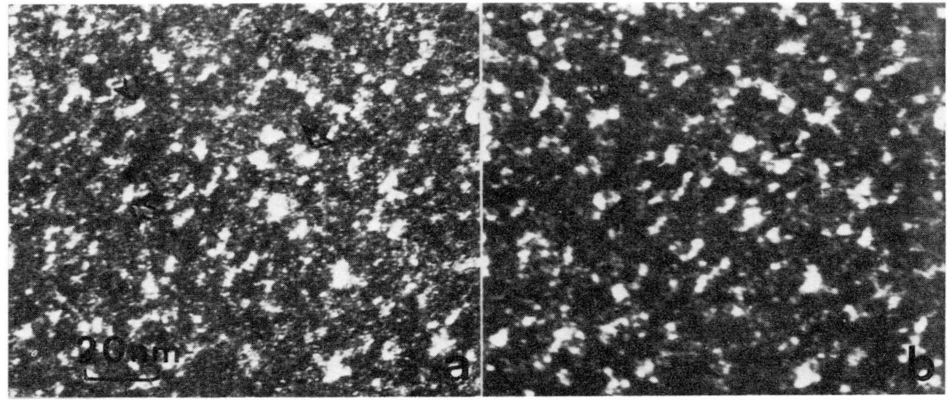

Figure 3.

a) $\Delta Z = 0$ nm ; b) $\Delta Z = -680$ nm

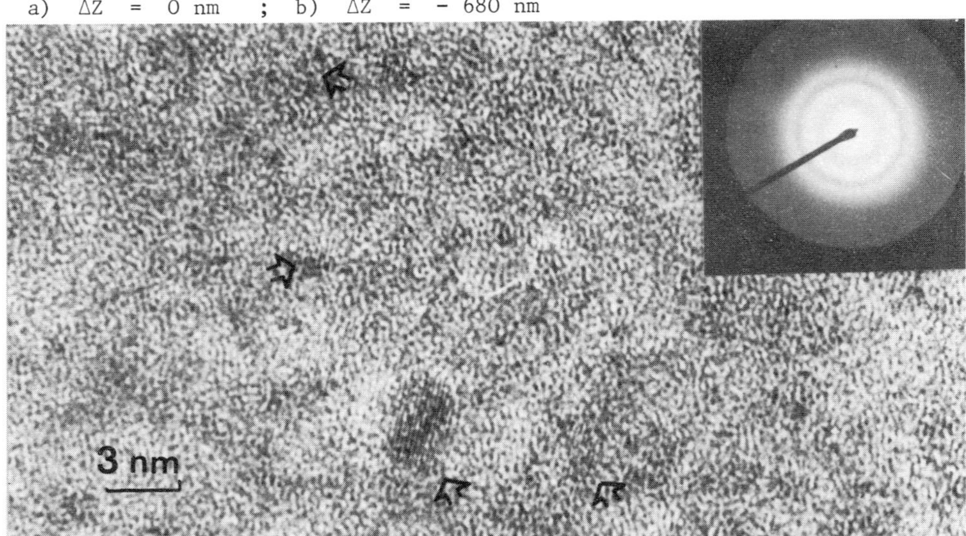

Figure 4.

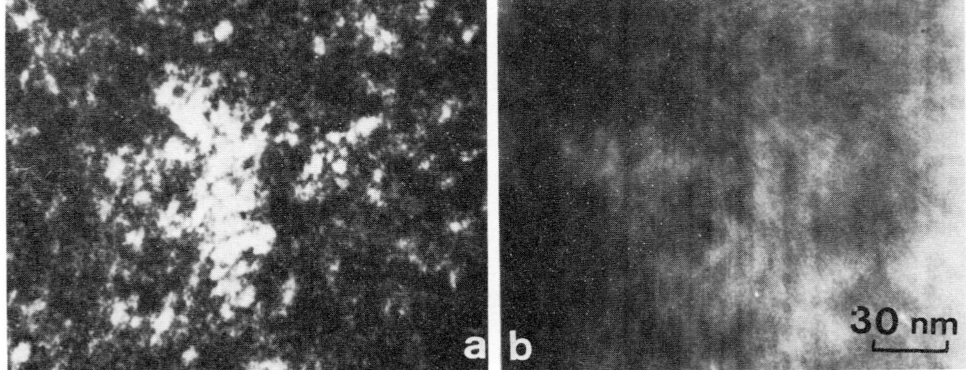

Figure 5.

a) Dark field ; b) Annular dark field

The structure of amorphous alumina

S M El-Mashri and A J Forty

Department of Physics, University of Warwick, Coventry.

L A Freeman and David J Smith

High Resolution Electron Microscope, University of Cambridge, Cambridge.

1. Introduction

Despite the important role of anodic oxidation in practical applications, such as the formation of protective coatings and the epoxy–resin bonding of metal sheets, the structure of amorphous alumina is not well-known. Wilsdorf (1951), and later Kerr (1956), accounted for the haloes in the electron diffraction pattern using a molecular model with a basic unit of Al_4O_6 in the form of an octahedron of O^{2-} with Al^{3+} bonded to three O^{2-} on four of the faces (Fig. 1a). Later studies by Oka et al (1979) showed that a model based on a disordered γ-Al_2O_3 crystalline analogue can give a good fit with the X-ray scattering pattern. In a recent EXAFS study Norman et al (1981) found that the Al^{3+} cations in the anodic layers are both octahedrally and tetrahedrally co-ordinated to O^{2-}. This suggests a model for amorphous alumina (Forty and El-Mashri, 1981) wherein Al_4O_6 molecular units are arranged in sheets (Fig. 1b) which, when stacked (Fig. 1c), give the appropriate admixture of octahedral and tetrahedral sites occupied by Al^{3+}. This is similar to the model proposed by Oka et al but with a different spacing of the Al^{3+} sheets: the amorphous state is thought to be a result of distortions of the Al–O bonds with accompanying distortions of the sheet structure.

The high-resolution studies described here have been undertaken in order to examine the structure of amorphous alumina, in particular to establish the possible existence of any quasi-crystalline regions which might be related to these various structural models. Furthermore, given the problems in epoxy–resin bonding arising from the tendency of alumina layers to become hydrated to a form of boehmite (Al O OH) – which is also amorphous but relatively weak mechanically – it was also of interest to examine the effect of the structure of amorphous alumina on the hydration reaction.

2. Experimental

The most important requirement for HREM studies is the preparation of very thin specimens, preferably having uniform morphology and composition. The results described below have been obtained from films approximately 60Å thick, formed by anodic polarisation at 2V in sodium tartrate solution on 6N purity aluminium which was initially polished mechanically and argon ion bombarded to give a clean smooth surface. The films are stripped from the aluminium in mercuric chloride solution and mounted on holey carbon grids.

The carbon film serves to reduce beam-charging of the alumina and also assists in selection of objective lens defocus. The effect of hydration has been examined using similar films which have been immersed in hot water ($\sim 80^\circ$C) for several minutes. Observations have all been at 500kV primarily with bright-field axial illumination, using the Cambridge University HREM (Nixon et al, 1977; Cosslett, 1980). A standard optical bench was sometimes used to record optical diffractograms from features of interest, and, particularly, to measure lattice fringe spacings accurately.

3. Observations

The most striking feature of our observations of the original "amorphous" samples has been the considerable variation in morphology seen from area to area, even in a single specimen. Large regions of apparently amorphous material were visible - see Fig. 2 - as indicated by high-resolution bright-field images (no fringe "packets" or regular structure), tilted illumination dark-field images (limited random speckle) and electron diffraction patterns (broad, diffuse rings). Yet, even in such "amorphous" regions, there would occasionally be packets of fringes visible, sometimes as much as 30–50Å across - see Fig. 3 - with fringe spacings primarily of 3.0Å and 3.7Å. Note, however, that these spacings could not be seen in any electron diffraction patterns. In other regions, considerable crystallinity was apparent, as indicated both by diffraction contrast as the objective lens focus position was altered, and by the presence of lattice fringes in high-resolution micrographs - see Fig. 4. The most common spacings observed in these regions were 1.4Å and 2.0Å, although some other spacings were occasionally seen. Electron diffraction patterns from such areas also revealed sharp rings superimposed on the diffuse haloes, verifying the presence of crystalline phases, and sometimes individual spot patterns appeared. Table I provides a summary of lattice spacings observed, together with a tentative identification of the particular phase where possible.

(a) Amorphous Alumina

```
D.P.              -    -    -   2.52  -    -    -    -   1.90 1.64 1.38 1.24
L.F.           3.71 3.03 2.62 2.56 2.30 2.07 2.00 1.98  -    -   1.38  -
F.and El-M     3.66 3.02 2.61 2.52 2.05            -    -   1.91  -    -   1.20
```

(b) Hydrated Alumina

```
D.P.              -    -    -   2.55  -    -   2.00  -    -    -   1.39 1.20
L.F.              -   2.86  -    -    -   2.00  -   1.71 1.46       -
```

Table I. a) Lattice plane spacings (Å) for amorphous alumina according to diffraction pattern, lattice fringes and the Forty/El-Mashri model. The spacings at 1.98 and 1.38Å are present in γ-Al$_2$O$_3$, 2.56Å in α-Al$_2$O$_3$ b) Lattice plane spacings for hydrated alumina. The spacings at 1.38, 2.86 and 2.0Å are present in γ-Al$_2$O$_3$, 2.55Å in α-Al$_2$O$_3$ and 1.39 in aluminium oxide hydrate.

Our characterisation of the hydrated samples is only at an early stage but it is already clear that a variety of structural changes are induced by the hydration treatment. In particular, more extensive crystalline regions are seen - Fig. 5, for example, shows a small patch with crossed 2.0Å lattice fringes, plus a further set of 2.86Å inclined at 45° to these. This configuration matches closely with that expected from (400) and (220) lattices of γ-Al$_2$O$_3$. Furthermore, electron diffraction patterns indicate

the occurrence of other spacings, some of which can be identified as arising
from γ–alumina and aluminium oxide hydrates.

4. Discussion

Despite the problem of specimen charging, which was alleviated to some
extent by the holey carbon support film, a considerable amount of information relevant tc the structure of amorphous alumina, as prepared by
anodic oxidation, has already been obtained. Thus, large expanses appear
to be truly amorphous except for the occasional small domain of fringes
the presence of which was only established by direct examination of high-
resolution micrographs – the occurrence of such ordered domains is consistent with the structural model for amorphous alumina recently proposed by
Forty and El–Mashri (1981). Although some of the observed lattice spacings
confirm the presence of known crystalline phases, for example γ–Al_2O_3 which
has been previously reported, there seems to be clear evidence of spacings
corresponding closely to the model of Forty and El–Mashri.

The Cambridge University 600kV HREM was constructed as a joint project
between the Cavendish Laboratory and the Department of Engineering with
major financial support from the Science Research Council. Continuing
support from SERC is gratefully acknowledged. S. M. El–Mashri wishes to
thank the Libyan Government for financial support.

References

Cosslett V E 1980 Proc. Roy. Soc. A370 1–18
Forty A J and El–Mashri S M 1981 Submitted to Phil. Mag.
Kerr I S 1956 Acta Cryst. 9 879
Nixon W C, Ahmed H, Catto C J D, Cleaver J R A, Smith K C A, Timbs A E,
    Turner P W and Ross P M 1977 Inst. Phys. Conf. Ser. 36 13–16
Norman D, Brennan S, Joeger R and Stohr J 1981 Surf.Sci.Letts, in press
Oka Y, Takahashi T, Okada K and Iwai S 1979 J. Non–Cryst. Solids 30 349
Wilsdorf H 1951 Nature 168 600

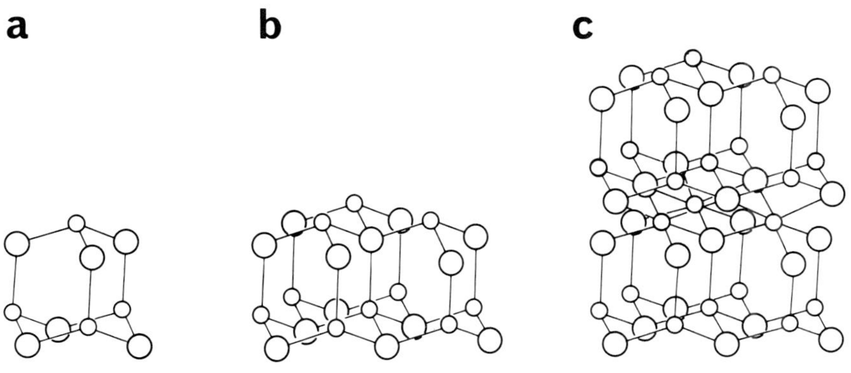

Fig. 1. The proposed structure of amorphous alumina showing a single Al_4O_6
    group, a sheet of these and a stack of sheets.

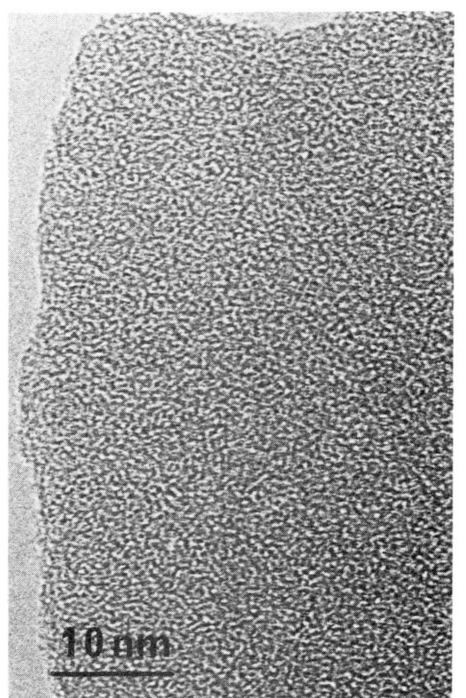

Fig.2.Area of amorphous alumina.

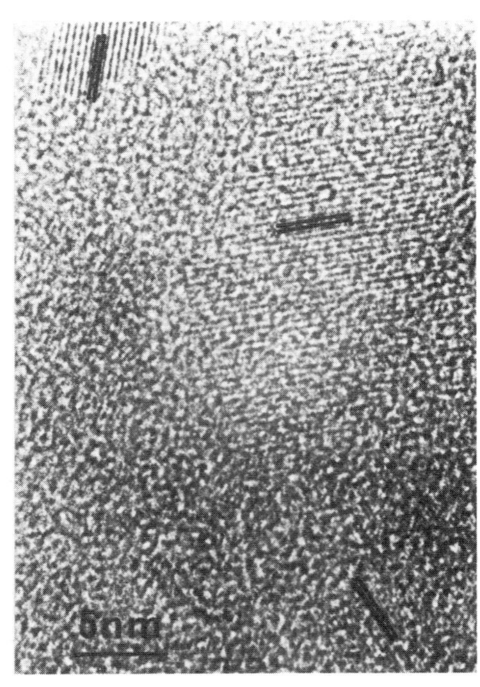

Fig.3."Amorphous" region with fringes
visible:spacings 3.0 and 3.7Å.

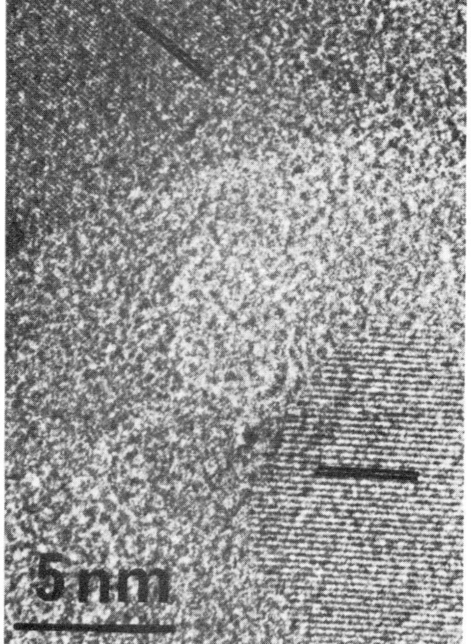

Fig.4.Region of extensive crystall-
inity:1.38Å and 1.98Å lattice
fringes of γ-Al_2O_3 visible.

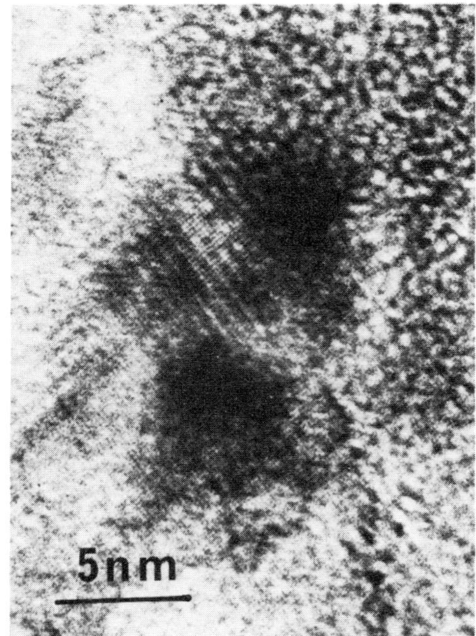

Fig.5.Hydrated sample showing crossed
2.0Å fringes, plus faint 2.86Å
fringes at 45°.

Imaging the various 'stages' in graphite–iron chloride intercalates

G.R. Millward   J.M. Thomas   W Jones and R Schlögl*

Department of Physical Chemistry, University of Cambridge, UK
*Institut fur Anorganische Chemie der Universitat Munchen, FDR

Intercalated graphites are of considerable interest because of their aniso-
tropic physical and electronic characteristics and also their catalytic
properties. In particular iron chloride-graphites are of consequence in
relation to catalysis of reactions such as Fischer-Tropsch synthesis (conversion of synthesis gas from coal into liquid hydrocarbons).

Early attempts by Evans and Thomas (1975) and Thomas et al (1976) to image,
and locate directly, intercalated species between graphite layers were
informative but not entirely satisfactory because of inadequate electron
optical specification for the electron microscope used - as discussed
by Millward and Jefferson (1978).

The results presented here were obtained using a JEOL 200-CX electron
microscope with a side-entry stage (C_s = 2.5 mm). For this instrument at
an accelerating potential of 200 kV the optimized Scherzer transfer
function (underfocus 941Å) extends out to a resolution of 2.9Å and is more
than adequate for resolving graphite-intercalate layered systems viewed
in a direction perpendicular to the c -axis where the significant lattice
periodicities are 3.365Å or larger - see Millward and Thomas (1979) and
Thomas et al (1981). This point is justified in Fig. 1 where an experimental image is compared with a computer simulated image calculated using
the   methods described by Jefferson et al (1976).

It is unfortunate for the present purposes that graphite particles, because
of their morphology, tend to lie on specimen grids with their c-axes
parallel, rather than (as required) perpendicular, to the direction of
viewing. In order to view the layers "edge-on" we rely on finding a small
region at the edge of a particle which has been twisted into an appropriate
orientation. By using an AF graphite of small particle size (nominally 1μm)
as the starting material for intercalate preparation, the chances of
finding suitable areas are improved over the more usual extended graphites.
The iron chloride was intercalated into the graphite using the novel photo-
chemical/solvent techniques reported by Schlögl and Böehm (1980) rather
than conventional vapour phase methods.

In an $FeCl_2$-graphite, characterized by X-ray diffraction and chemical
methods as a second-stage intercalate compound (i.e. two layers of graphite
between every layer of $FeCl_2$), we have found that the staging is rarely
perfectly regular, Fig. 2 being typical of the specimen. A mixture of
second with third staging is clearly shown. The statistical nature of the
stage-ordering is further indicated by rarer examples of regular first
stage (Fig.3) and regions where the staging is quite irregular (Fig.4).

It is by no means unusual to find terminating $FeCl_2$ layers of the type shown in Figs. 5 and 6, although the more complicated interpenetrating layered-assembly of Fig. 7 is a less commonly observed feature. Such observations provide some direct justification for the type of mechanism proposed by Daumas and Hérold for interconversion of one degree of staging to another; it is suggested that the staging changes by growth, or contraction, of interpenetrating differently staged domains.

A rare and unexpected observation is shown in Fig. 8, where two contiguous sheets of $FeCl_2$ can be seen intercalated into the graphite lattice. This interpretation is supported by computer simulation calculation as shown.

$FeCl_3$ intercalated graphite is much less air-stable than the $FeCl_2$ preparations, the $FeCl_3$ tending to de-intercalate from the graphite.[2] Fig.9 shows a typical image of a nominally first-staged $FeCl_3$ -graphite compound prepared for the electron microscope from suspension in ether using normal methods. The intercalation observed is, clearly, by no means first-stage. At the position marked A, in Fig. 9, fringes of periodicity 5.8Å, we think, represent $FeCl_3$ that has been extruded from the underlying graphite and become associated in crystalline or near crystalline form. Notwithstanding, these remarks, extended areas of first-stage $FeCl_3$ intercalated graphite, similar to that indicated in Fig. 3, may be observed provided that the preparations are stored dry (under argon), mounted directly onto electron microscope grids from dry powder (again in an argon environment) and subsequently transferred to the microscope within a few seconds.

The results described here form a part of a more general investigation of iron chloride intercalated graphite compounds including, as discussed by Bowen et al (1981), such appropriate experimental methods as Mössbauer spectroscopy and X-ray diffraction.

The authors are grateful for support from the Science Research Council, the National Coal Board and the Fonds der Chemischen Industrie.

References

Bowen P, Jones W, Thomas J M and Schlögl R and Böehm H P 1981 J C S Chem. Comm. - in press
Daumas N and Hérold A 1969 C.R. Acad. Sci. C268 373
Evans E L and Thomas J M 1975 J. Solid State Chem. 14 99
Jefferson D A, Millward G R and Thomas J M 1976 Acta. Cryst. A32 823
Millward G R and Jefferson D J 1978 Chemistry and Physics of Carbon 14 1
Millward G R and Thomas J M 1979 Carbon 17 1
Schlögl R and Boehm H P 1980 Carbon Conference, Baden-Baden, preprints
Thomas J M, Millward G R, Davies N C and Evans E L 1976 J. Chem. Soc. (Dalton) 23 2443
Thomas J M, Millward G R and Bursill L A 1981 Phil. Trans. Roy. Soc. A300 43.

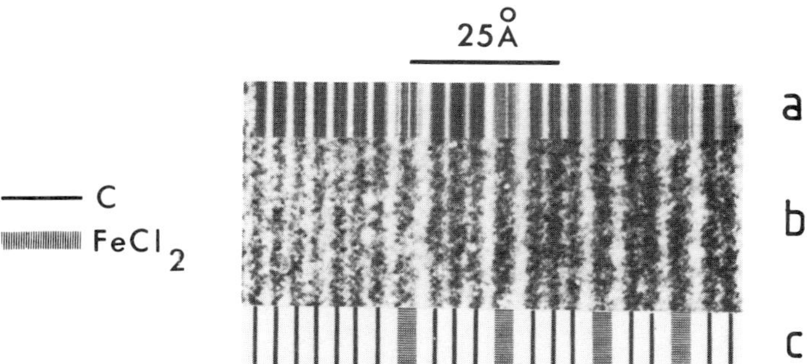

Fig. 1   FeCl$_2$ graphite viewed normal to the c-axis. Images are (a) computer simulated (b) experimental (c) schematic. Assumptions for (a) were: thickness 50Å, C$_s$ = 2.5 mm, underfocus 941Å, 200 K eV electrons, aperture limited resolution 2.9Å.

Fig. 2   Mixed 2nd and 3rd staging in a nominal "2nd stage" FeCl$_2$- graphite.

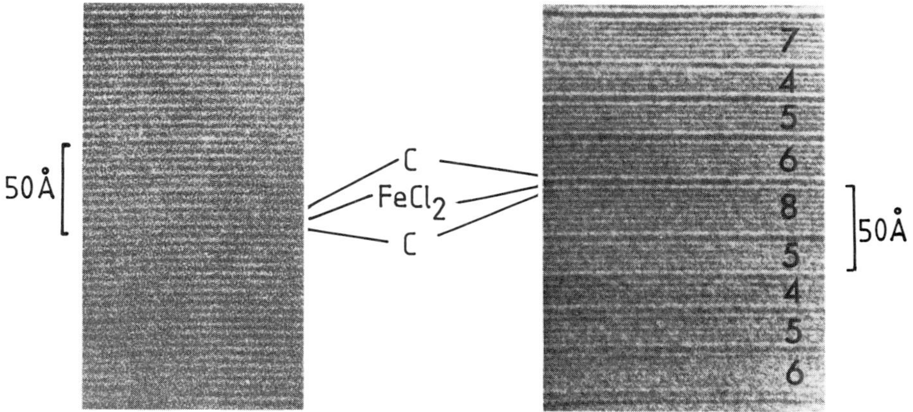

Fig. 3   Regular 1st staged intercalation in a nominal "2nd
stage" FeCl$_2$ -graphite.

Fig. 4. Variable staging in a
nominal "2nd stage"
FeCl$_2$ - graphite.

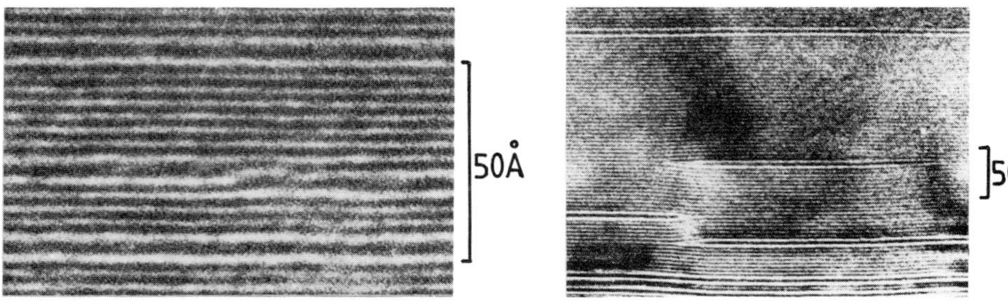

Fig. 5   Terminating $FeCl_2$ layers in $FeCl_2$ – graphite.

Fig. 6   Terminating $FeCl_2$ layers in $FeCl_2$ – graphite.

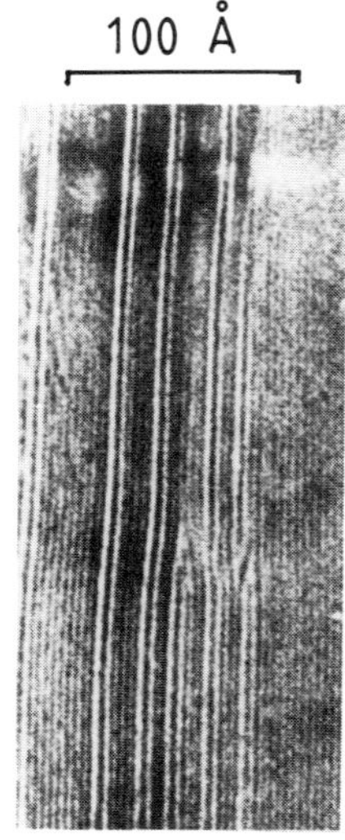

Fig. 7

Terminating $FeCl_2$ layers in $FeCl_2$ – graphite.

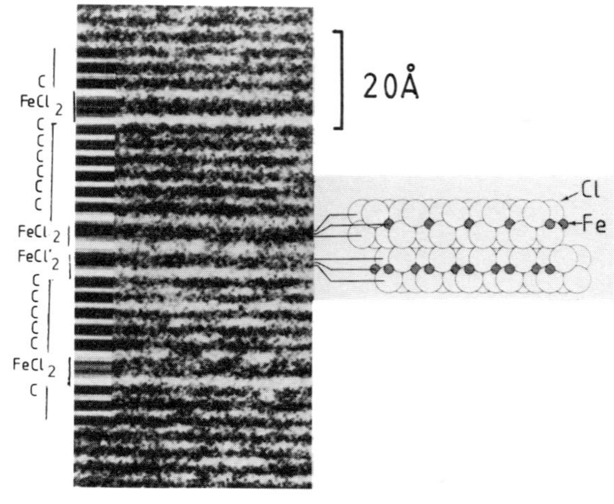

Fig. 8   Two adjacent layers of $FeCl_2$ intercalated into the graphite lattice, with two further individually intercalated $FeCl_2$ layers.

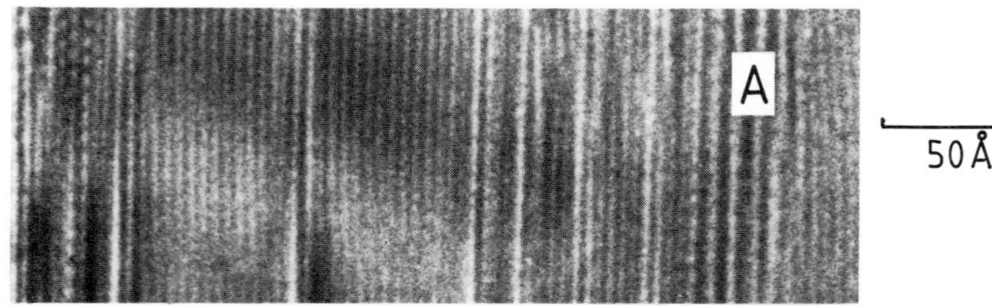

Fig. 9   Nominal "1st stage" $FeCl_3$ graphite. Some de-intercalation has taken place.

Fine structure in platinum catalyst particles on an alumina support

D. White, T. Baird, J.R. Fryer and David J. Smith*

Department of Chemistry, University of Glasgow, Glasgow

*High Resolution Electron microscope, University of Cambridge, Cambridge

1. Introduction

Transmission electron microscopy has provided useful information on the
size, structure, morphology and dispersion of metal particles on metal
oxide supports (Howie, 1980), although high support contrast and instrumental aberrations have been cited as resolution-limiting factors (Flynn
et al., 1974; Treacy and Howie, 1980). Model systems, specifically
adapted for TEM study, can reveal high resolution detail not easily
accessible in real catalyst systems (Baker et al., 1979). In addition,
high voltage electron microscopy provides the opportunity for enhanced
resolution below the 0.3nm level. The combination of model system plus
high voltage TEM has allowed direct structural characterisation of small
metal particles (Marks et al., 1979). This paper describes the high
resolution characterisation of platinum particles on a thin-film alumina
support. Lattice imaging in both fresh and sintered model catalysts has
revealed structural aspects of the platinum particles which may be of
catalytic importance.

2. Experimental

Thin support films were prepared by vacuum evaporation of γ-Al_2O_3 onto
cleaved rocksalt (100) crystals at room temperature, with the platinum
subsequently deposited on the alumina by evaporation of platinum wire.
After mounting on holey carbon films to minimise charging in the electron
beam, the samples were examined in the Cambridge University HREM (Nixon
et al., 1977; Cosslett, 1980) at 500kV using axial illumination. Careful
illumination alignment and a small incident angle ($\approx$0.2m Rad) ensured
adequate beam coherence and lattice fringes were visible in both particles
and substrates over a considerable focus range.

Samples of the model catalyst, prior to backing with a carbon film, were
treated in three sets of atmosphere/temperature conditions (White et al.,
1980):

i   reducing conditions – H_2 at 600°C
ii   oxidising conditions – 3^v/o O_2 in N_2 at 600°C
iii   oxidising conditions – 3^v/o O_2 in N_2 at 300°C with CCl_4
                             vapour traces added.

3. Results

The typical appearance of the initial model catalyst before undergoing
treatment is shown in Fig.1. The platinum islands are randomly oriented
and of 0.5-3nm diameter. Fine lattice structure was evident in most particles - marked in Fig.1 is a particle of around 2nm diameter clearly
showing the 0.227nm(111) lattice fringes. The evaporated alumina support
film consisted of microcrystalline domains of γ-Al_2O_3 embedded in an amorphous alumina matrix. This was suggested by electron diffraction and confirmed by lattice imaging (Fig.2) with 0.228nm(222) and 0.198nm(400)
lattice fringes being clearly resolved. The size and appearance of these
domains did not appear to change extensively during the heat treatments,
although no quantitative estimation has yet been attempted. The domains
also exhibited other crystallographic features such as twinning, shown in
Fig.2, and evidenced in the accompanying optical diffractogram. The presence of the platinum particles had no effect on the domain size or frequency in these films.

In each of the systems subsequently studied, particles of various sizes
were observed, often in the form of single crystals, lamellar twins and,
occasionally, multiply-twinned particles (MTPs), which could be identified
from their lattice spacings as being of metallic platinum. The results of
heating the model catalyst in various atmospheres were:

<u>i</u> H_2 reduction at 600°C - very little particle sintering was apparent
and the particle size distribution (PSD) was quite narrow, having a mean
$\bar{d}$ = 4nm, with a variance $\bar{\sigma}$ = 1nm. About 30% of the particles could be
identified as crystalline platinum from lattice fringes visible, of which
$^1/_3$rd were clearly twinned. The other particles might reasonably be pre-
sumed to be crystalline but incorrectly oriented to give visible lattice
fringes.

<u>ii</u>   Oxidation at 600°C - generally much larger particles were produced
with $\bar{d}$ = 43nm, $\bar{\sigma}$ = 17nm. Twinning again appeared to be extensive with the
same proportion of the crystalline particles exhibiting the characteristic
structure. A small number (~1%) appeared to be multiply-twinned giving
approximately decahedral particles (see Fig.3).   The twin structure was
usually accompanied by re-entrant surface structure (see Marks et al.,
1979) - Fig.4 shows the lattice structure visible at the boundary between
two tetrahedral segments.

<u>iii</u> Oxidation at 300°C - this produced a PSD with $\bar{d}$ = 3.6nm and $\bar{\sigma}$ = 0.8nm,
i.e. very similar to that from the 600°C H_2-reduction treatment. The total
proportion of particles visibly twinned was again approximately 30%.
The platinum particles did not appear to exhibit any bulk defects apart
from twinning. Fig.5, for example, shows a perfect single crystal with
the (111) and (200) planes imaged. Such particles normally appeared round,
and might reasonably be assumed to be quasi-spherical whereas twinned
particles displayed considerable surface variation (Fig.6). There was no
evidence for structural interaction between the platinum and its support.

4. Discussion

The present study has demonstrated that the very small supported metal
particles of a model catalyst can be well-characterised by direct lattice
imaging in an HREM. It was shown that such particles can have either

single crystal or twinned f.c.c. structure and, furthermore, that sintering is greater in 'oxidising' than in 'reducing' environments. Some
irregular f.c.c. structures (MTPs) occur but not apparently in significant
numbers.

The structure of the twin boundaries did not show any noticeable discontinuities or defects that might lead to increased surface area of the
particles. However, oxidative treatments are employed in the industrial
re-generation of platinum catalysts to regain metal surface area (e.g.
Yates, 1976). Our observations establish that there is no re-dispersion
under these conditions so that the structure associated with the twinned
particles may be critical to catalytic activity.

The Cambridge University 600kV HREM was constructed as a joint project
between the Cavendish Laboratory and the Department of Engineering with
major financial support from SRC. We are pleased to acknowledge continuing support from SERC and we are grateful to SERC and ICI Ltd. for
support to DW in this work.

References

Baker RTK, Prestridge EB and Garten RL 1979 J. Catal. 56 390
Cosslett VE 1980 Proc. R. Soc. London A370 1
Flynn PC, Wanke SE and Turner PS 1974 J. Catal. 33 233
Howie A 1980 Characterisation of Catalysts ed. JM Thomas
    and RM Lambert (New York : Wiley) pp 89-104
Marks LD, Howie A and Smith DJ EMAG 1979 397
Nixon WC, Ahmed H, Catto CJD, Cleaver JRA, Smith KCA,
    Timbs AE, Turner PW and Ross PM EMAG 1977 13
Treacy MMJ and Howie A 1980 J. Catal. 63 265
White D, Baird T and Fryer JR Electron Microscopy 1980 1, 220
Yates DJC 1976 US Patent Specification 1433864.

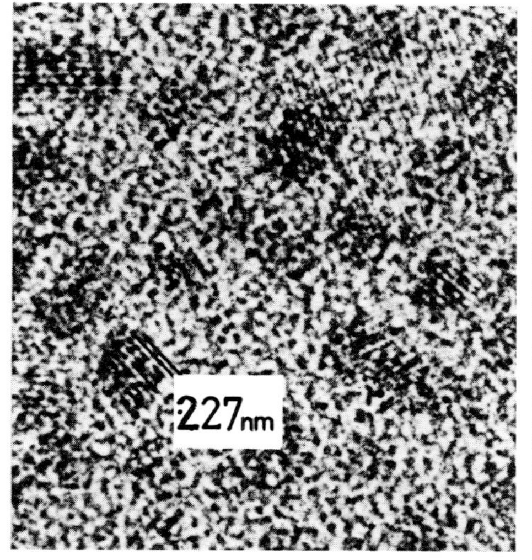

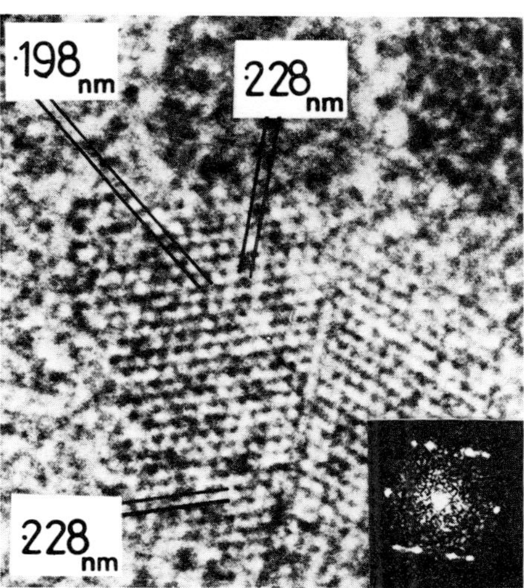

Fig.1: Model catalyst before treatment
       showing platinum islands with
       0.227nm(111) lattice fringes.

Fig.2: γ-Al$_2$O$_3$ micro-twin with
       0.228nm(222) and 0.198(400)
       lattice structure; optical
       transform inset.

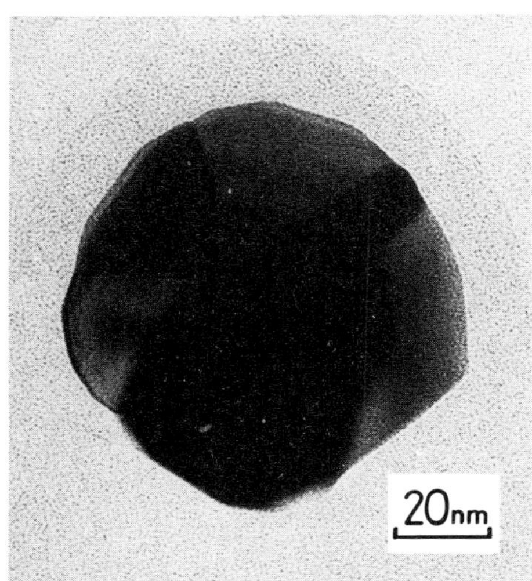

Fig.3: Modified "decahedral" MTP from oxidative sintering at 600°C.

Fig.4: Lattice structure at re-entrant twin boundary (see Fig.3).

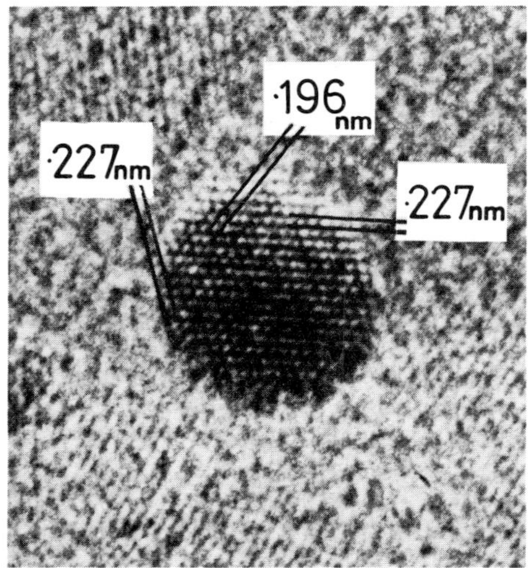

Fig.5: Platinum single crystal with 0.227nm(111) and 0.196nm(200) lattice fringes.

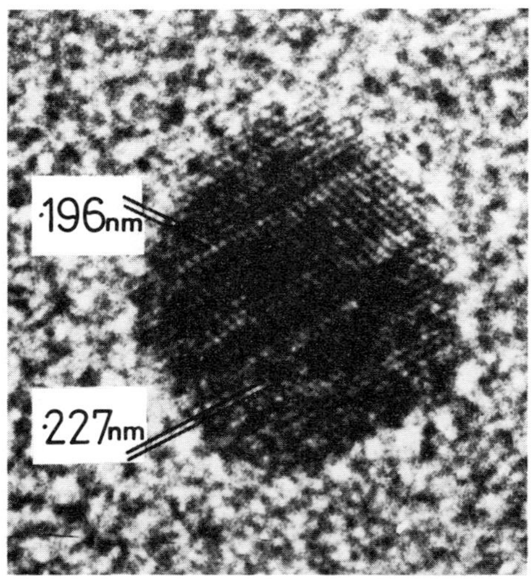

Fig.6: Platinum crystallite with twin lamellae and irregular surface structure.

Structure imaging and orientation relations of dislocation-free talc–tremolite interfaces

H-U Nissen, J L Hutchison*, A Kumao, R Wessicken and J Ylä-Jääski

Laboratory of Solid State Physics, ETH, CH-8093 Zürich, Switzerland
*Department of Metallurgy and Science of Materials, University of Oxford, Parks Road, Oxford OX1 3PH, United Kingdom

1. Introduction

The white-green nephrite from Scortaseo, Posciavo Valley, Swiss Alps consists of a crypto-crystalline intergrowth of tremolite, $Ca_2Mg_3Si_4O_{10}(OH)_2$ i.e. a chain silicate, and talc, $Mg_3Si_4O_{10}(OH)_2$, a sheet silicate (Dietrich et al 1968). Many examples of boundaries between tremolite and talc have previously been found to be strictly planar and oriented parallel to the (010)-plane of both phases (Hutchison et al 1979), and no dislocations have been observed in these planar boundaries. We have studied this material by high resolution electron microscopy at 100 kV and 200 kV in order to obtain structure images of the two phases and of the phase boundary area. The orientation relations between the tremolite and the talc have been analyzed on the basis of electron diffraction patterns and of optical diffraction patterns made from structure images of the two phases. Further, a possible reaction from tremolite to talc is discussed.

2. Structure imaging of tremolite

A new method of obtaining a through-focus series of structure images has been applied to tremolite. The series is taken from one single exposure of an area along the crystal margin having approximately constant thickness. The contrast of these images along the margin is presented in Fig. 1 together with dynamical multislice calculations using the EMSYS program system (Skarnulis 1979). Cell parameters of the pure phase given by Papike et al (1969) have been used. In the calculation, the number of excited beams was 419, the number of beams inside the aperture 55, the slice thickness 4.9 Å, and the specimen thickness 59 Å. The only variable parameter was the defocus value. Therefore the image contrast changes must be due to differences in the defocus position along the crystal margin. This originates from inclination and/or bending of that margin near a hole in the holey carbon film.

3. Orientation relations and reaction mechanism

The orientation relation between talc and tremolite has been investigated for cases in which the phase boundary is parallel to the (010)-plane of both minerals, i.e. the majority of cases. Three different orientation relations for phase boundaries in (010) of both tremolite and talc were

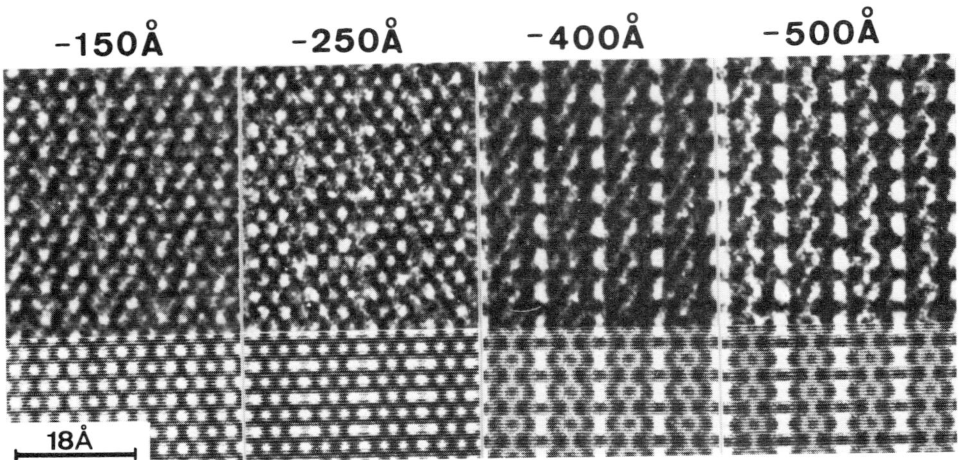

Fig. 1 Upper part shows structure image details along a crystallite margin of tremolite obtained from one micrograph. Lower part shows dynamical multislice calculations with defocus as parameter. 200 kV, [100]-projection.

found and are listed in Table 1. Case 1a from this table is presented in Fig. 2 which shows two phases, i.e. talc and tremolite separated by a phase boundary which is straight on the atomic scale and free of dislocations. The right hand part of Fig. 2 schematically shows the diffraction pattern of the boundary area. It is a superposition of the pattern of talc and that of tremolite and can serve to find the precise orientation relations between both phases (cf. Stemple and Brindley 1960). The three orientation relations are shown in Fig. 3. The plane of the figure is the (010)-plane for both structures. It may be assumed that the orientation relations represent positions of best geometric fit (cf. Bollmann and Nissen 1968) as well as minima of phase boundary energy.

After a cell transformation of the tremolite to a bodycentered I2/m cell (biopyribole cell, Thompson 1978) the geometrical similiarity of the (010)-planes of talc and tremolite is apparent. It is increased if the cell parameters of tremolite at 400°C (Sueno et al 1973) are used in the comparison (those of talc are not known as yet), and it appears possible that the reaction from tremolite to talc within one bicrystal took place at the temperature for which the coincidence of the two phases in the (010)-plane is complete. This presupposes that the bicrystals are a product of a metamorphic reaction during the change from amphibolite facies to greenschist facies conditions in the presence of water and CO_2. This reaction can be written as

$$Ca_2Mg_5Si_8O_{22}(OH)_2 + CaMg(CO_3)_2 + H_2O + CO_2 \rightarrow 2\ Mg_3Si_4O_{10}(OH_2) + 3\ CaCO_3$$

       tremolite   dolomite   talc   calcite

Calcite as possible reaction product is found in the nephrite material,and the nephrite is surrounded by dolomite rock (Dietrich et al 1968). Chain defects including 1-chain (pyroxene) and 3-chain defects are frequently observed in the tremolite phase. It can be seen from Fig. 4 that

Table 1 Orientation relations of talc and tremolite with the phase boundary in (010) for talc and tremolite

	electron beam direction		axes in diffraction pattern	
	tremolite	talc	tremolite	talc
1a	[101]	[001]	$010 \perp 10\bar{1}$	$010 \perp 100$
1b	[201]	[101]	$010 \perp 10\bar{2}$	$010 \perp 10\bar{1}$
2	[10$\bar{1}$]	[20$\bar{1}$]	$010 \perp 101$	$010 \perp 102$
3	[10$\bar{1}$]	[101]	$010 \perp 10\bar{1}$	$010 \perp 10\bar{1}$

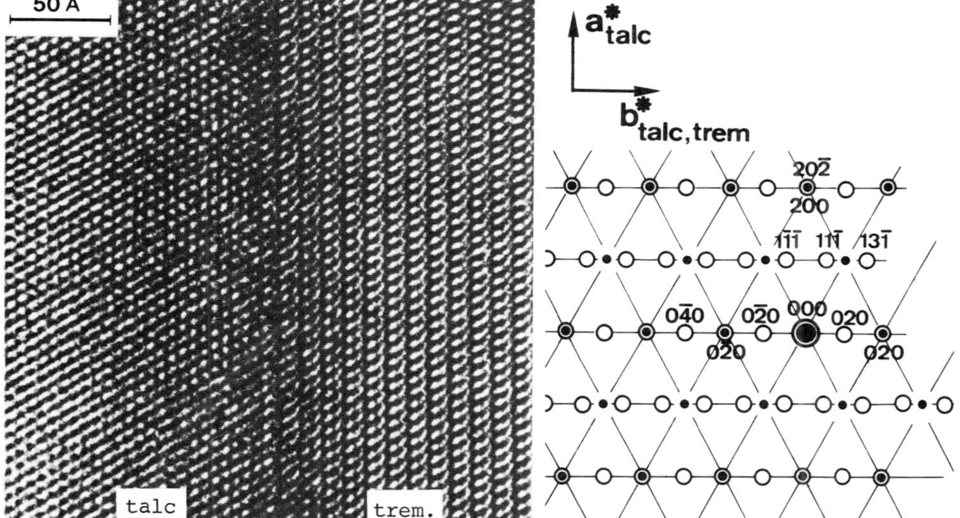

Fig. 2(a) Structure image of the phase boundary corresponding to case 1a of Table 1. (b) Schematic superposed diffraction pattern of talc and tremolite for that case

a single 1-chain defect in tremolite produces a shift of the entire structure by 1/2 c. In cases where the phase boundary is not parallel to (010) of tremolite, chain defects were observed to pass over into the talc structure where they produce a disturbance in the periodic contrast; this indicates that the phase boundary represents the front of a solid state reaction which has come to a stop before completion rather than a stable intergrowth boundary. The same can be deduced from the cases in which the talc-tremolite boundary shows irregular steps.

4. Conclusions

We have obtained a through-focus series of structure images of a thin crystal from a single micrograph. This is advantageous for cases in which the specimen damage does not allow enough time to take a through-focus series with sufficient exposure time. Also, we have taken high resolution structure images of two phases in the phase boundary region in nephrite, where one phase is talc which rapidly becomes damaged in the electron beam. This case has been studied by selected area diffraction as well as by optical diffraction patterns made from structure images in order to obtain

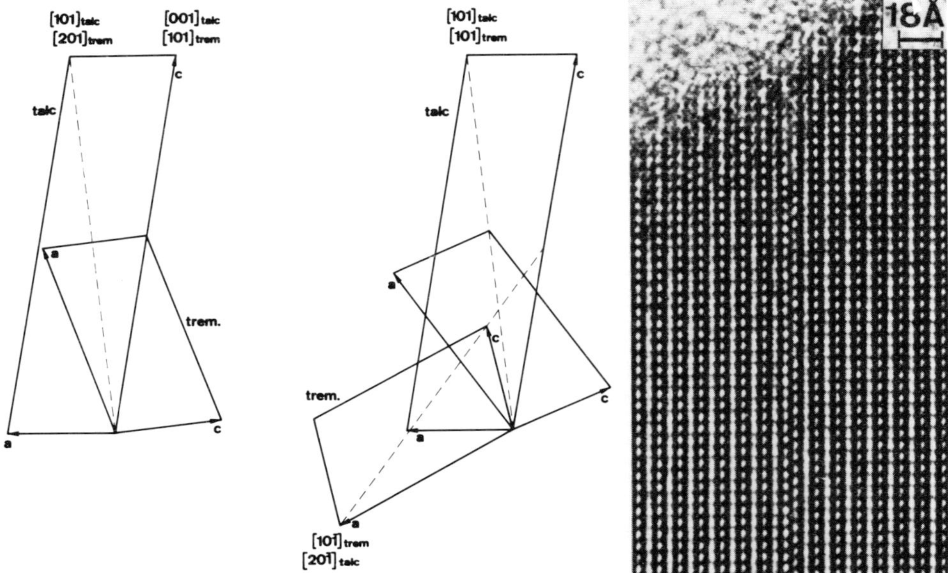

Fig. 3 Orientation relations of talc and tremolite. The projected plane is the (010)-plane of both structures. Indices represent beam directions for different experiments

Fig. 4 Single chain defect in tremolite showing relative shift of contrast pattern by c/2 along defect

diffraction information for each phase separately in the vicinity of the phase boundary. The total of this diffraction information gives the precise orientation relations of both phases which is prerequisite for an understanding on the atomic scale of chemical reactions along phase boundaries.

Acknowledgements

We thank Mrs. M. Brügger-Trombini for excellent photographic work. This work was in part supported by Schweizerischer Nationalfonds (A.K.) and by Schweiz. Volkswirtschaftsstiftung (J.Y.).

References

Bollmann W and Nissen H-U 1968 Acta Cryst. A24 546-557
Dietrich V, de Quervain F and Gross G 1968 Beiträge zur Geologie der Schweiz, Geotechn. Serie 46 1-78
Hutchison J L, Nissen H-U and Wessicken R 1979 Phys. Chem. Minerals 4 275-280
Papike J J, Ross M and Clark J R 1969 Mineral. Soc. America Spec. Paper 2 117-136
Skarnulis A J 1979 J. Appl. Cryst. 12, 636-638
Stemple I S and Brindley G W 1960 J. Amer. Ceramic Soc. 43 34-42
Sueno S, Cameron M, Papike J J and Prewitt C T 1973 Amer. Mineralogist 58 649-664
Thompson J B 1978 Amer. Mineralogist 63 239-249

Oxidation–reduction reactions in natural spinels

R. Frost [*] and P.L. Gai [+]

[*] Department of Geology and Mineralogy, Oxford, OX1 3PR
[+] Department of Metallurgy and Science of Materials, University of Oxford, Parks Road, Oxford, OX1 3PH, U.K.

The occurrence and compositional interrelationships of oxides - including spinels in natural gabbroic rocks are of considerable interest because of their industrial applications. However, investigations have been limited to mainly simple synthetic equivalents of these. It is therefore hoped that experiments using the natural minerals as starting materials in a controlled gas reaction cell (Swann & Tighe, 1971) fitted to a high voltage electron microscope (HVEM) together with high resolution EM(HREM), may provide a better understanding of the mechanisms of specific natural processes. Iron-titanium oxide spinels with small amounts of coexisting magnesium-iron-aluminium spinel pleonaste, $(Mg,Fe)Al_2O_4$, from Skaergaard intrusion ferrogabbro (East Greenland) were prepared for EM by ion thinning. The compositions were analysed by Philips EM400 and the microprobe. In-situ studies were conducted in an AEI-EM7 HVEM at 1 MeV to examine phase relationships between spinels and rhombohedral oxides. Oxidation was conducted in oxygen gas at temp. of 323K-1100K employing higher experi-mental O_2 fugacities to obtain faster reaction rates and for reduction 10% H_2-He gas was used. For HREM, a JEOL 200CX EM (Boyes et al 1980) was used, at 200kV.

Fig 1(a) shows the natural spinel with cubes of magnetite (Fe_3O_4) at M and lamellae of ulvöspinel (Fe_2TiO_4) at U with 2 sets of edge dislocations (Burgers Vector $\underline{b} = \frac{1}{2}[101]$ type) at their boundaries (Frost and Gai 1980), and (b) the $(1\overline{1}0)$ diffraction pattern (d.p.). 1(c) shows the direct structure image of M in (100). HREM confirmed the presence of disloca-tions at the interfaces. Rhombohedral ilmenite lamellae $(FeTiO_3)$ were found in some natural spinels (Fig.2) indicating partial oxidation of some ulvöspinels. Attempts were therefore, made to simulate the reactions in the spinels using the gas cell with 100torr O_2, at <900K to 1100K. Needles of ilmenite were observed to grow from ∿900K confirmed by d.p., analysis and HREM. Fig 3(a) shows a high resolution image from such an oxidised sample showing ilmenite lamellae of 30-300A° with a growth habit of (111) spinel $||$ (0001) ilmenite. The corresponding d.p. in (b) shows superlattice rhombohedral spots and some weak spots believed to be from an orthorhombic type of Ti rich cation deficient spinel (similar to that observed in chrome spinels) shown e.g. at C at the ilmenite junctions in Fig 3(a). More commonly, cation deficient spinels with primitive cubic structure (titanomaghemite, γ-FeTiO_3) were observed in the experiments. Reduction of the oxidised samples was conducted at <570K to 870K to examine the extent of reversibility. Titanomaghemite formed in oxidation at 850K (Fig 4(a) and (b), exhibiting primitive cubic reflections indica-tive of a cation deficient cubic spinel,superimposed on the spinel

Fd3m spots in the d.p.) could be reduced to ulvöspinel as shown in Fig 4(c) containing no superlattice reflections.

It seems likely from the HREM images that the boundary relationships between spinels and rhombohedral oxides are more complicated at atomic level involving several rows of atoms rather than the minor variations suggested previously on the basis of their similar lattice spacings and atomic arrangements (e.g. Porter, 1976; Withers & Bursill, 1980).   During oxidation pleonaste was also observed to form a cation deficient cubic spinel with the diffraction similar to titanomaghemite.

We thank the NERC and SERC (U.K.) for financial support.

References
Boyes et al 1980, Inst.Phys.Conf.Ser.52, 445
Frost R. and Gai P.L. 1980, Acta.Cryst.A36, 678
Porter J.R. 1976, Ph.D. thesis, London Univ.
Swann P.R. and Tighe N. 1971, J.Ann., 155, 497.
Withers R. and Bursill L.A., 1980, J.Appl.Cryst. 13, 346

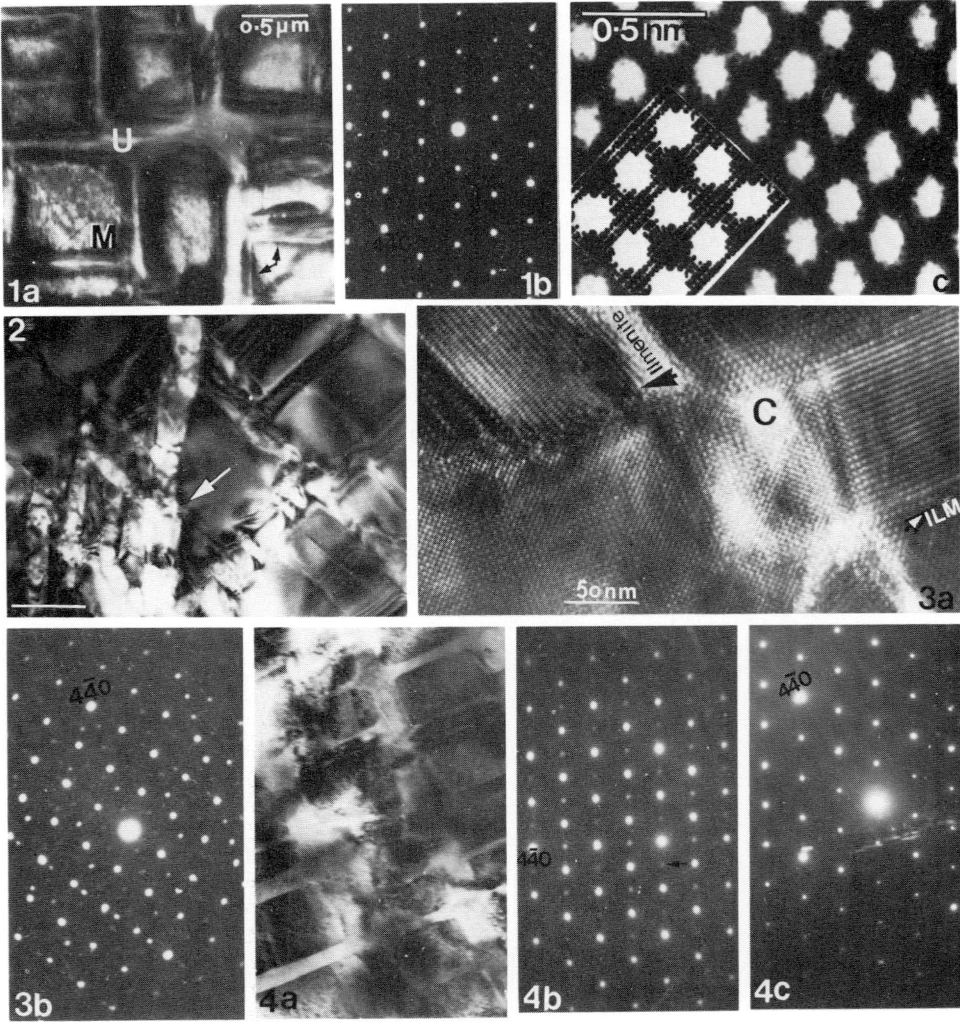

A lattice imaging study of gold–cadmium martensite

K M Knowles

Department of Metallurgy and Science of Materials, University of Oxford

1. Introduction

The remarkable rubber-like properties of the martensitic orthorhombic
phase of Au-47.5 at% Cd were discovered nearly 50 years ago by Ölander in
1932.   A fully transformed specimen of this composition can be severely
deformed with strains typically of the order of 8%, yet on removal of the
stress causing the deformation the specimen springs back to its original
shape.   Examination of the microstructure of this martensite by
conventional TEM shows that it consists of many different variants each
internally twinned on a fine scale and occasionally possessing faults
(Toth and Sato, 1968; Tadaki and Shimizu, 1977).   The rubber-like
behaviour can therefore be explained in terms of the reversible movements
of twin boundaries in the martensite (Lieberman et al., 1975).   These
twin boundary movements can be accomplished by glide of twinning dislocations, which are steps with dislocation character at the twin interfaces
(Christian, 1975).   It might be expected that such twinning dislocations
would be readily observable using the high resolution TEM technique of two
dimensional lattice imaging.   In this paper we report such observations
and also present direct evidence of the nature of the faulting in the
martensite.

2. Experimental Methods

Alloy ingots with nominal compositions of Au-47.5 at% Cd were prepared by
melting in evacuated and sealed quartz tubing and grown by a modified
Bridgman method.   The ingots were then ground to 2.3mm diameter rods and
annealed at 200°C for 2 hrs.   Discs were cut from the rods and electropolished into thin foils using conventional procedures (Tadaki and Shimizu,
1977).   The foils were examined at 200kV in a JEM 200CX electron microscope equipped with a high resolution objective lens and a double-tilt
(±12°) goniometer holder.   For suitable specimens, the limiting instrumental point-to-point resolution of this microscope is 1.8Å (Boyes et al.,
1980).   Two dimensional lattice fringe images were then obtained from
low index zones of the martensite using axial illumination.   Some of the
experimental difficulties encountered in obtaining lattice images from
metallic foils on this microscope are discussed elsewhere (Knowles, 1981).

3. Results and Discussion

Fig.1 is a typical example of the twinned microstructure of the orthorhombic
martensite observed in the thin foils.   Previous studies of Au-Cd have
shown that the twinning planes are {111} (Lieberman et al., 1955; Toth and

Sato, 1968; Tadaki and Shimizu, 1977) and this was confirmed in the present work. Fig.2 is a diffraction pattern from the twins containing the 111 twin reflection. Taking the lattice parameters of the ortho-rhombic martensite to be a = 3.150Å, b = 4.861Å and c = 4.755Å and the atomic positions to be $(0,0,0)$, $(0,\frac{5}{8},\frac{1}{2})$ for Au and $(\frac{1}{2},\frac{1}{2},0)$, $(\frac{1}{2},\frac{1}{8},\frac{1}{2})$ for Cd (Ölander, 1932), this diffraction pattern can be labelled either [1$\bar{1}$0] or [10$\bar{1}$] from the measured interplanar spacings and angles. The ambiguity arises from the near-tetragonality of the unit cell. However, by considering the spot intensities and allowing for the effect of double diffraction, this pattern is consistent with a [1$\bar{1}$0] rather than a [10$\bar{1}$] twinned diffraction pattern. The influence of double diffraction implies that the foils were relatively thick (at least for lattice images) even near the foil edges. It is therefore not surprising that the images obtained could not be interpreted in terms of the projected potential distributions in the twins along the beam direction. In particular, the 001 spots, which are those nearest the incident electron beam, are kinematically forbidden, but double diffraction permits them to have finite intensities and this is reflected in the 4.76Å periodicity observed in the lattice images.

Fig.3 is an example of a two dimensional lattice fringe image taken at a nominal defocus of $\Delta F = -500$Å of a portion of a twin interface with the diffraction conditions shown in Fig.2. Two features are immediately apparent - the twin interface is not accurately parallel to (111) and there is tentative evidence of the faulting on (001) planes found by previous workers (Toth and Sato, 1968). In addition, the pattern of the lattice images in each twin varies with position across the micrograph; this can be caused by changes in defocus, thickness and foil orientation across the area of interest. Close inspection of the interfacial region reveals that the deviation of the twin interface away from (111) is accomplished by three steps, two of height $2d_{111}$ perpendicular to the (111) plane at A and B and one of height $3d_{111}$ perpendicular to the (111) planes at C, where d_{111} = 2.310Å is the interplanar spacing of the (111) planes. In all cases the steps appear to be almost completely parallel to the electron beam and lie on (001) planes of the lower twin. It is particularly interesting that the step heights are multiples of the unit step height d_{111}, which one might expect from a twinning dislocation. For this particular twinning mode in Au-Cd martensite, shuffles are needed in addition to the twinning shear to produce the twinned structures, and so multiple step heights are therefore not necessarily unfavourable (Christian, 1975). It is, however, interesting that the steps observed do not all have the same height, and this requires further investigation.

Fig.4 is a lattice image of an (001) stacking fault taken at a defocus $\Delta F = -500$Å and also with the diffraction conditions shown in Fig.2. This clearly shows that the fault is very narrow and is at the most 2 (002) planes wide. Such faulting may be explained by considering the stacking sequence of the (002) planes and the shuffles required by the phenomeno-logical theory of martensite. For an ideal crystal of Au-Cd martensite, the stacking sequence parallel to the (002) planes can be represented as ..ABABAB.. where layer A has an Au atom at (0,0) and layer B one at $(0,\frac{3}{8})$. Layer B can then be considered to have had a shuffle of $(0,\frac{1}{8})$ from the atom positions in the parent cubic phase of Au-Cd. If, however, a shuffle of $(0,-\frac{1}{8})$ takes place, a new sequence ..AB'AB'AB'.. arises which equally well represents the orthorhombic structure. The fault shown in Fig.4 might then be a consequence of a stacking sequence such as ..ABAB'AB.. . However, it is impossible to confirm this without taking a through focal series of lattice images in thin ($\sim$ <100Å) foils and comparing the images with computer simulations using the dynamical theory of electron diffraction.

4. <u>Conclusions</u>

Lattice images have been obtained from Au-Cd martensite which demonstrate the presence of steps in the twin boundaries and reveal the fine faulting in the martensite directly. Further work is required to fully characterise the nature of both the faults and the interfacial steps.

<u>Acknowledgements</u>

I would like to thank Professor Sir Peter Hirsch FRS for the provision of laboratory facilities and the Science Research Council for financial support in the form of a Research Fellowship.

<u>References</u>

E.D. Boyes, E. Watanabe, A.J. Skarnulis, J.L. Hutchison, P.L. Gai, M.L. Jenkins and M. Naruse, 1980, Inst.Phys.Conf.Ser.<u>52</u>,445

J.W. Christian, 1975, The Theory of Transformations in Metals and Alloys (Pergamon Press, Oxford)pp 286-7

K.M. Knowles, submitted to Proc.Roy.Soc.(Lond.)A

D.S. Lieberman, M.S. Wechsler and T.A. Read, 1955, J.Appl.Phys. <u>26</u>, 473

D.S. Lieberman, M.A. Schmerling and R.S. Karz, 1975, in "Shape Memory Effects in Alloys" ed.J. Perkins (New York, Plenum Press)p.203

A. Ölander, 1932, Z.Krist.<u>83A</u>, 145

T. Tadaki and K. Shimizu, 1977, Trans.J.I.M. <u>18</u>, 735

R.S. Toth and H. Sato, 1968, Acta metall.<u>16</u>, 413

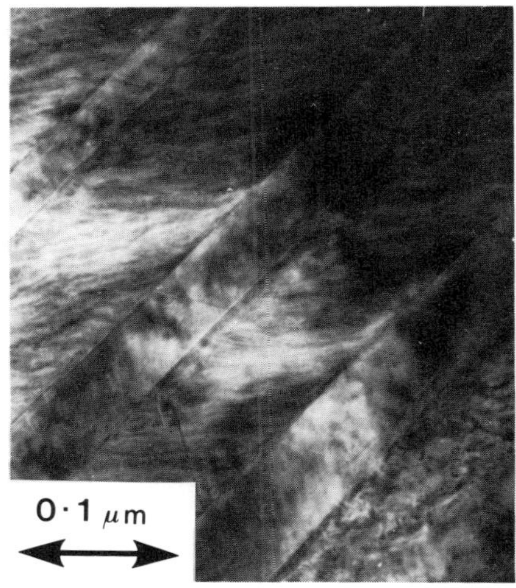

Fig.1 An electron micrograph of the twinned microstructure of the orthorhombic Au-Cd martensite. The twin plane is(111).

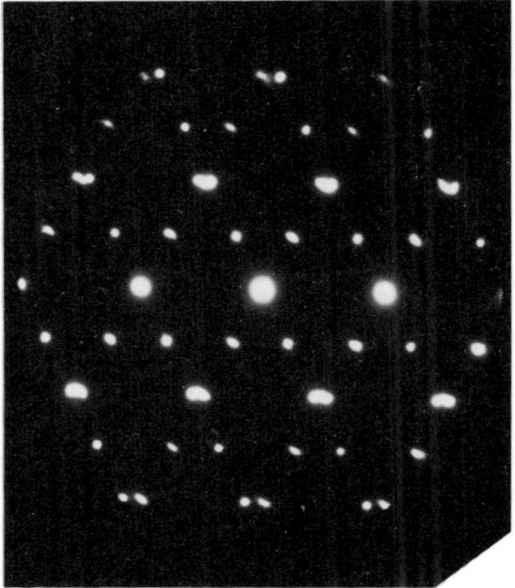

Fig.2 A twinned electron diffraction pattern from the martensite. [1$\bar{1}$0] zones.

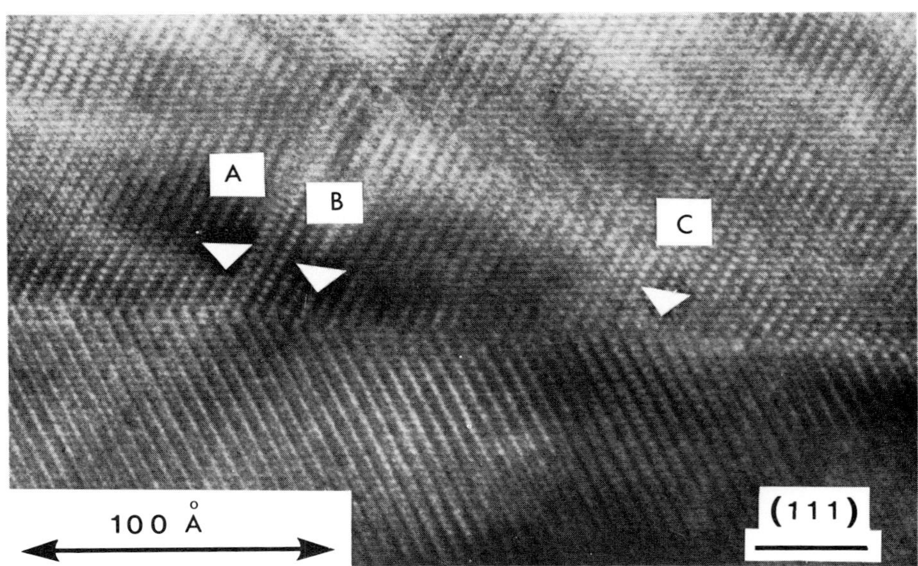

Fig.3   A two dimensional lattice fringe image of a twin interface showing steps at A, B and C. $\Delta F = -500\text{Å}$.

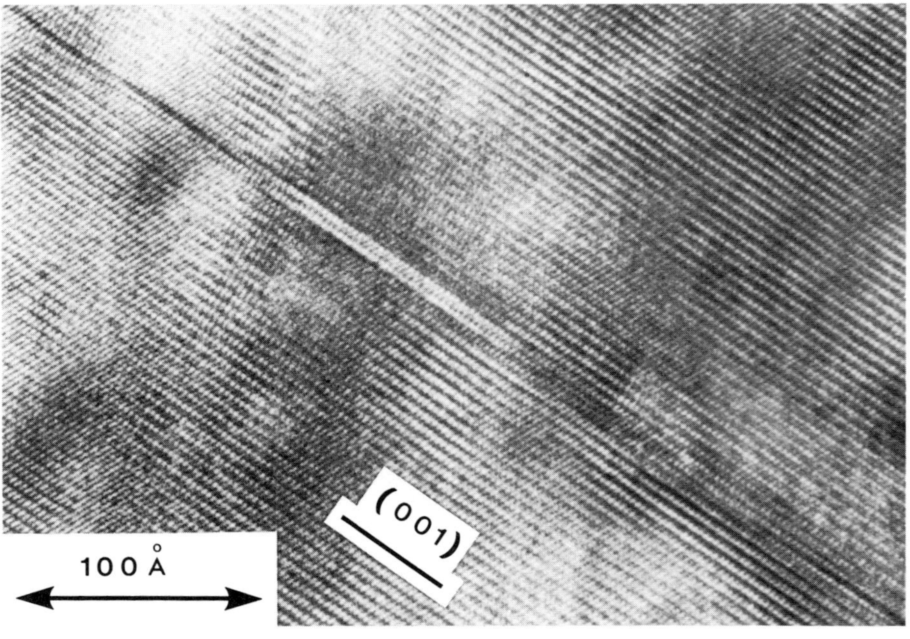

Fig.4   A two dimensional lattice fringe image of an (001) stacking fault. $\Delta F = -500\text{Å}$.

HREM of boundaries in a martensitic CuZnAl alloy

J M Cook, David J Smith* and W M Stobbs
Department of Metallurgy and Materials Science, University of Cambridge.
* High Resolution Electron Microscope, University of Cambridge.

The major deformation mechanism in martensitic beta-brass alloys is the motion of boundaries between different orientations, or variants, of martensite (Schroeder and Wayman, 1977). In order to understand this process, which is associated with the shape memory effect, it is important to obtain information on the structures of the boundaries by a variety of techniques. Two techniques considered here are firstly, the use of elastic side-band (ESB) imaging to observe periodic structures in the boundaries and secondly, direct HREM imaging.

ESB images have been obtained for the two important boundary types in a CuZnAl alloy, and are shown in Figs. 1 and 2. This alloy has a monoclinic structure derived from ordered f.c.c., with a stacking fault every third $(111)_{fcc}$ plane (i.e. the stacking order is ABCBCACAB). Both the boundaries are twin boundaries, one lying parallel to the $(1\bar{1}4)$ plane of the martensite and highly planar (Fig.1), the other being non-planar and separating two variants twinned by reflection in (2 0 10) (Fig.2) (Cook, 1980). Intensity scattered by the boundaries has been found at the positions and orientations indicated in the diffraction patterns, along lines perpendicular to the twinning plane normals. It is not yet possible to say whether these positions correspond to real periodicities in the boundaries, or whether the intensity there arises by double diffraction from other areas of the reciprocal lattice. Further experiments are required, with careful tilting of thin specimens to satisfy the Bragg condition for each aperture position used.

High resolution images were obtained using the Cambridge University HREM. The (2 0 10) non-planar boundary is shown in Fig.3. The boundary becomes less planar at the very edge of the foil, and its structure there is probably not characteristic of the bulk alloy. Thus, although the image at the edge is relatively simply interpretable, it will be necessary to analyse images from thicker regions, which are definitely not directly related to the structure, by simulation of images from structural models. A structurally interpretable image of a thin region away from a boundary is shown in Fig.4; the stacking sequence can be read directly from images of this type.

The major difficulties in obtaining the high resolution images are:
a) the tendency of correctly oriented regions of the foil to reorient spontaneously at the foil edge; b) the presence of internal stresses associated with the martensitic transformation which cause considerable buckling of the foil, and c) in particular, the need for very exact specimen and optic axis orientation. The lines of closely-spaced reflections through the (000) beam in Figs. 1 and 2 are forbidden when the

beam direction lies exactly in the twin plane. They appear, however, at small misorientations and give rise to strong 0.6nm fringes (corresponding to the spacing of the periodic stacking faults) which obscure features at higher resolution.

Future work will include examination of the effects of non-periodic faults on the two kinds of image, and attempts to correlate changes in the plane of the (2 0 10) boundary with changes in the boundary structure.

The authors are grateful to SERC for financial support, and to Delta Memory Metals Ltd. Ipswich, for financial support and provision of specimens.

References

Cook J M Ph D Thesis, Cambridge (1980).
Schroeder T A and Wayman C M Acta Met 25 1375 (1977).

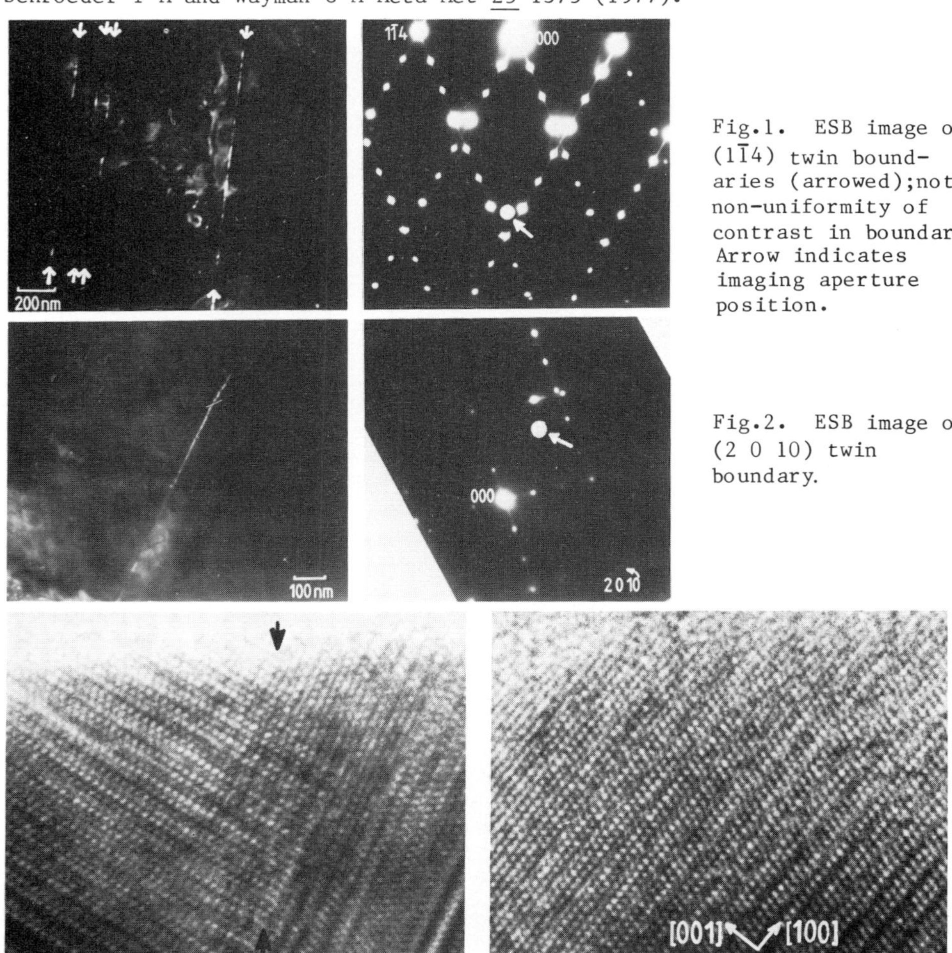

Fig.1. ESB image of (1$\bar{1}$4) twin boundaries (arrowed);note non-uniformity of contrast in boundary. Arrow indicates imaging aperture position.

Fig.2. ESB image of (2 0 10) twin boundary.

Fig.3. High resolution image of (2 0 10) boundary at foil edge; beam direction [010]. Arrows indicate line of boundary.

Fig.4. High resolution image showing stacking sequence; beam direction [010].

Electron microscopy of surface structure and reactions

A Howie

Cavendish Laboratory, Madingley Road, Cambridge CB3 0HE

1. Introduction

The study of structure and reactions on well characterised surfaces demands clean vacuum conditions as well as sample preparation and microanalytical facilities rarely available in electron microscopes and not easily compatible with high resolution performance. Nevertheless the faith of electron microscopists in the value of real space images has inspired a variety of compromise solutions to these difficulties and has in several cases been amply justified by the results obtained. We ignore the important contributions of field ion microscopy as well as SEM at modest resolution but which now may be carried out in UHV conditions and concentrate here on TEM. A useful and more complete account covering both EM approaches which now show some signs of convergence has been given by Venables (1981).

2. Improved vacuum modifications and environmental cells

Modified transmission instruments with improved vacuum specimen chambers or environmental cells have been constructed (Mills and Moodie 1968, Valdrè et al.1970, Baker and Harris 1972, Flower et al.1974, Heinemann et al.1975, Tatlock et al.1980) and applied to provide basic and striking information about epitaxy, particle growth and sintering, oxidation and catalysis. Fig.1 shows the misfit dislocation trigons nucleated by climb to relieve the tensile stress in (111) films of Pd grown epitaxially on Au. Similar defects may relieve the internal stresses in multiply twinned particles which have interesting facets and notches (Marks 1980) shown in fig.2 and whose (111) surfaces are under a comparable tensile stress. Fig.3 shows the process of Ostwald ripening for small Ag particles under 1 atmosphere O_2 at 220°C (the chamber being evacuated for each image exposure). Simple measurements of particle size distribution or of changes in mean particle size are much less reliable than detailed observations of many individual particles for identifying particle coarsening mechanisms in catalysts. Studies of this kind as well as other reactions relevant to catalysis (Gai and Goringe 1981) can be carried out very conveniently in high voltage microscope environmental cells which have been employed for a very wide range of in-situ work (Butler 1979).

3. Post-mortem techniques and transfer devices

A great deal of valuable information can be obtained in situations where the specimen can be transferred to an electron microscope with the surface reaction frozen at a particular stage. An outstanding recent example of this is the use of ion milling techniques to cut thin sections suitable

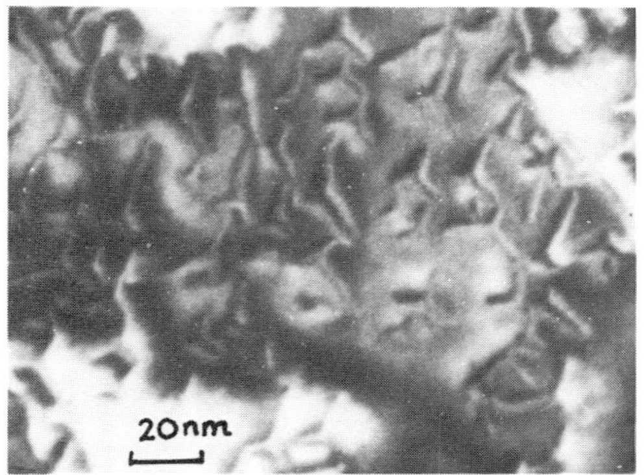

Fig.1 Misfit dislocation trigons in (111) epitaxial Pd (Cherns and Stowell 1975)

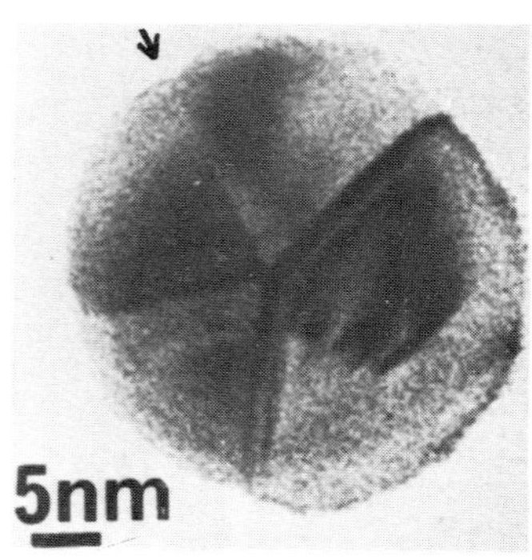

Fig.2 Decahedral Ag MTP with notch arrowed (Marks 1980)

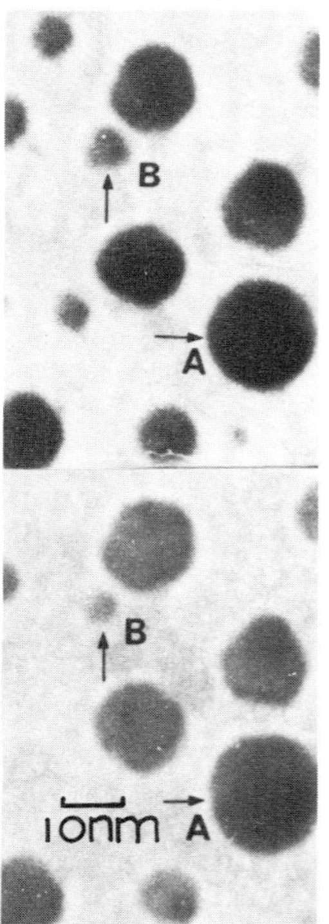

Fig.3 Ostwald ripening of Ag particles (T G Cunningham - to be published). Note growth A, shrinkage B and disappearance of particles after 10 hours (lower picture)

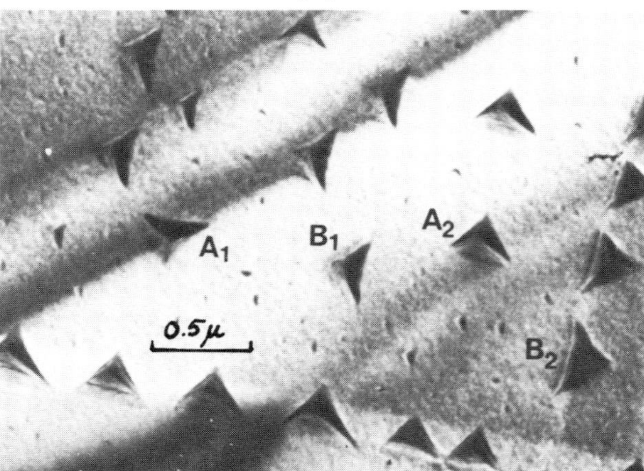

Fig.4 Different families of Cu_2O islands on (100) Cu (Levitt 1979)

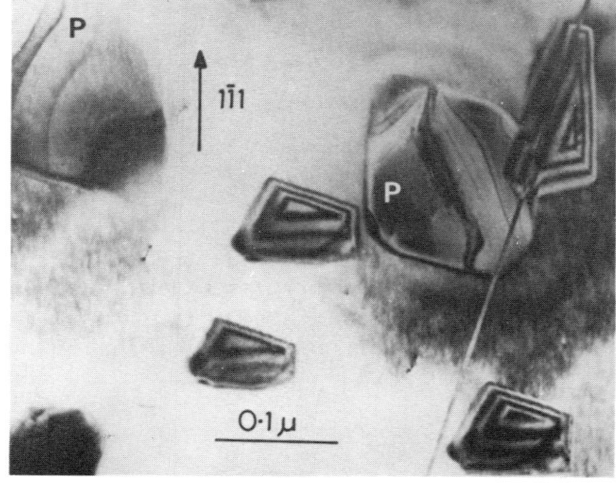

Fig.5 Cu_2O islands on (110) Cu with islands P on lower surface (Milne 1981)

for high resolution microscopy normal to epitaxial or oxide interfaces
(e.g. Krivanek et al. 1978).

An alternative technique employs a transfer device (Ambrose 1976) to convey
the specimen to a high resolution TEM from a UHV chamber fitted with all
the usual surface cleaning and analytical facilities. Genuine in-situ
studies are not possible with this system and return to the UHV chamber is
usually vitiated by exposure to the microscope vacuum. However, its usefulness in extending previous work (Goulden 1976) on the growth of oxide
nuclei has been demonstrated (Ambrose 1980, Milne and Howie 1980). Fig.4
shows the 4 families of Cu_2O islands growing on the (100) face of Cu and
the influence on the nucleation process of (111) planar defects, i.e.
large slip steps or twins (Levitt 1979). Similar defects in (111) Cu
specimens did not affect nucleation nor did those lying at 35° to the surface of (110) specimens; nucleation was affected in (110) by defects lying
on (111) planes normal to the surface. RHEED patterns indicated in some
cases a thin layer of CuO on top of the Cu_2O islands and this was subsequently made visible by weak beam microscopy (Levitt 1979). The importance of good specimen cleaning is particularly acute on the (110) Cu
surface. Fig.5 shows the sharp Cu_2O islands on the upper surface of the
sample which had been cleaned by several cycles of ion bombardment and
annealing (Milne 1981). The more rounded islands with a different epitaxy
are growing on the lower surface inaccessible to ion bombardment and are
similar to those observed in most previous studies. From observations on
a number of crystal faces (Milne 1981) it is clear that the oxide is predominantly related to the Cu by a 180° rotation about a common [111] axis
chosen closest to the surface normal and a plausible explanation of this
can be given at the atomic level (Howie and Milne, to be published).
Oxygen atoms occupying octahedral interstitial sites between (111) close
packed A and B Cu layers induce the B layer to diffuse away (thus providing most of the extra volume required for Cu_2O formation). The remaining expansion in and contraction normal to the (111) planes is predominantly a shear and leaves the O atoms correctly placed between A and C
layers of Cu to give a nucleus of Cu_2O in the observed twin orientation.

The usefulness of transfer devices seems likely to increase rather than
diminish in connection with more sophisticated microscopes like the STEM,
since it is difficult to tie up such expensive machines for several days
on a single experiment. It could also clearly be fruitful to devise systems to transfer catalyst and other samples direct from chemical reactors
to an environmental stage or pressure cell in the microscope.

4. Surface structure imaging techniques

Fig.6 shows an image of cleavage
step structures in graphite obtained in glancing angle scanning mode (Craven 1975) at
20 keV using a field emission
gun and secondary electron detector built into a standard
UHV chamber. The angle of incidence varies across the
field of view. Special techniques of this kind, including
also weak beam microscopy, are

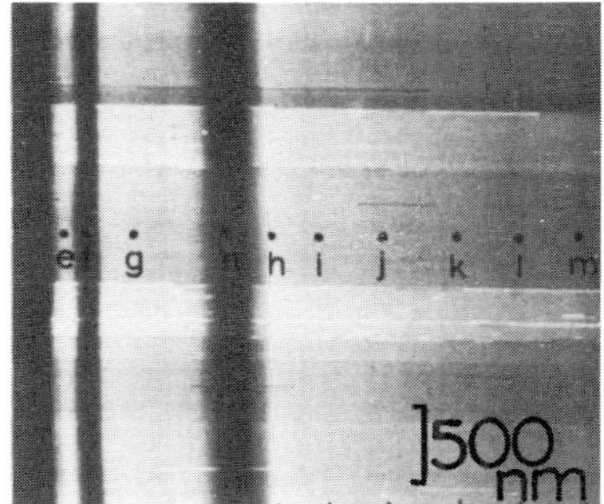

Fig.6 Surface structure on a
graphite crystal (Craven 1975)

essential in all but the thinnest transmission samples to enhance surface
relative to bulk scattering so that surface fine structure can be revealed
with adequate contrast. Despite the considerable foreshortening effect in
the images, the glancing angle scanning mode (using both secondary electrons and diffracted primaries) has been successfully employed by Højlund
Nielsen and Cowley (1976). Spectacular images, recently obtained by
Osakabe et al.(1980) using a UHV conventional microscope at glancing
angles, show atomic steps on the Si (111) surface and their influence on
the nucleation of the 7 x 7 surface reconstruction (see fig.7). The same
machine, operating at higher resolution in transmission mode (Yagi et al.
1979), was able to reveal details of surface reconstruction in extremely
thin (111) Au crystals. By comparison, the weak beam, dark field method
of Cherns (1974) which employed 'forbidden' reflections from partially
completed unit cells at the surface to reveal atomic steps on ∿ 10 nm
thick (111) Au crystals has developed rather slowly. Although overlapping
effects occur from both surfaces, these can usually be sorted out in
practice. Contamination effects in the conventional microscope are
probably a more serious difficulty judging from the lack of success so far
in extending this promising method to other materials and turning it into
a really useful surface structure technique.

Fig.7 Dark regions of 7 x 7
structure nucleating and propa-
gating from the outside edge of
surface steps on the (111) Si
surface (Osakabe et al.1980)

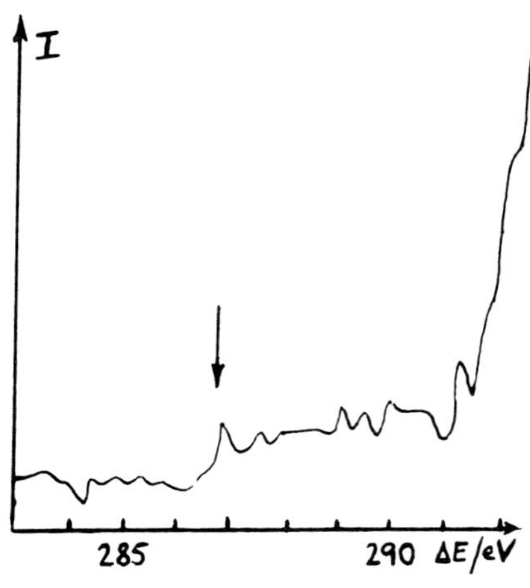

Fig.8 EELS surface edge structure (arrowed) below the K shell
edge in diamond (Pennycook 1981)

Comparative freedom from contamination is one of the advantages offered by
the field emission STEM where the vacuum conditions, even when the specimen is in the high resolution immersed position, are much better than in
most other microscopes. Electron energy loss spectroscopy (EELS) can also
be used, not only to provide filtered weak beam transmission images of
higher contrast or low loss images (Wells 1971) in the reflection mode
(Treacy et al.1981), but also to select surface-sensitive features in the
energy loss spectrum. Fig. 8 for example, taken near the K shell absorption edge of a 30 nm diamond crystal (Pennycook 1981), shows an additional
absorption edge whose height is independent of crystal thickness. It
could arise from two 0.25 nm surface layers of carbon or perhaps from a

modified surface structure including a bound surface state. Observations
of this kind could be greatly strengthened by the incorporation of an
Auger analyser in the system which would provide microanalytical and perhaps also binding information from the atoms at one of the film surfaces.
There is enough probe current for Auger spectroscopy, but it is not clear
at present whether the specimen would have to be placed outside the objective lens field or whether with some modified analyser, such as an axial
Wien filter, it might be possible to retain the immersed high resolution
position and use the field focusing action to increase the collection
angle for the Auger electrons.

The microdiffraction and energy loss facilities of the STEM, combined with
high resolution imaging, have been exploited by Cowley (1981) to study the
surfaces of MgO smoke particles and to display some of the complex elastic
and inelastic scattering processes involved in glancing angle conditions.
Similar observations by Marks (1981) of the low energy region of the loss
spectrum show that quite strong surface losses can be excited even when
the probe is at some distance from the smoke particle. These results are
in qualitative agreement with classical dielectric theory which also predicts that in glancing angle reflection conditions, even at 100 keV, most
electrons will suffer some inelastic scattering when outside the sample.
It is clear therefore that, although the glancing angle mode may prove to
be one of the most useful surface imaging techniques, any quantitative
theory of the contrast will be quite complex. The role of phase contrast
effects in revealing surface steps (fig.7) in the presence of strong
inelastic scattering is an interesting point but is already familiar in
transmission microscopy. The study of surface plasmons and related excitations could perhaps be a useful technique for surface characterisation
in the STEM but is unlikely to provide high spatial resolution unless
sufficient signal can be collected from electrons scattered at appreciable
angles where the interaction should be more localised.

Conclusions

Recent instrumental developments have opened up a number of opportunities
for high resolution imaging and local characterisation of clean surfaces.
A great deal of valuable work, particularly on well chosen surface
reactions, remains possible however on existing equipment. Transfer
techniques or other post mortem methods are still likely to be extremely
useful in certain cases.

Acknowledgments

I am grateful to the various colleagues mentioned in the text who have
kindly allowed me to use their material for illustrations. The work in
the Cavendish Laboratory, on which a good deal of this article is based,
was supported by the S.E.R.C.

References

Ambrose B K   1976   J. Phys. E $\underline{9}$ 382
Ambrose B K   1980   Electron Microscopy and Analysis 1979   ed. T Mulvey
    (London: Inst. of Phys.) pp 43-46
Baker R T K and Harris P S   1972   J. Phys. E $\underline{5}$ 793
Butler E P   1979   Re. Prog. Phys. $\underline{42}$ 833
Cherns D   1974   Phil. Mag. $\underline{30}$ 549

Cherns D and Stowell M J   1975   Thin Solid Films <u>29</u> 127
Cowley J M   1981   Ultramicroscopy (in press)
Craven A J   1975   Ph.D. Thesis, University of Cambridge
Flower H M, Tighe N J and Swann P R   1974   High Voltage Electron
  Microscopy, eds. Swann et al. (London: Academic Press) pp 383-395
Gai P L and Goringe M J   1981   Proc. EMSA Meeting (Baton-Rouge: Claitors)
Goulden D A   1976   Phil. Mag. <u>33</u> 393
Heinemann K, Bhogeswara Rao D and Douglas D L   1975   Oxid. Met. <u>9</u> 379
Højlund Nielsen P E and Cowley J M   1976   Surf. Sci. <u>54</u> 340
Levitt J P F   1979   Ph.D. thesis, University of Cambridge
Krivanek O L, Sheng T T and Tsui D C   1978   Appl. Phys. Lett. <u>32</u> 437
Marks L D   1980   Ph.D. Thesis, University of Cambridge
Marks L D   1981   paper in this volume
Mills J C and Moodie A F   1968   Rev. Sci. Instr. <u>39</u> 962
Milne R H and Howie A   1980   Proc. Eur. Conf. on Electron Microsc. <u>1</u> 262
Milne R H   1981 Ph.D. Thesis, University of Cambridge
Osakabe N, Yagi K and Honjo G   1980   Jap. J. Appl. Phys. <u>19</u>, L309
Pennycook S J   1981   Ultramicroscopy (in press)
Tatlock G J, Spain J, Raynerd G, Sinnock A C and Venables J A   1980
  Electron Microscopy and Analysis 1979, ed. T Mulvey, pp 39-42
Treacy M M J, Krakow W, Smith D A and Trafas G   1981   Appl. Phys. Lett.
  <u>39</u> 341
Valdrè U, Robinson E A, Pashley D W, Stowell M J and Law T J   1970
  J. Phys. E <u>3</u> 501
Venables J A 1981   Ultramicroscopy (in press)
Wells O C   1971   Appl. Phys. Lett. <u>19</u> 232
Yagi K, Takayanagi K, Kobayashi K, Osakabe N, Tanishiro Y and Honjo G
  1979   Surf. Sci. <u>86</u> 174

New developments in electron spectroscopy for bulk and surface microscopy and microanalysis

J. CAZAUX, D. GRAMARI, D. MOUZE, J. PERRIN, X. THOMAS
Laboratoire de Spectroscopie des Electrons, UER Sciences 51062 REIMS France

1. Introduction

We present here some recent results obtained with a new multi-purpose apparatus (Cazaux et al 1979) which allows us to apply the following micro-analysis techniques : surface microanalysis (information depth $\simeq$ 1 nm) by X ray photoelectron microprobe analysis (XPMA = ESCA microprobe) and Auger electron spectroscopy (electron induced : e$^-$AES and X ray induced : XAES) ; bulk microanalysis by electron microprobe analysis (EMA : inf. depth $\simeq 1\mu$m) and X ray fluorescence spectroscopy (XFS : inf. depth $\simeq 3$ μm up to 300 μm). For each technique, the processing facilities are the following ones : N(E) spectra obtained in the counting mode, first and second derivatives obtained by the use cf a lock-in amplifier for point analysis. Line analysis and two dimensional images can also be obtained in the scanning mode (cf. Fig. 6) and the last case includes Scanning X ray photoelectron microscopy (SXPM) and Scanning X ray radiography (SXR).

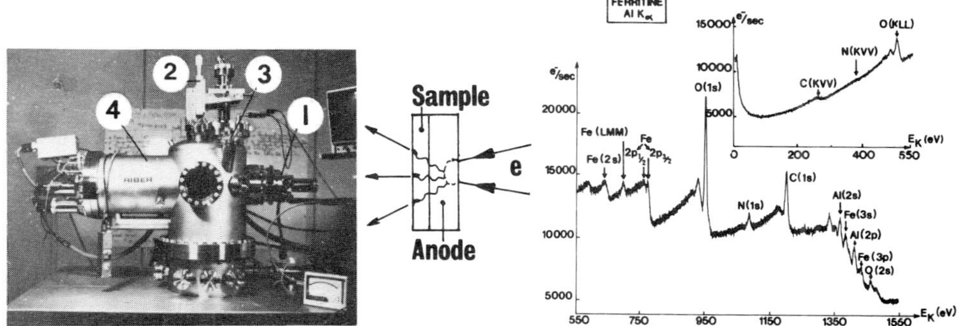

Fig. 1 View of the apparatus ①: scannable electron gun, ②: sample holder, ③: ion gun (for etching),④: CMA (and inside it a scannable Auger e$^-$gun).

Fig. 2 Schematic drawing of the experimental arrangement for X ray photoelectron microprobe analysis (XPMA) and scanning X ray photoelectron microscopy (SXPM).

Fig. 3 XPM analysis of a ferritin sample. Aluminium anode 5µm thick, V_O = 10 keV, $I_O \simeq$ 4µA (e$^-$ gun ① voltage and intensity). Spectrum obtained in 40 minuts ; analyzed area diameter $\phi \simeq$ 20 um.

2. EXPERIMENTS

The apparatus, shown on the Fig. 1, is composed of two opposite scannable electron guns, one of them being inside a CMA (for eAES only), and a composite sample is placed between these guns. In XPMA the sample in a thin film form is attached on the back of an aluminium anode (Fig. 2) in a form of a sheet. Rather thick sample like an alga 15 µm thick has been analyzed and Fig. 3 presents point analysis of another biological object (ferritin sample) ; SXPM images have be shown elsewhere (Cazaux, Mouze, Perrin 1980). Successive e^-AES and XPMA of the same sample with **approx.**the same spatial resolution (about 15-20 µm at the present state) allows us to compare the sensitivity of each technique (Fig. 4). By the measurement of intensities ratio related with a given Auger line - i.e MgKLL - obtained in XAES and e^-AES, it is possible to evaluate the X ray flux Φ_O which illuminates the sample :

$$I(e^-AES)/I(XAES) = \left| I_p(1+r)\ \sigma^e_{Mg} \right| \left| \Phi_O(1+c)\ \sigma^x_{Mg}(h\nu) \right|^{-1} \quad (1)$$

and we obtain $\Phi_O \simeq 3.10^9$ phX sec^{-1} by using the compilated cross-sections for electrons (Gryzinski 1965) and X rays (Scofield 1976) and by neglecting the back scattering coefficient r (in e^-AES) and bremsthralung effect c (in XAES). For bulk microanalyses, X ray analysis by electron spectroscopy is used : knowing the binding energy E_B of surface atoms A (converter) the photon energy $h\nu$ is deduced from the measurement of photoelectron kinetic energy E_K : $h\nu = E^A_B + E^A_K$ (2).
The corresponding intensity for a photoline is given by :

$$I(p.e) \propto \Phi_O.\beta^A(h\nu).T \quad (3)$$

where β^A is the quantum yield for photoelectron production - $\beta^A = N^A.\sigma^x_A(h\nu).\lambda^A_K$ -and T the C.M.A. transparency. An alloyed Al/Mg anode - Fig. 5 - has been analyzed in this way (Cazaux, Mouze, Perrin, Thomas, 1981 a). By setting a sample in sandwich between an anode and a converter it was possible, to obtain for the first time SXR images (Fig. 6) without the help of the synchrotron radiation (Ash 1980) but simply by means of a classical X ray source ; in this case the emerging X ray flux Φ_O is related to the X ray flux created in the anode material Φ_i by the relation $\Phi_O = (1/2).\Phi_i.E(\mu h)$ where $E(\mu h)$ represents the X ray absorption through the anode, the converter and the sample (Cazaux, Mouze, Perrin 1981**b**). At last thin film analysis by XFS (with corresponding X rays detected by electron spectroscopy) was also applied to detect the presence of a 3 µm thick Mg foil set under an Al sheet 5 µm thick (Fig. 7 and Cazaux, Mouze, Perrin, Thomas 1981**c**).

3. PERFORMANCES AND FURTHER DEVELOPMENTS

The apparatus used is a modified classical Auger spectrometer which allows to apply all the above described techniques. The corresponding **features** of our arrangement are the following ones (compared with the conventional apparatus) : i)it's possible in XPS to reduce the analyzed area down to the 100 μm^2 range (instead of the sq. mm range) and so 10^8 (10^{-15} g) surface atoms have been detected by this technique and the corresponding scanning images obtained.

ii) X ray analysis by electron spectroscopy is very interesting for light element X ray detection because the energy resolution $\Delta E \simeq 1eV$ corresponds to the convolution of the natural width of a photoline and the CMA energy resolution ; this technique can also be applied for heavier element's detection (by choosing a converter having a binding energy E_B which confers an easily measurable kinetic energy on the photoelectrons ($50 < E_K < 2000$ eV).

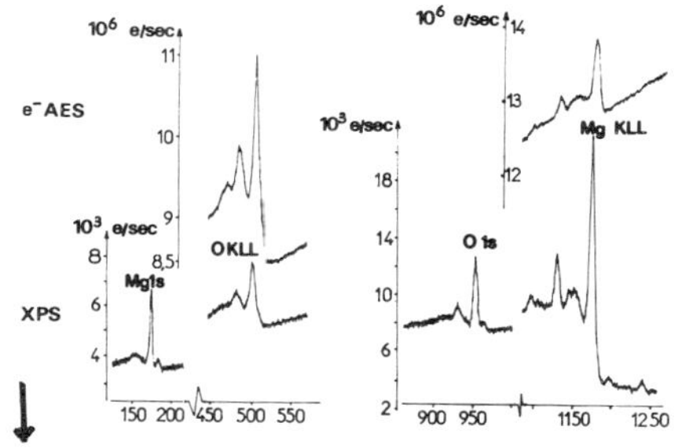

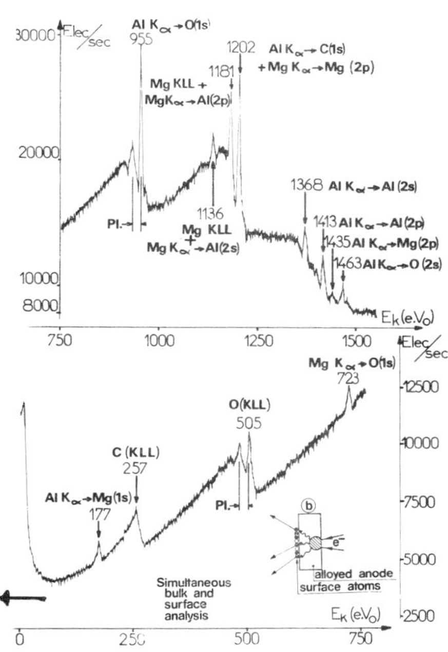

Fig. 4 Direct comparison (for the same sample and **almost** the same analysed area) between eAES(top curves) and XPMA (with the corresponding X ray induced AES lines - bottom curve). Note in each case the signal and background intensities. MgO thin film on an Al anode 5μm thick : Auger gun 5keV,0.3μA XPMA gun : 10 keV, 6μA.

Fig. 5 Bulk analysis (X ray microprobe) of an Al/Mg alloyed anode 5μm thick ; X rays are analyzed by X.P.S. and the surface atoms in front of the CMA act as converter**s**.

Fig. 6 Scanning X ray radiography :
a) Schematic drawing of the experimen-

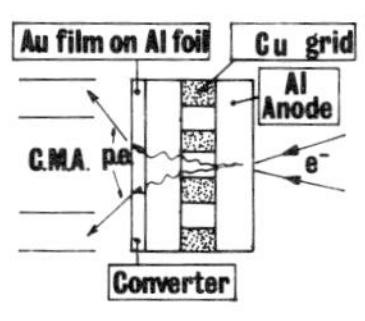

tal arrangement :
a copper grid
(200 meshes) in sand-
wich between 2 Al
foils,the thin gold
film in front of the
CMA is the converter
b) X ray radiographic negative image
c) line analysis (using Au N6,7 photoline excited by AlKα radiation.
d) corresponding point analysis in A
and B.

Fig. 7 X ray fluorescence microanalysis of thin films

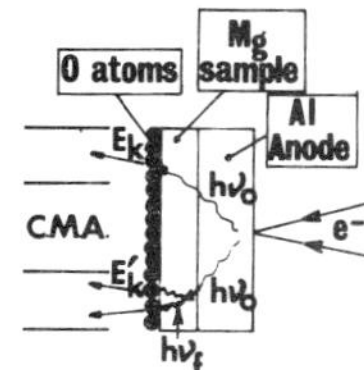

a) Schematic arran-
gement : primary X
radiation hν_O crea-
ted in the Al anode
(5μm thick) excites
the fluorescence ra-
diation hν_F of a Mg
foil (3μm thick).
Oxygen surface atoms play the converter role.
b) Corresponding photoelectron
spectrum.

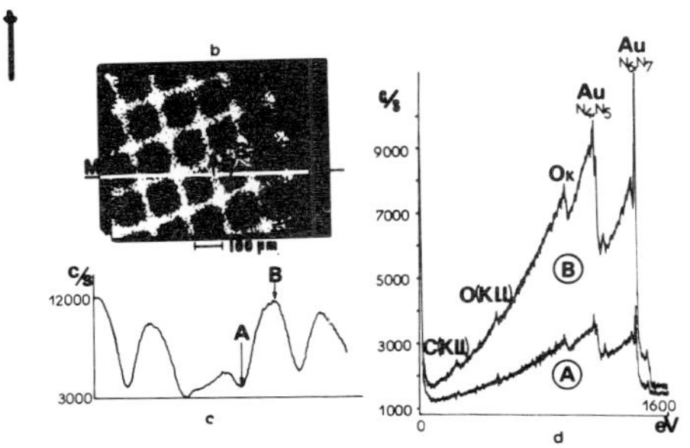

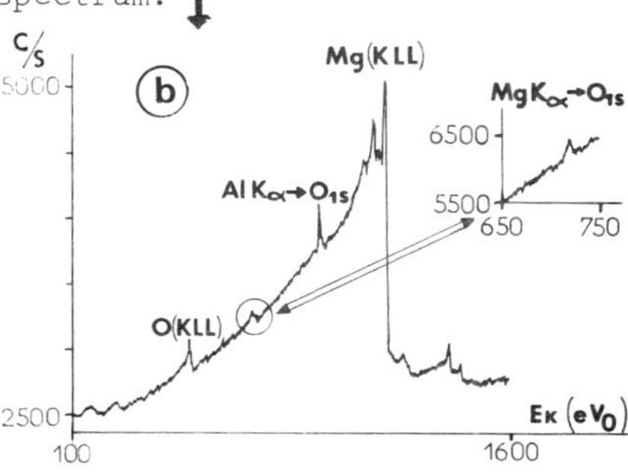

iii) Scanning X ray radiography can be performed in such an apparatus as well as in a simply modified scanning electron microscope with spatial resolution of about 10 microns.

iiii) The analyzed volume in XFS is several orders of range smaller than the volume analyzed in conventional XFS apparatus.

These performances are due to the high density X ray photon flux ($\simeq 10^7$ phX sec^{-1} μm^{-2}) obtained on the sample in our experimental arrangement, it is two orders of range higher than those delivered by synchrotron radiation. In each case the spatial resolution depends only on the size of the X ray source and the distance between this source and the sample (consequently in EMA, SXPM, SXR it is independent of the distance between the sample and the converter. In the future, improvements of performances will be related with a reduction of the X ray source size down to the micron range (actually 10 μm range due to the size of our incident electron probe). In this way we can expect a gain of two orders of range for the analyzed area in XPMA and the corresponding improvements in XPMA sensitivity, spatial resolution in SXR and analyzed volume reduction in XFS.

There is no doubt that such performances open new possibilities in thin film analysis.

Applications in various domains : physics, metallurgy, geology, biology can be easily established (for example : scanning X ray radiography of integrated circuits in semiconductor technology or XFS analysis of very small quantities of biological objects).

This work was supported by D.G.R.S.T. (contrat IM 80/563) and Ministère de l'Industrie (N° 29 3656/1979).

E.A. ASH 1980 Scanned Image Microscopy Academic Press.

J.   CAZAUX, P. COLLARD, D. GRAMARI, D. MOUZE, J. PERRIN 1979 J. Microsc. Spectrosc. Electron. 4 319.

J.   CAZAUX, D. MOUZE, J. PERRIN 1980 Proceed. 8th Int. Vac. Cong. Cannes Le Vide les couches minces 201 Sup. I 291.

J.   CAZAUX, D. MOUZE, J. PERRIN, X. THOMAS 1981a Appl. Phys. Lett. 38 1021.

J.   CAZAUX, D. MOUZE, J. PERRIN 1981b Appl. Phys. Lett. (submitted to)

J.   CAZAUX, D. MOUZE, J. PERRIN, X. THOMAS 1981c Journal of Physics D (submitted to).

M.   GRYZINSKI 1965 Phys. Rev. 1 138 336.

J.M. SCOFIELD 1971 J. Electron Spectros . Relat. Phenom. 8 129.

Low-loss surface imaging of thin films in the STEM

M.M.J. Treacy , D.A. Smith, W. Krakow and G. Trafas

IBM, T.J. Watson Research Center, Yorktown Hts, New York 10598

Low-loss surface images are usually formed in the SEM by collecting with
an energy filter those backscattered electrons which have lost little or
no energy. Such electrons have a shallow mean penetration depth, and when
collected at low emergence angles from a sample inclined to the beam,
strong topographic contrast can result (Wells 1979). This technique has
been used to study the surface structure of defects in thick samples (Morin
et al. 1979) and in thin films on thick supports (Krakow and Broers 1980).
In each case the electron collection geometry did not allow transmission
images to be obtained. A variant of this technique has been used in a VG
HB5 STEM, equipped with a high excitation objective lens and a tilting
stage, which allows low-loss and transmission images to be obtained from
the same region of a thin film (Treacy et al 1981).

The thin film sample is mounted on a tongued grid as shown in fig.1. The
grid can be made from standard 3mm EM grids which have been trimmed and
bent to form an upright tongue with a semi-circular base. The base is made
opaque to electrons by covering with a thick (> 100μm) Cu or Al foil. The
specimens are placed on the outer facing surface of the grid. To obtain
low-loss images the sample is tilted about the XY axis (fig 1) so that the
transmitted electron beam is blocked by the opaque semi-circular base (ray
1 in fig 2a).Electrons backscattered through less than some critical angle
θ_c with respect to the optic axis are then bent by the specimen post-field
of the high excitation lens onto the annular and bright field detectors
(ray 2 in fig 2a)(Wells 1973). Electrons scattered beyond θ_c either strike
the specimen cartridge or are sent back up the column (rays 3 and 4). It
has been shown by Broers and Sokolowski (1975) that for specimens tilted
close to $\hat{\theta}_c$ so that only shallow take-off angles are collected, there is
no observable difference between the filtered and unfiltered images. This
is because the large energy-loss electrons emerge predominantly at large
take-off angles (Wells 1979). Furthermore, in thin films the emerging
electrons will not have explored sufficient specimen volume to lose significant energy. Thus the unfiltered
annular detector signal can be used as the
low-loss signal.

For transmission images the specimen is
tilted back through the vertical (fig 2b)
thereby allowing the transmitted electrons
to form transmission images in the usual way.

Figure 3 shows an example of comparison low-
loss and bright field images taken of a 20nm

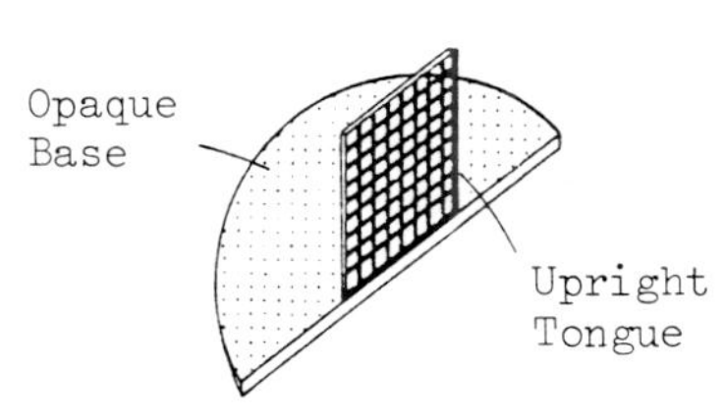

FIG. 1   Tongued grid

Present address: CNET, 196 Rue de Paris, 92220 Bagneux, France

thick (100) oriented gold film grown epitaxially on (100) rock salt, The surface topography visible in the low-loss image (fig 3a) can be unambig- uously identified with the microtwins visible in the bright field image (fig 3b). For comparison purposes figure 4 shows the STEM secondary electron image of a similar region. The surface structure is visible but shows poorer contrast and resolution.

The authors are grateful to J. Lynch of IFP France for supplying figure 4.

Broers A N and Sokolowski J, Proc of 33rd Ann, EMSA Meeting, 148, (1975)
Krakow W and Broers A N, Proc. of 38th Ann, EMSA Meeting, 182, (1980)
Morin P, Pitaval M, Vicario E and Fontaine G, Scanning, 2, 217, (1979)
Wells O C, Scanning, 2, 199, (1979)
Wells O C, Broers A N and Bremer C G, Appl. Phys. Lett., 23, 353, (1973)
Treacy M M J, Krakow W, Smith D A and Trafas G, Appl. Phys. Lett., 38, 341, (1980)

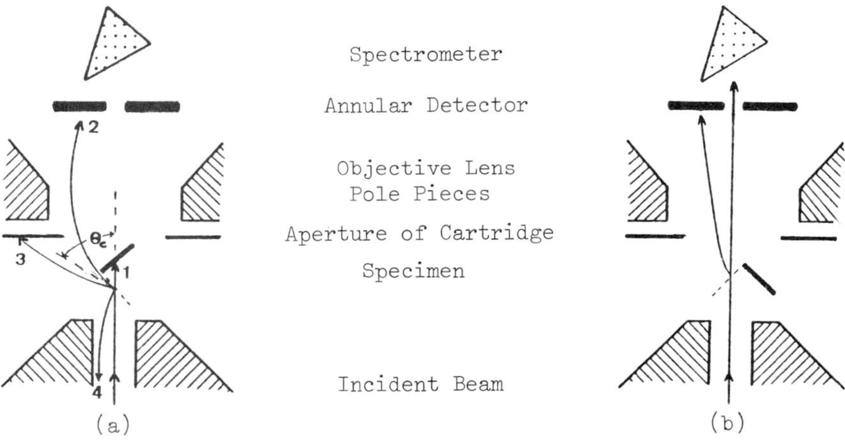

FIGURE 2: (a) Imaging with the low-loss electrons. The transmitted beam (1) is blocked by the opaque base of the tongued grid, whereas the shallow back- scattered low-loss electron trajectories (2) are bent by the objective lens specimen post-field onto the detectors, Backscattered trajectories (3) and (4) are lost to the detectors. (b) Imaging with transmitted electrons.

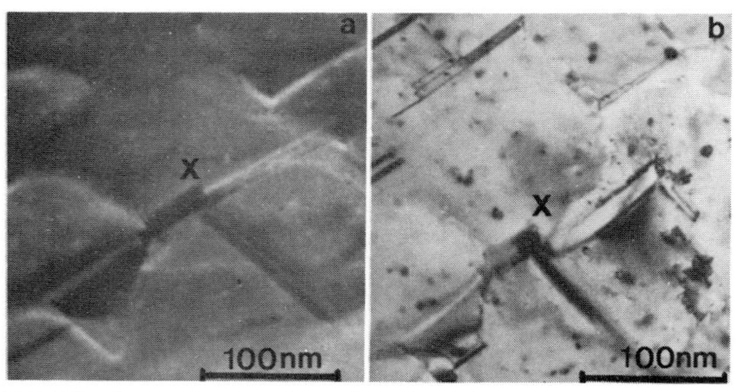

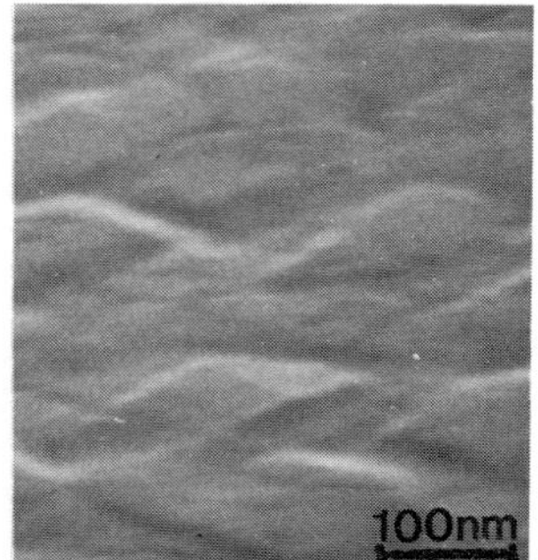

FIGURE 3:   20 nm thick (100) oriented Au film. (a) STEM low-loss image with inci- dent beam angle of 55°. (b) STEM bright field image. Specimen tilted at -65°.

FIGURE 4:
STEM Secondary electron image. Specimen tilt 30°

Dynamic studies of surface processes in the hot-stage UHV-SEM

A.M. Brown and M.P. Hill

Central Electricity Research Laboratories, Leatherhead, Surrey

1. Introduction

An ultra-high vacuum field emission source SEM equipped with a hot stage
and environmental gas cell was described at EMAG 79. The microscope can,
under normal operating conditions, resolve 50 Å for specimen temperatures
up to $800^{\circ}C$, but during observations of reactions in the gas cell, this is
reduced to 250 Å due to probe size and signal collection limitations
(Brown and Hill, 1979). Auger and EDX analysis can be used on heated
specimens when the gas cell is not in use. In this paper the experimental
problem of temperature measurement is discussed together with the pressure
limitations during in-situ observations of gas solid reactions. Three
examples of different types of work carried out with the instrument are
presented.

2. Experimental

Specimens are normally in the form of small coupons, typically
6 mm × 6 mm, mounted on electrically heated, 200 μm tungsten wires as
described previously (Brown and Hill, 1979). This method minimises heat
dissipation and therefore outgassing from surrounding structures. A
75 μm Pt − 13% Rh/Pt thermocouple, spot welded near a corner of the
specimen, provides a guide to specimen temperature but cooling at the point
attachment prevents accurate measurements.

Pyrometry is therefore used for temperature calibration. A conventional
disappearing filament micro-pyrometer (Pyrowerk), working at visible
wavelengths, is used for $T > 650^{\circ}C$ and an infra-red Pyrometry system
(Land) is mounted on another specimen chamber port equipped with a re-
entrant quartz window. It has a fixed focal distance of 12 cm. The field
of view at the specimen is either 4 mm or 1.5 mm diameter for the two
ranges in use, which are $300^{\circ}C$–$700^{\circ}C$ and $500^{\circ}C$–$1200^{\circ}C$ respectively.
Agreement between the two pyrometer systems is good.

The specimen chamber, field emission gun and column can be separately
pumped during gas cell use (Fig. 1). For typical gas/metal reactions in
methane, the maximum cell pressure is 0.2 Torr to maintain 5×10^{-11} Torr
in the field emission gun region. Pressures up to 50 Torr in the gas cell
have been used but above 2 Torr the electron gun must be turned off thus
losing the facility of continuous observation during a reaction. Between
0.2 Torr and 2 Torr there is a risk of damaging the emitter (depending on
the gas) and the image brightness is generally reduced.

3. High Temperature Behaviour of an Amorphous Cr-B Film

The stability of Cr-B films for possible use as protective coatings on

iron containing substrates, has been examined. A 500 Å thick film of
nominal composition Cr/25 At% B was prepared by co-evaporation onto
liquid N_2 cooled KCl crystals and floated off onto an iron foil substrate
for examination. Traces of KCl remained attached to the film after washing,
but were removed by heating to $500^{\circ}C$. The film was then heated in $50^{\circ}C$
steps each of 30 min. and the onset of crystallisation was detected at
$680^{\circ}C$ (Fig. 2) which compares with values of $410^{\circ}C$ and $800^{\circ}C$ for films
containing 8% and 35% B. Continued heating to $780^{\circ}C$ resulted in the
entire film transforming (Fig. 2b), yielding crystallites in a size range
up to 0.25 μm. Auger analysis of the 'as prepared' film, using a 300 Å
electron probe from the field emission source, showed evidence of boron
enrichment at the surface and carbon contamination (analysis: 28% Cr,
54% B, 18% C). After crystallisation and heating at $790^{\circ}C$ for 1 hour there
was a loss of Cr and C and a transfer of Fe, N and O to the film (analysis
49% B, 26% N, 9% O, 12% Fe). Further heating increased the iron content
to 23% with boron being reduced to 38%. The results show that although
the Cr-B film has a relatively high crystallisation temperature, it is not
stable on an iron substrate at temperatures where iron atom mobility is
possible.

4.   Segregation in a Nb-Stabilised 20% Cr/25% Ni Stainless Steel

The steel, which is used for high temperature service (300°-$850^{\circ}C$) in
nuclear reactors, contains the following minor constituents, 0.7 wt% Nb,
0.6 wt% Si, 0.6 wt% Mn 0.04-0.08 wt% C and N. Variations in the surface
composition and structure induced by heating in vacuum have been followed.
The specimen was initially ion etched to remove the air-formed oxide film
and other residual contaminants. A typical set of composition versus
temperature curves were obtained with heating steps of $50^{\circ}C$ every 15 mins.
up to $1020^{\circ}C$ (Fig. 3) using, for analysis, the Varian CMA integral gun
which excites areas of 500 μm^2. In the early stages, oxygen is seen to
concentrate at the surface causing chromium enrichment. Above $600^{\circ}C$
several minor constituents segregate to the surface and small particles of
a chromium rich carbide ($\sim$0.25 μm) develop. These redissolve at
$T > 800^{\circ}C$ and by recording the process on videotape, the dissolution
process (part shown in Fig. 4) can be measured to establish the kinetics.
Particle dissolution follows a $t^{\frac{1}{2}}$ rate law indicative of a rate controlled
by diffusion. Sulphur segregates above $600^{\circ}C$ and is stable on the
surface to $1015^{\circ}C$ suggesting that a sulphide phase is formed. However, no
evidence of sulphide crystallites could be detected and Auger analysis
showed that sulphur was distributed evenly over the surface.

5.   Mobility in Iron-Carbon Layers

The formation of carbon deposits by cooling-induced precipitation of
carbon dissolved in γ-Fe has been studied previously by in-situ SEM
(Brown and Hill, 1981a). The deposition of carbon from 50 Torr CH_4 onto
polycrystalline Fe at $750^{\circ}C$ has been followed in the gas cell using
imaging in the gas at 0.15 Torr pressure. Deposition of carbon is rapid
and a mobile scale containing both Fe and C develops. With continued
reaction the scale breaks up, (Fig. 5), to yield bright particulate
deposits. These particles also have sufficient mobility to coalesce
(Fig. 6), eventually forming extended networks of crystalline material
(Brown and Hill, 1981b). The deposits initially have a high (80%) carbon
content but with light ion etching this reduces to 50% which suggests the
presence of an iron carbide. Ion bombardment may partially dissociate
the compound and the true composition is therefore in doubt. If

precipitated carbon/carbide deposits (Brown and Hill, 1981a) are redissolved in γ-Fe by heating to 750°, mobility is again observed. For example the carbon covered mound in Fig. 7a exhibits a region at the mound/substrate interface where the mobility on dissolution gives the appearance of a molten zone (see Fig. 7b, taken from a video recording). The mobility occurs at ∿100°C below the Tamman temperature for iron. During the dissolution it is unlikely that long range crystallographic order exists and the Fe/C deposit may be behaving as an amorphous phase. The hot stage SEM has thus revealed new aspects of Fe/C interactions which would not have been observed in conventional experiments where specimens are cooled to ambient temperature before transfer to the SEM for examination.

6. <u>References</u>

Brown, A.M. and Hill, M.P., 1979, (EMAG 1979) Inst. Phys. Conf. Ser. <u>52</u>, 379-382

Brown, A.M. and Hill, M.P., 1981a, Carbon, <u>19</u>, 51-59

Brown, A.M. and Hill, M.P., 1981b, 15th American Carbon Conf., (p.400) Philadelphia

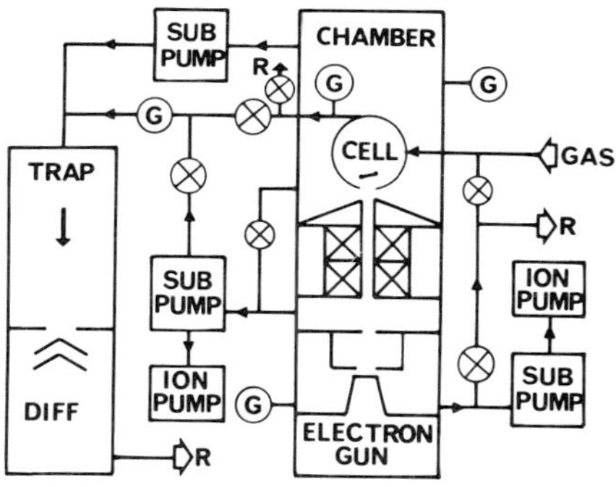

Fig. 1. Pumping arrangements for the gas cell SEM.

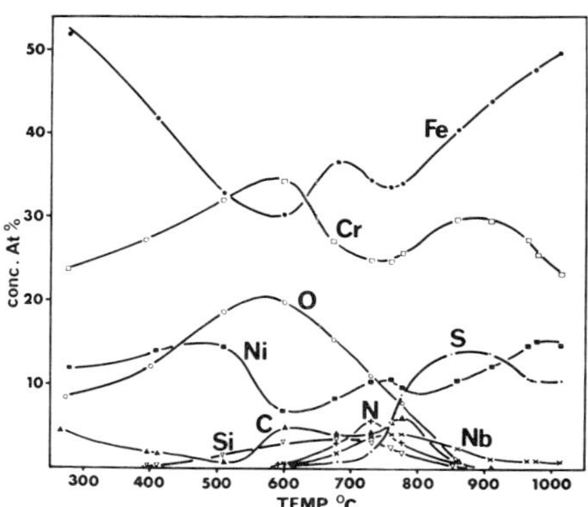

Fig. 3. Elemental segregation curves for a 20%Cr-25%Ni steel.

Fig. 2. Crystallisation of an amorphous Cr-B film.
(a) initiation at 680°C.
(b) development after 30 min at 780°C.

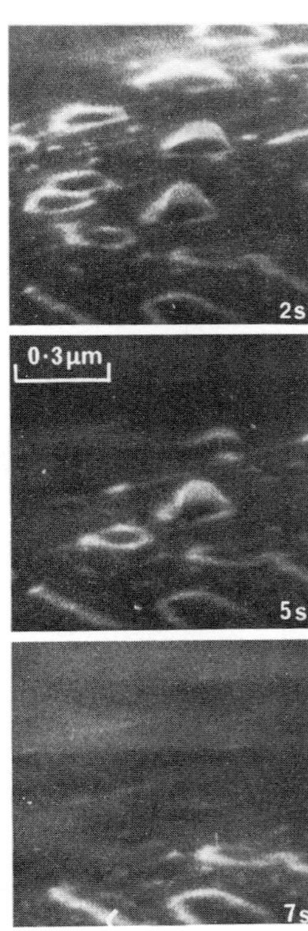

Fig. 4. Sequence showing carbide particle dissolution. (Taken from video recording)

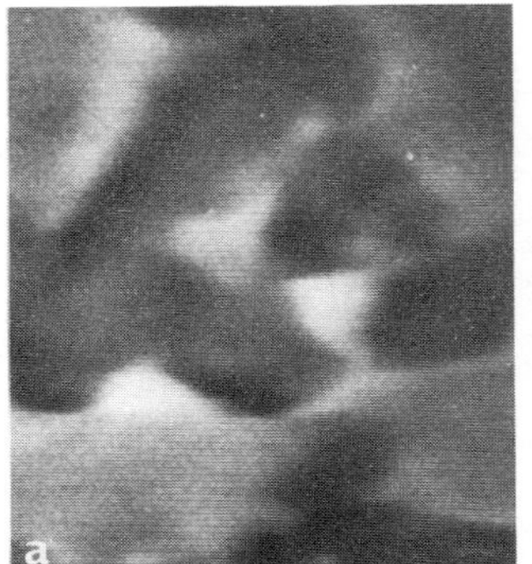
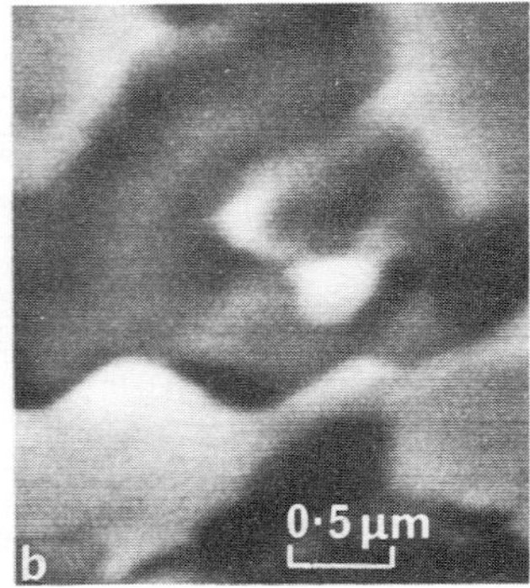

Fig. 5 Mobile Fe-C deposit showing separation into particles. (Imaged in 0.2 Torr CH_4 at 750°C)

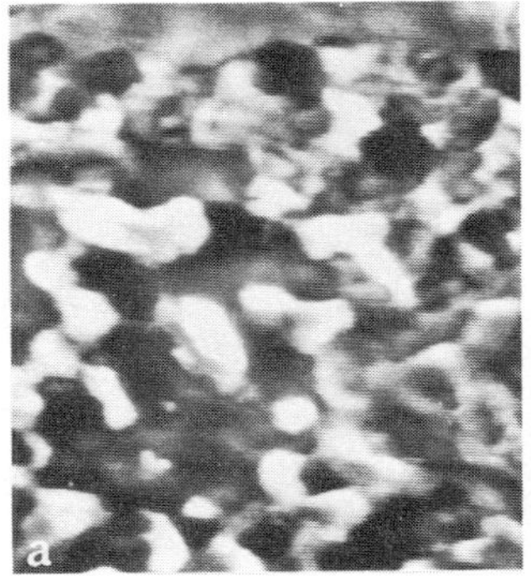
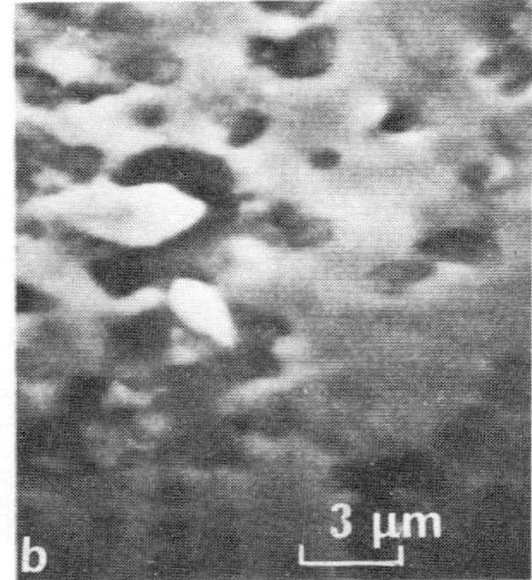

Fig. 6 Coalescence effects in a mobile Fe-C deposit. (Imaged in 0.2 Torr CH_4 at 750°C)

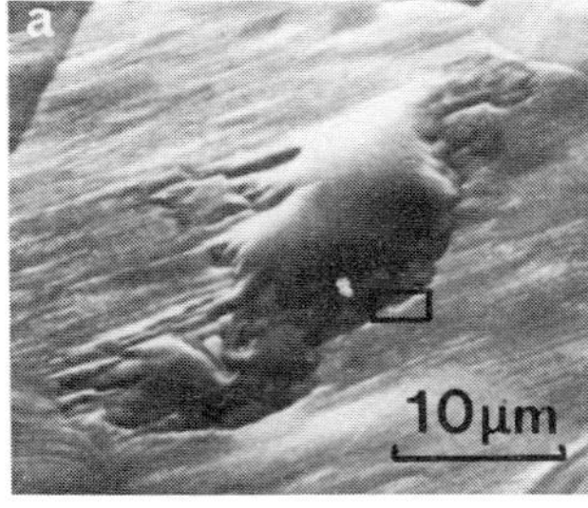
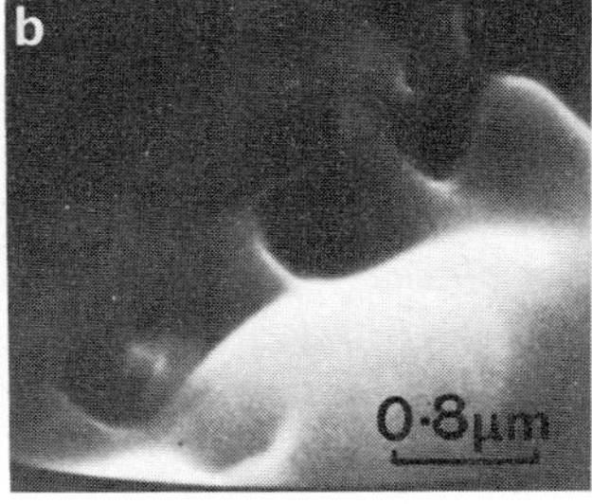
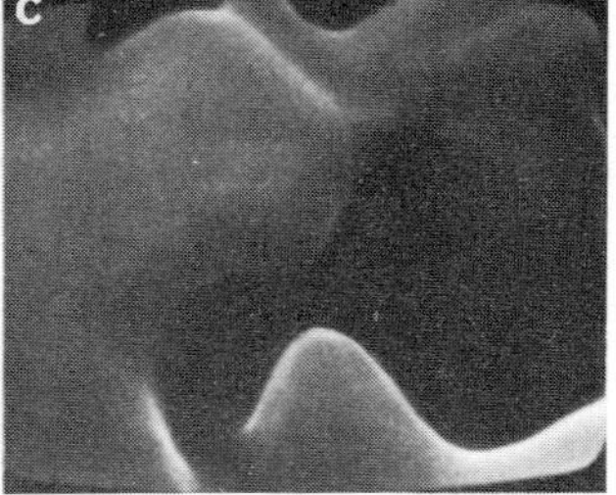

Fig. 7. Dissolution of a carbon mound in γFe at 750°C.
(a) mound, showing region of interest.
(b) onset of mobile activity (as seen on video recording)
(c) same field as (b) after 330 sec.

7.   Acknowledgements

The authors thank Dr P.J. Grundy of the Dept. of Physics, University of Salford for preparing the amorphous metal films. The work was carried out at the Central Electricity Research Laboratories and is published by permission of the Central Electricity Generating Board.

On the optimum incident beam voltage for Auger electron spectroscopy

H.E. Bishop

Materials Development Division, AERE Harwell, Didcot, Oxon, OX11 ORA

Abstract. The optimum conditions for AES are those that give the best signal to noise ratio subject to any constraints set by either the instrument or the specimen. In the direct counting mode of operation a good measure of the signal to noise is the $P/\sqrt{B}$ ratio. This ratio has been measured experimentally for the major peaks of Cu and Ta over a range of beam voltages and analyser resolutions. If spatial resolution is not important 10 keV represents a reasonable maximum working voltage. However the higher source brightness available favours the use of greater beam energies if high spatial resolution is desired.

Auger Electron Spectroscopy (AES) has evolved from being primarily a surface analytical tool into the microanalytical technique with arguably the highest potential spatial resolution on a solid sample, with the notable exception of the atom probe. (This transformation has been achieved by the move from adapted cathode ray tube guns to specially designed magnetic lens systems.) Auger microscopes with an electron imaging performance approaching that of a conventional scanning electron microscope are now commercially available. The beam voltage used in AES was initially limited to a few keV but, with improved design, beam voltages up to 10 keV have become common. Some of the new electromagnetic systems use beam energies up to 30 keV, or even to 60 keV in the case of the V.G. Scientific HB50. This wider choice of beam energy raises the question; what is the optimum beam voltage for AES?

The optimum conditions for any analytical technique are those that achieve the maximum signal to noise (S/N) ratio subject to any constraints set by either the instrument or the specimen. In AES one or more of the following factors may limit the sensitivity:

(1) the need to achieve a high spatial resolution,

(2) the maximum temperature rise the specimen can tolerate,

(3) the minimum electron dose that will induce significant changes in the specimen surface composition – either radiation damage, beam induced diffusion, adsorption or desorption,

(4) specimen charging.

In the latter case beam energy and angle of beam incidence have to be adjusted to try to eliminate charging.

Before considering these points it is necessary to know how S/N varies with beam energy for a fixed current. Although the ionization cross section is known sufficiently well to be able to calculate the variation of Auger yield with beam energy, there is not the corresponding information on the electron background variation to allow the S/N ratio to be calculated with any certainty. It was therefore decided to make an exploratory set of measurements in order to define the overall behaviour of the S/N ratio over the range of energies (2-30 keV) covered by our instrument, the V.G. Scientific MA 500. This instrument has a magnetic electron optical column with a hemispherical electron spectrometer. The angle between the electron optical axis of the microscope column and analyser input lens is 60°. Data for the lower beam currents used in high resolution work are recorded in the counting mode where it is convenient to take the standard deviation for the noise, i.e. the square root of the counts recorded, rather than the root mean square value conventionally quoted, since the difference between the two is negligible. (Incidentally quoting absolute counting rates obtained in this mode of operation should facilitate the fair comparison of different instruments.)

Two elements, Cu and Ta, were chosen for this study as between them they have peaks covering the main regions of the Auger spectrum. The specimens were cleaned by ion bombardment and tilted 45° to the incident beam so that the analyser axis was at 15° to the specimen normal. A fixed beam current of 10 nA (measured in a Faraday cup) was used in all cases and the beam was rastered over a 200 um area during measurements to average out any microtopographical effects. In most cases the peak height was taken as the difference between the counts at the peak and those on the high energy side of the peak, the parameter used in practice for Auger imaging. However for low beam energies and for the 60 eV copper peak, where there was a large background slope, the value for the background was interpolated.

The results of the measurements are set out in Tables 1 and 2. Table 1 demonstrates the effect of varying beam energy using a fixed analyser retard ratio corresponding to a nominal resolution of 0.5% similar to that used in most conventional AES work. Table 2 shows the effect of varying the spectrometer resolution. The results are all scaled to 1 sec counting time and 10 nA beam currents. Values for different counting times or beam currents may be readily calculated from these data. As expected from the ionization cross sections, the peak heights pass through a maximum and then decrease with increasing beam energy whilst the backgrounds decrease uniformly with energy. The peak to background ratios however increase with beam energy, eventually reaching a saturation value. The S/N ratio as measured by $P/\sqrt{B}$ ratio passes through a broad maximum. Varying the analyser resolution show that, for the particular peaks measured, the peak to background ratio is degraded by reducing the resolution but that this is more than offset by the increase in current when determining the signal to noise ratio. Thus for the highest sensitivity the lowest available resolution compatible with discrimination between peaks should be employed.

If none of the special factors outlined above applies, a beam voltage giving the best compromise S/N ratio for all energies should be chosen. An examination of Table 1 shows that a maximum energy of 10 keV is adequate under this circumstance. If however high spatial resolution is required the optimum S/N that can be obtained from a given area rather than from a probe current must be calculated. The maximum current that can be delivered into a given area by an electron optical system is determined by

Table 1

Beam Energy keV	Cu 57 eV	Cu 912 eV	Ta 164 eV	Ta 1670 eV
Peak heights (c/s)				
2	7500	3150	7500	–
5	5050	7500	5400	1950
10	3000	6100	3950	5150
20	1600	4150	2400	3800
30	1300	3350	1350	3500
Corresponding background (c/s)				
2	60000	93000	46500	–
5	31500	24000	29000	60000
10	18000	10500	18500	23500
20	10200	5200	10500	11300
30	7700	3550	8000	8000
Peak/background				
2	.125	.03	.16	–
5	.16	.32	.19	.03
10	.17	.58	.21	.22
20	.16	.80	.23	.34
30	.16	.93	.23	.44
Peak/(background)$^{\frac{1}{2}}$				
2	31	10	35	–
5	28	49	32	7.9
10	22	59	29	33
20	16	57	24	36
30	15	56	21	39

Table 2

Retard ratio	Nominal Resolution %	Cu 912 eV (10 keV)	Ta 164 eV
Peak heights (c/s)			
1	2.0	23000	24000
2	1.0	14500	10500
4	0.5	6100	3950
10	0.2	1750	850
20	0.1	650	–
Corresponding background (c/s)			
1	2.0	67000	130000
2	1.0	31000	57000
4	0.5	10700	18500
10	0.2	2150	3500
20	0.1	630	–
Peak/background			
1	2.0	.34	.18
2	1.0	.47	.18
4	0.5	.60	.21
10	0.2	.82	.24
20	0.1	1.02	–
Peak/(background)$^{\frac{1}{2}}$			
1	2.0	89	66
2	1.0	82	44
4	0.5	59	29
10	0.2	38	14
20	0.1	26	–

the brightness of the source which is in turn proportional to the beam voltage. To correspond to fixed probe size conditions the $P/\sqrt{B}$ figures should therefore be scaled by $E^{1/2}$ which then shows a clear advantage in chosing as high a probe voltage as possible. It must however be borne in mind that for very small probes the argument is complicated by the spread of the beam in the specimen.

In the case of a temperature sensitive specimen there will, for a given probe size, be a maximum tolerable power input. In contrast to the above situation a lower beam energy is now favourable provided the electron optical system can deliver sufficient current into the required probe, since an equivalent drop in beam current would adversely affect the S/N ratio.

The situation in which beam induced effects are present is more complicated. Bauer and Seiler (1980) have shown that for a given detection limit there is a minimum probe size that may be achieved depending on the sensitivity of the material to electron dose. The radiation damage in the surface layer is proportional to the energy lost in this layer which is approximately proportional to $\bar{E}^{0.5}$. For a given surface damage rate there may therefore be a slight advantage in using a higher beam voltage.

Reference

Bauer, H.E. and Seiler, H., Electron Microscopy 1980, <u>3</u>, 214 (Seventh European Congress on Electron Microscopy, Foundation, Leiden, 1980).

Backscattering effects in high resolution Auger electron microscopy

M M El Gomati, D C Peacock and M Prutton

Department of Physics, University of York, Heslington, York YO1 5DD, UK

1.   Introduction

Scanning Auger electron microscopy (SAM) is now established as a technique for high resolution mapping of the distribution of a chosen chemical element on a surface. Recently, considerable effort has been devoted to the development of submicron spot sizes in SAM because the resolution is determined mainly by the size of the incident electron beam (El Gomati and Prutton 1978; El Gomati et al 1979). It has been predicted that the direction of scanning of the electron beam with respect to a chemical edge affects the edge resolution (El Gomati et al 1979). In addition, the influence of high energy secondaries on LMM, as compared with KLL, Auger transitions has been reported by Shimizu et al (1978). This paper reports on experimentally measured edge resolutions for the same edge scanned in two different directions and comments on the effects of high energy secondaries upon the low energy Auger transitions.

2.   The experiment

The engineering compromises involved in designing and constructing a SAM often lead to the use of off-normal electron beam incidence at the specimen. Such constraints result in two extreme geometries for the detection of a chemical edge. Figure 1(a) shows the perpendicular mode, in which the plane of incidence and the scanning direction are both normal to the chemical edge, and figure 1(b) the parallel mode, in which the plane of incidence is parallel to the chemical edge but the scanning direction is normal to it. For the preliminary experiments presented here the sample used was a microcircuit with an SiO_2 overlayer, estimated from the SEM pictures to be 2 μm thick, on a silicon substrate. Although a thinner overlayer would have been preferable for quantitative experimentation we have not seen gross interference from topographical effects during this experiment due to the effective normalisation used (Prutton et al 1981) and the sloping profile of the steps cut through the overlayer to reveal the Si. Figure 2 shows the N(E) spectra of the two materials, Si (SiO_2) $L_{2,3}VV$ and

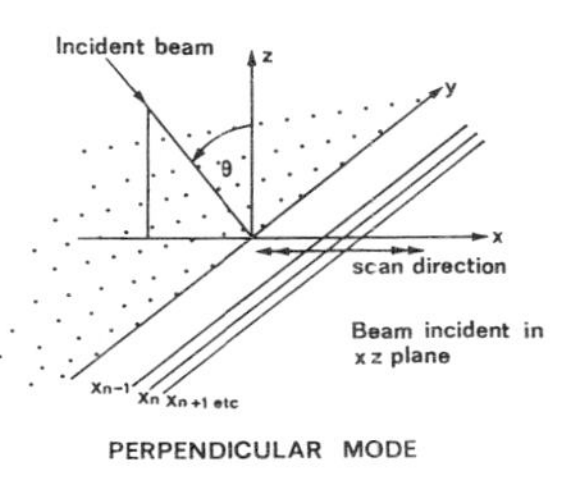

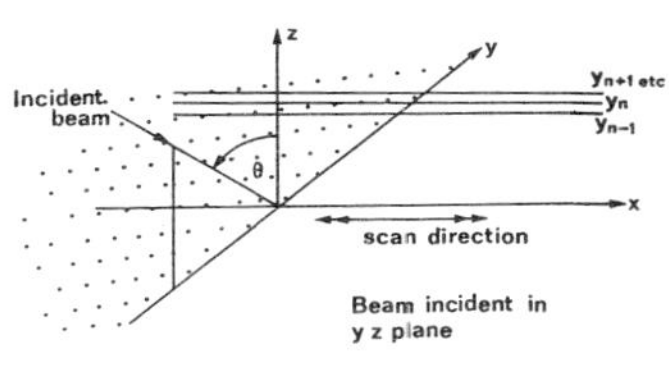

Fig 1:   The geometry of the two scanning modes

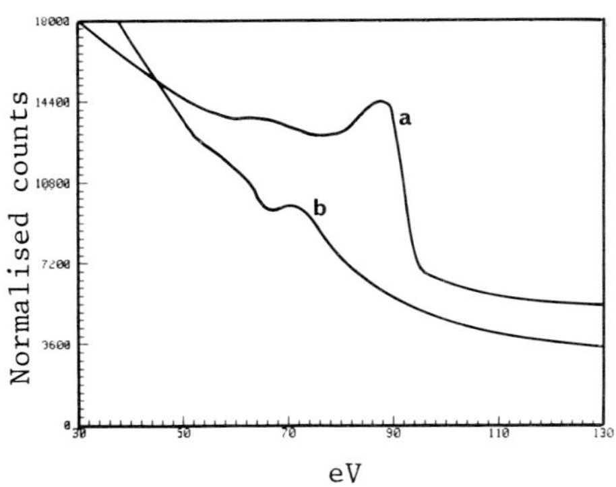

Fig 2: N(E) spectra of the specimen showing (a) Si $L_{2,3}VV$, and (b) Si (SiO_2) $L_{2,3}VV$ respectively

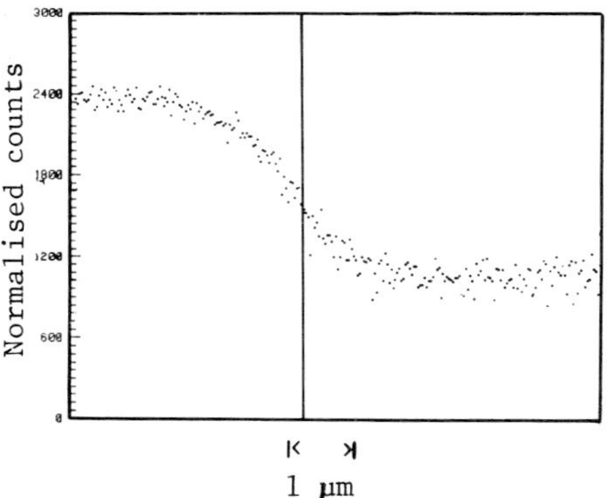

Fig 3: Measured edge profile in the perpendicular mode. Dwell time is 200 m sec per point, for 256 points. The profile is the average of 2 scans

Si $L_{2,3}VV$, after the sample was cleaned for 45 minutes with an argon ion dose of about 8×10^{-5} A/cm^2. Assuming a Gaussian beam profile the field emission column, with deliberately large apertures and operated in a thermally assisted mode, gave a full width at half maximum (FWHM), σ, of about 300 nm for a primary energy of 20 keV and a beam current of 2 nA stable over ten hours of continuous operation (El Gomati et al 1981).

3. Results

Figure 3 shows the measured edge profile in the perpendicular mode. The linescans presented are for the Si Auger signal which has a greater peak height above background, and therefore better linescan signal to noise (S/N) ratio, than the Si (SiO_2) although the same results were obtained when the latter signal was used. The electron beam made an angle of incidence, θ, of 45° to the sample surface normal but the chemical edge was in the same geometrical orientation as depicted in figure 1(a). The energy analyser (Browning 1981) was modulated at 1 kHz between two energies with a dwell time of 200 ms per line scan point. Two counts, N_1, at the Si $L_{2,3}VV$ Auger peak in the N(E) spectrum, and N_2, at an energy above the peak, were made. The difference $(N_1 - N_2)$ was used as the signal. To normalise against topographical effects and

beam current fluctuations $(N_1 - N_2)$ was divided by $(N_1 + N_2)$ as suggested by Prutton et al (1981). As expected, the edge profile is asymetric about the step centre in the perpendicular mode (Figure 3). This is because, with oblique incidence, the proportion of backscattered electrons reaching the escape depth for Auger electron emission from the surface is forward peaked. The computer simulation of electron trajectories (Figure 4) illustrates this effect for the neighbouring element, aluminium. In figure 3, as the electron beam approaches the edge the forward peaked backscattered electrons cease to contribute to the Si Auger signal at the detector. The signal in this area is due to primaries and, for this low energy Auger transition, the high energy secondaries (Shimizu et al 1978). In addition, the electron beam intersects the surface in an ellipse of major axis σ sec θ which is the effective beam diameter for this scan direction. Using the same criterion for the measurement of edge resolution, Δ_{50}, as adopted by El Gomati and Prutton (1978) we find $\Delta_{50} \cong 950$ nm in figure 3. For higher energy Auger transitions, El Gomati and Prutton (1978) predict a Δ_{50} of nearly 740 nm assuming an Auger backscattering factor, r, of 1.4 and using,

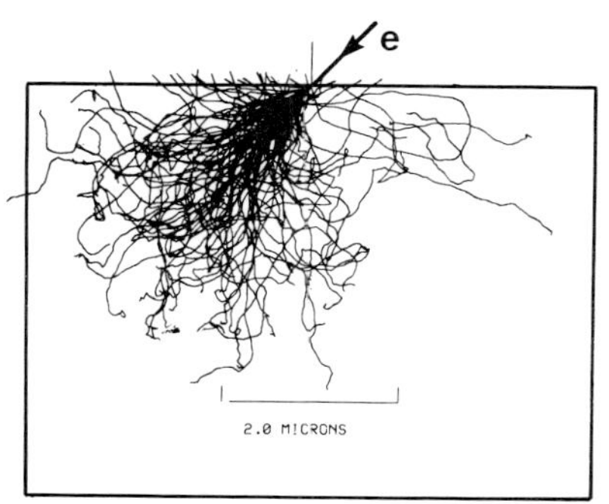

Fig 4:   Computer simulations of the trajectories of 100 electrons incident upon an aluminium target at θ = 45°

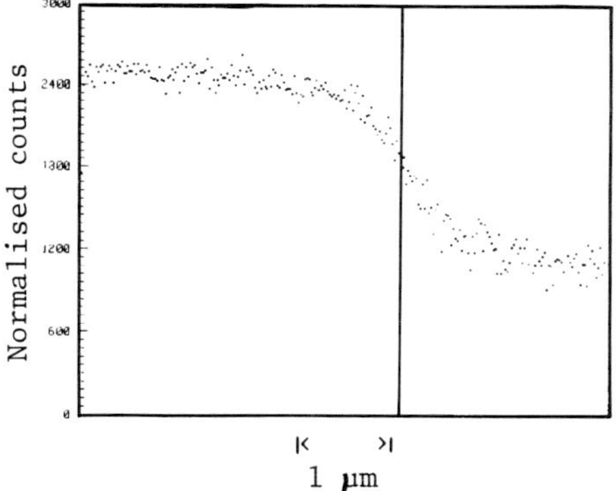

Fig 5:   Measured edge profile in the parallel mode same conditions as in Figure 3

$$\Delta_{50} \cong 1.25 \; r \; \sigma \; \sec \theta \; .$$

However, this theory does not take into account high energy secondaries which may broaden the spatial distribution of low energy Auger electrons. Also, the broadening effects of the forward peaking of the backscattering contribution for off-normal incidence in this mode may increase the value of Δ_{50}.

By rotating the sample in situ through 90° about the azimuth, a unique capability of the present instrument, it was possible to scan the same edge as in Figure 3 but in the parallel mode to give the edge profile shown in Figure 5. In this case the effective beam diameter is the minor axis of an ellipse, σ. Figure 5 shows a symmetric profile about the origin as predicted by El Gomati et al (1979). In addition, a Δ_{50} of nearly 600 nm is found from Figure 5. The theory suggests $\Delta_{50} \cong 530$ nm. This discrepancy might be due to the contribution from energetic secondaries.

4.   <u>Conclusion</u>

The present experiment has clearly shown that the orientation of a chemical edge with respect to the scan direction has an important influence on the spatial resolution of a SAM and must be taken into account when interpreting Auger images. Further, one advantage of being able to rotate a specimen about its surface normal has been demonstrated. Moreover, in order to obtain more quantitative estimates of the effects of high energy secondaries a shallower topographical step at the chemical edge and a beam diameter of less than 100 nm will be used. In order to compare such an experiment with theory the existing Monte Carlo calculations will need to be modified.

References

Browning R (1981) J Phys E $\underline{14}$ 58

El Gomati M M, Browning R and Prutton M (1981) this proceedings

El Gomati M M, Janssen A P, Prutton M and Venables J A (1979) Surf Sci $\underline{85}$ 309

El Gomati M M and Prutton M (1978) Surf Sci $\underline{72}$ 485

Prutton M, El Gomati M M, Peacock D C, Larson L A and Poppa H (1981) this proceedings

Shimizu R, Everhart T, MacDonald N and Hovland C (1978) Appl Phys Lett $\underline{33}$ 549

Normalisation of Auger electron images for beam current fluctuations and sample topography

M Prutton, D C Peacock and M M El Gomati

Department of Physics, University of York, Heslington, York, YO1 5DD, UK

and

L A Larson and H Poppa

NASA/Stanford Joint Institute for Surface and Microstructure Research, NASA Ames, Moffett Field, 230-3, Ca 94035, USA

1. Introduction

Fluctuations in the incident electron beam current and in the sample topography lead to undesirable variations in the Auger signal used in forming an Auger electron micrograph or linescan. Beam current fluctuations become most significant in high spatial resolution (< 100 nm) instruments which use field emission sources. They can be reduced in importance by normalisation to a signal proportional to the beam current which is measured at the same time as the Auger signal. Sample topography is more complex because it has several effects. For instance, the Auger yield can vary from place to place because: (a) the angle of incidence varies, causing a changing number of ionisations within the Auger escape depth; (b) the analyser samples varying fractions of the emitted Auger electron distribution – which is not, in general, isotropic; (c) the surface roughness can obstruct the incident beam and/or shield the analyser from the emitted electrons.

The simplest normalisation technique, which is usually readily available, is to divide the Auger signal by the conventional SEM signal. This corrects well for beam current fluctuations (Browning et al, 1977) in spectroscopy at a fixed place. It is not satisfactory for linescans and images because the collection geometry of the secondary electron detector is usually different from that of the electron energy analyser and because chemical effects contribute to the SEM signal.

Another method is to form a ratio derived from the electron count, N_1, at an Auger peak in the spectrum $N(E)$ or $EN(E)$ and a second count, N_2, at some energy just above the peak. This note compares normalisation methods using $(N_1 - N_2)/N_2$ (a similar quantity to that used by Todd and Poppa 1978) and $(N_1 - N_2)/(N_1 + N_2)$ (a quantity related to the logarithmic derivative used by Janssen et al 1977).

2. Method

The samples used in the comparison were anisotropically etched Si(001) slices kindly provided by Dr K Bean, Texas Instruments Inc (Bean 1978).

They had polygonal pyramids terminated in 4{111}, 8{331} and 1(001) faces protruding about 40 µm high from the slice surface. These samples were coated with about 1000 Å of either Au, Al or Ag for normalisation studies in which the objective was to find the method which was most successful in removing topographical contrast. By mounting the sample with a mean angle of incidence of 30° the 13 facets presented a well spaced set of angles of incidence θ, between 20° and 90° which could be imaged or linescanned within the field of view of the spectrometer. Figure 1 shows a view of one of the polygons used in this work. Numerical data on the Au NVV peak at 65 eV and the Ag MVV peak at 352 eV are reported here from experiments on the Stanford SAM (Todd et al 1979) and the conclusions were tested on the York SAM (Browning et al 1977).

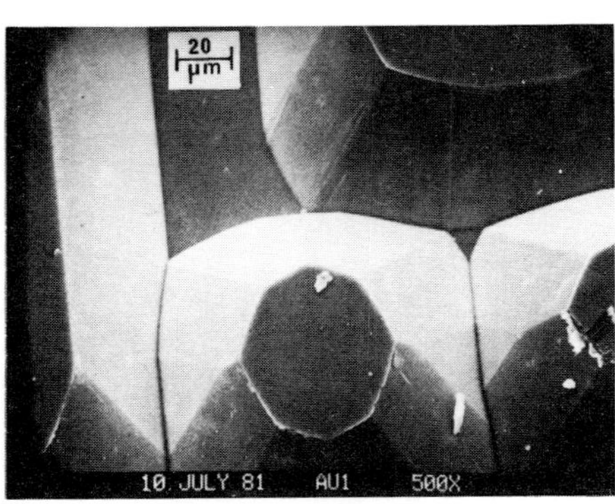

Fig 1:   SEM image of Au coated polygon

3.   Results

The ratios $(N_1 - N_2)/N_2$ and $(N_1 - N_2)/(N_1 + N_2)$ measured for Ag MVV at 352 and 365 eV and for Au NVV at 65 and 85 eV are compared in figures 2 and 3. All the surfaces studied showed both ratios increasing with θ which means that the Auger yield is usually a faster function of θ than is the background just above the peak. $(N_1 - N_2)/(N_1 + N_2)$ was always less sensitive to θ than was $(N_1 - N_2)/N_2$.

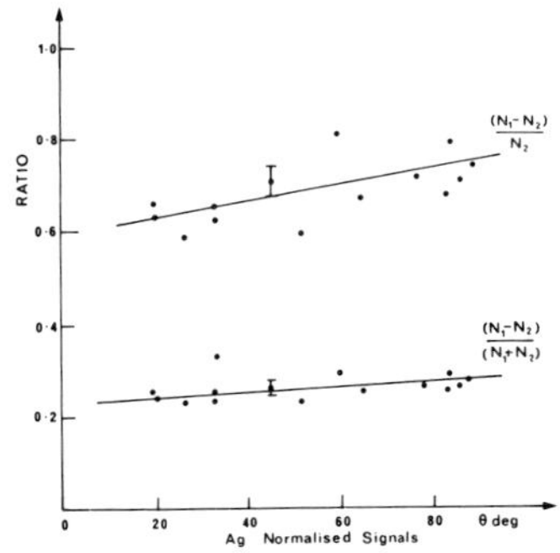

Fig 2:   The two ratios for Ag MVV

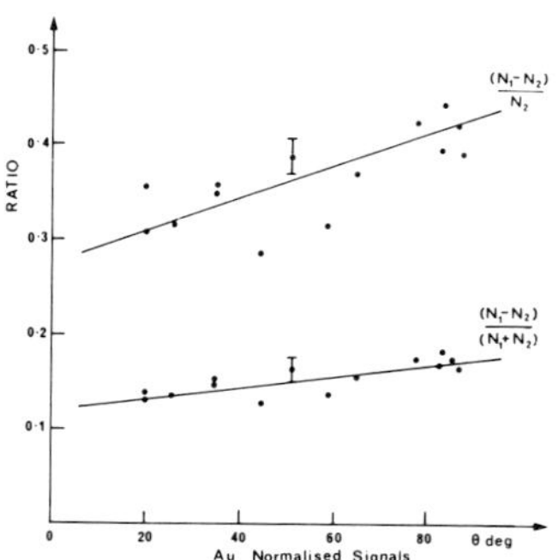

Fig 3:   The two ratios for Au NVV

If a randomly rough elemental surface had been used for these measurements then the topography would have caused random signal variations which could be measured by an rms deviation from some average value. The fractional rms deviations derived from the measurements for Au, Ag and the uncoated Si are summarised in Table 1 where they are compared with the ratios of twice the rms noise count in the spectra to the height of the Auger peak at θ = 45°. This latter ratio gives a measure of the rms deviation that might have been achieved for an ideal topographical correction.

It can be seen from the data in Table 1 that $(N_1 - N_2)/(N_1 + N_2)$ is the better ratio for topography correction although some systematic variation with θ is still present. The figures for Si are not so low as for Ag and

	$(N_1 - N_2)/N_2$	$(N_1 - N_2)/(N_1 + N_2)$	Noise ratio
Au	0.129	0.111	0.035
Ag	0.093	0.068	0.027
Si/SiO_2	0.210	0.200	0.050

Table 1: Fractional rms deviations of the two ratios and the corresponding
noise ratio

Au because Si and SiO_2 were present on the surface in fluctuating amounts
but the topographical contrast is still largely eliminated. These
results have been tested on the York SAM where $(N_1 - N_2)/(N_1 + N_2)$ has
also been found to be the better correction.

This result is demonstrated in the linescans of figure 4 which have been
taken across the polygon shown in figure 1. The ratio $(N_1 - N_2)/(N_1 + N_2)$

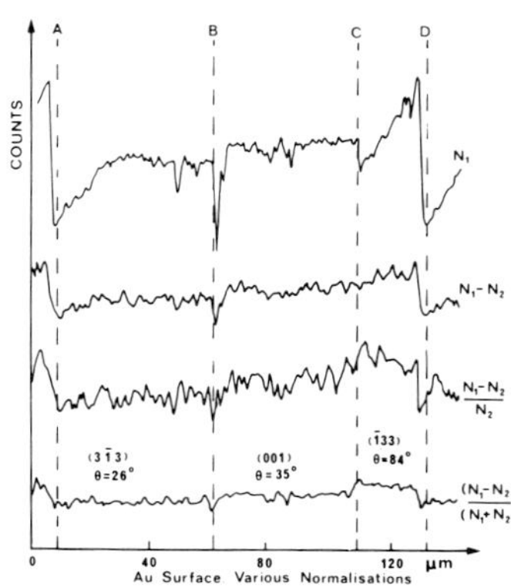

Fig 4:   Linescan from X to Y across polygon of figure 1

shows the best elimination of the topographical effects clearly visible at
A, B, C and D in the linescan using N_1 only. The details of the variation
of N_1 with position can be explained in terms of shadowing of the analyser
entrance aperture by the surface topography.

<u>References</u>

Bean K E, 1978, IEEE Transactions on Electronic Devices <u>ED-25</u> 1185
Browning R, Bassett P J, El Gomati M M and Prutton M, 1977, Proc Roy Soc
    Lond <u>A357</u> 213
Janssen A P, Harland C J and Venables J A, 1977, Surf Sci <u>62</u> 277
Todd G and Poppa H, 1978, J Vac Sci Tech <u>15</u> 672
Todd G, Poppa H and Veneklasen L H, Thin Solid Films <u>57</u> 213 (1979)

The structure of interphase interfaces

G.C. Weatherly

Department of Metallurgy and Materials Science, University of Toronto,
Toronto, Canada M5S 1A4

1. Introduction

In contrast to the extensive literature on the structure of grain
boundaries (much of which comes from electron microscopy), there have
been few definitive studies on the structure of interphase interfaces
other than those types where the two lattices are related by a simple
transformation strain. The problems associated with characterizing the
interface structure result not only from experimental difficulties in
unambiguously identifying the "defects" at the interface but also in
understanding how they can be related to a plausible dislocation model
that could account for the transformation strain connecting the two
lattices. This problem becomes increasingly difficult where the two
phases have different crystal structures and the transformation strain
connecting the two lattices is not uniquely defined. Examples of problems
of this type are considered in this paper.

2. The dislocation content of a boundary

It is convenient to distinguish those cases where the boundary structure
can be described in terms of dislocations having lattice translation
Burgers vectors (of either crystal) from those where they may not. In the
former case (and the only one considered here) the starting point for any
discussion of the dislocation content of a boundary is the equation

$$\underline{b} = (D_+^{-1} - I)\,\underline{p} \tag{1}$$

due first to Frank (1950), where $\underline{b}$ is the total Burgers vector inter-
sected by a vector $\underline{p}$ lying in the interface between two crystals (+) and
(−), I is the identity matrix, and the deformation of the (+) lattice
to form the (−) lattice is described by the matrix D_+. In the 0− lattice
formulation of Bollman (1970), an almost identical relationship is used,

$$\text{viz.} \quad \underline{x}_i^o = (D_+^{-1} - I)^{-1}\underline{b}_i \tag{2}$$

but now the equation defines the 0− lattice vectors $\underline{x}_i^o$ in terms of the
translation vectors $\underline{b}_i$ of the reference crystal (−). Equation (1) can
also be modified to include situations where the plane of the boundary or
the misorientation between the two phases changes slightly. For example,
a small change in the orientation of the boundary plane leads to a change
$\underline{\Delta b}$ in the dislocation content of the interface given by the equation

$$\underline{\Delta b} = (D_+^{-1} - 1)\ (R - I)\ \underline{p} \qquad (3)$$

where R represents the rotation of the boundary
plane (Sargent et al, 1980). Two simple examples,
both of which have been verified by electron
microscopy demonstrate this approach. If the
plane of misorientation of a simple tilt boundary
rotates in the sense shown in Fig. 1a, extra edge
dislocations lying parallel to the original dislocations of the boundary but having orthogonal
Burgers vectors must be introduced (Sargent et al,
1980). Due to the large torque associated with
the tilt boundary the misorientation of the
boundary becomes localized at these extra dislocations i.e. a step structure develops at the
boundary (see Fig. 1b). In general such steps
do not enhance the rate of boundary migration,
but they do allow the boundary to move without
rotation from its cusp position.

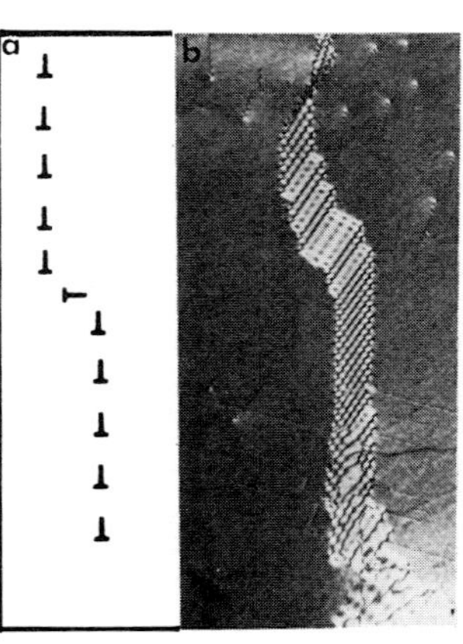

Fig. 1 The "ledge"
structure of a tilt
boundary

As a second example, we consider the interfacial microstructure of {100}
twist interphase boundaries in a thin film couple such as Au-Pd studied
by Hwang et al (1980). In this situation, there is both a misfit and
misorientation to account for. If the normal to the boundary is [001],
equation (1) gives for this case

$$\underline{b} = \begin{bmatrix} \delta & \phi & 0 \\ -\phi & \delta & 0 \\ 0 & 0 & \delta \end{bmatrix} \underline{p} \qquad (4)$$

where δ is the difference in the lattice parameters of Au and Pd (assuming
little or no interdiffusion during thin film production at Z = 0) and ϕ
is the angle of rotation (both δ and $\phi \ll 1$). An ambiguity now arises in
applying eq. (4). The misfit δ could be accommodated by an array of
epitaxial edge dislocations and the misorientation ϕ by a separate array
of screw dislocations. However the energy of such an array would be
greater than an orthogonal array of mixed dislocations which simultaneously accommodate the misfit and misorientation, and indeed a mixed array is
observed (Hwang et al, 1980). From eq. (4) we can show that the angle
between $\underline{b}$ and the dislocation line is $\tan^{-1} (\phi/\delta)$, which agrees both with
the result of Jesser and Kuhlmann-Wilsdorf (1967) when ϕ, $\delta \ll 1$, and with
the experimental observations.

The twist interphase boundary illustrates that even where the form of D_+
can be sensibly defined, one still has to appeal to other considerations
(in this case the energy of the boundary) to determine the optimum dislocation array. In many situations, even the correct choice for D_+ is not
immediately obvious; e.g. the precipitation of the interstitial oxygen T_6
phase (Fe_6Ti_6O, which has a F.C.C. unit cell, lattice parameter = 11.3 Å)
in the ordered intermetallic compound FeTi (B.C.C., a = 2.97 Å) is
interesting. A single set of edge dislocations, spacing $\sim$ 15 mn, is
observed at the interface (Fig. 2a) and the corresponding diffraction
pattern (Fig. 2b) shows an excellent fit between the two lattices, with

$(002)_{FeTi} \parallel (426)_T$ and

$(211)_{FeTi} \parallel (428)_T$.

A reasonable interpretation of the dislocations is that they are accommodating the residual misfit of 1.6% between the $(426)_T$ and $(002)_{FeTi}$ planes with $\underline{b} = a[001]$, (which is observed shortest lattice translation vector in the ordered FeTi structure) While this still does not permit a complete determination of D_+ and other dislocation arrays (below the resolution limit of the microscope) may be present, much useful information can be gained in this way about the structure of the interface.

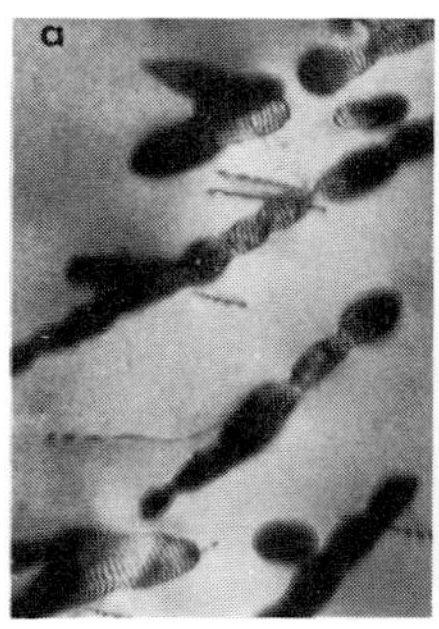

Fig. 2   (a) The interface dislocation structure and (b) the orientation relationship between FeTi and Fe_6Ti_6O

As a final example of the permutations that can be rung on this theme, the α(H.C.P.) - β(B.C.C.) interface in a Zr-2.5 wt% Nb alloy, heat treated to produce a coarse grained 2 phase structure is considered. The orientation relationship between the phases is Pitsch-Schrader i.e. $[011]_\beta \parallel [0001]_\alpha$, $[01\bar{1}]_\beta \parallel [01\bar{1}0]_\alpha$ and $[100]_\beta \parallel [2\bar{1}\bar{1}0]_\alpha$, and $(D_+^{-1} - I)$ is probably given by the relationship suggested by Kelly and Groves (1970). Using values for the lattice parameters of α-Zr and retained β(Zr-20 wt% Nb) we find that

$$D_+^{-1} - I = \begin{bmatrix} -0.085 & - & - \\ - & +0.116 & - \\ - & - & +0.03 \end{bmatrix} \qquad (5)$$

(Note again that there is no uniqueness in the selection of the components of this matrix. They have been chosen on the basis that all the strains involved are relatively small, but additional shuffles are required to move some of the atoms to their correct final positions). The optimum interface plane (i.e. of lowest strain energy) could be selected by using eq. (2) to define $\underline{x}_i^o$ (and hence the spacing of the dislocations at the interface) and then following either the procedure proposed by Bollmann and Nissen (1968) or by Ecob and Ralph (1981). An additional problem occurs in interpreting the experimental results in this particular example because both the plane of the interface and the misorientation across the boundary can vary (Perovic and Weatherly, 1981). Following the arguments presented earlier, both factors must lead to the incorporation of extra dislocations at the interface (see eq. (3)) which can be pure edge, screw or mixed depending on the sense of the rotation of the interface plane and misorientation of the two crystals from the exact Pitsch-Schrader relationship. Some examples of the structures observed at the interface in this alloy and their interpretation are given in Fig. 3.

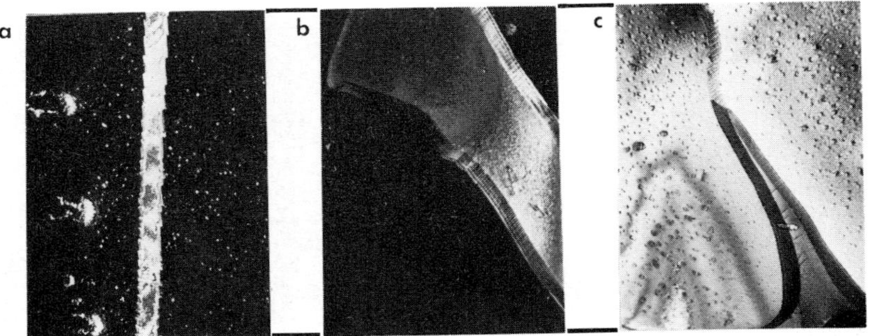

Fig. 3 The α(HCP) - β(BCC) interface in a Zr-2.5 Wt% Nb alloy
showing (a) ledges (b) [000C] misfit dislocations and (c) a
sub-boundary - interface dislocation reaction.

3.   Ledges at interfaces

Ledges are commonly observed at many interfaces (see e.g. Fig. 3a).
As in sub-boundaries they permit the interface to remain in a low energy
cusp position as the boundary migrates, but they also play an important
kinetic role in providing sites for atom attachment. Unfortunately
neither their size nor the diffraction contrast observed at ledges can be
explained in the framework of the approach based on eqs. (1) - (3). The
most widely studied example is the Al-θ' (CuAl$_2$) system (Weatherly, 1971,
Stobbs and Purdy, 1978) but similar effects are observed at plate shaped
particles in Al-Au (Sankaran and Laird, 1974), eutectic interfaces
(Garmong and Rhodes, 1974) and eutectoid structures in Zr-Nb (Perovic
and Weatherly, 1981). In all these systems ledges 2 nm or greater in
height are observed, having an effective Burgers vector normal to the
plane of the interface. While the direction of this vector ($\Delta\underline{b}$ in eq.(3))
can be predicted from the analysis, the observed strain field results
from forced _elastic_ coherence between the two phases. For Al-θ', Stobbs
and Purdy 1978 have suggested that the elastic strain energy at ledges
is minimized when 3½ θ' unit cells are juxtaposed with 5 Al unit cells,
with a small residual elastic strain of 0.5% bringing every 10th (002)$_{Al}$
plane into exact coincidence with every 7th (002)$_{θ'}$ plane. The ledge
height would then be 2.025 nm or multiples thereof. There is clearly
a relationship between this suggestion and the 0- lattice description of
boundary structure (Bollmann, 1970); in this case a relatively small
elastic strain allows a coincidence boundary to form. The interface
structure at the ledge would thus consist of an array of dislocations,
spacing 2.025 nm (as given by eq. (2)) superimposed on which is a longer
range elastic strain field associated with the co-incidence requirement.
Electron diffraction studies "see" only this latter strain field.

Clearly great care has to be taken to distinguish between the strain
field of such ledges and of dislocations lying at the interface which rotate the two lattices from their exact orientation relationship.
Contrary to the suggestion of Garmong and Rhodes, 1974, there is no
rotation of the two lattices associated with the migration of the pure
ledges described above.

4. Martensitic or displacive transformations

The analysis presented in sections
2 and 3 is concerned with those
situations where the deformation,
D_+, is given by the product of a pure
rotation and a lattice strain which
may be a combination of shears and a
dilation. In many martensitic
transformations, the requirement that
the habit plane be an invariant plane
necessitates in addition that one or
more lattice invariant shears occur.
Thus we have the possibility that if
these lattice invariant shears
occur by slip, the dislocations

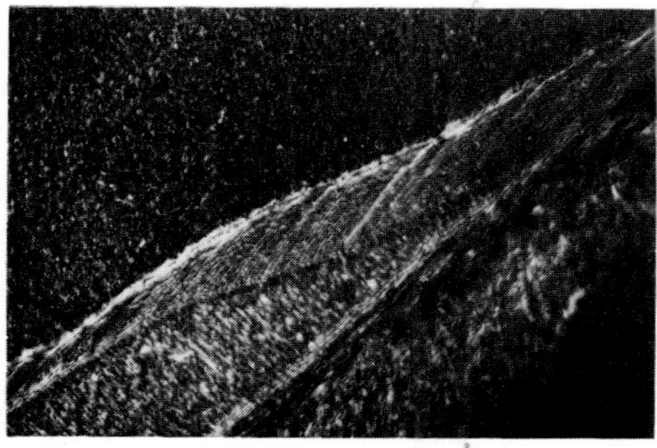

Fig. 4 An array of $\frac{a}{3}\left[10\bar{1}0\right]$
transformation dislocations at the
tapered end of a ZrH plate in Zr

lying at the habit plane interface can be resolved and characterized.
Furthermore, in certain situations it should be possible to resolve the
transformation dislocations (associated with D_+) at the tapering ends of
martensite plates. The direct lattice resolution studies of Sinclair
and Mohammed, 1978 on TiNi provided striking confirmation of this, while
our own studies on γ-hydride precipitation in Zr-H, (which is a diffusion
controlled martensitic transformation), have resolved the arrays of
$\frac{a}{3}\left[10\bar{1}0\right]_{Zr}$ transformation dislocations (see Fig. 4), as well as dislocations associated with a lattice invariant shear (Weatherly, 1981).

5. The visibility of interface dislocations

Considerable progress has been made in characterizing the dislocation
structure of grain boundaries using either the simple $\underline{g}.\underline{b}$ criterion
or the computer matching technique pioneered by Head et al, 1973. In a
parallel development, the improved resolution of modern T.E.M.'s has lead
to the fairly routine observation of dislocation arrays at interfaces
having spacings of 2-3 nm, see e.g. Rigsbee and Aaronson, 1979. It
has yet to be shown either theoretically or experimentally that the much
smaller spacings, $\sim$ 1 nm, predicted for many F.C.C./B.C.C. interfaces
(Ecob and Ralph, 1981) can be resolved, or indeed whether such
a dislocation description has any physical significance. Column
approximations of the type used by Weatherly and Mok, 1972 are clearly
inadequate in this context to predict the diffraction contrast from such
closely spaced defects.

References

Bollmann, W., 1970, Crystal Defects and Crystalline Interfaces,
(Springer-Verlag).

Bollmann, W. and Nissen, H-U., 1968, Acta Cryst., __24__, 546.

Ecob, R.C. and Ralph, B., 1981, Acta Met, __29__, 1037.

Frank, F.C., 1950, Symposium on the Plastic Deformation of Crystalline
Solids, p. 150, O.N.R. Pittsburg.

Garmong, G. and Rhodes, C.G., 1974, Acta Met., $\underline{22}$, 1373.

Head, A.K., Humble, P., Clareborough, L.M., Morton, A.J. and Forwood, C.T., 1973, Defects in Crystalline Solids, $\underline{7}$, (North Holland).

Hwang, M., Laughlin, D.E. and Bernstein, I.M., 1980, Acta Met., $\underline{28}$, 621.

Jesser, W.A. and Kuhlmann-Wilsdorf, D., 1967, Phys. Stat. Sol., $\underline{31}$, 533.

Kelly, A. and Groves, G.W., 1970, Crystallography and Crystal Defects (Longman).

Perovic, V. and Weatherly, G.C., 1981 unpublished research.

Rigsbee, J.M. and Aaronson, H.I., 1979, Acta Met., $\underline{27}$, 365.

Sankaran, R., and Laird, C., 1974, Acta Met., $\underline{22}$, 957.

Sargent, C.M., Purdy, G.R. and Weatherly, G.C., 1980, Scripta Met., $\underline{14}$, 725.

Sinclair, R., and Mohamed, H.A., 1978, Acta Met., $\underline{26}$, 623.

Stobbs, W.M. and Purdy, G.R., 1978, Acta Met., $\underline{26}$, 1069.

Weatherly, G.C., 1971, Acta Met., $\underline{19}$, 181.

Weatherly, G.C., and Mok, T.D., 1972, Surface Science, $\underline{31}$, 355.

Weatherly, G.C., 1981, Acta Met., $\underline{29}$, 501.

A methodology for the TEM characterisation of grain boundary films in ceramics

N.W. Jepps, W.M. Stobbs and T.F. Page
Department of Metallurgy and Materials Science, Pembroke Street, Cambridge

As a function of their composition and mode of fabrication, engineering ceramics exhibit a wide variety of grain boundary structures. In relation to the strong influence which such boundaries can have on mechanical properties, thin glassy films (typically 1-10nm in width) are of particular interest and a number of studies of them have been made using both lattice and dark-field imaging methods (e.g. Kirn et al.1978; Clarke & Thomas 1977; Lou et al. 1978; Krivanek et al. 1979; Clarke 1979). The glassy region presumably consists of a number of 'jumbled', but linked, coordination polyhedra providing a 'transition region' between the directed bond arrays of the two adjacent crystals. The polyhedra may be those of the parent crystal or they might be modified to varying extents as a function of sintering additives etc. Further, if the boundary transition region is very thin, then the arrangement of the polyhedra might be expected to be constrained to exhibit some degree of order (Spaepen 1978). Thus, a complete characterisation of grain boundary films in ceramics requires information concerning not only width but also structure and composition.

Our approach to the study of such thin grain boundary films in high performance ceramics (e.g. SiC, Si_3N_4 etc.) has been to explore the information available from the distribution of intensity in diffraction patterns from boundaries viewed 'edge-on' in thin foils. The main features expected in such a reciprocal lattice section are shown schematically in figure 1 and comprise various refraction and diffraction phenomena. However, since boundary regions may be both thin and of small specific volume, the boundary-related features shown in figure 1 are weak, broadened and difficult to observe directly in diffraction patterns. Thus, the intensity distribution has been examined using the indirect technique of dark-field imaging with a small objective aperture. The size of the aperture used in the search is critical since it must be small enough to allow feature selection from the diffraction pattern but not so small as to broaden the image features beyond recognition. The selection of aperture positions for exploring varicus boundary characteristics is discussed below.

Two SiC-based fine-grained, hot-pressed, materials have been used for initial study, namely Norton NC203 and an experimental sample from Prof. H. Suzuki (Tokyo) containing (1%B + 1%C). Thin foils were obtained by sectioning, polishing and ion-beam thinning (e.g. Sawyer & Page 1978) and were examined using a JEOL 120CX TEM operating at 100kV.

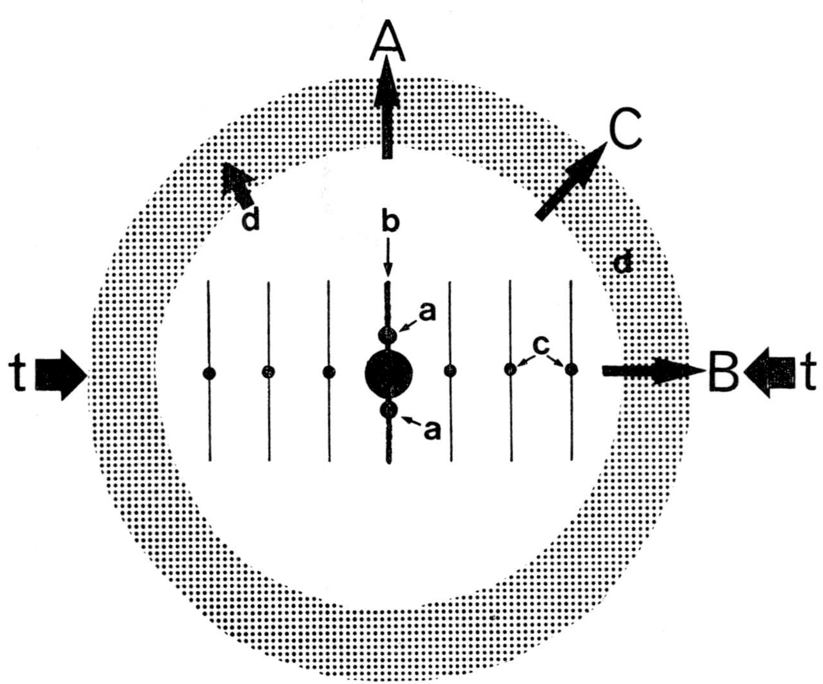

Figure 1.

Schematic features in the reciprocal space intensity distribution from a region containing an edge-on grain boundary (trace tt). The diffraction patterns from the crystals abutting the boundary have been omitted.
a) Refraction spot whose position is dependent both on the difference in refractive indices of the unperturbed crystal and the boundary and also on the exact orientation of the boundary to the incident beam. Combinations of diffraction and refraction events can result in a more complex distribution of refraction spots (Turner and Cowley 1981).
b) Fresnel diffraction streak whose intensity profile is dependent on the boundary thickness (t) with a first minimum at 1/t.
c) Reflections due to short-period order along the boundary trace (tt). The spots show Fresnel streaks in a direction perpendicular to the boundary plane (e.g. Carter et al. 1980).
d) Diffuse rings due to the presence of amorphous material in the boundary. Their shape is modified by the width of the boundary, being broader in a direction perpendicular to the boundary plane. The position of the rings depends on the composition of the glass (if this alters the radial distribution function).
Dark-field imaging with an objective aperture set along directions A,B and C can be used to sample different aspects of the information available.

The symmetry of the Fresnel fringes across the boundary region was used as
a critical test of boundary orientation; slight misorientations ($\lesssim 1^{\circ}$)
from exact edge-on alignment result in noticeable asymmetry – see figures
2a-2c. With the boundary edge-on, dark-field images were obtained with a
small ($10\,\mu m$) objective aperture whose position was first moved from the
optic axis in a direction perpendicular to the boundary trace (direction A
in fig.1). For small shifts in this direction, and with a necessarily
finite beam convergence, the main intensity in the dark-field image comes
from refraction at the two crystal-boundary interfaces (see (a) in fig.1).
However, because the difference in mean inner potentials (and hence
refractive indices) of the crystal and boundary regions is small, the
refraction spots lie very close to the optic axis ($\lesssim 10^{-4}$ radians).
Thus, for boundaries in the thickness range 0.5-3nm, moving the aperture
further out along A results in the image arising mainly from the Fresnel
diffraction streak ((b) in fig.1) – see figure 3. In principle, it should
be possible to estimate the boundary width from the positions of the first
minimum and maximum in the streak profile noted by the changes in
intensity of the dark-field boundary image with aperture position.
However, the intensity recorded depends on the position of the Ewald
sphere. Thus, in order to make intensity comparisons, $\bar{s}$ should be set to
zero for each successive aperture position. This is difficult.

Any short period order in the boundary (giving rise to (c) features in
figure 1) can be investigated by moving the aperture along a direction
parallel to the boundary trace (i.e. along B in fig.1). Rather
surprisingly, given the discussion above, no such phenomena have as yet
been detected unambiguously. This is possibly because any periodicity
present did not lie parallel to the boundary trace.

In order to avoid the diffraction and refraction phenomena along
directions A and B – and thus to detect amorphous regions in boundaries
unambiguously – it is necessary to move the objective aperture at 45° to
the boundary trace (i.e. C in fig.1). The position of the first, most
intense, amorphous halo ((d) in fig.1) may be determined as that giving
the brightest grain boundary image (see fig.4). Clearly glasses of two
different compositions (and thus different radial distribution functions)
may be differentiated in this way. However, for grain boundary glasses of
composition close to that of the parent crystalline material, amorphous
material on the surface of the specimen (resulting from ion-beam thinning
damage) may also be detected and thus obscure boundary information.

Our initial results show that both materials studied contain thin (~1nm)
grain boundary regions. However, while the Suzuki material displayed a
boundary composition based on disordered SiC units, the NC203 displayed an
amorphous film of different composition together with crystalline grain
boundary particles.

In conclusion, it is expected that various features of the width,
composition and structure of grain boundary films may be determined by
carefully sampling the diffracted intensity in different directions with
respect to the trace of the boundary when viewed edge-on. Complementary
information can also be derived from other techniques (e.g. EELS for
composition). However, the difficulty of estimating film widths of less
than 1nm remains and other techniques are currently being explored for
this. For example, we have found the variation of Fresnel fringe spacing
with defocus to give more accurate results than either a single lattice or
dark-field image and results obtained in this way will be reported
elsewhere.

Carter C B, Kohlstedt D L and Sass S L 1980 J Am Ceram Soc 63 623.
Clarke D R 1979 Ultramicrosc. 4 33.
Clarke D R and Thomas G 1977 J Am Ceram Soc 60 491.
Kirn M, Ruhle M, Schmid H and Gauckler L J 1978 Proc Ninth Int Cong E M
    (Toronto) 302.
Krivanek O L, Shaw T M and Thomas G 1979 J Appl Phys 50 4223.
Lou L K V, Mitchell T E and Heuer A H 1978 J Am Ceram Soc 61 392, 462.
Sawyer G R and Page T F 1978 J Mat Sci 13 885.
Spaepen F 1978 Acta Met 26 1167.
Turner P S and Cowley J M 1981 Ultramicrosc 6 125.

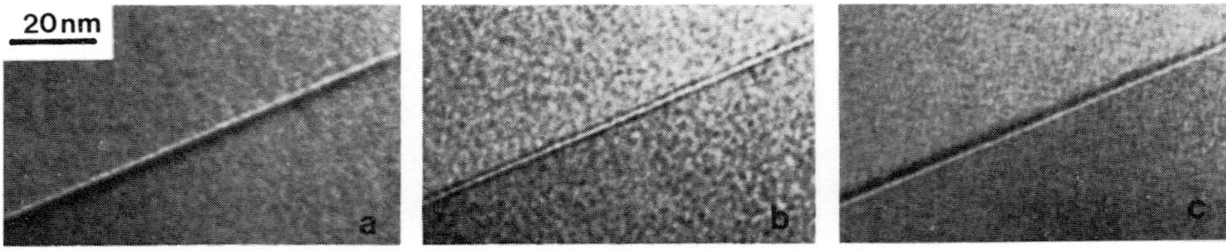

Fig.2(a-c). The changing appearance of Fresnel fringes as the boundary is
moved through the exact edge-on orientation in (b) (see Clarke 1979).

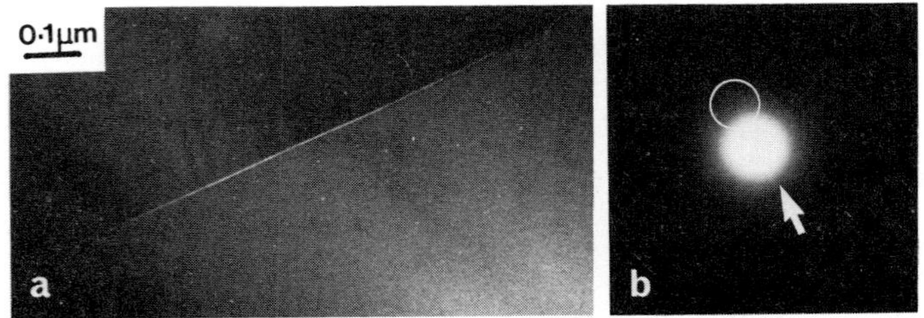

Fig.3. An edge-on grain boundary from a specimen of (1%B + 1%C) hot-
pressed SiC. a) Dark-field image, with aperture moved along direction A
in Fig.1. The bright line image is aperture size limited and measures
-1nm.

        b) Diffraction pattern showing Fresnel streak and aperture
position.

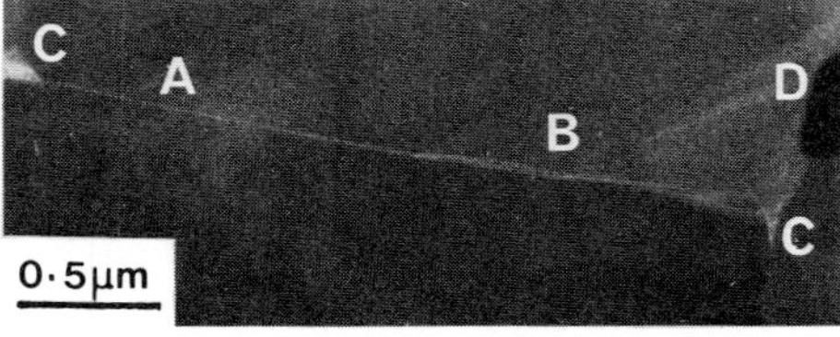

Fig.4. A dark-field image from a specimen of NC203.   The objective
aperture was placed on the first amorphous halo. The illumination of the
region does not depend on boundary orientation, though the projected width
does. The boundary is vertical at A and inclined at B. Features C are
glassy regions at triple points and D is a crystalline particle.

The nature of orientation relationships associated with precipitation on interphase boundaries

R.A. Ricks*, P.R. Howell+ and W.M. Stobbs*

*Department of Metallurgy and Materials Science, University of Cambridge
+Department of Materials Science and Engineering, Pennsylvania State Univ

1. Introduction

In general, orientation relationships between parent and product phases are quoted in precise crystallographic terms which express the parallelism of reasonably low index planes and directions. However few orientation relationships are obeyed precisely, and an orientation scatter is often observed. In the present investigation, a through focal electron microscopical technique has been employed to examine the spread in orientation of ε-Cu precipitates which have nucleated on mobile austenite/ferrite interfaces in an Fe-2%Cu-2%Ni alloy.

2. Results

Due to the spherical abberation of the objective lens, electron beams will be brought to focus at different points in space if they are not equiangularly displaced from the optic axis of the lens. This in turn leads to image displacements (y) which can be quantified via the following expression:

$$y = C_s \alpha^3 \mp D\alpha \quad \text{(Hirsch et al. (1965))}$$

where C_s is the spherical abberation coefficient of the objective lens, α is the angular displacement of the electron beam from the optic axis and D is the degree of defocus. Figures 1a,b plot the measured average image displacements for interfacially nucleated ε-Cu precipitates as a function of α, with D constant, (α was varied by displacing the chosen diffracted beam from the optic axis using the beam tilt controls) and as a function of D, with α constant (figures 1a and 1b respectively).

This image movement, as a function of defocus, can be used to investigate the spread in orientation within a precipitate colony. If a precipitate centred dark field image is obtained, it follows that, if any spread in the orientation exists, not all of the diffracted electron beams from individual precipitates can be accurately centred on the optic axis. Thus if the objective lens current is varied, the image movement observed will be directly related to the angular spread for that $\underline{g}$ vector. By taking doubly exposed photographic plates at various amounts of defocus, and measuring the movements of the individual precipitates, a plot of image displacement as a function of defocus can be obtained. The gradient of the curve in such a plot is the value of α, for that particular precipitate. By employing several $\underline{g}$ vectors, the angular spread in orientation can then be accurately defined.

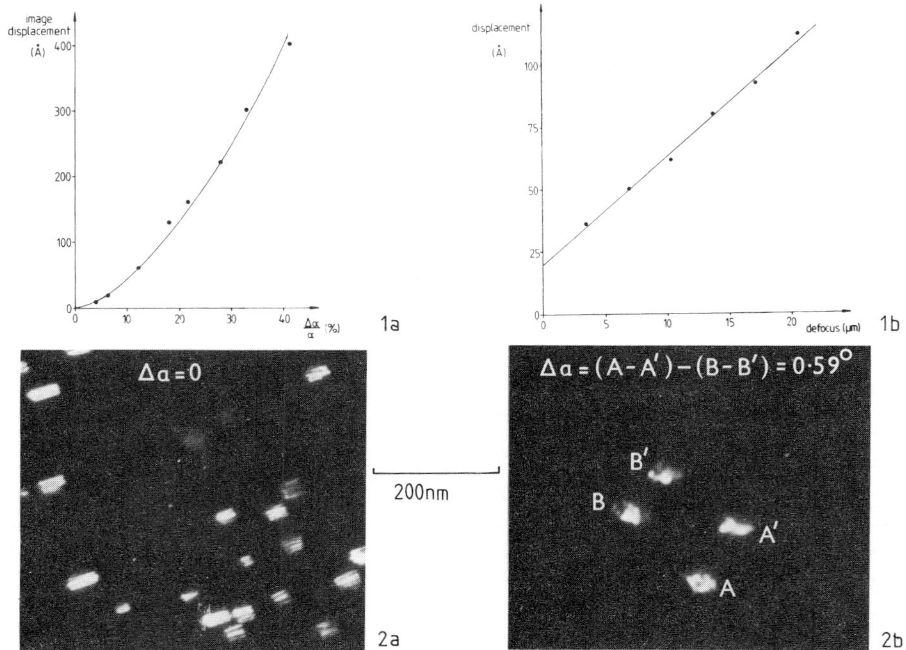

Fig.1   Image displacements as a function of α (a) and D (b)
Fig.2   Doubly exposed dark field micrographs showing no image displacement (a) and considerable image displacement (b).*

Nucleation of ε-Cu can occur in association with ferrite which is rationally related (by the Kurdjumov-Sachs (1930) orientation) or unrelated to the abutting austenite. In the former case, the ε-Cu can be K-S related to the ferrite and cube/cube related to the austenite such that a stringent three phase crystallography can obtain. Figure 2 a is a doubly exposed precipitate centred dark field image at different positions of defocus (41.74μm apart) of such a precipitate colony. As can be seen very little image movement is observed which indicates that the precipitates conform to a singular orientation with little orientation spread. In contrast, for precipitation on unrelated γ/α interfaces (where the ε-Cu can only be rationally oriented w.r.t. to ferrite) a considerable orientation scatter was observed as illustrated in figure 2b. Hence, the nature of the nucleation site can control, to a large extent, the orientation spread of the product phase.

In summary, the application of $2^{1}/2$D imaging is potentially a very powerful tool in analysing orientation scatter and in determining the effect of nucleation site on the strictness with which the orientation relationship is obeyed.

References

Hirsch P B, Howie A, Nicholson R B, Pashley D W and Whelan M J (1965) "Electron Microscopy of Thin Films", Butterworths.
Kurdjumov G and Sachs G (1930) Z. Phys. 64 325.

*It should be noted that the operating reflection used to form the images in figure 2b was deliberately displaced from the optic axis of the microscope to produce an additional constant displacement of the image. The orientation spread is then seen as a perturbation on this fixed shift.

Antiphase domain boundary tubes in plastically deformed, ordered Fe–30.5at%Al alloy

C T Chou and P B Hirsch

Department of Metallurgy and Science of Materials, University of Oxford, Parks Road, Oxford OX1 3PH

1. Introduction

Vidoz and Brown (1962) and Vasilyev and Orlov (1963) suggested that the increase of workhardening of an alloy with long range order relative to that of the same alloy in the disordered state may be due to the generation of antiphase domain boundary (APB) tubes formed by glide of jogs on super-dislocations which are not aligned along the Burgers vector direction (fig. 1). The superdislocation in fig. 1a consists of two partial dislocations, ABC and DEF, each with Burgers vector b_p, separated by a ribbon of APB. If the jogs B, E, produced by intersection with a forest dislocation are not aligned along the Burgers vector direction, a tube of APBs is generated (fig. 1b) when the dislocation moves and the jogs move conservatively. The tube is parallel to the Burgers vector direction, and if the jogs are pro-duced by intersection with superdislocations, the minimum thickness of the tube is the height of an elementary jog (in the superlattice) produced by intersection by a dislocation with Burgers vector $2b_p$. Tubes of larger thickness would arise from conservative motion of non-aligned superjogs. The width of the tubes would be controlled by the detailed configuration of the dislocations during the intersection mechanism. Tubes of APB could also be formed by the annihilation of superdislocations of screw character by cross-slip of the individual partial dislocations along different planes (see fig. 2).

So far there is only limited evidence for the existence of APB tubes. Kear (1964) reported the existence of segments of APB parallel to the Burgers vector direction in deformed ordered Cu_3Au crystals. Crawford (1976) observed faint lines along the Burgers vector direction on micrographs taken in superlattice reflections and streaks on diffraction patterns, of heavily deformed single crystals of ordered Fe-35.5 at % Al; these effects were consistent with the existence of APB tubes of varying thicknesses and widths.

2. Experimental Results

ections parallel to (011)from specimens of Fe 30.5 at % Al with B2 long range order ($b_p = \frac{1}{2}$<111>), deformed in compression at room temperature to about 4% shear strain, were examined using the "weak beam" technique with superlattice and fundamental reflections, in a JEOL JEM 100B microscope. Fig. 3a shows faint lines parallel to [11$\bar{1}$] in a dark field micrograph taken under weak beam conditions in the $\underline{g}$ = $\bar{1}$11 superlattice reflection, with ∿ 3g satisfied; the deviation parameter ω = s ξ_g ∿ 45, exposure time

$\sim$ 30 secs. Fig. 3b shows the same area taken in the fundamental $2\bar{1}\bar{1}$ reflection (exposure time $\sim$ 8 secs); the lines along [$1\bar{1}1$] are not visible. Most of the dislocations in fig. 3b are edge dislocation dipoles, which show the characteristic inside-outside contrast change when imaged in $\underline{g}$ and - $\underline{g}$. The lines along [$1\bar{1}1$] in fig. 3a have widths between $\sim$ 30Å and $\sim$ 100Å; some are observed to terminate at the dislocations, others apparently change their widths there. Figs. 4a, b show faint lines parallel to the projection of [$1\bar{1}1$] directions in a dark field micrograph taken in the $11\bar{1}$ reflection (fig. 4a, ω = 0) and under weak beam conditions (fig. 4b dark field image in $\underline{g}$ = $11\bar{1}$, with $\sim$ 1.8 $\underline{g}$ satisfied for reflection, $\omega \sim$ 20). While the lines show oscillatory contrast under weak beam conditions (fig. 4b), the contrast is uniform when ω = 0 (fig. 4a).

3. Discussion

The contrast shown by the lines along [111] can be summarised as follows:-
1) The lines are along <111>, the Burgers vector directions. 2) The defects are visible only when superlattice reflections are excited. They are not screw dislocations; the latter are imaged strongly as pairs of partials using the fundamental reflection. 3) The oscillatory contrast in fig. 4b shows that the defects are line defects. 4) The fringes in dark field are symmetrical about the centre of the specimen under two beam conditions, and no fringes occur when ω = 0.

1, 2 and 3 are as expected from APB tubes. In order to interpret the contrast in detail, and in particular 4, two beam and n-beam image contrast calculations were carried out using the method of scattering matrices (Hirsch, Howie, Nicholson, Pashley and Whelan 1965) for two overlapping APBs, i.e. two stacking faults, each with phase angle $\alpha = \pi$ (van Landuyt, Gevers and Amelinckx 1964). Because the anomalous absorption parameter ξ_g' is expected to be very large compared to the crystal thickness t, anomalous absorption effects can be neglected, and provided the distance (along the column) between the APBs (t_2) is small compared to the extinction distance of the superlattice reflection (ξ_g), it can be shown that under two-beam conditions to a good approximation

$$I_g(t)/I_o(0) = \exp\left(- \frac{2\pi t}{\xi_o'}\right) \sin^2\beta \; \{\sin^2(\pi\Delta kt) - 4\sin(\pi\Delta kt_2)\sin(\pi\Delta kt)$$

$$\chi \; [\sin^2\beta \cos\pi\Delta k(t-t_2) + \cos^2\beta \cos(2\pi\Delta kt')] + 4\sin^2(\pi\Delta kt_2)\cos^2\beta\} \qquad (1)$$

where $I_g(t)$ is the dark field intensity at thickness t of crystal, t' is the distance of the middle of the pair of overlapping APBs (along the column) from the middle of the crystal (see fig. 5), ξ_o' is the mean absorption length, $\beta = \cot^{-1}\omega$, $\Delta k = \sqrt{1+\omega^2}/\xi_g$. Results of n-beam calculations show this formula to reproduce the contrast to a good approximation. Equation (1) shows that the oscillatory contrast from an inclined defect arises from the $\cos^2\beta \cos(2\pi\Delta kt')$ term which is symmetrical about the centre of the foil, as observed. At the reflecting position $\beta = \pi/2$ and the oscillatory term disappears; thus uniform contrast is predicted, as observed. The disappearance of the oscillations at ω = 0 is characteristic of the contrast for two overlapping faults each with $\alpha = \pi$; the overlapping APBs have the same effect as a change in thickness $2t_2$. This is because at ω = 0 the two Bloch waves change branches at the first APB, travel with the "wrong" velocity till the second APB, where they switch back again to their original branch. The two Bloch waves therefore suffer phase changes of $\pm (2\pi\Delta kt_2)$ respectively, equivalent to a change in thickness of $2t_2$.

The thickness t_2 of a tube can be estimated from the intensity of the tube contrast relative to that of the neighbouring perfect crystal, using equation (1) for the interpretation. A number of tubes parallel to the (011) sections normal to the electron beam in wedge shaped specimens has been investigated, under weak beam conditions. The procedure is as follows: 1) Intensity profiles are obtained by densitometry across the tubes at points at which the tube contrast is a maximum. The peak corresponds approximately to $|\cos(2\pi\Delta kt')| = 1$. 2) The intensities at the peak (I_p) and in the neighbouring perfect crystal (I_b) are measured. 3) The intensities are corrected for diffuse incoherent background determined by displacing the diffraction aperture to the side of $\underline{g}$ in the weak beam image. 4) The ratio of the corrected ($I_p/I_b - 1$) depends only on Δk, t, t_2 and β, according to (1). (The $\sin^2\beta \, \cos\pi\Delta k(t-t_2)$ term can be neglected for large deviation parameters); t is measured from thickness fringes using super-lattice or fundamental reflections. The calculated values of ξ_g appropriate to the alloy are used. Where possible checks are made from projected lengths of inclined defects. 5) Δk is determined from the number of fringes up to thickness t under the weak beam conditions utilised; this also gives β. Knowing Δk, t, and β, t_2 can be calculated from ($I_p/I_b - 1$), using equation (1). The values of ω used are very large, (~ 40), and the intensities are typically 50-100 times lower than for normal weak beam intensities from dislocations using fundamental reflections. Preliminary estimates indicate that the thickness of the tubes giving weakest contrast is of the order of 5-6Å, i.e. corresponding to the elementary thickness of a tube.

4. Conclusions

APB tubes have been observed in deformed ordered Fe 30.5 at % Al alloy. In a few cases they appear to be attached to individual superdislocations, as expected from the original mechanism of formation by the conservative glide of non-aligned jogs. Many tubes are found to be associated with edge dislocation dipoles suggesting that these may be formed by cross-slip and annihilation of screws of opposite sign.

References

Crawford R C 1976 Phil. Mag. 33 529
Hirsch P B, Howie A, Nicholson R B, Pashley D W and Whelan M J 1965
   Electron Microscopy of Thin Crystals (Butterworths)
Kear B H 1964 Acta Met. 12 555
van Landuyt J, Gevers R and Amelinckx S 1964 Physica Status Solidi 7 519
Vasilyev L I and Orlov A N 1963 Phys. Metals Metall. 15 1
Vidoz A E and Brown L M 1962 Phil. Mag. 7 1167

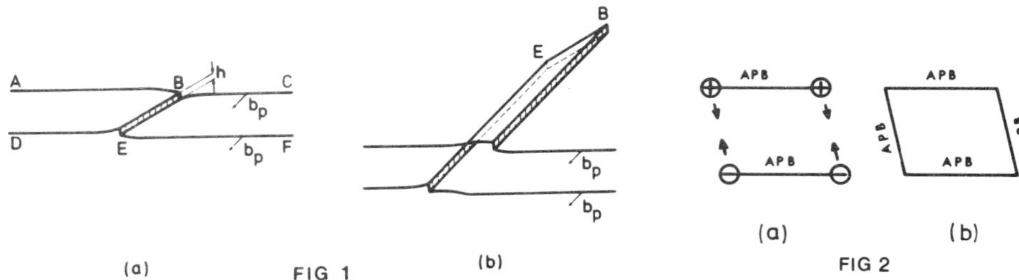

Fig. 1a) Superdislocation consisting of partials ABC, DEF, with two jogs B, E not aligned along the Burgers vector direction, b_p. b) Formation of APB tube by glide of non-aligned jogs B, E.

Fig. 2 Formation of APB tube by cross-slip and annihilation of screws.

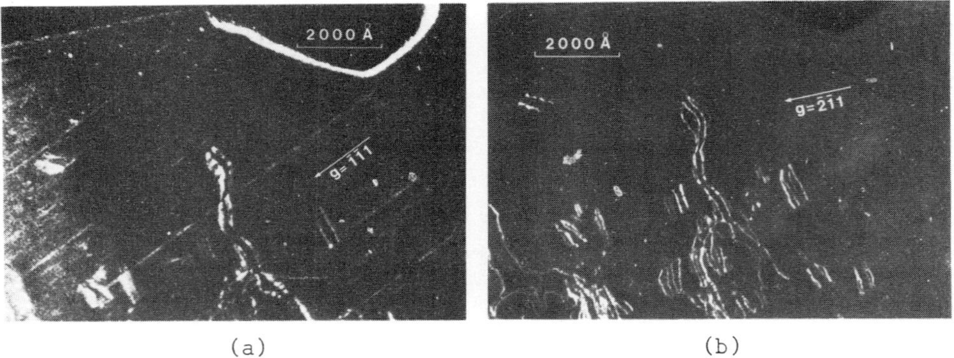

Fig. 3 a) $\bar{1}\bar{1}1$ weak beam image of specimen parallel (011) of deformed B2 ordered Fe $\sim$ 30 at % Al alloy showing APB tube images along $\bar{1}\bar{1}1$. b) $\bar{2}\bar{1}1$ weak beam image of same area showing only dislocations.

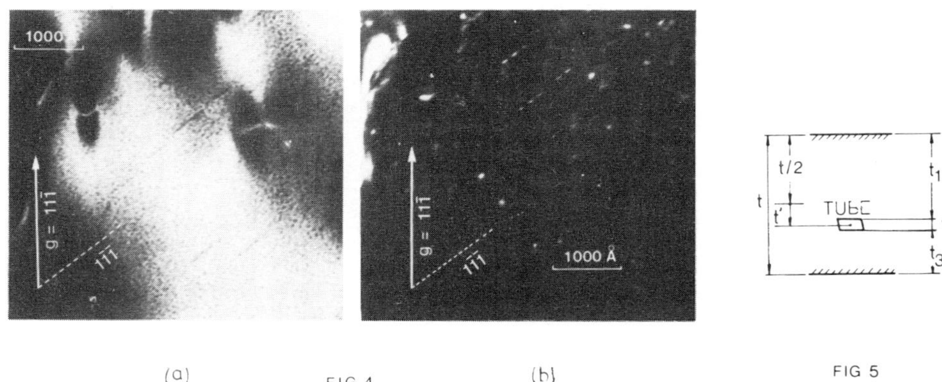

Fig. 4 a) $11\bar{1}$ dark field image of specimen parallel (011) of deformed B2 ordered Fe 30.5 at % Al alloy. Faint lines are along projections of [$1\bar{1}\bar{1}$]. b) $11\bar{1}$ weak beam image of same specimen, showing oscillatory contrast along APB tube defect.

Fig. 5 Notation for theory of contrast from APB tube of thickness t_2.

TEM investigation of the interaction of shear bands with crystals in partially crystallised metallic glasses

P.E. Donovan and W.M. Stobbs
Department of Metallurgy & Materials Science, University of Cambridge.

At room temperature, metallic glasses undergo very localised deformation, for reasons which are not fully understood. As part of an investigation of this behaviour we have observed shear bands in partially crystallised glass. The appearance of shear bands in completely amorphous material has been reported previously (Donovan and Stobbs 1981). The deformed glass produces very little contrast in the TEM, and to locate the shear bands we found it necessary to use a special method for preparing specimens whereby the surface steps associated with the shear bands are retained on one side of the foil. A shear band is then visible both because of the abrupt change in specimen thickness at the step and because of localised structural changes in the shear band in the immediate vicinity of the originally stress free surface.

Three glasses were studied which were known to remain ductile when partially crystallised: $Cu_{50}Zr_{50}$ (Freed and Vander Sande 1978), $Ni_{80}B_{20}$ (Lewis et al. 1979) and $Pd_{80}Si_{20}$ (Masumoto and Maddin 1971). $Cu_{50}Zr_{50}$ and $Ni_{80}B_{20}$ form large crystals, ($\sim$ 1μm), while $Pd_{80}Si_{20}$ transforms uniformly to a microcrystalline ($\sim$5nm) state.

$Cu_{50}Zr_{50}$ proved to be difficult to polish, and hence it was impossible to make detailed observations of the contrast. Only the gross morphology of the shear bands could be studied. Figure 1 shows a typical interaction between a crystal and a shear band in this material; the shear band has been deflected locally by the crystal, passing round it and leaving it undeformed (Freed and Vander Sande 1979).

In $Ni_{80}B_{20}$ a different form of interaction was observed. Here the crystals may be deformed with the glass. Figure 2a shows the interaction of two shear bands with a crystal in this material; the crystal has been sheared and the bands have been permanently deflected from their original planes. Elastic side-band images of the points where shear bands reach the crystals suggest that the interfaces are extremely well-bonded, as shown in figure 2b. There is no sign of any additional bright speckle contrast which might have indicated the presence of extra voids or particularly disordered material. Long-range interactions also occur in $Ni_{80}B_{20}$, as in figure 2c where a shear band has been abruptly deflected by a crystal some distance away.

Perhaps the most interesting behaviour is that of $Pd_{80}Si_{20}$. This material undergoes localised deformation like a glass even when it is microcrystalline. Figure 3a is a bright-field image of a shear band, AB, and a crack, C, both of which appear to have propagated between the crystallites so that their edges are bumpy. In figure 3b, an elastic side band image, bright speckle suggests that voids are present in the plastic zone around the crack, but no voids can be seen in the shear band in this image.

The similarity in behaviour between amorphous and microcrystalline material was unexpected. In particular, the apparent width of the deformed region is similar in both cases, and broader than the average crystal diameter. This helps to differentiate between different models for the shear process in both classes of material.

Donovan P E, Stobbs W M, Acta Metall. <u>29</u> 1419 (1981)
Freed R L, Vander Sande J B, J. Non.Crys.Sol. <u>27</u> 9 (1978)
Freed R L, Vander Sande J B, Met. Trans. <u>10A</u> 1621 (1979)
Lewis B G, Davies H A, Ward K D, Scripta Metall. <u>13</u> 313 (1979)
Masumoto T, Maddin R, Acta Metall. <u>19</u> 725 (1971)

We are grateful to Professor R.W.K. Honeycombe for provision of laboratory facilities, to Dr. P.H. Gaskell for the $Pd_{80}Si_{20}$ ribbon, to Professor J. Vander Sande for advice and to the SERC for financial support.

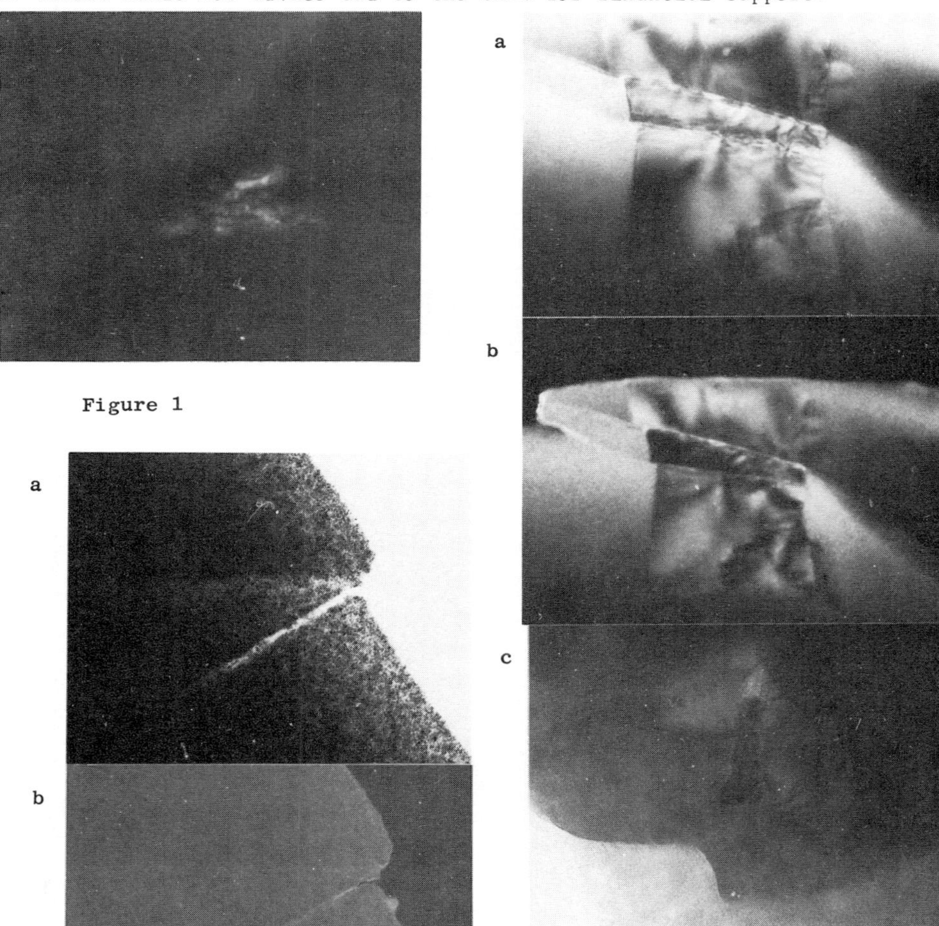

Figure 1

Figure 2

Figure 3

A combined CTEM, STEM, SEM study of grain boundary precipitation in Co–20Fe

Shirley Payne and Paul Howell*
University of Oxford, Department of Metallurgy and Science of Materials,
Parks Road, Oxford OX1 3PH
*Department of Materials Science and Engineering, Pennsylvania State
University, PA 16802

Introduction

The rate of growth of grain boundary precipitates has in previous studies
frequently been characterized by studying the growth rate of particular
precipitates (often the largest). The system Co–20%Fe has been the subject
of a number of precipitation studies (eg. Ryder et al. 1967) since the high
temperature f.c.c. phase (γ) partially transforms on cooling to a low
temperature b.c.c. phase (α). The nature of α precipitation at γ grain
boundaries is observed to be strongly boundary dependent as shown in fig.1.
In this study backscattered electron imaging was used to trace the development of the size distribution of boundary precipitates along boundaries
whose misorientations were determined from electron channeling patterns
(ECPs) and serial sectioning. On a finer scale the interfaces between the
precipitates and the adjacent grains were studied using TEM, and STEM
analyses were used to study the composition profiles around them.

Sample treatment and experimental procedure

Co–19.6 w/o Fe specimens in the form of thin strips were solution treated
at 1200°C for 30 mins, aged for various lengths of time at 730°C and then
water quenched. Specimens for TEM were prepared in a twin-jet electropolisher and examined in a Philips EM400 TEM/STEM equipped with an EDAX
X-ray analyser. For microanalysis the specimen was tilted to an angle of
45° in a low background beryllium holder and the resultant spectrum was
analysed using the standard EDAX "EN" program. Specimens for SEM were jet
electropolished and examined in an ISI 100B using an annular Schotky
barrier detector for both the backscattered imaging and the ECP recording.
Precipitate lengths were measured from enlarged prints using a Quantimet
720 automatic image analysing computer and were corrected for sectioning
effect assuming an approximate shape of abutting spherical caps.

TEM observations

All α precipitates grew with $(110)_\alpha||(111)_\gamma$ with respect to at least one of
the abutting grains but with a range of orientations between $[1\bar{1}1]_\alpha||[1\bar{1}0]_\gamma$
(K–S) and $[001]_\alpha||[0\bar{1}1]_\gamma$ (N–W). On some boundaries all precipitates
nucleated with exactly the same orientation while in others adjacent precipitates showed slightly differing orientations, giving rise to low angle
boundaries where they impinged. A typical precipitate in a specimen aged
6 hrs 730°C is shown in fig.2. A well developed facet lies along the
'coherent' plane and ledges are visible in the region of boundary curving
away from this.

STEM observations

Analysis at precipitate free boundaries showed no detectable segregation
and analysis of the parent grains was unaffected by the presence or absence
of any quench-aided intragranular transformation. The results of analyses
performed over the area shown in fig.3 are shown in table 1 and fig.4.
These consistently give the iron content of the untransformed γ well away
from the precipitates as 21.6 a/o, slightly higher than the overall composition of 19.6 w/o given by wet analysis. The measured compositions cannot
therefore be taken as absolute, but reflect relative compositional variations in this area. All the α compositions lay within 2 standard deviations
of the mean, as did all γ compositions measured at points >0.5µm from the
nearest interface. The compositions close to the precipitates varied from
interface to interface and with position. The compositions shown in fig.4
are taken from within 0.5µm of interfaces marked a,b,c,d but excluding
analyses close to the precipitate ends.

Class no.	source of analyses	mean a/o Co	standard devn. (a/o Co)	no. in class
1	α	74.7	0.3	30
2	γ >0.5µm from interfaces	78.3	0.3	24
3	γ <0.5µm from interfaces	81.6	2.8	37
4	<0.5µm from interface b	81.5	2.1	6
5	<0.5µm from interface d	80.9	2.2	7
6	as (4) excluding measurement at ppt end.(b')	80.7	0.8	5
7	as (5) excluding measurement at ppt end.(d')	80.1	0.4	5
8	<0.5µm from interface e	82.5	3.3	6
9	<0.5µm from interface a	78.7	0.9	5

Table 1

Close to the planar interface 'a' the bulk composition is maintained
virtually to the interface, whereas at the curved and ledged interfaces
the γ is depleted in Fe. This effect is more marked at the curved interface c than at the mixed ledged and facetted interfaces b and d. Signifi-
cence tests gave a 98% confidence level for the difference between the
compositions close to 'a' and those close to the other boundaries.

SEM growth measurements

During the earlier stages of ageing the precipitate sizes followed a log-
normal distribution as shown in fig.5. Prolonged ageing produced a skew
towards larger precipitates (fig.6), indicating coarsening. A summary of
the results obtained from analysis of the measured size distributions is
shown in fig.7.

Discussion of results

The STEM analyses show growth inhibition at the low energy facets. At the
mobile curved interface c growth proceeds at a rate limited by the diffusion of iron to the precipitate but when the interface lies along a low
energy plane as at a its position is metastable. Growth is then limited
by the rate of nucleation of new ledges. This slow growth causes little
deviation of the composition of the surrounding matrix from its bulk value.
The slope of a log/log plot of precipitate length against time gives the
exponent n for any power law growth, indicating the rate controlling
mechanism. As seen in fig.7 the arithmetic and geometric mean precipitate
sizes increase steadily with time giving n $\approx$ 0.26. This agrees with the

value $0.25 \lesssim n \lesssim 0.37$ given by Brailsford and Aaron's (1969) treatment of the collector plate mechanism and this mechanism was confirmed by other STEM analyses (Howell and Payne 1981).

The sizes of the largest precipitates show a different behaviour from that of the means. Initially, the maximum precipitate size increases more rapidly than the mean with $n \approx 0.6$ but later increases more slowly with $n \approx 0.22$. This reflects the development of the size distribution from log-normal during growth to that typical of coarsening. A coarsening distribution shows an increase of frequency with size up to a cut off at a small multiple of the size of the precipitate which is neither growing nor dissolving. This multiple depends on the rate controlling mechanism of coarsening. For the 113hr aged material $\ell_{max} \approx 2.0\ \bar{\ell}$, in good agreement with the value predicted for the cut off in precipitate size when coarsening is controlled by interfacial diffusion.

Conclusion

Where there are several possible growth mechanisms the overall development of the precipitate size distribution must be studied, rather than either the mean or maximum size alone, to determine unambiguously which growth or coarsening mode is dominant and which mechanism is rate controlling. Grain boundary allotriomorphs in Co-20Fe aged at 730°C grow by a boundary dependent collector plate mechanism. As the precipitates grow, low energy facets develop which can move only by the propogation of ledges. Inhibition of thickening leads to interface controlled growth but the boundary diffusion fields soon overlap and coarsening occurs with interfacial control.

References

Brailsford A D and Aaron H B 1969 J.Appl.Phys. **40** 1702
Howell P R and Payne S M in preparation.
Ryder P L, Pitsch W and Mehl R F 1967 Acta Met. **15** 1431

FIG.1. Channeling contrast micrograph showing wide variation of grainboundary precipitation in Co-20Fe aged 113hrs 730°C.

FIG.2. Facetted grain boundary precipitate in Co-20Fe aged 6hrs 730°C.

FIG.3. TEM montage of area over which STEM analyses performed.

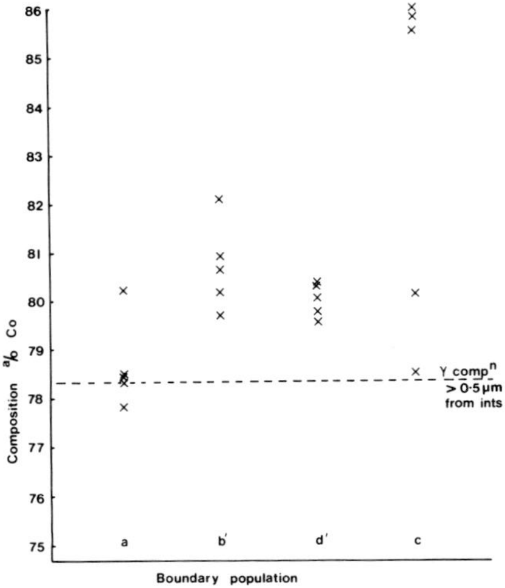

FIG.4. Results of spot analyses made adjacent to interfaces marked in fig.3.

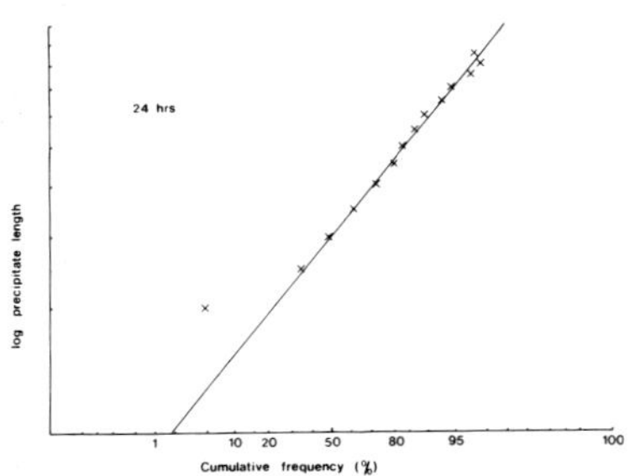

FIG.5. Log-normal plot of precipitate size distribution for Co-20Fe aged 24hrs. 730°C.

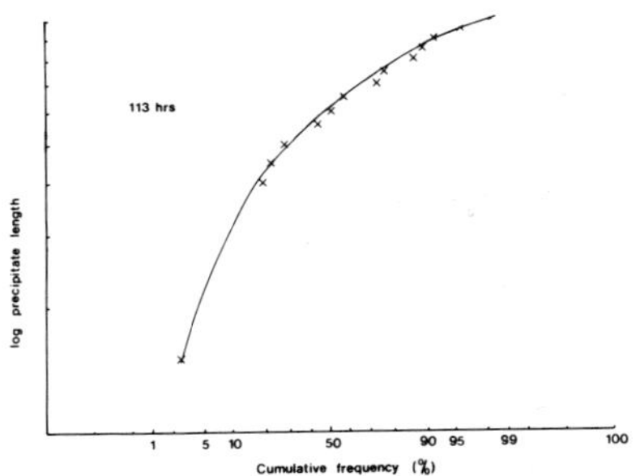

FIG.6. Precipitate size distribution for Co-20Fe aged 113hrs. 730°C.

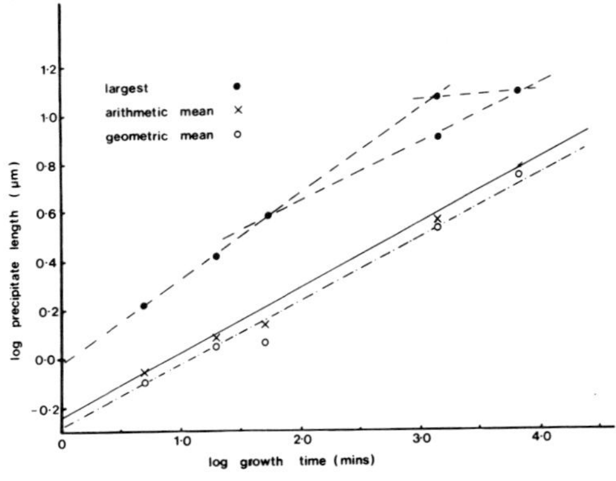

FIG.7. Changes in parameters characterising Co-20Fe precipitate size distribution with ageing time.

Dislocation movements in twin boundaries in copper

M.E. Welland and D.J. Dingley

H.H. Wills Physics Laboratory, University of Bristol, Bristol BS8 1TL

1.   Introduction

In the last decade much advance has been made in interpreting and
defining the structure of grain boundaries of polycrystalline metals
using transmission electron microscopy (Smith (1975) and Carter and Sass
(1980)). Here we present the first results of a detailed investigation
of the morphology of gb's in copper and copper-bismuth alloy. They are
concerned with grain boundary dislocation (gbd) behaviour and boundary
movements during recrystallization of deformed Cu with special emphasis
on the growth of Σ 3 twins. Although gbd's in Σ 3 twins have been
investigated before, for example Mori and Tangri (1979), the importance
of twinning during recrystallization warrants further study to elucidate
the many varied mechanisms.

The effects of annealing time on bulk specimens, beam heating on thin
foils, and repeated annealing of thin foils have been examined.

2.   Experimental

The material used was industrial grade pure copper sheet which was cold
rolled to 90% reduction and annealed at 220°C for times varying from
five minutes to eight hours. Rapid warm up and cooling down times
allowed for effective freezing-in of the partially recrystallized state.
Thin foils were prepared and examined in a Phillips 301 and latterly a
JEOL 100 CX TEM. In those specimens which had only been annealed for
short periods, it was observed that the electron beam stimulated dislocation movement. These foils were subsequently re-annealed so as to
follow further structural changes. Due to the lack of a heating stage
for in-situ annealing experiments, the annealed samples were removed and
annealed in a vacuum furnace at a pressure of $\sim 10^{-6}$ Torr and temperature
of 220°C. In order to reduce oxidation to a minimum the furnace tube and
pumping system were initially purged with argon. This enabled up to five
or six repeated heat treatments without so severely oxidizing the
specimen as to render it opaque in the TEM. The problem of accidentally
damaging thin foils during these operations was overcome by mounting them
in a small brass cup which could be inserted directly into the specimen
holder of the TEM.

3.   Results

3.1 Electron beam effects.

Fig 1 shows a Σ3 twin in the early stages of recrystallization with well
formed coherent gb   planes and a heavily dislocated region where the
incoherent boundary might be expected to form.   During initial investigation of this region it was noticed that slight focussing of the beam in
the region of the coherent boundary produced some gbd movement.   Figs 2a
and b.   A fully focussed probe stimulated rapid movement although certain
dislocations were observed to be sessile.   The beam was then selectively
condensed on different regions of the twin and any dislocation movement
recorded.

The coherent boundary beam became
invisible in a simultaneous twin (T)
- matrix (M) {200} reflection, and
the (311)$_T$, (113)$_M$ (s_g=0)   diffraction spots were superimposed.
This is consistent with a perfectly
coincident Σ3 relationship and an
exact (111) twin boundary.   Tilting
experiments identified the glissile
dislocations as having Burgers vec-
tor $\underline{b}$ = ^{a}/6 [1$\bar{2}$1] . Gliding of
such dislocations in the coherent
boundary is accompanied by an interfacial shift of ^{a}/3 [111] . Movement of dislocations in the deformed, "incoherent" region was also
seen although in no apparent direction and at a much reduced rate.
These dislocations could not be
identified.

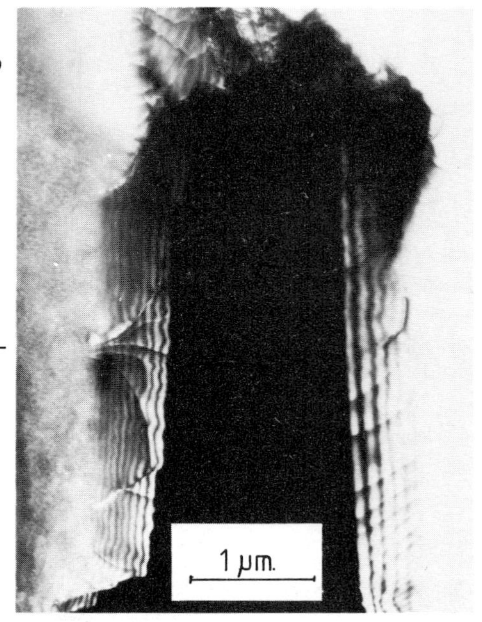

Fig 1.   Dark field g = [1$\bar{1}$1]$_M$.

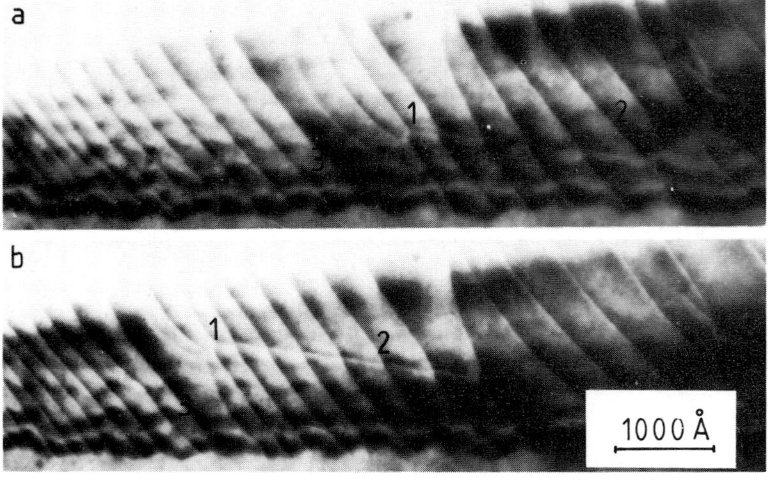

Figs 2a and b.   Numbers refer to identical dislocations in both
pictures.   g = [1$\bar{1}$1]$_M$.

3.2 Annealing experiments.

Fig 3a shows an isolated twin in material which had been recrystallized for 15 minutes. As in the previous section one end of the twin is seen to be dislocated. One of the coherent boundaries is seen to contain a fault. After tilting experiments to categorise the orientation and boundary plane the specimen was removed from the microscope and further annealed in steps of 15 minutes, as described above, up to an anneal time of 60 minutes. Changes after this became very slow and degradation in image quality due to contamination rendered study impossible. Figs 3b, c and d were taken after 30, 45 and 60 minutes annealing respectively, and in identical dark field $\{111\}_M$ diffracting conditions as in Fig 3a. It is observed that the fault in the boundary gradually becomes smaller and eventually disappears in Fig 3d, and that the deformed region of the twin rapidly becomes less distorted and planar. This micrograph also shows new dislocations in the coherent interface which were stationary under the electron beam, and had clearly been generated during the heat treatments. Further heat treatment resulted in movement of these dislocations which were identified as having Burgers vector $\underline{b} = {}^a/6\,[1\bar{2}1]$.

Fig 3a.　15 minutes anneal.

Fig 3b.　30 minutes anneal.

Fig 3c.　45 minutes anneal.

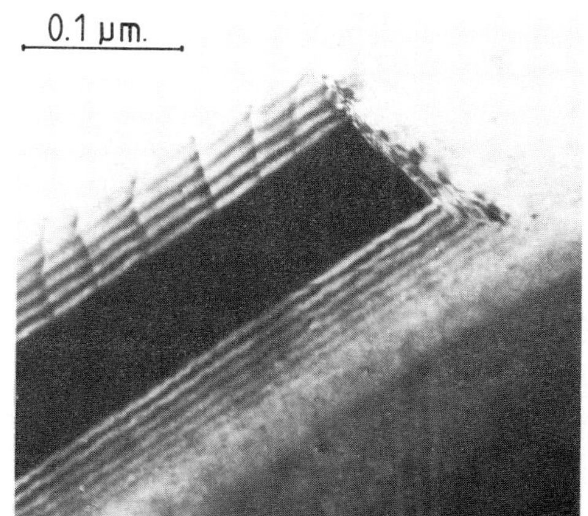

Fig 3d.　60 minutes anneal.

4. <u>Discussion</u>

It is clear that in the early stages of recrystallization, the incoherent portion of the growing twin is very poorly defined. This region consists of a mass of entangled dislocations with no well formed interfacial plane. As twin growth proceeds $^a/6 \langle 121 \rangle$ gbd's appear to be forced from this region down along the coherent boundary (producing lateral growth of the twin). From the observation that these dislocations are more clearly spaced after the short annealing times, and become more widely spaced at longer annealing times, it would seem that the process of generation of the $^a/6 \langle 121 \rangle$ gbd's slows down as the incoherent region takes on a more defined structure. There were no observations of $^a/2 \langle 110 \rangle$ type dislocations in the coherent boundaries or of $^a/6 \langle 112 \rangle$ type which were likely to have been generated by the disassociation of $\langle 110 \rangle$ lattice dislocations, as observed by Mori and Tangri (1979) in aluminium.

The fault seen in Figs 3a - 3c is likely to impede the motion of the $^a/6 \langle 121 \rangle$ gbd's and that is probably why few were seen after short annealing time. However, once the fault had disappeared, such dislocations were seen to move along the length of the coherent boundary. It is conceivable that the fault was eliminated by absorption of these dislocations, although no direct evidence for this was observed. It is curious that no $^a/6 \langle 121 \rangle$ dislocations were observed in the opposite coherent boundary in Fig 3.

In this case, beam induced dislocation movement is believed to be due to a heating effect, rather than a contamination induced strain (Gale and Hale (1961) and Pashley and Presland (1961)), as movement started as soon as the beam was focussed in the boundary region and the contamination rate in the JEOL 100CX was very low.

5. <u>Conclusion</u>

$^a/6 \langle 121 \rangle$ dislocations were observed, in the coherent boundary of partially recrystallized Cu, to increase in spacing with annealing time. It is concluded that the $^a/6 \langle 121 \rangle$ dislocations were generated as the incoherent region became better defined, and moved along the coherent boundary producing lateral growth and possibly eliminating a faulted region in it.

6. <u>Acknowledgements</u>

M.E. Welland was funded through an SERC Postgraduate Grant.

7. <u>References</u>

Carter C.B. and Sass S.L. (1980) MSC   Report 4232, Cornell University, Ithaca, USA.

Gale B. and Hale K.F. (1961) Brit.J.Appl.Phys. <u>14</u> 218.

Mori T. and Tangri K. (1979) Met.Trans.A. <u>10A</u> 733.

Pashley D.W. and Presland A.E.B. (1961) Phil.Mag. <u>6</u> 1003.

Smith D.A. (1975) Journal de Physique pp C4.1-C4.15

Correlation of dislocation structure with cathodoluminescence at hardness indents in MgO

S D Berger and L M Brown

Cavendish Laboratory, Madingley Road, Cambridge CB3 OHE

1. Introduction

Two previous studies of the cathodoluminescence (CL) from hardness indents
in MgO have been made. Veledintskaya et al.(1975) explained the change in
CL contrast seen after annealing of the indents by assuming that only point
defects act as recombination sites. Pennycook and Brown (1979) however
attribute much of the CL to dislocations and explain the change in contrast
in terms of changes in dislocation density. The observed phenomenon is
shown in fig. 1. Briefly, the centre of the indent is nonluminescent at
room temperature, but an anneal at 500°C renders it highly luminescent.
To explain this more fully a knowledge is required of the exact nature of
the damage caused to the material directly below the indenter, but this is
not available at present. However, a comparison of CL spectra from indents
before and after annealing should indicate whether any new defects become
active as a result of the anneal, or whether the phenomenon is to be explained solely in terms of defects present at room temperature.

2. Experiment and Results

Two single crystals of 4N purity MgO were indented with a 100 g load
Vickers microhardness indenter at room temperature. One of the crystals
was then annealed in air at 500°C for half an hour. CL micrographs of the
two crystals are shown in fig. 1. Spectra were taken from single indents
under identical system conditions at a temperature of T = 6.4 K. They
were obtained using a liquid helium CL stage designed for a JEOL JSM–35 SEM
at Imperial College London. The spectra are shown in fig. 2. Spectrum A
is from the unannealed indent and spectrum B is from the annealed indent.
As expected, the spectra taken at liquid helium temperature show much
greater detail than spectra taken at room temperature (Datta et al. 1979,
Pennycook and Brown 1979), but the spectra still take the form of two
broad bands. When comparing the spectra taken before and after the anneal,
little difference is seen in the shape, though there is a small shift
($\sim$ 10 nm) in the peak positions. This suggests that no new CL centres have
been created during the heat treatment. There is, however, a change in
the relative intensity of the two bands. It can be seen that after the
anneal the intensity of the long wavelength band is approximately double
what it was before, while the intensity of the short wavelength band has
slightly decreased.

3. Interpretation

Under the action of the indenter, both point defects and dislocations are
produced and the results of Chen et al.(1975) and Pennycook et al.(1980)

suggest that the blue CL seen in fig. 2 is due to recombination at these sites. Datta et al.(1980) attribute the red band to iron impurities. Keh (1960) showed that annealed dislocations around a hardness indent become trapped by the diffusion of impurities from the surrounding crystal. The increase in the red CL in spectrum B can therefore be ascribed to a local increase in the impurity content. The point defects also become mobile at the annealing temperature (Sibley et al.1969) and tend to diffuse away from the deformed region, whereas the dislocations are in the main unaffected by the heat treatment. Thus the slight decrease seen in the short wavelength band of spectrum B confirms the observation of Pennycook et al. (1980) that most of the original blue CL is due to the dislocations, although the evidence is still rather indirect.

Acknowledgments

We wish to thank Dr D B Holt and Mr P Grant of Imperial College London for help in obtaining the spectra; Dr S Datta for many useful discussions; Professor A H Cook for provision of laboratory facilities, and the Science Research Council for financial support.

References

Chen Y, Abraham M M, Turner T J and Nelson C M 1975 Phil. Mag. 32 99
Datta S, Boswarva I M and Holt D B 1979 J. Phys. Chem. Solids 40 567
Datta S, Boswarva I M and Holt D B 1980 J. de Physique 41 C6-522
Keh A S 1960 J. Appl. Phys. 32 1538
Pennycook S J and Brown L M 1979 J. Luminescence 18/19 905
Pennycook S J, Brown L M and McGovern S 1980 Inst. Phys. Conf. Ser. No. 52
  Ch 3 161
Sibley W A, Kolopus J C and Mallard W C 1969 Phys. Stat. Sol. 31 223
Velednitskaya M A, Rozhanskii V N, Comolova L F, Saparin G V, Schreiber J
  and Brummer O 1975 Phys. Stat. Sol.(a) 32 123

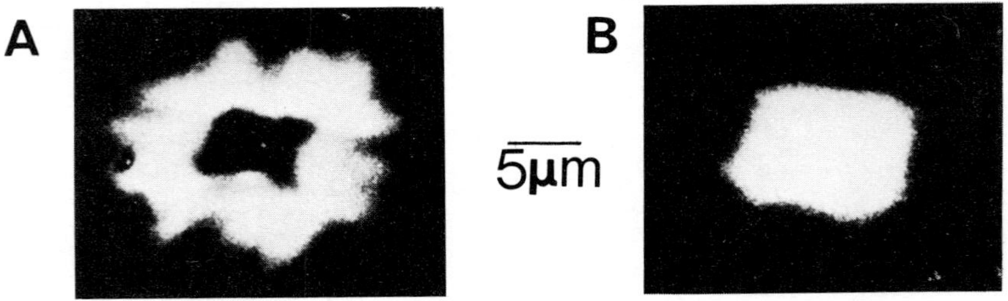

Fig.1   CL from 100 g hardness indents in MgO
        A: from an un-annealed crystal;
        B: from a crystal annealed in air at 500°C for 30 minutes.

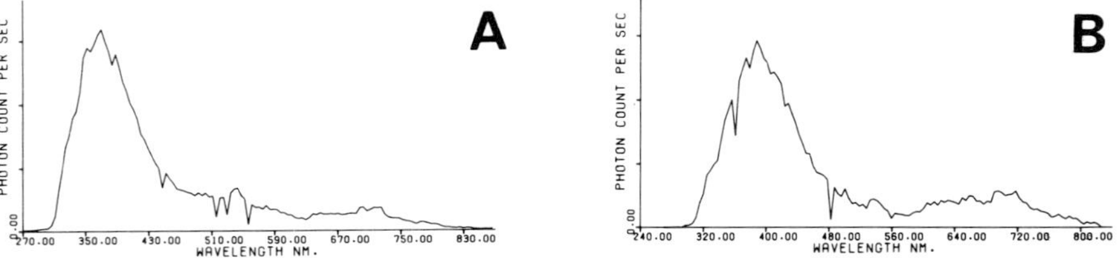

Fig.2   Spectra taken at T = 6.4K . A: from the un-annealed crystal; B: from
        the annealed crystal, Spectra are not corrected for instrumental
        response. (Corrected spectra will be presented at a later date.)

Defects in polydiacetylene single crystals

R J Young, R T Read and J Petermann*

Department of Materials, Queen Mary College, Mile End Road, London E1 4NS
*Werkstoffphysik und Werkstofftechnologie, University of the Saarland,
Bau 2, 6600 Saarbrücken, West Germany

Abstract.   The structure of single crystals of a substituted
diacetylene polymer (pTS) has been investigated using transmission
electron microscopy. It has been found that edge dislocations with
Burgers vector parallel to the chain direction are present at a density
of up to 10^{13} m^{-2}. It is suggested that the dislocations may be present
first of all in the monomer crystals and become locked into the monomer
structure during polymerization. A common defect in the crystals was a
stacking fault; using dark-field microscopy it has been found to have a
displacement vector of $\frac{1}{2}$ [$1\bar{2}1$]. It has been shown that such a stacking
fault can be accommodated without any disruption to either the molecular
backbone or the relatively large side-groups on the molecule.

1. Introduction

All crystals of materials are known to contain defects such as dislocations,
stacking faults and vacancies but knowledge of such defects in polymer
crystals is somewhat limited. This is rather unfortunate because there are
several reasons why interest should be shown in defects in macromolecular
crystals, for example:

(a) From a fundamental view-point it is essential to know what types of
defects exist in polymer crystals and how they compare with defects in
crystals of other materials;

(b) It is desirable to know if there is any correlation between the
presence of defects and the mechanical or electrical properties of polymer
crystals.

The study of defects in metal crystals has been undertaken using electron
microscopy for over 25 years and a high level of understanding of metal
defects has now been reached. In contrast, relatively little is known
about the types of defects that can exist in polymer crystals. There are
two main reasons for this. The first is that single crystals of
conventional polymers are normally in the form of microscopic chain-folded
lamellae with the molecules orientated approximately perpendicular to the
large crystal surface [1]. This lamellar morphology imposes limitations
upon the orientations in which the crystals can be viewed.

The second reason is that conventional polymers tend to suffer radiation
damage in the electron beam [2]. This means that the time available to
study defects in any particular crystal is severely limited as the crystals
rapidly become damaged by the radiation. In addition, defects can be

induced by the radiation itself.

There has been some notable success in identifying dislocations in lamellar polyethylene single crystal: edge dislocations caused by the termination of fold-planes within the crystals have been identified using Moire techniques [3], interfacial dislocations have been seen in overlapping single crystals [4] and screw dislocations with Burgers vectors parallel to the chain direction have been identified through their associated surface relaxations [5]. However, the chain-folded crystal morphology restricts investigations to viewing the crystals in the electron microscope parallel to the chain direction and defects which can only be observed by viewing crystals in directions perpendicular to the chain direction therefore cannot be investigated. This particular problem has now been overcome by the preparation of large, relatively perfect polymer single-crystals by solid-state polymerization [6]. This technique involves the growth of crystals of the monomer in the form of single crystal and the subsequent polymerizat-ion reaction takes place in the solid state [6]. Suitable monomers include sulphur nitride [7] and certain substituted diacetylenes [8]. Polymers from these monomers can be obtained in the form of good single crystals with the molecules in a chain-extended conformation. Previous investigat-ions by the authors using transmission electron microscopy have shown that polymeric sulphur nitride $(SN)_x$ [9, 10] and polydiacetylenes [11, 12] can be obtained in the form of thin single-crystal films in which the molecules are extended in the plane of the film.

2. <u>Experimental Procedure</u>

Monomer crystals prepared by conventional techniques, described elsewhere [8], were supplied by Dr D Bloor and Mr D Ando of Queen Mary College. The pTS polymer crystals were prepared by the solid-state polymerization of monomer crystals grown from solution on a carbon film, as described in detail elsewhere [17, 18].

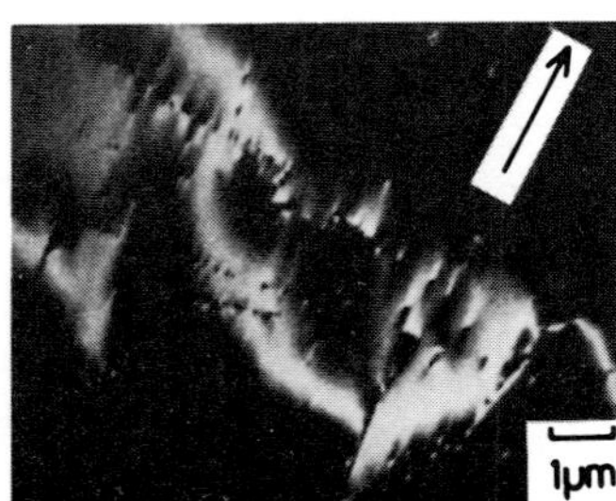

Fig 1. Edge dislocations in pTS single crystal ($\underline{g}$ = [2$\bar{1}$1]).

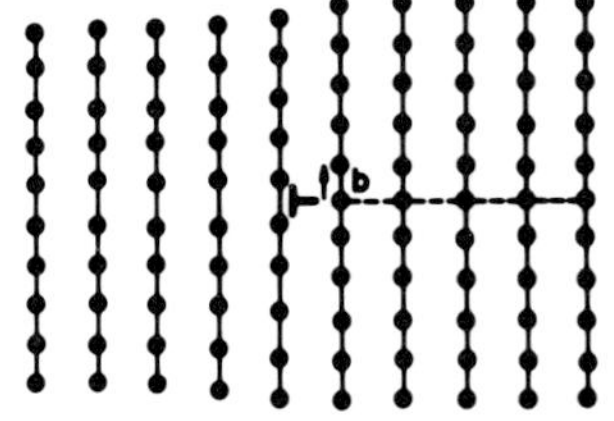

Fig 2. Schematic diagram of edge dislocation in pTS.

3. <u>Dislocations</u>

Detailed examination of pTS crystals by centred dark-field electron microscopy showed that in certain areas of the crystal in the Bragg con-dition there were small regions that exhibited characteristic black/white contrast patches. An example of such an area is shown in Figure 1. These regions are reminiscent of the black/white patches observed for screw dislocations in lamellar polyethylene single crystals [5, 13, 14] which were shown to be similar to the type of contrast observed for screw dislocations seen end-on in thin foils of metals by transmission electron microscopy [15, 16]. The similarity between the contrast due to surface relaxation [16] observed from the dislocations in the aluminium foil and

polyethylene lamellae and the contrast in the pTS crystals suggests strongly
that the pTS crystals contain dislocations.

Tilting of the specimen holder stage in the electron microscope by up to
$\pm 40^{\circ}$ did not allow any dislocation line to be imaged. This implied that
in the relatively thin ($\sim$ 1000 Å) pTS crystal foils the dislocation lines
are approximately perpendicular to the foil surface. Since the chain
direction of the crystals lies in the planes of the foils [11] it must be
concluded that the dislocation lines are perpendicular to the chain direction, c. The conventional method [15] of determining the Burgers vector of
a dislocation is to view the crystal in different Bragg conditions and see
whether or not the dislocation lines are visible or invisible. However,
for dislocations seen end-on, as in this present investigation, it is not
such a simple matter but the situation has been analysed at least for
isotropic solids by Tunstall, Hirsch and Steeds [16]. Their analysis
allows the identification of both edge and screw dislocations viewed end-
on by using different imaging conditions. It is found that the patches in
Figure 1 have strong contrast with g $=[2\bar{1}1]$ and g $=[002]$ but are very weak in
g $=[2\bar{1}0]$ (where g is the scattering vector used to generate the dark-field
micrograph). Tunstall et al [16] have predicted the nature of the contrast
expected for both screw and edge dislocations viewed end-on. The expected
contrast due to surface relaxation from screw dislocaitons in isotropic
materials is in the form of black/white patches and it is further predicted
that the line of no contrast between the two patches should rotate when the
imaging conditions are changed such that this line is always parallel to
the scattering vector, g. The situation is more complex in the case of
edge dislocations. Contrast is predicted to be very weak when g.b = O when
the pattern always consists of a single patch. When g is parallel to b
(g.b $\neq$ O) the contrast is expected to be stronger, and as for the screw
dislocations, again consists of black/white patches with the line of no
contrast parallel to g.

As far as the pTS crystals used in this present study are concerned, some
general conclusions can be formulated about the dislocations from the
nature of the observed contrast. The contrast is weak when g $=[2\bar{1}0]$ and
strong for g $=[002]$ and $[2\bar{1}1]$. This is consistent with the dislocations
having their Burgers vector parallel to the chain direction, c. Since it
has been shown that the line of the dislocations is perpendicular to the
chain direction it follows that the dislocation in the pTS film must be of
mainly edge character. The density of dislocations in the crystal is found
to vary from crystal to crystal. The areas with high densities contain
up to 10^{13} m^{-2} (10^9 cm^{-2}). The dislocation is sketched schematically in
Figure 2 for a pTS crystal containing a row of molecules lying in the (010)
crystal plane and hence projected parallel to the dislocation line. The
extra half-plane of the dislocation corresponds to an extra half-plane of
chain segments which appear as an extra half-row of segments in the projection in Figure 2. The structure of the dislocations will be discussed in a
forth-coming publication [17].

4. Stacking Faults

Figure 3 is a bright-field electron micrograph of a relatively thick wedge-
shaped pTS crystal obtained using g = $[2\bar{1}1]$. The main features that can be
seen are five defects which lie parallel to the chain direction and become
narrower as they approach the specimen edge. The defects each display
characteristic black/white fringes and closer examination shows that the
contrast is reversed between the dark- and bright-field images. The defects

in Figure 3 are very similar in appearance to stacking faults, such as
those which are found in metals (compare with Figure 10.11a in [15]).

The defects are also found to give either strong or weak contrast depending
upon the g used to image them in the dark-field. It is found that the
defects give strong contrast for g = [2$\bar{1}$1] and g = [4$\bar{2}$1] but appear only
weakly for g = [2$\bar{1}$0]. Examination of many examples of these defects in
different imaging conditions has indicated that they are stacking faults
with a displacement vector of R = ½ [1$\bar{2}$1] as shown in a recent publication
[18].

It is of interest to examine what is meant by a stacking fault with a
displacement vector of R = ½ [1$\bar{2}$1] in terms of the perturbation produced in
the molecular stacking. The plane of the stacking fault can be calculated
from the line of the fault and the displacement vector, R. The plane of
the fault must contain both the line, [001] and R,½[1$\bar{2}$1]: the only plane
which can contain these two directions is (210). The crystal structure of
pTS [19] containing the stacking fault is represented schematically in
Figure 4 as a projection parallel to the chain axis. There are two molecular segments per unit cell and the molecules tend to be planar with the
side groups lying approximately in the (210) planes. The stacking fault
can be accommodated without any disruption to the molecular backbone or to
the relatively large side-groups on the molecule [18].

References

1.   A Keller, Rep Prog Phys 31 (1968(623.
2.   D T Grubb, J Mater Sci 9 (1974) 1715.
3.   P H Lindenmeyer, J Polymer Sci C 15 (1966) 109.
4.   V F Holland and P H Lindenmeyer, J Appl Phys 36 (1965) 3049.
5.   J Petermann and H Gleiter, Phil Mag 25 (1972) 813.
6.   G Wegner, Pure and Appl Chem 49 (1977) 443.
7.   R H Baughman, R R Chance and M J Cohen, J Chem Phys 64 (1976) 1869.
8.   G Wegner, Z Naturforsch 24b (1969) 824.
9.   J Petermann and J M Schultz, J Mater Sci 14 (1979) 891.
10. J M Schultz and J Petermann, Phil Mag A40 (1979) 27.
11. R T Read and R J Young, J Mater Sci 14 (1979) 1968.
12. R J Young, R T Read, D Bloor and D J Ando, Faraday Discussion 68 (1979)
     510.
13. J Petermann and H Gleiter, J Polym Sci A-2 10 (1972) 731.
14. E L Thomas, S L Sass and E J Kramer, Phil Mag 30 (1974) 335.
15. P B Hirsch, A Howie, R B Nicholson, D W Pashley and M J Whelan,
     Electron Microscopy of Thin Crystals, Butterworths London 1965.
16. W J Tunstall, P B Hirsch and J Steeds, Phil Mag 9 (1964) 99.
17. R J Young and J Petermann, J Polym Sci Polym Phys Ed (to be published).
18. R J Young, R T Read and J Petermann, J Mater Sci 16 (1981) 1835.
19. D Kobelt and E F Paulus, Acta Cryst B30 (1974) 232.

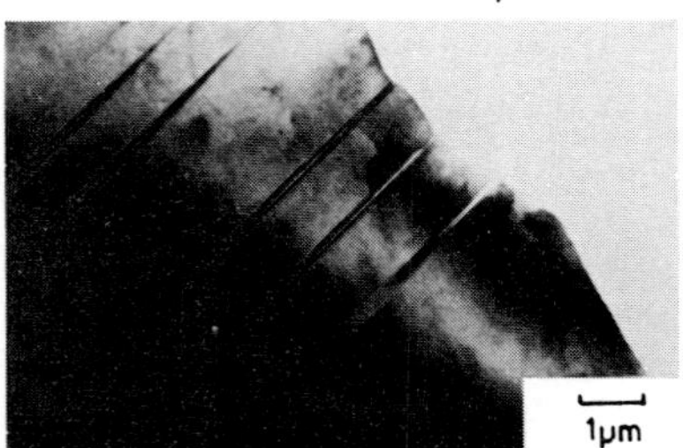

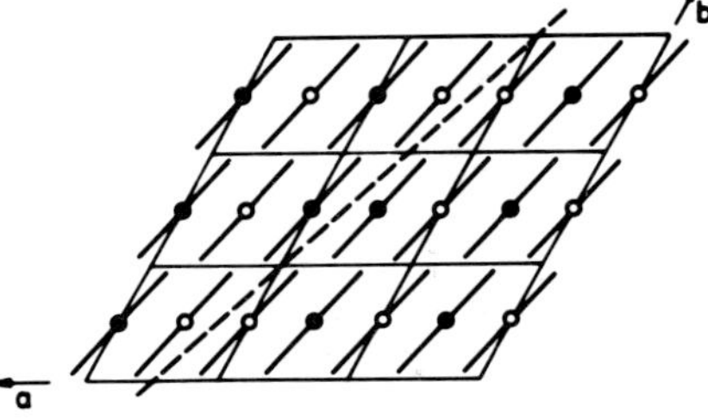

Fig. 3. Stacking faults in pTS crystal   Fig. 4. Schematic diagram of fault

Edge-on observations of metal–oxide interfaces

S.B. Newcomb and W.M. Stobbs

Department of Metallurgy and Materials Science,
University of Cambridge, Cambridge CB2 3QZ, U.K.

1. Introduction

The problem of preparing thin foils from metal–oxide composites has considerably reduced tne impact of transmission electron microscopy on the study of oxidation. Most of the TEM work reported in the literature has been confined to the examination of stripped oxide films (Lloyd et al. 1977) or of nucleation in the plane of the foil (Sanderson and Scully 1969). Clearly the oxidation process could be better understood if the progressive changes in structure were to be examined edge-on.

Here we describe a method that we have developed, related to that described by Manning and Rowlands (1980), for the TEM observation of both the retained metal–oxide interface and the in-situ variations in oxide morphology, structure and chemistry ahead of this interface. We also give some preliminary results of an investigation, the overall aim of which is to compare the oxidation behaviour of a range of austenitic and ferritic chromium steels as well as pure iron.

2. Experimental Technique

A combination of nickel plating, electrochemical polishing and ion-beam thinning has been used in the preparation of metal–oxide interface foils.

Initially a cylinder of the metal or alloy is machined and the material then cut down the middle. The faces of the two halves are given a pre-oxidation treatment of either abrading or electropolishing. After oxidation both halves are nickel plated to protect the fragile oxide in later grinding and polishing. The two protected semi-cylinders are then clamped together at each end and again plated but this time around the outside. Specimens are cut off and ground down to 35–50μm, the interface lying across the centre of the 3mm diameter discs.

These are then polished on a gravity-feed electropolisher to reduce the specimen thickness to 20μm or less and to ensure that the pre-ion-beam thinned surface is flat. However, during ion-beam thinning preferential removal of the oxide occurs without suitable masking. Shadowing the oxide with a mask means that the ion-beam can only strike the oxide from the direction of the metal (Manning and Rowlands, 1980). This technique produces appropriate areas of near uniform thickness suitable for examination in the electron microscope.

Figure 1 illustrates the range of alloys that are being studied and also shows whether the alloys are austenitic or ferritic. Oxidations are beinng carried out at $600^{o}C$ in three environments a) air, b) steam and c) 1% CO/CO_2 for upto 1000 hours. Pure iron has also been included for comparison but is oxidized at $550^{o}C$.

3. <u>Experimental Results</u>

A metal, such as iron, with a variable valency can form different oxides and may produce a sequence of layers depending on the oxidant pressure. The oxide scale formed when pure iron was oxidized at $550^{o}C$ in flowing CO_2 for three hours is shown in figure 2. It is magnetite (Fe_3O_4) of duplex form with a thinner inner layer than outer. The oxide grains in both layers are columnar but in the inner layer are much smaller and extend across the layer. Oxide growth is likely to follow a conventional mechanism (Hussey et al. 1977) with the pre-oxidized metal surface lying at the inner and outer oxide boundary.

The oxidation of electropolished iron surfaces is currently being investigated to determine the crystallographic orientation relationships between the metal and oxide.

Examination of the oxide scales formed on an austenitic and a ferritic steel reveal clear morphological and chemical differences between the oxides formed in each case indicative of the difference in oxidation mechanisms involved. For example, oxidation of the ferritic Fe–10Cr–34Ni alloy in air produced a two–layered scale (figure 3) which was made up of an inner fine–grained oxide layer of mainly Cr_2O_3 bounded by an outer spinel. Both these layers thickened considerably after oxidation for 1000 hours. However, an austenitic alloy of higher chromium content, Fe–20Cr–34Ni, showed both a thinner inner oxide layer for a given oxidation period and less development of this inner layer with time as demonstrated by figure 4.

EDX and EELS can be used with the interface foil to obtain data about the oxide compositions. EDX analysis of the oxide scale formed on the Fe–10Cr–34Ni alloy (figure 5) showed that the inner layer was nearly all chromium oxide with small additions of both iron and nickel, while the outer layer contained high concentrations of iron. EDX is also being used to quantify significant alloy depletions behind the metal–oxide interface.

One problemm of analysing these oxide scales has been the sputtering of material during ion–beam thinning onto the thin areas of the foil. Methods are being developed to prevent sputter damage, but this limitation has made quantification of EELS spectra very difficult. The problem is compounded by the fact that EELS analysis can only be done on the thinnest parts of the oxide where there is relatively more sputter damage.

High resolution lattice imaging is being used to obtain structural details about the metal–oxide system which should give a better understanding of the oxidation mechanism. A lattice image of an inner layer Fe_3O_4–Fe interface is shown in figure 6. This technique is also being used to study the oxide deficit structures. An example of the observations is shown in figure 7 where defects are revealed as local intensity changes in the areas indicated.

Acknowledgements

We wish to acknowledge financial support from the S.R.C. and the C.E.R.L. (Leatherhead).

References

Hussey R J, Sproule G I, Caplan D and Graham M J 1977, Oxidation of Metals 11 65-79.

Lloyd G O, Saunders S R J, Kent B and Fursey A 1977 Corrosion Science 17 269-299.

Manning M I and Rowlands P C 1980 British Corrosion Journal 15 184-189.

Sanderson M D and Scully J C 1969 Corrosion 25 291-299.

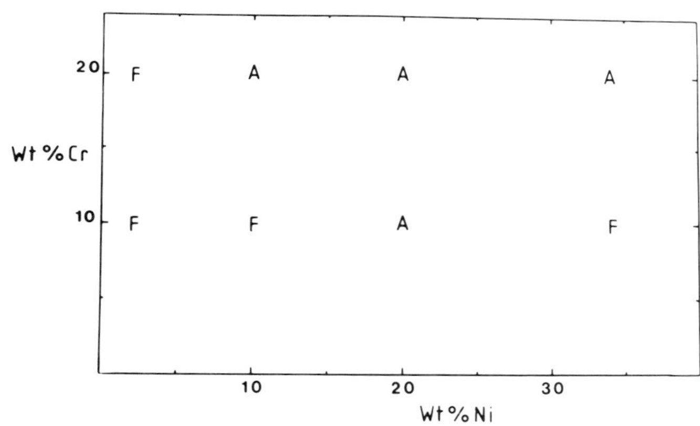

Fig.1. Range of austenitic (A) and ferritic (F) steels being studied showing the variation in Ni and Cr content.

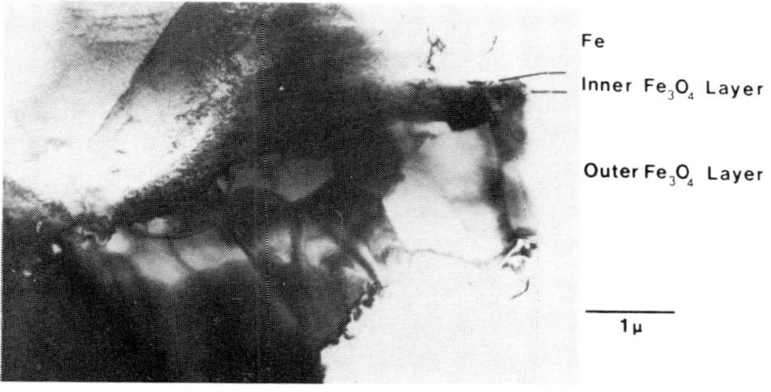

Fig.2. Abraded iron oxidized in CO_2 ($\sim$1atm) at 550°C for 3 hours.

Fig.3. Fe-10Cr-34Ni oxidized in air at 600°C for 40 hours.

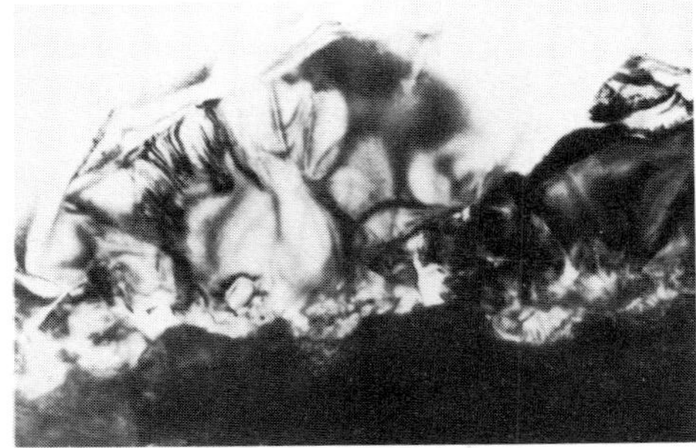

Fig.4.   Fe-20Cr-34Ni oxidized in air at 600°C for 1000 hours.

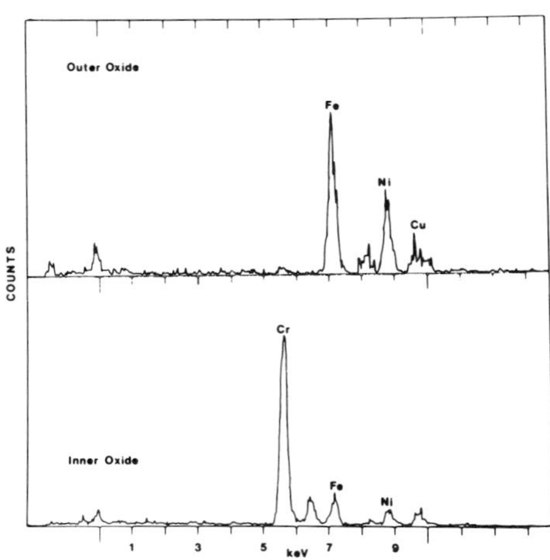

Fig.5.   EDX spectrum of specific areas of the inner and outer oxide layers formed on Fe-10Cr-34Ni in air at 600ºC (40 hours).

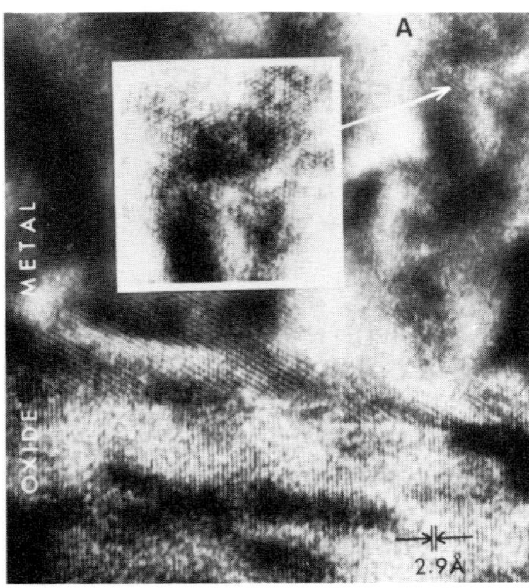

Fig.6.   Lattice imaging of Fe-Fe$_3$O$_4$, interface. The metal is near (1$\bar{1}$0) as can be seen at A, but there are indications that the interface is inclined.

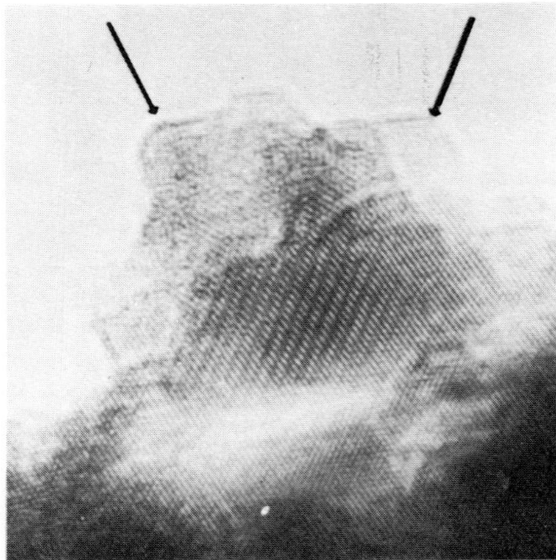

Fig.7.   Spalled magnetite from electropolished iron oxidized at 550ºC. Note the apparent imperfection of the oxide, as arrowed in the thin regions, cannot be the result of foil damage, ion-beam techniques not having been employed.

Oxide formation during the corrosion of silver–gold alloys in nitric acid

David J. Smith and L.A. Freeman

High Resolution Electron Microscope, University of Cambridge, Cambridge.

P. Durkin and A.J. Forty

Department of Physics, University of Warwick, Coventry.

1. Introduction

Recent studies have shown that silver–gold alloys corroded in nitric acid
develop characteristic morphologies, with gold-rich islands distributed
over the corroded surface (Forty and Durkin, 1980). It has also been
established that the selective dissolution of silver from the silver–gold
alloys leads to an oxidation of the residual gold (Durkin and Forty, 1981).
The oxide grows in surface domains in three distinct but related orientations, with a well-defined epitaxy on the alloy. It is meta-stable,
decomposing to gold which then tends to cover the surface of the alloy by
island growth; these islands are then important in the subsequent
corrosion behaviour of these alloys. Clearly, knowledge of the oxide
structure is required for a full understanding of the corrosion mechanism.
Consequently, we have undertaken a detailed high-resolution examination
of this phase, in particular its epitaxial relationship to the alloy
substrate, in order to confirm and augment the previous electron diffraction and lower resolution studies (Durkin and Forty, 1981).

2. Experimental

Thin films of silver–gold alloys of various compositions, as well as pure
gold, have been prepared by vacuum evaporation onto heated single-crystal
substrates of (111)-oriented silver. These composite specimens were then
carefully stripped in nitric acid (see Forty and Durkin (1980) for details).
After corrosion in nitric acid solutions of various strengths, the films
were mounted on microscope grids and then examined in the Cambridge
University 600kV high resolution electron microscope (Nixon et al, 1977;
Cosslett, 1980). Use of a double-tilt goniometer allowed an orientation
close to (111) to be obtained, with the pole-figures associated with bend
contours providing precise local orientations. Micrographs were recorded
with two different objective aperture sizes: one excluded the six $(2\bar{2}0)$-
beams of the substrate to give only oxide lattice fringes and oxide-
substrate moire fringes; the other included the $(2\bar{2}0)$-beams so that the
oxide-substrate epitaxial relationships could be established directly
from the angles between the oxide lattice fringes and those from the
substrate. A standard optical bench with a 10mW He-Ne laser was used
to provide optical diffractograms from the experimental micrographs.

3. Results

Following corrosion, electron diffraction patterns from the alloys are
transformed from the simple hexagonal pattern characteristic of a (111)-
oriented face-centred-cubic alloy to quite complicated arrays, such as
that shown later as an inset of Fig. 1.   The strongest extra spots appear
as clusters, symmetrically disposed about the central beam, whilst the
others seem to be geometrically related to these and the fundamental alloy
reflections.   Some sort of epitaxial relationship with the alloy thus
seems indicated; indeed, it can be shown (see Fig. 2) that all aspects of
the diffraction pattern can be accounted for by assuming an overgrowth of
gold (I) oxide on the substrate.

Observations at moderate magnifications (say 100,000X), with the smaller
objective aperture, show that, as well as the characteristic island
morphology, the alloys are also covered with small crystalline regions
ranging upwards in size from perhaps 25Å to as much as 200Å in some cases.
Direct fringe measurement, as well as use of the optical bench, showed
basic lattice spacings of 2.45 and 2.82Å which correspond to the (200)
and (111) oxide spacings, as well as a variety of larger-spacing moire
fringes.

High resolution observations established that the gold (220) lattice
fringes, spacing 1.44Å, could not be resolved even in samples precisely
oriented into the (111)-orientation; this was not surprising, considering
that the typical foil thickness was in the region of 500Å.   Consequently,
in order to investigate the epitaxial oxide-substrate relationship in more
detail, samples of pure gold with a nominal thickness of 300Å were also
examined after corrosion treatment.   Extensive regions of gold lattice
fringes were then easily obtained, together with the oxide particle
fringes observed previously.   A typical field of view, with a small
cluster of oxide particles, is shown in Fig. 1, whilst Fig. 3 shows an
isolated one.   An optical diffractogram has been recorded from the latter
area (see Fig. 4); the diffraction spots arising from the gold lattice
fringes (G) provide a calibration standard for measurement of both oxide
fringes (O) and the larger spacing moires (M), as well as establishing the
epitaxial angular relationships.   The gold lattice fringes here appear
to be $\frac{1}{2}$ x (2.35Å): their origin is being investigated.

4. Discussion

Dark field micrographs of specimens containing the oxide phase have
previously shown that the oxide exists in three distinct crystallographic
orientations or domains, with each domain giving rise to its own set of
diffracted beams (Durkin and Forty, 1981).   Fig. 2 provides a diagram
illustrating the structure and epitaxial relationship of the oxide and
gold substrate expected in reciprocal space.   This indicates that the
diffraction pattern can be accounted for by an overgrowth of a simple
cubic structure on a (111)-oriented fcc structure, with the epitaxial
relationship between these structures being defined by
$(110) \ [1\bar{1}0]_{\text{oxide}} \ || \ (111) \ [1\bar{1}0]_{\text{alloy}}$.

Our high resolution observations have directly confirmed this interpretation of the domain structure in terms of an oxide phase epitaxed with
the alloy substrate; three distinct orientations of the lattice fringes
from the oxide relative to those of the substrate were observed.   Moreover,
accurate measurements of lattice spacings from the optical diffractograms
indicated an oxide lattice parameter of a_o = 4.90Å, in very close agree-

ment with measurements made direct from the electron diffraction pattern.

Although an explanation for the observed corrosion morphology of the
alloys was deduced originally from electron diffraction and low resolution microscopy it would have been difficult to achieve this for the
pure gold specimens which have a much smaller amount of oxide. Our
observations indicate the potential for real-space investigations of other
systems where the amount of surface phase present is too small to be
studied by conventional diffraction techniques. Direct lattice imaging
combined with optical diffraction should then be invaluable.

The Cambridge University 600kV HREM was built as a joint project between
the Cavendish Laboratory and the Department of Engineering with major
financial support from SRC. DJS and LAF are grateful for continuing
support from SERC; AJF and PD are grateful for financial assistance
provided by AERE, Harwell.

References

Cosslett V E 1980. Proc. R. Soc. A370 1-18
Durkin P and Forty A J 1981 accepted for publication in Phil. Mag.
Forty A J and Durkin P 1980 Phil. Mag. A42 295
Nixon W C, Ahmed H, Catto C J D, Cleaver J R A, Smith K C A, Timbs A E,
    Turner P W and Ross P M EMAG 1977 pp 18-16

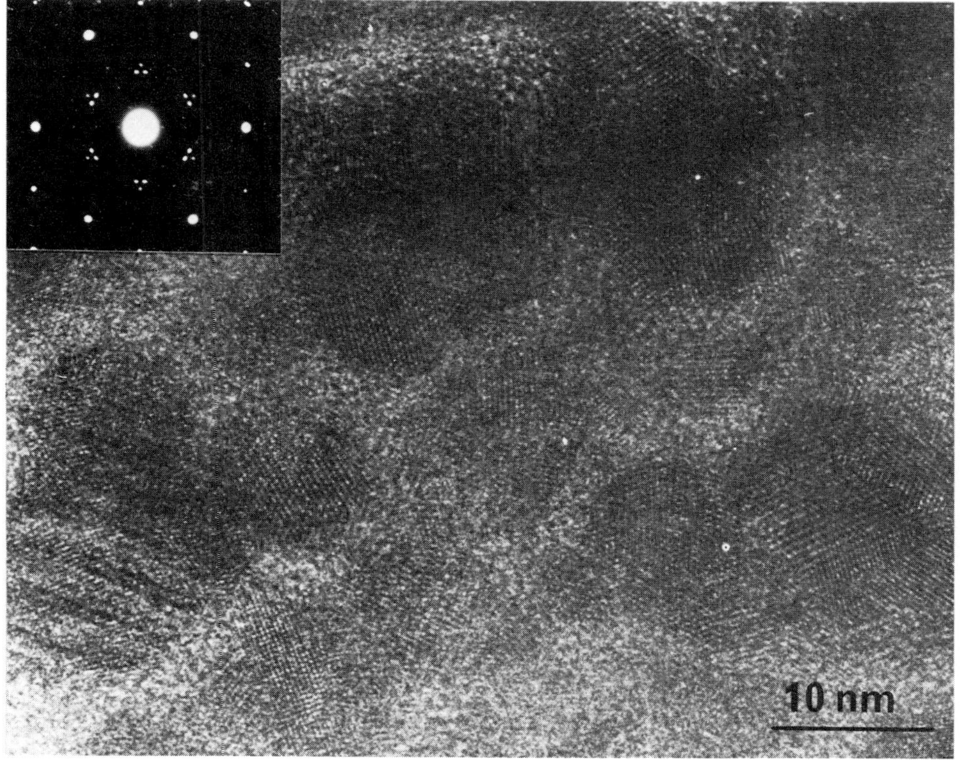

Fig.1. Cluster of gold oxide particles showing oxide lattice and oxide-
substrate moire fringes: electron diffraction pattern inset.

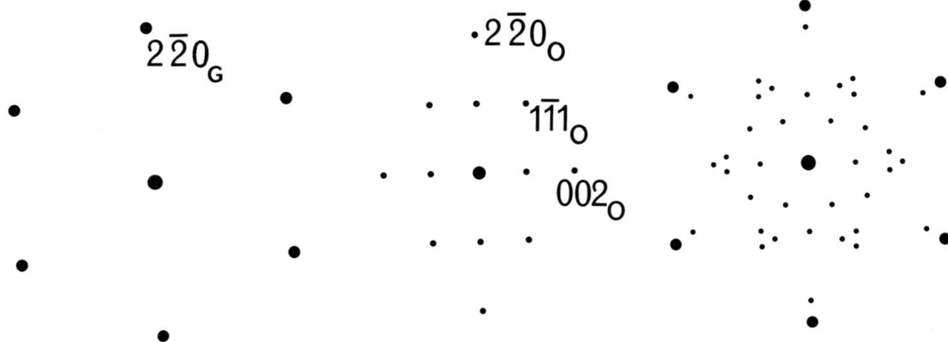

Fig.2. Showing, from left to right, the reciprocal lattice for gold (111), gold oxide (110) and gold with an oxide overlayer (110) $[1\bar{1}0]_{oxide} || ^{\ell}$ (111) $[1\bar{1}0]_{gold}$ using the three possible $<1\bar{1}0>_{gold}$ directions. In real space the oxide overlayer grows in three distinct domains corresponding to these three epitaxial relationships.

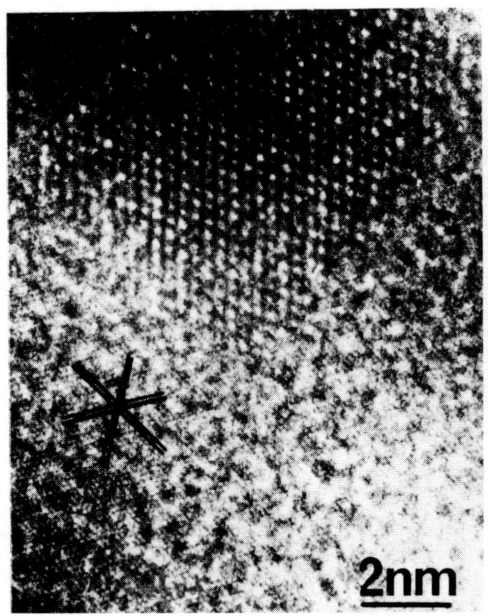

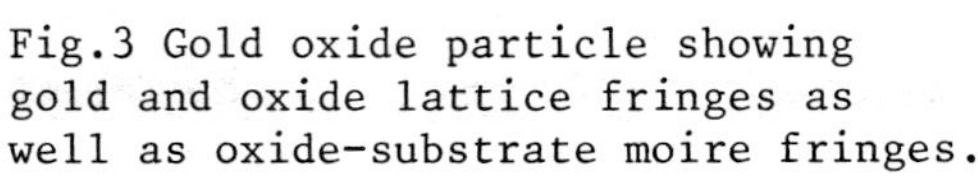

Fig.3 Gold oxide particle showing gold and oxide lattice fringes as well as oxide-substrate moire fringes.

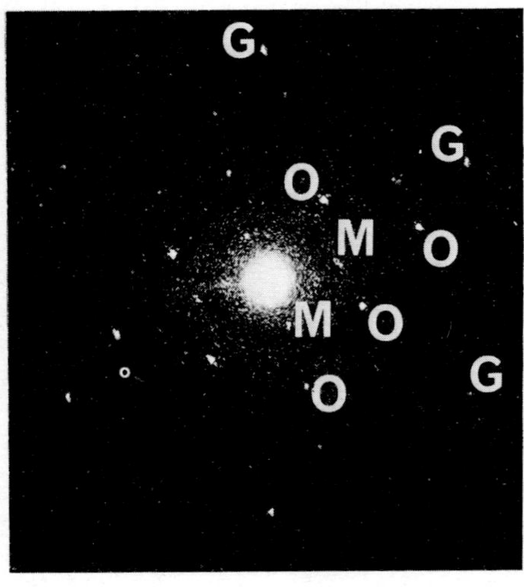

Fig.4 Optical diffractogram from region of Fig.3: G≡gold; O≡oxide; M≡moire fringes.

High resolution electron microscope studies of passive oxide films on iron in aqueous solutions

J. Kassim, T. Baird and J.R. Fryer

Department of Chemistry, University of Glasgow, Glasgow, G12 8QQ.

1. Introduction

The nature of thin protective films formed on iron in aqueous solutions
has been extensively studied and the films shown to contain $\gamma\text{-}Fe_2O_3$ which
has a pseudospinel structure closely related to that of Fe_3O_4 (Itaka,
1937; Mayne and Pryor, 1949; Cohen, 1952; Davies and Evans, 1956; Foley et
al., 1967). X-ray and electron diffraction data from these oxides displayed all the reflections of cubic Fe_3O_4 and thus a cubic structure
similar to Fe_3O_4 was assigned to $\gamma\text{-}Fe_2O_3$ (Welo and Baudisch, 1925; Verwey,
1935; Hägg, 1935; Haul and Schoon, 1939). However, some preparations
(Oosterhout and Rooijmans, 1958; Oosterhout, 1960; Schrader and Buttner,
1963) exhibited faint extra reflections that could be indexed in accordance with a tetragonal unit cell with $c_O = 3a_O$, where $a_O = 0.833$nm and is
equivalent to the cubic unit cell of Fe_3O_4.

The present high resolution studies showed that $\gamma\text{-}Fe_2O_3$ has a tetragonal
structure and provided a clear distinction between $\gamma\text{-}Fe_2O_3$ and Fe_3O_4 by
the use of lattice imaging techniques and optical diffractograms.

2. Experimental

Passive-oxide films were obtained on pure iron sheet (50x20x0.5mm) by
reaction in deionised water, 0.1M $NaNO_2$, 0.1M NaOH, 0.1M $NaNO_3$, and
0.1M Na_2CO_3 for several days depending upon the thickness of the oxide
layer required. Visible oxide films were stripped from the sheet using
the methanol/iodine method (Mayne and Pryor, 1949; Vernon et al., 1939) or
carefully scraped from the surface using a thin glass edge and then ultra-
sonically dispersed in propanol and mounted on carbon covered grids.

Synthetic $\gamma\text{-}Fe_2O_3$ and Fe_3O_4 were prepared as follows:

i   Fine particles of magnetite (Fe_3O_4) prepared by the method of Swaddle
and Olthmann 1980, were oxidised in air or in oxygen/water vapour
mixtures at 220°C for 10 hours to give $\gamma\text{-}Fe_2O_3$.

ii   Polycrystalline α-Fe prepared by evaporation of pure Fe onto a
heated carbon substrate at 350°C in vacuo (10^{-5} torr), was oxidised
either in air, or in a stream of oxygen/water vapour, or in water,
for various times to give Fe_3O_4 or $\gamma\text{-}Fe_2O_3$.

iii α-FeOOH preparations (Gildawie, 1979) were dehydrated by heating in
air at 250°C for 10 hours. The red $\alpha\text{-}Fe_2O_3$ formed was then reduced

to black Fe_3O_4 in dry hydrogen at 350°C for 10 hours and this was carefully oxidised to reddish-brown $\gamma-Fe_2O_3$ in air or oxygen/water vapour at 220°C for 10 hours.

All the oxidation and reduction stages were carried out in a Pt boat in a glass system. Samples were examined in a Siemens Elmiskop 1A or in a JEM 100C at 80 and 100kV respectively.

3. Results and Discussion

The stripped oxide films were shown by microscopy to consist of an extensive homogeneous layer of particles of a size ranging from 5-50nm diameter depending on the reaction conditions. Films formed in sodium nitrite and sodium hydroxide solutions were much denser than those formed in water, sodium carbonate or nitrate solutions. Diffraction patterns consisted of continuous rings and the spacings agreed well with either $\gamma-Fe_2O_3$ or Fe_3O_4. Some patterns showed extra reflections at 0.43nm, 0.372nm, 0.343nm and 0.272nm which could be identified with tetragonal $\gamma-Fe_2O_3$ (Fig.1). Regions of the less dense films often contained large particles such as shown in Fig.2 where lattice fringes of 0.34nm and 0.32nm are evident. Diffraction patterns from this material gave reflections at 0.595, 0.340, 0.325, 0.296, 0.276 and 0.23nm which agree with X-ray data for tetragonal $\gamma-Fe_2O_3$ (Schrader and Buttner, 1963).

Close examination with a laser diffractometer showed the cross lattices in Fig.3 to have spacings of 0.695nm(01$\bar{2}$) and 0.43nm(105) at an angle of 61.6° and 0.695nm(012) and 0.482nm(1$\bar{1}$3) at 99.2° in accord with tetragonal $\gamma-Fe_2O_3$. Further evidence of a tetragonal structure is given in Fig.4 which shows crossed lattice fringes of 0.43nm(105) and 0.415nm(220) at 59°. The (200) planes at 0.415nm in a cubic structure would be at 90°.

The oxidation products from cubic Fe_3O_4 consisted of reddish-brown rounded particles comparable in size to the original oxide material. Extra reflections at 0.703, 0.625, 0.428, 0.352 and 0.338nm are consistent with tetragonal $\gamma-Fe_2O_3$. Fig.5 shows crossed lattices at 0.595, 0.34 and 0.296nm again in agreement with a tetragonal unit cell.

Single crystal data (Fig.6) from $\gamma-Fe_2O_3$ obtained by successive reduction and oxidation of α-FeOOH indicated a tetragonal unit cell with $c_o = 3a_o$.

Finally, oxidation of α-Fe in air in the presence of water vapour for 6 hours, or in water for 2 hours, produced small rounded particles of $\sim$5nm diameter with pronounced extra reflections at 0.54, 0.43, 0.41, 0.377 and 0.35nm and is consistent with tetragonal $\gamma-Fe_2O_3$. Oxidation for 12 hours in water resulted in needle-like crystals of γ-FeOOH rather than $\gamma-Fe_2O_3$.

4. Conclusions

The oxide films formed on iron under the various experimental conditions used in this work all consisted of $\gamma-Fe_2O_3$. These films and synthetic $\gamma-Fe_2O_3$ were shown by high resolution microscopy and electron diffraction techniques to have a tetragonal structure with unit cell dimensions of $c_o = 3a_o$ where $a_o = 0.833$nm and is equivalent to the cubic unit cell of Fe_3O_4.

References

Cohen M 1952 J. Phys. Chem. 56 451
Davies D E and Evans U R 1956 J. Chem. Soc. 4373
Foley C L, Kruger J and Bechtoldt C J 1967 J. Electrochem. Soc. 114 994
Gildawie A M 1979 PhD Thesis University of Glasgow
Hägg G 1935 Z. Phys. Chem. B29 95
Haul R and Schoon Th 1939 Z. Phys. Chem. 44 216
Itaka I, Miyake S and Imori T 1937 Nature 139 156
Mayne J E O and Pryor M J 1949 J. Chem. Soc. 1831
Oosterhout G W 1960 Acta Cryst. 13 932
Oosterhout G W and Rooijmans C J M 1958 Nature 181 44
Schrader R and Buttner G 1963 Z. anorg. allg. chemie 320 209
Swaddle T W and Olthmann P 1980 Can. J. Chem. 58 1763
Vernon W H J, Wormell F and Nurse T J 1939 J. Chem. Soc. 621
Verwey E J W 1935 Z. Krist. 81 65
Welo L A and Baudisch 1925 Phil. Mag. 50 399

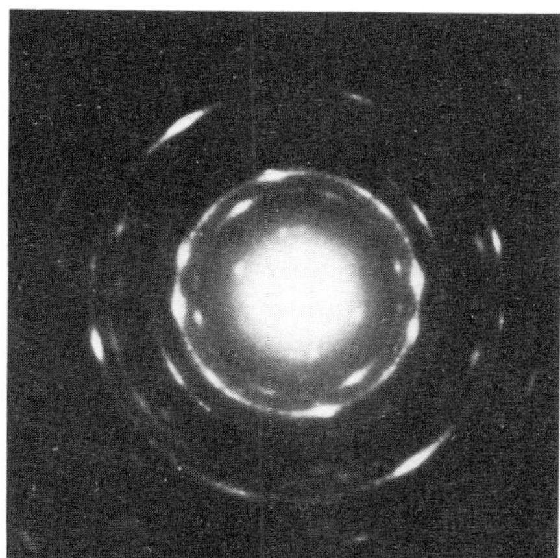

Fig.1 Electron diffraction pattern from iron oxide (γ-Fe_2O_3) formed in 0.1M $NaNO_2$.

Fig. 2 Passive oxide film formed in water showing 0.34nm(213) and 0.32nm (205) lattice structure.

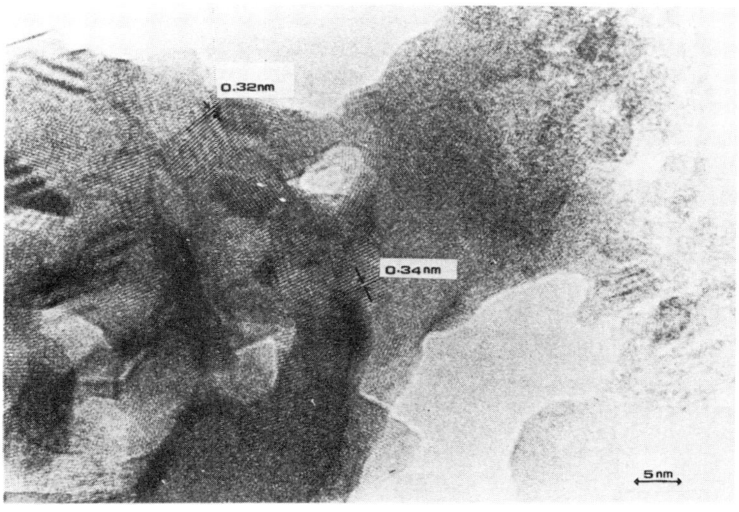

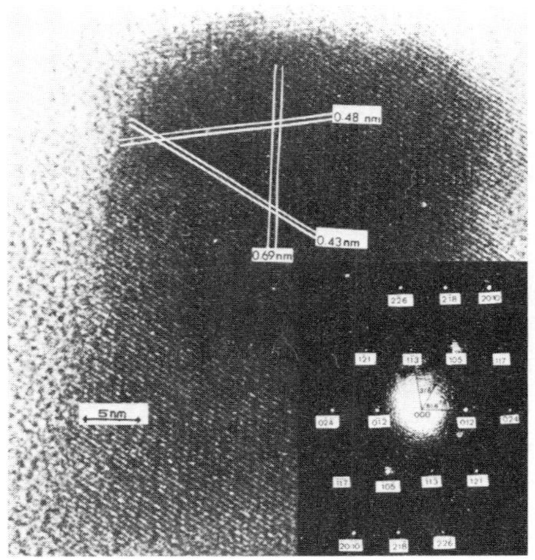

Fig. 3 Large single crystal of
γ-Fe$_2$O$_3$ showing crossed lattice
structure. Inset: optical
diffractogram.

Fig.4 Single crystal of γ-Fe$_2$O$_3$
with 0.43nm(105) and 0.415nm($\bar{2}\bar{2}\bar{0}$)
lattice fringes evident.

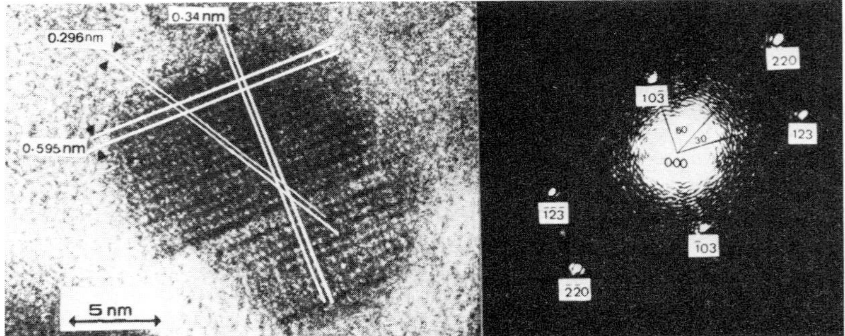

Fig. 5
Twinned crystal of γ-Fe$_2$O$_3$ formed by oxidation of Fe$_3$O$_4$ showing
lattice structure. The optical diffractogram from this crystal
is indexed in accord with a tetragonal unit cell.

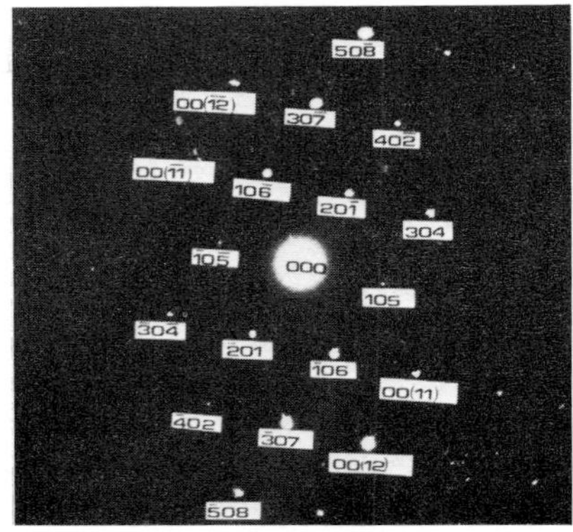

Fig.6 Electron diffraction
pattern from γ-Fe$_2$O$_3$ obtained
from α-FeOOH. The pattern is
indexed based on a tetragonal unit
cell.

Applications of convergent-beam diffraction to nickel super-alloys

R Vincent and T A Bielicki

Physics Department, University of Bristol, Bristol BS8 1TL

1. The γ'' Phase

The established methods for analysis of convergent-beam diffraction (CBD) patterns (Steeds 1979) are used to investigate the structure of precipitates extracted from two commercial Ni-base super-alloys. The first example is the tetragonal γ'' strengthening phase $NbNi_3$ in the alloy Inconel 718. In comparison to the more common cubic γ' super-lattice, which has the same nominal MN_3 composition, the M atoms in γ'' are arranged on a sub-set of sites in the fcc matrix with lower symmetry. CBD patterns from γ'' (space-group I4/mmm) along zone-axes equivalent to <111> fcc contain only a single vertical mirror plane (Fig 1). The presence of super-lattice reflections reduces the projection symmetry from 6mm for fcc to 2mm, and then the higher-order Laue zone (HOLZ) interactions, which are clearly visible in the bright-field disc (Figs 1 and 3a), reduce the symmetry even in the zero layer from 2mm to m. In a fcc crystal with 3m whole pattern symmetry along <111> , the mirror planes would bisect the three segments of the second HOLZ ring in Fig 2, which are centred about points 120° apart. Now only the lowest segment is bisected by the mirror in Fig 1; the other two segments contain an asymmetric distribution of super-lattice reflections and are related by the same vertical mirror. In addition, the prominent Kikuchi lines near the centre of each segment are associated with matrix reflections but show a similar loss of symmetry, where the vertices in the two clusters related by the vertical mirror contain tangential off-sets relative to the third set bisected by the mirror. These shifts are linked to a tetragonal distortion of the lattice previously noted in X-ray measurements (Paulonis et al 1969), which is consistent with, but not required by the point-group symmetry of the crystal.

A direct estimate of the tetragonal c/a ratio can be obtained from the deficiency lines associated with HOLZ reflections in the bright-field disc (Fig 3a), where the symmetry of the matrix lines is reduced from 3m to m by the tetragonal distortion of the lattice (cf Figs 3b and c). The aim of the calculations was limited to checking the c/a ratio and to comparing the sensitivity against X-ray methods. The patterns drawn in Fig 3 are based on a kinematic approximation, where the electron wave vector K and the c/a ratio are treated as two independent parameters which determine the scale and distortion of the pattern. The best fit to the observed pattern (Fig 3b) is obtained from a c/a ratio consistent with the X-ray figures, but the experimental accuracy ($\sim 0.2\%$) is limited because the HOLZ lines are inevitably broadened in thin crystals. Any small improvement gained from a complete dynamical calculation is likely to be marginal.

2. M_3B_2 borides

The second example is a boride, M_3B_2 (space-group P4/mbm;Beattie 1958)where
the detection of this phase can be linked, at least indirectly, to the
presence of significant levels of boron in the matrix. However, the X-ray
spectrum is not always a reliable guide to the phase identity. In Inconel
718, two M_3B_2 borides are found with differing metallic components. The
normal type contains principally Nb and Cr, whereas a few large precipitates are relatively Mo-rich. The metal elements in similar borides extrac-
-ted from the alloy Astroloy are essentially Mo and Cr. Although the
presence of boron has been established in energy-loss spectra (Fig 4) ob-
-tained from borides mounted directly on a 2000-mesh grid, the boron K
edge is not always visible when spectra are taken through the carbon film
in extraction replicas. Generally, CBD patterns are the most efficient
and direct guide to the crystal structure. In particular, the direction
of the tetrad axis in the M_3B_2 reciprocal lattice is easily recognised
because the layer spacing along [001] is unusually large. The [001] CBD
pattern has 4mm symmetry and alternate reflections along the hOO and OkO
rows contain dark bars due to the glide planes and associated screw diads
(Fig 5a). When the crystal is tilted to [011] (Fig 5b), the dark bars along
the vertical and horizontal axes are linked to the (100) glides and [100]
screw diads, respectively. The zero-layer pattern contains only a vertical
mirror due to HOLZ interactions in the bright-field disc. Finally, the
(010) glides lie in the plane of the [010] pattern (Fig 5c) and vertical
rows of $h0\ell$ reflections for h odd are absent.

3.   Titanium Carbo-sulphide

The final phase considered here is a tabular, intergranular precipitate in
Astroloy which shows strong Ti and S peaks and a weak Zr peak in EDX
spectra. The phase is hexagonal with 6mm symmetry along [0001], but the
presence of a mirror normal to the hexad axis is less clear, expecially in
faulted crystals. The point-group symmetry is 6mm or 6/mmm, and the pres-
-ence of {1010} glides (Fig 6) limits the space-group to P6$_3$/mmc or P6$_3$mc.
The lattice constants are consistent with a carbo-sulphide, $(Ti,Zr)_4C_2S_2$
(Kudielka and Rohde 1960), where the carbon and metal atoms form local ABC
stacking sequences equivalent to a MC carbide. If the sulphur is replaced
with metal atoms, the lattice is iso-structural with the H-phases,hexagonal
analogues of the perovskite structure. The phase is uncommon in alloys
and may be confused with Ti sulphides, in particular Ti$_3$S$_4$(P6$_3$mc), which
has similar lattice constants. Evidence in favour of the carbo-sulphide
is listed here: (i) the c/a ratio differs by 5% from any Ti sulphide;
(ii) the EDX spectrum is definitely Ti-rich relative to sulphur, and
(iii) the weak 2nd HOLZ ring (Fig 6) is consistent with the carbo-sulphide
structure factors. Also, we have observed the carbon K edge in energy-loss
spectra from particles mounted directly on a grid. Interpretation is inevi
-tably uncertain due to the possibility of contamination but the boride
spectrum (Fig 4) shows no similar edge.

References

Beattie H J 1958 Acta Crystallogr. $\underline{11}$ 607
Kudielka H and Rohde H 1960 Z.Kristallogr.$\underline{114}$ 447
Leapman R d, Rez P and Mayers D F 1980 J.Chem.Phys. $\underline{72}$ 1232
Paulonis D F, Oblak J M and Duvall D S 1969 Trans.ASM $\underline{62}$ 611
Steeds J W 1979 Introduction to Analytical Electron Microscopy ed J J Hren,
  J I Goldstein and D C Joy (New York:Plenum) pp 387-422

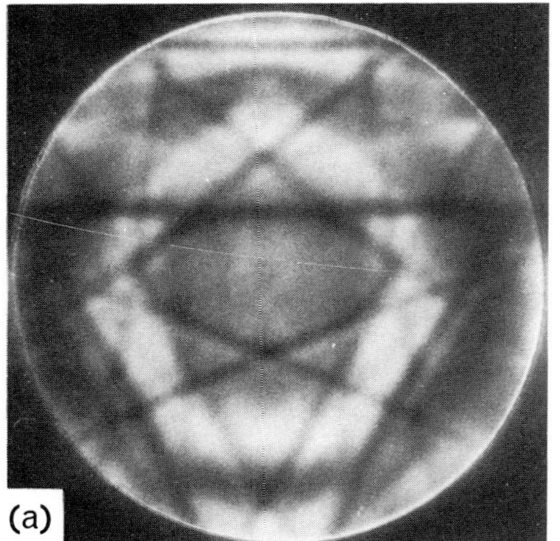

Fig.1 Zero layer of the γ" CBD
pattern along < 111 > fcc.

Fig.2 Segments of the second HOLZ
ring associated with Fig 1 centred
on points 120° apart.

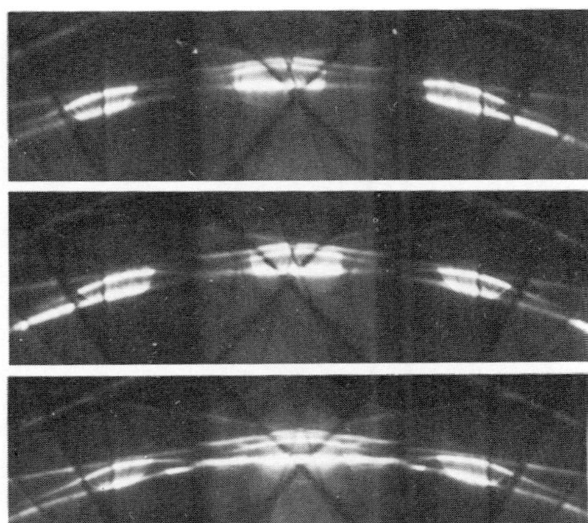

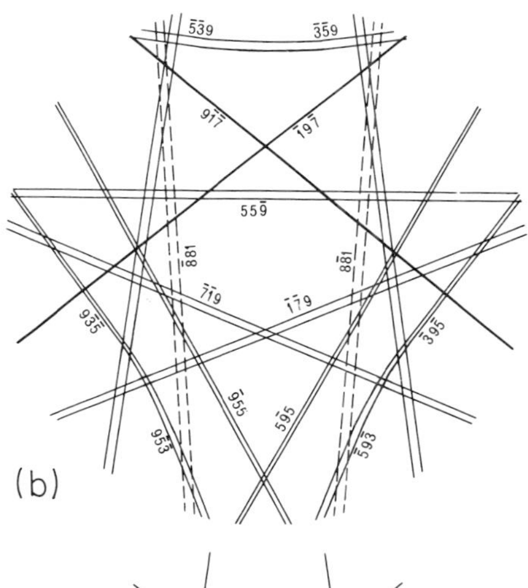

(a)

Fig.3 Pairs of lines in (b) are equi-
-valent to an estimated error of ± 0.1%
and represent the best fit to (a),where
a=3.620±004Å and the volume of the
unit cell, a^2c is fixed at 97.27Å^3. The
overall scale is matched by adjusting K
to fit the diagonal lines which are
least sensitive to changes in a. Lines
are indexed on the side facing the HOLZ
reflection, where curved lines labell-
-ed with two sets of indices represent
dynamical interactions and dashed lines
represent super-lattice reflections. In
(c),the tetragonal distortion is remo-
-ved (a=3.650Å) and the matrix lines
have 3m symmetry.

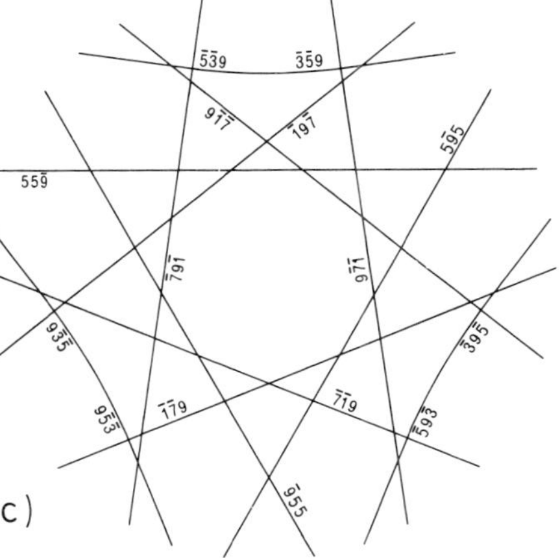

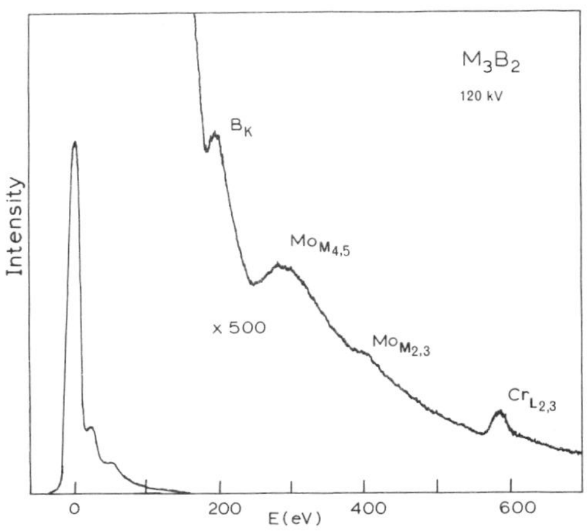

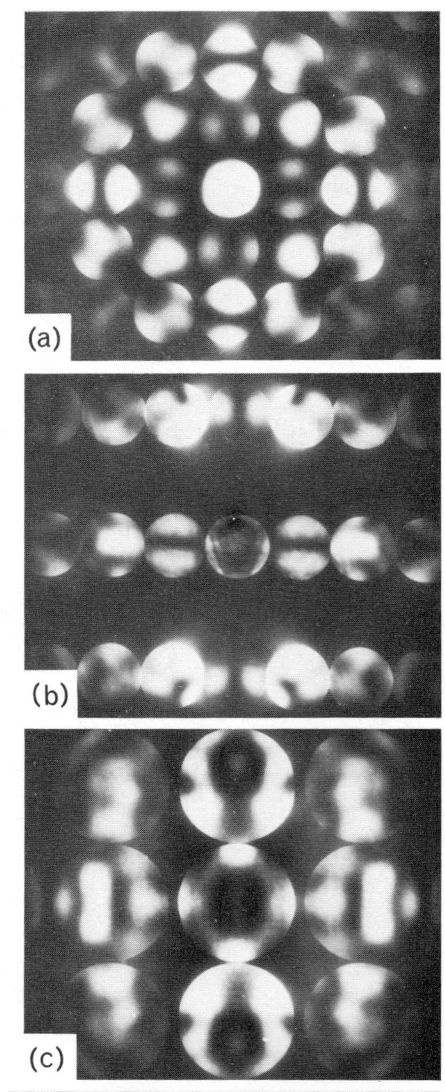

Fig.4 (above) Energy-loss spectrum for a M$_3$B$_2$ boride extracted from Astroloy. The chromium and boron edges coincide with their expected positions, but the maximum cross-section for the prominent molybdenum M$_{4,5}$ peak does not occur at the onset of the edge (230 eV), but is shifted towards higher energies (Leap--man et al 1980). The weak molybdenum M$_{2,3}$ edge at 400 eV is distinguished from the nitrogen K edge by comparison with a standard Mo spectrum.

Fig.5 (opposite) Standard CBD patterns from a M$_3$B$_2$ boride where the crystal is tilted 90° about a horizontal [100] axis from [001] (a) through [011] (b) to [010] (c). All three patterns contain at least a vertical mirror.

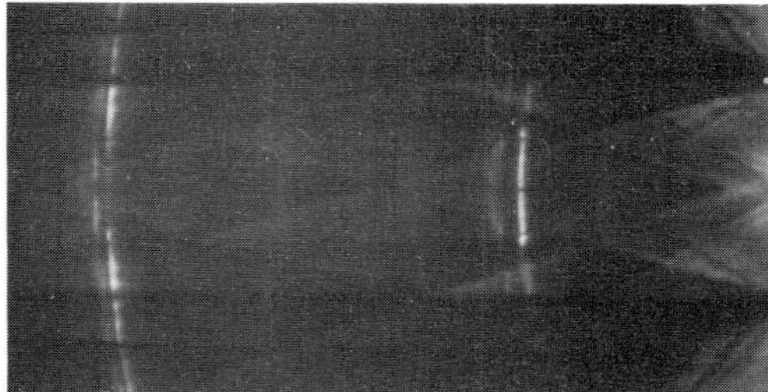

Fig.6 Section parallel to [11$\bar{2}$0] from the [0001] CBD pattern for Ti$_4$C$_2$S$_2$ (printed from two negatives). The 2nd HOLZ ring is almost invisible and the 6 6 $\bar{1}$2 3 reflection on the mirror line in the 3rd ring contains a dark bar due to the glide planes.

Deformation twinning in aluminium

R C Pond and L M F Garcia-Garcia

The University of Liverpool, Department of Metallurgy and Materials Science,
P.O. Box 147, Liverpool L69 3BX.

1. Introduction

Deformation twinning is an important mode of plastic deformation, particularly in bcc metals, and is also known to occur in a variety of fcc metals
and alloys (Mahajan and Williams, 1973). In the present work the dislocation process leading to accommodation of the stress at the tip of deformation twins in high purity aluminium has been studied using transmission
electron microscopy. This is of interest because such accommodation has
not been studied microscopically in fcc metals previously and deformation
twinning has been regarded hitherto as not occurring in aluminium (Venables,
1964). Thin foils of annealed aluminium were prepared by the window
technique of electropolishing, and specimens were often found to contain
short cracks running approximately perpendicular to the specimen edges.
These cracks are thought to arise from tensile stresses caused by hydro-
dynamic pressure waves when the specimen is moved in the electrolyte.
Since the stress state is essentially plane stress, mode III type cracks
are expected to propagate at approximately 45° to the specimen surface.
Deformation twins were observed to have grown from the tips of these cracks
in such a way that the twinning shear was close to the direction of maximum shear.

2. Observations

Fig.1 shows a deformation twin viewed along the $[1\bar{1}0]_{M/T}$ direction common
to the twin and matrix, i.e. with the coherent twin planes seen edge-on.
Note the buckling of the foil as indicated by the bend contours and that
the tip of the twin, A, is blunt. Using a travelling microscope it was
found that one coherent interface was essentially planar while the other;
was curved in a way consistent with steps introduced by a pile-up of twin-
ing dislocations. By tilting the specimen it was possible to resolve the
individual twinning dislocations in regions away from the twin tip. The
specimen's crystallography is shown in the stereogram, fig.2, and the
traces of the twin plane, $(1\bar{1}1)_{M}/(1\bar{1}1)_{T}$, and the plane and direction of the
maximum shear stress, σ, are indicated. Fig.3 depicts schematically the
twin emanating from the crack tip, B. In addition to the twinning dislocations, two other types of dislocations were observed. Close to the tip
of the twin a coarsly spaced arrangement of dislocations was seen as shown
in fig.4(a) and (b). Contrast experiments were used to establish that the
Burgers vectors, $\underline{b}$, of these dislocations, hereafter called residual dislocations,were $\frac{1}{6}[1\bar{2}1]_{M}$, $\frac{1}{6}[21\bar{1}]_{M}$ and $\frac{1}{6}[\bar{1}12]_{M}$

0305-2346/82/0061-0495$01.50   © 1982 The Institute of Physics

Note that these are glissile perfect interfacial dislocations in a $(1\bar{1}1)_M$ twin interface. The final type of dislocations observed were perfect dislocations in the matrix ahead of the twin having slip traces which extended from the twin tip. These are referred to as emissary dislocations, and emission from the twin tip was actually observed in the microscope. Fig. 5(a) and (b) show part of a Burgers vector determination of an emissary dislocation, C; the twin tip is at A and $\underline{b}$ for this emissary was found to be $\frac{1}{2}[0\bar{1}1]_M$.

3. Discussion

The observations are consistent with the following sequence of events. The thin foil specimen was subjected to a stress pulse sufficiently intense to tear its edges. The stress in the region of the crack tip was sufficiently high to lead to the homogeneous nucleation of a deformation twin by the emission of screw-type $\frac{1}{6}[12\bar{1}]_M$ partial dislocations on successive $(1\bar{1}1)_M$ planes. The shear stress in the foil was great enough to sustain growth of the twin into material where the enhancement of stress due to the crack was insignificant. At an instant when the stress at the twin tip fell below a critical value (see below) the twin became blunt without striking an obstacle (the foil was essentially dislocation free since it was annealed). The blunting can be considered to occur by the following mechanism in which the leading three twinning dislocations react to form emissary and residual dislocations:-

$$3 \times \frac{1}{6}[12\bar{1}]_M \rightarrow \frac{1}{6}[12\bar{1}]_M + \frac{1}{6}[21\bar{1}]_M + \frac{1}{2}[0\bar{1}1]_M$$
$$\qquad\qquad\qquad\qquad \text{residuals} \qquad\qquad \text{emissary}$$

$$\text{or} \qquad 3 \times \frac{1}{6}[12\bar{1}]_M \rightarrow \frac{1}{6}[12\bar{1}]_M + \frac{1}{6}[\bar{1}12]_M + \frac{1}{2}[\bar{1}1\bar{0}]_M$$

Due to mutual repulsion the emissary dislocation is ejected into the matrix and the residual dislocations glide back along the planar coherent interface. Further blunting can occur by coalescence of subsequent triplets of twinning dislocations which then cross-slip across the newly created incoherent interface and dissociate as described above. Blunting will cease when the local stress is insufficient to overcome the mutual repulsion of the component dislocations of a triplet. As the stress decreases further nucleation will also stop, and eventually many of the piled-up twinning dislocations will be able to glide back to the crack tip and out of the foil leaving a long narrow blunted twin as observed. The thickness of the blunt tip, expressed as a number of $(1\bar{1}1)_M$ planes, should be equal to 3/2 times the number of residual dislocations observed (or 3 x times the number of emissary dislocations), and this is in complete accord with observation.

It has been shown by Vitek (1970) that single and double layer faults on {211} in bcc materials are unstable. Thus, the leading three dislocations in bcc deformation twins are probably best regarded as a single dislocation spread over three planes and stabilised by the applied stress. In fcc materials intrinsic and extrinsic stacking faults are stable. However, if the interfacial tension forces acting on each twinning dislocation are considered, it is found that a stress acts pulling the leading three dislocations together. With reference to fig. 6 it is seen that if γ is the stacking-fault energy (for both intrinsic, I, and extrinsic, E, faults) and the twin boundary, TB, energy is about $\gamma/2$, then an interfacial tension of about $3/2\,\gamma$ tends to cause coalescence of the leading three dislocations.

In the case of aluminium (high γ) this corresponds to a stress sufficiently
great to overcome the mutual repulsion of these dislocations (based on
elasticity theory). Thus, deformation twins in aluminium may not require
an obstacle in order to become blunt since the leading triplet of twinning
dislocations may become unstable.

References

Mahajan S. and Williams, D.F., 1973, Int. Met. Rev., <u>18</u>, 43.

Venables, J.A., 1964, 'Deformation Twinning', New York(Gordon and Breach).

Vitek, V., 1970, Scripta Met., <u>4</u>, 725.

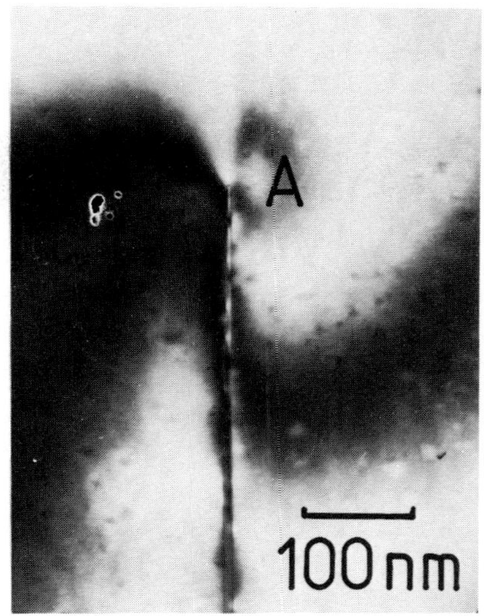

Fig.1. Deformation twin in Al
viewed along $[1\bar{1}0]_{M/T}$

Fig.2. Stereogram showing the orientation
of the matrix and twin in the untilted
orientation.

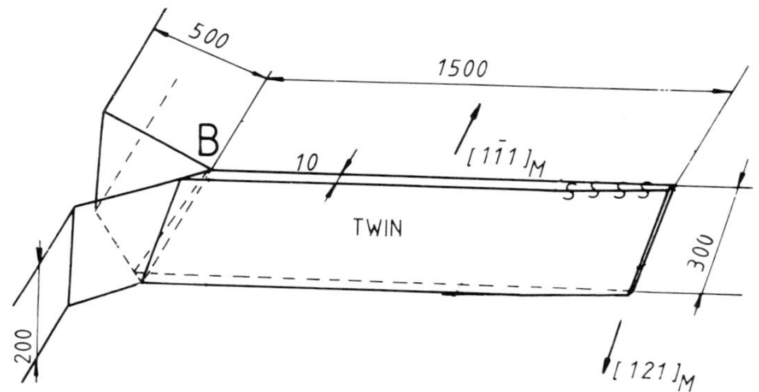

Fig.3. Schematic illustration of deformation twin emanating from a crack at
the edge of a thin foil. All dimensions shown are in nanometers.

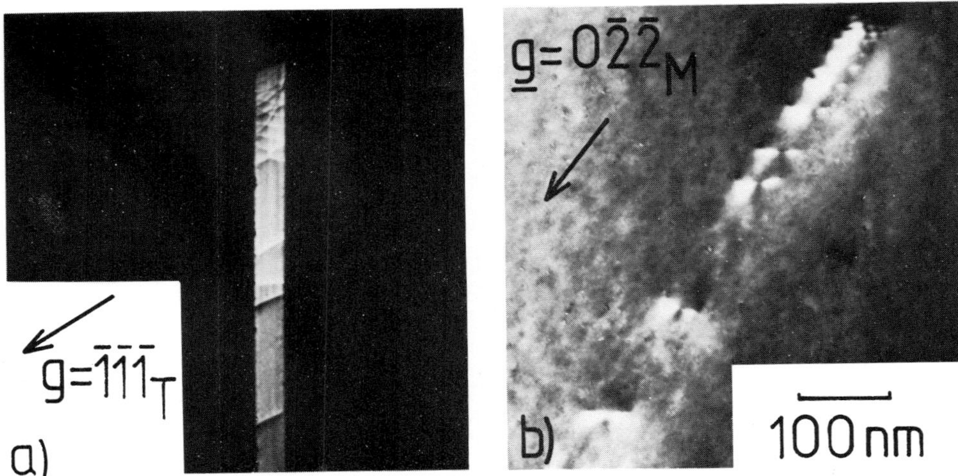

Fig.4. Dark field images a) using a twin reflection, and b) using a common reflection. Note that the residual dislocations with $\underline{b} = \frac{1}{6}[21\bar{1}]_M$ are invisible in (b).

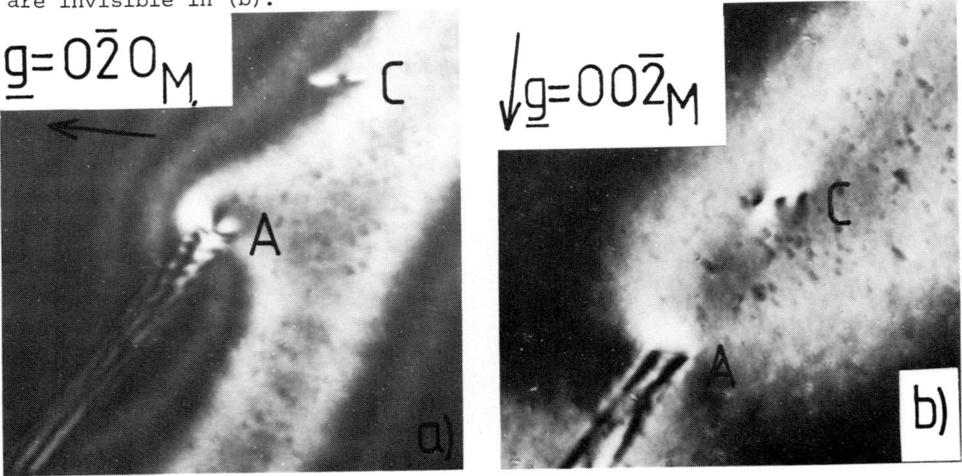

Fig.5. Burgers vector determination of an emissary dislocation. The dislocation was invisible using $\underline{g} = 0\bar{2}2_M$.

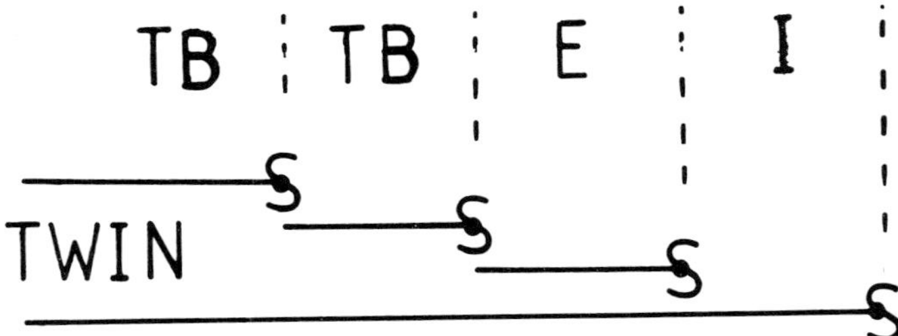

Fig. 6. Schematic illustration of the twinning dislocations at the tip of a deformation twin. The faults separating individual dislocations are indicated.

The effect of ion beam charge composition on sputtering in oxides

P M Marquis[†] and J Franks[°]

[†]Centre for Materials Science, University of Birmingham, P.O. Box 363,
 Birmingham B15 2TT

[°]Ion-Tech Limited, 2 Park Street, Teddington, Middlesex TW11 OLT

1. Introduction

Ion-beam etching is a well established method for the preparation of
thin specimens for transmission electron microscopy. The main objective
of the technique is to produce a damage free specimen of uniform cross-
section in the shortest possible time. The nature of the ion-beam
damage that can be introduced during etching depends on many factors,
including the nature of the beam and specimen composition. For example,
multi-element materials may undergo changes in surface composition due
to preferential etching. This effect is of particular relevance in
oxides where oxygen is preferentially sputtered. The creation of an
oxygen deficient surface region can also distort the surface charge
balance. Most ion-beams, produced using commercial ion-beam thinners,
consist of a combination of positive ions, neutrals and electrons.
Electrostatic interactions between the beam and sample can therefore
influence ion-beam sputtering behaviour. In this paper the results of a
study of the nature of ion-beam damage and sputtering rates in oxides is
reported.

2. Experimental

The results reported in this paper were obtained using Ion-Tech B11
water cooled ion guns. The plasma chamber in one of the guns (B11N) was
modified to produce an increase in the neutral ion content of the beam
to 90%, compared to a B11 neutral content of 30%.

Single crystal specimens of Ni, Mg, NiO, MgO and Al_2O_3 were prepared by
slitting from the bulk followed by mechanical polishing. The amount of
material sputtered during ion-bombardment was determined from weight
loss measurements. An electro-microbalance was used with a sensitivity
of 1 µg. For the experiments reported in this paper a fixed applied
voltage of 6 kV was used. The gas flow rate was then fixed for each ion
gun to produce the same thinning rate for the pure metal standards.
Comparative thinning rates were then determined for the oxide specimens.
The specimens were positioned 0.5 cm from the exit cathode of the ion
gun and the angle of incidence for the ion-beam was fixed at 60°. Each
specimen was thinned for 90 mins. Cross sections of the oxide samples
were prepared and variations in the metallic ion concentrations
determined using scanning electron microscopy combined with energy-
dispersive X-ray analysis.

3. Results and Discussion

The results of the X-ray analysis have shown that ion-bombardment, for the conditions studied, can produce localised deviations from specimen stoichiometry. This behaviour is illustrated in figure 1. It has been found that thin oxide foils when sputtered from one side develop a cation rich region at the unthinned surface. The extent of the effect has been found to vary with beam conditions and in particular it has been noted that the amount of cation enrichment is substantially reduced for beams having a high percentage of neutral ions, B11N gun. In addition the B11N gun also results in a 10 - 25% increase in oxide sputtering rate, compared to the B11.

It is proposed that the positively charged ions in the beam exert electrostatic forces on ions in the sample. The attractive forces on the anions will produce an increase in the preferential sputtering of the oxygen ions with an associated creation of a positively charged surface region, for ionic solids. The electrostatic repulsive forces from both the beam and surface will therefore produce a driving force for cation diffusion away from the thinned surface. Increasing the neutral ion content of the ion-beam will reduce the driving force but since a reduced amount of preferential oxygen sputtering will still occur the positive surface charge will still be produced. The conditions studied, in the experiments reported in this paper, are not those normally used for ion-beam thinning for electron microscopy. However, it is apparent that ion-beam thinning of ionically bonded oxides can create a non-equilibrium distribution of defects and that optimum thinning rates and minimal damage are associated with ion beams of high neutral ion content.

4. Acknowledgements

The authors would like to thank Dr. D.W. Jones for the provision of laboratory facilities.

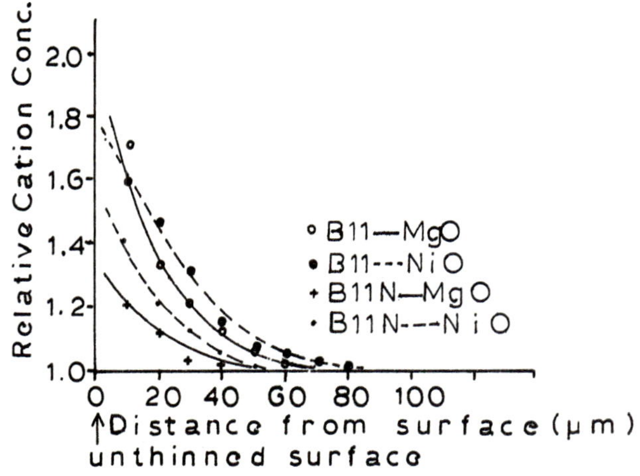

Fig.1. *Energy dispersive X-ray analysis traces of cation composition across cross-sections of ion-bombarded oxide samples.*

STEM spectroscopy of semiconductors

P. M. Petroff

Bell Laboratories, Murray Hill, NJ 07974

1. Introduction

With the development of instruments for scanning transmission electron microscopy (STEM), new spectroscopic techniques can now be applied to small individual defects and interfaces. Most of these techniques which aim at understanding the electronic properties of the material or defects in the probed volume were initiated in the field of scanning electron microscopy (SEM). Among the most important are: charge collection and luminescence spectroscopy.   For more details on the physical concepts involved in these techniques the reader is referred to the recent reviews by Leamy 1981, Varker 1979, Holt 1974 and Kimerling 1981. The major contribution of STEM spectroscopy resides in the possibility of simultaneously probing the structural properties of individual defects and their electronic properties with a marked improvement in spatial res-olution. This type of analysis requires a thin film sample and an electron beam energy significantly higher (50 to 200 KeV) than that of the SEM (1-40 KeV). The resulting experimental conditions can significantly affect the resolution and sensitivity of the spectroscopy and introduce new defects   in the sample. In this paper we discuss for the charge collection and luminescence spectroscopy, the differences between the SEM and STEM analysis. Applications of STEM spectroscopy are presented and the techniques are discussed.

The instrumentation for implementing this type of measurement is based on a modified 200 KeV STEM instrument which has been described in detail by Petroff et al. 1978. A single tilt liquid helium stage which allows for low temperature (18°K), high spatial resolution (30 Å) analysis of the sample in the STEM and spectroscopic mode has been added.

2. The Charge Collection and Luminescence Spectroscopy in STEM

The excess minority carrier produced by the electron beam is collected as a current providing there exists a built in potential V_{bi}, in the sample. The film thicknesses, t, required for structural analysis range between .01 µm and 1.5 µm for 100 or 200 KeV electrons are close to the width of the depleted layer under 0 bias conditions for GaAs or Si doped in the range $10^{15} \leq N_D \leq 10^{17} cm^{-3}$. Thus a very efficient carrier collection is expected and the contribution for carrier diffusion to the collected signal will be negligible. The absence of a diffusion signal is one of the reasons for the small signals $(10^{-7}$ to 10^{-10} Amp) in charge collec-tion STEM spectroscopy.

The charge is collected as a current commonly referred as the electron beam induced current I_{EBIC}, given by

$$I_{EBIC} \simeq \eta \, I_0 (1-f) \frac{E_0}{E_A} \int_0^t \phi \left(\frac{dE}{dZ}\right) dZ \tag{1}$$

where η is the carrier collection efficiency; E_0 and I_0 are the incident beam energy and current; f is the fraction of elastically backscattered electrons; E_A is the threshold energy for electron-hole pair formation; $\phi(dE/dZ)$ is the energy loss function along the z direction and t the film thickness. η is a function of the junction field, minority carrier diffusion length and carrier loss processes through recombination during diffusion or at defects and surfaces. To eliminate minority carrier losses to the thin film surfaces it is recommended to include on the surfaces carrier confining layers consisting of a larger band gap material. We also note that in thin films, the defects electronic properties are detected through recombination processes taking place in the depletion layer where non-uniform high electric fields (10^4 to 10^6 V/cm) can strongly affect these processes. Therefore it is preferable to analyze defects located in the low field or neutral region of the thin film. Equation (1) further indicates that smaller signals are expected in the STEM-EBIC mode because of the smaller generation volume compared to the SEM case.

The concepts presented above also apply to the luminescence spectroscopy commonly referred to as cathodoluminescence (CL). However η in equation (1) becomes the radiative quantum efficiency of the material and the collected signal is in the form of a current, I_{CL}, originating from a photon detector. This digitized signal is computer processed to yield CL spectra which are corrected for the transfer function of the optical system. We note the less restrictive nature of STEM --CL spectroscopy since the presence of a depletion layer in the sample is not required. Therefore the interpretation of radiative or non-radiative recombination at defects is simpler than in the STEM-EBIC case.

3. <u>Resolution and Spectral Sensitivity in STEM Spectroscopy</u>

The shape and dimensions of the excited volume V(t), the minority carrier diffusion length and the defect position affect the resolution of the EBIC and CL analysis. For a high energy beam, Monte Carlo calculations (Soum et al. 1979) suggest that a good approximation to the shape of the excited volume is a truncated cone as shown in Fig. 1. The beam spreading b, for small film thicknesses, t, can be computed assuming elastic scattering (Goldstein et al. 1977) and a small (30Å) electron source

$$b \simeq \frac{Z}{E_0} (\rho/A)^{1/2} t^{3/2} \tag{2}$$

where Z and A are the material atomic number and atomic weight and ρ its density. Figure 1 gives a lower limit of the beam spreading as a function of E_0 for thin films of Si with thicknesses 1 and 2 μm. Equation (2) provides a lower limit of the resolution in STEM spectroscopy since it does not take into account the minority carrier diffusion. Nevertheless the improvement in resolution due to a reduction of the beam spreading when working with thin films and higher beam energies will be significant when

probing VLSI structures. In practice, the observed resolution is smaller
than the minority carrier diffusion length by as much as an order of magnitude. Several factors may be responsible for this effect: a) the influence of surface recombination (Donolato 1979) b) a negligible amount of
carrier diffusion takes place when the width of the depletion layer approaches the film thickness and c) the electron-hole plasma within the
generation cone may produce carrier screening and a reduced effective carrier diffusion length. The variations in image resolution with E_o have
previously been discussed (Petroff et al. 1978). We note that with the
increase in the electron beam range (Fig. 1) fewer carriers are generated
in the thin film and thus a "trade off" between spatial resolution and
sensitivity will direct the choice of the optimal beam energy.

4. <u>STEM-EBIC Applications to the Study of Recombination Processes at
Defects</u>

The STEM-EBIC method has been used for analyzing carrier recombination at
individual dislocations (Fathy 1980, Petroff 1980a). Fig. 2 gives an
example of   carrier recombination at individual dislocation and dislocation nodes introduced by the misfit strain at a $Ga_{1-x}Al_xAs_{1-y}P_y$-GaAs
interface. A dislocation marked D and several dislocation nodes ($N_1,N_2,$
N_3) are associated with an absence of carrier recombination. The STEM
analysis of these defects indicates a correspondence between the edge
character of the dislocations or dislocation nodes and the absence of
carrier recombination. A core reconstruction model in which dangling bonds
are eliminated to minimize the core energy has been proposed to explain
the electrical behavior of the edge "Lomer" dislocation (Petroff et al.
1980a).

The transient charge collection regime (Kimerling 1979) or scanning deep
level transient spectroscopy (SDLTS) (Petroff et al. 1977, 1978)
are also used to measure the energy levels of deep states and provide a
visual display of their distribution in the material. The activation
energy for thermal emission of a defect state which is filled by the e-
beam injection pulse is measured by varying the emission rate window as in
the standard capacitance DLTS (Lang 1974). A comparison with deep level
data previously obtained by capacitance DLTS on large diodes is necessary
to complete the defect identification. Applications of the SDLTS technique to defect analysis in thin films are difficult because of the
small current transients (10^{-12} - 10^{-14} Amp). The detection of SDLTS
signals thus will require a small time constant for the measuring electronics (including the sample). An application of SDLTS to an analysis
of optically induced damage in $Ga_{1-x}Al_xAs$ double heterostructure (DH) is
given in Fig. 3. The two SDLTS maps at -40°C and -150°C respectively display the distribution of a deep donor level (E_t = 0.28 eV) also called the
"DX" center (Lang et al. 1977) and of a shallower level in the regions A
and B of the sample which were optically degraded at room temperature.
The corresponding EBIC micrographs are maps of the junction collection
efficiency. These are used in the deconvolution of the SDLTS signals
(Petroff et al. 1979). The decrease in the "DX center" concentration produced during the dark line formation (indicated in A) is detected by a
quantitative analysis of the SDLTS (T = -40°C) and EBIC (T=-35°C) signals.
The optical degradation process also produces the shallow level detected
in A and B in the SDLTS and EBIC maps at -150°C. STEM analysis of these
areas does not show any strain contrast associated with the B region of
the film.

5.   <u>STEM-CL Applications to Defects and Interfaces</u>

The qualitative chemical analysis of thin films and detection and identifications of small concentrations of impurities are the most common applications. The STEM-CL analysis of interfaces between $Ga_{1-x}Al_xAs$ and GaAs films grown by liquid phase epitaxy (Fig. 4) has shown unambiguously the existence of a non uniform transition layer at these interfaces (Petroff et al. 1980b). Quantitative measurements of the effects of interfaces on the GaAs luminescence have also been carried out by STEM-CL analysis (Petroff et al. 1981). Double heterostructures (DH) and multiquantum well (MQW) structures grown by molecular beam epitaxy with nearly identical thicknesses of GaAs (Fig. 5) are compared. Analysis of the neutral donor bound exciton (1.512 eV) CL intensity in the (DH) structure and that of the free exciton (1.523 eV) in the MQW structure (Fig. 6) indicates a luminescence increase (up to 15 times) associated with multiple interfaces in the structure. Recent results support the idea of a gettering mechanism of non radiative centers by interfaces in MQW structures.

6.   <u>Anomalous Effects in STEM Spectroscopy</u>

Anomalous effects in STEM spectroscopy are always present because of instrumental requirements(high energy beam, small gap objective pole piece). The effects of backscattered electrons and diffracted electrons on the x-ray yield from thin films have already been discussed. Identical effects have been reported in quantative electron energy loss spectroscopic analysis (Zaluzek et al. 1980). In the case of interest here, the semiconductor sample to be investigated acts as a detector not only for the excess minority carriers generated by the incident electron beam but also for carriers generated by secondary ionizing radiation. Such radiation is produced by the interaction of backscattered or diffracted electrons with the sample surroundings. The schematic diagrams in Fig. 7 illustrates this process for the case of STEM and SEM analysis. In the case of a STEM analysis for example the introduction of an objective aperture which intercepts some of the diffracted beams will result in an increase of ionizing radiation flooding the sample. These in turn generate electron hole pairs in the sample and are detected together with those generated by the incident electron beam. Primary backscattered electrons will also produce a secondary anomalous signal identical to that generated by the diffracted beam. The total signal, I_T, detected is a convolution of the signal, I_p, generated by the incident beam and a signal, I_s, generated by secondary radiation:

$$I_T = I_p \times I_s \qquad\qquad (3)$$

In cases for which $\Delta I_p/I_p << \Delta I_s/I_s$ the collected signal yields a contrast given by: $\Delta I_T/I_T \sim \Delta I_s/I_s$. Thus the EBIC or CL signals could correspond to a dark field STEM or to a dark field backscattered signal of the thin film. Figure 8 illustrates this effect in a silicon MOSFET device which was chemically thinned under the gate area. The gate is floated and the EBIC signal is collected through the drain. The channeling pattern in the STEM image is also detected in the EBIC image. The sign of the EBIC contrast is given by the presence of microcracks in the film (dark Bands). The excess minority carriers generated in the regions of reduced channeling account for the bright contrast of the channeling pattern in the EBIC image. The strength of this signal is markedly increased ($\sim$ 5 times) when the objective aperture is introduced below the sample. The second

example is taken from CL imaging of   misfit dislocationsat the interface
between a GaAs and a $Ga_{1-x}Al_xAs_{1-y}P_y$ film. The near bandgap and below
bandgap monochromatic CL micrographs in Fig. 9 show different background
contrast (white bands) which are more visible in the below bandgap images.
The contrast of these bands is complementary to that seen in the STEM
image and is consistent with an increased carrier generation due to backscattered electrons in the region of reduced channeling. These anomalous
secondary signals are very important when small variations in the contrast
are used in quantitative EBIC or CL analysis of individual defects. The
SEM spectroscopy suffers from the same problem.

7. <u>Radiation Damage Effects in STEM Spectroscopy</u>

The usefullness of the STEM spectroscopy can be offset by beam induced
radiation damage in the analyzed material. Two types of radiation damage
may be introduced: a) the damage associated with the formation of Frenkel
pairs in the bombarded thin film (Fathy et al. 1980) which takes place at
high beam energies for most semiconductors (E_O > 200 KeV) and b) the
damage produced by beam induced ionization of impurities in the thin film.
This so called subthreshold damage which occurs at very low beam energies
(E_O << 50 KeV), produces deep levels or interface states which usually
correspond to non-radiative recombination centers. In the presence of an
electric field, this ionization induced charge can become mobile and
change the device characteristics. The beam induced damage effects are shown
(Fig. 10)in a MOSFET device which has been chemically thinned to a thickness
of less than .l µm under the gate area. The device I-V characteristics
are unchanged by the thinning operation. The gate area of the device is
imaged in the STEM instrument operated at 100 KeV or 200 KeV. The drain
to source current is used for forming the EBIC image. The EBIC contrast
in Fig. 10 shows recombination centers (dark lines) introduced by focussing the beam at high magnification and moving it over the surface to create
the fine line pattern (.3 µm). This type of damage which is not detected
by STEM analysis, is introduced rapidly (seconds) and is located in the
SiO_2-Si interface as indicated by the changes in the I-V curve. The other
type of damage is illustrated in Fig. 11. Misfit dislocations visible in
the STEM image are seen as bright lines of recombination centers in the
corresponding EBIC micrograph. In addition recombination centers
visible as small white spots in the EBIC image appear as a function of
exposure to the electron beam for a beam energy of 200 KeV. This effect
does not occur at a beam energy E_O ≤ 150 KeV. The nature of the damage
and the high threshold energy   E ≥ 150 KeV   leads to the conclusion that the damage is associated with production and subsequent clustering
of Frenkel pairs.

8. <u>Conclusions</u>

This survey of the STEM spectroscopy techniques point out their usefullness
for: a) defect analysis on a submicron scale, b) spectroscopy of multilay-
ered structures and c) rapid correlation between electronic and structural
properties of defects. However spurious effects associated with multiple
scattering of the primary beam, radiation damage, surfaces and depletion
layers in thin films should be further investigated before useful quantitative information can be routinely extracted from these methods.

<u>References</u>

C. Donolato, Appl. Phys. Lett. <u>34</u>, 80 (1979)
D. Fathy, T. G. Sparrow and U. Valdre, J. of Microsc. 118, 263 (1980).
A. George, J. M. Fournier, J. P. Gonchond and D. Bois in "Scanning Electron
  Microscopy" (SEM, Inc. AMF O'Hare Chicago, IL 60666 USA) 4, 69, 1980.
J. I. Goldstein, J. L. Costley, G. W. Lorimer and S. J. Reed, SEM, Inc.
  O'Hare Chicago, IL 60666 USA, 315 (1977).
D. B. Holt, "Quantitative Scanning Electron Microscopy", D. B. Holt,
  M. B. Muir, P. R. Grant eds. (Academic Press, New York) p355,(1974).
L. C. Kimerling in Physics of Semiconductors 1978, Inst. Phys. Conf. No.43
  p. 113-122 (1979).
L. C. Kimerling, "Defects in Semiconductors" J. Narayan and T. Y. Tan eds
  (North-Holland Inc., New York) p. 85, (1981).
D. V. Lang, J. Appl. Phys. <u>45</u>, 7, 3023 (1974).
D. V. Lang and R. A. Logan, Phys. Rev. Lett. <u>39</u>, 635 (1977).
H. J. Leamy, "Physical Electron Microscopy", O. C. Wells, K. F. J. Heinrich
  AND D. E. Newberry eds. (Van Nostrand Reinhold New York) (1981).
P. M. Petroff and D. V. Lang, Appl. Phys. Lett. <u>31</u>, 60, (1977).
P. M. Petroff, D. V. Lang, J. L. Strudel and R. A. Logan, SEM Inc. (AMF
  O'Hare Illinois 60666 USA) 325 (1978).
P. M. Petroff, D. V. Lang, R. A. Logan and W. D. Johnston, Jr., Inst. Phys.
  Conf. Ser No 46, London, 427 (1979).
P. M. Petroff, R. A. Logan and A. Savage, Phys. Rev. Lett. <u>44</u>, 4,287(1980a).
P. M. Petroff, and R. A. Logan J. Vac. Sc. Tech. 17(5) 1113 (1980b).
P. M. Petroff, C. Weisbuch, R. Dingle, A. C. Gossard and W. Wiegmann, Appl.
  Phys. Lett. 38, 12, 965 (1981).
G. Soum, F. Arnal, J. L. Ballladore, B. Jouffrey and P. Verdier, Ultramic-
  roscopy 451, (1979).
C. J. Varker, "Nondestructive Evaluation of Semiconductor Materials and
  Devices, J. N. Zemel ed. (Plenum, New York) (1979).
N. J. Zaluzec, H. J. Hren and R. W. Carpenter EMSA Proceedings ed.
  G. W. Bailey (Claitor's Publishing Baton Rouge LA 70821) 114 (1980).

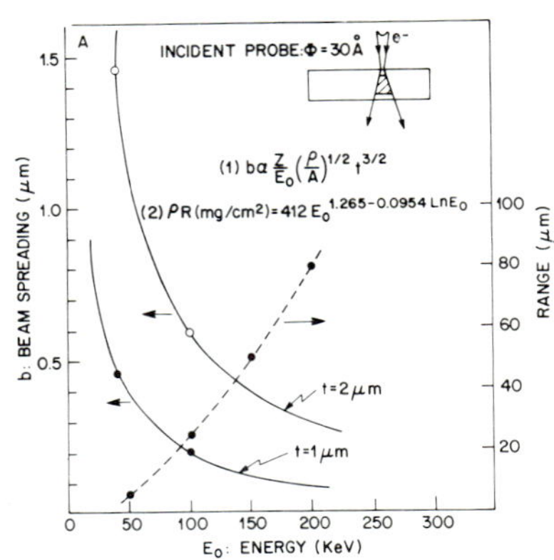

Fig. 1: Beam spreading (b) and range
(R) as a function of beam energy E_0.

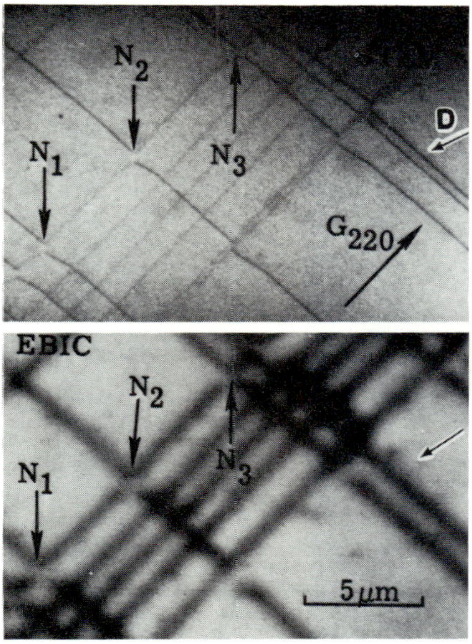

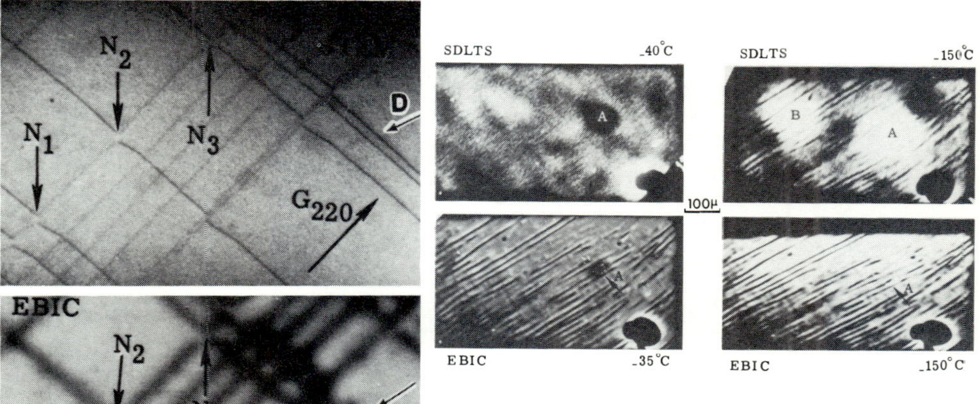

Fig.3: SDLTS-EBIC micrographs of optically degraded GaAs-Ga$_{1-x}$Al$_x$As double hetero-structures. The emission rates for both SDLTS maps is 500 nsec.

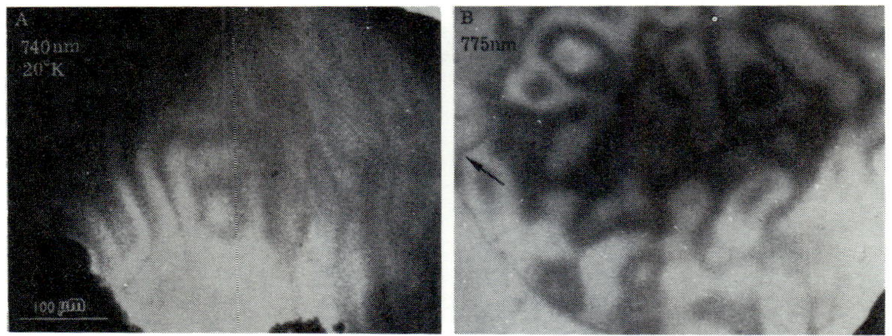

Fig.2: STEM-EBIC micrographs of misfit dislocations. Dislocation D and parts of dislocation nodes N$_1$, N$_2$, N$_3$ do not exhibit carrier recombination effects.

Fig.4: CL monochromatic micrographs of interfaces in a Ga$_{1-x}$Al$_x$As-GaAs double heterostructure.

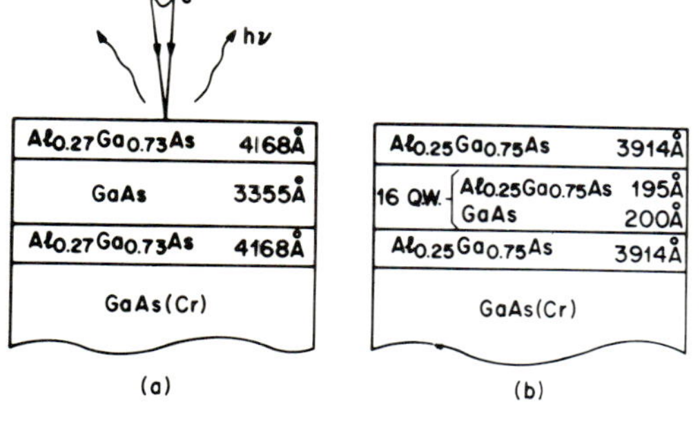

Fig.5: Schematic of the double heterostructure and multiquantum well structure analyzed in Fig.6.

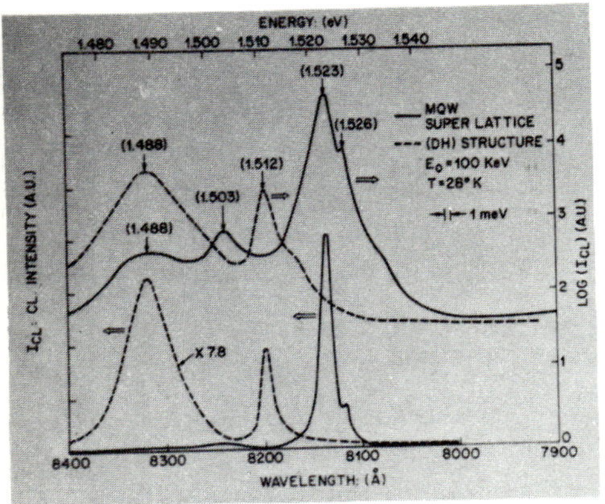

Fig.6: CL spectrum of the structures shown in Fig. 5.

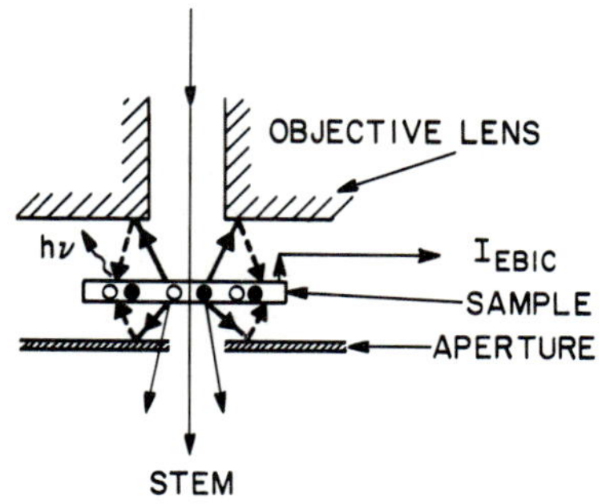

Fig.7: Schematic of the spurrious signal generation processes in STEM analysis.

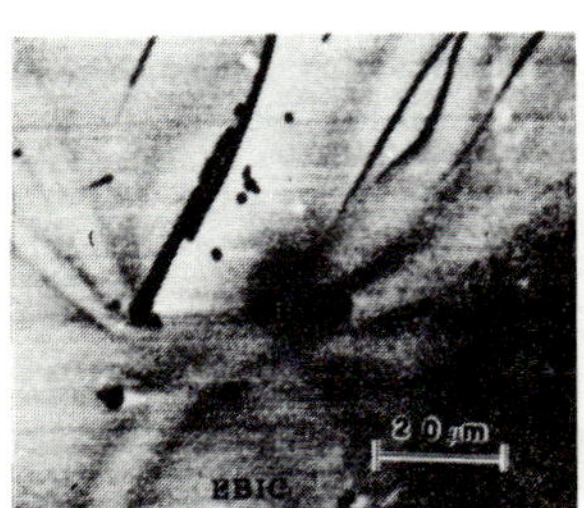

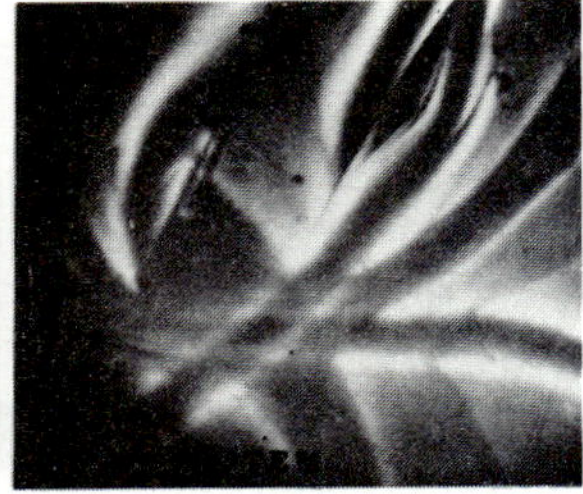

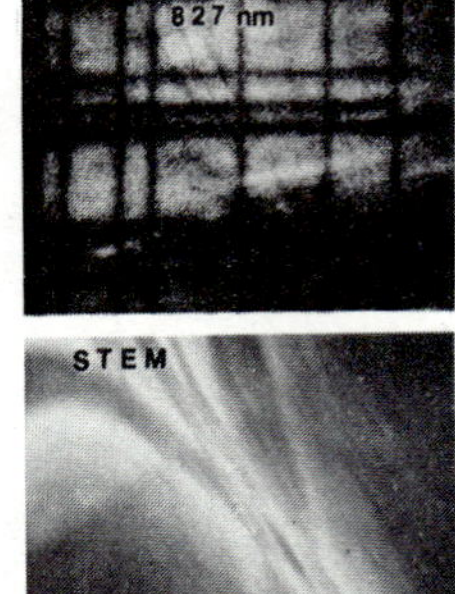

Fig.8: STEM-EBIC analysis of the gate area in a MOSFET Si device.

Fig.9:   STEM-CL micrographs a thin film with misfit dislocations.

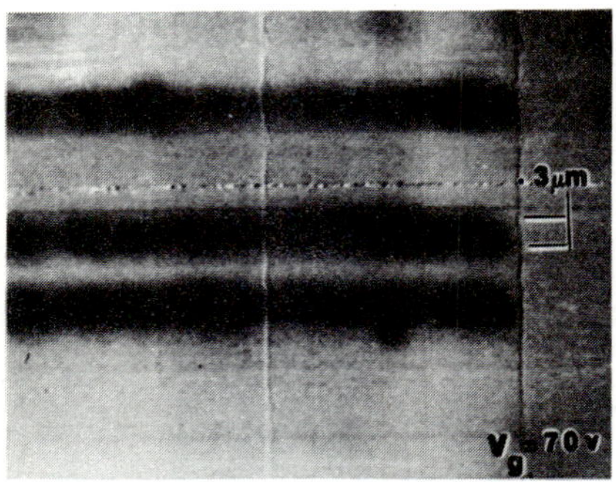

Fig.10: STEM-EBIC micrograph of a pattern (dark lines) due to beam induced damage in a MOSFET Si device.

Fig.11: EBIC micrograph of near threshold radiation damage in a GaAs-Ga$_{1-x}$Al$_x$As$_{1-y}$P$_y$ thin film containing misfit dislocations (white contrast correspond to recombination centers).

Microanalytical electron microscopy of semiconductor devices

D Fathy and L M Brown

Cavendish Laboratory, University of Cambridge, Madingley Road,
Cambridge, CB3 OHE.

1. Introduction

A considerable amount of work has been carried out on the electrical
properties of defects near p-n junctions in Si based devices. However,
although the position of the depletion layer can be determined accurately,
to date it has not been possible to measure the concentration of dopant
across the junction. Such concentration profiles have had to be
estimated from indirect evidence: see for example Brown and Fathy (1981).
It is true that the technique of secondary ion mass spectroscopy can
give excellent impurity profiles for ion implanted or diffused impurities
by gradual erosion of the surface (Zinner 1980), but such techniques
cannot be used for devices with junctions perpendicular to the surface.
However, thin film x-ray microanalysis (STEM/EDX) offers improved
sensitivity of detection and spatial resolution over techniques based
on bulk samples, and it is of some interest to see how well the impurity
profiles can be characterised in a standard device.

2. Experimental

The silicon power transistor devices used in this work were thinned
from the substrate side (collector) until transparency near the step S
(Fig. 1) was obtained (Fathy and Valdre 1980). The details of the
devices as supplied were: the phosphorus doping in the n-Si side was
$\sim 2 \times 10^{19}$ atoms/cc and the boron doping on the p side was $\sim 8 \times 10^{16}$ atoms/cc,
giving an estimate of the number of phosphorus atoms of the order of
0.001 p.p.m in the depletion region of the device. For this work, a
fixed cartridge with a graphite specimen holder was used in a V.G.
HB 5 STEM.

Fig. 1 shows a schematic section of the device and the x-rays generated
by the beam and collected by the Si (Li) detector positioned at a take-
off angle of approximately 12° with respect to the specimen. Figs. 2a,
b, c are bright field images obtained close to the surface step. Previous
experiments had shown that the position of the bright band B in Fig. 2a,b
coincides with the depletion region of the device. The x-ray spectra
were obtained along a line marked on Fig. 2b which covers a distance
from the n-Si to p-Si doped region.

The spectra were taken using a scanned probe at high magnification. In
this condition the diameter of the area illuminated by the beam was $\simeq$
2nm. The acquisition time was chosen such that the peak value for Si

reached the 65K counts maximum
available in the Link system used.
In this way also all the spectra
were normalised, revealing only the
spatial variation in doping. With
the combination of the two apertures
used, i.e. probe forming and selected
area (the former stops x-rays from
the gun and the latter excludes x-rays
generated by the first aperture), the
hole count was less than 1 count/sec.
as compared to count rates of 2000–
3000 per sec. over the whole spectrum
generally used.

3.Discussion and Results

When thin foils are studied at
higher beam energies than those
commonly used in SEM, the following
advantages are apparent: (a) the
excitation volume of x-ray generation
is small, and can be assumed to be
a cylindrical volume with a diameter comparable to that of the
probe; (b) the Bremsstrahlung radiation produced by inelastic interaction is a great deal smaller,
being nearly linearly dependent on
thickness and inversely proportional
to the accelerating voltage (Reed
1975).
Fig. 3a, b shows two typical spectra
obtained over n-Si and p-Si region
respectively, showing the presence
of phosphorous kα and kβ in the
former and not the latter. The n-
Si and p-Si regions were about 300
nm thick (as measured by the number
of mean-free path of plasmons
excited). For the spectra shown
here, the background, about 1 count
per second per 20 ev channel, is
to within a factor of two of what
one expects from Bremsstrahlung.
Thus the minimum detectable number
of atoms in the probe is determined
mainly by the ability to detect
characteristic radiation in a
background of continuous radiation;
it is not determined by instrumental
factors, and one should, in a very
thin film, be able to achieve close
to single atom detection (Brown
(1981). However, in the thicker
film, which was of necessity used

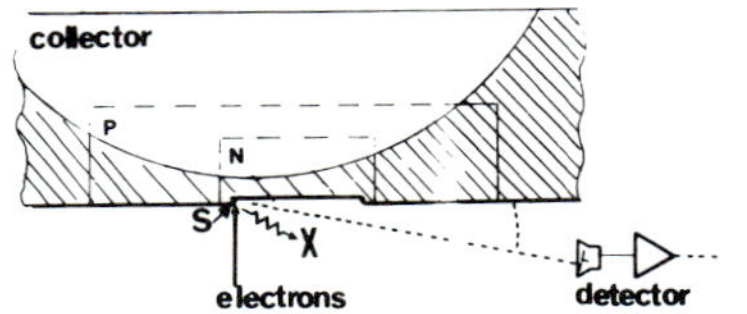

Fig. 1 Schematic drawing of the
device, showing the position of
the step S with respect to the
electrical junction and the
relative position of Si (Li)
detector.

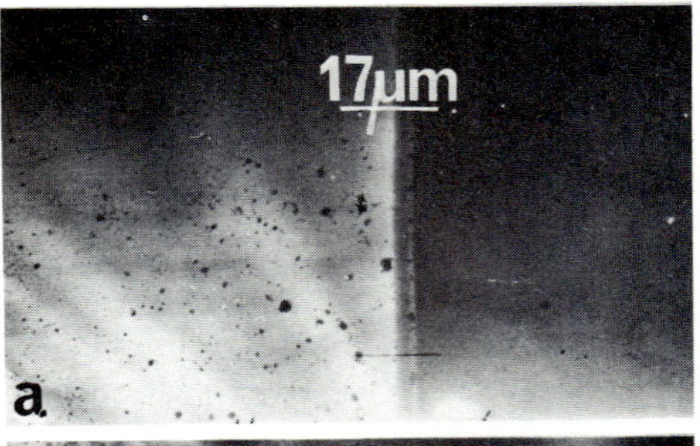

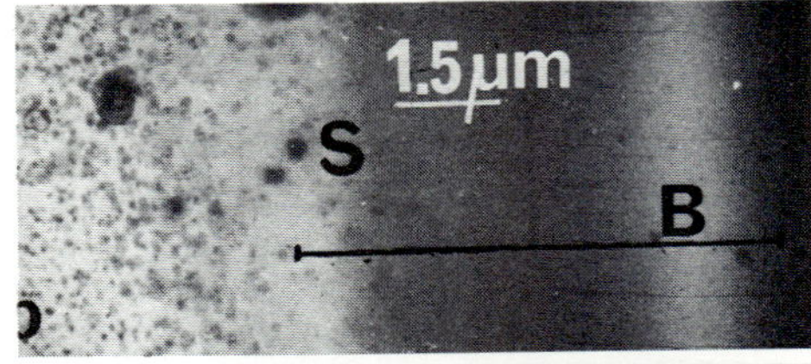

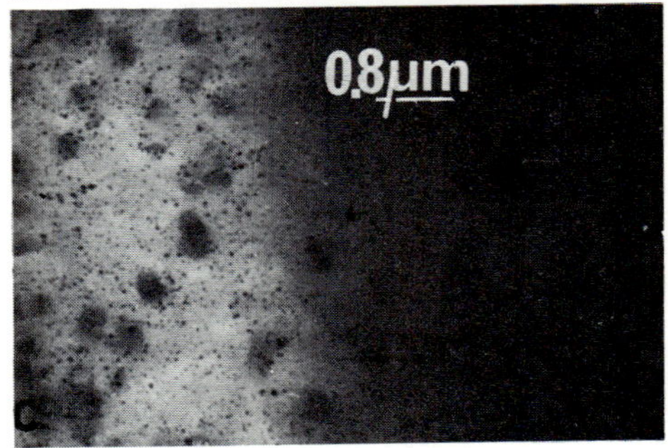

Fig. 2a,b,c Bright field images
obtained in STEM. In Fig. 2b, the
letter B shows the position of the
depletion region. The line
indicates the extent of the area
where the spectra were obtained.

here, the minimum number of phosphorus atoms detectable is increased
because of the increased Bremsstrahlung; of course, the spatial resolution
is also degraded in this case to about 200 nm (Hutchings et al 1979).

Fig. 4 shows the results obtained for the variation of phosphorus
doping across the p–n junction. The detection sensitivity is about
0.02% or an equivalent detection of 80 atoms in
     the scanned area.

4. <u>Conclusion</u>

With improved experimental
conditions, it is possible
to profile impurities in semiconductors, either for the
lateral spreading of impurities
or where heavy ion implantation may lead to the formation of very shallow junctions
$\sim$1000 Å deep, enabling the
measurement of impurity variations
down to 10^{18} atoms/cc.

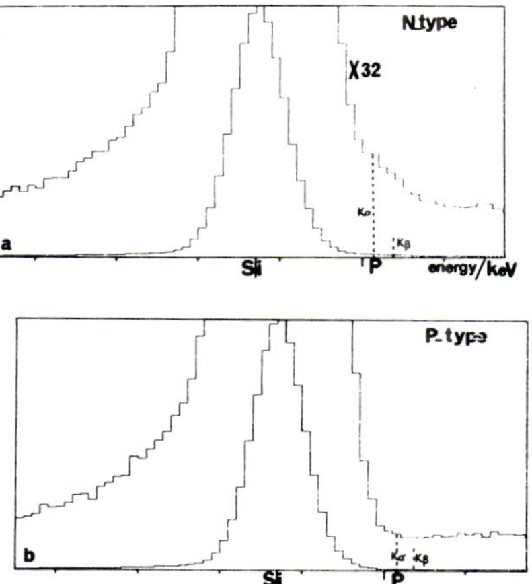

Fig. 3a,b The spectra obtained
from n and p region of the device
respectively.

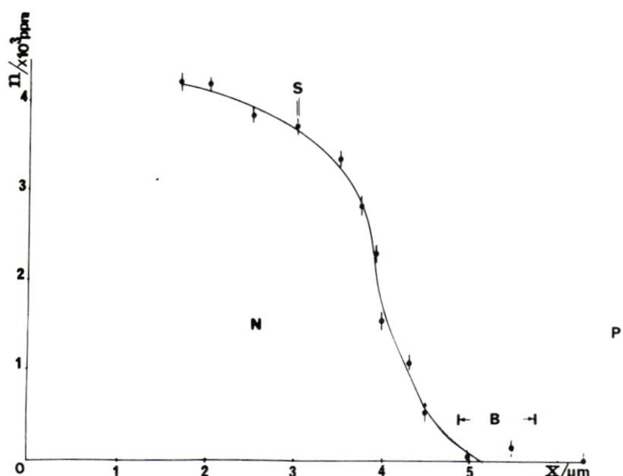

Fig. 4 The profile of the spatial variation
of phosphorous in the range marked in Fig. 2b.
The spectrum was corrected for the background
and the contribution of the Si peak in the
region of the phosphorus peak.

References

Brown L M 1981 J. Phy.F 11 1
Brown L M and Fathy D 1981 Phil. Mag. 43 4 715
Fathy D and Valdre U 1980 J. Micro. Spectrosc. Electron. 5 175
Hutchings R, Loretto M H, Jones I P and Smallman R E 1979 Ultramicroscopy
   3 401
Reed S J 1975 Electron Probe Analysis (Cambridge: Cambridge University
   Press)
Zinner E 1980 Scanning 3 57

X-ray composition analysis of semiconductor alloys with high spatial resolution

Lars-Åke Ledebo

Department of Solid State Physics, Lund University, 220 07 Lund, Sweden

Abstract. Low voltage electron beam excited X-ray analysis is very useful to characterize semiconductor alloys with multiple epitaxial layers. The resolution may be as good as 0.2 μm at an excitation voltage of 5 kV. For the analysis to be practical it is essential to use a rapidly repetitive line-scan with signal averaging.

1. Introduction

Electronic devices based on multiple layers of semiconductor alloys are becoming increasingly important in semiconductor technology. The hetero-junction laser continuously radiating at room-temperature is a well-known example of a device based on multiple layers of the ternary alloy $Al_xGa_{1-x}As$ (0<x<1). Much progress has been made recently within IR optical communication using complex structures of the quarternary alloy $In_xGa_{1-x}As_yP_{1-y}$ (0<x,y<1). Some other important applications of semiconductor alloys are light-emitting diodes of $GaAs_xP_{1-x}$ and bandgap-tuned infrared detectors of $Hg_xCd_{1-x}Te$. Successful growth of the material for such devices requires careful analysis of alloy composition, layer thickness and material homogeneity.

In conventional electron microprobe analysis at high excitation voltage, the spatial resolution is limited due to the comparatively large volume in which X-rays are generated by the electron beam. A number of other methods have been suggested to obtain a profile of the composition of semiconductor structures. The ion microprobe is an obvious possibility which however is not very accurate in quantitative analysis. Monemar (1978) suggested profiling by monitoring the spectral distribution of the photo-luminescence from a wedge-etched cross-section of the sample. Goodfellow et al (1978) developed a contact-resistance profiling method, also on wedge-shaped cross-sections. A group at the Post Office Research Department, UK, have developed an electrochemical technique for profiling (Ambridge and Faktor 1975). Even though this method is very powerful for the profiling of carrier concentrations, it is less straight-forward for composition analysis. All these methods have the common disadvantage that they are destructive and require large volumes of material which are not always available. In this paper it is demonstrated that for most semiconductor alloys it is possible to make a highly competitive, rapid and non-destructive analysis with good spatial resolution by using the L-lines for the heavier elements (Ga, In ...) and the K-lines for the lighter elements (Al,P ...). Data are presented here for $Al_xGa_{1-x}As$, but similar results have been obtained for $GaAs_xP_{1-x}$, ZnS_xSe_{1-x}, $In_xGa_{1-x}As_yP_{1-y}$ and $In_2S_xSe_yTe_z$ (x+y+z=3).

2. Data acquisition

The equipment used in this investigation is a Cambridge Stereoscan S4
scanning electron microscope in combination with a Tracor Northern
TN-2000 X-ray analyzer. The peak half width of the detecting system at
the As-L peak at 1.28 keV is about 110 eV. In the quantitative analysis
the spectrum is deconvoluted and corrected for matrix effects by using
the commercial Tracor software with either pure or mixed standards.
Semiconductor alloys provide a particularly efficient means to test the
accuracy of the quantitative analysis, as most alloys are known to
deviate very little from exact stoichiometry. In $Al_xGa_{1-x}As$, for example,
the As content is 50 atomic percent. In the range of composition
from GaAs to AlAs, the As-concentration determined by the system in
everyday routine measurements is normally within 2 atomic percent units
of the accurate value. Larger deviation may be caused by inadequate
geometry, dirty final aperture or decomposition of the semiconductor
surface.

Complex epitaxial structures are easily analyzed by profiling the cross-
section obtained by cleaving the wafer perpendicularly to the surface. A
very useful feature of the analyzing system in this context is the time-
scan facility. In our particular system it has been found practical to
run the microscope beam in a line-mode over a selected region, and to
trigger the X-ray analyzer from the microscope. In each time interval, the
number of energy quanta within a given energy interval are counted by the
analyzer and added to a channel corresponding to the beam position. By
running the line sweep rapidly repetitive it is easy to monitor the data
collection. A sweep time of 0.4 sec is normally used. After the start of
the data collection, it is more or less immediately obvious whether the
region selected is well chosen, and whether the number of channels is
convenient. Furthermore, it is easy to judge when the statistical quality
of the data is sufficient, whereupon the collection can be interrupted
and the data plotted or recorded. A typical result is shown in fig 1,
where the number of energy quanta within the Al-K peak are presented.

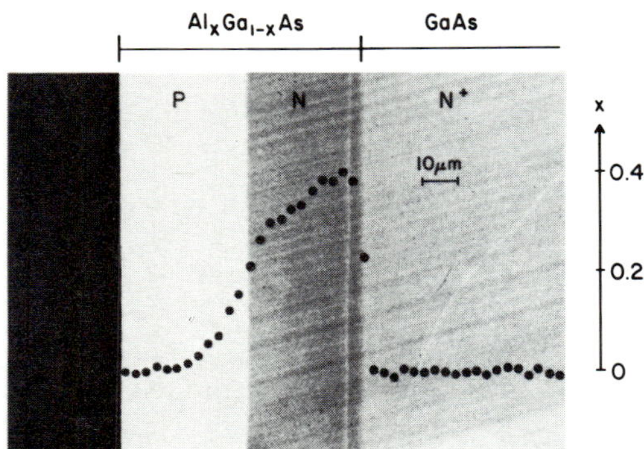

Fig 1. A time-scan superimposed on a normal photomicrograph
of an epitaxial structure grown by liquid phase epitaxy. The
accelerating voltage was 10 kV, and the acquisition time 8
minutes. The absolute value of x has been determined by
quantitative analysis in one point in the middle of the N-type
layer.

The background has been assumed approximately constant and is subtracted
in figure 1. As the matrix correction is also linear in x to a good
approximation, it is sufficient to make the full quantitative analysis
in only one or a few points in order to calibrate the spatial distribution
in absolute Al content.

Fig 2. A time-scan of the
number of pulses within the
Al-K peak obtained at 5 kV.
The acquisition time is
4 minutes.

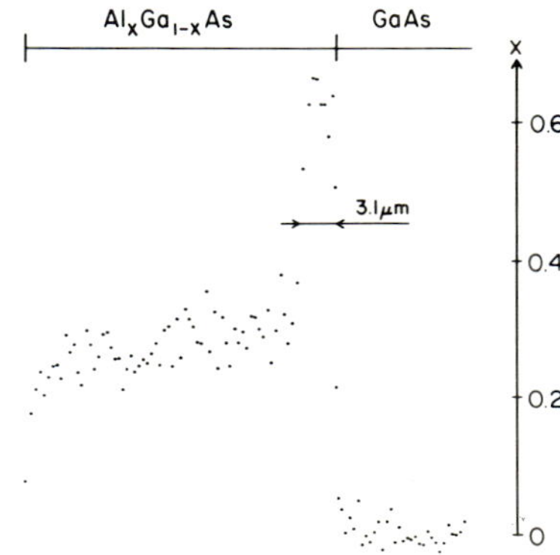

The above figure is taken for an epitaxial structure where a relatively
thin layer acts as a buffer with high Al content between the substrate
and a thicker layer in which a pn-junction is to be made. In comparison
to figure 1 the number of points have been increased, and the accelerating voltage has been decreased in order to improve the resolution. This
illustrates a difficulty with the constant sweep-rate, as the increased
number of points gives additional statistical noise, which is annoying
particularly in the thicker layer where the high resolution is not needed.
This problem could be solved by using a digitally controlled beam with
step-sizes determined from the data being collected. In this manner the
step-size could be kept small in regions with rapid changes in composition
and larger where the composition only changes slowly. The hardware needed
is readily available, but it seems that the software is not yet fully
developed for this particular task.

3. Spatial resolution

The spatial resolution has been measured as a function of the excitation
voltage for a GaAs-$Al_{0.7}Ga_{0.3}As$ step, made by liquid phase epitaxy. Such
steps are known to be abrupt on a more or less atomic scale. It
was found practical to use the width from 10 % to 90 % as a definition of
resolution. Fig 3 demonstrates the result together with the signal-to-
background ratio for the Al peak. The probe function has been determined
for the same step, and is displayed in fig 4 for 5 kV and 20 kV excitation
potentials. The full lines are a fit to the numerical data, which contain
some statistical noise. It is quite obvious that a reduction of the excitation potential to 5 kV considerably improves the result. A further
decrease of the voltage beyond 5 kV will normally not improve the resolution further, as the beam current has to be increased substantially to
compensate for the less efficient X-ray generation. This in combination
with the low excitation voltage will in most microscopes lead to a resolution limitation due to loss of definition of the electron beam.

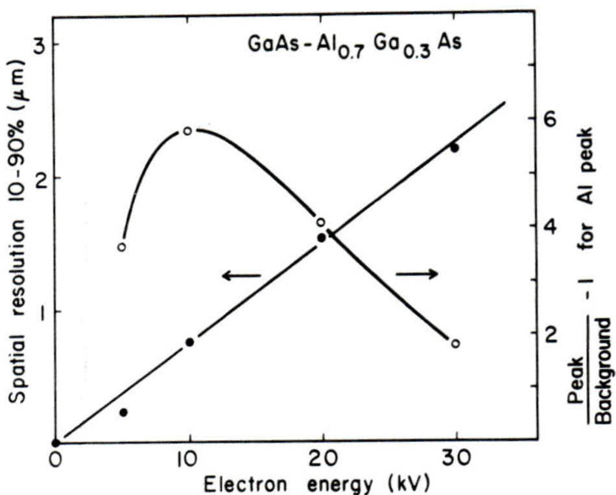

Fig 3. Spatial resolution and peak-to-background ratio for
the Al-K peak as a function of excitation voltage.

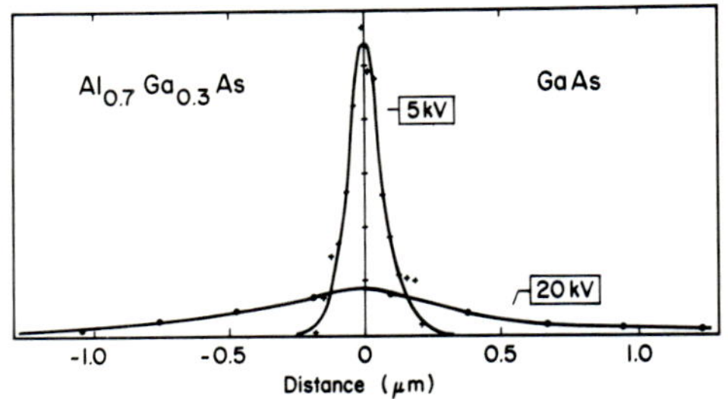

Fig 4. The probe function at 5 and 20 kV. The curves have
been normalized to constant area. The full curves are fits
to the data points, which are indicated in the figure.

For optimum resolution it is important that the epitaxial layers are
oriented so that they lie in the plane defined by electron beam, detector
and sample. By using a probe stage with rotation, tilt and tilt-rotation
in addition to the x,y and z movements, it is possible to make this
orientation and profiling also with mounted and tested devices. This is
of great value in error analysis, as for example when degradation behaviour is to be correlated with layer properties. Such characterization
has been found to be particularly successful when simultaneously combined
with measurements of EBIC and cathodoluminescence.

4. References

Ambridge T and Faktor M M 1975 Inst. Phys. Conf. Ser No. 24, p 320
Goodfellow R C, Carter A C, Davis R and Hill C 1978 Electronics Letts.
   14 328
Monemar B 1978 J Appl Phys 49 2922

A system for EBIC analysis of defects using controlled low temperatures in the SEM

E D Boyes, G Dixon-Brown and A Ourmazd

Department of Metallurgy & Science of Materials, University of Oxford

An analysis of the temperature dependence of EBIC contrast from defects over a suitable temperature range can be used to determine the positions of the energy levels associated with individual defects in semiconductor materials and devices. When used in conjunction with other characterisation techniques, such as transmission electron microscopy, this can provide a powerful diagnostic technique for the study of the electrical properties of individual defects, including dislocations[1].

A system based on a JSM-35X scanning microscope has been developed, which includes a high resolution liquid nitrogen cooled 90K cold stage. The UHV-flanged demountable cryostat is fitted to one of the large spectrometer ports in the specimen chamber of the microscope and is connected to the teflon insulated (0.57 J-sec^{-1} cm^{-1} at 300-76K)[2] stage assembly by means of a thick but flexible high conductivity copper braid (930 J-sec^{-1} cm^{-1} at 300-76K). Other temperatures can be achieved accurately (±2K) by using a small heater and a thermocouple[3] or CLTS[4] device mounted very close to the specimen. Mechanical control and X-Y movements with a limited amount of tilt are provided by the standard goniometer stage, to which the specimen mount is attached by means of a thermally isolated sub-stage, providing working distances (WD) of 25-0mm, Fig.1. At small WD the lens aberrations are reduced and the probe current density correspondingly increased. The airlock mechanism is retained and samples may be exchanged and the necessary electrical connections engaged with a minimal specimen cool-down time (<3mins).

By changing the sample temperature and thus the position of the Fermi level in the band-gap, the charge state of a dislocation can be altered. Since the EBIC contrast is in part determined by the dislocation charge, this presents the possibility of determining the energy position of a dislocation level by monitoring its EBIC contrast as the Fermi level is swept across the gap. Fig.4 shows the T-dependence of the EBIC contrast from an individual screw dislocation in n-type Si, Silicon (P; 10^{16}/cm^3), similar to those shown in figure 3. The reduction in the EBIC contrast as T becomes greater than 180K provides an indication that the level associated with this dislocation is at 0.1eV from the conduction band-edge. The results were obtained with the phase-sensitive detection system described elsewhere in these proceedings (Ourmazd, Wilshaw and Cripps, 1981)[5].

Preliminary results have shown that:

(i) the resolution of the microscope is not significantly reduced by the cold stage (Fig.2),
(ii) contamination, usually a severe problem at low temperatures, has not been unacceptably high and that the microscope vacuum of 5x10^{-7} torr is adequate. (The specimen is the least cold part of the cryo-system, which provides some additional cryo-pumping),
(iii) analysis of EBIC contrast as a function of temperature can be used to study the electrical properties of individual dislocations of different character.

References

1. Ourmazd A, Weber E, Gottschalk M, Booker G R and Alexander H 1981 Inst.
     Phys. Conf. Ser. No. 60 (eds Cullis & Joy) p 63
2. Croft A J, Cryogenic Laboratory Equipment 1970 Plenum Press
3. Chromel/Alumel is useful down to ~70K
4. CLTS - Oxford Instruments Ltd, Oxford
5. Ourmazd A, Wilshaw P R and Cripps R M, The proceedings of this conference

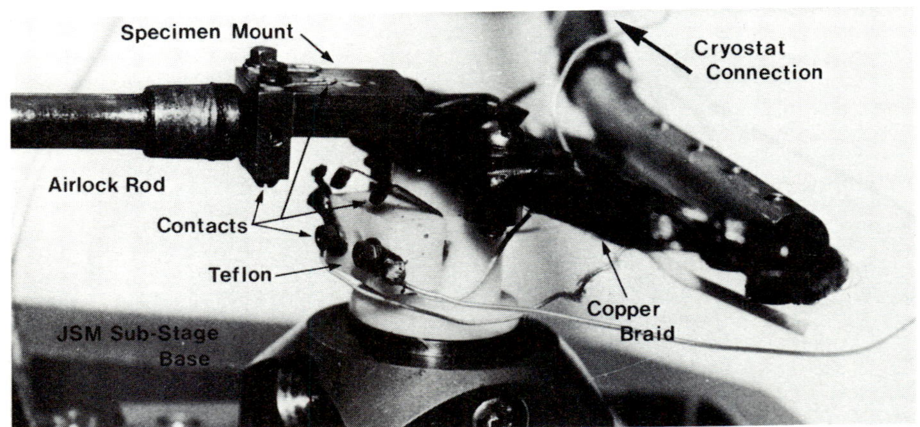

Fig.1. Cold sub-stage system for JSM-35X goniometer stage.

Fig.2. Image resolution retained
     with the cold stage.

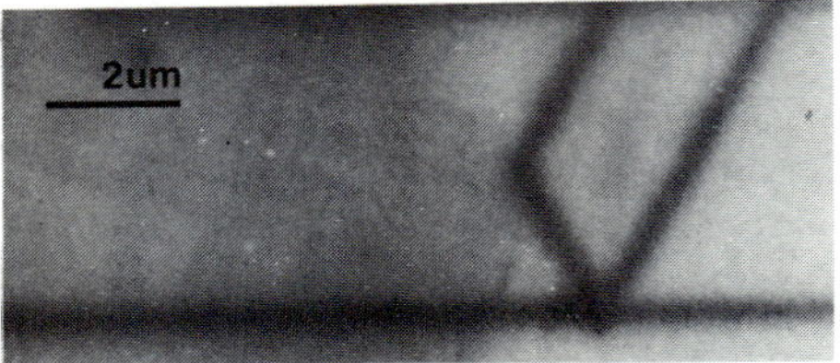

Fig.3. EBIC image of dislocations in
     n-type silicon.

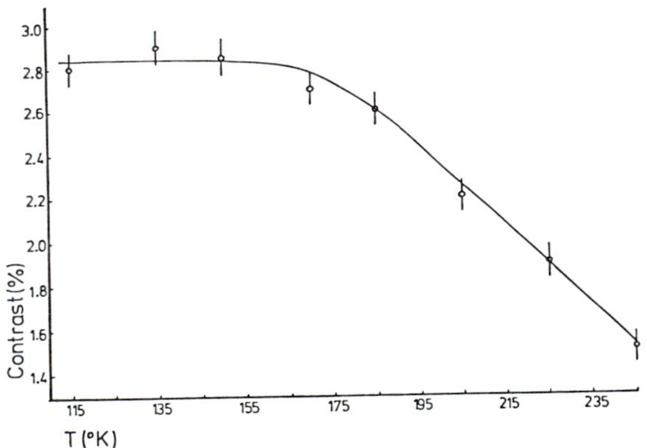

Fig.4. Temperature dependence of EBIC contrast from screw dislocations in
     an n-type silicon.

Measurement of contrast from individual dislocations by lock-in EBIC

A Ourmazd, P R Wilshaw and R M Cripps

Department of Metallurgy & Science of Materials,
University of Oxford, Parks Road, Oxford OX1 3PH

1. Introduction

The greater understanding of the macroscopic properties of semiconducting
materials on the one hand, and the trend towards the miniaturisation of
electronic devices on the other, have increased the importance of investigating the physics of individual microscopic systems such as defects.
Electron microscopical techniques, with their high spatial resolutions, are
ideally suited to such investigations. Thus the structural properties of
defects such as dislocations can now be determined down to the 2Å level
(Anstis et al. 1981). The Electron Beam Induced Current (EBIC) mode of the
SEM can, in principle, yield electrical information on a localised scale of
less than 1μm. This technique, however, has been mainly used as a method
of revealing the presence of electrically active microscopic defects and,
until recently, has not been used to obtain quantitative information about
the physics of electronic processes at defects.

The EBIC technique utilises the electron beam of a SEM (or STEM) to create
electron-hole pairs in a region of the semiconductor known as the generation volume. The excited minority carriers are then collected by a p-n
junction or Schottky barrier to produce a signal. Any electrically active
defect is revealed by a reduction in the collected current when it lies in
the vicinity of the generation volume. The spatial resolution of the
technique is in part determined by the size of the generation volume, which
in turn depends upon the energy of the incident electrons as $E^{1.75}$. Thus
to achieve high spatial resolution (<1μm) using the SEM, it is necessary to
use low incident electron energies. Under such circumstances the EBIC
signal is also small, not only since fewer electron-hole pairs are created,
but also because the excess carriers are generated closer to the surface,
where recombination can further reduce their equilibrium concentration. In
many instances it is also necessary to use low incident electron beam
currents to ensure the establishment of the low excitation regime, thus
simplifying the interpretation of the data. Consequently, obtaining high
spatial resolution, low excitation regime EBIC data generally implies very
low EBIC signal intensities.

The lock-in technique is widely used in many branches of physics to recover
small signals. It is the purpose of this paper to (i) outline the principle of lock-in detection as applied to the EBIC mode of the SEM, (ii)
describe a system utilising this principle and (iii) illustrate the possible applications of such a system by presenting preliminary results, which
yield detailed information about the electronic processes involved in
carrier capture at individual dislocations in Si.

2. The Lock-in Technique

The lock-in (or phase-sensitive) detection technique, which is widely used
in a variety of systems, relies on encoding the signal, which can then be
distinguished from the background noise by the so-called phase-sensitive
detector (PSD). The signal is encoded by the modulation of the source
intensity; thus in the case of EBIC, the incident electron beam is repete-
tively chopped, or the beam intensity is sine-modulated. The PSD is given
the chopping or modulation frequency as a reference and is thus able to
select the encoded signal from the background noise.

A PSD can be viewed as a normal amplifier with a certain band-width, which
is determined by its risetime. But it is different from a normal amplifier
as the position of this band-width in the frequency domain is centred at
the reference (i.e. modulation) frequency. Fig.1 is the noise spectrum
present in a typical EBIC system. If the source intensity is not modulated
at a certain frequency, the signal is accompanied by the part of the noise
spectrum falling inside the amplifier band-width, which is centred at, or
near, zero frequency. By modulating the source intensity, however, the
signal can be shifted to a different position in the frequency domain. It
should be clear that, for a given amplifier band-width, unless the noise
spectrum is flat (white noise), the signal to noise ratio can be improved
by shifting the signal and the PSD detection window to that part of the
frequency domain where the noise is a minimum. Depending on the particular
circumstances, a substantial improvement in the signal to noise ratio can
be made by a suitable choice of the modulation frequency.

3. A Lock-in EBIC System

The system utilised by the authors consists of a JSM-35X SEM, whose electron beam can be periodically chopped by means of an electromagnetic
chopping system installed above the first lens of the SEM. A signal
generator provides a square-wave, which is amplified to produce both the
chopping pulse and the reference for the PSD. The EBIC signal is amplified
by a current pre-amplifier and fed into the PSD, whose output can be displayed on the SEM video screen or recorded on a chart recorder. The
chopping system can operate over the frequency range dc-10MHz, while the
PSD can process signals with modulation frequencies in the range 0.5Hz -
100kHz. Care has to be taken to minimise interference from the chopping
coils; any noise from the chopping system is in phase with the signal and
can produce a substantial reduction in the signal to noise ratio. In cases
where the signal is very small, a small current pre-amplifier installed inside the microscope chamber can be used.

The signal-to-noise improvement is determined not only by the chopping frequency and its position in the noise spectrum, but also by the band-width,
or equivalently, the risetime of the PSD. The risetime of the PSD used in
the present work can be varied over the range 20μs-300sec. It is obvious,
however, that the larger the risetime of the amplifier is chosen to be, the
slower the electron beam must be scanned across the feature of interest or
the image-width etc. will be determined by the amplifier response time.
Thus it must always be ensured that the time taken by the electron beam to
scan across a feature of interest is much longer than the particular amplifier risetime setting used. This consideration imposes an upper limit on
the risetime of the PSD. For practical purposes, this limit is 10ms when a
good quality EBIC micrograph (≳1000 lines/frame) is to be produced in a
reasonable time (∿500 sec.).

4. Applications

4.1 Signal to noise and resolution improvement and the
    study of the behaviour of individual defects

Fig.2 is an EBIC micrograph of a deformation-induced dislocation network in
Si. The EBIC signal was produced with an accelerating voltage of 7kV and a
beam current of 10^{-11} A and collected by a Au-Pd Schottky barrier on the
top surface. The beam was chopped at a frequency of 10kHz and the PSD
risetime used was 1ms. The resolution, defined as the full width of a dislocation image is ∿0.3μm. It must be pointed out that the "resolution" of
an EBIC micrograph depends not only on "system" parameters such as the
incident beam energy, but also the particular sample geometry employed. It
is nevertheless apparent that the use of low beam energies made possible by
lock-in detection can produce significant improvements in resolution for a
given sample geometry.

As an example of the signal to noise improvement possible using a lock-in
EBIC system, figs.3(a) and (b) are EBIC line profiles using an ordinary
EBIC system, consisting of a Keithley 427A current amplifier, and the
present lock-in system respectively. These profiles indicate an improvement in the signal to noise ratio of more than 15 dB.

Using the system described in section 3 it has been possible to study, for
the first time, the electrical behaviour of individual screw and 60° dislocations in plastically deformed Si (Ourmazd et al. 1981). For example, by
changing the sample temperature by means of a nitrogen-cooled cold stage
(Boyes et al. 1981), the Fermi level has been swept over part of the upper
half of the band gap and the temperature dependence of EBIC contrast from
individual screw and 60° dislocations has been studied (fig.4a). Such
measurements can, in principle, yield the positions of the energy levels
associated with individual defects. Thus the curves of fig.4a naively
interpreted, indicate an energy level at 0.1eV from the conduction band
edge to be associated with screw dislocations, and the level associated with
the 60° dislocation to be deeper than 0.2eV from the conduction band edge.

4.2 The study of time-dependent phenomena

It is well-known that if a process with a characteristic time τ is to be
studied, the system must not be disturbed at a rate faster than $1/\tau$. This
presents the possibility of studying the different steps involved in the
capture and recombination of carriers at individual defects. As each step
has a characteristic time associated with it, by sweeping the electron
beam chopping frequency from dc upwards, one can "freeze out' the various
capture/recombination steps one at a time.

Fig 4b shows the frequency dependence of EBIC contrast from screw and 60°
individual dislocations in n-type Si, in samples where no dangling bond EPR
centres are associated with the dislocations (Ourmazd et al. 1981). Screw
and 60° dislocations exhibit different frequency dependences. Fig 4c however, shows the frequency dependence of EBIC contrast from screw and 60°
dislocations in n-type Si, when dangling bond EPR centres are present at
the dislocations. These frequency dependences are the same, indicating
that the EPR centres, being the dominant recombination centres, have masked
the behaviour characteristic of the dislocations. Fig 4d shows the frequency
dependence of a 60° dislocation in p-type Si, where again dangling bond EPR
centres are present.

Using data such as those outlined above, it should be possible to obtain
detailed information about the electrical behaviour of individual dislocations in Si and indeed other semiconducting elements and compounds.

The authors wish to express their gratitude to Dr. E.D. Boyes and

Mr. G. Dixon-Brown of Department of Metallurgy & Science of Materials,
Oxford for their invaluable assistance in installing the beam-blanking
equipment and to Prof. H. Alexander and co-workers of Köln University,
F.R.G. for the plastically deformed Si samples.

<u>References</u>

Anstis G R, Hirsch P B, Humphreys C J, Hutchison J L and Ourmazd A 1981
  Inst. Phys. Conf. Ser. in press
Ourmazd A, Weber E, Gottschalk H, Booker G R and Alexander H 1981 Inst.
  Phys. Conf. Ser. in press
Boyes E D, Dixon-Brown G and Ourmazd A this volume

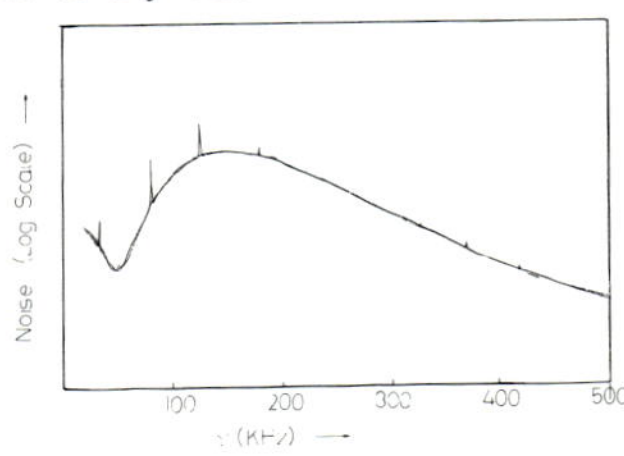

Fig.1 Noise present in a
typical EBIC system.

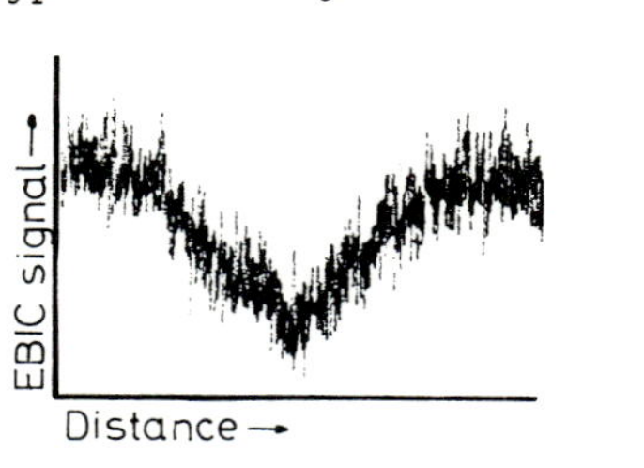

Fig.2 EBIC micrograph of
dislocations using lock-in
amplifier.

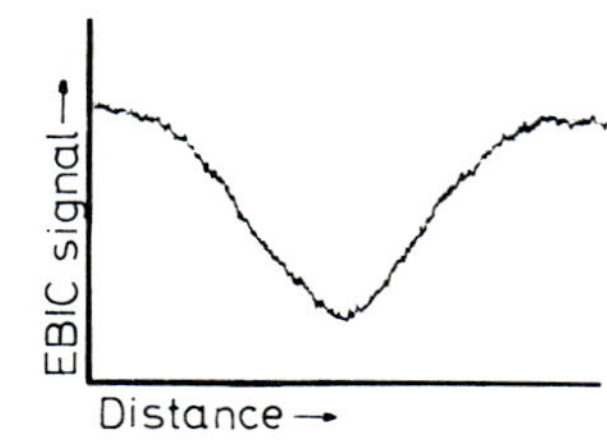

(a)   (b)

Fig.3 EBIC profile of a dislocation a) using a Keithley 427A, risetime
       30mS b) using lock-in amplifier, risetime 30mS.

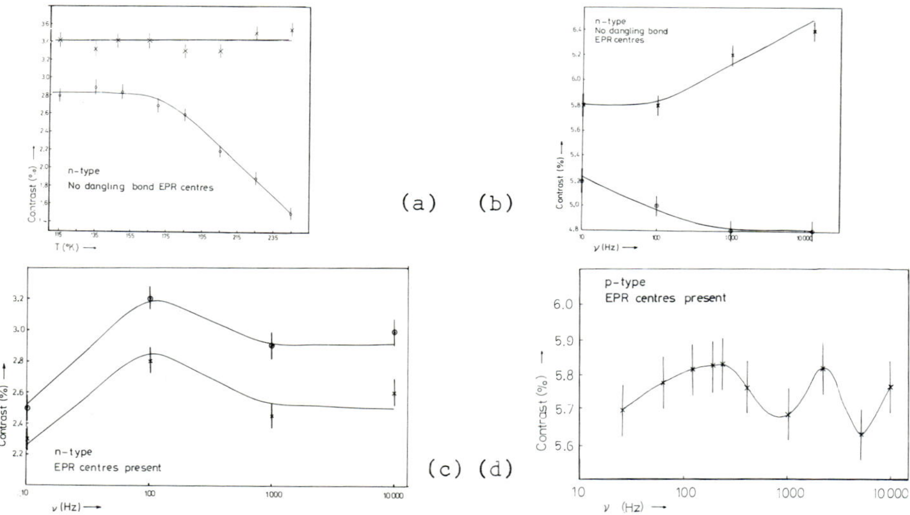

(a)   (b)

(c)   (d)

Fig.4 Temperature and frequency dependence of EBIC contrast from 60°
       (×) and screw (o) dislocations in silicon.

Correlation between process-induced defects and imaging performance of a charge-coupled device

S Blythe

Cavendish Laboratory, Madingley Road, Cambridge CB3 0HE

1. Introduction

It is well established that certain of the processes used in silicon semiconductor device manufacture can produce electrically active crystallographic defects which impair device performance. Usually this departure from designed characteristics is a consequence of the contributions from a large number of defects acting together and can be related to a macroscopic defect density. Only when this is comparable to the density of discrete components in the device can the effects of individual defects be considered. A device comprising a large-area array of identical discrete cells, such as a solid-state memory or image sensor, and containing a small, non-uniform defect distribution enables correlations to be made between measured electrical characteristics and microscopic imperfections. Such correlations have been made for a 385x576 pixel charge-coupled device area image sensor, in which the horizontal and vertical spacings between adjacent photosensitive elements is 22 microns.

2. Experimental

The devices have been commercially manufactured on p-type (100) float-zoned silicon using buried n-channel m.o.s. technology. The channel was ion implanted with phosphorus, the source and drain regions were phosphorus diffused and the inter-channel isolations were formed by boron diffusion. The gate structure consisted of a dry, thermal oxy-nitride dielectric with three levels of overlapping, doped poly-silicon electrodes which, not contributing to performance defects, will not be considered further. Other oxidations were performed in a steam ambient at 1100 degrees Centigrade. The essential features are shown in a schematic horizontal cross-section through one channel in Fig.1.

The performance of completed devices was assessed, in part, by observing on a television monitor the output of the device operating in the dark. Such an image is shown in Fig.2, where it can be seen that there is a non-uniform distribution of dark current spikes. For the purposes of this investigation, the operating conditions were chosen to optimise the visibility of these imaging defects. Additionally there was observed a background distribution of much weaker spikes. The cross-hatch was electronically generated to aid defect location.

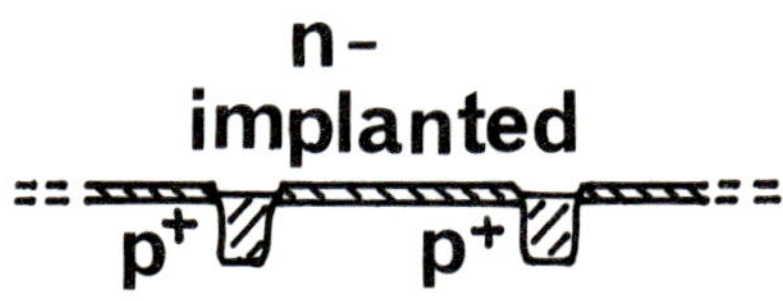

Fig.1 Schematic cross-section of
CCD cell structure.Not to scale.

Fig.2 Dark-current performance
showing more intense spikes.

For microscopic observation, the gate structure was stripped and
selected devices were scribed and broken into 2mm squares. These were
chemically thinned to perforation using 15% HF/HNO_3 solution. The
exact location of the thin region was determined by optical
microscopy and was compared to the positions of defective elements in
the television image. Most of the subsequent analyses were performed
on a V.G. Microscopes HB5 S.T.E.M., equipped with energy-dispersive
X-ray and electron energy-loss spectrometers.

3. <u>Results and Discussion</u>

Initially, electron-transparent regions of the device were located where
dark-current spike generation had been observed. Such a region is
shown in the annular detector region of Fig.3. The horizontal stripes
correspond to the preferentially thinned region of the inter-channel
isolation diffusions. The origin and nature of the feature labelled A
in the figure are unknown at present and only those of type B will
be considered here. A correlation was found between the dark-current
spikes and device cells incorporating one or more B-type defects
close to the channel-isolation diffusion. An example of this is shown
at higher magnification in Fig.4.

Previous investigations of charge-coupled device micro-defects had
revealed a large density of stacking faults lying either along or
across the channel isolation. Although these had usually been observed
to be undecorated, the regularity of the features shown in Fig.4
suggested that they may have been impurity decoration of a fault.
Tilting experiments in a high-voltage (500 keV)TEM revealed no such
structure, but better resolved the core of the type B defects, as shown
in Fig.5. The elemental composition of these structures has been
investigated by comparing X-ray and electron energy-loss spectra from
the central core with those of the surrounding matrix·

The X-ray spectra from both regions showed no peaks other than silicon.
Electron energy-loss spectra from the two regions are shown in Fig.6
and Fig.7. respectively. The latter shows two silicon edges at 100eV
and 149eV.   The edge at just below 200eV

in Fig. 6 may be attributed to the presence of either boron or phosphorus.
If the latter were present, it would give rise to another edge at 135eV
and should be detectable by X-ray techniques. Since neither of these
obtained, it may be concluded that the defects are boron-containing
precipitates, which is consistent with the introduction of a high
density of boron to the inter-channel isolation.

Fig.3 STEM micrograph of
electrically active region

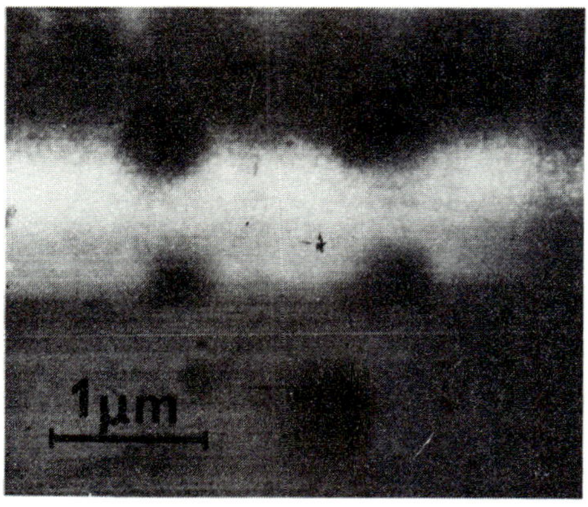

Fig.4 STEM micrograph of
precipitates near p-n region
of inter-channel isolation

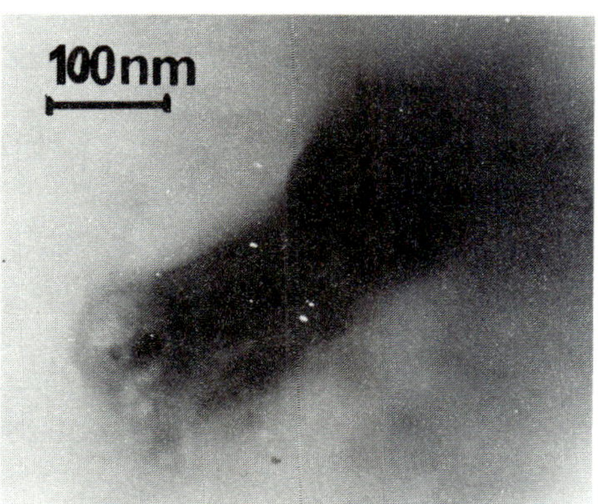

Fig.5 HVEM micrograph of
precipitate core structure.

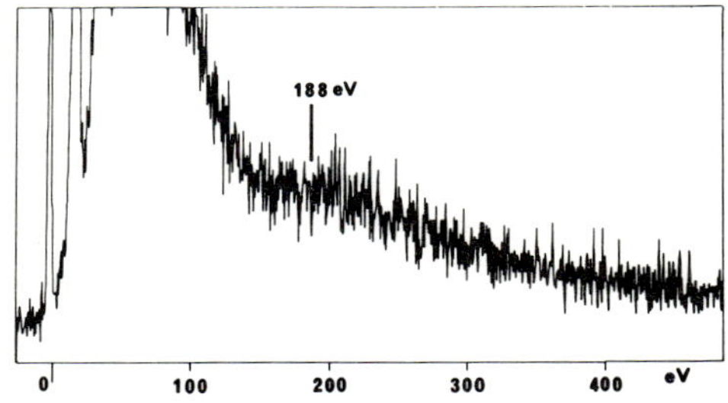

Fig.6 Electron energy-loss spectrum of precipitate

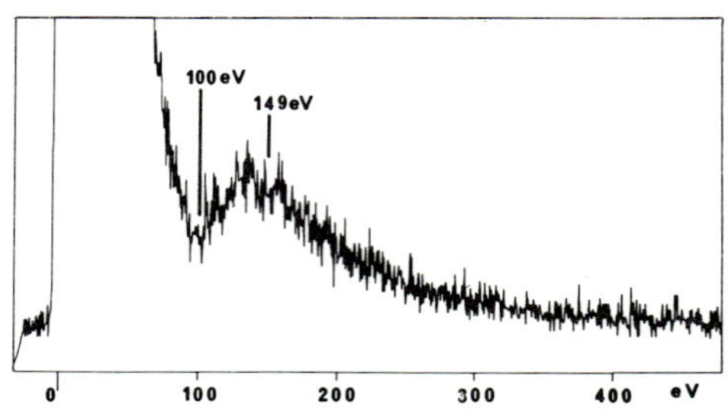

Fig.7 Electron energy-loss spectrum of silicon matrix

4. Acknowledgements

I acknowledge the financial support of the Science Research Council and G.E.C. Hirst Research Centre for supplying the devices.

Quantitative voltage contrast in the SEM

A.R. Dinnis, A. Khursheed & P.D. Nye

Electrical Engineering Department, Edinburgh University,
King's Buildings, Edinburgh EH9 3JL

Specimen Preparation

The best specimen is one with clean metal surfaces, such as an integrated circuit which has completed all processing steps up to the production of the aluminium interconnections but has not been covered with a coating of insulating 'passivation' material. If the circuit is not available in this form, then several possibilities are available:
(i) Strip the passivation off, preferably using dry chemical plasma etching.
(ii) Examine the specimen as it stands, with the passivation in place. The passivation is likely to charge up in this case, but by suitable choice of accelerating voltage (around 7 kV), the layer becomes sufficiently conductive for voltage contrast to be observed. Spatial resolution in this case is fairly poor, and no great confidence can be placed in the precision of quantitative measurements made in such a situation.
(iii) Coat the surface of the passivation with a slighly conductive coating, in a similar manner to that normally used in the SEM for non-conducting specimens. The coating must obviously not be so thick that it completely shields all electric fields from the beam, nor must it be allowed to short-circuit the bond pads, which are not covered with the passivation. In practice, the resulting voltage contrast is good, as shown in Fig.1, and appears at a wide range of accelerating voltages. The effect on the operation of the circuit should not be significant, though this aspect requires further investigation.

Electron Collector System

This must, firstly, extract a good fraction of the secondaries from the specimen and then perform an energy filtering operation on these electrons. In view of the fact that electric fields in excess of 1 MV/m can exist at the surface of an integrated circuit, a fairly strong extraction field is needed. These surface fields can also give the electrons very substantial transverse velocities, so that the energy filter must be able to cope with quite widely divergent electron trajectories. For the present work, a very simple simple form of extractor/filter has been used, consisting of an extractor grid maintained at between 400 to 1000 V which is a few mm. above the specimen surface, followed by a second, retarding, grid which performs the energy high-pass filtering function. The grids obstruct viewing of the specimen to some extent, but by aligning the grid bars and removing some bars from the central area, it is possible to view an area of about 1 mm. square and maintain adequate energy resolution. The electrons passing through the filter grid are accelerated and deflected into a conventional scintillator cage. Figure 2 shows the computed electric field and electron trajectories for this arrangement. The computer programs for this purpose have only recently been completed, and it is expected that with the aid of these, modifications to the geometry of the electrodes will considerably improve the transport efficiency of the deflector.

Quantitative Voltage Contrast

The characteristic which separates voltage contrast from other contrast mechanisms is that voltage contrast results from the difference in the _energy distribution_ of the secondary electrons with surface potential, rather than in the total quantity of secondaries emitted. It is therefore necessary to connect some electronic system to the energy filter so that the necessary information can be extracted as rapidly and accurately as feasible. This is at present achieved by applying a d.c. bias plus a superimposed sinusoidal voltage to the filter grid. The resulting signal from the head amplifier on the SEM is then suitably processed to produce quantitative results. The accuracy of the results depends largely on the degree of sophistication of this electronic processing. Using a very simple analogue system, a precision of 0.1 volts can be obtained in about 10 ms for steady voltages.

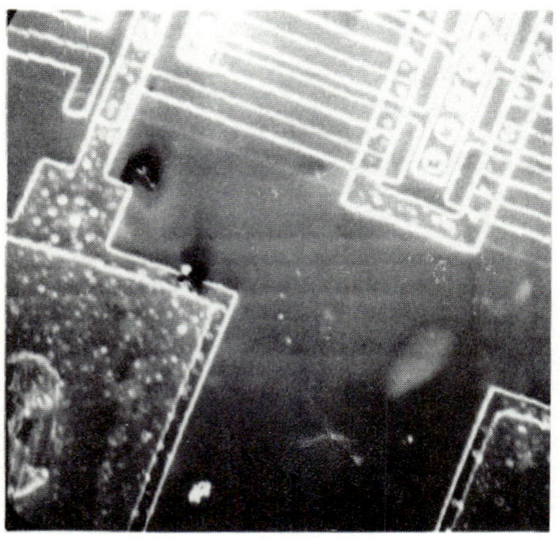 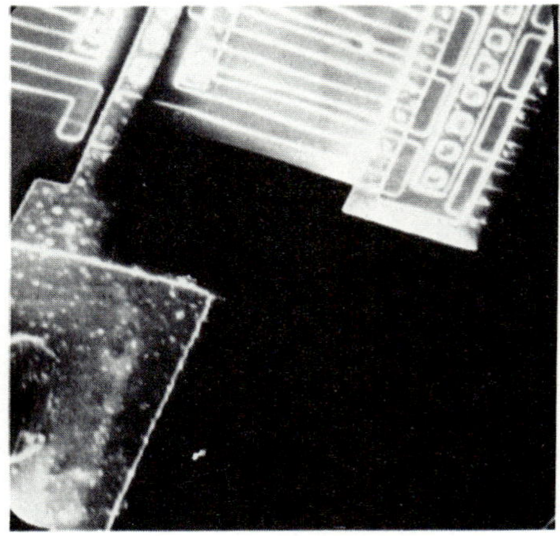

(a) No applied bias

(b) Left-hand pad at 0 V.
Right-hand pad at +5 V.

Figure 1.   IC chip with conductive coating over passivation.
EHT = 5 kV.   Picture width = 150 μm.

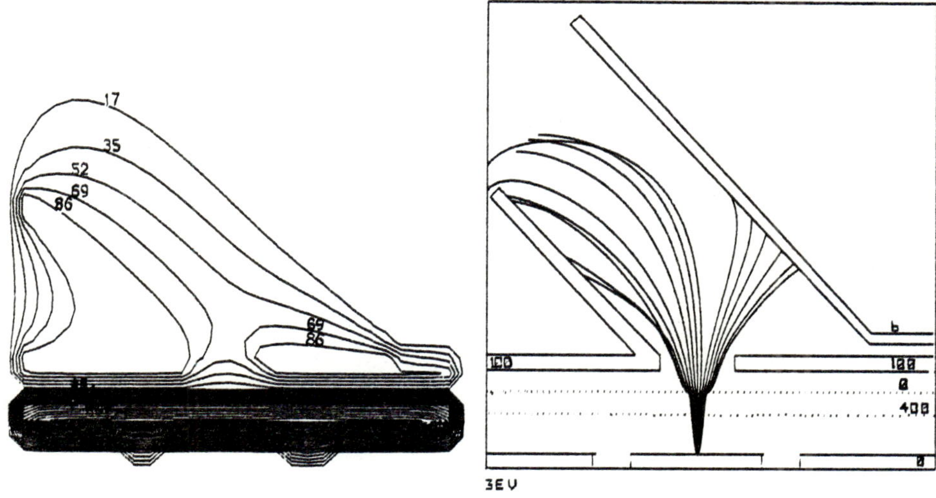

(a) Equipotentials

(b) Trajectories of electrons leaving
specimen with 3 eV energy.
Filter grid at 0 V.

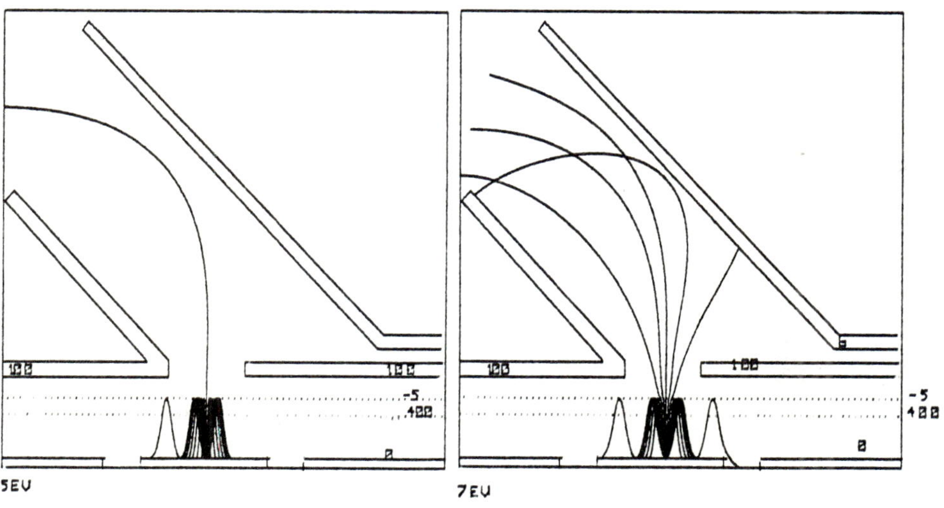

(c) Electrons emitted at 5 eV.
Filter grid at -5V.

(d) Electrons emitted at 7 eV.
Filter grid at -5 V.

Figure 2.   Computer analysis of electron extractor/filter.
A conventional scintillator with a cage bias of
+250 V is placed just to the left of the system.

Microcomputer Control

A computer system has been developed which can control all elements of the SEM, assimilate the data obtained, process it and present it to the operator. This can be done automatically or with a varying degree of interaction with the operator. The centre of the system is a 'Superbrain' microcomputer. This is connected to the many functions of the SEM using the IEEE 488 General Purpose Interface Bus (GPIB). It is also connected by a serial (RS 232) link to the Department's PDP 11/60 computer which has facilities for bulk storage and can perform large scale data processing if required.

The specimen stage is driven in the X & Y directions by stepper motors which are controlled by a local dedicated microprocessor, enabling either remote control from the 'Superbrain' or simple push-button positioning. This relieves the central computer of the task of keeping track of stage position as it can simply feed coordinates to the stepper processor.

A digital scan generator has been constructed which allows very flexible control of the scanning or positioning of the electron beam either from the front panel or remotely via the GPIB. This too has its own microprocessor which allows 'spot', 'line', Record or Visual scans; 11 scan rates from TV to 60 s under local operation and virtually unlimited under remote control. For interactive use, a bright-line cursor is provided which is moved by thumbwheels so that when 'spot mode' is selected, the beam goes to the position which was indicated by the cursor. At the same time a trigger output starts the voltage measuring electronics. When the measurement is complete, the scan generator is automatically returned to scan mode by a 'ready' signal from the analyser. This minimises damage and contamination buildup caused by the beam dwelling at one point for too long.

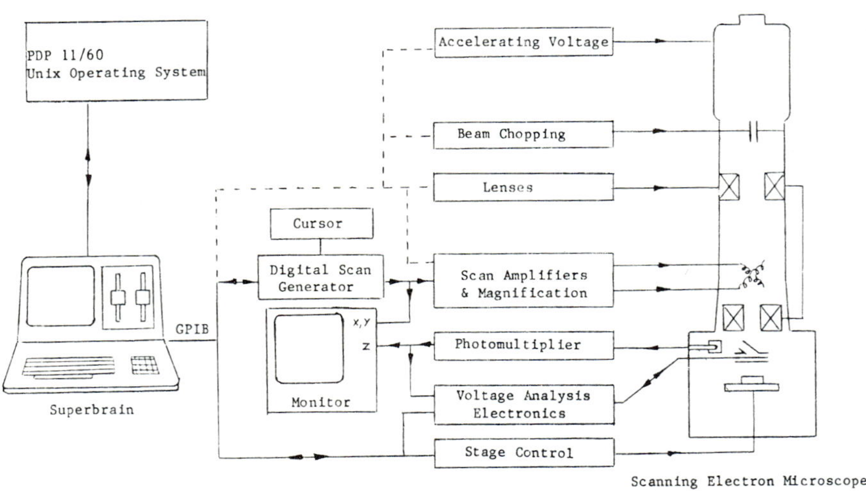

Figure 3.   Control system for SEM.

A structure image study of defects in cadmium sulphide

P. Pirouz[a], J. Echigoya[b] and J. W. Edington[b]

(a) Department of Metallurgy and Science of Materials, University of
    Oxford, Parks Road, OXFORD OX1 3PH, England
(b) Materials Durability Division, College of Engineering, University of
    Delaware, NEWARK, Delaware 19711, U.S.A.

1. Introduction

A thin film polycrystalline solar cell based on cadmium sulphide is a most
promising candidate as a low cost terrestrial cell reaching a conversion
efficiency of at least 10%.   Both low and high angle grain boundaries
are a natural component of polycrystalline CdS solar cells and they both
significantly reduce the solar conversion efficiency.   The low performance of such cells, as compared with single crystal cells of the same material is largely caused by lower short circuit currents arising from carrier
recombination at grain boundaries.   However, all boundaries do not influence photovoltaic performance by the same amount.   Hence it is important
to correlate the photovoltaic properties to the grain boundary and defect
structure in CdS.   In fact, there have been relatively few studies of the
defect structure in the wurtzite form of CdS.   The limited evidence available indicate that growth faults can occur on (0001) and $(11\bar{2}0)$ in vapour
deposited single crystal platelets that grow with $(11\bar{2}0)$ parallel to the
large area surface.   Grown-in perfect dislocations in such material were
reported to have a $[0001]$ Burgers vector by Mohling and Heydenreich (1968)
whereas Yoshiie et al. (1980) and Osipyan et al. (1980) demonstrated that
the Burgers vector of slip dislocations was $1/3<11\bar{2}0>$.   Recently, Cockayne
et al. (1980) observed the dissociation of basal dislocations in plastically deformed single crystal CdS and dislocation glide in the dissociated
form.   This paper presents some preliminary results of an electron microscopy study of defects in single and polycrystalline CdS using the lattice
imaging approach.

2. Experimental Details

Single crystals of CdS were grown by the gas transport technique in the
form of $(11\bar{2}0)$ platelets.   The resistivity of the material was found to
be $> 10^6$ ohm cm indicating high purity.   Polycrystalline thin films were
vapour deposited in vacuo on a copper substrate.   The thickness of the
CdS film was about 25–30 μm.   The through-thickness resistivity was 1–10
ohm cm and the impurity content was < 650 ppm.

Specimens for TEM were produced by ion beam thinning.   In the case of
polycrystalline CdS, the copper substrate was removed in concentrated
ammonium persulphate solution before thinning.   The specimens were then
examined in a high resolution Philips EM 400 operating at 120 kV.   The
microscope was equipped with a LaB_6 filament and the coefficients of

spherical and chromatic aberration were 1.1 mm and 1.2 mm respectively.
In each case a through focus series of micrographs were taken with axial
illumination in areas having thicknesses in the range < 10 nm.

3. <u>Results</u>

The single crystals of CdS were found to be quite perfect with a very low
density of defects.   Both single perfect and also dissociated dislocations
and basal and prismatic stacking faults were occasionally observed.   Fig.
1a shows an image from a region of a $(1\bar{2}10)$ plane of a CdS single crystal
showing a perfect dislocation.   The shortest perfect Burgers vector in the
wurtzite form of CdS is 1/3 $<11\bar{2}0>$.   Fig. 1b shows the viewing geometry
and since the electron beam direction is $<11\bar{2}0>$, we interpret the image as
a 60° perfect dislocation.   The thickness of this region of the specimen
was estimated from the thickness fringes to be about 4 nm.   As our calculations show, each bright spot in the image corresponds to a projection
of two columns of Cd and S atoms in the crystal.   The separation of the
latter is 0.146 nm which is beyond the resolving power of the microscope.

To check the validity of our interpretation, image simulations were carried
out using the multi-slice method (Skarnulis, 1979).   Images of both perfect crystal and 60°-dislocation were calculated using the operating parameters of the microscope, (120 kV, C_s = 1.1 mm, Δf = 4 nm, divergence angle
= 0.002 rads).   The dislocation image was calculated using the method of
periodic continuation (Cowley, 1975).   A large unit cell (73 atoms) containing the 60°-dislocation was constructed in which the positions of the
atoms were determined from the isotropic elasticity theory.   Figures 2
and 3 show the calculated images of the perfect crystal and the 60°-dislo-
cation at a defocus of −100 nm.   At this defocus, we found the best matching between experimental and calculated images for both perfect crystal
and the dislocation.   Comparison of the simulated images with the projected potential distributions showed that each bright spot in the image
corresponded to a Cd-S atom pair.

Fig. 4a shows a basal stacking fault in a different region of the crystal
close to the specimen edge.   The geometry of viewing the partial and
stacking fault is shown in Fig 4b.   The dissociation takes place according
to the following reaction:

$$1/3 \ [11\bar{2}0] \ = \ 1/3 \ [10\bar{1}0] \ + \ 1/3 \ [01\bar{1}0]$$

There is a two-fold axis of symmetry falling on the line which passes
through the row of bright spots in the stacking fault.   Thus, the bright
spots in the image correspond to projections of atomic columns rather than
tunnels, (Olsen and Spence, 1981).   The stacking fault seems to end just
at the edge of the specimen.   In this case we obtain a separation of
about 8.6 nm between the partials giving a stacking fault energy of about
79 mJ/mm^2.   However, this value may not be representative due to surface
effects.   Due to insufficient resolution, it is not possible to say anything in support of the shuffle or glide models of dislocation dissociation
in CdS from our work.

Prismatic stacking faults were more frequently observed than the basal ones
particularly in the polycrystalline specimens.   An example of such a fault
is shown in Fig. 5 where the viewing direction is [0001].

In the polycrystalline material both low and high angle boundaries were

observed.   Fig. 6a shows an image of a low-angle grain boundary consisting
of a regular array of dislocations.   The viewing direction in this case
is [0001].   The misorientation, as measured from the image, is about 10°
and the spacing between the dislocations is about 2 nm.   The boundary may
be interpreted in terms of the classical Read-Shockley model which predicts
a misorientation of 10.275°.   A model of the boundary was constructed
with the aim of minimising the number of dangling bonds.   This is shown in
Fig. 6b.

Finally, Fig. 7 shows a high angle boundary in a CdS polycrystal.   The
important point to note here is the continuity at the boundary.   Under
the conditions under which the CdS was deposited, this corresponds to zone
1 (Thornton, 1977), namely a high deposition rate and a low substrate temperature with little surface diffusion.   Contrary to the general belief
that such deposition conditions lead to gaps at grain boundaries and subsequent local variations in electronic properties, we did not observe such
gaps in any of a number of high angle boundaries studied here.

Acknowledgements

We are grateful to Professor R. Sinclair of Stanford University for the
use of the microscope and to Mr. T. Yamashita for technical assistance
and to Dr. G. R. Anstis for useful discussions concerning image calculations.   This work was partly supported by the Solar Energy Research Institute under contract DE-AC01-79ET23105.

References

Blank H, Delavignette P and Amelinckx S 1962 Phys. Stat. Sol. 2 1660
Cockayne D J H, Hons A and Spence J C H 1980 Phil. Mag. A 42 773
Cowley J M 1975 Diffraction Physics (Amsterdam: North Holland)
Mohling W and Heydenreich J 1968 Phys. Stat. Sol. 7 155
Olsen A and Spence J C H 1981 Phil. Mag. A 43 945
Osipyan Yu A, Petrenko N F, Strukova G K and Khodos I I 1980 Phys. Stat.
  Sol. A57 477
Read W T and Shockley W 1950 Phys. Rev. 78 275
Skarnulis J M 1979 J. Appl. Crystallog. 12 636
Thornton J A 1977 Ann. Rev. Mat. Sci. 7 239
Yoshiie T, Iwanaga H and Shibata N 1980 Phys. Stat. Sol. A59 K9

(a)

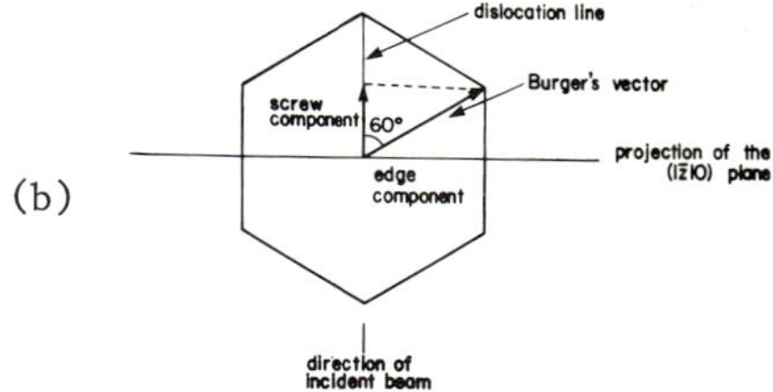

(b)

Fig. 1a)   Image of a perfect 60°-dislocation
        in CdS
     b) Viewing geometry

Fig. 2) Calculated image of a
perfect crystal 4 nm thick along
the <11$\bar{2}$0> at a defocus of
-100 nm.

Fig. 3) Calculated image of a 60°-
dislocation in a 4 nm thick
crystal at a defocus of -100 nm

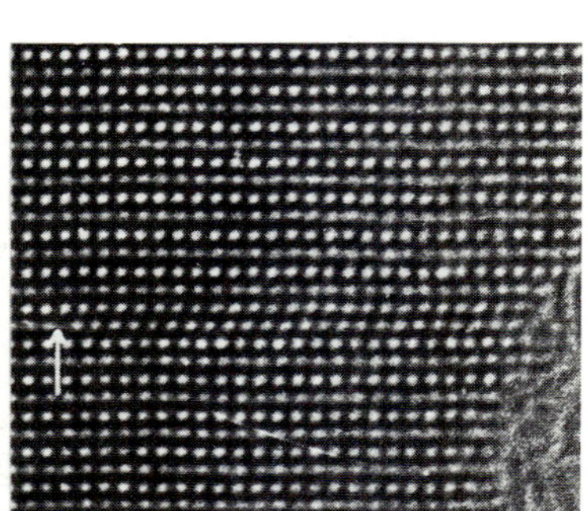

Fig. 4a) A basal stacking
fault. The partial on the
left is arrowed. Viewing
direction is <11$\bar{2}$0>

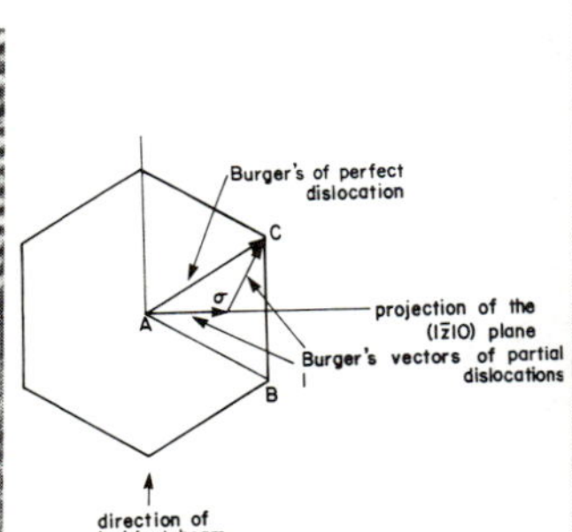

Fig. 4b) The viewing
geometry

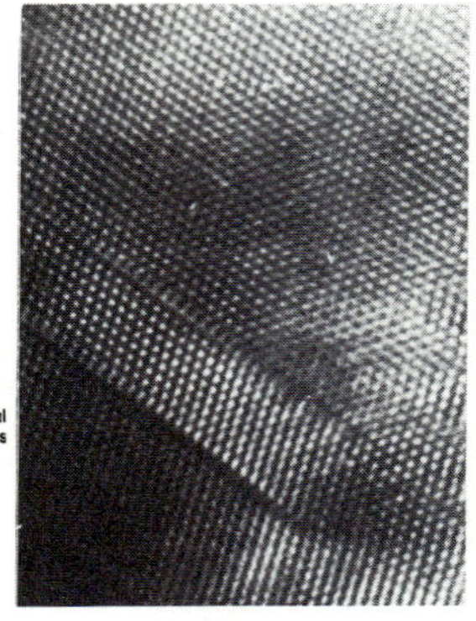

Fig. 5) Prismatic
stacking faults in CdS
Viewing direction is
[0001]

Fig. 6a) A low angle near-
tilt grain boundary in CdS.
Viewing direction is [0001]

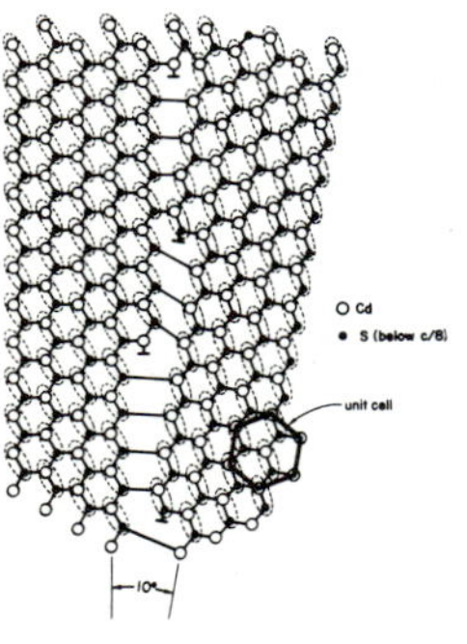

Fig. 6b) Proposed
model of the low
angle grain boun-
dary shown in Fig.
6a

Fig. 7) A high angle grain
boundary in CdS.
Viewing direction is
[0001]

Electron microscope studies of the semiconductor to metal phase changes in epitaxial layers of elemental tin

A G Cullis, R F C Farrow, N G Chew and G M Williams

Royal Signals and Radar Establishment, Malvern, Worcs. WR14 3PS

1. Introduction

It has been known for some time that the usual metallic β-phase of the element Sn can undergo a phase transformation at reduced temperatures to the α-allotrope which has the diamond cubic structure and exhibits semi-conducting properties (Busch and Kern, 1961). Although early attempts to prepare thin films of cubic Sn did not succeed, recent improvements in vacuum technology coupled with metal beam deposition techniques (Farrow et al 1979) have now made this possible, as demonstrated in the present article and as described by Farrow et al (1981). We have chosen to grow the α-Sn films on substrates of the semiconductor InSb since the latter (a_o = 6.4798Å at 25°C) is a close lattice match to α-Sn (a_o = 6.489Å at 25°C). Furthermore, due to the small band gap of α-Sn, semiconductor structures of the type fabricated in this work may find application for the photovoltaic detection of long wavelength infra-red radiation (Farrow and Robertson, 1981).

2. Experimental

Deposition of Sn was carried out in an ion and titanium sublimation pumped ultrahigh vacuum chamber with a base pressure of $\sim 10^{-10}$ Torr. A beam of metal atoms was obtained from a Knudsen effusion oven containing double zone refined Sn of purity > 99.999%. The beam composition was monitored using modulated beam mass spectrometry. Substrates were usually wafers of (001) InSb (n-type, $n_D - n_A \sim 10^{14} cm^{-3}$) which had been mechanically and then chemically polished, the latter with a proprietary oxidizing solution (MCP Electronic Materials Ltd.). Some substrates were in the form of discs which had been thinned from the back to electron transparency by use of HNO_3/HF chemical reagent. These discs were suitable for direct insertion into the transmission electron microscope (TEM) after layer deposition. Substrate surfaces were cleaned *in vacuo* by sequential Ar^+ ion bombardment and annealing (200°C) treatments, after which Sn deposition took place with a substrate temperature of $\sim 25°C$.

TEM examinations which employed 120kV electrons were carried out using a double-tilting heating holder so that *in situ* annealing experiments could be performed. Some specimens were studied by scanning microscopy using secondary electrons.

3. Results and Discussion

When relatively thin (~ 2000Å) Sn films were examined by scanning microscopy they were found to be uniform down to the scale of a few hundred Angstroms

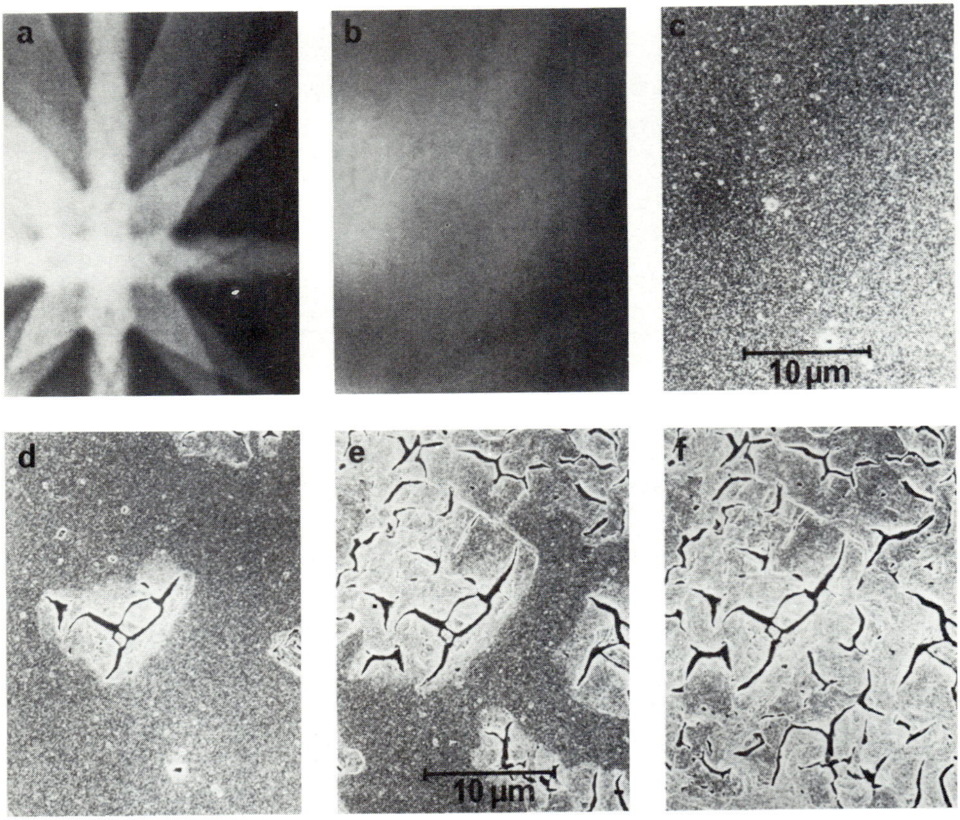

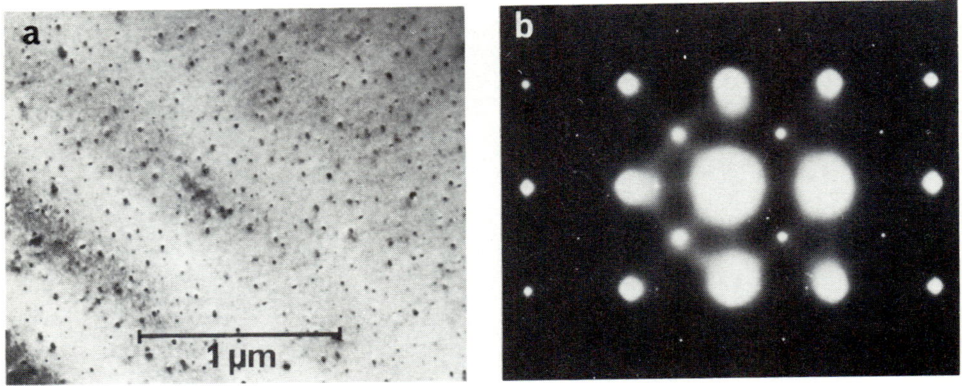

Fig. 1. Electron channelling patterns obtained from a) as-deposited α-Sn and b) transformed metallic β-Sn. Scanning electron images show the same specimen area, c) as-deposited, d), e) and f) after progressive transformation at ∿75°C (brighter areas are β-Sn).

Fig. 2. Transmission electron image (strong beam) and [001] pole diffraction pattern obtained from as-deposited α-Sn film on (001) InSb.

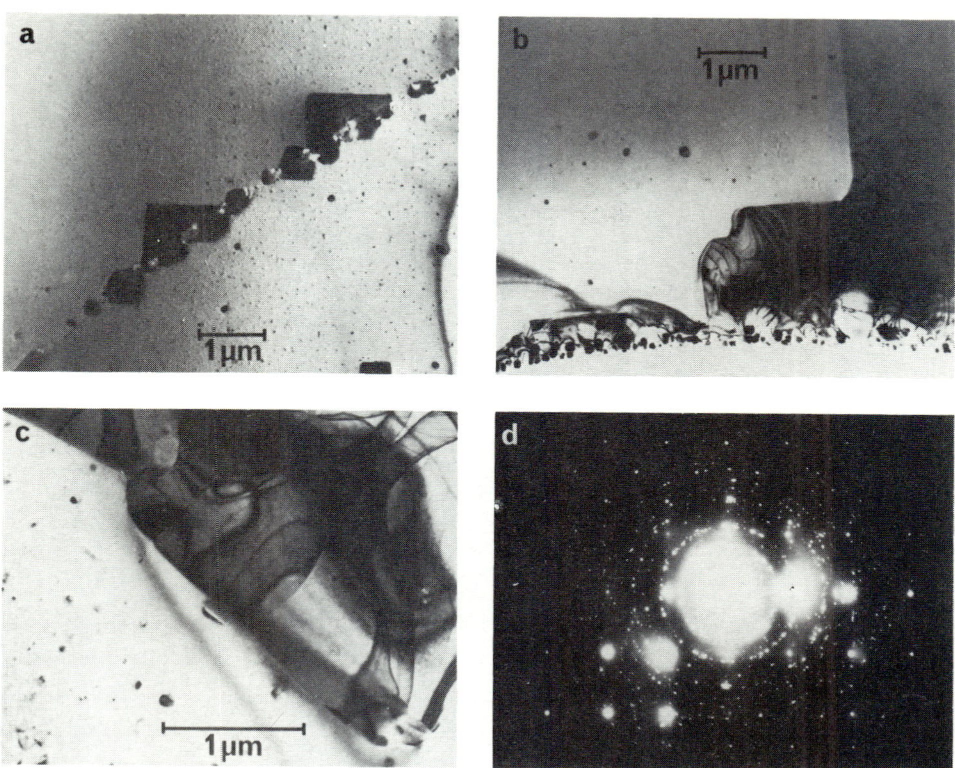

Fig. 3. Transmission electron images showing a) nucleation of β-phase
Sn, b) and c) boundaries of the β-phase advancing from the right.
d) Transmission diffraction pattern showing β-phase rings.

and electron channelling patterns (Fig. 1a) demonstrated that they had the
cubic α-Sn structure which was aligned with the substrate lattice (that is,
$(001)_{Sn} /\!/ (001)_{InSb}$). It is to be noted that α-Sn is a metastable phase at
room temperature and above, since the temperature of bulk transformation
to the β-phase is 13.2°C. It is, therefore, of special interest to study
this metastability and to examine details of the transformation when it
does occur. Accordingly, specimens were heated slowly *in situ* and a corresponding series of secondary electron images is shown in Fig.1. The initial
as-deposited surface showed few initial features (Fig. 1c) and little
change was observed until a temperature of ⪆70°C was achieved. At ~75°C
the β-phase was seen to nucleate in local areas (Fig. 1d) which then expanded into the remaining α-Sn film (Fig. 1e,f). The newly formed β-Sn
appeared brighter than the original film and contained many microcracks
due to its greatly increased density. Furthermore, electron channelling
patterns obtained from the transformed film (Fig. 1b) showed few features
indicating that its crystal perfection was poor.

When an as-deposited α-Sn film on a thinned substrate was examined in the
TEM it was found to be high quality single crystal with few large-scale
extended defects (Fig. 2). The main structural feature was a fine dispersion of very small particles and point defect clusters mainly in the region
of the heteroepitaxial interface. The precision of the alignment between
the film and substrate lattices led to the virtual superposition of Bragg

diffraction spots. However, it is interesting to note that X-ray lattice parameter measurements (Farrow et al, 1981) indicated that such films exhibited in plane compression with symmetrical uniaxial dilatation along the [001] normal. When a film was slowly heated in the TEM to ∿75°C, the metallic β-phase was suddenly seen to form as scattered islands across the specimen. The new phase often nucleated at preexisting irregularities in the film as occurred, for example, along the line of a scratch shown in Fig. 3a. The β-phase islands were often bounded by straight edges aligned along <110> crystal directions in the α-phase. An advancing β-phase growth front is shown in Fig. 3b. The transformed material was heavily defective and consisted of interlocking polycrystals. This is demonstrated along a relatively straight interphase boundary in Fig. 3c where several β-phase grains are extending side by side. The polycrystallinity of the tetragonal β-phase material is also evident from the spot array in the diffraction pattern of Fig. 3d. With continued heating at ∿75°C, the β-phase material ultimately propagated throughout the deposited film.

Finally, it is important to note that electrical measurements showed that the as-deposited α-Sn films exhibited p-type conductivity. Heterojunction p-n photodiodes formed with the InSb substrate showed significant photo-voltaic response beyond the InSb long wavelength cut-off of 5.7μm at 77°K.

4. Conclusions

The close lattice match between InSb substrate material and α-Sn allows the latter to grow in thin film form as a metastable phase significantly above its bulk transformation temperature. This can be exceeded by more than 60°C before the transformation occurs in films ∿2000Å thick. The high temperature β-phase is preferentially nucleated at preexisting irregular-ities, and islands so formed spread rapidly across the deposited film. Nevertheless, the enhanced stability of the semiconducting α-Sn at room temperature is likely to be important for infra-red detection applications.

5. References

Busch G A and Kern R 1961 Solid State Physics 11 1
Farrow R F C, Cullis A G, Grant A J, Jones G R and Clampitt R 1979 Thin
    Solid Films 58 189
Farrow R F C and Robertson D S 1981 British Patent Application No. 81 09471
Farrow R F C, Robertson D S, Williams G M, Cullis A G, Jones G R, Young I M
    and Dennis P N J 1981 J. Crystal Growth 48

Transmission electron microscopy of Sn-implanted and laser-annealed GaAs

M A Shahid, B J Sealy* and K E Puttick

Department of Physics, *Department of Electronic and Electrical
Engineering, University of Surrey, Guildford, Surrey GU2 5XH, UK

There is considerable interest in using a high power pulsed laser beam in
order to remove the damage introduced in semi-conductors during ion-
implantation (Sealy 1979).   We have used transmission electron
microscopy for studying the various annealing stages of Sn-implanted GaAs.

Si(100) GaAs was implanted in a non-channelling direction to a dose of
5×10^{14} Sn^+/cm^2 and at ion energies of 100 and 300 KeV.   The implanted
substrates were cleaved into 2mm x 2mm squares which were subsequently
annealed in air using a single 25 ns pulse from a Q-switched ruby laser.
The laser beam energy densities (EDs) ranged from 0.12 to 1.5 J/cm^2.   An
L-shaped quartz rod was used as a diffuser to give a uniform energy
distribution over the area of the laser beam.   The thin foils for
electron transmission were prepared by chemically thinning these specimens
from the back untreated side.   A JEOL TEMSCAN 200CX was used to examine
the specimens.

The results are summarized in the following table:

Table

ED (J/cm^2)	100 KeV		300 KeV	
	M	SADP	M	SADP
0	FD	BR	FD	BR
0.12	MF,P	FR	MF,P	FR
0.2	MF,P,T	R	MF,P	FR
0.3	SC	ST	MF,P	FR
0.47	SC	ST	SF,ppt,D	ST
0.7	SC,ppt	ST	SC,ppt	ST
1.2	D,ppt	ST	D,ppt	ST
1.5	D,ppt	ST	D,ppt	ST

(key: M=micrograph, SADP=selected area diffraction pattern, FD=fine
damage, BR=broad rings, MF=Moire fringes, P=polycrystalline, FR=fine
rings, T=dislocations in tangles, R=broken rings, SC=single crystal
having clear background, ST=spots, SF=stacking faults, ppt=precipitates
and D=dislocations)

A glance at the table gives an idea about the sequence of events taking
place during laser annealing.   In the case of 100 KeV implants, the
thickness of the damage layer is $\sim$ 1000 Å and an ED of 0.3 J/cm^2 is
sufficient to push the melt front through to the crystalline substrate so
that recrystallization takes place in the liquid phase (Baeri et al 1979,
Auston et al 1978).   On the other hand, for ED < 0.3 J/cm^2 the melt

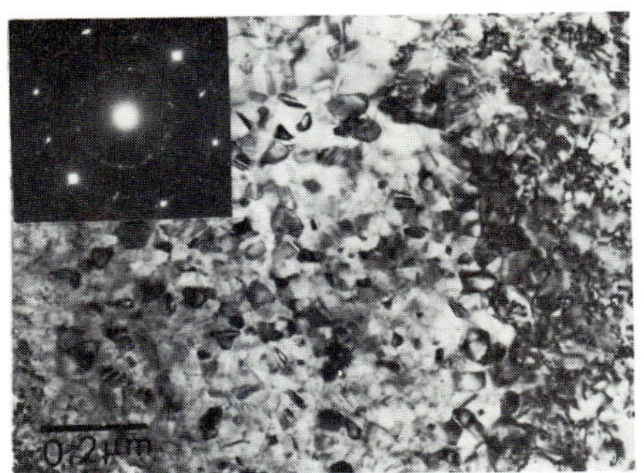

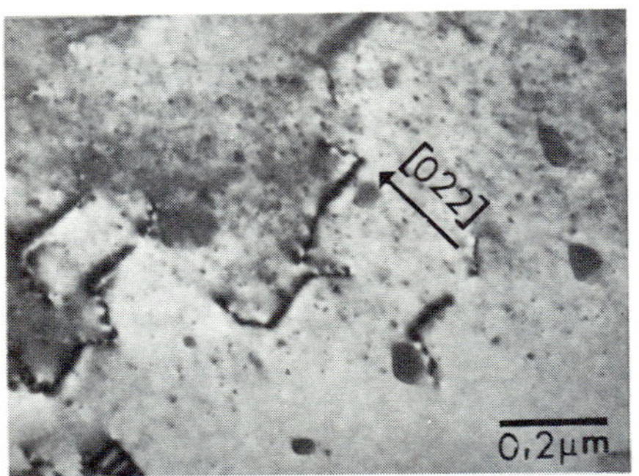

Fig.1   A bright field transmission electron micrograph of Sn-implanted GaAs showing a polycrystalline surface layer after laser annealing at an ED of 0.2 J/cm^2.   Inset shows the SADP.
ion dose=5x10^{14} Sn$^+$/cm^2
ion energy=100 KeV

Fig.2   A bright field transmission electron micrograph of Sn-implanted GaAs showing residual damage after laser annealing at an ED of 0.47 J/cm^2.
ion dose=5x10^{14} Sn$^+$/cm^2
ion energy=300 KeV

front does not penetrate the damage layer so that a polycrystalline material results in the regions of the surface (Fig.1).   In the case of 300 KeV implants (a damage layer thickness of $\sim$ 2000 Å), recrystallization takes place at an ED of about 0.47 J/cm^2.   At this ED there is a lot of residual damage in the form of stacking faults, dislocations and Ga-rich precipitates due to evaporation of As (Fig.2).   For EDs between 0.47 and 1.0 J/cm^2 good crystalline material is obtained with a very low dislocation density.   But for ED $\geqslant$ 1.2 J/cm^2, a high density of dislocations along with severe surface roughness (as seen by optical microscopy) results.

It is evident that there is a threshold ED for recrystallization of the amorphised implanted layer.   For a given ion dose, this threshold ED depends on the ion energy and is $\sim$ 0.3 J/cm^2 for 100 KeV and $\sim$ 0.47 J/cm^2 for 300 KeV Sn implants into (100) GaAs.   Moreover, there is an energy window in the middle range within which good quality regrowth can be achieved without affecting the surface topography significantly.   The present work suggests that the Sn implanted layers in GaAs can be completely recrystallized at an ED of 0.6 J/cm^2 irrespective of the ion energies up to 300 KeV.

The authors wish to thank the SERC for financial support and the Microstructural Studies Unit, University of Surrey, for technical assistance.

References:

Auston D H, Golovchenko J A, Smith P R, Surko C M and Venkatesan T N C 1978 Appl. Phys. Lett. 33 539
Baeri P, Campisano S U, Foti G and Rimini E 1979 J Appl. Phys. 50 788
Sealy B J 1979 J. Crys. Gro. 48 655

EBSP from semiconductor materials

D.J. Dingley, L. Baker and L. Hunnings

H.H. Wills Physics Laboratory, University of Bristol, Bristol BS8 1TL

1. Introduction

The two SEM micro diffraction techniques of electron back scattering, Venables et al (1976) and micro Kossel X-ray diffraction, Dingley (1978) have been used for the investigation of semi-conductor materials and devices. The latter of the two methods provides very accurate lattice parameter values with precision of the order 1 part in 20000 and example patterns from semi-conductor materials are shown in Fig 1. The sample volume needed to obtain these patterns was several microns cubed. They are, however, very weak and if the sample volume is made smaller, e.g. a thin film device, then the patterns become undetectable. Venables et al (1981), however, have shown that good electron back scattering patterns, EBSPs, can be obtained in these circumstances and so we have added this technique to complement that of Kossel diffraction.

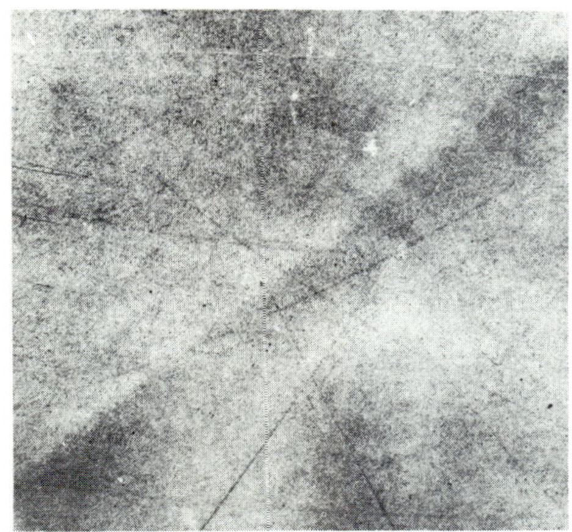

Fig 1a.

Back reflection Kossel diffraction
pattern from GaP.

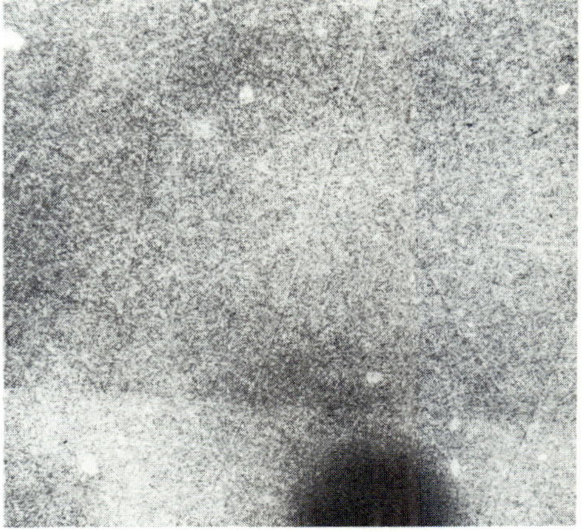

Fig 1b.

Transmission Kossel diffraction
pattern from GaAs.

2. Apparatus

Following Venables' original design (1973), we observe the EBSP directly in a Cambridge S4 SEM by placing a phosphor screen within 20 mm of the specimen. An RT Laboratories low light level TV camera was used to view this screen so that the diffraction pattern could be imaged on a

monitor under normal operating conditions of the microscope. The gain of
this camera was in excess of 3000 times and we found it possible to
observe diffraction patterns directly with specimen currents as low as
0.1×10^{-9} amps.

Our technique differs from that of Venables in that the diffraction
pattern is recorded on 35 mm film placed in the specimen chamber of the
microscope. The film is loaded in cassettes and drawn in front of the
phosphor screen when a pattern is to be recorded. Holes are cut at
intervals along the film so that when these are in front of the phosphor
screen, the diffraction pattern can be observed as described above. The
apparatus is shown in Fig 2. The EBSP camera/phosphor screen assembly is
mounted in the rear of the specimen chamber with the Kossel camera,
energy dispersive and secondary electron detectors fitted to the other
ports.

An example of an EBSP obtained by direct photography from GaP is shown
in Fig 3.

Fig 2.

Photograph of EBSP camera with
phosphor screen incorporated.

3. Experimental investigation of EBSPs

Most experiments have been carried out on GaAs or GaP single crystals, although patterns have been obtained from many other materials.

3.1 Pattern dependence on specimen current.

EBSPs were obtained at 18 kV from GaAs on Ilford FP4 film for specimen currents ranging from 100ρA down to 5ρA. The exposure time over this range, to obtain patterns of equal intensity, varied from 75 secs to 175 secs. The specimen to film distance was 40 mm. This is 20 mm greater than we now use. The calculated electron beam spot size on the specimen for the pattern obtained at 5nA is approximately 800 Å. Further reduction in the spot size was impractical because of contamination to the specimen surface. Even at 5nA the beam had to be moved from time to time to a clean area of the specimen surface during exposure to avoid this problem.

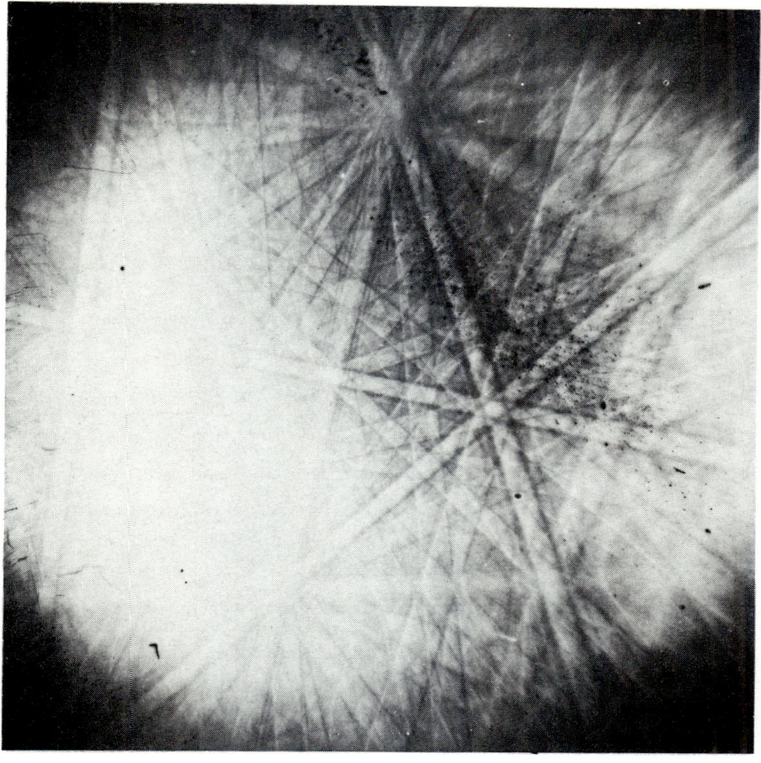

Fig 3.

EBSP recorded directly on
FP4 film from GaP.

3.2 Pattern dependence on beam voltage.

Using the RT Laboratories TV image intensifier system, it was observed that equally good patterns could be observed on the monitor down to 4kV which was the minimum excitation energy of the phosphor. When recording the patterns on FP4 film, the specimen current was maintained at 50ρA and patterns were recorded down to 7kV. Towards this voltage the patterns became very poor and high order lines were not imaged.

3.3 Pattern dependence on specimen inclination.

In all the specimens observed, good patterns could only be obtained if the specimen was inclined at angles less than 20° to the incident beam. The quality improved as this angle was decreased, but investigation becomes impractical for angles less than 5°.

4. <u>Effects of surface layers</u>.

A GaP single crystal 5 mm long was coated with aluminium in such a way that part was left uncoated – part had one coat and part had two coats. The amount of aluminium evaporated was the same in each case and vacuum conditions during evaporation maintained constant. The crystal was un- heated. The estimated layer thickness was 80$\overset{\circ}{A}$ for the single layer and 160$\overset{\circ}{A}$ for the double layer. It was appreciated that for such small coat- ings the film is probably not continuous and due care was accordingly taken during recordings of EBSPs. At 30kV patterns could be obtained from the GaAs beneath the singly coated portion of the specimen, but not from the doubly coated portion. This was independent of the probe current. At 14kV, whereas good patterns were still recorded from the uncoated region, they could not be detected even from the singly coated zone.

5. <u>Conclusions</u>

It has been shown that Venables' technique for obtaining EBSPs in the SEM can be used in a routine fashion for investigating semi-conductor materials and that the pattern can be obtained under a wide range of experimental conditions. The minimum selected area for this technique, using an SEM of average vacuum , appears to be about 1000$\overset{\circ}{A}$ and the minimum practical voltage $\sim$10kV.

6. <u>Acknowledgements</u>

The authors would like to thank Mr. G. Burns and Mr. M. Gabb for help in constructing the EBSP camera, the RT Laboratories, Wells, for the loan of their low light level TV camera and Professor J. Ziman for permitting part of this work to be carried out as an undergraduate research programme.

7. <u>References</u>

Dingley, D.J. 1978,   Scanning <u>1</u> 79.

Venables, J.A. and Harland, C.J. 1973 Phil. Mag. <u>27</u> 11.

Venables, J.A., Harland, C.J. and Akhter, P. 1981, J.Phys.E, <u>14</u> 175.

Venables, J.A., Harland, C.J. and Bin-Jaya, 1976, in Developments in
     Electron Microscopy and Analysis, London Academic Press, 101.

Microstructural investigation of AuGeNi contacts on InP

R.J.Graham, J.F.Mansfield, G.M.Rackham
H.H.Wills Physics Laboratory, University of Bristol
Royal Fort, Tyndall Avenue, Bristol BS8 1TL

1. Introduction

AuGeNi contacts on InP are prepared by sequentially evaporating Ge,Au and
Ni on to (100) InP and heat treating to temperatures of up to 400°C.
We have made extraction replicas of the interface, prepared after removal
of the metal contact with mercury, for examination in transmission.
In scanning we have been looking at the metal face of the interface with
the InP removed by HCℓ, and the InP face with the metal contact removed
with mercury.

2. AuGeNi contacts on InP

Firing temperature seems to affect the appearance of the contact as seen
in scanning and transmission. After firing at 350°C a fairly continuous
sheet of particles containing Au,Ge,Ni and P is dotted with polycrystalline
lumps of a ternary phase containing Au, Ge and P which is under further
investigation. For a 375°C firing the contact region is much thicker and
continuous, consisting predominantly of Au and P with some Ge. The 400°C
fired contact exhibits large thick polycrystalline lumps containing Au, P
and Ge. Regions of amorphous Ge are also present as are a large number
of round pits in the InP, as seen in scanning, where Au or other mercury
soluble phases have intruded into the InP.

The doping level also appears to affect the contact. For a constant firing
temperature (375°C) a reduction in doping level by a factor of 1000 produ-
-ces a much thicker contact layer.

The InP was given several different types of cleaning prior to metallisa-
-tion. Solvent cleaning produced contacts with round areas more or less
void of any particles, known as 'white spots'. The origin of these is not
clear. Carbon replicas of the solvent cleaned InP surface showed shallow
crater-like depressions with small protrusions in their interior. These
correspond to the pits in the gold occurring at the centre of white spot
as seen in scanning of the contact with the InP removed (Figs.1,2). All
white spot features are absent and pitting in the InP is reduced if the
wash prior to metallisation is KOH based.

A phase containing Ni,Ge and P has been seen on contacts fired at 350°C.
Convergent beam microscopy in a Philips EM400 has been used to determine
the structure as hexagonal a = 5.94 Å c = 4.91 Å. Using the tables of
Buxton et al (1976), the point group 6/mmm is indicated from the convergent
beam symmetries, with no absences detected. Small crystals of inferior

quality make positive space group identification difficult.

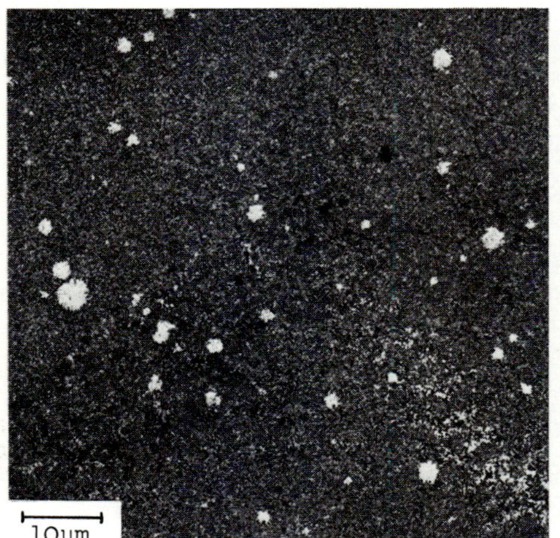

Fig.1. White spot on extraction replica (375°C firing).

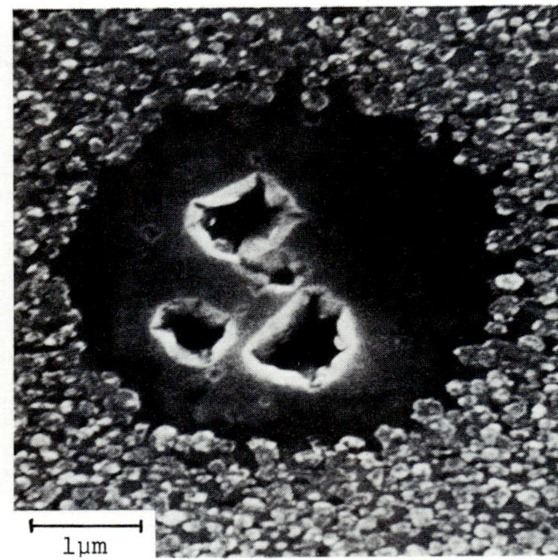

Fig.2. White spot on gold face of interface (375°C firing).

3. EDX calibration

To determine the atomic percentages of the constituents of the NiGeP phase by EDX, a calibration of the KEVEX X-ray detector on the Philips EM400 was carried out. For thin specimens primary X-ray absorption and fluo--rescence is small, so that the intensity of X-rays characteristic of a particular element is proportional to its concentration (Cliff and Lorimer 1972). Thin mineral specimens of known stoichiometry have been used as calibration standards by Cliff and Lorimer (1975). We have used the crystallised forms of microdroplets of solutions of various salts including nitrates and sulphates. These were generally about 4 µm in diameter, polycrystalline and sufficiently thin so that the peak to back-ground ratio of the radiation was constant (Cliff and Lorimer 1972). The droplets were illuminated by focussed beam and did not degrade. Detector efficiency and X-ray production efficiency is determined as a function of atomic number for Kα radiation. This has given the proportions of Ge:Ni:P as 2:6:5 (±3%) in the ternary phase.

4. Conclusion

The nature of the contacts varies greatly depending on materials prepara--tion and processing. Work is in hand to correlate the presence of the new phases and other features with electrical measurements and device performance. We wish to acknowledge the support of RSRE Malvern for supplying samples for our investigations. One of us (RJG) holds a CASE studentship with RSRE Malvern and another (GMR) received support from DCVD (M.O.D.P.E.)

References

Buxton B F, Eades J A, Steeds J W and Rackham G M 1976 Phil.Trans.Roy.
   Soc. 281 pp 171-194
Cliff G and Lorimer G W 1972 Proc.5th Eur.Cong.on EM pp 140-141
Cliff G and Lorimer G W 1975 J.Microsc.103 pp 203-207

Workshop on STEM

A group discussion on the current and future state of the art, organised under several headings. What follows is an edited version of the tape-recorded discussion; editor L M Brown (Cambridge).

Can STEM be justified to a large industry?

This topic was opened by Dr P Flewitt (Central Electricity Generating Board, Scientific Services Department, Gravesend, Kent) who described how analytical microscopy is used by the CEGB to assess both the short-term and the long-term integrity of operating plant, sometimes pushing the useful life of a plant beyond the design lifetime. The main problems are associated with impurity distributions and precipitates in all types of steels and in welds. Fracture in boiler components is very often associated with the opening of voids around precipitates, some types of which can be identified as having particularly deleterious effects.

The combination of microdiffraction and EDX analysis has proved very powerful in this context. Most impressive is the size of the electricity generating plant involved, and the financial rewards for avoiding long outage times justify the use of costly STEM equipment. In this work it is important to develop effective methods of specimen preparation to produce uniform contamination-free thin films. Studies of oxides produced in the steam environment are also very important. Dr Jeanne Silcock (CEGB Central Research Labs, Leatherhead) pointed out that although for many problems the relationship between integrity of plant and STEM results might be in question, nevertheless the STEM in the laboratory was very heavily used and provided much information of great interest to the CEGB. Dr Ricks (Cambridge) questioned whether STEM was the best way to tackle these problems, because techniques of convergent-beam diffraction permitted direct crystallographic identification of metallic carbide and nitrides in steels. However Dr Flewitt pointed out the importance of identifying the metallic elements in these precipitates, and all seemed to agree that a combination of analytical outputs can produce solutions to these problems much more effectively than a single output. Dr Hugh Bishop (Harwell) reminded the audience that an SEM or microprobe equipped with a STEM attachment is a much cheaper way of getting the answers, especially for the larger particles, and that wavelength-dispersive x-ray analysis can be used for lighter elements. The important question of the determination of carbon in ferrous solution was raised, and Dr Scott (Bath University) pointed out that carbon sensitivity levels of about .05% (by weight) had been claimed for many years using soap-film crystals as the dispersive element in wavelength - dispersive analysis. He suggested that attention paid to the development of better crystals might pay dividends. He recommended however, that EELS is a more profitable line. Dr Stobbs (Cambridge) said that quantitative analysis of carbon in solution is extremely difficult by EELS; the preliminary results obtained by his group seemed not to agree very well with those obtained by other methods such as the atom probe. It was recognised that the case of carbon in iron is particularly difficult because the carbon edge appears on a high background from the iron M losses. With the proviso that carbon is ferrous solution posed considerable difficulty for all techniques, the

tone of the discussion was optimistic and all participants looked forward
to increasing application of STEM in industry.

<u>Cathoduluminescence and EBIC* outputs: What do they tell us?</u>

This question was taken up by Dr Pierre Petroff (Bell Labs, Murray Hill,
N.J.). He pointed out that these techniques were essentially part of
the range of spectroscopic techniques already employed in solid-state
physics but unique in their ability to correlate the spectrum directly
with microstructure; he predicted increasing applications to biological
and geological samples, as well as to semiconductors, where most recent
work has been done. On semiconductors, as components get smaller, the
importance of defects and of interface structure increases, and one needs
to monitor band-structure and electrical activity on an ever finer scale.
In the study of devices, the determination of carrier lifetimes and
mobility and the analysis of device failure are prominent applications.

Dr Petroff went on to discuss the problems of radiation damage. He
showed a picture of lines written by the probe in a high-voltage STEM;
these were lines of radiation damage (i.e. charged states) at the oxide/Si
interface in a MOSFET†. Other problems associated with radiation damage
were mentioned, in particular damage in GaAs and other III/IV compounds
at electron energies well below the threshold energy needed to cause
atomic displacements. The lines of radiation damage had a width
controlled by beam-broadening, which is appreciable. Dr Petroff indicated
that in a very large scale integrated devices a probe size less than
0.1 μm was necessary and therefore one must use thin films.

Finally Dr Petroff pointed out the very narrow wavelength range allowed
by existing photomultiplier systems, about 0.9 μm to 0.3 μm, and indicated
how difficult it is to decide whether transitions observed in
cathodoluminescence are radiative or not because they may radiate outside
the detectable waveband. In these studies, there is a need for low
temperature stages as well as a need to extend the waveband.

The question which of STEM or SEM is better was hotly taken up by Dr
Ourmazd (Oxford) who pointed out the devastating effect of surface
inversion layers in thin films and of high voltage electrons in pinning
dislocations, and in producing other types of damage. He felt that since
specimens prepared for STEM were not in any case thin enough for weak-
beam studies, SEM followed by thinning and CTEM was the more attractive
technique. After some lively discussion, both Dr Petroff and Dr Ourmazd
agreed that one significant advantage of STEM is the availability of a
wide range of facilities both for analysis and imaging in one instrument.

Dr Stan Davidson (GEC Hirst Lab) interjected that when it comes to the
analysis of thin-film MOSFET devices STEM is the only possible technique.
He indicated that modern technology, in attempting to produce more closely-
spaced devices, is trying to use 'worse' silicon, for example, that grown
on sapphire. The films, of thickness typically 0.5 μm, are loaded with
defects, many of which are known to be electrically active. Dr Davidson
asserted that the economic return resulting from an improvement in MOSFET
technology would amply justify STEM. He wondered whether
cathodoluminescence could be refined to achieve the same level of

* Electron beam induced current

† Metal oxide semiconductor field effect transistor

sensitivity to impurities as photoluminescence techniques, where one part in 10^{10} of optically active impurities could be detected. Current STEM techniques using EDX or EELS are really only sensitive at the 1000 ppm level; what was needed was a technique at the 1 ppm level, or better. This problem was particularly acute for transition metals, trace amounts of which could affect minority carrier lifetimes and hence device leakage currents. Dr Brown (Cambridge) indicated that at such levels, the small electron probe could contain zero impurities, or occasionally one impurity atom and questioned whether STEM could ever be applicable under such circumstances; Dr Davidson agreed.

Dr Petroff appealed to the manufacturers to try to produce a STEM which is better adapted to semiconductor work: capable of taking chips which are 1cm square, and equipped with really efficient optical detection systems.

How can we extract better microdiffraction patterns?

Dr John Spence (Arizona State University) started the discussion by emphasizing the power of diffraction techniques; many structural features encountered by a 1nm probe could be interpreted with a ruler - i.e. crystal structure and interplanar spacings - and there was no need for elaborate theoretical calculations except in special circumstances. He then stipulated the design requirements for a effective microdiffraction output: (i) parallel readout; (ii) high angular acceptance; (iii) small probe size; and (iv) the system must not compromise the use of the electron spectrometer. He described briefly the system at Arizona State, and showed examples of its operation. He pointed out that the Grigson* system of serial readout is markedly inferior, in that many interesting effects due to the coherent probe are not observable because of the long recording time coupled with radiation damage and specimen drift. To obtain higher-order Laue zones, which allow three-dimensional crystal structure to be determined, it is essential to have large angular collection which can be achieved by post-specimen lenses as at Glasgow University or by the post-specimen field of a high-excitation objective lens. At Arizona State, the microdiffraction pattern is extracted through a fast transmission phosphor placed after the electron spectrometer: the phosphor is imaged by an image intensifier and T.V. camera. Such a system is not possible using a window-frame spectrometer of the Gatan type now being installed in many instruments. Dr Spence showed a remarkable example taken by Prof Cowley of microdiffraction patterns from a probe whose size is smaller than the unit cell of the crystal. Under such circumstances, the pattern symmetry is that sampled by the probe, and it may become possible to work out the symmetry of selected points within the unit cell directly, without the need for the usual analysis by structure factors. For these applications, as well as for patterns from highly inhomogeneous materials such as multiply-twinned particles, detailed theoretical analysis is essential but the methods for carrying it out are now available.

Dr Derek Houghton (McMaster University) enquired about the cost of the system at Arizona State University. After some discussion in which large variations in prices for the necessary equipment were noted, it was agreed that a 'consumer guide' of available T.V. cameras, image

* In which the diffraction pattern is scanned past a collector aperture
  by post-specimen deflection coils.

intensifiers, and frame stores would be highly useful. (Perhaps appropriate for the EMAG newsletter, ed.). Dr Lynch (Institute Francais du Pétrole, Paris) emphasized the utility of such video displays, especially since if they are properly designed with a hole in the phosphor screen they allow simultaneous viewing of the image and the diffraction pattern, which is very important for characterisation of small, possibly radiation-sensitive, catalyst particles. This point was taken up enthusiastically by Dr Spence (Arizona) and by Dr Boyes (Oxford) who indicated that with parallel recording of a microdiffraction pattern, the first few frames of a video tape are observed to contain patterns from undamaged crystal, and thus this is an extremely efficaceous way of getting information from beam-sensitive materials.

Dr McMullan (Cambridge) described a system under design in his laboratory in which photographic film is used to record the pattern, but the film is protected from the U.H.V. environment by a thin electron-transparent window, made of mica or of plastic. Such a system should allow fast recording of high-quality patterns, but it must be placed before the electron spectrometer if the spectrometer is of the window-frame type and it could be designed to be swung out of the path of the beam when not required. It should be used in association with a fast video-display of the microdiffraction pattern for routine use.

Dr Craven (Glasgow) and Dr Boyes (Oxford) pointed out that direct camera recording of a phosphor screen through a transparent port is very inefficient, and that Dr McMullan's suggestion, which is based on established techniques in astronomy, should be much better. Dr Spence wondered whether gelatin-free soft x-ray film might be more compatible with U.H.V. systems than traditional emulsions.

A final useful point was made by Dr Veseley (Brunel) who stressed that for each diffraction problem there was an optimum probe and that there is little point in pushing for even smaller probe sizes unless there is good reason. The act of optimising the probe size is the key to successful use of microdiffraction in STEM.

<u>The EDX output</u>

Dr G.W. Lorimer (Manchester) stressed the need for adequate standards to establish the "K-factors" (essentially ratios of x-ray intensity produced in the STEM by pairs of elements present in equal mass). He indicated that mineral standards seemed best from many points of view. Several members of the audience wondered whether theoretical expressions for cross-sections combined with a small number of standards to choose the best values for parameters in the theory could produce equally good results with less labour. Dr Lorimer replied that if an accuracy of $\pm$ 5% in ratios were sought, the use of standards is unavoidable. Dr Peter Statham (Link Systems) suggested that a good test of the x-ray detection system might be the ratio of K to L production by a metal foil. Such a technique gives an indication of (for example) the cleanliness of the Beryllium window over the detector. After some discussion, this point was agreed, always bearing in mind that it is not possible to get "K-factors" this way.

Dr Loretto (Birmingham University) introduced the principle of coincidence counting to reduce the effect of beam-spreading in the x-ray signal. The idea is that only those x-ray pulses should be counted which coincide with the electrons received by the spectrometer in a certain

angular range and energy window above the edge of interest. In this way,
the electrons which produce the counted x-rays would be ones known to have
sampled a very limited volume of foil, because their trajectory is confined
to a small forward scattering angle. Several members of the audience
expressed scepticism, because it was felt that the count-rates are much
too high. Dr Loretto pointed out that the technique would be used only
for trace elements with K-edges of small cross-section, so the count-
rates should be acceptable.

Dr Lorimer introduced the subject of beam broadening by writing out the
famous formula of Stephen Reed* which predicts a beam diameter at the
electron exit surface proportional to the three-halves power of the foil
thickness and inversely proportional to the accelerating voltage. Dr
Stephen Reed (Cambridge) gave a lucid account of the derivation of the
formula, which is based on Rutherford scattering of electrons. He
pointed out that the assumption of single-scattering is a good one,
because a given deflection is accomplished with greater probability by
one scattering event than it is by two smaller ones, as Rutherford
originally said. He mentioned an important limitation of the formula,
which is that it does not give an accurate indication of the current
distribution resulting from an incident probe of finite width. Dr Stobbs
(Cambridge) said that in his group's experience, the apparent shape of an
impurity profile at say a grain boundary is rather insensitive to the
assumed probe shape, and wondered whether the use of a large probe with
a well-defined current distribution might lead to more reliable results.
Dr P Doig (CEGB) supported Dr Stobbs, but said that a Gaussian probe
shape is in any event the result of multiple scattering, and allows easy
allowance for beam broadening with depth by simply adding mean square
widths: a point with which Dr Reed agreed. Mr S Collett (Cambridge)
asked whether Reed's derivation properly included the Poisson scattering
distribution. Dr Reed replied that this is unnecessary for thin films.
(Of course, this problem is made complicated because in many instances
it involves a foil thick enough for there to be a small number of elastic
scattering events but not enough for statistical treatment – Ed.) Dr
Loretto (Birmingham) supported Monte Carlo methods, but Dr Boyes (Oxford)
pointed out that they could not evolve into an interactive interpretative
technique for inhomogeneous samples, because of restrictions on computing
time. Dr Loretto reported that the Monte Carlo methods were in principle
better than those based on analytical models, especially in the range of
foil thickness of interest.

An important problem is the incident probe shape resulting from the use
of too large a condenser aperture, a shape like a "witches hat" (to use
Dr Lorimer's phrase) with a very wide brim. Both Dr Van der Maast
(Phillips) and Dr Bovey (V.G. Microscopes) pointed out that with a double
condensor system, this problem can be avoided by correct setting of the
instrument. According to Dr Bovey, the probe shape is most easily
assessed in a dark-field image.

<u>Towards analytical electron microscopy in surface physics</u>

Dr A Howie (Cambridge) introduced the discussion by distinguishing
between high-resolution instruments in which the specimen is immersed in
the lens, and instruments of resolution perhaps 3 nm in which the specimen
is illuminated at a working distance of several millimetres. Vacuum

* Goldstein, J, Costley, J.L, Lorimer, C.W, and Reed, S.J.B, 1977 10th
  symposium on scanning electron microscopy (Chicago IITRI) P 315.

conditions better than 3 x 10^{-10} torr could be guaranteed for the latter,
but for the former the effective vacuum at the specimen is unknown. X-
ray detectors are not compatible with surface physics equipment because
of the need for baking. However EELS should easily be available, and Auger
spectroscopy in the thin foil for the non-immersed specimen, and possibly
for the immersed specimen if use can be made of the magnetic field lines
around which the electrons will spiral out of the lens. Such instruments
promise a marriage between conventional surface physics and electron
microscopy. The problem of surface preparation is best solved by a
transfer device, which could enable specimens to be cleaned without ·
tying up the expensive major equipment. Dr Howie emphasised the use that
can be made of various low-loss emaging modes to reveal surface structure
by selecting electrons which are scattered very close to the surface.
Dr Boyes (Oxford) pointed out that A.C. fields would limit electron
collection by spiralling. He emphasised the importance of being able to
view the specimen in an environoment cell while reactions were taking
place, and of learning about the specimen surface from the secondary
image. Dr Gordon Tatlock (Liverpool) stressed that the experimental
approach to be taken depends very much on the type of problem to be
studied. He was in favour of a dummy sample chamber with a transfer
device, if at all possible. Dr Roy Willis (Cambridge) spoke for the
surface physicists, who "work flying blind above the woods, trying to
determine the position of the trees by feeling the leaves". He suggested
that a microscope with as low a resolution as 1 μm would be a considerable
advance on the current 1 mm areas under study. The lifetime of a clean
iron surface at pressure of 2 x 10^{-10} torr is about 1 hr, so operating
with a transfer device under these circumstances is very difficult.
However, he felt certain that the next step in instrumental development
is to combine the surface physics techniques with microscopy.

Towards Nanometre Lithography

This final topic for discussion was introduced by Dr Haroun Ahmed
(Cambridge) who described experience with electron beam lithography in
the Engineering Department. He pointed out that by making existing
semi conductor devices smaller, one could improve their performance;
if devices could be constructed below 10 nm size, new device concepts
would be possible and of course interesting physics experiments such as
electrical conduction in nanometre wires or electron interference
experiments could be contemplated. The instrument in use in the
Engineering Department is a TEM/STEM with a lanthanum hexaboride gun.
It has two modifications; one is the inclusion of a diagnostics chamber
to monitor electron dose and the other is a microprocessor based pattern
generator which drives the scan coils. Most of the work has been done
on electron resists of PMMA, or with double resists to control the line
shape; the double resist is a thin film of PMMA on top of which is a
more heavily cross-linked polymer. Dr Ahmed said that calculations of
beam-broadening within the resist were carried out by Monte Carlo methods
- a fact no doubt of interest to microscopists. The STEM/TEM currently in
use maintains current in the beam stable to ± .1% over short periods,
but since many devices in practice involve complex patterns, this
stability must also be maintained over a period of several hours.
Impressive examples were shown of lines of width 15-16 nm intended for
superconducting devices. Dr Ahmed briefly described the next
generation of writing instruments which would be based on ion sources;
liquid metal sources developed at Culham Laboratories were bright enough
when combined with electrostatic optics to do the job. Ion beams should

display much reduced broadening in the resist, and because they are more
damaging should enable much less sensitive resists to be employed. Dr
Chapman (Glasgow) said that to judge from work jointly undertaken by the
Electrical Engineering Department and the Natural Philosphy Department a
new resist was needed; in PMMA it seemed that an intrinsic granularity
prevented the production of lines finer than about 10 nm. Dr Bovey
(V.G. Microscopes) showed a picture from Professor Isaacson's group at
Cornell University in which lines of 2 nm width had been achieved using
the HB5. The work was done using very thin substrates; the resist was
Nacl. The problem undoubtedly was how to transfer these structures to
useful substrates such as Si. Dr Ahmed commented that perhaps the
method of x-ray exposure as used at Glasgow could provide an answer.
He wondered what sorts of physics experiments could be done with such
devices. Dr Nixon (Cambridge) said that x-ray zone plates, already used
to focus radiation from fusion reactions and for x-ray astronomy may yet
provide us with the interference x-ray microscopy so long contemplated
but not yet available. He reminded the audience of the paper by
Professor Burge (Queen Elizabeth College, London) in the conference
proceedings.

The discussion was on the whole optimistic and forward-looking, and
Dr Chapman (Glasgow) asserted that more work on the materials problems
associated with electron bema lithogrpahy seems well worthwhile.

Author Index †

Subject Index†

† Page numbers refer to the first pages of the papers in which the citations occur.